The Battle of the Little Big Horn has had more books written about it, by far, than any other battle in the Western Hemisphere, and that includes Gettysburg. Why? Because there are mysteries galore surrounding this epic confrontation between Custer's vaunted Seventh Cavalry and the greatest American Indian fighting force ever assembled on the western plains. The fascinating cast of characters alone is unparalleled: Custer, his officers and troopers, and a fascinating mix of white and American Indian scouts on one side, and Sitting Bull, Crazy Horse, Gall, and many other great warriors. The authors of this exhaustively researched and well-written book have examined the mysteries and produced fresh and often intriguing answers—and a new battle timeline—based on primary sources, not legend or hearsay. This is simply one of the best books on the battle I have ever read.

-- Jim Donovan, author of *A Terrible Glory*

Custer's Last Stand Demystified provides an exciting new interpretation of the events surrounding one of the most famous battles in American history. Clearly written and based on prodigious research, this book enhances our understanding of the Custer fight while challenging many standard assumptions about the battle. While I did not agree with everything the authors wrote, I nevertheless was enthralled with the book, as will all readers captivated as I am by the Little Big Horn. This now joins a short list of must-read books on the battle.

-- Paul Andrew Hutton, author of *The Undiscovered Country*

The Battle of the Little Bighorn has been mired in controversy and contradiction for one hundred and fifty years. But clarity reigns in *Custer's Last Stand Demystified: The Story of an Epic Defeat* by columnist and award-winning author Phil Guarnieri with seasoned Civil War and Little Bighorn scholar and living historian, Bill Rini. The authors present a fast-paced countdown of the spiraling events of June 25, 1876. Layering together a multitude of source material, decades of research, and brilliant, fresh insight, Bill Rini and Phil Guarnieri delineate and unknot many of the tangled paths leading to that blood-soaked battlefield. A Little Bighorn library unto itself, *Custer's Last Stand Demystified* is brimming with incredible maps, eyewitness testimony from both US soldiers and Lakota and Cheyenne participants, battle timelines, and appendices. It is a must-have for every serious student of the Battle of the Little Bighorn.

--Siobhan Fallon, author of *The Confusion of Languages* and creator of the Seventh Cavalry YouTube channel *@SiobhanFallon7*

“I love your book. It is electrifying.... I can't tell you how many times I was exclaiming YES to myself as I read. I am working on a novel inspired by the Reno Court of Inquiry. I must have read the courtroom accounts, from the original version, to Col. Graham's transcript, to the *Chicago Times* and *Pioneer Press* accounts, at least ten times.... I agree wholeheartedly with your findings.... I was getting goosebumps as I read *Demystified*, seeing how you provided evidence for so many things that I and others have intuited to have happened. No one has laid it all out the way you gentlemen have and explained it with such rigor and clarity, which is why there seems to be at least one "eureka" in every chapter. I know I won't be the only reader to feel this way. I enjoyed all the literary twists and allusions in your book as well. I can't wait for your book to be out in the world. It is amazing.”

-- Excerpts from an email from Siobhan Fallon, author of *The Confusion of Languages* and creator of the Seventh Cavalry YouTube channel *@SiobhanFallon7*

For those of us who have studied the battle since childhood, reading *Custer's Last Stand Demystified,* it is not like looking through a prism to the past but rather revisiting a familiar memory that parallels our own lives, particularly if, like Bill Rini, we have been involved with reenacting on the ground that he and Phil Guarnieri write about. I was taken first by their prose, written in a most literal and yet poetic style. The concise sentences that tie it all together are thoughtful, authentic, and profound, respecting both the amateur reader and professional historian of this emotional and moving event. Bill Rini and Phil Guarnieri are no strangers to this subject, and their wide concept and scope of literature and world history combine to bring the reader along on a ride like no other recounting of this dramatic retelling of our western history.

A myriad of books have been printed about this segment of our past, but very few capture the flavor, excitement, and cerebral approach Bill and Phil have committed to binding between two covers. To me, this tome is not a labor of reading but rather a journey for the senses. I was once again in the saddle, absorbing, analyzing the sights, sounds, smells, and sensations one can only experience along the banks of the Greasy Grass.

This is no 'Johnny come lately' take on the Little Big Horn. It's a culmination of truth and the passion of thoughtful minds that have lived this story and can tell it in depth, proving to the reader that it was more than a skirmish between cultures on the American Frontier. Bill and Phil know what they are writing about.

-- Steve Alexander "Foremost Custer Living Historian" and author of: *G. A. Custer to the Little Big Horn, Believe in the Bold: Custer and the Gettysburg Campaign, Dandy General Custer's Little Brown Horse,* and *Boots and Bruises My Life as Steve Alexander by General Custer.*

There are many books on the Battle of the Little Bighorn, but new material and learned interpretations continue to come to light, to be woven into the material presented by others in the past. That is what Phil Guarnieri and Bill Rini have done with *Custer's Last Stand Demystified*. Using a variety of primary sources, archaeology, and personal knowledge of the battlefield, they have presented us with an incredible book that sifts through both legends and facts. I have not had the pleasure of meeting Mr. Guarnieri, but I have known Bill Rini for 20-plus years, so I can speak from first-hand knowledge of his profound dedication to the preservation of historical truth and unbiased accuracy. Bill has great respect for the Native views of the battle and presents those accounts with the same unbiased zeal he has for all others. A true Man of Honor, and therefore, I assume with confidence, that Phil is as well.

-- Michael Badhand, Native American historian, author, and lecturer. December 2025.

Bill Rini and Phil Guarnieri's long-awaited publication answers sixteen of the most often-asked questions about the battle in June 1876 along the Little Bighorn River in the then Montana Territory. The questions started immediately after the gunfire stopped. Underpinning the extensive bibliography are Rini's three-plus decades of moderating the LBHA's Discussion board. Thirty-one maps illustrate the narratives. Their explanations will no doubt generate much discussion. *Custer's Last Stand Demystified* belongs on the shelves of every student of this battle, the people involved, as well as the history of the American West.

-- Colonel Rodney G. Thomas, U.S. Army, retired, author of *Rubbing Out Long Hair* and *Biilaachia-White Swan.*

Custer's Last Stand Demystified

The Story of an Epic Defeat

A Tactical & Timing Analysis of the Little Big Horn Battle

Custer's Last Stand Demystified

The Story of an Epic Defeat

A Tactical & Timing Analysis of the Little Big Horn Battle

Phil Guarnieri and Bill Rini

Garryowen Publications
Roslyn, New York

Cover painting "A Day Of Legend - When Courage And Destiny Meet, 1982 ,"
by Joe Grandee (1929–2021)
by permission of Nancy Pinson

Maps by Geoff Lackey

Library Of Congress Cataloging-In-Publication Data

Rini, Bill. 1955-
Guarnieri, Phil. 1957-
Custer's Last Stand Demystified: The Story of an Epic Defeat: A Tactical & Timing Analysis of the Little Big Horn Battle / Phil Guarnieri and Bill Rini.

First Edition.
Includes bibliographical references and index.

1. Little Bighorn, Battle of the, Mont.,1876. 2. Custer, George Armstrong, 1839–1876. 3. Indians of North America 0151Government relations 01511869–1934. 4. West (U.S.)—History, Military—19th century. 5. United States, Army—History—19th century. 6. United States—Military policy—19th century. 7. United States Army, Cavalry Regiments, 7th. 8. Command of troops—Case studies.

Library of Congress Control Number: 2025924715
Hardcover ISBN: 979-8-993905-70-9
Paperback ISBN: 979-8-993905-71-6

The paper used in this publication meets the minimum requirements of American Standard for Information Sciences—Permanence of Paper for Printed Library Material.
ANSI Z39.48-1992.

Printed in the United States of America
First edition, 2025

Dedication

To my wife, Sonia:

If I could write the beauty of your eyes
And in fresh numbers number all your graces,
The age to come would say "This poet lies;
Such heavenly touches ne'er touched earthly faces."

–Shakespeare Sonnet XVII

Phil Guarnieri

<<<❋>>>

Dedicated to the memory of all the officers, soldiers, scouts, and civilians of the 7th U.S. Cavalry, along with the Native American warriors who fought at the Little Big Horn on June 25, 1876.

Bill Rini

The Little Big Horn Battlefield – North End

The Little Big Horn Battlefield – South End

Table of Contents

Table of Maps

Custer's Last Stand Demystified

The Story of an Epic Defeat

A Tactical & Timing Analysis of the Little Big Horn Battle

Prologue

"Man who plants feet too deep into dogma, have hard time cleaning moccasins."

How someone analyzes and dissects the anatomy of a disaster may say more about that individual's personal agenda, disposition, and outright prejudices than about the true causes of an unanticipated cataclysm. While the Battle of the Little Big Horn provides a fertile field for analysis and investigation, it is also treacherous territory. The most challenging stories to tell are those where humans kill each other. This, in part, is because it is always hard to see clearly through a smoke-clouded battlefield, especially one that has been marked with such controversy and disputation as that of the Little Big Horn. That is particularly true with this story, as all who rode with Custer on that fateful afternoon never lived to see the sunset fade over the Montana horizon. Never would they bear witness to what happened and why.

So, the tale must be told by indigenous people of another culture who expressed their life experiences in ways unfamiliar to white Europeans. The Indians spoke their own language and expressed their culture through customs and perspectives unique to their conventions and habits. In other words, the Plains Indians, mainly the Sioux and Cheyenne nations, were a world apart from those soldiers they vanquished that day and then those who later attempted to recreate the battle in letters, newspapers, and books. Since the Indians were the only survivors, telling this epic story would be shaped by their memories, impressions, and imaginations. Over the years, historians, battle students, linguists, archivists, and archeologists have all tried to fill in the narrative's lingering gaps. Over the years, a mountain of testimony has been accumulated, and researchers have been combing through it to reach reasonable conclusions. Yet, tantalizing mysteries remain, and we still search for those exasperatingly elusive answers.

The past remains hidden until it is dug out, dissected, analyzed, questioned, and expressed. The poet Walt Whitman once said about another great conflict on American soil that "the real war will never get in the books." We have come to find this pessimistic sentiment truer of the Little Big Horn than any other battle in modern times. In the three hours from Custer's retreat up Deep Coulee and Butler Ridge, until the last soldiers were killed at various points from Finley Ridge, Deep Ravine, and Last Stand Hill, there have always been more questions than answers. Perhaps it will always be so. Still, we must work with what has come down to us, what we can read, examine, compare, and test. The milieu of history, messy as it can be, allows us to revisit and reevaluate sources and the evidence they furnish. The novelist William Faulkner once famously said, "The past is never dead. It's not even past."

Custer's Last Stand Demystified

This book asks sixteen questions in an honest attempt to uncover the truth about what happened on June 25, 1876, in southeast Montana when General George Armstrong Custer's 7th Cavalry attacked Sioux, Cheyenne, and a smattering of other tribes gathered in an extraordinarily large village beside the Little Big Horn River. This attack has become one of modern history's most inexplicable and debated disasters. For the American public, the shocking news of not only the defeat of the U.S. Army's most famed regiment by tribes still mired in the Stone Age but also the annihilation of all five companies commanded by a true-blue Civil War hero and the nation's most celebrated Indian fighter was as electrifying as it was unbelievable. The question of why such sudden ruin and disaster befell the regiment has been pondered and debated ever since. Everyone seems to have their own pet theory about the cause of the catastrophe—Custer's vainglorious pursuit, the failure of his subordinate commanders, or the overwhelming number of Indian warriors that converged on Custer's troops.

Because of the mystery surrounding it, the bloody and irredeemable drama has excited wonder and curiosity, which, despite the passing of 150 years, breathes new life into its undying controversies and the passions surrounding them. Even mystics have waded into its perplexing vortex, trying to untie the Gordian knot of how Custer's command so dramatically and completely unraveled. With so few facts, at least at first blush, the epic encounter took flight on the wings of lurid imaginations and unwarranted supposition.

That could be expected since mystery always breeds a minefield of perpetuating perplexities, second-guessing, and a host of legends. With time and patience, those who endeavor with the forensic zeal of irrepressible detectives to piece together clues of a haunting crime scene, the ambiguity clouding the blood-stained battlefield has come into better focus. While we now understand what happened and the chain of events that precipitated the 7th's collapse, gnawing gaps in the story remain. Why certain command decisions were made and why others were not acted upon, or acted upon belatedly or even half-heartedly, is a question that remains as tantalizing and hotly contentious as ever.

The only thing to do was go back to the mountainous documentation to hunt down the facts, no matter how elusive they might be. Yet, with a forest of paper already dedicated to analyzing this battle, the natural question is: why another book? The answer is personal and plain simple: We believe there is something new to say and something old to affirm. This is not a big-picture book that provides a general overview but rather one that seeks the devil in the details and the demon in the minutia. Neither bound by orthodoxy nor seduced by novelty, we vow faithfulness to the facts, rendering to the best of our capability the most thorough and forthright interpretation attainable. By applying logic, common sense, and an understanding of 19th-century cavalry tactics combined with virtually all the extant Indian accounts about the battle, we state our opinions as clearly as possible and as far as the evidence will allow and no further. Where we are forced to speculate, we tend to acknowledge that we are speculating.

In exploring this boundless subject, we have moved in and out of the shadows of the great valley of the Little Big Horn, endeavoring to find answers wherever they might lie or lead. Anyone embarking upon this subject should never doubt that this is an enormously complex battle. No single volume could serve as an encyclopedic survey of the Battle of Little Big Horn, and any detailed diagnosis of its causes and results must be weighed against that reality. This is an epic story in which the personalities of its three central characters, George Armstrong Custer, Marcus Reno, and Frederic Benteen, resonate so loudly amid strife, smoke, and emotions that have charged, but also clouded, debates over whose shoulders the preponderance of blame should rest.

This was an unqualified and humiliating disaster for the U.S. Army on the eve of the nation's much-awaited Centennial celebration. But so riveting and indelible were the events of the battle that, henceforward, the lives and reputations of those who led the 7th would principally be defined within the parameters of one Sunday afternoon in June as all three commanders, like Thomas Carlyle's Dr. Guillotin, were doomed to forever wander, "as disconsolate ghosts on the wrong side of the Styx and Lethe."

After years of scouring all available Little Big Horn records on both sides, including primary accounts, letters by participants, newspaper articles, government documents, diary entries, and memoirs, we have earnestly endeavored to furnish for the reader a sage and reasonable evaluation of the controversies that have colored, sometimes darkly, a saga that has been aptly called the Western Iliad. Writing about a subject that has aroused so much interest, this book will excite the wrath of more than a few students and scholars of the battle. It would be a rash presumption and the height of folly to state that every question asked and answered in this book unlocks the enduring mystery of Custer's last movements and motivations as his battalion was seen confidently galloping toward Medicine Tail Coulee. Hence, we make no such claim nor invite any such pretension. History does not work with such immediacy, nor does it ever employ the language of finality. No matter how well-mined a subject becomes, there is always something more to be discovered or said. We are fortunate to be the beneficiary of so much research and analysis. Due to the inexhaustible persistence of those who have studied this subject so scrupulously and tirelessly, we wish to acknowledge our everlasting debt and gratitude.

At this point, it is tempting to give the usual obligatory assurance that this is a non-partisan book, completely objective, where all prejudices and predispositions that plague and define other efforts and interpretations have been eviscerated. Of course, we did our best to be as detached and unbiased as possible. But to say we have cleared the loom of every warp and woof of impartiality would deceive the reader and ourselves. No account of the Battle of the Little Big Horn can be utterly unbiased since conjectures inevitably intrude into this multilayered and intricate narrative. Ultimately, one alternative must be chosen over another, whatever the conclusions of some of its more notable and distinguished chroniclers. No matter how perspicacious and deeply researched a book might be on the Little Big Horn, the subject is so kaleidoscopically layered by tangled webs woven by perplexing variables that all the unanswered questions can never be definitively resolved, especially since those who could have spoken were forever silenced in the ravines, coulees, and hillocks of a landscape once swept with unremitting merciless violence.

Sometimes, when confronted with the absence of solid evidence, a writer must declare his position about specific points of contention. This involves some educated guesswork that can easily lead to snares and pitfalls. Whatever our perspectives and conclusions, we can honestly say we were guided by what we dare to call an informed bias, bolstered by the source material and documentation of the U.S. Army, and the voluminous scholarship of innumerable historians and researchers who have studied the fine print of this battle. But we have also employed in this analysis the copious testimony of native tribes, which we have painstakingly reviewed, some from unpublished sources (see endnotes), and which have afforded us access to the most extensive and comprehensive Indian accounts ever committed to paper. This does not make our judgments infallible, for no one can speak ex-cathedra regarding this battle's timeless controversies. Still, we believe it does introduce an added perspective and a higher perch from which to view and judge the evidence.

Always exercising caution, we do not hesitate to state what we think likely happened without pronouncing upon the matter apodictically or with ontological certainty. Like every thesis, our questions and answers await the winnowing crucible of time and peer review to separate the wheat from the chaff—judgments bound to ripen over time. In the meantime, we will allow the proverbial chips to fall where they may, knowing that more than a few highly respected battle scholars will likely disagree with some of our conclusions. While we remain confident in our weighted conclusions, we will listen attentively to dissenting voices.

As for our modus operandi in searching out these narratives, we believe the best way to find answers comes from the old Socratic method of asking questions and plenty of them. The more serious and crucial inquiries, the better understanding of what happened results—haunting questions that have been asked since the ghostly white bodies of Custer's command were discovered, prostrate, naked, and mutilated over a wide berth of dusty terrain on the east side of a meandering, previously nondescript river flowing through a valley of death and destruction. We have adopted Francis Bacon's precept that "if a man will begin with certainties, he shall end with doubts, but if he will be content to begin with doubts, he shall end in certainties." Hence, we have operated more from the dark mines of doubt than the sunlit uplands of certainty.

For students and researchers who want to go deeper into the minutia of the battle, the authors have also provided in Appendix A (Participant Timeline of the Battle of the Little Big Horn) a tactical and timing analysis of the battle—a general overview of when the battle began, its key events, and when it ended. This includes a sourced Continuum supporting these conclusions and furnishing a range of time estimates approximating when events occurred.

The book is divided into chapters, each posing a different question to answer. It is a format we believe lends convenience and clarity to the subject matter, beginning with the readiness of the 7th Cavalry. Sometimes, a particular event was referenced again when related to another subject or question in a different chapter. This book is an inquiry into the basic contours of the battle and all its accompanying controversies, and we attempt to provide enough detailed narrative to give a fuller picture, even discussing in each chapter some of the more minor attendant questions that have continued to buzz about, but still interesting and relevant enough to cover. This is not a book about the moral questions belabored in today's political and social discourse: The so-called Western imperialism, often called

Manifest Destiny, where the U.S. waged an unprovoked war on the Great Plains against native tribes that long preceded the arrival of Europeans.

We still wrestle and agonize over the fate of these indigenous people, knowing the steep price they paid for the westward expansion of American civilization, even though total war, as practiced by William Tecumseh Sherman and Phillip Sheridan, almost pales in comparison to native tribes in their internecine conflicts over the centuries. Moralists and ethicists can use other texts and materials to explore those essential dimensions of the story. We have purposely limited ourselves to the battle proper. In so doing, we have learned that in this riveting saga, nothing was inevitable about the events that occurred on that hot, harrowing, sanguinary day in June.

Indeed, having thoroughly researched the subject matter, we found it at once a familiar story and one that startles with its maze of twists and turns. It has been said that there are six million stories in the Naked City, but we have also learned that more than a few crowd the accounts of the American Frontier. This book is an invitation to take a step back and analyze one of the more gripping puzzles in military history, so epic in scope and character that it has never released its grip on the public imagination. American anthropologist Raymond J. DeMallie once wrote, “Just as we are outsiders to other cultures, we are also outsiders to the past.” Yet, that foreign country still beckons, and we diminish who we are by refusing to step over that threshold for another look into that always nettlesome but fascinating chapter of what once was and why it came to be.

Introduction

The Battle of the Little Big Horn is the most told, written about, and debated saga in America's long romance with the American West. What began with a stark ultimatum rapidly turned into a human tragedy with an unforeseen climax that has never released its grip on this country's collective imagination. After a century of expansion and conquest, the U.S. government vigorously undertook the eradication of the last, refractory, nomadic tribes of the Great Plains. The inexorable push westward by European Americans across the continent was devastating to the indigenous population. By 1876, these tribes had all the earmarks of a dying race, their way of life and sustenance near extinction. In a cold prophecy, the 19th-century historian Francis Parkman wrote that if the American Indian would not assimilate to the ways of the white man, then he and his forest must perish. But in a hot summer a century and a half ago, no capitulation would be forthcoming from these free-roaming tribes, no surrendering of their ancient ways and customs.

For decades, the U.S. government had been perplexed about how to deal with what was gloomily referred to as the "Indian problem." Because of war and disease, the number of Native Americans dwindled to a pitiful few while those of European descent multiplied unchecked. Nonetheless, the spirit of these dwindling tribal people remained strong and unbroken. The Second Treaty of Fort Laramie, signed in 1868, guaranteed the Lakota Sioux exclusive possession of the Dakota territory west of the Missouri River. That seemed to be a tolerable stopgap to an insoluble problem.

However, the Panic of 1873 had plummeted the U.S. economy into the nation's worst depression yet. Then, in 1874, gold was unearthed in the Black Hills. and much like the California Gold Rush a quarter of a century earlier, hordes of white miners, in barefaced violation of the Fort Laramie Treaty, trekked westward, trespassing upon Indian lands, greedy to make their fortunes. Heavily armed and as stubborn as the mules they brought, these transgressors were not about to abide by any treaties or take orders from the authorities. Nor was the U.S. army eager to enforce the treaty with the Indians. Not yet ten years had elapsed since a fratricidal war had torn the entire nation apart. Americans killed other Americans by the hundreds of thousands. Presiding over a still tenuous and tattered Union, the U.S. government was unwilling to risk more bloodshed by forcibly removing these prospectors. Nor, in turn, could they convince the Lakota to sell their lands where the last remnants of a once vast buffalo herd, the Indians' source of sustenance and lifestyle, still roamed freely. The northern plains tribes would never acquiesce to a life reduced to farmers scratching at the earth with worn-out tools; they were proud warriors, and to surrender these hunting grounds and the very earth where their ancestors were buried was unthinkable.

Custer's Last Stand Demystified

A peaceful solution to the crisis seemed impossible. President Grant had endeavored to embark on a more humane policy of moving the tribes onto reservations away from westward expansion. However, louder voices in the U.S. government argued that more muscle than diplomacy was needed to deal with the Indians. Moreover, the government never fully understood that many chiefs who agreed to treaties with the whites never represented or spoke for the tribes at large. Faced with this conundrum and to attempt to avoid widespread bloodshed, the government recanted on the signed agreements and acquiesced to its errant citizens invading the sacred lands of these indigenous people. The government issued an order mandating that all Indians return to the carefully demarcated Sioux reservations by January 31, 1876, or be considered hostile. Manifest Destiny—the doctrine that maintained Divine Providence willed European Americans to overspread the North American continent—had, by decree of the U.S. government, become armed, militarized, and sent out to march against whatever pocket of resistance they came up against. In effect, it gave license to the army to wage war on those Native Americans who refused to yield to progress and civilization.

That such a dire message could be communicated and disseminated to the warrior/hunter culture during the bitter and harsh Dakota winter of 1875–1876 was beyond credulous. This improbability, coupled with the native tribes' innate antipathy to being domesticated, made the final clash between two diametrically opposed cultures inevitable. Indians were vulnerable in the winter, but the U.S. army could also be humbled and hobbled by freezing temperatures and mountainous snow drifts that could slow columns and ravage supply trains. This was no longer the mighty Army of the Potomac, crusading triumphantly through the Confederacy at the end of the Civil War. The combined frontier forces now numbered scarcely three thousand soldiers scattered thinly across the plains, often part of ill-trained and poorly equipped garrisons. After one feeble attempt to embark upon a winter campaign against the tribes, the army opted to march in milder climes later in the spring. Meanwhile, rather than being cowed by the army's ultimatum, warriors from various tribes gathered in unprecedented numbers along the banks of a meandering river in a valley they called "Greasy Grass," known today as the Little Big Horn. An unexpected victory a week earlier along the Rosebud against General George Crook's column, the massive size of their village, and the strange and prophetic visions of a mystical holy man called Sitting Bull all combined to create for these tribal people a false sense of security and empowerment. At the Little Big Horn, Indians saw their camp as too big and their medicine too strong to be seriously threatened by any enemy force.

But fate had other designs. Unbeknownst to them, a threat stealthily snowballed toward their village on the morning of June 25, 1876. The fabled 7th Cavalry, with twelve companies of soldiers and scouts, in all 655 men, approached the undiscerning campsite of multiple Indian tribes. The 7th was led by the ostentatiously brash, colorful, supremely confident, albeit controversial, George Armstrong Custer, known for his mettle in war. His command was now intrepidly galloping into history in hot pursuit of Custer's self-appointed destiny, only to encounter a fate he could have never imagined for himself and his beloved regiment. Though graduating last in his class at West Point, Custer soon became a favorite of General Phillip Sheridan, having emerged as a dashing and flamboyant cavalier, a dandy whose dress, long yellow curls, and daring exploits during the Civil War won him the newspaper headlines that made him a national hero. Leading scores of hellbent charges over blood-soaked, internecine battlefields, brazenly pitching into anything that had the temerity to oppose him, the "Boy General" (a name he was denominated with because of his youth) was not only outrageously successful as a cavalry commander but had miraculously managed to come out of the mayhem relatively unscathed aside from a minor wound in the fall of 1863, despite having eleven horses shot from under him. These exploits feathered his growing reputation, and his meteoric rise through the ranks seemed as providential as it was breathtaking: a Brigadier General at age 23, one of the youngest in American military history, and a Major General at age 25, he was both at Gettysburg thwarting the legendary Jeb Stuart and with Grant at Lee's surrender at Appomattox.

This thunderbolt on horseback with an overwhelmingly pugnacious bent had an instinctive knack for being in the right place at the right time and making the most of it. His many detractors scoffed at what they called 'Custer's luck.' But, in truth, no one could be that lucky. Custer may have been arrogant, overconfident, and an incorrigible glory hunter, but there is no gainsaying that he was a magnificent soldier: fearless, quick-thinking, a marvelous horseman, and possessing an unerring instinct for the attack. If the future cannot be ruled or even coaxed, then it can, Custer believed, be persuaded, if not shaped with the single-minded purposefulness he brought to every campaign. By 1876, despite a roller coaster post-Civil War career marred by a court-martial and suspension of rank and command, Custer had nonetheless carved out a reputation as arguably the most famous and formidable Indian fighter on the Northern Plains. General Alfred Terry, commander of the Montana expedition, who was ordered to force the tribes back onto their reservations, demonstrated his unwavering confidence in his subordinate by giving him the task of finding the Indians as well as the latitude to use his own discretion to prevent their escaping by any means necessary.

Starved for publicity, Custer relished the assignment, for it was said that he was only truly happy, truly alive, in war. The sound of the guns, the flash of sabers, and the sanguinary drama of mortal conflict were an enchantment, a sorcery, an adrenaline-fueled intoxication for him. Leading a search and destroy mission was part of his natural makeup, where thought and action fused. If anyone could sleuth out the hostiles over a vast and daunting geographical area, it was him. He knew tribal customs and once supposedly confided to a few friends that he could even be an Indian. He remained confident and undaunted despite the maddening maze of coulees, valleys, hollows, and intervening ravines that dotted the great, unceded western territories. By early morning light, June 25th, after four days of dogged pursuit, on an inconspicuous promontory known as the Crow's Nest, some fifteen miles from his designated target, razor-sharp eyes scanned the emerging horizon as night's darkness faded. Custer would make good on his boast, locating his elusive prey without alerting them. Though the village domiciled nearly twice the number of warriors that he had anticipated, curiously, this was not what worried him. America would triumphantly celebrate its centennial anniversary in barely more than a week. The West, or so it was believed, had been won. Laurels and the national limelight propitiously awaited the conqueror of the last of the Indians. Militarily, this assignment amounted to little more than a mop-up operation against a large body of intransigent hostiles, who, while capable of inflicting terror and devastation upon wagon trains, mining camps, and civilian settlements, always crumbled in the face of focused, tactical cavalry charges. Custer remained vexed only at the possibility of being discovered by the Indians, who, he knew, would scatter fissiparously into the endless expanse of a desolate Montana landscape like a host of apparitions.

Now, having sighted the village in the hazy distance, nestled peacefully alongside the Little Big Horn River and shadowed by steeply rising bluffs, the once Boy General was sure that the moment of glory he had so thirsted for was finally at hand against what had proved to be a surprisingly complex, resourceful, and frustrating enemy. The time to act was now; fortune favored the bold. As the hours relentlessly marched toward the fateful hour of 1:00 p.m., Custer, bristling with confidence and anticipation, had already divided his regiment once. Now, mistakenly informed by his scouts that the village was alerted and running, he turned sharply upon his saddle to address his second in command, Major Marcus Reno, and forthrightly gave Reno the fateful order to attack...

Chapter 1: Was the 7th Prepared for Battle?

On June 22, 1876, Montana Territory, along an indiscrete bank of the Yellowstone River, George Armstrong Custer's 7th Cavalry Regiment, having downed a hearty breakfast washed down with steaming black coffee, wheeled its finely muscled horses sharply to the left to form columns of four on damp and muddy ground from a stormy deluge the evening before. With regimental trumpeters on their white mounts lightheartedly playing the fight song and anthem of the 7th, the spirited "Garryowen," all twelve companies of the command, each carefully choreographed to ride its own-colored horse in a splash of aesthetic panache, marched in tempo over a rude and jutting landscape covered with sparse grasses and dotted with sagebrush all under the watchful and penetrating gaze of Department Commander General Alfred Terry and his reviewing party. It was, by all accounts, a magnificent display of martial ardor, stirred by a gentle breeze riffling through the air, guidons fluttered as if they were butterflies fountaining in a spray of spring. Two sergeant color bearers were behind Terry; one carried the regimental flag of the 7th, and the other carried Custer's personal battle flag of the 3rd Cavalry Division, which he proudly bore in the last year of the Civil War. It was just the kind of parade review that Custer relished.

For weeks, reports of large swaths of Indians that the government considered hostile were leaving their reservations to freely roam their old hunting grounds in the way of their ancestors. The authorities demanded they return to the agencies where they had been assigned, but the Indians, in a disobliging mood, refused to be quartered and fenced in. According to popular sentiment of the day, something needed to be done to clear up the last vestiges of hostile resistance lingering in the western territories. It was felt in some circles that these free-roaming natives impeded progress, development, and civilization. Thus, Custer was among the chosen to pursue and subdue them, bringing Indian warfare on the Great Plains to a close.

Without missing a beat, Custer galloped toward the reviewing party, looking every inch the frontier cavalier in his bone-white, broad-brimmed felt hat, smartly appareled in his fringed white buckskin jacket and polished knee-high black boots that comfortably rested astride the flanks of Dandy, his favorite fifteen-and-a-half-hand tall Morgan bay. General Terry and General Gibson were duly impressed by the 655-man regiment's esprit de corps and splendor. For a frontier army, it was an impressive procession of martial aplomb, with 175 pack mules assembled directly behind the troops. Rationed for two weeks, the packs supplied the regiment with 24,000 spare rounds of ammunition. Each man carried one hundred rounds of carbine ammunition in addition to his standard issue sidearm, primarily the now legendary, fully loaded Colt .45 revolver, with an additional twenty-four revolver cartridges on their person. Morale in the 7th appeared high, and confidence was unflagging.

Custer's Last Stand Demystified

Custer had called the '7th Horse' the best cavalry in Uncle Sam's service. At a luncheon in New York City, Custer boasted to his dining companions that the 7th could whip all the Indians on the plains. A mildly irritated General G. M. Dodge, who had fought the Sioux, cautioned Custer about his conceit. Still, the confident cavalier brushed the criticism aside with an audible demonstration of lightheartedness that signified an unshakable self-assurance.

In some sense, Custer was a walking resume for war. It was his natural milieu—feeding and nourishing him—for there was no greater elixir than a saber-clashing mounted charge into enemy lines. One officer acutely remembered that Custer was relied on for great things because he was dependable. Take the engagement he fought at Williamsburg on May 5, 1862. Dashing ahead of the infantry, he single-handedly captured six rebels and a large silk flag—the first colors taken by the Army of the Potomac. Twenty days later, he overran an enemy picket post at Chickahominy. Lieutenant Custer was not only the first to cross the stream, but he was also the first to open fire and the last to leave the field. Just a typical performance in a series of Custer's exploits during the Civil War.

Out of envy and the counsels of prudence, others saw things differently. Custer's ostentatious dress and manner may have captivated his admirers, but it outraged his detractors even more. During the war, he sported a black velveteen jacket with five gold loops on each side, a blue sailor shirt with an outsized collar, a garish red tie, black boots that reached just over the knees, and a soft, felt Confederate slouch hat from which his shoulder-length hair, with its yellow curls, attracted the attention of the press. Being noticed was Custer's objective.

Much of this stylish apparel remained with him during the Plains Indian wars. Major James Brisbin, commanding a battalion of the 2nd Cavalry, and an adversary of Custer, was in the reviewing party on June 22, and he certainly needed no convincing of what he considered the Custer's irresponsible rashness, going so far as to express astonishment that Terry was sending this wild man and insufferable ass down Rosebud Creek in pursuit of a large body of hostiles. Custer even managed to give ammunition to his critics when they heard General Gibbon say, "Now, Custer, don't be greedy! Wait for us!" Laughing, Custer replied, "No, I will not!"—a play on words that left his intentions an open question and made him, after the full tragic circumstances of the Little Big Horn campaign, a wide-ranging target for his enemies and critics who pointed to this brazen remark as but one more indication of his untethered arrogance and unbridled grandiosity.

Since his death, there has been a strong tendency to portray Custer as a maniacal Ahab of the plains, haunted by and in relentless pursuit of his Great White Whale, hunting down the enemy no matter the cost or consequences, even to himself. Many historians and the broader popular culture have embraced his supposed megalomania, as the restless zeal of the single-minded individual fascinates and makes for good copy. But Custer's hunger for distinction and glory was not necessarily shadowed by delusion or a sense of personal omnipotence. Whatever his weaknesses, they were largely a byproduct of his great strengths. Major General William Ludlow said Custer "was the personification of bravery and dash." But Ludlow also recognized the shortcomings of such qualities. "If he had only added discretion to his valor," Ludlow sadly lamented, "he would have been a perfect soldier. His impetuosity very often ran away with his judgment. He was impatient of control. He liked to act independently of others and take all the risk and all the glory to himself. He frequently got himself into trouble by assuming more authority than really belonged to his rank."

Yet, despite these tendencies, Custer was a born soldier and leader. General A. B. Nettleton remarked on Custer's innate *coup d'oeil*, his "instantaneous quickness of his eye and his lightening [lightning] like formation and execution of successive, correct judgments in a rapidly shifting situation that were attributes of a successful combat officer."

Even more effusive was Lieutenant Colonel Jacob Greene, who served with Custer during the Civil War and in a letter to General Edward Godfrey dated September 1, 1904, recalled:

> He was remarkable for his keenness and accuracy in observation, for his swift divination of the military significance of every element of a situation, for his ability to make an instant and sound decision…exhaustless energy which he everlastingly drove home his attack. And the swiftness and relentless power of his stroke were great elements in the correctness of his decisions as well as in the success of his operations. He was wise and safe in undertaking that in which a man slower in observation, insight and decision, and slower and less instant in action, would have judged wrongly and failed.

If there is an element of hyperbole in that description, it did not miss the mark by much. As anyone who saw him in action and under fire readily attested, Custer's determination, resolve, and fortitude in battle were overwhelmingly acknowledged. So how was it that he would become the principal character in what is arguably the most decisive military defeat in the Plains Indian War at the hands of irregular tribal forces—untrained, undisciplined, and unorganized as a military entity?

Is it possible that the resources the Army afforded to Custer's regiment in the Little Big Horn campaign were inadequate to subdue the hostile Lakota Sioux and Cheyenne? It is a subject worth considering since it has often been suggested that a major contributing factor to Custer's cataclysmic defeat, occurring only three days after he departed from the Yellowstone River, was that the 7th was ill-prepared, ill-equipped, and plagued on the day of the battle by a disproportionate number of mounts breaking down from fatigue.

General Terry certainly understood Custer's critical role in the campaign, even producing a new plan on the *Far West*, a steamer carrying army supplies and anchored along the Yellowstone River's banks, but it was scarcely different from his first blueprint. Its strategy was still the soul of simplicity: Find the Indians, defeat the warriors, and destroy their village and means of sustenance. It was assumed that the Indians would probably be found at the headwaters of either the Rosebud or the Little Big Horn River. Terry had meticulously stuck pins in the map spread out on a table in the cabin of the *Far West* to show Custer his line of march. Because Terry was nearsighted, Major Brisbin highlighted the trajectory with his blue pencil. The new plan was based on intelligence from Major Marcus Reno, who was conspicuously absent from this gathering due to his failure to follow orders. During his scout up the

Rosebud, Reno had unaccountably departed from Terry's orders. Still, despite all the contretemps associated with his botched mission, he did manage to provide a more likely location of where the Indians were, narrowing it down to somewhere between the Rosebud and the Little Big Horn.

On this fresh information, Terry decided to launch a pursuit and attack plan utilizing two independent columns. Custer would lead the faster-moving strike force of twelve companies of cavalry, and Gibbon's column, consisting primarily of slower-moving infantry, would theoretically block the Indians' escape to the north after Custer struck the village from the south. The objective was to catch and trap the Indians between two separate forces to achieve victory and prevent their escape, assuming the hostiles were in the Little Big Horn Valley, as they could have turned in any direction.

Although Terry had dabbled in military history and tactics, even visiting famous battlefields in Europe, the revised plan also had its defects. The plan looked good on paper, but trying to ensnare a highly mobile enemy by coordinated troop movements hundreds of miles away is like lassoing the wind itself.

Although Terry and Custer never settled on a specific date to join up, it is clear that Terry cared more about letting Custer know where he (Terry) might be found rather than coordinating an attack in concert with him. Terry, for all intents and purposes, anticipated that Custer would strike first. In other words, for all practical purposes, Terry saw Custer as the lone executioner in this operation. Custer undoubtedly saw Brisbin's penciled blue line as one that signaled him forward, not a red one he could not cross. As far as he was concerned, the die had been cast—he was the favored one, and what was done could not be undone. Lieutenant James Bradley, Gibbon's Chief of Scouts, got to the matter's nub: "It is understood that if Custer arrives first, he is at liberty to attack at once if he deems prudent. We have little hope of being in at the death, as Custer will undoubtedly exert himself to the utmost to get there first and win all the laurels for himself and his regiment."

The 7th Cavalry that marched so cocksure and bullish from Terry's burning scrutiny the next morning was anything but impaired, fearful, or marred by doubts about meeting the enemy. As one of the junior officers later wrote, "There was never a more eager command who had been ordered after the Indians. There was nothing that could save the Red Man except flight." Nor would most of the 7th's mounts, as they marched up the Rosebud, be seriously compromised over the next seventy-two hours before the battle, hoping to corner their unsuspecting prey. A contingent did emerge that would later swear to horses being played out. But these voices were on a fishing expedition, trying to hook onto anything that would explain why the regiment was so shatteringly defeated. First Sergeant John Ryan described the campaign as more of a cowboy roundup than a military clash between two opposing sides of relative military strength. No one seriously believed the 7th was walking into a lion's den. The regiment, Ryan noted, never looked better, as all the men were in good spirits. The troops marched in columns of platoons with their guidons flying and their horses prancing, all to the stirring notes of "Garryowen."

Much of the Philippics surrounding broken-down horses emanated from the belief that Custer, in a rush to glory, set a too-punishing pace for his mounts in sweltering weather and over rugged terrain.

Chapter 1: Was the 7th Prepared for Battle?

It was a tune not hummed but sung with all the power and virtuosity of an operatic aria. Lieutenant Gibson from Benteen's H Company was unstinting in stating that their horses were jaded, and Private William Slaper blamed the forced night march and the lack of grain for the "worn condition of our horses during the battle." Lieutenant Godfrey testified that their horses were beginning to peter out, many struggling at the rear. He would later qualify his statement by saying that the general condition of the horses was good, but that those with Custer were "very much ridden down, compared to the rest of the stock of the other companies." Lieutenant George Wallace of G Company testified that the men were tired, and the horses were worn out when they reached Ford A before hostilities commenced.

However, these remembrances and their concomitant implications do not square with the known facts. Custer stated that he expected the regiment to travel twenty-five to thirty miles a day, and not a single protest was heard from any officer that this pace would overtax the horses. Even on June 24th, the day before the fateful encounter with the Sioux and Cheyenne, when the command rode for some twenty-eight miles, the march, although starting rapidly, slowed considerably, with many halts and a four-hour rest between 1 and 5 p.m. A gait that could hardly be described as blistering. Even Captain Frederick Benteen, one of the 7th's battalion commanders and no friend of Custer, wondered why there were so many halts. Benteen admitted that Reno's mounts were only "blown" [exhausted] when they reached a stream in the high bluffs after a frantic chase by an Indian war party, which showed the horses were rested enough and in good condition. The command stopped at Busby Bend around dusk for an additional two-and-a-half-hour rest before moving out at 11 p.m. Headquarters time. They halted again on the 25th, the day of the battle, at about 2:15 a.m. and rested until eight in the morning, when they saddled up only to halt again at about 9:30 a.m. for another hour or so break.

While some have made much of the poor condition of Custer's mounts, Sergeant Finkle and troopers Peter Thompson and James Watson are the only documented cases to have dropped out of the column because of trouble with their horses. Both water and grass along the trail appeared plentiful for grazing. On June 22, before Custer took leave from the Yellowstone River, Colonel Gibbon noted the number of fine horses in the regiment and was told there was not a single sore-backed horse amongst them. In his "Annual Engineer Report," Lieutenant E. Maguire wrote, "Both men and horses were in excellent condition when they set out on the 22nd." Fred Gerard, an army interpreter and part of Custer's command, did not notice any fatigue among the horses in the regiment, further stating that he could ride his horse "40 or 50 miles the next day." In his memoirs, Charles A. Windolph from H Company wrote, "...our mounts were doing fairly well. Most of them were strong, young cavalry horses." Riding down the valley with Reno at the opening of the battle, the experienced scout George said, "The horses had galloped in good style."

So, precisely what does one make of these conflicting testimonies? Looking at all the facts, rather than focusing on a few isolated incidents, it would be a stretch to maintain that the horses, as a rule, were overtaxed in this campaign. A few of Lieutenant Francis Gibson's mounts were sent off the main trail with Benteen on a scouting assignment, and over rough ground, they may have become fatigued on this sweltering day. It was Benteen's horses, and not those of the regiment at large, that Lieutenant Godfrey, quoted above, likely referred to. As for Lieutenant Charles Varnum, he stated the horses were played out only after they raced out of the timber, crossed the stream, and ascended the one-hundred-foot-high bluffs to Reno Hill in the oppressive heat. He said nothing at all about their mounts being spent before this.

When the testimony is carefully broken down and parsed, the weight of the evidence is that at the onset of the battle, the horses were not overly ridden. Private Slaper was one of the very few enlisted men with anything negative to say about running down the horses. He also noted that Benteen's command was scrupulous in riding around rather than over the hills to not fatigue the horses. As for the officers, Lieutenant Wallace, who was appointed Reno's adjutant, was known to be outwardly biased in protecting Reno's reputation in the battle's aftermath and seized upon anything that would redirect blame from the Major's performance over which, to this day, a dark, brooding, accusatory finger still points. Doctor Henry Porter, an assistant surgeon in the regiment who had no reason to be laden with bias or partiality, offered a more sober observation: "The horses were in pretty good condition and ready to run."

The accusation that some troopers from Custer's command had broken off from him and were found on Reno Hill, some four miles away, is not indicative, as some have argued, of a general breakdown of horseflesh. All the mounts of Custer's five companies had been used in Reno's earlier ten-day scouting effort, which, in part, accounts for the estimated eleven soldiers from Custer's command that eventually found themselves on Reno Hill, and a few of those, it is generally agreed, were stragglers. Although there were soldiers whose horses were played out, Custer had at least 210 mounts that were not, despite the insistence of Colonel Elwood Nye, who, in an article for the Army's *Veterinary Bulletin* wrote seventy-five years later, bluntly, stated, "The whole story of his (Custer's) military life shows a brutal disregard of the well-being of his men and animals. There can be no doubt that this contributed, in part, to the tragic result beside the Greasy Grass [the Lakota name for the Little Big Horn River]."

Although Nye was a veterinarian, it is hard to see how he could have made a proper diagnosis about the condition of the horses at the Little Big Horn three-quarters of a century later. While acknowledging the broken terrain and heat under which the 7th labored and fought, Private Edward Davern testified at the Reno Court that the horses were in tolerable condition. This sounds about right since the Indians, who were connoisseurs of horseflesh, were mostly silent about the horses being in bad shape, which strongly implies that there could not have been more than a few. Three Indians interviewed by Lieutenant Oscar Long on June 27, 1878, just two years after the battle, contradicted the notion of played-out horses in the regiment. These Indians, White Bull (also known as Ice), Brave Wolf, and Hump, speaking of the 7th that day, said that neither the horses nor the men seemed to be tired.

Most soldiers in the 7th had good riding skills, having spent as many hours on horseback as most cowboys on the range. For those city boys who came out west and had never been on a horse before, this inexperience was quickly remedied by enlisting with the mounted cavalry and living out west where such a skill as horse riding was necessary for a soldier and for survival. In other words, these inexperienced soldiers learned fast—they had to. Cavalry recruits spent hours learning to steer horses in the direction the rider intended. When Major Reno's battalion barreled down the valley toward the village, the troopers kept the horses in line formation, which was not only a highly proficient task but the very essence of horsemanship, with several horses bolting past Reno's dismounting point—another indication that the horses, far from being overtired, were still packed with verve and vitality.

A critical analysis of Custer's itinerary underscores that most of the horses were not driven inordinately hard. There were, as noted earlier, frequent and lengthy halts, even though they were on a fresh Indian trail. These breaks were taken to rest the horses as well as to afford the scouts time to discuss and interpret the signs of the hostiles' campsites. The regiment had traveled about one hundred miles in nearly four days—not an inordinate distance. There are also other factors to consider. General Fry, who later would be one of Custer's pallbearers when his remains were interred at West Point, stated that while being on the offensive is more fatiguing than the defensive, the loss of a battle is seldom, if ever, due to the weariness of the attacking force. Fry points to the outbreak of Indian hostilities at the Pine Ridge Agency in 1890, where a battalion of the 9th Cavalry, under Colonel Henry, marched 102 miles in thirty hours to fight a successful skirmish against the hostiles. Henry had traveled almost the same distance as Custer but in less than half the time. So, it is logical to conclude that the notion that Custer's glory-seeking march overburdened the cavalry mounts and the men who rode them, a notion so fetching among Custer's detractors, is unsupported by the evidence.

What about the accusation that the 7th's firearms were not commensurate with those of the enemy? Scandalously, the hostiles' most steady and reliable source in providing them with weaponry was the U.S. Government's Indian Department and the Indian traders who supplied them with repeating rifles and ammunition. But what about the quality of the regiment's arms? More than a few critics have blamed Custer's defeat on the inferior and second-rate weapons his soldiers carried. They alleged that many of the guns were defective. Expended shells became jammed and had to be pried out with knives, which consumed valuable time and energy critical in a firefight. But, this allegation breaks down under analysis. According to battlefield archeologists Dr. Doug Scott and Dr. Richard Fox, extraction failure, though occurring occasionally, was not instrumental in the 7th's monumental collapse in the Little Big Horn Valley. The argument that the Indian tribes were better armed than the soldiers does not hold up under scrutiny.

The 7th, which included all 607 uniformed personnel, was armed with single-shot .45 Springfield carbines, which exceeded by 250 yards or more the range of the two hundred or so repeating rifles the warriors had. The Springfield carbine functioned optimally against the warriors by keeping the latter far beyond the effective range of their own weaponry. This relieved battlefield stress by giving the soldiers time to reload. While the Springfield carbine, in competent hands, could fire four shots a minute, such a pace of fire tended to rapidly deplete one's supply of ammunition. With the cavalry employing the tactics of a mounted army on terrain of their choosing, the Springfield carbines were the optimal weapons to fight Indians whose fighting style centered more on the caprice and temperament of the individual warrior rather than an organized, collective expression of warfare. Such fighting usually left the warriors bereft of command and control, in stark contrast to the U.S. Army's regimented and tactical thoroughness, permitting integration and coordination among disciplined firing lines during combat.

Each soldier, as mentioned elsewhere, was ordered to carry a hundred rounds of carbine ammunition, in addition to twenty-four rounds for their revolvers. The pack trains supplied an additional 24,000 rounds. Although the army would make a few minor adjustments to weaponry after the calamity at the Little Big Horn, the Springfield carbine remained the standard firearm of the U.S. Cavalry until 1893—seventeen years after the 7th's staggering defeat. That continuity demonstrates that not even the army seriously believed the deficiency of their arms was the reason for Custer's loss. True, the military suffered budget cuts during the post-Civil War period, the by-product of a dramatic and precipitous de-escalation from a fully mobilized wartime footing to meet an existential crisis. The cuts reduced the average allotment per soldier to twenty rounds of ammunition for practice. There is no denying that the overall attenuation of ammunition, considerably limiting opportunities for target practice, may have compromised the marksmanship soldiers had enjoyed during the war. While a diminution in target practice was inhibiting, it would be a wild overstatement to say it was fatal.

Nor can Custer's refusal to take Gatling guns be blamed for the defeat. He had little or no confidence in that weapon's use in fighting a war against Indians who used guerrilla tactics. The Gatling guns had severely inhibited Reno in his earlier scouting expedition, and they would undoubtedly slow a pursuing force when speed is paramount in chasing down a mobile and elusive enemy. General Nelson Miles, who had plenty of experience with these weapons, considered the Gatling gun useless when fighting Indians. Rather than attacking en masse, the warriors dispersed and used the terrain to their advantage, nullifying the gun's effectiveness.

As for the physical condition of the soldiers themselves, there seemed to be few issues adversely affecting their battle readiness. While there is some anthropological evidence suggesting that several soldiers of the 7th were slightly malnourished, most were lean and robust, much more so than what is typically believed by those intent on probing for the regiment's soft underbelly. The 7th was comprised of young men, averaging twenty-seven years of age, and living anything but a sedentary life. The open frontier during the 1870s, with its harsh elements and unrelenting physical challenges, was not where weaklings thrived or where those with delicate constitutions sought employment. Army life on the open frontier was physically and psychologically demanding, and a soldier had to be resilient to cope with such a rugged environment.

Spartan military conditions made these men hardier, largely inuring them from the debilitating effects of fatigue and hardship. They carried no illusion of the mortal dangers accompanying a military outpost. Benteen, king of the one-liner, told recruits that the government paid them to get shot at. Captain Moylan later testified that some of the men were very well drilled and knew their duty. Those who were not well drilled had not been in the service long. Yet, by the time of the Little Big Horn, more than thirty percent of the enlisted men benefitted by having more than five years' experience in the military. The inexperienced were mostly relegated to guarding the pack train or left to guard the Powder River supply depot. Familiarity with guns, as with horseback riding, was a necessity on the frontier. While 19th-century frontier soldiers were not as well trained as today's U.S. Army, most of them learned the rudiments sufficiently to fire a gun from astride a horse.

Chapter 1: Was the 7th Prepared for Battle?

Sergeant John Ryan of M Company wrote in his memoirs, "a great many of the men in the 7th had seen service in the Union army during the Civil War, with a few in the Confederate service and were familiar with squad drill, setting-up drill, dismounted drill, saddling and unsaddling the horses, preparing to mount and all the tactics pertaining to a mounted trooper. After we got so that we could drill pretty well, we had mounted drills by companies, platoons, and finally battalion drills." In other instances, Ryan seemed to harbor a more negative opinion when he said the regular cavalry was comprised of mostly green recruits so unreliable that Custer dared not fight them mounted.

Yet, when Custer was reinstated following his one-year suspension in September of 1868, he wasted no time getting the 7th combat ready for the campaign against the Cheyenne. Training days were spent ironing out the kinks, with an intensive regimen of morning and afternoon skirmish drills and two-week marches. Because Custer was a martinet and a no-nonsense taskmaster, the 7th was probably at its peak of readiness in 1868. Undoubtedly, his presence in the 1876 campaign, as had been true in the Washita campaign, had reenergized the regiment, and Ryan was not the only one noting the 7th's commitment to training and battle readiness. Major Reno, a good drill master, wrote about "how the regiment would undertake one hour of daily target practice although later reduced to a half hour," so the idea that the troops were not trained to use their firearms properly—at least by the military standards of the day—appears overstated.

In addition, the detractors asserted that the troops' performance on the frontier was subpar. This was most critically reported in a periodical in the final quarter of the 19th century. The publication noted that all the valuable survivors of the volunteer army had returned to civilian life, "Only the malingers, the bounty jumpers, the draft snakes and only the worthless remained in the army after the Civil War...these with the scum of the cities and the frontier settlements, constituted more than half of the rank and file on the plains." However, that was true only in the broadest sense. The discipline and training soldiers received in the army, though not up to the standards of the Army of the Potomac during the Civil War, still vastly improved the standard and readiness of the enlisted man.

The battle evidence shows that the army performed exceptionally well at Washita in 1868, and the two battles fought on the Yellowstone River in 1873. Just one year after the Little Big Horn debacle, Lieutenant Colonel Nelson Miles—whose command included elements of the 7th—won every battle he fought against the Indians. General Crook's forces performed well in the field, even at the Battle of the Rosebud, although they were taken by surprise and suffered a strategic setback. The same is true with Lieutenant Colonel Ranald Mackenzie, who pursued the Indians after the defeat at the Little Big Horn and brought most of the hostile tribes under control. These soldiers were far from the group of tyros who first entered the service. To portray the enlisted men as a military version of the movies' Keystone Kops was another tactic to minimize and shield the surviving officers at Little Big Horn from blame.

The truth is that the soldiers of the 7th were reasonably versed in the military arts and were often referred to as the "flower of the cavalry." As author Frederic Wagner correctly noted, the men of the 7th were professional soldiers, and they viewed their duties strictly through the prism of a military operation. The same can be said of many non-commissioned officers, some of whom were career cavalrymen and served in the Civil War. The regiment's percentage of raw recruits was no higher than the norm of that era—about ten percent of the outfit. Only nine of them had served less than six months; six of the nine were in Captain Tom McDougall's B Company, the rear guard of the pack mule train that trailed well behind most of the regiment on June 25th. At least 75 percent had a year or more in the army, and seventeen men had fifteen years or more dedicated to military service. Inexperience was not a major problem for the 7th Horse, much less the salient cause of their monumental defeat.

Whatever rust or lack of training that plagued the outfit, forty days in the field and riding on a dusty, rugged trail for more than 350 miles before the battle helped shore up and toughen the soldiers. Wagner concluded that before entering military service, young men in those days were far more likely to have ridden a horse or fired a rifle than young recruits today would be in driving a tank or firing a rocket launcher.

Even those who had never experienced combat still had the virtue of becoming soldiers. They became acclimated to regimentation, including following orders and accepting disciplined instruction. They genuinely understood command structure and tactical operations. This differed markedly from that which characterized tribal warfare, with its premium on individual honor and what was called 'counting coup' and making bravery runs, few of which served any useful tactical military purpose. Similarly, men in the 7th were proud to belong to such a famed and regaled outfit. One soldier stated, "What Indians would dare fight the 7th Cavalry? ...it was the finest cavalry regiment in the world. It just did not seem like anything could ever happen to it."

Indeed, many knowledgeable military personnel would agree that Custer's regiment was the best in the U.S. Army, although Mackenzie's 4th Cavalry in the Southwest might challenge that claim. Listen to Private Charles Windolph, who beamed with pride at serving in the 7th: "You felt like you were somebody when you were on a good horse, with a carbine dangling from its small leather ring socket on your McClellan saddle and a Colt army revolver strapped on your hip and a hundred rounds of ammunition in your belt and saddle packets. You were part of a proud outfit with a fighting reputation and were ready for a fight or a frolic." Even today, the reputation of the 7th Cavalry is the stuff of legend, and its very name reverberates with the sounds of glory and the aura of romance.

Moreover, the regiment was also heavily bolstered by highly capable and experienced scouts. Accompanying the 7th was the half-Sioux, half-French scout Mitch Boyer, who had been with General Gibbon on the long march from the Fort Ellis country and was widely judged, aside from Boyer's mentor, the legendary Jim Bridger, as the best guide in the country. George Herendeen, a remarkably versatile frontiersman who knew the territory like few others, also scouted in the campaign. A black scout named Isaiah Dorman, a former slave who had married a Lakota Sioux woman and served as an interpreter and courier was renowned for knowing his business and being a rugged campaigner. Also joining the 7th was Lonesome Charley Reynolds, dubbed "Lucky Man" by the Indians and considered the best white scout on the Northern Plains. He was a dead shot, fearlessly determined, and a marvelous hunter. Although Reynolds had developed a painful felon on his hand and was advised by Dr. Porter to sit out the campaign, the scout brushed off the doctor's misgivings, determinedly asserting that he waited two years for this expedition and that he would rather be dead than miss it.

Others rounded out the regiment, like the sunburnt thirty-seven Arikaras, often called Rees. Still in the prime of life, these veterans averaged about twenty-nine years of age and were remarkable physical specimens. Arikara means horns, referring to how they wore their hair with two pieces of bone standing up on each side of the crest. These Arikaras had enlisted in three prior army campaigns totaling at least nineteen months in the field before saddling up with Custer. The 7th Horse would also be buttressed in the campaign by the invaluable Crow scouts. If any party was motivated during this campaign against the Sioux, it was the Crows. The Crows had a blood score to settle with Sitting Bull and his tribe, who had previously and pitilessly driven them off these very hunting grounds. The Crow knew the territory now occupied by the Sioux better than anyone. Custer himself noted: "This employment of Indian allies exploiting ancient hatreds was a great weapon—a powerful auxiliary."

As for the officers of the 7th who took part in what the historian John S. Gray would later call the "Centennial Campaign," there was little to carp about in terms of their qualifications. At least seventeen were trained at West Point, the most distinguished military institution in the United States. Several other officers were trained in other highly reputable and well-regarded military establishments. Indeed, many of the non-commissioned officers had fought in the Civil War, where they faced not only gunfire from a trained, professional army but also the fearsome artillery of the enemy. Captain Frederick Benteen was widely saluted as being a soldier's soldier. Highly decorated for his actions in the Civil War, despite a difficult and misanthropic personality, Benteen was recognized by his peers—including Custer, who had more than a few dustups with him—as a good combat officer who could be counted on to be calm and collected under even the most stressful and contentious battlefield conditions.

Even Major Marcus Reno, the other battalion commander, though seen by many today as the weak link of the three highest-ranking officers, was at least considered competent, albeit colorlessly so, and had pursued Indians (but did not fight them) as early as 1860. Reno had been stationed in the Oregon Territory and fought in the Civil War, in which he was cited for bravery. However, as events unfolded in the upcoming campaign, it became clear that he was more suited for a supporting role rather than one in leadership. Myles Keogh, George Yates, Thomas Weir, Thomas French, Myles Moylan, Edward Godfrey, Edward Mathey, Thomas Custer, and Thomas McDougall had all served previously in the American Civil War, with Tom Custer (the General's brother), the recipient of two Medals of Honor.

Overshadowing them all was George Armstrong Custer, who had a stellar career in the late war. With a flair for publicity and showmanship, he had an uncanny knack for being in the right place at the right time. Custer first impressed General George McClellan, Commander of the Army of the Potomac, when, accompanied by McClellan's Chief Engineer, General John Barnard, he rode up to the bank of the Chickahominy River wherein Barnard said, "Jump in!" Custer, with all the sangfroid of a seasoned performer, jumped in, waded across the river to the opposite bank, and made a reconnaissance of the enemy position, thus earning him an invitation to serve as a Captain on McClellan's staff. But it was not all about showmanship. No less a figure than General Philip Sheridan, one of the Union Army's three most celebrated Generals, thought Custer's gallantry was virtually unparalleled during that war, and he believed that Custer's heroic exploits helped bring the Confederacy to its knees. Sheridan even gifted Elizabeth "Libbie" Custer the table at Appomattox, where General Robert E. Lee officially signed the surrender papers to General Ulysses S. Grant. He said it was a token of his and the nation's gratitude for her husband's devotion to preserving the Union.

Custer's meteoric rise landed him the rank of Brigadier General at the tender age of twenty-three and Major General at twenty-five, one of the youngest in American military history. He fought in numerous battles to preserve the Union, including Gettysburg, where he led his Wolverines (as he called his Michigan brigadet) in a successful headlong charge into Jeb Stuart's invincible cavalry, thwarting Stuart's attempt to circle around the Union army's rear and hold the field. Charging headlong into the enemy was a Custer specialty. He led multiple times from the front, where eleven horses were shot from under him, with Custer barely receiving so much as a scratch. 'Custer's Luck,' it was called. True or not, he believed in it and in himself—an indispensable quality for those who lead men. Dramatic and daring, Custer was drawn to war as few warriors had ever been. His military exploits, flamboyant image, and willingness to risk everything under fire captured the popular imagination and the hearts of many who served in his command. Part of his admiring society was the public at large, who became enthralled by his success on the battlefield.

At least on paper, lack of leadership was not the reason for the 7th's defeat that day. Some have hypothesized that the lack of reliable intelligence about the enemy and the scouts misinterpreting some of the more ominous signs going into the battle could explain why the Indians were so severely underestimated, leading to overconfidence and catastrophe. But even this explanation falls short. Whatever failures and shortsightedness hindered the campaign's planning and the actions leading up to the battle certainly were not game changers as far as the outcome was concerned. While the intelligence Custer received was not exactly precise in terms of warrior numbers and location, it was still good enough. The regiment, after all, did locate the village in the light of day while remaining, incredibly enough, undetected by the enemy. The earliest intelligence estimated eight hundred warriors. However, Custer believed that another two to seven hundred warriors could drift over from the agencies, increasing their number to as many as 1,500 warriors. Other estimates were significantly higher.

Frederic Francis Gerard, an interpreter for the Rees, believed there could be as many as twenty-five hundred to three thousand warriors, and he even once confided to Terry that if all the warriors who left the reservations united, the number could be as high as four thousand. The 7th Horse, despite a present-day inclination to report lower numbers on June 25, 1876, probably confronted more warriors than many of today's official tallies are usually willing to concede. A later chapter goes into warrior numbers more deeply, but for now, suffice it to say that had Custer known the exact numerical strength of the warriors beforehand, it would likely not have deterred him or changed his plans.

Custer believed that the right tactics, coordinated movement between his battalions, and the element of surprise were enough to bring the campaign to a successful conclusion regardless of the warrior's numerical strength. The soldiers seemed to know that a big fight lay ahead. On June 24, they passed several abandoned Indian camps. The trails migrated, causing confusion and misinterpretation such that many believed the camp was breaking up and the trails were diverging instead of converging, which was not the case. Varnum even thought that it was one continuous camp. Unbeknownst to the 7th, one of the camps was the scene of the Lakota Sun Dance earlier that month, conducted by a holy man known as Sitting Bull, who proved a magnet for the Indian free roamers captivated by his spiritual leadership.

The size of the abandoned camps indicated thousands of inhabitants. Custer mused over this, telling his orderly, Private John Burkman, that there were more warriors than he anticipated. Nevertheless, this did not dampen the Custer's bravado, when he quickly followed his cautionary remark with, "Not too many to lick, though." That the hostile village was unaware of the encroaching presence of the 7th Horse was not accidental or the residue of pure luck. Precautions to conceal themselves were strictly observed as the regiment silenced their trumpets and built only small fires that could be quickly extinguished by smothering them with dirt. Lieutenant William W. Cooke ordered each troop to ride a little off from the preceding companies to avoid raising large dust clouds, which could reveal their presence to the enemy. They did not bring along sabers to reduce the weight on their horses and for fear that the enemy might hear their loud clattering. Again, the predominant concern was preventing detection and the hostiles fleeing, not the strength of the Indian encampment.

Students unfamiliar with the battle and the nature of Plains Indian warfare become preoccupied with warrior numbers since it is the one barometer by which they can measure, understand, and compare. Experience on the Great Plains had amply demonstrated that warriors dispersed in the face of cavalry charges, which meant that a smaller, regular military force with disciplined lines of fighting, speed, and maneuvering could expect to handle a much larger irregular warrior force that far exceeded their own. During the march to the Little Big Horn, the civilian scout George Herendeen pointed out a location on the west side of the divide between the Rosebud and the Little Big Horn Valley, where he and 150 civilians held off over six hundred warriors. It was reasoned that if civilians were capable of such mighty feats against an Indian force many times their number, what could 607 professional soldiers, abetted by dozens of experienced scouts, do? So, numbers, or even potential numbers, as far as Custer was concerned, would not be a major factor in this battle, provided the battalions of his command acted concertedly and decisively.

The scouts did not entirely comprehend all the signs leading up to the Rosebud and toward the village. Most of them thought the abandoned camps along the trail were made, at least during the approach to the village, by single camps rather than multiple ones converging. Fifteen years after the battle, a startling admission by Lieutenant Edward Godfrey surfaced—the scouts had completely misinterpreted the signs. Be that as it may, nothing of significance would have changed even if all the signs were read flawlessly. In this campaign, it was clear to everyone in the regiment that Custer intended to attack the village, regardless of its size.

While much is made about stories of the 7th's premonitions of death and defeat, these things were retrospectively overplayed and overemphasized because of the terrible and unexpected fate that befell the regiment. The leader of the Crow scouts, Half Yellow Face, portentously told him that if he went down into that valley, they would both go home by a road they did not know or not see the sunrise or something similar. Other Indian scouts, like Red Star, less quoted in popular history, told Custer, upon discovering the enemy encampment, to attack the village immediately while the prospect of surprise was still possible. Custer confidently replied, "Yes, it shall be as you say."

Even though Custer did not yet understand the full meaning of the converging trails, he knew that the fresh tracks of the enemy meant the agency Indians were now joining the hostile encampment. The Crows he had sent out ahead reported that the Indian trail they were following had gone west, off the Rosebud, and crossed the divide into the Little Big Horn valley. The freshness of the trail meant the village was much closer and on the lower part of the river, not farther up the valley as Terry had originally speculated. As the village was perhaps a day's march away, anticipation swept through the 7th Horse. What mattered now was not following the precise wording of Terry's instructions but surprising and defeating a village of hostiles now much closer than anyone previously anticipated. The time of pursuit was coming to a close, and the time to execute was rapidly approaching.

Just six weeks after the crushing defeat at Little Big Horn, General Crook's force approached General Terry's on the Rosebud. First thought to be an enemy force, Captain Charles King of the 5th Cavalry from Crook's command could not help but notice how the 7th expertly fanned out, looking as ready and efficient and as prepared for action as any military unit could appear:

> Another minute and I have reached the bluff, and there got a good view of the coming host. Indians! ... Scores of them, darting about in equal excitement to our own. But no Indians are they who keep in close column along that fringe of trees; no Indians are they whose compact squadrons are moving diagonally out across the broad plains, taking equal intervals, then coming squarely towards us at a rapid trot. Then look! Each company, as it comes forward, opens out like the fan of a practiced coquette, and a sheaf of skirmishes is launched to the front. Something in the snap and style of the whole movement stamps them at once. There is no need of fluttering the guidon and stirring trumpet call to identify them: I know the 7th Cavalry at a glance, and swing my campaign hat in delighted welcome.

This passage from an experienced army officer reflects the meticulous mounted training in the field for which the 7th Horse was noted. In general, their marksmanship might have been somewhat lacking, but the accepted method of defeating Plains warriors was not by producing a high body count derived from expert shooting but rather by destroying their village, infrastructure, and horse herds, forcing them to survive on the harsh realities of nature away from their camp and hunting grounds. This was something the 7th was more than capable of accomplishing. Private John Burkman, Custer's orderly, pridefully noted that when General Terry and Gibbon reviewed the 7th Horse on the morning of June 22, "there had never been a regiment that looked as fit as we did that day."

Chapter 1: Was the 7th Prepared for Battle?

Winfield Edgerly, who had fought in the Little Big Horn battle and examined the battlefield afterward, would write and reflect on the debacle for the rest of his life. He would also rise in the following years and decades to the rank of Brigadier General in the U.S. Army. Edgerly would be one of the twenty-three men called to appear before the Reno Court of Inquiry. Edgerly would provide, among all the witnesses, one of the most extensive testimonies of the battle in which he attested to the fiber of its troopers and officers. The following is a sample of the Court's questions and Edgerly's responses that touch on the fitness of Custer's regiment, contradicting, along with Captain King's observations noted above, that the 7th Horse was ill-trained or ill-prepared for combat that day:

Q. State what, to your mind, were the evidences of the struggle, whether that of a desperate struggle or of a panic or a rout without much resistance?

Edgerly: Knowing the men as I did I have no doubt that they fought desperately for a few minutes.

Q. From your knowledge of the character of General Custer and of the officers and men under his command, and from the evidences of the manner in which his entire command was annihilated, state whether or not, in your opinion, General Custer could have fled the field with a portion of his command by abandoning the others to their fate?

Edgerly: I believe he fought very desperately.

Q. State if there were any evidence of the company commanders abandoning their companies and leaving the men to their fate?

Edgerly: I don't think there were any such evidences. There was no evidences of their abandoning their companies in a disgraceful or cowardly manner.

Q. Were there not in that command a number of soldiers who had experience in actual warfare?

Edgerly: Yes, sir.

Q. Do you think they needed any special instructions?

Edgerly: I think not.

Q. Do you think Captain Benteen and many other officers needed special instructions?

Edgerly: I think not.

The defeat at Little Big Horn on June 25, 1876, cannot be attributed to inexperienced troops led by inexperienced officers. There is little evidence that these soldiers fought with any less zeal and skill because of a lack of training. Poor army intelligence before the battle is another canard, even if a few signs regarding the Indian trail were misread and misinterpreted.

As for weapons, some expended shells indeed had to be pried from guns, but far too few to be consequential. Only six of Reno's 340 guns on the hill were found to have jammed, and there is no reason not to extrapolate that small percentage to the entire regiment. The 7th was already among the army's best-outfitted and trained cavalry units. It is reasonable to conclude that after forty days in the field, they were, in modern parlance, 'ready to rumble.' Reasons for their stunning and total defeat must be found elsewhere.

Chapter 2: Did Custer Disobey Orders?

Few questions surrounding this battle have been as hotly debated as whether Custer disobeyed General Terry's written orders and, in doing so, hastened the onset of the battle and its disastrous consequences. With army intelligence strongly indicating that the preponderance of Indians was most likely on the upper reaches of the Rosebud or the Little Big Horn, a view bolstered by the fact that this valley was a prized hunting ground of the Lakota, Custer was directed to pursue the trail of the Indians up the Rosebud Creek. Custer was all too willing to grab the reins of command and use whatever authority he was given to pursue the enemy. This chapter examines the nature and substance of these orders and how closely Custer adhered to or diverged from Terry's orders. Knowing how emphatic the ideologues on either side of this question can be, it is worthwhile to look at the subject anew and see what others have concluded from it over the years.

Brigadier General Terry, a lawyer, literate, and an inherently cautious man, believed that in Custer, he had a highly capable cavalry officer and a man of action who would not only be keenly aware of the tactical possibilities of any battle situation but utterly decisive in its execution. Once on a trail, Custer had the nose of a bloodhound and felt the pulse of battle like few others. No one understood this better than General Terry. His written orders to Custer would be thoughtful, careful, and deferential, expressing a mixture of confidence and caution—mostly confidence—regarding the latitude he granted Custer's independent command. Unbeknown to Terry, the wording of these orders would become a lightning rod and a Rosetta Stone, not only among future historians and scholars but also his contemporaries who, in the aftermath of the battle, struggled to come to terms with the reasons for the 7th being so thoroughly vanquished, with Custer and his five companies annihilated. Terry's orders to Custer became fair game to reflect upon and pick apart.

In the annals of American military literature, perhaps no other set of orders has been so thoroughly dissected, analyzed, broken down, parsed, and agonized over as the 407 words Terry crafted for Custer. Context is always critical, and no doubt Custer interpreted his orders with as much latitude as possible, which became apparent in a last-minute dispatch he composed for the *New York Herald* newspaper stating that he would take up the Indian trail and "follow the Indians as long and as far as horse flesh and human endurance could carry the command to the point of eating horsemeat before we're through." He also informed Lieutenant Godfrey and Captain Moylan, as if to underscore his determination, that the men better carry an extra supply of salt because they might have to live on horsemeat before they got through.

Pursuing Indians entailed taking chances, and Custer was prepared to move his command far from their supply base and supply steamer, the *Far West*, which had been navigating the Yellowstone River in concert with the movement of the columns. With this as a background, rather than quoting excerpts and cherry-picking Terry's orders, we will lay out Terry's famous missive in its entirety without abridgment, condensation, or editorial assistance:

Camp at Mouth of Rosebud River
Montana, Territory, June 22, 1876
Lieut. Col. Custer, 7th Cavalry

Colonel:

The brigadier general commanding directs that as soon as your regiment can be made ready for the march, you will proceed up the Rosebud in pursuit of the Indians whose trail was discovered by Major Reno a few days since. It is, of course, impossible to give you any definite instruction in regard to this movement, and were it not impossible to do so, the Department Commander places too much confidence in your zeal, energy, and ability to wish to impose upon you precise orders which might hamper your action when nearly in contact with the enemy. He will, however, indicate to you his own views of what your action should be, and he desires that you should conform to them unless you should see sufficient reason for departing from them. He thinks you should proceed up the Rosebud until you ascertain the direction in which the trail above spoken of leads. Should it be found (as it appears almost certain that it will be found) to turn towards the Little Horn, he thinks that you should still proceed southward, perhaps as far as the headwaters of the Tongue, and then turn towards the Little Horn, feeling constantly, however, to your left, to preclude the possibility of escape of the Indians to the south or southeast by passing around your left flank. The column of Colonel Gibbon is now in motion for the mouth of the Big Horn. As soon as it reaches that point, it will cross the Yellowstone and move up at least as far as the forks of the Big and Little Horns. Of course, its future movements must be controlled by circumstances as they arise, but it is hoped that the Indians, if upon the Little Horn, may be so nearly enclosed by the two columns that their escape will be impossible.

The Department Commander desires that on your way up the Rosebud, you should thoroughly examine the upper part of Tullock's Creek and that you should send a scout through to Colonel Gibbon's column, with information of the result of your examination. A detachment from Colonel Gibbon's command will examine the lower part of the creek. The supply steamer will be pushed up the Big Horn as far as the forks if the river is found to be navigable for that distance, and the Department Commander, who will accompany the column of Colonel Gibbon, desires you report to him there not later than the expiration of the time for which your troops are rationed unless in the meantime you receive further orders.

Very respectfully, Your obedient servant,
E. W. Smith, Captain, 18th Inf.,
Acting Assistant Adjutant General

The intent is important, and exactly what Terry intended has also been debated. According to Mary Adams, Custer's African American cook who, along with her sister Maria, worked in the Custer household, recalled overhearing Terry saying to Armstrong: "Use your judgment and do what you think is best if you strike the trail ..." But exactly what Mary Adams heard has also been the subject of controversy and debate. Many believe she misheard or misconstrued what Terry said. She had always been loyal to the General and his wife Libbie, and it may have been not so much what Terry said, but in the post-disaster milieu of charged emotions, what she heard may have come down to what she wanted to hear. Though a credible source, Mary Adams's recollection of that day should not stand alone. We must look elsewhere to obtain a fuller and more concrete understanding of the language of Terry's orders and the intent of his words.

Based on the language Terry used, more than a few battle students, including those with military backgrounds, insist that Custer disobeyed his orders in pursuit of the Indian village. They do admit that Terry's verbiage permits a degree of discretion and command decision. Therefore, those embodied inflections and nuances clearly in the orders' body deserve special attention. Terry's key phrases, such as "impossible to give you any definite instructions in regard to this movement," and "if it were not impossible to do so, the Department Commander places too much confidence in your zeal, energy, and ability to impose upon you precise orders," and again, "... the Department Commander desires that you should conform to his view unless you should see *sufficient reason* for departing from them." [Emphasis added.]

Terry's letter is also punctuated by phrases such as the commander's "desires" or "wishes" for this to be done, which seems less forceful and demanding than what would normally be characterized as "orders." Some have argued that Terry's "desires" and "wishes" have the same emphasis and weight as "commands" and "orders"—that although the document did not use declarative language, it was still a written document, weightier than mere verbal instructions, and therefore, should carry the full gravity of an order. This is where the element of subjectivity emerges and where other factors, such as custom and previous usage, come into play. It is important to interpret the written word within the context of its time—in this case, how orders were delivered and understood in 1876. While it is true that what Terry wrote would carry the weight of an order today, a century and a half earlier, his language would have been interpreted differently. For this reason, General Fry, a contemporary of General Terry, when analyzing the contents of this letter a decade after the battle, called Terry's missive a "list of instructions" rather than orders. Again, words matter—and so does context. It makes a difference if one reads Moses's Decalogue as the "Ten Commandments" or interprets the instrument more like "Ten Suggestions." A literal interpretation is always more binding, inescapable, and obligatory, while a looser reading takes on a more elastic and malleable context.

Terminological exactitude is critical, especially when issuing military orders. There is no substitute for clarity. But, while mathematical plainness in language is important, ambiguity also has its purpose. When Terry wrote these instructions to Custer, he did not know the location of the hostile tribes, their true strength, or the circumstances in which they would be found. All he knew was that Gibbon's scouts had recently reported seeing smoke on the Little Big Horn, from which they inferred that Indians were nearby. Terry was also unaware that five days earlier, the hostiles had successfully engaged General Crook's larger army. Uninformed about these developments, Terry wrote in a particular way, and, based on what he then knew, it made sense to allow Custer some flexibility in following his orders.

By comparison, Terry's language in his orders to Reno, who had embarked on a scouting foray twelve days earlier, is markedly different. Always cautious and circumspect, Terry wanted a scouting party to search all waterways south of the Yellowstone and east of the Rosebud. Major Reno became the surprise choice to lead the party rather than Custer. Some believe Terry's decision to go with Reno may have emanated from the General's wariness about the Indians' whereabouts, and other historians have speculated that Terry was afraid that if Custer commanded the scouting party, he would head straight to the Indian village on his own. The newspaper correspondent Mark Kellogg, however, has the simpler and likelier explanation, which was that Custer had, in fact, declined Terry's offer to lead the scout, convinced that there were no Indians east of the Rosebud. The task then fell to Reno, and from the wording he used, it is clear that Terry did not place the same confidence in Reno's ability and wherewithal as he did in Custer's. Accordingly, his language leaves no room to negotiate any discretionary action or afford any unnecessary latitude on Reno's part. Here is the full order in which the differences in his wording and expression are clearly discernible:

Headquarters Department of the Dakota

In the Field

Special Field Orders # 11 Camp on Powder River June 10, 1876

Major M.A. Reno, 7th Cavalry, with six companies (the right wing) of his regiment and one gun from the Gatling battery, *will* [emphasis added] proceed at the earliest practical moment, to make a reconnaissance of the Powder River from the present camp to the mouth of the Little Powder. From the last-named point, he *will* cross to the headwaters of Mizpah Creek and descend that creek to its junction with Powder River, thence he *will* cross the Pumpkin Creek and Tongue River and descend the Tongue to its junction with the Yellowstone—where he may expect to meet with the remaining companies of the 7th Cavalry and supplies of subsistence and forage. Major Reno's command will be supplied with subsistence for 12 days, and with forage for the same period, at the rate of two pounds of grain per day for each animal. The guide Mitch Boyer, and eight Indian scouts to be detailed by Lieutenant Col. Custer, will report to Major Reno for duty with his column.

Acting Asst. Surgeon J.E. Porter is detailed for duty with Major Reno.

By Command of Brigadier General Terry

With these orders, there is no latitude, license, or elbow room. There is no ambiguity. Nor is there any of the softer, pliable, negotiable terminology like "desires," "wishes," or "should" that populate the order to Custer. That explains why Terry was so angered when he learned that Reno veered from his orders despite securing some valuable and even face-saving intelligence. In a letter to his sisters, a smoldering General Terry bitterly complained that the Major ignored his instructions to stay away from the Rosebud and thereby risked alerting the Indians of his presence, placing the entire campaign in jeopardy.

When comparing what Terry wrote to Reno and what he wrote to Custer, it seems obvious that Terry's orders to Custer were largely composed within the context of what were primarily instructions. He only occasionally veered into the stronger language used in the Reno missive when he wished to emphasize certain specifics. Those orders were embodied in the opening sentences of the June 22 Custer order: "The brigadier general commanding *directs* that as soon as your regiment can be ready for the march, you *will* proceed up the Rosebud *in pursuit* of the Indians whose trail was discovered by Major Reno a few days since..." That would constitute an order to Custer—a closed circle with no escape clause for command discretion. The remainder is a series of instructions and more of an expression of Terry's desires and wishes, except when he pointedly refers to Tullock's Creek, which will be discussed later regarding Custer's obligation to scout that area for the presence of Indians. Only in that instance, and nowhere else, can a serious argument be made that Custer disobeyed an order. In fact, Custer's Chief of Scouts, Lieutenant Charles Varnum, saw Terry's order to Custer as carte blanche, permitting him to act on his own initiative at his own discretion.

But while words matter in a campaign, so does the allotment of resources. The latter spoke volumes about what was expected of Reno and what Terry expected of Custer. General Terry furnished Custer with more than twice the force that Reno had. It was plain that if the occasion arose, he wanted Custer to strike the Indian encampment and not, like Reno, merely report its location. It is also the reason why Custer was initially offered four additional companies of the 2nd Cavalry (some 160 men) and Gatling guns, which he refused for reasons of speed, unity of command, and the assurance that only he alone would be calling the shots and not doing so in conjunction with the commander of another regiment. General Gibbon and his Chief of Scouts, Lieutenant Bradley, understood Custer's prerogative perfectly, saying that he was at liberty to attack. Such aggressiveness is exactly what was expected of Custer by his commanding officer. Only after the catastrophe struck did everyone associated with the campaign begin changing their story that would best insulate them from criticism and incrimination.

So, Custer, pursuant to his orders, proceeds up the Rosebud amid the cottonwoods, wild rosebushes, tributaries, and streams. The command would encounter multiple abandoned campsites of notable size where some of the campfires still smoldered, indicating a recent departure by the hostiles. Soon enough, the scouts ascertained the direction in which the trail led, but when it turned west, despite Terry's instructions to continue south to the headwaters of the Tongue, Custer purposely turned toward the Little Big Horn following their trail. Terry, having feared that the Indians would escape to the Little Big Horn Mountains, planned for Custer to move southward at a faster pace than the hostiles to cut off their path, all the while feeling constantly to his left to prevent the escape of the Indians who might slip around his left flank. The overriding concern, General Gibbon remembered, was to "prevent the escape of the Indians, which was the idea pervading the minds of all of us." Custer became so obsessed over the matter that when Lieutenant Godfrey reported a trail leaving the valley that led southeast, fearing that the Indians might escape in that direction, he ordered his already exhausted Chief of Scouts, Lieutenant Charles Varnum, who had been in the saddle twenty-four hours "to see that no trail led out of the one we were following"—making sure that he was in full accord with Terry's wishes. The divergent trail returned to the main lodge pole trail several miles farther on.

From the Rosebud, Custer, according to Terry's wishes, would cross over to the headwaters of the Little Big Horn, move north and attack from the south, hoping to sandwich the hostile village between his and Gibbon's command holding at the junction of the Little and Big Horn Rivers. Because the trail was so fresh (a sure sign that the 7th was rapidly closing in on the Indians), abandoning it to continue south, as the instructions stated, meant that he could no longer emphasize the larger intent of Terry's orders essentially leaving Custer no alternative but to pursue the Indians, attack their main encampment, and destroy any means or vestige of resistance. The decision to follow the trail as it moved westward into the Little Big Horn Valley would mark the first of two instances in which Custer would depart from Terry's instructions. Custer's decision to follow the trail rather than continue moving south was justified under the rubric of Terry's escape clause of "sufficient reason"—the unexpected freshness of the trail—which, indeed, seems a reasonable assumption.

No doubt, Custer was also influenced by the army's recent failures to track down the hostiles and subjugate the enemy. This failure, in conjunction with Terry's orders, must have also swayed his thinking about the tactics he would employ. Other military commanders had been recently criticized for being too reluctant to pursue the Indians aggressively and bring them to battle. Reno had been harshly reprimanded by Custer for exercising the same forbearance just days earlier. No doubt, this was heavily on Custer's mind when he arrived at a deserted Indian village site, since he then turned to Varnum and pointed out that this is where Reno made the mistake of his life, noting that he (Reno) would have made a name for himself if he pushed on. Custer believed that inhibitions, whatever their provenance, thwarted the army's effectiveness in dealing with the Indian problem decisively. Given a chance against the hostiles, Custer was determined not to hesitate or delay in his duty to pursue and, if necessary, bring the matter to a bloody close.

The whole reason for this campaign was that the Indians were not going to voluntarily return to their reservations or capitulate to the military authorities. They would have to be forced, and someone had to force them. With the reins of command now firmly in his hands, Custer was determined that there would be no turning back once he picked up the scent of the enemy. He was committed to following the Indian trail even if it took him, as he had noted more than once, to the agencies on the Missouri and in Nebraska hundreds of miles away. He didn't earn the cognomen "hard ass" (his nearly superhuman ability to stay in the saddle for interminable stretches of time) because he was easily deterred and turned back in the face of hardship and adversity.

When, on the evening of June 24th, Custer learned from Gibbon's Crow scouts of the village's location, he needed no compass or counsel about what to do. Custer immediately understood that continuing south to satisfy a technicality in his commanding officer's written orders made no sense when the village was only a short distance from him. To have proceeded, as Terry prescribed, would have taken Custer to the end of the Rosebud Valley, some twenty miles away from the site of the Indian encampment, and jeopardized the command by greatly heightening the prospect of being spotted by hostile scouts. This would jettison the element of surprise, and the Indians, being adaptable and mobile, would inevitably scatter in multiple directions, melting into the great western abyss, resulting in yet another humiliating failure for the army.

Being faithful to the larger intent of Terry's instructions required Custer, as intelligence of the Indians' whereabouts congealed into certainty, to eschew fidelity to those peripheral details. That outcome was, at best, unlikely, if not implausible. General Terry, after all, insisted that his written instructions should *not* be read as an imposition "which might hamper your action when nearly in contact with the enemy." In effect, Custer obeyed the essence and intent, though not necessarily the particulars of every aspect of Terry's instructions. For Custer, this was enough to justify violating the letter of Terry's wishes to continue south rather than follow the fresh Indian trail to the west, where the Indian encampment was now located .

Given the historic defeat Custer suffered, it is natural, in retrospect, to conclude that Custer made the wrong decision. But after having scrupulously studied all the elements of this campaign, General Miles would later weigh in that once General Terry divided his command, taking part of it with him up the Yellowstone and sending General Custer with the other portion far out into the Indian country, putting as much as seventy-five to a hundred miles between the two commands, it imposed upon Custer, whatever the wording of the instructions, the responsibility of deciding to act on the defensive or offensive. It is hard to see how one can effectively dispute General Miles's conclusion.

Indeed, as it turned out, Terry and Gibbon had decided to stop at the mouth of the Little Big Horn and Big Horn River—some eighteen miles from the battlefield. They continued only after the Crow Indian scouts informed them of the mind-bending news that the 7th had unaccountably met with disaster in the Valley of the Little Big Horn. How likely was it, General Miles mused afterward, that the tribes would have permitted two columns (Terry and Gibbon) to move around the country with infantry, cavalry, and Gatling guns and then patiently wait for them to converge upon their encampment on both sides? That notion borders on the ludicrous, and such a scenario only has merit in a fantastical imagination.

Another important factor to consider is weighing the unexpected. In every campaign there are knowns, unknowns, and unknown unknowns. No one, least of all Terry, thought Custer going up the Rosebud would overtake the Indians so quickly (three days), considering the regiment was rationed for fifteen days. But sooner than even he expected, Custer began to see the pieces falling into place as the Indians, lured by herds of antelope grazing in the Little Big Horn Valley, moved slowly and deliberately up the Rosebud thence over the divide and into the Little Big Horn valley where the battle would eventually be fought. It is impossible to dismiss the impact that chance and circumstance had on shaping events when the soldiers began closing in on the village. Once the Indians reached the Little Big Horn Valley, they would uncharacteristically settle down for five bountiful and leisurely days, hunting and gathering by the banks of the river, seemingly without a care or a concern for any danger.

Besides good hunting, something else drew the Indians to this location. The spirits of Sitting Bull's warriors had been buoyed by having so recently thwarted Crook's nearly 1,200-man army that easily outnumbered them. Crook was forced to withdraw from the field after directly engaging with them. The mystic supplications of Sitting Bull's Sun Dance preceding the Indians' engagement with Crook gave rise to feelings of invincibility and an unwavering belief that their medicine was powerful. They knew of no army in the area besides Crook's, which they had just forced to retreat to Goose Creek near where Sheridan, Wyoming, is today, and Gibbon's column along the Yellowstone. Sitting Bull's warriors felt secure and unthreatened in this fruitful valley. They were in no hurry to leave what turned out to be their fool's paradise, much less move south where Crook's still formidable army threateningly lingered.

The Indian accounts are quite clear that they were not in the least anticipating an attack on their village. Pretty White Buffalo Woman, a Hunkpapa, remembered they were in camp with no danger of a battle: "I have seen my people prepare for battle many times and this I know: that the Sioux that morning had no thought of fighting." Gall, a heavy set Hunkpapa Lakota war chief, noted, "Buffalo scouts brought word to us late at night on June 24th where soldiers were camped." He found this surprising, if not unbelievable because they did not expect any other soldiers unless they came from the south, where they had defeated General Crook a week earlier, or from Gibbon's column along the Yellowstone.

The reluctance of the Indian village to move from such fertile environs along with their false sense of security from impending attack, combined with the nimble facility and unadulterated aggressiveness that Custer displayed in a military campaign, enabled the 7th to reach their prey much faster than anyone had anticipated. This left Lieutenant Godfrey to further speculate on the implications of following Terry's instructions to the letter by utilizing a thought experiment of "what if" to underscore his point:

> Now suppose the Indians had been located on the headwaters of the Rosebud, and the Tongue, or Powder, and not the Little Big Horn, should Custer have backed away, sent a scout through to Terry, made a detour to get to the south side to obey the very letter of Terry's order?

It seems, as we consider Godfrey's hypothetical, that it would take a real impoverishment of the imagination or an inflexibly stubborn streak of partisanship to believe that, if Terry had known all that Custer knew, he would have decided upon anything other than ordering Custer to attack.

There was also something else that struck Lieutenant Godfrey. It was not just the Indians' ignorance of approaching danger but, as mentioned above, Custer's grit and tenacity that became critical in catching the village by total surprise, a situation that the 7th and its command did not yet realize, even when they were on the precipice of charging the village. The consequences of this are enormous, especially because of the firestorm of criticism unleashed over Custer's supposed impetuosity to strike the encampment too quickly, covering himself with glory. To paraphrase Godfrey, if the attack had been delayed even a few hours, Custer would not have found all the Indians in the village. But, because Custer moved unhesitatingly and decisively, the hostiles had no time to strike the tepees and steal away. Their pony herds were still out grazing, a few warriors were out hunting, and others were sleeping and unconscious of any impending threat.

Chief Sitting Bull had been caught so off guard by the attack that he was ready to negotiate with the soldiers if he had a chance. Yet more than a few critics of Custer's actions that day embraced the quaint notion that it would have been best to forego an immediate attack in return for a better opportunity. That kind of wishful thinking was often a losing proposition when fighting Indians. A surprised enemy often translates into a demoralized one. Seizing the day was the only effective strategy for subduing the tribal warriors. The only way to do that was to locate the Indian encampment, prevent their escape, and then attack with everything the soldiers had.

Surprise was the key element in defeating the Indians, and against almost all odds, Custer had accomplished this feat in broad daylight. That is why author Bruce Liddic titled his superb study on the battle *Vanishing Victory*—since catching Indians off guard was tantamount to almost certain victory. With the camp napping away directly in Custer's sights, all that was required was a concerted and spirited effort by the battalions of the 7th to tighten the noose. Why this didn't happen and how nearly a certain victory morphed into a crushing defeat is a complicated story that will be discussed and dissected in the forthcoming chapters.

Despite all the preliminary plans and preconceived ideas that guided the army's strategy before the battle, once it was learned that the 7th was badly defeated and Custer's entire command destroyed, as stated elsewhere, the proverbial wagons quickly circled around everyone who had anything to do with the campaign. History may be the polemic of the victor, but in this battle, the vanquished would also insist on having their say. Once the news of the massacre was headlined in all the major newspapers, a paroxysm of accusations, alibis, and rationalizations regarding the campaign erupted, and this would include the "true meaning" of Terry's orders, resulting in a melee of finger-pointing and recriminations about who was responsible. The blame game was off and running at full speed. Tongues wagged. but those silenced on the battlefield were unable to mount a defense to the charges leveled at Custer's command. More than a few animadversions were colored by blatant inconsistency and the self-preserving instinct to protect one's reputation at all costs.

Custer's Last Stand Demystified

In 1892, more than a decade and a half after the battle, General James Brisbin shrewdly argued that Terry's orders were deliberately composed in such a way as to restrain Custer's untethered zeal until Gibbon and Terry could arrive on the scene. This view mirrored exactly that of General Gibbon in his report to the Secretary of War, which stated that Terry had strongly cautioned about marching too rapidly and that Custer, by not objecting, gave at least the pretense of understanding this. Captain Benteen went as far as to say that had Custer carried out the orders he received from General Terry, "the commands would have formed at a junction exactly at the village and have captured the whole outfit of tepees and probably any quantity of squaws, papooses, but Custer disobeyed his orders from the fact of not wanting any other command or body to have a finger in the pie and thereby lost his life."

As far as Brisbin was concerned, Custer was the antithesis of Terry's oft-expressed motto, "Zeal without discretion only does harm." Brisbin joined with Gibbon in censuring Custer and punctuated his scathing attack on Custer's impulsiveness in a letter to General Edward Godfrey dated January 1, 1892, laced with language loaded with dynamite, saying, "If not in letter than in spirit, and I think and have ever thought, in letter as well as spirit..." Then, writing later on that, "If this doesn't constitute disobedience," he would indignantly and self-righteously exclaim, "I do not see how it is possible for the charge of disobedience of orders to be held against any man, under any circumstances, when away from his superior."

It is a powerful indictment, but one also fraught with an agenda of personal resentment. Just three days after the battle, on June 28, 1876, when memories were far more vivid, it was this same James Brisbin who, in an account published in the *New York Herald*, forthrightly stated that General Terry had said Custer "would strike the blow and that General Gibbon and his men received that information without a murmur." This begs the question: if it was believed that Custer was too rash, impulsive, and headstrong to follow orders, then why did General Terry, who marched behind Gibbon, not take the lead instead of delegating the central thrust of the mission to Custer?

Terry, however, understood his limitations as a field general and recognized Custer's greater experience fighting Indians. Still, critics argued that if Terry had accompanied Custer, the command would have held more closely together when cooperation was most needed. But it was not in Terry's grain to act in such a unilateral manner, something that Lieutenant Godfrey and other leading officers of the 7th carped about. In his diary, Godfrey lamented that there must be something wrong with General Terry—that he could not hold control of Cavalry and Infantry without having merely nominal control. Custer, however, had no reservations about his ability to do it all. That is one reason, among others, he rejected any assistance from the 2nd Cavalry. "The 7th," he asserted confidently, "can handle anything it meets." He told Lieutenant John Carland, "This is to be a 7th Cavalry battle, and I want all the glory for the 7th Cavalry there is in it."

No doubt there was a downside to Custer's virtues as a dashing and aggressive cavalryman. His combativeness courted both opportunity and danger. But Terry was willing to risk that because of Custer's natural acumen for soldiering, his past military successes, and a decade of experience on the Great Plains. While Terry was a highly respected officer, he was more of an armchair General, more comfortable in administrative matters, planning operations, and charting maps for lines of marches than he was in leading forces to engage with armed nomadic tribes that he knew little about and had even less experience fighting.

Moreover, anyone familiar with Indian warfare would have realized that Indians could never be caught by the slow-moving infantry that Terry and Gibbon commanded. With their speed and mobility, only the cavalry could corner a fleeing Indian village. No acutely discerning military intelligence or a conspiratorial turn of mind is necessary to realize that Custer's instructions were worded in a way, though perhaps not consciously on Terry's part, that would make Custer responsible for allowing the Indians to escape or being reckless in attacking them. It was a 'damned if you do' or 'damned if you don't' scenario. Anyone who knew anything about Custer, and Terry knew a great deal, understood that he would rather be damned doing than condemned for not doing.

Inevitably, this refocused everyone's attention on Terry's oft-analyzed epistle, the logomachy of which was stretched every which way, including inside out, by a slew of semantical sleuths belaboring over every syllable. Their adjudications varied immensely depending on their inclination to be either condemnatory or exculpatory. First Lieutenant Edward Godfrey, to cite one example, took exception to those who dogmatically stated that Terry's language of wishes and desires had the full force of a direct order. Godfrey interpreted Terry's missive in the same vein as General Fry, "Orders given by a commander, in which he utilizes 'wishes' and 'desires' are not equivalent to positive commands in all circumstances. When a commanding officer furnishes written orders, the words "command" and "direct" or, for that matter, any use of the imperative should leave no ambiguity about his intentions. That Terry does not exercise these word choices leads one to think that vagueness was purposeful and intentional."

The closing paragraph in Terry's orders refers to sending a scout to Gibbon's column to relay intelligence obtained on the upper part of Tullock's Creek. That would mark the second occasion of what some saw as Custer's putative disobedience. On June 16, Gibbon's scouts noticed large columns of smoke near Tullock's Fork, which indicated hostile tribes in the area. Terry had specified in his written orders that he desired Tullock's Creek to be scouted, and not a few battle students have claimed that this order was not formally adhered to since Custer's Chief of Scouts, Charles Varnum, had said that he never heard anyone mention Tullock's Creek.

Yet, we know Custer had this area scouted because when alerted that smoke was seen coming from the direction of the Tullock, Custer dismissed the report, stating that he had scouts on that side and if Indians had been present in that location, the scouts would have reported it to him. As it turned out, the "smoke" seen from Tullock's Creek was more likely fleecy clouds on the horizon, an illusion created by heat rising from the hills on that side. Just one of the many bewitching mirages emerging from the dry air of the haunting and mystical West.

On the other hand, there is no denying that, after passing the divide, a scout did not depart to General Terry to inform him about Tullock's Creek since the instructions were to advise Terry of his findings about whether Indians were present. It is true that earlier on June 24th, Custer released George Herendeen, who would be paid a handsome $200 for risking his neck, to ride to Terry but was dissuaded by Herendeen himself. Mitch Boyer joined with Herendeen in advising Custer that there was a better departure point for the scout up ahead, to which Custer replied, "All right. I could wait."

Yet, when Herendeen, on the morning of June 25th, opined to Custer about the necessity of immediately departing for Terry, Custer chose not to reply. Although Custer had, in effect, already released Herendeen from the command, and the scout was free to leave at any time, some battle students have pointed to this as evidence of Custer leaving Terry out of the equation precluding Gibbon from arriving on the battlefield so that he alone could win all the plaudits of victory. Although Herendeen agreed that there was no longer any need to scout Tullock's Fork, the scout could not help but sense that "Custer wished to fight the Indians with the 7th alone and that on the trail, he was making every effort to do so." Be that as it may, it was no longer feasible to investigate the Upper Tullock since the location of the Indian village had already been determined to be in the Little Big Horn Valley, and this was a cold, hard fact, whatever visions of glory might have been marinating in Custer's mind.

While divining the motivations of others is always problematic, Herendeen's observation about Custer reserving all the praise for the 7th is perhaps accurate, but it is certainly not the whole story. The notion that Custer made his night march on June 24 to strike the village before Gibbon could arrive to participate—the usual argument for Custer's undue hastiness—falls flat. It omits the fact that on two separate occasions, one of which he called the officers together on the night of June 24, where he reiterated his previous statement that he planned to attack the morning of June 26, and that June 25 would be a day of reconnaissance, demonstrating that Custer fully intended to approach the attack methodically and without undue haste.

But fast-moving and unforeseen events quickly overtook rapidly devised plans and long-held intentions, not only leaving Herendeen to his own devices but compelling Custer to attack the village a day earlier than he planned. As Custer reached the divide between the Rosebud and the Little Big Horn valleys, he was informed that the regiment had been discovered, and the Indians were running. That changed everything, including conducting a reconnaissance and sending Herendeen to Terry. There were now more important things to worry about. The reported discovery of his regiment would precipitate a scattering of the tribes, likely dooming the campaign to an ignominious failure.

As for Herendeen, if Custer even gave him a passing thought amid the clutter of rapidly unfolding events, how would sending him now to deliver intelligence about Tullock's Creek serve Terry? Presumably, having been alerted to the soldiers' presence, the Indians here were running. Additionally, they had a clear view of the valley of Tullock's Creek from the Crow's Nest. Gerard said,"…we could see all over that part of the country, and as no trail led that way, that there were no Indians in that part of the country." Indeed, Lieutenant Winfield Edgerly, reflecting on the Herendeen incident, stated that "he didn't believe any good officer, under those circumstances, would have obeyed that order at that time regarding the scouting of Tullock's Creek." And even if Herendeen had left on his mission on the morning of June 25th, it would have made not a whit of difference to General Terry, nor would it have resulted in General Terry arriving at his objective—the junction of the Little and Big Horn Rivers any earlier than he actually did. By the time Herendeen could have reached Terry, Custer's fight on June 25th would have been effectively over."

Another key point to remember is that Herendeen was a civilian and not subject to Custer's command. Thus, he had no obligations other than to himself. Custer had already granted Herendeen permission to leave on June 24th, so he did not need to repeat it on the morning of June 25. The bottom line is that Herendeen, on his own volition, chose not to go. Custer did not request him to stay.

Nevertheless, there is no denying that a sense of urgency was rapidly accelerating throughout the command. Custer once wrote, quoting Cervantes, that a "Faint heart never won fair lady." The famous line may have subtly served as a military maxim for the ambitious cavalier, for neither did a faint heart ever pursue and overtake a hostile Indian village. The unquenchable fire that burned within Custer never dwindled—the winds of war drove him. There never seemed to be a time when doubt or inhibition prevailed on whatever battlefield he found himself: "I was always fortunate," he whimsically recalled, "since the hour of my birth and always would be—my guardian angel has always clung to my side since the day I left the cradle." One cannot help but note the wistful optimism of the man, a faith touched by an almost fiery sense of mission. That a soldier who so often brazenly courted death and yet was capable of such breezy and self-assured sentiments underscores Custer's invincible optimism. But it might also reflect a nature that would not permit or even entertain how cruel and unpredictable the vicissitudes of fortune can be, as well as someone who could not be restrained when victory was within striking distance.

Throughout his rocky yet colorful military career, Custer was known to seize the day with a surge of overconfidence, if not imperious certainty, at every opportunity that came his way. Often disregarding the odds, he was driven to gallop full stride onto the field of strife wherever it beckoned. Such was the wellspring of his fame and celebrity. Why should things be different now as he stood on the promontory of what promised to be a great victory? While still mindful of Terry's orders, he would duly deploy Reno and Benteen, as well as the pack train contingent—seven companies in all—to attack the encampment from the south to be in line with Terry's wishes. But even where danger loomed largest, he never lost his relish for rolling the dice in pursuit of victory. After all, Custer possessed that one quality that Napoleon admired in his Cavalry officers: *"L'audace, l'audace, toujours de l'audace!"* (Audacity, audacity; always audacity).

Driven by the fates, this left Custer wide open to charges by his critics, ever poised to strike at his Achilles' heel, that his one overpowering desire, overshadowing all other considerations, was a rendezvous with glory. If things went badly, as they overwhelmingly did at Little Big Horn, then parsing the words of Terry's written order in his favor would not redeem him. Instead, his critics' charges of insubordination and recklessness would be forever affixed to his restless ghost, as gnawing in death as it was in life. Custer, mounted on his steed, would forever be seen through the prism of faint light and flickering shadows from a long-ago battle, on a faraway frontier, that refuses to release its icy grip upon our collective imaginations.

Chapter 3: Should Custer Have Divided the Regiment?

Heading to the divide on the morning of June 25, the 7th Horse had no way of knowing they had taken their first fateful step on the road to Armageddon. The large, abandoned campsites they passed as they followed the lodgepole ruts were telltale signs that they were on the Indians' trail. Despite these portending signs, the 7th gave not the least indication of uneasiness or apprehension—just the opposite. The men joked and played practical jokes on one another. According to the Indian scout William Jackson, the scarred trail was three hundred yards wide, indicating some fifteen hundred lodges. While this may have been an exaggeration, it also visually and palpably conveyed to the outfit that an extraordinarily large number of warriors lay ahead.

Riding in fine fettle, the mounted soldiers must have welcomed the crisp morning mountain air, knowing that as the sun rose, the heat would become oppressive, making action in the high country challenging. Trampling over fresh pony tracks and overgrazed grass, they gazed into the distance where the burgeoning gray horizon seemed expansive enough to encompass the whole earth. Ahead, rising from the plains were ascending bluffs whose heights screened the sought-after but still-hidden village quietly and inconspicuously nestled deep in the bosom of a verdant valley. In these moments of reverie, Custer once remarked that the monotony on the Plains changed upon seeing the mountains, leaving one in speechless veneration.

Custer was in good spirits, having recently written to Libbie, "I feel hopeful of accomplishing great things." Lingering doubts faded when the tracks they followed pointed squarely in the direction the Indians were heading. Impatient to pinpoint his quarry, he ordered his scouts to locate the Sioux camp. They were to sacrifice sleep and, if necessary, travel all night to find it. On the evening of the 24th, six Arikara and five Crow scouts, Lieutenant Varnum, Charley Reynolds, and Mitch Boyer, had scrambled to a perch high on the Wolf Mountains, later dubbed the Crow's Nest, where they could see for a considerable distance into the Little Big Horn Valley. But the darkness of the night gravely limited their visibility. Guided by the scarcest starlight, they waited impatiently for dawn's first light, searching for a glimmer of illumination in the verdant valley below.

As the curtain of darkness slowly receded, signs of an encampment began to come into view. Not everyone could distinguish its outline. Lieutenant Varnum, rubbing sleep-deprived, bloodshot eyes, could discern nothing in the pre-dawn shadows. Custer's Chief of Scouts had been in the saddle for twenty-four hours, had ridden seventy miles the previous day, and now exhaustion had seeped into every sinew and bone of his body. But Varnum knew that the eyesight of the Indian scouts who accompanied him was sharper than even a white man aided by binoculars. Soon, the scouts pointed to a wisp of smoke rising from the Montana landscape.

Although no tepees could be detected because the high ridge hid them, black specks populating the landscape indicated that a massive pony herd was grazing on the herbage of the western foothills. Even in the shadowy predawn, the Indian scouts did not doubt that down in the valley was the biggest village they had ever seen. Close by, the always reliable and keenly observant Charley Reynolds nodded affirmatively to what Red Star was describing. That was all Varnum needed to hear. He wasted no time dispatching the Arikara scouts, Red Star and Bull, with a message to inform General Custer of the sighting at the halt where he anxiously waited.

Spurring his mount on, Red Star outdistanced Bull. He arrived at the halt, first meeting the scout Isaiah Dorman, who immediately escorted him to Custer's bivouac, whereupon he encountered Custer's striker, John Burkman, telling him there was important news for the General. This was the news Custer was anxiously waiting for. Leaping upon his favorite mount, an army-issued Morgan named Dandy, Custer dug his spurs into the horse's flanks and rode to the base of the Crow's Nest, followed by a small contingent that included the Arikara scout Bloody Knife and interpreter Fred Gerard. Shortly before 9 a.m. Headquarters time on June 25, Custer's party approached the hill. Varnum met him at the base of the hill, excitedly confirming that an Indian village had been sighted. Scaling the grassy incline, Custer reached the precipice of a crest that overlooked the valley. In that muggy silence of a Montana morning, Custer gazed outward with mounting expectation.

Custer's eyes strained through the shimmering heat that wavered over the deceptive stillness of a boundless landscape, making its observers appear insignificant. Custer prided himself on his acute eyesight, but like Varnum, he could see nothing of an Indian encampment through the haze that shrouded the earth for miles. The Indian scouts, tense, frustrated, and impatient, told Custer what they repeatedly told Varnum: Look for worms moving in the grass. The scouts anxiously moved their fingers in a crawling motion to signify an enormous Indian pony herd grazing along the western foothills, numbering perhaps as many as twenty thousand.

Because of the demand for grass and game that sustained them, a village and herd of such size could not remain in one place very long. Nevertheless, the enormous herd and the traces of smoke from what were presumed to be hundreds of campfires were evidence enough that the Sioux and Cheyenne had not yet left for better forage. The longer the scouts peered into the valley, the more convinced they became that what they saw spanning the valley was a village containing more Indians than any of them had ever seen in one place. Custer, however, remained skeptical. According to the Ree Interpreter, Gerard, Custer insisted there was no hostile village in the valley as he could see no sign of it through the morning haze.

On his return from the Crow's Nest, Custer called for his last Officers' Call shortly after 10 a.m. Headquarters time, when he solemnly informed his troop commanders that signs of the hostile village had been spotted by his scouts in the Little Big Horn Valley roughly fifteen miles ahead. At about the same time, it had also been reported that enemy scouts had spotted his command. This necessitated an immediate 'advance-to-contact,' that is, to strike the village before the Indians became aware of the soldiers' approach, thereby preventing them from having any chance to scatter and escape. Hence,

Custer's original plan to scout the area and attack on the morning of the 26th would be discarded and hurriedly replaced by an impromptu advance.

Possibly because Custer had not seen the village for himself, doubts over the hostiles' exact whereabouts might have continued to linger in his mind. This uneasiness compelled him to take another good, hard look into the distant Little Big Horn Valley. Varnum insisted that Custer did not re-ascend the Crow's Nest to take a second look, yet Lieutenant De Rudio stated that Custer had, in fact, taken a second look into the valley and had even borrowed his binoculars to do so. Lieutenant Hare confirmed the account of the binoculars.

Custer did not return to the Crow's Nest as there was no time for that; however, he likely stared through De Rudio's glasses more intensely down into the valley from the top of the divide. At this point, Custer was now convinced that his prey would be found in the Little Big Horn Valley about 15 miles away.

After a long trek down the Rosebud and passing one Indian campsite after another, Custer must have felt great satisfaction that he had finally zeroed in on his prey. With the hard part over, his expectations of success soared. Every soldier thought that finding Indians was much harder than fighting them—or so they had come to believe. The time for waiting and planning had come to an end.

This Indian encampment promised to be even bigger than anticipated, large enough perhaps to constitute a small city. Here was Custer's long-awaited chance for redemption. His military career had been in a downward spiral just over a month earlier, due to Custer's criticism and congressional public testimony against members of President Ulysses S. Grant's administration for taking kickbacks in the widespread fraud that had plagued the Bureau of Indian Affairs. The scandal implicated not only his Secretary of War, William W. Belknap, but the President's brother, Orvil Grant. Incorruptible himself, Grant was stone-blind to the faults of family and friends to whom he was loyal to a fault, with some profiting from their positions of public trust and their relationship with the President. The resulting political fallout from Custer's testimony made him persona non grata with the President. He was thus removed from commanding the 1876 campaign and from his beloved 7th Horse and was replaced by the Division of Dakota commander, General Alfred Terry.

Only the timely intervention of Generals Sherman and Sheridan, along with the press, which pilloried Grant's vindictiveness toward Custer, rescued the once 'Boy General' from oblivion, giving him a new lease on life serving under Terry's command. Now, Custer was leading the regiment in the hunt for Sitting Bull. He was not about to squander this second chance with the tentativeness that crippled Reno during his scout—for his entire military career was characterized by one who foreswore the caution that would enervate decisiveness. The old 'Custer's Luck' was still smiling broadly.

With the contretemps of politics behind him, Custer was back in his element. He had found the hostiles, but reservations over reports of the regiment being discovered quickly tempered his hopes of surprising the enemy. Custer had been so careful at concealing his approach that he rebelled against the idea of being detected. But that was before his brother Tom informed him that Sergeant Curtiss' detail, which had backtracked on their trail looking for missing personal items, unexpectedly came across several Indians rifling through a box of hardtack. Shots were fired before the hostiles disappeared. It was surmised (incorrectly, as it turned out) that these warriors would race back to warn the village that the bluecoats had suddenly arrived at the threshold of their encampment. Instead, these Indians did nothing more than tail Custer's troops and circumnavigated the enormous encampment, arriving in the village at the battle's conclusion.

Custer, however, was unaware of their maneuvering away from, not toward, the Indian encampment, assuming that these warriors would be setting off alarms to alert the camp that it was in mortal danger. He knew he must strike immediately if any surprise was to be achieved. He had pursued native hostiles for the better part of a decade with a degree of notable success, leading to his reputation as the Army's most renowned Indian fighter. Custer's old plan to attack the next morning, June 26, after a thorough surveillance of the enemy encampment was now scrapped.

There was no time available to fulfill Terry's orders to scout the upper reaches of Tullock's Creek, now clearly visible from the divide, despite scout Herendeen having been authorized to do so a day earlier. Custer and Herendeen both recognized that the situation had changed, rendering such a scout a waste of time since they could clearly see no Indians there. Believing he was discovered, the only thing to do was attack without delay. The army had been criticized for failing to bring hostiles to battle, but Custer was determined to change that and end the Plains Indian Wars for good.

Custer rode back from the divide to his regiment. His mind was made up, choosing what the Prussian General and military theorist Von Clausewitz called the 'bloody decision.' There would be no negotiation, bartering, or ententes with the enemy. This would be no tentative strike, no feeling-out process, or sniping at the enemy from safe distances, but a blitzkrieg aimed at the heart of the Indian infrastructure, bringing ruin to the hostile tribes by destroying their village, cutting off their food supply, scattering and capturing their pony herds, and crushing all remnants of opposition.

In the brazen manner of frontier warfare, the army's idea of total war often paled compared to the well-known brutality of the Sioux and the Cheyenne on the western Plains. In 1867, General Winfield Scott Hancock ordered Captain Myles Keogh, "without regard to age, sex, or condition, to kill all the Indians you may encounter belonging to the Sioux and Cheyenne except if they belong to bands of friendly Indians." At the Little Big Horn, Custer would give scarcely a thought to the potential size of the village. Knowing that speed was a great equalizer, he could turn a numerical disadvantage into a victory, provided he used those critical quantities of time and space to maximum advantage. Custer understood that boldness, initiative, and establishing a favorable offensive position counted in every war and every battle. The only question was how he would orchestrate this attack. That answer would come soon enough.

As the regiment descended to the lower terrain on this blistering, cloudless day, they presented an atypical but poignant picture, uniforms dusty, wrinkled, and even torn after nearly forty days in the field. The Hunkpapa, Little Soldier, remembered that the Washichu (whites) wore campaign hats, coats of light blue (likely due to the dust coating them), trousers of a lighter blue with yellow stripes, black boots with flat heels, and spurs with rowels. Lieutenant Garlington, who had joined the 7th a month after Little Big Horn, stated that no two soldiers or officers in the outfit were dressed alike. There were all kinds of hats, and nearly every soldier had reinforced canvas for trousers. One or two of the officers wore black hats, and some wore white, with a small smattering wearing straw hats, and the same could be said of the troopers' apparel, though none sported straw hats. A few others (including Custer) wore buckskin blouses as they rode astride their powerful mounts in military parade arrayed with mathematical precision in columns of four—but here, on rough and broken ground, they asymmetrically pounded the beaten grass of the western slopes.

Like a rose in a grotto, Custer, as usual, stood out. In a letter from Captain Edward Godfrey to Edgar S. Paxson (who would later render a famous painting of the "Last Stand"), Custer was described as wearing a broad-brimmed, whitish-gray hat with a low crown and a fringed white elk buckskin suit, the blouse of which was double-breasted with military buttons. He carried a Remington Sporting rifle with an octagonal barrel, two self-cocking English, white-handled pistols with a ring in the butt for a lanyard, a hunting knife in a beaded sheath, and a canvas cartridge belt.

Despite the heterogeneity in their appearance, each mount was impressively fitted with a McClellan saddle, and each trooper was armed with a model 1873 Springfield carbine. Most carried a six-shot Colt revolver. Sergeant Jack Vickory was given the distinct honor of carrying the regimental colors and, while there was no regimental band to rouse the troops with strains of "Garryowen" or "The Girl I Left Behind Me" (Terry had refused the 7th permission to take the band beyond the Powder River) the guidons of the 7th fluttered proudly whenever they caught a slight breeze. By now, the temperature of the regiment, revving up with anticipation, was running as hot as the weather, and everyone seemed to know that they were in for a big fight. Every trooper bristled with jaunty confidence—"the idea of the 7th being licked," remembered John Burkman, "never entered our minds."

Laden with thoughts of an imminent engagement, anticipation accelerated their pace. The intervals between the columns of four grew longer and wider apart until Custer rode up to Captain Benteen, now leading the advance, and cautioned him to slow down. It was one of the more curious ironies of the campaign, considering that dawdling is what Benteen would later be most criticized for. A gentle admonition but one more notch in the belt of the ongoing, stormy relationship between Custer and Benteen. It has been said that friendships need history—so do enmities. The relationship between Custer and Benteen was like bottled lightning. There are few things more certain, amid all the uncertainties of this battle, than the simmering, combustible emotions between the two. This ongoing friction would, at least unconsciously, affect how the upcoming battle unfolded. Any understanding of what happened at Little Big Horn and why some actions were taken while others were somewhat neglected might well be explained by how the relationship between the regimental commander and one of his chief subordinates fatally played out.

The command did not travel much beyond the divide (the summit ridge between the Rosebud and the Little Big Horn valleys) when it came to a sudden halt near the headwaters of what was then known to the Crows as Ash Creek. Surrounded by hills dotting the rugged landscape, Custer and his adjutant,

Lieutenant William Cooke, spritely trotted their mounts to an isolated spot just out of earshot where the two were seen conversing. Cooke busily scribbled on a piece of paper torn from his notebook. Clearly, something was up. Without knowing the size of the village, whether it ran upstream or downstream, or the enemy's true strength, Custer decided to divide the regiment into several battalions.

That was the same tactic he used at the Washita when he attacked Chief Black Kettle's sleeping village from different points, rolling the enemy up and quickly shattering all resistance. Custer was looking to replicate that success. Riding close to Custer, Lieutenant Edgerly vividly recalled watching the General and Adjutant Cooke make divisions into battalions. As Edgerly would later testify, Major Reno would be assigned Companies A, G, and M, while Captain Benteen would have Companies K, D, and H. One battalion was given to Captain Keogh and another to Captain Yates. Captain McDougall, with one company, was to be the rear guard.

Striking Indian encampments from different points stunned and intimidated the enemy, eviscerating their fighting spirit and throwing them into chaos and panic. He wrestled with the officer assignments, carefully assessing the needs of each battalion. He took Captain French from Benteen's column and assigned him to Major Reno. He took Captain Weir from Reno and placed him in Benteen's battalion. In making these changes, he might have thought French, one of the most capable combat officers in the regiment, would compensate for Reno's inexperience in engaging Plains Indians. Reno would write in his official report that little attention was paid to him, and only when he assumed command of the companies without definitive orders did he move forward.

As for Weir, there had been friction between him and Reno. But Custer, not trusting Benteen and being closer to Weir, knew that bringing the two together might give him a measure of control over the difficult and thorny captain. Dividing a regiment resulted in more exposure, and to compensate for the added risk, Custer certainly would have wanted to strengthen the leadership of each column for reasons of command and control. There was no better way to prevent an Indian village from scattering than to strike from different points to hem them in. Operating in the clear daylight rather than in the semi-darkness of Washita and against an encampment with far greater numbers than Black Kettle's compact village presented its own challenges. Yet Custer remained unvexed by the odds, knowing that a big village's fighting spirit could be just as easily broken as a small one if the same elements of surprise and attack were in place.

As he rode toward the valley, Custer was still operating as a 'reconnaissance in force,' with no definite plan except possibly to attack at the earliest opportunity. Because of the irregular ways of tribal life, intelligence gathering in Indian warfare was often excruciatingly incremental, perplexingly inexact, largely indeterminate, often in flux, and fraught with unreliability. Army scouts did their best to pick up bits and pieces of information and then try to piece it together like a jigsaw puzzle, but that was rarely enough to tell the full story. Custer must have intuitively known that the campaign's success would come down to his regiment, with neither Terry nor Gibbon playing a role, at least at the outset. Although Custer kept his own counsel when plans were discussed on the *Far West*, he likely realized that it was sheer fantasy to expect Terry's infantry to march through the wide-open country, find a mobile and transient

Indian population in the Little Big Horn valley, and then catch them between two forces—Custer from the south and Terry from the north.

Meanwhile, as set down by Adjutant Cooke, Custer's orders directed that Major Reno be detailed three companies to ride alongside his five down Ash Creek and toward the village. Until then, Reno was merely accompanying the regiment, so he was startled to find himself thrust into what might prove to be an active combat situation with the hostiles. At the same time, an irked Benteen was relieved of leading the advance. His three companies were ordered to make a left oblique and proceed to a line of bluffs to the south and west, a maneuver known as a flank march. Eight years earlier, at Washita, Custer had neglected to reconnoiter the surrounding area thoroughly and had been surprised by a thousand or more warriors from several large villages downstream with nearly disastrous consequences. Presumably, Benteen's march and scout for satellite villages would prevent a repeat of such an unwelcome surprise.

More than a few battle students believe that in partitioning his forces, Custer had, in both theory and practice, committed a cardinal sin that directly led to the regiment's catastrophic defeat. Despite the risk, dividing forces in the face of the enemy had become a sound tactic in mounted warfare against Plains Indians, provided that there was a willingness among the regiment's disparate columns to move in concert, with celerity, and in full cooperation among battalions in the theatre of operations. This meant an enveloping attack in conjunction with a frontal one, playing not for safety but fighting to win at every opportunity. Custer had successfully divided his forces during the Civil War and again, against an irregular, mobile enemy on the Plains.

However, war is never a zero-sum game; every action has a reaction, and there is no telling when things might rapidly spin out of control, especially if battalion commanders fail to act decisively and in a timely, collaborative manner. Dividing the command can sometimes weaken the regiment by creating a situation where a mobilized and determined foe can overpower the smaller, separate columns in detail, especially if the time and distance needed to bring the respective commands back together is miscalculated or underestimated. Even physically divided, the attacking force must, in essence, act as one unified entity, collectively committed to coming together and acting in concert. This requires top-down, self-assured leadership with shared goals of timely cooperation between battalions on different parts of the field to shatter whatever cohesion the enemy might possess. A miscue of any of these elements could produce perilous consequences. In war, the line between victory and defeat can be notoriously slender.

Given the tension and animosity between himself and his two battalion commanders, Reno and Benteen, the question arises as to why Custer would split his forces in such a manner. Concerns bubbled up among some of the scouts about whether there were enough soldiers or too many Indians to fight. .

Earlier, Benteen had lamented that Custer, rebuffing the offer of four companies from the 2nd Cavalry, would be something the 7th would come to rue. "I think we will miss them," he said within earshot of his commanding officer. Custer had his reasons. He never got along with Major James Brisbin, the 2nd Cavalry's commander, considering his contentious relationship with Benteen and Reno, adding Major Brisbin's 2nd Cavalry to the 7th Horse would also come with the caveat of replacing Custer as the field commander in place of General Terry, as Brisbin made it clear he would not consent to the 2nd Cavalry's participation in the field without the presence of General Terry. Custer saw subduing the warriors as a job for the 7th alone. Nor was he the only one. This confident attitude was a hallmark of the 7th. As Private Windolph stated in what amounted to another collective boast, "There wasn't a man who didn't believe it was the greatest cavalry outfit in the entire U.S. army."

Nevertheless, skepticism persisted among some of Custer's key personnel. On the Crow's Nest, the half-breed Mitch Boyer, Custer's most reliable scout, plainly told Custer, "If you don't find more Indians in that valley than you ever saw before you can hang me." Half Yellow Face (in a quote often erroneously attributed to Bloody Knife) ominously told Custer they would both be going home by a road they did not know. Lonesome Charley Reynolds, known for his reticence, voiced his apprehensiveness about surviving the upcoming fight. Benteen, not surprisingly, allegedly objected, supposedly asking Custer, "Hadn't we better keep the regiment together, General? If this camp is as big as they say, we'll need every man we have." Custer had had enough, brusquely firing back, "You have your orders, Captain Benteen."

At least that was Private Charles Windolph's story about the Benteen-Custer exchange, and he would relish the retelling of it for a long time, living nearly seventy-five years after the battle. As seductive as his account is, segueing as it does into the popular narrative of Custer's overconfidence and even carelessness, there is reason to doubt the story. Benteen never mentioned it at the Reno Court of Inquiry, an army inquest charged with investigating the conduct and decisions of Major Reno at the battle. Nor did he broach the subject in any of his lengthy post-battle correspondence. Benteen spent the rest of his life justifying his actions and blaming Custer's rashness for the disaster at Little Big Horn, so it is highly unlikely that he would omit a story that put him in such a prescient, favorable light and Custer in such a myopic and shortsighted one.

Wherever the truth lay, Custer was unmoved by the trepidations of others. Indian scouts were, at times, prone to exaggeration and hyperbole. He was also convinced that martial organization, top-down leadership, tactics, and strategy trumped superior numbers when they did not operate under similar command and control. His primary concern remained the Indians escaping. On June 24, Custer became so preoccupied with hostiles eluding his net that he ordered flankers to the right and left to provide a sharp lookout for renegade satellite villages. He was determined that none of the hostiles should leave the main trail without his knowledge.

Meanwhile, five or six troopers would be taken from each of the twelve companies and assigned to Lieutenant Mathey for the protection of the mule train carrying 24,000 rounds of reserve ammunition. Many combat students believe that having a full 20 percent of the command guarding the slow-moving pack train did not make much tactical sense and was not an intelligent use of manpower. In retrospect, it is hard to find fault with that assessment. Custer does appear to be overcompensating for his error at Washita when he almost lost the entire pack train to the enemy. Hindsight, of course, is 20/20, but B Company would have been better employed supporting Custer's attack in the valley rather than guarding a pack train already adequately protected by the additional guard assigned to it, especially since no evidence indicated an enemy was tracking them at their vulnerable rear.

Another highly questionable strategy was ordering Benteen on a scout to the left and then directing the captain to engage anything he encountered while the remainder of the regiment was still fifteen miles away from the valley. True, there were legitimate concerns about Indian villages scattered along the Little Big Horn River. Striking them and driving them to the north into Terry's path would fit with the campaign's overall plan of trapping the Indians. But Terry was also worried about blocking the hostiles' escape to the south.

It was not a question of Benteen's scout to the left; it was the timing of the order that was injudicious. Expecting a cavalry battalion, as Benteen well understood, to advance twelve miles to the river over rough terrain with no trails or watercourses was foolhardy, fraught with danger, and inevitably impeded concerted action. Custer should have delayed Benteen's departure until they were closer to the Little Big Horn, perhaps at the junction of the South Fork, before subjecting horseflesh to such a grueling route cross-country. Keeping the command together would have allowed Custer to maintain control of the entire regiment longer, affording him the necessary latitude to gather more intelligence as things materialized and shortening the all-too-critical supporting distances between the respective columns.

Because of the timing of Benteen's departure from the regiment, there is reason to suspect that something else was at work other than Benteen scouting for Indians to the south. The human element of conflicting personalities can never be underestimated in the decision-making process. Friction between Custer and Benteen continued to grow, and recriminations flew between them like electric sparks. Custer was likely peeved by what he viewed as Benteen's duplicitous and self-serving actions, where, under seemingly false pretenses, the Captain had made the dubious claim that his company had fulfilled the prescribed allotments of ammunition, which, according to Custer's protocol, would earn Benteen the honor of leading the advance. It was clear to Custer that the edgy Captain could not have accounted for his company's ammunition so quickly. This may have prompted Custer to relieve Benteen of the advance by sending him on a scout to the left, which would undoubtedly make Benteen resentful.

This raises the question of whether Benteen's scouting assignment was more punitive than strategic, a way of demonstrating who was in charge, and that Custer was not gulled by his refractory Captain's deceptiveness and guile to claim the honor of the advance. In other words, was Custer showing his battalion commander the cost of artifice and duplicity by ordering him south, knowing there would be nothing to pitch into? Benteen never doubted that Custer was punishing him, sending him on a wild goose chase, since it became obvious that the ground to his left was too uneven and rugged for even Indians to traverse. Suppose Custer truly believed the main Indian encampment was at the South Fork of Ash Creek. Why was a surgeon not sent along with Benteen's three companies, since pitching into Indians invited the likelihood of casualties?

Would it not have been more sensible to send Benteen on a flank march some eight or more miles closer to the village, somewhere around the soon-to-be-discovered Lone Tepee, where he would be far more likely to intercept a fleeing village after an attack? While divining Custer's every motivation is impossible, it must be considered that his order to Benteen was not to continue on his reconnaissance no matter the circumstances, but to immediately return to the regiment on the main trail if no Indians were encountered. This was something that Benteen, to his discredit, later tried to conceal in a dishonorable attempt to make Custer look even more culpable for the shocking denouement at the Little Big Horn.

However, if Custer was engaging in a bit of one-upmanship, he failed to appreciate the law of unintended consequences, a game fraught with peril. Whether personal animus was involved, Benteen's flank march to the left of the regiment's advance down Ash/Reno Creek had graver consequences than Custer could have imagined, even if it were consistent with military tactics. As far as Benteen was concerned, he viewed his mission as 'valley hunting ad infinitum,' that is, looking in places where Indians would never go. Despite Custer's orders to return with dispatch if no hostiles were encountered, that order turned Benteen's contentious temperament into a smoldering volcano.

Lieutenant Charles Roe thought that by sending Benteen to the left, Custer aimed to keep Benteen out of the fight as the rest of the regiment was making a beeline toward the enemy's main encampment. This is important since no less a student of the battle than Walter Camp thought that Benteen's scout precipitated the division of the regiment and, hence, was the genesis of the disaster. Lieutenant Edgerly, however, had a different take, believing that Custer wanted Benteen to drive into the Valley of the Little Big Horn and either attack the Indian encampment from upstream or head them off if they fled from being attacked from the north. How much Benteen's later tardiness resulted from his resentment against his commanding officer has been a subject of unending and fractious debate that continues to divide battle scholars.

Granted, it is altogether possible that Benteen's march to the left was a legitimate order, considering what was then believed about where the Indians were and how far their satellite villages, if any, ran downstream. Additionally, probing south with a detachment was not necessarily Custer's intent but Terry's suggestion and perhaps the cause of Custer dividing his regiment more times than he would otherwise. Walter Camp concluded that Benteen's leftward movement materialized only because of Terry's order to stay to the left of the Indian trail by staying on the Rosebud.

That is feasible, but another possibility must be considered. When Custer was up on the divide, glassing the landscape, he might have mistaken the South Fork of Ash/Reno Creek as part of the valley he planned to attack. Because of the intervening bluffs, even sharp, trained eyes like his could easily conflate the two. Indeed, a small camp near the mouth of the South Fork was spotted from the divide. Might that indicate, like Washita, other unseen camps stretching along the valley? If so, they had to be dealt with to satisfy Terry's orders not to let any hostiles escape to the south. Custer was the designated attack force, the hammer, and the success of this campaign now rested solely on his shoulders. He did not want anything to slip by him. Consumed with the village breaking up, he continually ordered flankers right and left to preclude such an eventuality. The idea, however, of driving the Indians north up the valley and into the waiting embrace of the Terry-Gibbon column was nothing more than a textbook fantasy. This plan might look good on paper, but it had too many imponderables and uncertainties to afford it any practicality. Custer knew better than to have anything to do with such pipe dreams.

However, mistaking the South Fork of Ash Creek as being a part of the Little Big Horn Valley is a distinct possibility. Private Daniel Newell, a reasonable, reliable, and trustworthy witness, claimed he overheard Custer and Captain French talking about a small cluster of some twelve teepees that were seen in the South Fork area of Ash/Reno Creek. Newell picked up on Custer and French, agreeing that it was a small Indian village and that, as a result, victory might come quickly and easily. If there were indeed clusters of Indians along this route, it would be justification enough to address a Washita-like possibility since Indians were known to segregate themselves at such distances even for sanitation purposes alone. It would then make perfect sense for Custer to send Benteen on a flank march just when he did. Benteen's scout would then be consistent with the plan that was now developing, as he would not be pitching into the satellite village alone or in isolation but would be within supporting distance of other battalions following along the main trail.

On the other hand, would it not have been wiser for Custer to have sent a small scouting party ahead of the advance to investigate the South Fork rather than sending Benteen's three companies, which Dr. Charles Kuhlman, studying the battle decades later, aptly called a 'Reconnaissance in Force.' If Custer had dispatched a small scouting party rather than an entire battalion of soldiers, it would not only have gathered the same information, but it would have returned that intelligence more expeditiously than a battle-strong 'Reconnaissance in Force'—all the while maintaining the advantage of keeping the command fully intact. While many knowledgeable battle scholars maintain that three companies were required to defend themselves if Benteen's march encountered a large war party, an Indian scouting detail riding ahead of the advance served Crook's purposes in relatively similar circumstances at the Rosebud fight eight days earlier. Using the same strategy could have accomplished that for Custer.

Leaving aside Benteen's scout and returning to whether dividing the regiment under these circumstances was tactically sound, the spacious, undulating terrain, the enemy's guerrilla tactics, and their inevitable penchant to scatter in the face of a cavalry charge are also important factors in reaching an acceptable conclusion about whether Custer made the right decision. Despite the large number of hostiles, the effective strength of the warriors was deceptive since they were weakened by being too diffused over an area and with no formal mechanism to alert the village of impending threats or possessing limited command and control capabilities in reacting to an enemy assault. Astounded by Custer's brazen attack on the village in broad daylight, reinforced in their belief that Sitting Bull's sun dance insulated them from harm and lulled them into complacency after Crook withdrew from the field a week earlier, had placed the village in a precarious position.

Another explanation for not being more vigilant is that the warriors became convinced their village was too large and strong to be attacked. Low Dog and others later recalled that their unpreparedness and overconfidence were directly proportional to their illusions of invincibility. Despite the discrepancy in numbers, it cannot be emphasized strongly enough what a sizable advantage it was for the 7th to have caught nearly all the warriors and inhabitants by near total surprise. Many warriors were either asleep, away hunting, or bathing in the river during the attack. Many of the village's ponies that warriors would mount in the event of an attack were grazing away from the village, making the encampment particularly susceptible to being rolled up by mounted cavalry charging from different directions.

If each of Custer's battalion commanders resolved to carry out their orders with a minimum of wariness and hesitation, the chances of success were quite high. However, the vagaries of war can be a harsh and punishing mistress. Whatever errors Custer might have made during the campaign, and, again, he made his share, it is still obvious, or will later become obvious, that he was not well served in this battle by his chief subordinates, whether success was ultimately possible or not. This was not because the task Custer set for them was beyond their leadership capabilities (certainly hard to believe of Benteen) but more likely because they did not possess the motivation or resolve to do what needed to be done, even when a series of exigencies arose threatening not only the success of the entire campaign but, as it turned out, the life of the regiment itself.

This factor would be as true for Captain Benteen as it was for Major Reno, whose role as a commander of Companies M, A, and G will be discussed in a later chapter. Possibly, the most grievous error Custer made during the campaign was not in dividing the regiment, which admittedly was risky, but rather in misjudging either the fighting resolve or the loyalty of those he most depended upon in battle. As events turned out, Custer seemed to judge others by what he knew of himself and through the prism of his own abilities. It was the failure to consider those critical individual differences more than the fickleness of fortune or, for that matter, perhaps any other factor that ultimately led to Custer's unexpected and overwhelming defeat.

Chapter 4: Did Custer Have a Plan?

At least one renowned historian of the battle has stated that when Custer approached the valley, there was no plan of attack and that he was "flying by the seat of his pants." This is true, but only in the narrowest sense. Custer conducted a partial reconnaissance before the battle, but he had no clear-cut strategy on how the tribes would be defeated except by those same tactics that had worked against fighting hostiles in the past. When the Indian village was discovered around the pellucid waters of the winding Little Big Horn, Custer may have hoped to replicate his successful campaign at the Washita River by using the same multipronged attack that he used nearly eight years earlier in the Oklahoma territory. But that was not even remotely possible. He not only had fewer troops, but it became apparent early on that this village, although its exact dimensions were unknown, was far larger than Chief Black Kettle's campsite at Washita. The substantial pony herd on the Little Big Horn's western foothills made that more than abundantly clear.

At Washita, Custer had time to conduct a night scout, spending hours reconning the site and positioning his troops to surround it. This would not be the case at the Little Big Horn, where the village was too big to be encircled, and early intelligence indicated that the element of surprise may well have been lost. Seizing the initiative, striking suddenly and from multiple angles, dispersing the pony herd, coalescing the troops, and always erring on the side of aggression were integral to the plan, regardless of the guise it ultimately took.

"It is an immutable law of war," said Lord Woolsey, "that when there is no king to conquer, no capital to seize, no organized army to overthrow, no celebrated strongholds to capture, and no great centers of population to occupy"—then the first objective was to capture whatever the enemy values most. In the case of the Indians, this included their families, food supply, and pony herds. While there were no detailed orders regarding an assault on the village, from the outset, Custer's objective was to target the Indian village and every resource that sustained it, including those contiguous campsites feeding off the nest.

With this objective, the 7th doggedly trotted over the divide through the ever-narrowing valley, edging forward in pursuit of the enemy, destiny, and history itself. The ascending sun bathed the surrounding hills with a faint golden hue, stretching endlessly outward to the western horizon. It had been a long tour of duty, but despite five weeks in the field, Custer was in his element away from the comforts and sanctuary of home and hearth. He had left behind his faithful hounds, Blucher and Tuck, who noisily howled up a storm at being left with the pack train in the care of his personal orderly, John Burkman, who was as loyal to Custer as his canines. The General would tolerate no distractions or diversions. He had the command he wanted, all the scouts he needed, and his Headquarters Staff, which included his brother Tom and his adjutant, the outrageously whiskered Canadian William Cooke, whose bonhomie was a welcome relief to the cloying tension that always preceded a big battle.

Following in their train were two color bearers and a trumpeter named Giovanni Martini, now anglicized as John Martin, an Italian immigrant who was said to have been a drummer boy in Garibaldi's army back in the old country. He now found himself as far away from the Mediterranean coast's tranquil climate and its changing colors as imagination could conceive. Young Martin never dreamed he would be trekking about in the far reaches of the American West, fighting an enemy completely unknown in Europe. In America, just three years later, the bedraggled little trumpeter still became tongue-tied with the rudiments and nuances of English pronunciation, which he gamely wrestled with daily. Some of the men in the regiment found his malaprops a source of amusement, guffawing heartily when they referred to the Italian as "Dry Martini."

Now that Custer had separated from Benteen, with the pack train trudging behind, he had eight companies in all, plus a gallimaufry of experienced scouts, both Indian and white. But not all who rode with the 7th that day were soldiers or scouts. Custer's exuberant nineteen-year-old nephew, Autie Reed, would not hear of being left behind with the pack mules, brushing off John Burkman's admonitions about joining with the lead columns as he jauntily embarked on his first big campaign on the Indian frontier, hungry for adventure. Mark Kellogg, a newspaperman for the *Bismarck Tribune*, rode close to Custer, looking slightly comical on his long-eared mule, trying to negotiate the uneven, dusty trail. Kellogg's rugged determination and his literate pen nicely fit the bill for this campaign. From his halcyon Civil War days, Custer knew the value of publicity, and so did other celebrated officers in the army. General George Crook was known to take several correspondents on a campaign. Now thirty-six years old, Custer yearned for the kind of thrilling coverage that once won him the wreathed laurels from his superiors and the admiration of an adoring American public. He knew nothing fired the imagination of the nation's bustling and growing eastern cities than the specter of howling and bloodthirsty painted savages looking for scalps on the western frontier.

As for Kellogg, he was forty-two years old and a widower, corresponding regularly with his two daughters, and itching to chronicle what might be the last great clash between two opposing civilizations, one racing toward the future and the other in the past, gasping its last, desperate breaths on the vast North American continent. A hard-boiled reporter, handy with a Spencer carbine, Kellogg saw himself as an obituarist for the nomadic tribes, predicting for them a grim fate at the hands of Custer. He said more than he knew when he opined that he longed to be in at the death, which ironically turned out to be his own demise along with the entire Custer wing.

As a precaution, Custer ordered the pack train and its 160 mules to follow off the main trail to cut down on the dust. He still held onto hope that their presence had yet to be discovered. Despite these efforts to remain undetected, the pack train, preceded a good distance by the ammunition mules, methodically pounded the dry earth, inevitably raising dusty clouds that alert Indian eyes might see. Major Marcus Reno and his three companies rode on the left side of the small, bubbling waterway the Crows called Ash Creek, while Custer, with five companies, hugged the right side, trotting along briskly and following the Indian trail. Custer believed the Indians were strung out along the Little Big Horn River, but he was still unsure whether they would be found upstream or downstream from the mouth of Ash Creek (See Map 1).

The ubiquitous scouts were buzzing about the regiment, frenetically moving back and forth in a hive of activity, gathering whatever intelligence they could muster about the concealed enemy camp. Apprehensiveness built into a steamy cauldron of anticipation. From the time Custer got his first look into the valley from the Crow's Nest earlier that morning, he fretted over whether the command had been discovered. But from his lookout atop the White Rocks, a series of elevations lying between Ash Creek and the North Fork of Ash Creek some four miles east of the river (See Map 2), Hairy Moccasin had scouted a portion of the big village and assured Custer, much as Varnum had earlier, that despite reports to the contrary, the Indian camp showed no sign of being aware of the approaching regiment. The village was intact with no scattering, and Custer savored the prospect of devouring it in one unexpected swoop.

This latest intelligence emboldened Custer. For whatever the numerical incongruities, he never wavered in his belief that if these recalcitrant tribes were caught unsuspecting, the 7th would shear them as if they were nothing more than a flock of sheep haplessly grazing in the countryside. As expectations of a confrontation grew, the 7th began to increase its gait, churning up chunks of earth to cover ground and make time. As was his habit, Custer would ride out front for a short distance, studying the surroundings and assessing things for himself. Trotting intermittently, splashing down the creek and toward the verdant valley, the 7th zeroed in on their target. But as hurried as the pace was, this was not a hell-bent-for-leather dash to glory. Running horses in hot weather burns them out. It made no sense to arrive at a village that could flee at any moment with spent and overheated mounts unable to chase their quarry.

As they rode to the huge encampment, no battle plan had been formulated. Nor should there have been one. Lieutenant Edgerly spoke for most of the regiment when he said that the general feeling was that there wouldn't be any hard fighting. Typically, Indians ran from the cavalry. Other than pursuing and cutting off escape routes, planning an attack would be profitless without knowing the location and layout of the village. Even Benteen admitted such plans would be premature until they knew where the Indians were and the extent of their encampment. The most pressing issue remained whether the Indians were aware of the approaching 7th. Everything depended on knowing that answer. If discovered, they would have no choice but to strike immediately before the Indians could scatter.

Custer's first objective was to steamroll over whatever satellite villages were in their path, small vassal-like entities near the main encampment, sweeping everything before them on their way to the main camp, primarily targeting the dozen or so lodges that had been sighted earlier from high on the Crow's Nest. Ready to pounce, the scouts leading the regiment began to accelerate toward this small campsite just under six miles from the sprawling village. When the advance arrived, the Indian scouts were crestfallen to find it abandoned. Campfire embers crackled, and utensils littered the grounds, indicating a precipitous and hasty departure. Some battle historians remain skeptical about the possibility that loitering Indians, either deliberately or by sheer chance, made a narrow escape in this vicinity. Still, there are too many attestations of a hurriedly abandoned satellite village to believe otherwise. At least eight eyewitnesses thought it important enough to mention. Indeed, Benteen's scout to the left, some twelve miles from the village, only makes sense if there were hostiles somewhere along Ash Creek that Custer intended to target.

The regiment discovered the first lone tepee at this campsite, just over three and a three-quarter miles from the river. Two miles farther west, about 1.7 miles from the river, they came across another abandoned, lone tepee (see Map 2). The soldiers found evidence of speedily struck tepees and a quick getaway by the site's former tenants. Many battle students see these two lone tepees as one and the same. But the best evidence, based on eyewitness accounts, is that two separate tepees were encountered before reaching the Little Big Horn River. With this campsite deserted and no Sioux warriors to kill, an Arikara scout, Young Hawk, angrily rode up to the first conical structure and plunged a sharp knife into its thin shell of skin, gutting it as if he were slaughtering a buffalo. Inside, on an elevated scaffold, they were surprised to find a dead Sans Arc warrior (a casualty of the still unknown Rosebud battle eight days earlier) surrounded by the relics of death as well as the concomitant victuals customarily left to sustain the departed in the afterlife. The Indians who fled so quickly may have been a residue of mourners left behind or, possibly, hunters tracking buffalo and antelope.

Custer rapidly followed the scouts into the campsite. If he was disappointed to find no warriors in this smaller camp, he showed his usual impatience that time was being wasted by the scouts lingering around an abandoned site when the great prize, the main Indian encampment itself, lay unmolested in the distance. Pointing to the lone tepee, he ordered it torched, and soon, it burned a bright flickering yellow. Heavy black smoke curled upward into the blue, cloudless sky—an ominous sign of war that could be sighted for miles on this clear, bright day. Exactly why this tepee was set aflame can only be guessed at since Custer had been so preoccupied with concealing the regiment's presence. Was he sending a message that the 7th, like the Israelites of old, had become the chosen instrument of death and destruction for the Canaanites? Or was it a signal to General Terry, somewhere north of the village, that he had arrived at the very den of the hostiles and the long-awaited battle was about to begin? Either is plausible, but the more likely explanation is that torching the tepee was Custer's way of dramatically signaling Captain Frederick Benteen, whom he had yet to hear from, that it was time to follow orders and return to the main trail to join the regiment.

As the tepee burned, Reno's three-company battalion emerged from around the hills bordering the creek. He had been riding parallel to Custer on the left side of Ash Creek, sometimes only fifty yards apart and at other times several hundred yards, depending on the contour of the land. Reno had fallen behind partly because he was traversing over hillier, rougher terrain and because Custer was riding as quickly as he could without exhausting his mounts, bearing down on his quarry, as was his wont, with a vengeance. When Reno caught up, Custer promptly beckoned him over to the north side of the creek. The General had big plans for Reno's battalion, whether the major was ready for them or not.

Chapter 4: Did Custer Have a Plan?

Major Marcus Reno, the 7th Cavalry's second-in-command, had graduated from West Point several years before Custer. He served in the Civil War, and in what would turn out to be one of the great ironies of his military career, considering what occurred at Little Big Horn, he was honorably cited for coolness under fire while covering a retreat at the Battle of Gaines Mill in 1862. The absence of level-headedness is exactly what he would be later crucified for at the Little Big Horn. He was assigned to the 7th on December 26, 1868, after the Washita battle. Thus, he was insulated from the bitterness and controversy that followed the battle's recriminating aftermath. When he first arrived, Custer warmly welcomed him to the outfit as a brother officer. But while relations were initially cordial, if not moderately friendly, they grew frostier and more strained with time. A widower whose sole child was being cared for by relatives, the desolation and loneliness of the frontier seemed to affect Reno emotionally more than most. Stolid, morose, and quarrelsome, he relied increasingly on the bottle to soothe his dark and brooding feelings. This saturnine disposition, along with a lack of aplomb, at least when contrasted to his dashing, death-defying commander, would cloud and complicate their relationship—and never more tragically than on the last day of Custer's life.

Reno's sullenness and sense of isolation were not helped by what he perceived as Custer's lack of support. When Reno discharged the Arikara interpreter, Fred Gerard, Custer promptly reinstated him. After Reno drafted a complaint for insubordination against Captain Thomas Weir, a then-favorite of the General, Custer dismissed the charges. Just days before the Little Big Horn battle, Custer had bitterly upbraided Reno for disobeying General Terry's orders on his scout up the Rosebud Creek. Dustups with others were not uncommon with Reno, and it is not surprising that his relationship with the other battalion commander of the regiment, Captain Frederick Benteen, was hardly a model of mutual respect and forbearance, with Benteen once slapping Reno across the face over a disagreement.

But despite their biting personal differences, Reno and Benteen had at least one thing in common: They both despised Custer more than they hated each other. These seeds of contention did nothing to improve morale among the 7th's high command. The far-reaching and weighty repercussions of these fractious and volatile relationships about who did what and when before and during the battle were destined to be forever debated and wrangled over by their respective supporters and detractors long after the smoke from the battlefield cleared and the dead had been buried.

Custer paused to survey the abandoned Indian camp and all its impedimenta. But any thoughts of a lost opportunity to strike a blow against the hostiles suddenly ended when Fred Gerard, the Indian interpreter, wheeled his horse around at a knoll some yards away, with large gestures, waved his hat in the direction of fleeing mounted warriors, and thundered: "There goes your Indians General, running like devils." Forty or so mounted Indians could be seen bolting from the surrounding environs, racing toward the main village, ostensibly to warn the tribes that soldiers had arrived. Herendeen and Hare, who were with the scouts at the right, almost simultaneously espied in the distance large dust clouds over the village, signifying, or so it was believed, that the main camp had already been alerted of the soldiers' approach and it was poised to make a run for it.

The sudden emergence of these dust clouds was mystifying, for it was hardly possible that those renegade horsemen Gerard spotted could have already warned the village about an approaching enemy. For reasons that to this day cannot be entirely explained, the village horse herders were somehow signaled in advance about the impending danger (although not the encampment at large, which stretched downstream for more than a mile and a half), and they began to move the pony herd, which accounted for the massive dust clouds seen by Custer and his scouts. The village was so large, so segregated, so achingly unprepared for any attack that it would take an inordinate amount of time to disseminate what the horse herders knew into alarms of an impending attack.

Custer did not immediately comprehend all of this and concluded that the village was embarking on a dead run, so he must cut off their escape. The need for concealment had passed, and the time for action had arrived. Following military protocol, Custer turned to his senior battalion commander and ordered Reno to proceed to the vast Sioux and Cheyenne village, which was not visible some three miles away, telling the Major he would support him. Since the divide, Reno seemed unnerved by the unexpected events and his assigned role, believing he was being ordered to do too much with too few. When Lieutenant Cooke, back at the divide, had assigned Reno his complement of companies without including Benteen's battalion, Reno could only watch Benteen heading south. Looking at his three remaining companies, he asked Cooke, "Is that all?" As disconcerting as the absence of Benteen's battalion might have been to Reno, there were compensations.

M Company, one of the strongest in the regiment, would be assigned to his battalion along with another thirty-five scouts. He was also afforded the services of Captain Thomas French, who was not initially in Reno's command but was included in his battalion when Custer reordered the officer assignments. Despite a portly physique and falsetto voice, French was a first-class fighting officer who measured up to a challenge with formidable military bearing: steady, brave, experienced in Indian warfare, and tactically sound. Additionally, Reno should not have felt abandoned since he received continuous assurances that the entire outfit would support him. But whatever redress was made to Reno for the departure of Benteen's battalion, he had no time to weigh it in the balance. Once Custer barked out the order to attack, events unfolded with such astonishing and unpredictable swiftness that it is difficult to describe its sequence with any concrete precision.

Plains Indians had been called the best light cavalry on the North American continent. The horse was at the center of their way of life and was indispensable in hunting and war. So necessary were they for the tribes on the western plains that after vanquishing Black Kettle's warriors, Custer, with brutal efficiency, cold-bloodedly slaughtered more than eight hundred of their ponies. When slitting their throats proved impractical, they were shot down en masse to undermine the Indians' means to make war on the plains. The cacophony of frightful cries echoing from these helpless, dying beasts sounded hauntingly human and only added to the horror of a very bloody, heartrending morning. It was but another gruesome reminder of the pitiless savagery of frontier warfare. Benteen later alluded to this wholesale slaughter of ponies. Never one to miss an opportunity to take another swipe at his favorite target, he caustically quipped that Custer, the so-called famous Indian fighter, never killed anything but horses.

Chapter 4: Did Custer Have a Plan?

Once the dust clouds were seen over the Little Big Horn, it became a race against the clock to reach the village before escape was possible. No one realized that the overwhelming majority of the encampment's inhabitants were either sleeping, cooling themselves in the river, or going about their daily duties, utterly unaware that by only the slightest strand did Damocles' sword hang over their heads. Meanwhile, Custer was still fuming that his Indian scouts stopped to investigate the tepee rather than follow the escaping warriors that Gerard had seen, even threatening to disarm and unhorse them until his interpreter smoothed over the misunderstanding. Once Custer got past these contretemps with the Rees, things started to move simultaneously, and Reno found himself directly in the eye of the maelstrom. Later testifying as to his exact orders, Reno stated, "Lieutenant Cooke came to me and said, General Custer directs you to take as rapid a gait as you think prudent and charge the village, and you will be supported by the whole outfit."

Others heard the same thing with slight variations. Private Edward Davern, Reno's orderly, recollected that Reno was directed to take his three companies and drive everything before him. He was reminded that Benteen would be on his left (returning from his scout to the south) with the exact instructions. Dr. Henry Porter concurred, remembering Adjutant Cooke telling Reno that the Indians were three miles ahead, and he was to charge them, and the whole outfit would support him. Gerard heard the same orders: "The major was to overtake the hostiles and bring them to battle." John Martin, Custer's orderly trumpeter, similarly understood that Reno would go down, cross the river, and attack the village.

Having been mildly assuaged by reassurances that Custer would support him, Reno resolutely moved forward. Although a reluctant warrior, at least when fighting Indians, Reno's initial reservation was entirely apparent. The task may have appeared to him overwhelming with just one battalion of his wing, some 175 men, from which nearly two dozen Indian scouts would be dispatched to capture or disperse the massive pony herd sprawled out on the western foothills. He would be plunging into unknown and unscouted territory against a warrior camp whose size, though unspecified, promised to contain a large number of warriors. The last words Reno ever spoke to Custer were an imploring: "Let us keep together," which Custer affirmatively acknowledged, as Reno later put it, in his jaunty way by lifting his broad-brimmed hat as if to say, "I hear you." But, according to Reno, Custer did not hear or at least did not heed him since he would end up waiting in vain for Custer's immediate support.

At least, that was Reno's story. There is a difference between close or immediate reinforcement and being supported. The dynamic between what Custer intended and what Reno expected and how each acted on those presumptions is one of the many controversies that have vexed and obsessed battle students since news of the disaster at Little Big Horn was first learned. A cavalry battalion can be supported in different ways, including flanking the enemy you seek to overtake. Reno, however, would later feign ignorance that Custer could support his command in other ways besides immediately following Reno directly into the valley. According to General Nelson Miles, in his autobiography, *Personal Recollections*: "Custer's order to Major Reno to move forward on the trail and attack the village, and that he would be supported by the other battalions, was a proper command and did not imply that the supports would follow immediately in his footsteps. An attack by the battalion on his right or on his left, or by both simultaneously, would have been the most effective support he could have had." Regardless, Reno would later allege that there was no agreed-upon plan, save a general assumption that Reno would charge into what was still believed to be a running village and Custer would immediately support him. According to Reno, that was, in substance, the only communication between them.

Yet, according to Sergeant Daniel Kanipe, just before being ordered to charge the village, he witnessed Custer and Reno riding side by side, conversing for two hundred yards, before the Major took leave and headed to the river crossing. We can only speculate on what these two officers discussed with each other. However, it is unlikely they were talking about the upcoming centennial celebration in Philadelphia. No matter what Kanipe concluded from seeing the two officers riding together, Custer's intent to support Reno in the valley can hardly be doubted, notwithstanding Reno's claims that Custer had abandoned him. That this support was not employed as Reno had expected does not mean he would not receive any support at all.

A good deal is known about Custer's mentality and battle tactics. When he ordered Reno's charge, he believed, based on Gerard's sighting of fleeing Indians and the dust clouds in the distance, that the village was breaking up and running. Otherwise, why would he have likely consented to a subordinate leading an attack against an enemy ready to stand and fight? Custer would rarely take a backseat or allow another to rob him of the honor and glory of leading an attack against the enemy, a testament to his willingness to share in its every danger.

If, however, the tribes were escaping, as he first believed, then Custer would indeed, as he had indicated, support Reno by following him into the valley and maneuver his forces according to where and in what direction the enemy was fleeing. The 7th did not have time to flank a running village or strike in another direction. The quickest and most optimum way to arrest their flight was to move directly into the valley and toward the village, with Reno's three companies as the first wave and Custer's five companies following as a second wave—a one–two punch, with the shortest distance between two points being a straight line. But after Reno's three companies left to attack, things abruptly changed as Custer, unbeknownst to Reno, received fresh intelligence that the Indians were not fleeing.

Shifting tides were rocking the boat, making it impossible to navigate a single course of action. A battle is not a static and inert phenomenon but more like a storm at sea, where the suddenness of winds and waves can toss events every which way, back and forth, upside down, with mesmerizing and dizzying swiftness. The chain of events, as best as they can be reconstructed, leading to Custer's change of plans not to follow Reno into the valley, can be summarized as follows:

No sooner was Reno off to the valley, believing, as Custer did, that he was chasing a running enemy, than Lieutenant Varnum came riding back to Custer. After asking Custer where Reno was going, he informed the General of a radically different set of circumstances from the one that resulted in Custer ordering Reno to charge a running village. From a high butte on the south side of Ash/Reno Creek, now known as Varnum's Lookout and just beyond where the river bends, the Lieutenant saw a portion of the village with more Indians than he had ever seen. But it was also evident to Varnum, contrary to what was most recently communicated, that the village, incredibly enough, was still anchored, immobile, and unaware of any threat. This was exactly what Hairy Moccasin had reported earlier. But, this intelligence was discarded as erroneous when Gerard sighted that band of Indians racing back to the main village.

What turned out to be a seesaw chain of events, Varnum's report again changed everything. Agreeing to Varnum's request to join Reno's advance, Custer must have thought that maybe his initial intuition was correct, and the village really was unmindful of any imminent threat. If so, Reno's order to charge a running village had been based on false information. This is another example that underscores the importance of reconnaissance before a battle and how it can liberate a military force from being a prisoner of moment-to-moment events, shifting innuendo, and anecdotal speculation about the enemy's whereabouts and intentions. However, time and circumstance conspired to make Custer repeatedly draw different conclusions at different times in close sequence to each other.

Just after Custer sent Reno, and just before receiving news from Varnum of a still stationary village, Custer had also sent Cooke and Keogh, two of the most trusted officers from his own right wing, to accompany Reno's command, perhaps to make certain he was carrying out his orders to overtake a running village, and for Keogh's squadron to prepare to follow Reno into the valley. Varnum remembered that when he left the main column, Custer's battalion was still moving at a walk as he followed Reno, no doubt to give Benteen time to catch up with the rest of the regiment. Custer understood that he needed all twelve companies to overpower the village, prevent escape, and defeat the warriors.

In the meantime, in his ever-smiling and good-natured way, Cooke informed Reno that he and Keogh were accompanying him to the river crossing But, while sending Cooke was appropriate and well within the duties of an adjutant in the field, Keogh was another matter. Having not heard from Benteen and perhaps projecting a lack of timeliness in Benteen reinforcing Reno from the left, there is a real chance that Custer might well have contemplated another strategy whereby he would send Keogh's C and I Companies to supplement Reno's undermanned battalion even if he was pursuing rather than battling tribal warriors determined to fight it out.

There is also the possibility that Keogh might have been sent to scout the crossing and the terrain in the likely event of his joining Reno. That will never be known for sure. What we do know is that Gerard had already crossed at Ford A. He then galloped back to the crossing to report—or at least attempt to report—that, contrary to everything that was believed, the Indians were not running but coming out in force to engage the battalion. Because of bad blood between the two, Reno had no use for Gerard and pointedly ignored him, but not his warning. Gerard, believing he was making no headway with his commanding officer, then rode across the river to relay the message to Custer, but met up with Cooke, informing him of this latest alarming development.

This blast of intelligence was a lightning bolt in a concatenation of swiftly changing events that once again altered the entire tactical picture. Receiving this news, Cooke and Keogh rode back to inform Custer of a fighting, not an escaping village. But Gerard, now returning to follow up Reno's advance down the valley, was completely mistaken about Indians coming out to fight, which again strikes at the heart of how easily perceptions can be distorted and misinterpreted under the stress of battle.

The valley fight, however, did not commence with Reno's charge down the valley toward the Indian village. It began with the Ree scouts, who went ahead of Reno's battalion, attempting to capture and stampede the Indian ponies grazing on the western foothills, a tactic designed to paralyze the village by stripping them of their best means of mobility. These scouts rode over two miles down the valley, just beyond the Garryowen Bend of the river, near the outskirts of the southern portion of the village, where the Hunkpapas were camped, and began firing. At this extreme southern end of the encampment, a few warriors, seeking to defend the village from this surprise attack, mounted their horses and fired back. As far as these warriors knew or suspected, they were not being attacked by soldiers but rather by their ancient enemies, the Arikara and the Crows, and they had no idea that the U.S. Cavalry was close behind.

During this altercation, a few of the 7th's Arikara (Ree) scouts may have overreacted to how strongly and quickly the village responded to the threat against their horses. They did not fully realize how unprepared the Sioux and Cheyenne were to meet any onslaught. It took time for the warriors in the village to realize what was happening and respond to it. But Gerard had as little understanding of what was occurring as did the Indian scouts, who misinterpreted the first signs of resistance as the enemy coming out in force. This misunderstanding proved to have legs—passed on from the Rees to Gerard to Reno to Cooke and finally back to Custer.

The information that the warriors were not fleeing but making a stand would soon be reinforced by two more messengers, Private Archibald McIlhargey and later Private John Mitchell, whom Reno had sent back to inform Custer what he had been told. For reasons that remain obscure, Custer either did not send a messenger back to Reno with further orders or if he did, Reno never disclosed it. Private Theodore Goldin always claimed that he had carried a written message from Custer to Reno, who had read it and placed it in his pocket without further comment. If true, other than Goldin's assertion that he delivered it, this order disappeared from history without a trace.

The story's veracity, however, has been questioned by some, as Goldin, just seventeen years old, proved an inveterate liar in stretching, distorting, and dissembling some of what he experienced that day. Even Goldin's presence is a mystery since no one knows when he joined Reno's command. As battle scholar Bruce Liddic stated, "If one believes all the tales Goldin told, then he did everything one could possibly do at the Little Big Horn except die with Custer." On the other hand, it would not have been out of character for the tight-lipped Custer to be unduly reticent with his officers about his intentions. That was a trait Custer shared with other notable military commanders such as Robert E. Lee, "Stonewall" Jackson, and the Duke of Wellington. Thus, he decided, upon learning of the warriors' aggressive disposition, not to issue new orders to Reno or inform him that he would not be following him into the valley. Custer, perhaps, expected an experienced military officer of the 7th Cavalry, which Reno was, to comprehend from such signs as seeing Custer's battalion riding high along the eastern bluffs as ample proof that he had changed gears and was maneuvering to 'fix and flank' the enemy. Custer waving his hat from the bluffs to Reno's charging troops in the valley would reinforce that understanding, and therefore, his battalion should act according to the new circumstances. In any event, we can speculate that the news Custer received that the village was standing, and not running away, did not warrant any change in Reno's original orders to advance to contact with his advance guard battalion.

Chapter 4: Did Custer Have a Plan?

After hearing from Varnum, Cooke (via Gerard), McIlhargey, and then Mitchell that the Indians were in Reno's front and on the offense, Custer must have believed that the tribes were either making a rare stand against a cavalry charge or, more plausibly, that it was the usual rear-guard action to allow women, children and other non-combatants time to escape the charging soldiers. Because of conflicting intelligence, Custer could not know the truth without ascending the nearly 100-foot bluffs southeast of the village to obtain a bird's eye view for himself. He had previously sent two Crow scouts, Half Yellow Face and White Swan, up to the bluffs, but misunderstanding Custer, they both diverted to instead accompany Reno's advance into the valley. He then sent Boyer and four Crow scouts, Goes Ahead, White Man Runs Him, Hairy Moccasin, and Curley, who knew the terrain better than anyone, up to the heights to serve as his eyes and ears as well as reliable guides should he decide to follow their ascent and orchestrate a flank attack or an enveloping maneuver as another way of supporting Reno.

Regardless of whether the village was fleeing or standing, it is obvious that Custer still intended Reno's battalion to act as an advance guard to bring the enemy to battle, develop the situation, act as a tool and a means to assess the strength of the enemy, its location and disposition, occupying or distracting them, allowing other attack elements of the regiment to strike either as a follow-up to Reno down the valley or attack stealthily from another angle or vantage point of the battlefield.

While some battle students positively recoil from the idea of Reno's battalion acting as an advance guard, it is hard to deny that not only did Reno's column exhibit all those classical earmarks but, on two separate occasions, Reno himself called his battalion at the Little Big Horn, an advance guard.

After ordering the charge, Custer followed Reno down Ash Creek at a distance. This was to create separation between the two since another function of an advance guard was to conceal the main column (Custer's) from the enemy. About a half mile from the river crossing, Custer ceased to follow Reno into the valley but, instead, made a sudden ninety-degree turn, ascending the high ground on the east side of the river, likely intending to ascertain and, from the heights, sort out contradictory intelligence about the Indians' actions in the valley.

Sergeant Kanipe, however, said Custer's unexpected detour was precipitated by the sighting of some sixty to seventy-five Indians on the bluffs. But this is highly unlikely and, even if true, this was hardly sufficient reason for Custer to ascend the bluffs in pursuit. Custer's foremost objective was to destroy the village, not to chase small bands of Indians to kingdom come. Shoed cavalry mounts, which easily carried more weight than Indian ponies, were not about to run down skillful, unencumbered horsemen of the plains. Custer himself noted that only a racehorse could run down Indian horsemen. It is worth noting that no one other than Kanipe saw these Indians on the bluffs, including Custer's eagle-eyed scouts. The scouts, who had preceded Custer up to the bluffs, would have signaled him if warriors were present.

Despite numerous accounts from interviewers and archivists like Walter Camp and others who laboriously pieced every hanging thread of the battle together over the succeeding decades, only Kanipe referred to these Indians on the bluff. Kanipe has always been a tough witness to vet since he was reasonably intelligent and seemingly sincere, and he got some important things right about this battle. Walter Camp also interviewed the Crow scout Curley, who shed some light on this story. Walter Camp had concluded that Custer went up the bluffs across the North Fork of Ash Creek. But Camp simply misunderstood Curley, who meant that he (Curley and the Crow scouts—not Custer) had gone up the bluffs by that route. Custer did not ascend the bluffs until within about half a mile of the river (see Map 3). There is more than a reasonable chance that Curley and his fellow scouts may well have been the Indians that Kanipe had seen on the bluffs. Camp's mistaken notion that Custer had ascended the bluffs at the same place his Crow scouts did also resulted in the unwarranted belief that he had watered his horses at the North Fork, which is nearly dry at that time of year and could have never slaked well over two hundred mounts that were with Custer's command. Whatever the truth, Custer likely thought that the flurry of reports he was receiving about the Indians coming out to confront Reno's battalion could not be ignored. This is why Custer charged up the bluffs with all five companies instead of sending a small reconnaissance force to report back to him, as he was now convinced that the village, with the warriors coming out to fight a rear guard action, had no time to get away and scatter.

By ascending the eastern bluffs, Custer could, once and for all, definitively assess the size of the village and determine whether the tribes were acting offensively or defensively. Depending on what he saw, he could either directly support Reno by following him into the valley or move around the southern end to strike the village from the north. Custer almost certainly believed that Reno would see his command on the bluffs, indeed making a visible effort to be seen by waving his hat at Reno's command in the valley, expecting Reno to conclude that he was on the bluffs because he was maneuvering his battalion to engage in a 'fix and flank'. From that same elevation, Custer could also see the dust from Benteen's battalion, indicating that he was back on the main trail and would soon reinforce Reno's forces.

If Custer had followed Reno into the valley, the battle would likely have ended in a stalemate similar to Crook's fight at the Rosebud or perhaps a limited victory due to the hostiles being caught completely by surprise. Ultimately, he would have brought all twelve companies into the valley, where they would have had the mutual support of all the battalions and supplied with all the ammunition they needed. Instead, five companies of the 7th ended up on the east side of the river, segregated and unsupported by the remainder of the regiment, thinly spread over a large expanse of relatively poor defensive ground where they faced a determined enemy that significantly outnumbered them. Yet, the victory that Custer may have achieved by following Reno into the valley may well have been a tarnished and incomplete one since many of the Indians facing a one-sided, frontal attack would have left wide open escape routes for the village to scatter to the north and west while the warriors fought a rear guard holding action, the very thing Custer was trying to avoid.

On the morning of June 25, Custer had some plan in mind, flexible and unexpressed, based on what he knew then. His plan was necessarily limited and constricting because of the early prevailing belief that the enemy had discovered the regiment. This precluded a proper and timely reconnaissance for fear that their foes, now aware of their presence, were already scattering and escaping. From then on, Custer's intelligence about the disposition of the Indian tribes and the actions of his own disparate columns in the regiment were always a beat behind, playing catch-up without ever catching up, preventing a consistent, orchestrated, and synchronized response against the village.

Like many 'Hussar'-type cavalry officers, delineating detailed plans to his subordinates was not how Custer typically operated, frustrating as that might have been to some officers under his command. Libbie Custer, who knew her husband better than anyone, once described his mental processes on the precipice of battle: "The General planned every military action with so much secrecy that we were left to divine what certain preliminary movements meant."

While it is true that Custer's taciturnity left much unsaid, there were likely enough indicators for seasoned commanders like Benteen and Reno, especially when their commanding officer had been observed on the bluffs riding north to the other end of the village rather than into the valley, to discern enough of what Custer intended that they understood more than they were later willing to admit. But, because of the circumstances that the 7th found itself operating under, as well as the unfortunate and contentious relationship between the lead battalion commanders (Reno and Benteen) and their commanding officer, neither Benteen nor Reno was willing to exercise independent command beyond their perimeter of engagement without explicit orders. The upshot was that Custer's subordinate battalion commanders interpreted their orders and the circumstances they found themselves in, according to their own lights, rather than adhering to the expectations of their commanding officer and the regiment as a whole. This largely precluded the timely reunification of the regiment, which was necessary for military success that day, and explains, perhaps more than any other reason, why the 7th failed so miserably and completely.

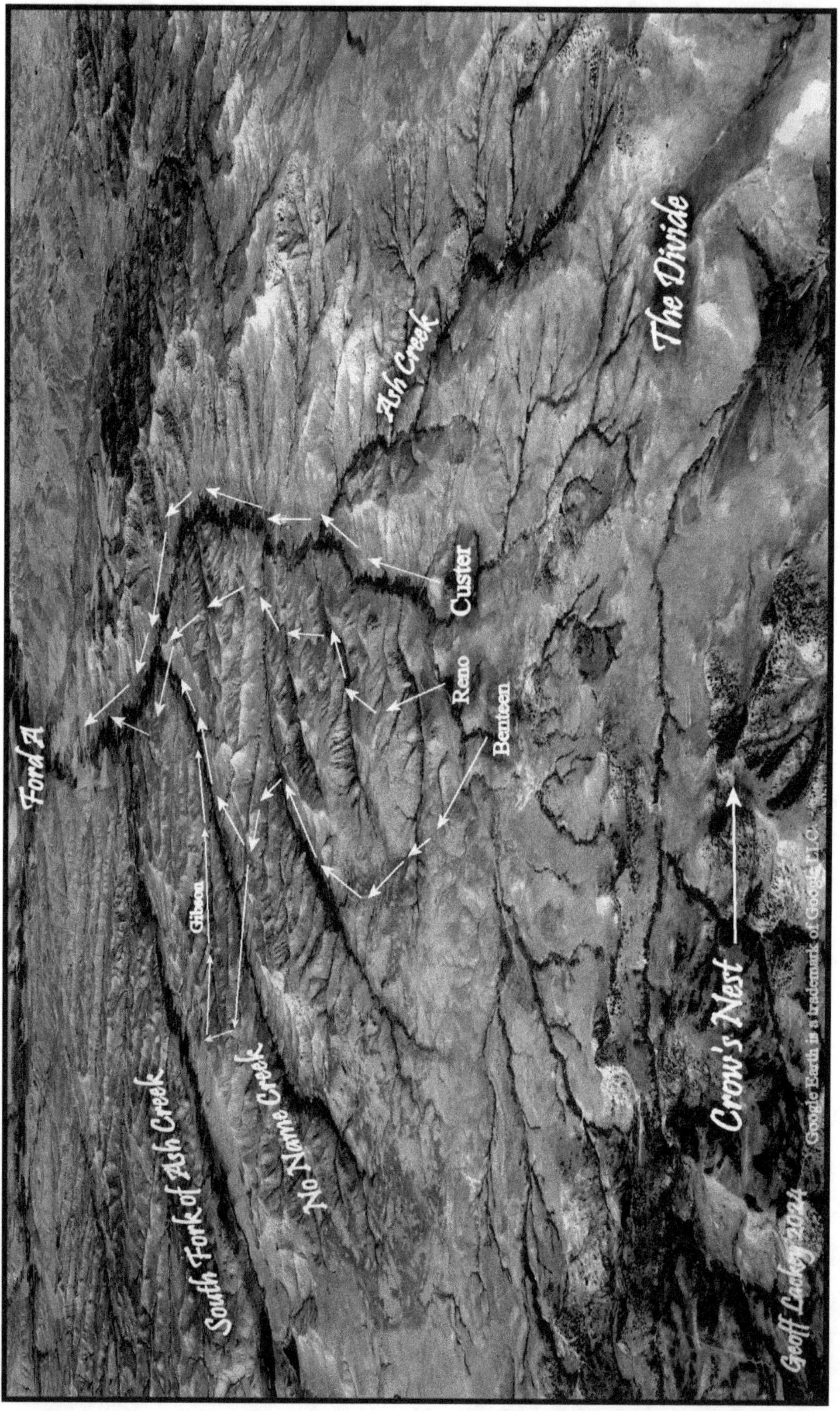

Map 1. The attack battalions of the 7th Cavalry advance toward the Little Big Horn Valley along Ash/Reno Creek after crossing the Divide.

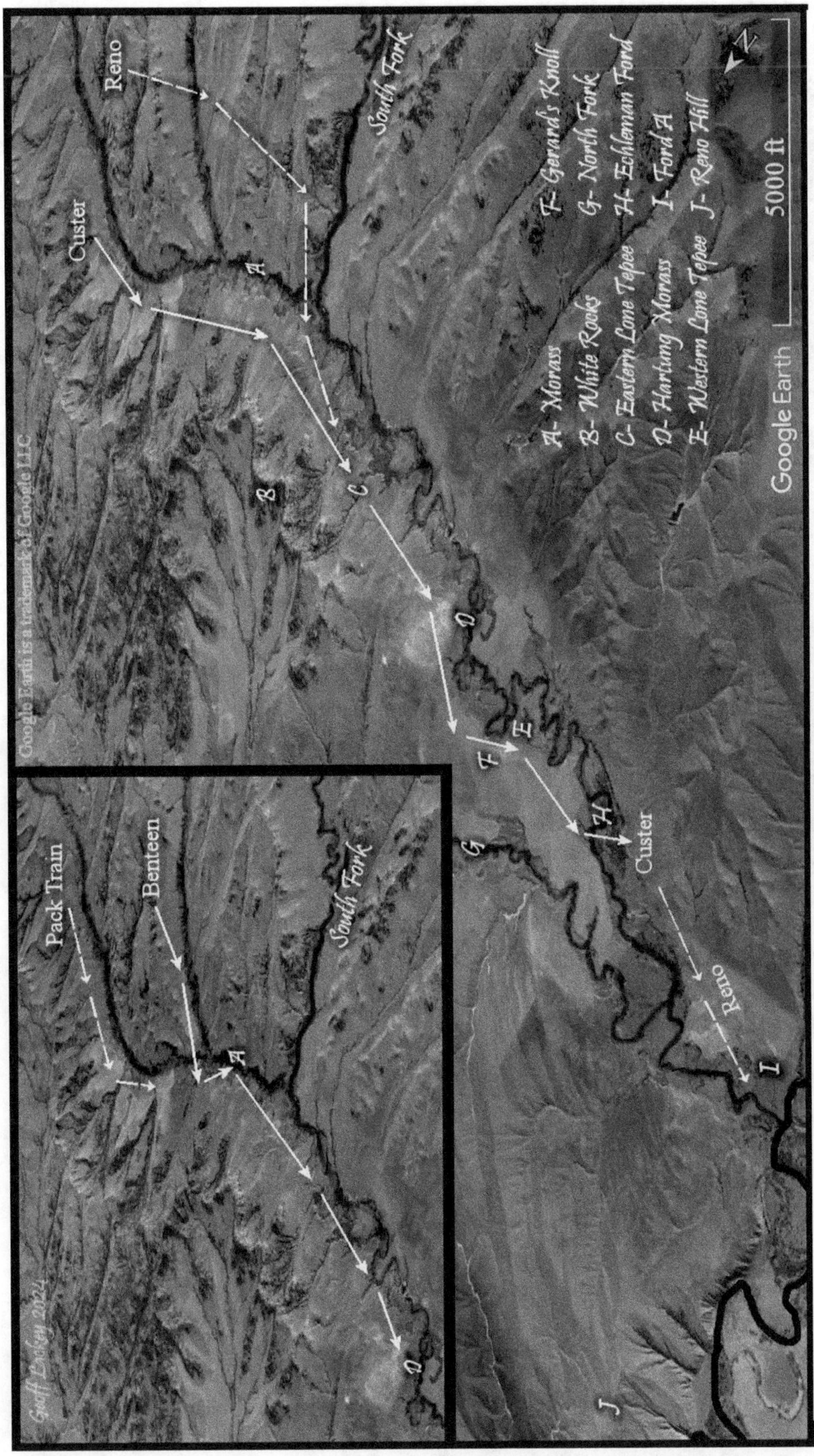

Map 2. The 7th Cavalry approaches the Little Big Horn Valley as they move along Ash/Reno Creek.

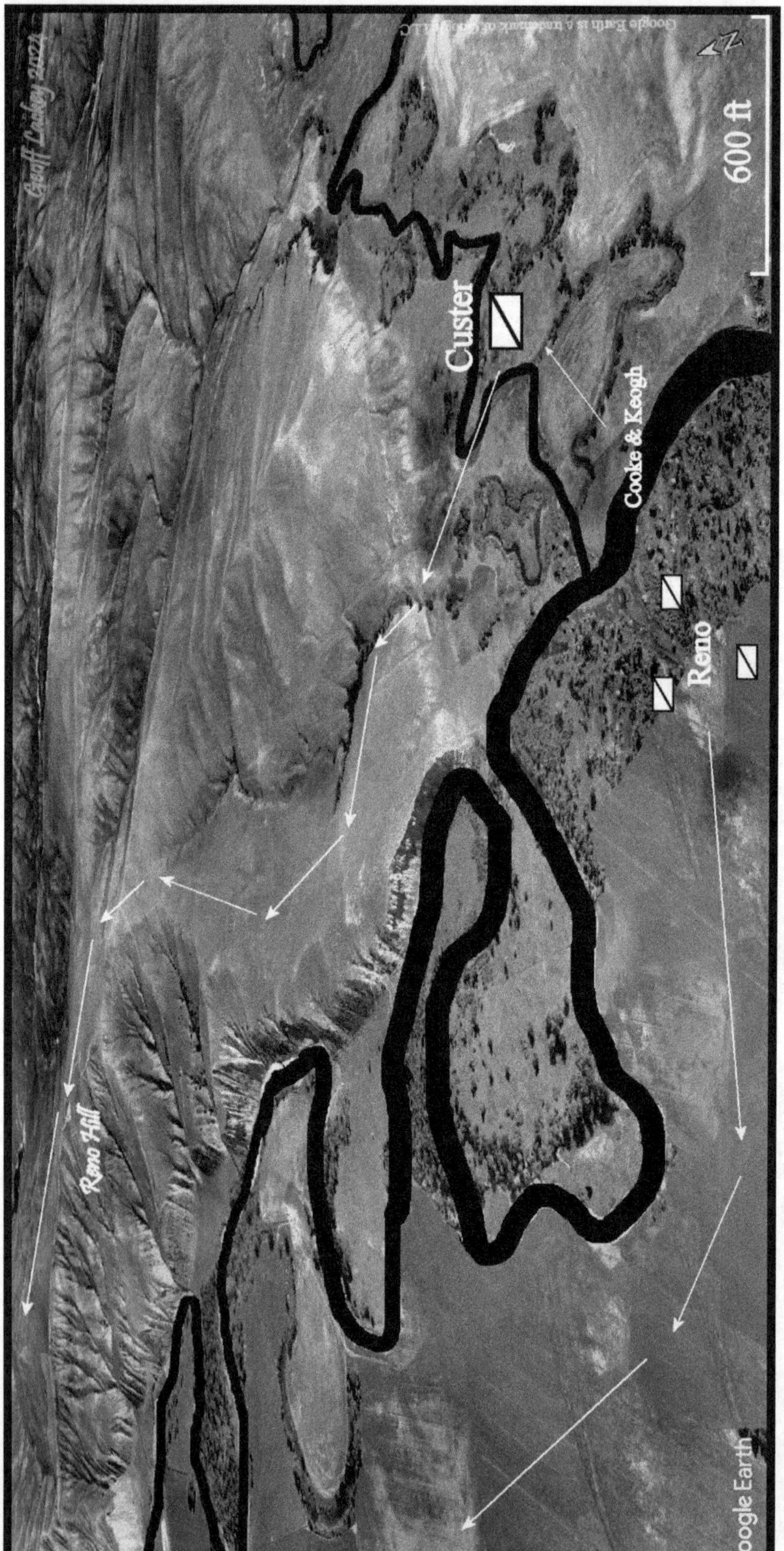

Map 3. Major Reno's battalion enters the Little Big Horn Valley while Custer's wing ascends the bluffs.

Chapter 5: Did Benteen Dawdle?

Captain Frederick Benteen was in a foul mood. Just minutes earlier, his company was leading the 7th's advance. Now, along with a battalion of approximately 117 troopers, he had been ordered off to the left to look for hostiles in places where they almost certainly were not. He watched as a portion of the Gray Horse troop, i.e., E Company, nearly two miles to the north, galloped away toward what the 7th expected to be a large encampment of the Sioux and Cheyenne nations. This might be the last great battle of the West, and it would soon be underway. Yet Benteen had been assigned to search for phantoms upstream instead of sharing in the regiment's dangers and glory. He seethed at his assignment and loathed his commanding officer, George Armstrong Custer.

A native Virginian, life had not been kind to Benteen and his wife, to whom he was utterly devoted. Four of their five children had succumbed to spinal meningitis before their first birthday, a disease which they had tragically inherited from him. This intense sorrow undoubtedly aggravated Benteen's already surly and sometimes vindictive disposition. He had defied the Confederacy and his father to fight for the Union in the American Civil War, in which he fought in numerous engagements and with distinction. Stouthearted and a highly capable cavalry officer, Benteen's sharp tongue cut like a saber.

Now, astride his favorite mount Dick, his charged emotions were evident as he rode with imperious indignation in a southwestern direction into what he considered a preposterous pursuit of hostiles. Riding over rough and hilly ground, Benteen's H Company held the advance, Captain Weir's D Company was second in the column, and Lieutenant Godfrey's K Company brought up the rear. To avoid fatiguing the horses of the entire battalion, Benteen, following Custer's directions, dispatched Lieutenant Francis Gibson, second in command of Company H, and six troopers to ride nearly a quarter mile ahead of Benteen, who rode some one hundred yards ahead of his battalion. It made no sense for the whole battalion to exhaust their mounts riding up and down hills when a small detail could do the same job more effectively. The uneven terrain was challenging, but Gibson dutifully rode to the top of each succeeding bluff that Custer had pointed out to Benteen earlier to see what lay in the valleys between the bluffs. Gibson scanned the open landscape with field glasses, and each time he surveyed the terrain, he saw no Indians

Twice, Custer had sent messages to Benteen, first by the regimental Chief Trumpeter Henry Voss and then some fifteen minutes later, by Sergeant Major William Sharrow. Each carried the same orders: to move to the next valley if no hostiles were seen. The objective was to block any attempt by the Indians to escape to the south and to avoid surprise from possible hostile camps that might be located up the valley. Blocking this escape route was critical since to the west lay Crow country, sworn and bitter enemies of the Sioux, and General Terry and Gibbon were up north. Running south would be the Indians' best hope to elude this wide-cast net, and Custer was determined to cut off that avenue and drive any hostile camps back toward where the main village was thought to be. Wending from one valley to the next, Gibson signaled from every ridgeline that the coast was clear of any Indians, a fact that did not surprise Benteen in the least. After riding several miles from where they left the main trail, Gibson finally reached a point where he glimpsed what he believed was the Little Big Horn Valley. Again, no sign of anything ominous—no camps, no hostiles, no critters. Scanning along the Little Big Horn Valley, Gibson noted what he saw:

> Up the valley I could see a long distance, but in the direction of the village was only a short one, owing to the turn in the valley and the broken character of the country. I hurried back to Benteen and told him it was no use going any further in that direction. Therefore, in compliance with General Custer's orders, in case no Indians were seen in the valley, we were to turn to the shortest route to the trail of the main command and follow it up.

With the mercury rising and the battalion having traversed through two separate valleys and a series of ridges and bluffs, a disgusted Benteen had had enough. With his mounts dusty and thirsty, he ordered a right oblique moving northwest down "No Name Creek" back to Custer's command after deciding, as he would testify some thirty months later at the Reno Court of Inquiry (RCOI), that he was engaged in doing nothing more than "valley hunting ad infinitum"—as likely to find Indians over this rolling and intractable terrain as he would unicorns.

And so, there is the crux of Benteen's dilemma. He despised his commanding officer and his orders but was obligated to submit to both. Historian Charles Kuhlman concluded that there is almost as great a difference of opinion concerning Benteen's scout south of the trail as about anything else connected with the battle. Students of the battle continue to disagree about the object and extent of Benteen's scout, which remains in dispute today. Did Custer order the scout in deference to Terry's suggestion that he should feel constantly to the left to prevent the hostiles from escaping around his left flank? Or was it, as Lieutenant Edgerly maintained at the Reno Court, to stop the hostiles from fleeing south if attacked from the north? Or did Godfrey come closest to the truth when he asserted that it was simply a reconnaissance in force—standard procedure when neither the strength nor disposition of the enemy was known? Perhaps it was all these things combined.

For the next two decades, Benteen insisted that he disobeyed Custer's orders by turning back, which he described as vague and senseless. But this was self-justifying nonsense. Benteen's orders were that if he did not see any Indians in the valley, he was to return to the main Indian trail on Ash/Reno Creek. As for his insistent cavils that his orders were vague or incomplete, Benteen was not an inexperienced, shave-tailed officer who had just fallen off the supply wagon and needed to have his shirt tucked in and his nose wiped at every turn. He was arguably one of the finest officers in the United States

Cavalry, and it was not necessary to furnish such an astute and experienced cavalry officer with a handbook of instructions explaining the nature of military tactics when deep in enemy territory.

Nevertheless, for the remainder of his life, Captain Benteen insisted that the whole exercise was a nonsensical and appalling waste of time. If he had not taken matters into his own hands, as he later informed anyone who would listen, his battalion could have traveled another twenty-five miles without finding the valley or any Indians. The truth, however, is more complicated and troubled. As time passed, Benteen became inconsistent in his actions and the reasons he gave to defend them. Later, Benteen would be of two different minds regarding what he said and thought about the events leading up to and through the battle. At the Reno Court of Inquiry, he testified, "When I received my orders from Custer to separate myself from the command, I had no instructions to unite with Reno or anyone else. There was no plan at all," he insisted. But on July 2, 1876, a week after the battle, Benteen sang from a different hymnal when he wrote his wife, "Custer divided the 7th Cavalry into three battalions, about fifteen miles from an Indian village, the whereabouts of which he did not know exactly. I was ordered…to the left for the purpose of hunting for the valley on the river—Indian camp or anything I could find."

As time passed, the implications of partitioning the regiment became more worrying and controversial. Once the regiment was partitioned, the potential for the respective units to spread too far apart increased dramatically. This separation would force each battalion to rely solely on its available resources and manpower against an enemy force that greatly outnumbered any of Custer's battalions and was, in reality, larger than the entire regiment. However, not everything can be evaluated at face value, and the deeper one probes, the more problematic these questions become.

The Little Big Horn is a mind-bending battle with many moving parts. Seeing how it fits together, including the scout upstream, is convoluted even when subjected to rigorous analysis. The sentiments Benteen expressed to his wife so soon after the battle likely reflected more closely what he understood then than anything he stated afterward. Benteen's recollections, however, are not the only ones that help piece together this tantalizing puzzle. Other accounts offer another perspective and perhaps present a more honest assessment of the nature of Benteen's orders and the importance of the regiments remaining within supporting distance of each other. For example, nine days after the battle, July 4, 1876, Lieutenant Gibson, writing to his wife, disclosed more about this scout than Benteen was willing to admit. The orders were not simply for Benteen to locate the Little Big Horn Valley but for his battalion to return to the command as soon as it was determined that there were no Indians in the direction he was heading.

This part of the Custer order tends to be neglected by battle analysts, and curiously, Gibson was never called upon to testify at the Reno Court despite his key role in the Benteen scout. Because of the early morning haze, Custer may not have realized the distance Benteen needed to travel to reach the Little Big Horn Valley, and from his vantage point up on the Crow's Nest, Custer most likely conflated the valley of the South Fork with that of the Little Big Horn Valley. But aside from that, Gibson's word that the battalion was to return to the main command and Benteen's immediate post-battle report stating the same thing does not harmonize with Benteen's testimony at the Reno Court, where he insisted that there was no order to return to the main trail and that he only did so on his own volition when the inanity of Custer's orders were thoroughly exposed.

This testimony would strike the Court as implausible because Custer's failure to order Benteen to return would deprive himself of three companies while facing an enemy that numerically dwarfed his own strength. Dissatisfied with Benteen's explanation under oath, the Recorder of the Reno Court pressed on: "Was there not anything in Custer's orders indicating that he was not to return to the main trail rather than continuing down to the valley?" It is only then that Benteen becomes more truthful, stating that Custer trusted his judgment to do whatever the situation required: "I don't think General Custer would have told me that. He knew I would have come up." And coming up is exactly what Benteen did when he found the succeeding valleys empty of hostiles—he followed those very orders he adamantly insisted were never issued, indeed, by returning to the main trail. Further corroboration of Benteen's orders is found in a second account given by Gibson to Lieutenant Edward Godfrey on August 8, 1908, attesting to the same narrative:

> Now as to my scout to the left to find the Little Big Horn Valley, I can state definitely that I did find and see it. Benteen sent me with a small detail… and gave me field glasses to take with me. I got some distance in advance. I crossed an insignificant stream (No Name Creek) running through a narrow valley I knew was not the Little Big Horn, so I kept to the high divide on the other side of it and from the top I could plainly see the Little Big Horn Valley (to the south) for a long distance with the aid of glasses … I saw not a living thing in it and I hurried back and reported to Benteen who then altered his course so as to pick up the [main Indian] trail and you know the rest.

Whether Gibson could see into the Little Big Horn Valley from his position on the ridge is irrelevant. The critical question is, what would Custer expect of Benteen if he found no Indians? In that same letter to his wife, Gibson's language is unmistakable in revealing what they understood as Custer's true intent: "We were to hurry and join the command as quickly as possible." Gibson also wrote that the horses were jaded after the scout. That would account for and justify Benteen moving to the main trail with less alacrity than was necessary. It would, except that no report exists of stragglers or mounts from Benteen's battalion falling out and being left behind, which happened to several mounts in Custer's battle wing. Nor does Lieutenant Winfield Edgerly of Benteen's command, who did testify at the Reno Court, mention anything about jaded horses. Gibson may not have been referring to the battalion's mounts in general but only those of the detail that had accompanied him up the steep ridges and were ridden hardest among Benteen's three companies. Undoubtedly, the horses were thirsty, not having watered since the evening before. But this would not account for much lost time, especially since no time was spent watering the mounts until they reached the morass on the main trail.

Determining the time expended and the distance traveled might provide some insight. But even here, memories vary. Godfrey thought the pace was rapid, but he exaggerated the distance, saying the scout was twelve to fifteen miles when it was actually several miles shorter and likely between eight and eleven. In contrast, Private Windolph thought the scout lasted two hours upon their return to the main trail. Whatever the time, distance, and exertion involved, there is no hard evidence that these mounts, other than being a bit wearied and parched, were in any way jaded, much less enervated, indicating that fatigued horses might have been a convenient excuse to use after the battle to explain away criticism of Benteen's purported dawdling.

The open and loaded question is how determined Benteen was to return from his scout or even communicate with his commanding officer. Having returned to the main Indian trail in about an hour and forty-five minutes, Benteen still had not sent a messenger reporting the findings of his scout, seeing his excursion to the left as amounting to nothing more than a spiteful attempt to keep him out of the battle. To be fair, there were those who supported Benteen's contention. Writing to Walter Camp, Lieutenant Roe stated that Custer's object in sending Benteen to the left was nothing less than to keep him out of the fight—a point Camp would never seriously entertain. In time, Custer would learn that there were no Indian camps upstream, not from Benteen but from his scouts (the Rees, who were reconnoitering to the south, and the Crows scouting to the north of the main trail) that had been sent out ahead of his command.

Since Benteen's orders were to send a messenger if Indians were discovered, technically, he disobeyed no orders, since they never encountered any hostiles. But his steadfast reticence is indicative that Benteen appeared more than satisfied to do the bare minimum until turning back and reaching the command. When Benteen reached the main Indian trail, he was barely a mile ahead of the snail-pace-moving pack train and some four miles behind Custer's eight companies. Still, up to this point, Benteen could not be reasonably accused of loitering on his scout because he had struck the trail and arrived ahead of the pack train, sluggish as it was.

Once back on the main Indian trail, Benteen was in no hurry as he took his time to water the horses at a morass, a marshy area about a quarter mile west of the junction at the main trail. The site was essentially a glorified mudhole, just a little short of five miles from the crystal-clear water of the river. The squalor at the so-called oasis was not an exaggeration, as Custer himself once noted that when water is found on the frontier, there are usually mosquitoes, gadflies, and other pests. But Benteen was not inclined to leave the grotesqueries any time soon to head for the nearby pristine water of the Little Big Horn River. Indian encampments were always situated near fresh water, so Benteen could hardly feign ignorance about knowing that the river was where he should be hastily heading. At the officers' call the night before, Custer made it clear that the Indian village was somewhere in the valley, and the whole purpose of the campaign was to destroy the village. That was the target—you go where the enemy is—and the Indians always camped near the banks of a river.

At the morass, the horses slurped up mouthfuls of the clouded liquid, but these waters were often alkaline and, as trooper Chip Creighton remembered, it was enough "to burn the hide off your tongue," but at least "it was wet." Some of the officers began to carp about why Benteen stayed at the morass for such a lengthy duration instead of ordering the battalion out ahead to the clear waters of the nearby river, where the Indian village must be. To no one in particular, one puzzled officer said, "I wonder why the old man [Benteen] is keeping us here for so long." One reason was that the morass was not large enough to accommodate the whole battalion. Each company had to take its turn watering their horses, all the while consuming precious minutes on the march. It would take Custer just ten minutes to water the horses of five companies at the Echleman Ford as opposed to Benteen's three companies watering for twenty-five minutes—an interval that matches closely with the timing of the ensuing events when using the conventional rates of marching for cavalry.

While the animals slaked their thirst, Boston Custer, riding a fresh mount that he picked up from the pack train, breezed past them, giving an enthusiastic shout-out and a cheery salute to Lieutenant Edgerly as he hustled to join his older brothers George and Tom for the upcoming fight. With the battalion mired down at the morass and Benteen showing little consternation about the delay, there was at least someone in the outfit who was moving to join the General.

But the sight of Boston jauntily loping by on horseback did nothing to uplift Benteen's spirit. He brooded over the blood relations in the regiment, the sanguinary ties that he believed had turned the 7th into a nest of nepotism. The regiment included not only the three Custer brothers but also the General's brother–in–law, the handsome Lieutenant Calhoun, who commanded L Company and was married to Custer's sister, Maggie, and the General's exuberant eighteen-year-old nephew, Autie Reed. Benteen disdainfully referred to the whole bunch as the Custer clan or the royal family. Nevertheless, Boston was well on his way to the General, and that is more than can be said for Mathey's pack train, from which Boston had come, which was stretched several miles from front to rear, and some of the mules, driven by thirst, bolted away from their handlers and sprinted to the morass, where they got mired in the mud. As Private Coleman noted, the mules were an ongoing headache that demanded multiple packings as equipment became loose and fell off.

With faint shots now being heard in the distance, Captain Thomas Weir bristled over the inordinate delay. Without receiving orders, he led Company D past a surprised Benteen. The gunshots Weir heard were probably from the Ree scouts who rode out ahead and were shooting Indians they encountered east of the river, killing a thirteen-year-old Indian boy named Deeds from the village, who was looking for lost horses. Deeds was probably the first casualty at Little Big Horn, run down by the Arikara Indian scouts as he desperately made a run for it, not far from the river. Among warring tribes, little or no discrimination was made between combatants and non-combatants. Some ten women and children were among the first victims of the battle, shot down in cold blood by these scouts south and east of the village. A burly, barrel-chested chief named Gall, later discovering that his two wives and three children were among the slain, said it made his heart bad and bitterly vowed to avenge them.

Weir's sudden departure, moving out ahead toward the sound of the gunfire, finally jarred Benteen out of his complacency. He followed Captain Weir, eventually overtaking him at a second watering hole, the Hartung morass, some two and a half miles from the river. Before reaching this second slough, Benteen arrived at the still-smoldering eastern lone tepee with a dead warrior inside about a mile from the morass. In typical Benteen fashion, he caustically remarked that he had no time to investigate the cause of this Sans Arc warrior being made a "good Indian"—some gallows humor playing on the oft-expressed line paraphrased from General Philip Sheridan, "The only good Indians I ever saw were dead."

Again, the horses relieved their thirst in the cool waters of the slough as Weir and, presumably, Benteen waited for the rest of the battalion to catch up. Just before this, the ubiquitous Sergeant Kanipe, riding back as either a messenger or due to a tired horse, saw Benteen's troopers approaching the Hartung morass. He rode over to tell the senior captain he had a message for the pack train. More than three decades later, Kanipe expanded on this message, claiming that he had told Benteen Captain Thomas Custer wanted Benteen and the ammunition packs to go to the front because the command had struck a large Indian encampment. But when he testified at the Reno Court, Benteen never wavered from saying that in his brief conversation with Kanipe, the sergeant curtly replied that the order did not pertain to Benteen but rather to Captain McDougall, who commanded the pack train. As Kanipe then headed over to McDougall, he was overheard saying, "We got them boys, they're on the run, and we're beating the stuffing out of them," or language to that effect. The distinct impression was that Custer had attacked and captured the village. Whatever Kanipe's exact wording, it left little doubt among those who heard it, including Benteen, that the battle had essentially been won, and all that remained was to congratulate Custer.

Benteen must have been infuriated, thinking that Custer had again seized the laurels and had reduced Benteen's battalion commander's role to a mop-up operation. It helps explain why, after Kanipe left, an uninspired Benteen proceeded toward the still audible gunfire at what even he described as a stiff walk. Why hurry? The battle had already been won. Barely twenty minutes later, another messenger could be seen in the distance making his way across to Benteen's three-company battalion. It was the Italian trumpeter John Martin. Recently assigned to Headquarters Staff, Martin had followed Custer's backtrail south, moving toward Benteen's dust cloud five miles to the southeast. At the Reno Court of Inquiry, Martin testified that he was dispatched after the command had executed the "Column Left" and they were a little below the ridge and almost to the head of the ravine. Looking back, Martin remembered seeing Custer's command galloping down into the ravine. Almost to the day he died, Martin would, in dramatic fashion, regale his spellbound listeners about his storied journey back to Benteen.

Cooke's message to fetch Benteen and the pack train saved the trumpeter's life. Riding along a high ridge, he encountered Boston Custer, who anxiously inquired where his brother's command was and if they had attacked yet. Martin directed Boston to the next ridge, warning him of nearby hostiles, sobering news that nonetheless left Boston unfazed as he unflinchingly proceeded on into enemy territory. During Martin's historic ride, he was shot at by the hostiles, and until it was later pointed out to him by Benteen, he did not even realize his horse had been wounded, although his mount's blood had spattered his blue blouse. Hearing shots, he glanced back and saw Reno's column engaged in the valley but did not stop to take in the particulars. He would come to mistakenly believe that he was hearing the first shots of the Custer fight east of the river, which turned out to be an important clue when creating a timeline for the opening of this stage of the battle.

After trotting back to Ash Creek, the trumpeter saw Benteen with his orderly William Ramell on the flats between North Fork and Ash Creek, a short distance west of the second lone tepee. They rode two or three hundred yards in front of the battalion. Waving his hat to attract the captain's attention, Martin rode up to Benteen, saluted him, and then handed him a message that had been written, or more precisely, scribbled on a sheet torn from Adjutant Cooke's notepad. Later, Benteen would condescendingly refer to Martin as "a thickheaded, dull-witted Italian, just as much cut out for a cavalryman as he was for a king." Be that as it may, from that day forward, Martin won a newfound respect from many in the regiment who admired his dauntless, solitary ride down the back trail where every ravine, ridge, or gully could have concealed an Indian intent on lifting his scalp.

The scribbled message torn from Cooke's notebook —the most famous military message in the history of the American frontier —was remarkably terse, considering the volumes of heated discussion and debate that its simple and declarative contents would arouse. Benteen would be the first among countless others to read its urgent appeal, even as he critically remarked (being Benteen, how could he help himself) that the note had no time notation. As immortal as it is controversial, the message contained just fourteen lean, spare, monosyllabic, and even abbreviated words:

Benteen
Come on. Big
Village. Be Quick.
Bring Pack.
WW Cooke
& bring pacs.

While Benteen, Captain Weir, and Lieutenant Edgerly pondered over the precise meaning of Cooke's message and how to respond to it, Martin left much the same impression as Kanipe, albeit in broken English, that the soldiers of Custer's command were sweeping over the Indian encampment, leaving destruction and death in their wake. He said Custer was about three miles away, but he clearly stated that Reno had attacked the village, routed the hostiles, and that they were killing Indians and squaws left and right. As with Kanipe, some of the men cheered at this news with disappointment tempered by relief that the battle was essentially over and that their anticipated participation had come to an uneventful and unceremonious end.

Much remains to be unpacked in the accounts of these two messengers, so it must be broken down step by step, beginning with Kanipe's dubious journey from Custer's battalion back to Benteen. At the time, and long afterward, no one questioned the sergeant's story. That he had diverted from his initial path to speak to Benteen, or so he maintained, afforded him some credibility, at least in retrospect. Later, skeptics noted that Kanipe spoke very little to Benteen, and perhaps for good reason. It would not have served his purpose to say much, even if the senior captain had pointed out that the order should be directed—not to himself, but to McDougall, who commanded the pack train. Still, the more closely the Kanipe accounts are examined, with their various and ever-changing details, the more questions are raised about him being a designated messenger from Custer's command.

Chapter 5: Did Benteen Dawdle?

One of the first skeptics of Kanipe's story was Little Big Horn archivist par excellence and author Frederic Wagner III, who astutely pointed out the problem in taking the sergeant's story at face value. At first, Kanipe told no one about his ride back to the pack train. Only after he learned Custer's command had been annihilated, and many years after the battle, did he tell the story of how Captain Tom Custer, pursuant to the general's directions, ordered him back to bring the packs straight across the high ground and told Benteen to 'come quick.' But it was odd, Wagner noted, that Kanipe would be sent back when no shots were fired at Custer's command and with the Reno fight in the valley having not yet begun.

Moreover, why would Custer send a sergeant when Kanipe's C company was already short of Non-Commissioned Officers (NCOs), and there was no shortage of orderlies to send? Why send a much-needed sergeant who could command a detachment of troops to Benteen when a military engagement against a much larger force was imminent? Why would Martin not have been sent back instead of Kanipe? Why were two messengers sent back within some twenty minutes of each other with essentially the same message? Why is Kanipe not ordered to return to the command like the other Custer messengers? Why did Martin, who was with Tom Custer and the Headquarters Staff, never mention in his numerous post-battle accounts (he lived to 1922) a single word about Kanipe being sent back when, as an orderly to Custer, Martin would have surely heard this order?

In a way, the questions become their own answers. While Kanipe continues to have his hard-core supporters, it seems more likely that Kanipe's horse began to play out rather than him being sent back as a messenger. Several other mounts from C Company had broken down, so this would not have stood out as something unusual or suspicious. Understandably, Kanipe would not have wanted to be left without a horse when he was standing at the precipice of battle. So, it is possible, if not probable, that the sergeant saw discretion as the better part of valor and opted to return to safer ground. As Wagner points out, "If the Custers had survived, it would have been easy enough for Kanipe to claim his horse broke down, and he was just doing his duty as any NCO would do to bring up the packs as he saw fit." So, there was not much risk for Kanipe to tell such a story and be able to cover himself.

To probe deeper invites even greater doubt regarding Kanipe's tale. While leading the pack train, neither Lieutenant Mathey nor Captain McDougall remembered a sergeant carrying orders to speed them up. Both were firm on this point and testified accordingly at the Reno Court of Inquiry. So, exactly what did Kanipe testify to at the Reno Court? The answer is that he did not testify, and for good reason. No one even knew about Kanipe's story until an entire generation had passed from the scene. After the battle, he all but vanished, so he was never called before the Reno Court. Indeed, Kanipe left the Yellowstone camp barely ten days after the battle from what suspiciously looked like a self-inflicted gunshot wound to his calf. Although he eventually married the widow of a fellow sergeant of his company who died with Custer, more than twenty-six years had passed before Kanipe finally recorded his story in the Montana Historical Society in 1903. By that time, Reno was dead, and so was Benteen. Few were left to contradict Kanipe, and so, quite possibly, he began to take some liberty with the truth, buttressed with the additional assurance that over time, memories either fade or disappear altogether.

There was little danger in deflecting anything McDougall or Mathey might say about the incident if they said anything at all. There is little question about Kanipe returning to the pack train after briefly talking with Benteen. Private Augustus DeVoto of McDougall's company stated that he met a soldier from Custer's command whose name he could not remember, but who was almost certainly Kanipe. This dovetails with Benteen's recollection that a mile or so from the "lone tepee with the dead Indian," he met a sergeant coming back with instructions to the commanding officer of the pack train to "hurry up the packs." Benteen brusquely told him the pack train was seven miles back (one to two miles to the advance pack and three to four miles to the remainder was closer to the truth), but that he could take the order back to Captain McDougall, who was in charge.

Even though Kanipe returned to the pack train, he transmitted no orders, making himself largely indiscreet. Years later, he played a prominent role in the feeding frenzy of a battle now being regaled as a stirring and epic historical event. A full two decades after the battle, details of who said what became muddied even further when Kanipe asked McDougall to furnish him with a character reference as he was seeking a position with the United States Postal Service. McDougall agreed, and in an astonishing refreshment of memory, completely contrary to what he testified twenty years earlier at the Reno Court when recollections were still green and fresh, he furnished Kanipe with the following testimonial:

> On the afternoon of June 25th, 1876, when the entire country was full of Indians, Sergeant Kanipe brought to me an order from General Custer to bring the pack train across the way where I found Major Reno ... I take great pleasure in giving him this small certificate of merit.

So, there it is—a favor from one old soldier to another who had shared the communal experience of participating in the West's most legendary and mythical battle. What is the harm, after all, of a gentle fabrication, a kindly, liberal stretch of the truth, if it serves to lend a helping hand to an old comrade in arms?

With the battle having passed into legend and lore as few others ever have, Kanipe clearly understood his newfound importance to a legion of historians, battle students, and curiosity seekers. No source is inerrant, and when accounts are coalesced and evaluated against each other, some of Kanipe's testimony creditably lends to the historical record a more concrete and seamless picture of what happened at the Little Big Horn. Still, it can be overwhelmingly seductive to accept at face value what a viable source like Kanipe says in an attempt to fill in all those yawning and exasperating gaps in the story and find answers to tantalizingly unanswered mysteries.

Even a battle scholar as insightful and discriminating as Bruce Liddic endorsed the core of Kanipe's story of carrying an order from Tom Custer. As Liddic surmises, had Kanipe not delivered this order, why did the pack train cease to follow the main trail and move to the right, cross country, towards the bluffs where he had last seen General Custer? However, Liddic makes a rare mistake in reaching this assumption because the pack train did not initially divert from the main trail but continued to follow it.

Only much later did it move off and toward the bluffs after Captain McDougall, now commanding the pack train, was briefed by a messenger from Lieutenant Mathey saying that the troops were consolidating on Reno Hill.

The belief that Kanipe was carrying this order spurred one of the great and enduring misunderstandings about the battle, which remains mostly unchallenged today. Because Kanipe maintained that Tom Custer ordered the packs to be brought to the high ground, it was thought that Cooke's "Be Quick, Bring Packs" note must have meant that Benteen was to meet up with Custer high on the bluffs, rather than with Reno down in the valley. That pernicious fiction stubbornly persists to this day. But such a notion is a categorical error that can easily throw even the most informed historians off the trail. Benteen and the packs were always expected to go to the valley, not to Custer, because that was where the "Big Village" was, and because, at the time, that was where all the attack battalions of the 7th were expected to go. The village was always the target, and the quickest and surest route to it was through the valley.

Once Custer took his five companies up on the bluffs, there was no one else to support Reno's three companies other than Benteen, who had yet to arrive. The idea of Custer ordering the excruciatingly slow-moving packs over the hundred-foot bluffs, even as he moved farther and farther away from the rest of the command, borders on the absurd. Such an order would have been an offense against the platitudes of common sense. That route would have taken the pack trains two hours to reach Custer's command on the east side of the river, leaving Reno alone in the valley to fend for himself against a village of perhaps several thousand warriors until Custer could make his way across the river with the packs to support him. Kanipe's fanciful ploy about bringing the packs to the high ground has, for generations, clouded the entire picture, distorting what Custer intended to be the destination for Benteen's battalion and McDougall's pack train. Benteen and the pack train's destination was the valley—always the valley.

Meanwhile, Benteen was never confused about the content of Cooke's note, at least insofar as to what was expected of him and where he was to go. While the note, written in great haste, was hardly a specimen of pristine clarity, it was clear enough to Benteen, despite his later laments to the contrary, that what Custer really wanted was for the thirteen mules carrying the ammunition in the advance group of pack mules, only a short distance behind Benteen, to head to the valley. After characteristically carping about the inherent contradiction in Cooke's note, Benteen easily unraveled the supposed conundrum of simultaneously being quick and bringing along the painfully slow pack train trailing him in a timely fashion. His statement, "if I am to be of any service to him [Custer], I think I better not wait for the packs," resolves whatever confusion there was regarding Benteen's understanding of the intent of Cooke's note. Also noteworthy is that none of his other officers objected to moving ahead without the packs. But what definitively clinches the argument that Benteen completely understood his role and what Custer intended is that he did not turn north to the bluffs where Kanipe and Martin had come from but headed straight toward the valley—exactly as Custer intended him to do.

The testimony of John Martin clearly recollected what Cooke said when he handed him the written order:

> Orderly, I want you to take this dispatch to Captain Benteen and go as fast as you can. Take the same trail we came down. If you have any time, and there is no danger, come back, but if there is any danger or Indians in the way stay with your company.

Notably, Martin was to return to the command if there was no danger of Indians blocking his return path, but in that case, he was ordered to stay with H Company of Benteen's battalion. There would be no reason to give Martin such a warning if he were riding back to Custer with Benteen and the packs, but only if, as Cooke says, he was coming back alone. Additionally, in following Custer's trail, Lieutenant Edgerly of D Company, now riding at the head of Benteen's battalion, had no hesitation about following the main Indian trail that led to the river crossing known as Ford A, the same ford Reno had earlier crossed into the valley. Benteen never asked Martin which side of the river Custer was on because it did not matter. Benteen was not going to Custer over the bluffs. It was in the valley that Benteen was expected to report to Custer—not at some undisclosed position east of the river that was never even alluded to when Benteen embarked on his scout upstream.

It is also important to acknowledge that Benteen made the right decision not to continue to the Little Big Horn Valley during his scout. Otherwise, Martin would have never reached him with Cooke's note. Custer never would have sent Martin if he did not know approximately where Benteen was. And knowing where he was would have been impossible had Benteen not returned to the main trail. The dust from his column and that generated by the pack train allowed his battalion's location to be generally pinpointed. This gave Custer confidence that Martin could find his senior captain's battalion. During his scout, Benteen correctly sensed that they were reaching the point where their respective columns could no longer be supportive, prompting him to return to his commanding officer.

In his celebrated timing analysis, Dr. John Gray concluded that Benteen, in returning from his scout, moved at a reasonable pace considering the heat and condition of his horses, arriving at the main trail on Ash Creek about four miles behind General Custer's wing. After watering his horses for an extended period of time, Benteen continued at a more leisurely pace, which he maintained for another three miles. As one experienced mounted cavalryman explained, the proper gait of a cavalry battalion is a trot whereby the horse uses less energy but covers the ground more rapidly. Deviating from this pace to negotiate an obstacle, climb a steep grade, or rest horses should occupy only twenty percent of the time.

Benteen's battalion was not trotting, at least not until after he received the message carried by Trumpeter Martin. Benteen testified that E troop was moving at a gallop when he last saw them, implying that such a pace resulted in a greater distance between Custer's command and his own. That would create the impression that Custer was leaving him behind and that this was the cause of his own tardiness. But there is little evidence that Custer began to gallop before he reached the bluffs. If Custer had galloped, he would have overtaken his scouts, which he never did.

What Benteen probably observed with E troop is what is known as the "accordion effect," a kind of optical illusion whereupon a gap or separation between the companies in the battalion results in the lagging group temporarily galloping forward to close the distance, giving the impression that the horses are moving faster than they really are. Even more importantly, based on the meeting location of Trumpeter Martin and Benteen, the latter was less than twenty minutes from joining Reno in the valley.

Some battle students have maintained that Benteen proceeded slowly and cautiously because he feared being ambushed or led into a trap, much like Captain William Fetterman was ten years earlier in what widely became known as the Fetterman massacre. But that line of reasoning is unpersuasive because Benteen's battalion was following a trail of eight companies of cavalry just a short distance ahead. After Martin arrived, Edgerly said the gait increased to a trot. There was no slowing down for fear of being jumped. Instead, the pace accelerated. Benteen knew that Custer had preceded him not only over the same ground but also the Ree scouts who were out ahead on the lookout for such snares and ambuscades. It is extremely unlikely that Benteen harbored any concern about falling into a trap, and he never once used that as an excuse to defend himself against persistent charges of lateness.

The primary reason for Benteen's delay was the prolonged watering of his horses at the first (eastern) morass when the rattling of gunfire was beginning to be heard, gunfire that was neither heavy nor sustained. While it is impossible to divine the motivations of others, it is hard not to believe that Benteen was still sulking at being removed from leading the advance to what he saw as a meaningless scout, which to him was an appalling waste of time. He was in no hurry to get anywhere anytime soon. This dilatory inclination began to change with the sudden and unexpected arrival of Martin. Until then, Benteen had been moving at a leisurely pace, but after reading Martin's note, the battalion's pace increased to a trot.

In fairness to Benteen, from the time he read Cooke's note, both Kanipe and Martin gave a definite impression that the hostiles were on the run and that the 7th Horse had already triumphed over the warriors. Being informed by two different sources that Custer was mopping up the Indian encampment in the valley, Benteen felt vindicated, thinking he had been purposely left out and his role marginalized. Moreover, neither Kanipe nor Martin, for some unknown reason or perhaps for no reason at all, bothered to inform Benteen that Custer had divided the regiment and separated from Reno. At the time, Benteen had no way of knowing that Reno's battalion was in dire peril. If he had, it can hardly be doubted that someone with his soldiery qualities, innate courage, and experience, notwithstanding his unflagging and visceral hatred for Custer, would have acted with more urgency and dispatch.

But even with Kanipe and Martin's attestations that Reno's battalion was wreaking havoc upon the hostiles, something was not quite right with the scenario that Kanipe and Martin described. Benteen must have sensed as much. Cooke's note said not once but twice to bring the packs. But why would the ammunition packs be needed if the battle was virtually over, with the hostiles suffering a total defeat? That alone should have raised some troubling questions about this received intelligence.

Over the years, there has been much discussion but little understanding regarding the pack train. To come to terms with this battle and the command decisions made on that long afternoon, it is necessary to have a working knowledge of the organization and function of the pack train. This understanding comes down to one question: When Cooke's note said to bring the packs, was he stating that Custer primarily wanted the ammunition packs or the whole pack train? To put it another way, did Adjutant Cooke, under time pressure, unintentionally omit the word 'ammunition,' leaving others to interpret that he meant the whole pack train? Dr. Charles Kuhlman argued in the affirmative that, for reasons based on time alone, Custer wanted only the ammunition packs and not the entire pack train. Kuhlman stuck to this interpretation until he was informed by General Hamilton Hawkins, an old-time cavalry warhorse, that in the argot of the army, "packs" always means the pack train—the entire pack train—unless otherwise stated.

That appeared to settle the issue. Or did it? What Hamilton said was true enough, providing the pack train operated as a single entity. But this was not the case at the Little Big Horn. What is often overlooked is that back at the divide, Custer had partitioned the pack train into two components, separating them by about twenty minutes, which, because of the more accelerated pace of the ammunition packs, compared to the pack train at large, lengthened the distance between them during the march, ultimately leaving a gap between them of some forty-five minutes. Of the nearly 175 mules that made up the pack train, twenty-five to thirty served as an advance group, of which approximately thirteen carried ammunition. As the faster-moving packs, they were positioned to lead the train, allowing them to arrive earlier at the front lines with precious ammunition.

Scarcely a battle student or historian has picked up on this important and deeply relevant distinction. As we shall see later, this fact became critical in bringing the pack trains to the Benteen/Reno command when they became, for reasons that continue to be furiously debated, bogged down at Reno Hill while the rest of the command, Custer's command, was fighting for its life in a losing battle on the east side of the river. It is more than likely that Custer wanted those thirteen mules with the ammunition packs pronto since waiting for the whole pack train would have been severely time-prohibitive in such dire and exigent circumstances.

Returning to Benteen, it is fair to say that from the time he received Martin's message to be quick until he first reached the tattered remnants of Reno's defeated and demoralized command, there is no compelling reason to attach any serious blame to Captain Benteen save that he spent too much time at the first waterhole. After Benteen reached the first, eastern morass, he developed, as President Lincoln once said of General George McClellan, a case of the slows. This is not necessarily to maintain that had Benteen, or Reno, for that matter, acted more energetically or decisively, they would have saved the day. No one can satisfactorily answer that question, and it is possible that had either Benteen or Reno moved more quickly, the 7th might have either suffered an even more debilitating defeat, or a stunning victory—one can only speculate.

Nevertheless, Benteen felt defensive over accusations of his dawdling back to the main trail. He bizarrely testified at the Reno Court that by the time Martin reached him, he believed Custer and his command had already perished at the hands of the hostiles. But how is it possible? According to both messengers, who had been with Custer just a short time before, the command was not only very much alive and unscathed but severely chastising the tribal warriors, shooting them down left and right. Even two days after the battle, when the regiment was rescued, and a teary-eyed General Alfred Terry informed Benteen of the disastrous fate of Custer's command, he still refused to believe the news, thinking that most of Custer's command was very much alive somewhere down the Little Big Horn River grazing their horses.

The interrogators at the Reno Court did not press the point about Benteen's false and ludicrous contention that he believed Custer's command was already dead before he had arrived on the battlefield. However, the court was not charged with investigating the actions of Benteen but rather Marcus Reno. No doubt Benteen concluded, perhaps even rationalized, that Custer had recklessly moved too far downstream, outdistancing his anticipated reinforcements. By doing that, no one other than he (George Armstrong Custer) was the primary author of his and his command's doom.

Plausible arguments have been posited to make this case. But Benteen's smokescreen of claiming Custer's command was destroyed by the time Martin reached him has been tried before the tribunal of history and found wanting. It is hard not to conclude that his explanation was a subterfuge to insulate himself from blame or protect a guilty conscience—maybe both. It was certainly contrary to what he believed at the time. Perhaps it was an unconscious admission that he knew he could have done more than he did to rescue Custer's command, even if the result had been the same or worse, as he himself would later admit. Undoubtedly, the notion that Custer and his command were destroyed by the time Martin reached Benteen, based on a documented timeline, was a fatuous and unwieldy attempt to argue that, with Custer dead, Benteen was no longer under any military obligation to follow Custer's orders to "be quick and bring packs." Sly maneuvering on Benteen's part for sure, but it does not change Benteen's pointed disingenuousness, if not outright fabrication, regardless of whether one believes Custer had outdistanced his reinforcements. It is one thing to lie to others but quite another to lie to oneself, as Benteen clearly and unrepentantly did.

Chapter 6: Were Reno's Actions in the Valley Fight Justified?

After a mile pursuit of some forty to fifty warriors heading west toward the big village that the scout Fred Gerard first spotted at a knoll back on the trail, Reno's battalion finally rested at a river crossing. Wearing a distinctive broad-brimmed straw hat and a red handkerchief tied loosely around his neck, Major Marcus A. Reno stood out among the three companies when his battalion first saw the Little Big Horn Valley. The mercury recorded a high of ninety-one degrees that Sunday afternoon, and men and horses sweltered under a torrid sun.

In 1876, there were no standardized time zones on the North American continent, as each town or city established its own time and affixed it to whatever measure provided a basis or standard of comparison. It would be another seven years before standardized time zones became the law of the land. There is good reason to believe the regiment's Little Big Horn Campaign watches were set on Saint Paul's Headquarters watch time. Frederick C. Wagner III, author of *Strategy of Defeat,* makes a convincing case that the watch time used by the campaign's commanding General, Alfred H. Terry, was based on his headquarters in Saint Paul, Minnesota, just under an hour later than local Montana sun time, rather than the Chicago standard, which ran about a half hour later than St. Paul's. Using this as our touchstone, Reno arrived at the ford crossing that led into the valley just before 1 p.m. Headquarters watch time, not 3 p.m. as most accounts have it (see Participant Timeline in Appendix B).

The natural ford, later referred to at the Reno Court of Inquiry as Ford A, just a little south of where Ash Creek flows into the Little Big Horn River, had already seen some use as a portion of the Indian village had descended from Ash Creek and crossed here on the morning of the 19th. The river ranged from thirty to fifty feet wide and three to four feet deep, or about belly high for horses. The river had a gravel bottom, and its water was clear and cold. Surveying the open plain ahead, Reno saw only a large swath of open prairie. Looking outward, the soldiers felt the twinge of excitement in the eerie stillness pervading the open, unfolding valley in the dry, motionless air. To the west, hills stood barren and mute, testimony to the silence of the ages looming over this ancient, crusted continent.

Perhaps feeling a ripple of anxiety, Reno took a swig from his alcohol-filled flask to settle a vexing uneasiness. Somewhere in the lingering unknown was a still-unsighted, large enemy encampment. Yet, no further word had arrived from the Ree scouts who had been sent to the valley to capture and stampede enemy horses, nor was there any sign from Lieutenant Hare or the other scouts who had already fanned out ahead into the valley. The quiet before the storm—no one knew.

Adjutant Cooke and Captain Keogh had dutifully accompanied Reno's battalion, hastening up to the bank of the river by the ford. The good-natured Cooke bantered back and forth with Major Reno in an apparent attempt to keep Reno in good spirits. According to Reno, Cooke, in a sort of jesting manner, stated, "we are all going with the advance and Myles Keogh is coming too." Since they had left the Yellowstone and up to the divide, Reno complained he had not been consulted about anything, and yet here he was, improbably leading an advance guard battalion into a valley containing a large hostile village of indeterminate size. Reno had pursued Indians, guarded them as captives, and dealt with them in the northwest before the Civil War, but he had never fought them. That was about to change as they soon would be thrown into an unimaginable whirlwind of chaos and tribulation.

As the soldiers in Companies M, A, and G crossed the river, Trumpeter William Hardy of A Company espied Cooke's steadfast and soldierly form poised on the eastern bank, intently watching Reno's progress. That Captain Myles Keogh also accompanied him might well mean that his squadron was to join Reno's command, adding to the battalion's strength and leadership. Riding to the ford allowed Keogh to review the troops, observe the crossing, and make certain the mounts got across in good order. In a streak of impetuosity, M Company haphazardly raced over to the other side, bringing a sharp rebuke from Cooke: "For God sakes men don't run your horses; you will be needing them later. We have hot work ahead of us."

By now, most of the scouts were well ahead of the battalion and moving down to the valley floor. It took Reno's battalion at least fifteen minutes to ford the river with many sweaty mounts gulping down the cool, refreshing waters. At the same time, more than a few troopers filled their canteens and laved the refreshing liquid over their hot faces. Arriving at the crossing in columns of four, Companies M, A, and G, in that order, traversed the stream, timber, and the troublesome undergrowth on the west side of the river ford. They dismounted and re-cinched their saddle girths before swinging back onto their mounts, where they confidently scanned the open, expansive ground. Reno squirmed in his saddle, counting the seconds, trying to buy time to allow Custer to close the gap between them. Reno's jitters were hardly pacified when he previously saw one of his Indian scouts pick up a handful of grass, slowly allowing the blades to drift to the ground and point down the valley. His meaning was unmistakably ominous: The warriors ahead were as thick as grass.

The pressure of command bore down on Reno as never before. Despite the majority of his battalion expressing excitement and confidence, no doubt tremors of foreboding wormed their way into the souls of some of the soldiers, for nothing is more unsettling than an enemy whose presence could be sensed but not seen. Reno lifted his flask and took a long swig. Later, he adamantly denied ever taking a drop of liquor until after the day's tragic events. But several witnesses saw him do exactly that several times. Like many who felt the burden of the frontier, Reno was a tippler, known to drown his demons with the palliative of intoxicating spirits.

He was observed drunk just three days earlier, on June 22, when the regiment had left General Terry just east of the mouth of the Rosebud. The only silver lining was that if Reno had serious problems with the bottle, he nonetheless gave every indication of being a high-functioning alcoholic. Had he been as inebriated that early afternoon as some of his detractors claimed, he likely could not have stayed on top of his horse, much less galloped down the valley. While Reno's drunkenness was not overtly obvious, a palpable excitability was. When Lieutenant Carlo De Rudio's headstrong mount plunged into the river near Reno, splashing the major, he snapped: "What are you trying to do, drown me before I am killed?" A sentiment hardly calculated to inspire confidence among those he led and an unmistakable indication that his nerves were taking a toll.

While Reno's three companies were at the crossing, up ahead, the Ree scouts shouted in their own Arikara language that the villagers were not running, as previously reported, but that the hostiles were coming out in force. The interpreter, Gerard, anxiously galloped back to relay this alarming news to the Major, who still lingered at the river. As noted before, Reno had no use for Gerard, having once tried to cashier him from the regiment, so he pointedly ignored the scout. Not having impressed Reno, Gerard raced over to inform Cooke. Both he and Keogh then promptly wheeled their horses around and rushed to advise Custer of this surprising development. Their sudden disappearance did not go unnoticed by some. Lieutenant George Wallace, speaking of Cooke and Keogh, stated, "They were with us when we started from General Custer and were with us when we crossed back to the left of the little stream [Ash Creek], which runs into the Little Big Horn. I thought at the time that they went into the fight with us. When they turned back, I don't know."

Keogh and Cooke had gone back to where Custer waited with the remaining five companies at a wide bend in the creek, roughly one quarter to one-half mile to the rear and even with the mouth of a broad coulee that offered immediate access to the bluffs directly to the north. That location was not picked by chance, and Reno would have understood the implications as it bore directly upon his role as the advance guard. By hanging back from the river at this spot, Custer was hedging his bets and keeping his options open. If the Indians were running as had been previously reported, he would immediately cross the river once Reno had cleared the ford. If they were not, as Cooke and Keogh were now ostensibly reporting, Custer would move north and strike the enemy in the flank. An officer of his caliber was not going to needlessly limit his options by committing his entire force to either course of action until Reno entered the valley and the enemy's reaction could be positively confirmed.

With Gerard's bulletin that the Indians were coming up the valley, there was a brief concern among Reno's command but soon settled into another pregnant stretch of torpid silence. Despite Gerard's warning, they saw no sign of warriors charging up the valley, heard no terrifying war cries echoing through the hills, nor observed any sign of armed aggression. Perplexed and uncertain, Reno continued to stare hard at the long, narrow, dusty plain stretching deep into the valley with an asymmetrical timberline bordering the western bank of the river to his right. He was supposed to be chasing a running village, not a fighting one, so the notion that warriors, as Gerard announced, would be challenging his advance came as something of a shock even though he had yet to see a hostile or hear a single war-whoop. Perhaps even more unsettling for Reno was that as the seconds and minutes ticked away, there was still no Custer—no Benteen—and no pack trains with 24,000 rounds of ammunition. Reno had yet to receive countermanding orders to forego his instructions to sweep everything before him into the valley. Now, even Cooke and Keogh had left him. He was alone.

Was Gerard seeing Indians where there were none—or at least misinterpreting what he heard from the Rees? No one could even hazard a guess, although there was a simple enough explanation for the confusion generated by Gerard's report. About fifteen minutes ahead of Reno's column, the Rees were attempting to capture and stampede the massive Indian pony herd when a small Sioux war party suddenly challenged them. The commotion the Rees raised had nothing to do with Indians coming out to confront Reno but rather with warriors challenging the regiment's Indian scouts who were trying to pilfer or run off prized Sioux ponies.

The false alarm signifying a direct attack on Reno's battalion nonetheless put Reno in motion. Focusing on carrying out his orders, he plodded through fifty yards of the fallen trees and timber populating the west side of the Little Big Horn contiguous to the crossing. Reno might not have been the most intrepid and gallant of soldiers, but he had never shown cowardice either during the Civil War, the Western plains, or thus far at the Little Big Horn. Notably, Reno's battalion was the first to cross the stream into enemy territory. It would also be the first to lead a charge against hostiles. Whatever else Reno is criticized for, including emotionally and psychologically fracturing when events reached a critical mass, up to this point, charges of cravenness were unjustified.

Reno reformed his command into columns of four with A Company and M Company in front and G Company in reserve—the proper formation for a cavalry charge. With Doctor Porter on his left and Reno's adjutant and close friend, Lieutenant Benny Hodgson, on his right, Reno hollered "forward," and 153 men, all but fifteen of them uniformed, spurred their mounts toward what was believed to be the location of the village. While riding down the valley, Reno, according to Private William Taylor, yelled "*charrrge*" with an emphasis on the slurring of the "r's," the private's not-too-subtle hint that Reno already had too much to drink.

It is critical to note that whether Custer was following Reno into the valley or flanking the enemy, Reno's battalion was acting as an advance guard. Using an advance guard was already an established offensive tactic by the time of Alexander the Great, and one Reno would have learned early at West Point and would have seen executed during the Civil War. The concept was simple, though not without its hazards. The advance guard moves ahead of the main body of troops, searching for the enemy and making contact with him. With sufficient strength to take care of itself for prolonged periods, one of two things will occur when contact is made. The first possibility is that the enemy will run. In this scenario, the advance guard will give chase. The other possibility is that the enemy will stand and fight. If so, the advance guard is to engage the enemy, maintain contact, attempt to fix him in place, and stay engaged for as long as necessary to allow the main body of troops to maneuver to the opposing force's front, flank or rear, and at the opportune moment, strike him there.

Put simply, Reno's job was to ride into the valley and pick a fight with anyone in it. If the Indians run, he is to hound them until they finally stop running. If they decide to fight, he is to draw all the warriors upon himself, grab hold of the front of their shirt, and hang on, no matter what, for however long it takes for Custer and Benteen to hit them where it will hurt most. In other words, to keep the warriors busy and pinned down while the rest of the command picks its spot and blindsides them from another direction. He is not there to conquer, he is there to hold them in place.

With the order to charge, Reno's column began to accelerate over the hard, ash-like soil, populated with sagebrush, an aromatic North American shrub that thrives in the West. They moved from a walk to a trot—then a canter—and then a near gallop heading menacingly toward the still indiscernible Indian encampment, which was hidden just around what later would be called the Garryowen Bend, the westernmost loop in the river. Pumped with adrenaline, the troopers were now in high dudgeon, yelling and cheering until Reno, insensible to the battalion's soaring spirits, admonished them to "Stop that noise." Private John Sivertsen, the lone dissenter to this narrative, later claimed Reno had indeed encouraged the bravado so necessary for esprit de corps. As is often the case, the truth comes down to who one chooses to believe.

So, the column sets out, with G Company on the left, M on the right, and A in the center. They rode not in a straight line but along the course of the river. Things were mostly a blur as they passed a spate of cottonwood trees, bullberry bushes, and scattered patches of sage and prairie grass. The village, three miles away in a straight line, is blocked from view by the timber along the Garryowen Bend. Any delay in the charge resulted from closing up the column, a natural occurrence when a trot ripens into a gallop. Rather than riding to the middle of the valley, Reno decided to keep the river as a screen, and his advance north was biased toward the river. At least ten flankers led by Sergeant John Ryan formed a mounted skirmish line to guard against any Indians in the brush along the river. Meanwhile, other scouts were up ahead on the far left, protecting an open flank. In short, solid, commendable military tactics were employed—strictly by the book.

The reverberating sound of hundreds of hooves pounding the powdered, dry earth provided the only early warning the Indian villagers had, save a few pairs of wide-open, watchful eyes that the tribe had situated on the western foothills gazing in awe at the charging soldiers as if the "Horsemen of the Apocalypse" themselves were thundering out of the fiery pits of hell. For this unsuspecting small city of Native Americans, it might as well have been.

Cavalry charges of this era bear little resemblance to what we are used to seeing on television and in movies, with riders stretched out in a very long line of one or two rows that strike the enemy like a crashing wave and then fight it out right there from horseback. In reality, a charge was meant for penetration and sowing panic by breaking the line and getting to the enemy's rear, causing all cohesion to fall apart. As the charge began, the reserve would be left in place to the rear as a support element to the front columns, the latter whose riders would be stacked in column of fours to form a narrow yet deep front, like a striking fist. The compact charging columns would drive through the enemy front and then loop back to the reserve and evaluate the result to determine the next course of action and manner of attack. The American cavalry of this day were actually dragoons rather than pure cavalry, being able to fight either mounted with saber and pistol or dismounted with their carbines. Aside from the charge, limited fighting was actually done while mounted because of the incredible difficulty of hitting anything beyond the range of a long stick from a moving horse. Only if the enemy broke into outright flight would the entire body continue on in the manner we typically think of cavalry behaving. It is when the enemy is in flight, with his back to the foe, that horsemen really start running up the body count as they chase down the now disorganized and vulnerable enemy.

Returning to the charge down the valley, scouts Charley Reynolds and Fred Gerard rode in the column's tail end, striving to catch up to the fast-moving battalion. In this stretch of the Little Big Horn River valley, the landscape broadened out between 0.75 and 1.25 miles across for a run of several miles encompassing the whole portion of the valley relevant to the battle, with bluffs along both sides. The river hugs the base of the high, eastern bluffs for most of two miles from where Reno crossed (see Map 4), then makes a sharp turn west into the middle of the valley, crossing more than half its width, only to sharply turn back upon itself and return to the eastern bluffs. This projection, known today as the Garryowen Bend after the small town and gas station once located near its bow, narrows the valley's width to less than half a mile, with the Indian village located just beyond.

As news of the oncoming danger traveled with grudging slowness from tribal camp to tribal camp, it became devastatingly apparent to its inhabitants that the village, despite banking on the false security of its intimidating size, was under a full-blown military attack. The entire village was thrown into utter pandemonium. When the alarm was first raised in the Hunkpapa camp, located at the most southerly portion of the encampment and first in line to be hit by a cavalry onslaught, shrieks of terror forewarned that the chargers, the long knives, as they referred to them (though that day the 7th carried no sabers) were coming with a vengeance to sweep over them like lava from an erupting volcano.

Most of the Indians had been oblivious to what was now shaping up as an end-of-the-world apocalypse. The inhabitants were either down by the river, leisurely swimming, fishing, hunting, watering their horses, digging for turnips, or sleeping in their tepees after engaging in a long, festive night. But now, in a heartbeat, life in the village had turned from an easy-going, quiescent morning into a torrent of hysteria and panic. Chaos and disorder reigned; everywhere, mothers frantically searched for their children while ululating the tremolo to inspire the young men to take up arms and defend them; oldsters sang death songs; and warriors caught so scandalously unaware desperately sought their horses, the vast majority of which were grazing outside the confines of the village.

Younger teenage males seized a few ponies and rode furiously to and fro on the perimeter of the village, churning up enormous dust clouds in the hope of impairing the enemy's vision, arresting the attack, and buying precious time for the non-combatants to evacuate and for warriors to mount some defense. Sitting Bull was astonished by the surprise attack. Lakota Eagle Ben said none of the Indians expected a fight. Bathing in the cool waters of the river when he first heard that the dreaded bluecoats were closing in, the Sioux's legendary warrior, Crazy Horse, breathlessly raced back to the village, exhorting his warriors to kill all the soldiers. The famous warrior now began his long preparation for the ritual of war, which he believed would make him impervious to the enemy's bullets.

Chapter 6: Were Reno's Actions in the Valley Fight Justified?

As Reno's battalion moved down the valley, perhaps a third of a mile west of Ford A, toward the now aroused camps of the Hunkpapas, Santee, Cheyenne, Minneconjou, Sans Arc, Oglala, Brulé, and the Blackfeet, Custer's battalion was spotted by his men high on the bluffs across the river. At least seven troopers saw a portion of the General's five companies riding rapidly to the north, conspicuously standing out against the broad, blue western skies. Whether he also saw Custer's troops on the bluffs or learned about it from others, Reno got his first intimation that he could expect no immediate support from his rear—an expectation he was fervently relying on. This prospect was profoundly unsettling, and Reno quickly sent another messenger, his cook, Private John Mitchell, with ostensibly another urgent plea to his commanding officer for support, perhaps fifteen minutes after he had sent Private McIlhargey at the Ford B river crossing.

To many in Reno's battalion, it appeared that Custer was endeavoring to get to the north side of the village in an attempt to envelop the enemy between the two commands, again, a tactic commonly used in military maneuvers and referred to as a 'fix and flank'. It became apparent to Hare, De Rudio, and Varnum that Custer's troops would attack the village's lower end. Still, some aficionados of this battle questioned exactly how much of this Major Reno, despite his military experience, understood. But there should be no mystery about what Reno comprehended that early afternoon. Ten days later, in his after-action report, he wrote, "It was evident to me that Custer intended to support me by moving down the stream and attacking the village in the flank."

Despite his stated reservations over Custer's intentions, up to now, Reno had just started the advance toward the village, paying attention that the columns were close together, protecting his flanks with skirmishers, and keeping the command in line with trumpet calls about the same time Custer was first spotted on the bluffs. Varnum, Wallace, Hare, and the Ree scouts were positioned in the left front flank of the advance while Reynolds, Herendeen, and Gerard rode in the rear. It was a solid positional arrangement, but the closer Reno came to his objective, the more doubts he had about plowing straight into a village of such unknown proportions. That the men in his battalion were still cheering indicated that their spirits were high and that few mounted warriors could be seen challenging them.

Shortly after the advance began, well before the halfway point, Reno ordered G Company up from the rear and placed it next to A Company, with the formation now being, from left to right, G, A, and M companies. Reno is no longer in a column formation, a natural precursor to a charge because a column is maneuverable. While impressive to see, the line formation is unwieldy to maneuver. Reno's decision to reshuffle the column into a line formation tells us something many historians overlook—Reno has taken the charge off the table before even setting eyes on the enemy. There is no longer a reserve in place, and no experienced officer, such as Reno, will charge a village in that manner, snarling the entire command in a forest of teepees, much like a stretched-out rope, which is exactly what a line formation is. In effect, Reno has committed the companies to dismounting ahead and making the fight on foot.

Indeed, the soldiers would not fire a shot until they dismounted. There was no attempt to shoot their way in since there was nothing to shoot at. Having recovered from their initial shock, the warriors could catch their breath and mount some sort of defense, even going on the offense. Incredibly enough, only a week after they encountered General Crook's forces, they were under attack—this time, not at the break of dawn but smack in the middle of a bright, sunny afternoon. Many believed, then and now, that the village was a proverbial sitting duck, including several Indians in the village who were certain that the cavalry, in a devastating tide of blue, would sweep over them.

Pretty White Buffalo Woman remembered that there "was no council the night before—there was no need for one ... our camp was not pitched anticipating a battle and were attacked from a direction we did not expect. The warriors would not have picked out such a place for a fight with white men, open to attack from both ends and from the west side." He Dog concurred, "When we got to the Little Big Horn, the Sioux thought there would be no more fighting. We had our wives and children with us and had to get buffalo meat for them and wished to be left alone." The camp was beginning to stir. Eagle Elk recalled that only a few of the old women were up, but "many were still asleep." Others, like Runs The Enemy, took comfort in the belief that the soldiers would never attack such a large encampment in the daytime. He paid no heed to the shouts and alarms he heard, the cacophony and the ensuing chaos around him, as he continued to sit in his tepee smoking, believing it a false alarm.

But this was no boy crying wolf, no exercise in artifice, no optical illusion—no false alarm. Even worse for the Indians, Reno's path was still largely unobstructed. However, ominous silhouettes of warriors were beginning to take shape down the valley, vaguely espied through a whirlwind of rising dust that the Indian boys were churning up with their ponies. Varnum said the valley was full of Indians riding in every direction, but the weight of the testimony indicates that most of the mounted warriors were not hindering Reno. Endless debates spanned a century and a half over whether Reno surrendered the initiative by not driving full steam into the horrified village, striking a concussive blow to the Hunkpapa camp, the first in a series of circles within the encampment. More than a few battle students believe this would have set off a chain reaction, a domino effect, in which the respective camps would have folded out of sheer terror, breaking the back of whatever resistance the warriors might have mustered. Several tribal inhabitants agreed, including Pretty White Buffalo Woman, who felt her people were ripe for the taking.

When it comes to rolling up Indian villages, having the right leadership, striking at the right time, and having the right disposition are much stronger factors for success than the size of the striking force. But these factors were not in play at the charge's culmination. As Lieutenant Edward Godfrey colorfully put it, "Reno had failed to hold a leg even if he couldn't skin"—meaning that, barring charging into the village, he could not even do the bare minimum—keep the Indians occupied by maintaining his defensive position in the valley. Whatever initial boldness Reno might have displayed charging down the valley, its denouement had no resemblance to Murat's charge at Eylau when Napoleon's storied cavalry commander furiously swept into the serried ranks of the Russian center. After studying the valley fight, Captain Charles King of the 5th Cavalry stated that a bold and dashing charge was required, but Reno had no dash, and the attack was spiritless and abortive.

Chapter 6: Were Reno's Actions in the Valley Fight Justified?

Others who were present believed that no matter what Reno did, no matter how bold or aggressive, if he had charged into the village, he would have been swallowed up by its roiling immensity—tantamount to a suicide mission, and only by aborting it was the command saved from annihilation. Lieutenant De Rudio, no admirer of Reno, expressed relief that the charge was halted, stating that another five hundred yards and the whole battalion would have been butchered. Although Varnum, looking over his shoulder, was unhappily surprised to see Reno's command halt and dismount, several other officers of Reno's battalion, with the exception of Captain Thomas French, believed that the Major had averted disaster by halting a direct assault upon the hostile village.

Reno's supporters continue to believe that he has been unfairly maligned—that his worst offense was not a failure to perform his duty but rather a failure to adequately and convincingly justify his actions, as was the case with his reconnaissance for Terry. They also maintain that by dismounting and forming a skirmish line, Reno could better defend his battalion than had he driven headlong into the village. The hastily formed skirmish line actually provided good support for Custer because the village was a fixed point, not a moving one. As such, by dismounting to fight on foot, Reno could still fulfill his role as an advance guard (see Map 4), whose purpose was not necessarily to rout the enemy as it was to occupy and distract him. In other words, establishing a solid skirmish line would not only pressure the village by endangering it but divert the warriors' attention from other possible avenues of attack, which Custer was clearly endeavoring to orchestrate from the north.

Far more damning than halting the charge was Reno's testimony that it was a ravine full of Indians that thwarted his advance and forced a dismount. He described the ravine as being ten feet wide and five feet deep, and, at least according to Reno, it seemed to be disgorging red devils as if the command had plunged into the bowels of Dante's Inferno. However, the notion that this ravine was an obstacle to the charge's trajectory was more a product of Reno's worried imagination than a physical reality. There is even a question of whether Reno could have seen this ravine, as it was a quarter mile in front of him, and everything ahead was clouded with dust. But for argument's sake, if Reno had seen it and Indians pouring out of it, could it have really been an obstacle? If hundreds of Indians can maneuver it, it is safe to say that the 7th could as well.

There was indeed a ravine called Shoulder Blade Creek, which ran in an east/west direction, near where the main horse herds were located, and the ravine served as a covered access from the western upper foothills down to the valley floor where the skirmish line would be deployed (see Map 4). But, this ravine was northwest of Reno's charge, about three hundred to four hundred yards in front of the most advanced point of Reno's dismounted skirmish line. It did not traverse the entire valley; thus, it could not cause Reno's dismount because Shoulder Blade Creek was not visible from where he had halted. Let Reno's own words speak for him on why he did not surge into the southernmost portion of the village when he had the chance:

> I, however, saw that I was being drawn into some trap, as they [the warriors] would certainly fight harder and especially as we were nearing their village that was standing; besides I could not see Custer or any other support, and at the same time, the very earth

> seemed to grow Indians. They were running toward me in swarms and in all directions. I saw that I must defend myself and give up the attack mounted. This I did.

Exactly how many Indians were confronting Reno at this point is also in dispute. Almost every account bears its own estimates. With respect to the ravine, Varnum, one of the lead scouts down in the valley, made no mention of any ravine or ditch obstructing their advance—it was all open prairie most of the way. He added that the terrain of the valley floor ahead of him was the same as that which they had just passed over. By this time, the Indians were kicking up so much dust that De Rudio said all you saw were shadows of warriors, but no one could physically count them. But, according to Reno, his battalion was not only fighting the entire Sioux nation but all the desperados, renegades, half-breeds, and squaw men between the Missouri and the Arkansas and east of the Rocky Mountains. The scout Herendeen said there were too few Indians in range to shoot at, which is closer to the truth. He estimated that there were two hundred warriors, but conceded that it could have been as many as three hundred, most of whom were well out of range.

Lieutenant Luther Hare perhaps gave the best answer when he said it was impossible to know for certain how many Indians there were because they were always on the move and never stayed in one place long enough for anyone to count. Both Herendeen and Hare agreed that once the battalion halted and the momentum dissipated, hundreds of Indians poured out of the ravines from the north and west.

Whatever the circumstances facing Reno, the uncertain number of warriors to his front, the lack of visibility, the village was stationary and not fleeing, and its formidable size are all legitimate reasons for Reno to give the order to halt and dismount. But a ravine replete with Indians directly in his path was a figment of fear's creative power. When considering Reno's aborted charge, the question remains: Should Reno have ordered the halt and dismount? The question, however, is academic. It is enough to acknowledge that for an officer of Reno's disposition and capability, this particular situation made the order to dismount and form a skirmish line defensible. Although Custer and, yes, Benteen would have undoubtedly shown more elan and daring in successfully charging the Hunkpapa camp. Nonetheless, three companies charging into a portion of a village of that size with impaired visibility and no support in sight is problematic enough to justify erring on the side of caution. So, Reno's battalion halted, dismounted about two miles from the river crossing at Ford A, and prepared to fight on foot (see Map 4). Save for a few rambunctious horses that could not be halted, the battalion had little difficulty adhering to the order. Still, those troopers who could not rein in their horses are often the subject of much innuendo and speculation among Little Big Horn aficionados. The mounts of six troopers, five from M Company and one from A Company, bolted and ran toward the village.

Private Sullivan from Company A returned safely, though he was later killed during the retreat crossing the Little Big Horn River. Trooper James Turley also could not bring his horse to heel as he rode toward the Indian line into which he disappeared. Sergeant John Ryan maintained that he never saw Turley again, but Private Morris and Private Slaper did. As it turned out, Turley was also killed in the retreat, found with a hunting knife driven to the hilt in one eye. Private Rutten's horse, which he claimed could smell Indians, became so agitated that it took off, carrying Rutten to the perimeter of the Hunkpapa camp. Fortuitously, Rutten finally turned his mount around and returned safely to the skirmish line. Trooper John Meier's horse also bolted through the Indian lines, but he miraculously shot his way back to his battalion with his six-shooter. Trooper William Meyer, who rode closer to the village than the others, somehow managed to circle back only to be killed in the valley retreat on the right bank. The capriciousness of war is such that the only trooper who did not make it back after inadvertently penetrating the hostile lines was the much-to-be-pitied Private George Smith, whose remains were later found in the village with his corpse described in various states of decomposition, his head split wide open by a war club.

Reports circulated that some troopers from M Company reached the Hunkpapa camp, the most southerly portion of the village, and torched some tepees. But the report was false. Some tepees did catch fire, but the Indians' burning campfires likely set off these small conflagrations. The Indian non-combatants may have inadvertently ignited them in their frenzy to dismantle the tepees and make a mad rush to escape with their lives. The closest soldiers were still at least six hundred yards away from the outer perimeter of the encampment, with its mass array of tepees that lay beyond.

As Reno approached the village, he commanded Companies M and A to halt, and Lieutenant Hodgson relayed the order to G Company on the extreme left. The battalion dismounted and deployed in skirmish order about a hundred yards in advance of the point of dismount, securing their right flank by the western edge of the Garryowen Bend of the river bend and reversing the position of M and G Companies from their advance down the valley to their deployment on foot. How close they came to the village is also disputed. Lieutenant Wallace thought they were just one hundred yards away, while Dr. Porter said about a mile. Others thought it was half a mile to a thousand yards to the closest tepees in the main village. The most educated guess remains somewhere between six hundred to eight hundred yards.

While events seemed to unfold in a mesmerizing, discombobulating moment, one thing was clear—the sleeping village was no longer dozing. Following the dismount, Sergeant John Ryan took ten troopers to clear out any Indians lurking in the timber on the right to prevent flank fire. One trooper for every four horses led the mounts into an old stream bed in the timber east of the skirmish line. Thus, the skirmish deployment's firing power was reduced by those holding horses. The thin blue line of scarcely more than ninety men advanced on foot another hundred yards closer to the village.

Some of the men fired from a prone position, others knelt. A large prairie dog town enabled some men to huddle behind mounds of dirt, providing a flimsy breastwork. The remainder of the battalion spread across the valley, facing an enemy force growing in strength but not yet reaching unmanageable proportions—although Sergeant Ryan thought there were probably five hundred warriors surging from the village. Obstructed by the timberline on the right, Lieutenant Varnum declared that only those in advance on the left could see the tops of a few tepees. But the one thing everyone seemed to realize was that they were near the boundary of a very large encampment.

Reno strode along the skirmish line, giving instructions while French boastfully called attention to his marksmanship after he unhorsed an Indian with his rifle. Initially, the troopers were restrained, if not jovial, methodically firing, loading, and then shooting again. Despite French's enviable accuracy, Trooper Thomas O'Neill said that few effective shots were fired from the skirmish line. But the men remained in good spirits, talking and laughing until the Sioux began to circle the line's left flank.

Estimates vary on the number of troopers initially on the line. But based on those missing—every fourth trooper holding horses, the company doctor, six orderlies, eleven officers on horseback, and one casualty—they probably numbered less than a hundred. Within a few minutes, that number would be steadily and markedly reduced. The line stretched out across the valley floor, measuring some 250 to 375 yards across—no more. Most observed the prescribed five-yard interval between troopers—although, when the danger increased, the soldiers bunched up closer for safety.

One of the more crucial questions about the valley fight—and among the most disputed—is the site of the skirmish line. Various sources give several different locations. The best evidence places it near the location of the old Garryowen Post Office by the extreme southwestern edge of the Garryowen Bend of the river. Of the twenty-seven primary sources that reference Major Reno's valley skirmish line, eighteen support that locale, and four support the old Pitsch Slough at the next loop of the river upstream from the Garryowen Bend. (Unfortunately, artifacts in that area purported to be evidence of the skirmish line are highly suspect and may lack reliable support.) The nine remaining sources are unclear about location. The strongest evidence indicates that the Reno valley skirmish line (though the precise angle of the line remains ambiguous) was situated at the Garryowen Bend of the river, which was not far from the timber position (see Map 4).

When attempting to assess the approximate location of the skirmish line, it is critical to develop a clear picture of who was fighting where and when. Even those erudite in the fundamentals of the battle and abetted by tools of science can go astray. The unearthing of army cartridges, for example, noted by the battle Archaeologist Dr. Douglas Scott, led to the conclusion that the line was situated at the Garryowen Bend and angled off in a general northwest to southeast direction. However, this analysis may have included evidence of the Army's Ree scouts firing that occurred before the skirmish line was established, possibly misleading investigators. Thus, the actual alignment of the skirmish line, with its right flank resting on the end of the Garryowen Bend, remains open to debate, as does the length of the skirmish line.

Much confusion and misunderstanding exist among battle students regarding this issue. Some say the line was no longer than 250 yards, while others have stated it was as long as eight-tenths of a mile, or approximately 1,400 yards. The latter measurement, based on Lieutenant Maguire's map of the battlefield, is wildly off and comes from a misunderstanding of Lieutenant Maguire's intent in drawing a dotted line across the valley floor to represent Reno's skirmish line. The line was drawn too long, and its direction and placement were inaccurate. At the RCOI, Maguire told the Court that the length of this dotted line did not represent the true length of the valley skirmish line. Perhaps the best description comes from Lieutenant George Wallace's RCOI testimony. The following questions were put to him by the Court, along with his very revealing answers, with our annotations in brackets:

> Question: Does this dotted line [on the official Maguire map used by the Reno Court] represent the position and angle [of the valley skirmish line] correctly?
>
> Lieutenant Wallace: No, sir ... The map is not correct; it does not represent the ground.
>
> Question: How would you represent the skirmish line?
>
> Lieutenant Wallace: By cutting off about 3/4 of it.
>
> Question: Was the angle [of the skirmish line] such as is represented there [on the map]?
>
> Lieutenant Wallace: The left [end of the skirmish] line was farther to the front; more towards the village.
>
> Question: Estimating the number of men and the distance, what was the length of the skirmish line?
>
> Lieutenant Wallace: He probably had 70 or 75 men on the skirmish line [after the horse-holders and G troop had been taken off the line and redeployed in the timber position].
>
> Question: What distance would that cover?
>
> Lieutenant Wallace: If he had 75 men, there would be 74 intervals of five yards each.

Lieutenant Wallace indicates that the dotted line representing Reno's valley skirmish line drawn on Lieutenant Maguire's map is inaccurate in terms of its position, angle, and length. He believed the line should have been reoriented with the left end facing more northerly towards the village location. This was how the line looked before a portion of it was redirected to face westward, as discussed later in this chapter. The upshot is that Major Reno's valley skirmish line was about 370 yards long, roughly 250 yards facing north toward the village and 120 yards long facing west toward the foothills after Captain French modified the left part of this line after warriors were seen riding around the left flank of Reno's skirmish line (see Map 5).

Three guidons were planted to mark each of the respective companies' positions. The troopers were encouraged by Lieutenant Benny Hodgson, popular with the men and given the nickname "Jack of Clubs"—a reference to his card-playing habits—to take their time and aim carefully, which, for the most part, they did at least early in the valley fight. At the outset, confidence was high because Plains Indians were rarely, if ever, able to penetrate a skirmish deployment with disciplined lines of fire. Those not familiar with 19th-century military tactics often overlook that a skirmish formation might at first appear to be nothing more than a thin elastic line. However, an appropriately organized skirmish line can be an astonishingly resilient and formidable battle deployment when directed by confident leadership and supported with ample ammunition. A well-formed skirmish line will not only hold a significant area on the battlefield but can also repel large, mounted forces that oppose it. It also possesses flexibility, whereby troopers can switch rapidly from defensive to offensive operations by bringing up rested horses to pursue a retreating enemy. In almost every case, certainly those involving Plains Indian warfare, the only way for warriors to overcome a skirmish line was to flank it.

At this juncture, Reno's skirmish line was effectively engaging the enemy. Even though the warriors were beginning to show signs of aggression, they were careful to stay beyond the range of the soldiers' breechloading Springfields. A few warriors, usually in groups of four or five, would emerge from the screen of dust and embark on bravery runs, testing their courage in the face of enemy gunfire. The Indians were mostly buying time for the non-combatants to escape, so the soldiers were still largely controlling the action. Despite Lieutenant Hodgson urging them to take their time, some were firing too often and failing to select a target. This was especially true of A Company, which may well have used significantly more ammunition than the others.

Approximately how long Reno's skirmish line remained intact is one of the minor controversies of the battle. But the weight of evidence indicates that it was not as long a time as might be expected. Varnum thought it stood in place for ten to fifteen minutes; others in the battalion thought it stood for a little longer. After the first five to ten minutes of fighting, Reno withdrew most of G Company into the timber position after mistakenly being informed that the hostiles were infiltrating and threatening the safety of his battalion's horses (see Map 5). Before another ten minutes, A Company, under Captain Myles Moylan, had suddenly evacuated its position after a group of warriors began to flank the left end of the skirmish line. A Company headed back toward the timber position, leaving M Company, under Captain Thomas French, alone on the skirmish line (see Map 6).

Shortly before 2 p.m. (Headquarters time), less than twenty minutes after exchanging fire with tribal warriors, Reno's line began to waver. Fears grew that the Indians were beginning to close around the rear of the deployed troopers, forming a complete circle. Why Reno's skirmish line seemed to disintegrate suddenly is perplexing since it was not under undue duress from the warriors. Almost inexplicably, it became caught up in a sort of gravitational pull that drew the battalion toward what was, at least unconsciously perceived by the troopers, the stronger and safer confines—an area to its right, toward the brow of the timber area where Sergeant John Ryan's flankers had earlier ensured was clear of Indians.

The lure of safety drew soldiers on the skirmish line toward the wooded area. The call to retreat to the timber may have appeared to have come from some unknown and unspecified source—but Little Big Horn author Larry Sklenar thought its provenance emanated from what he called a "disembodied voice," one that inaudibly communicated that Indians were penetrating the timber. Inaudibly, at least to others, because Major Reno, Sklenar caustically noted, seemed to be the only one who heard it. But that may not be entirely true. Lieutenant George Wallace also got word that the Indians were infiltrating the woods and the horses were in danger of being captured. Since their mounts were the sole means of their mobility and carried the troopers' extra ammunition, such an eventuality would have jeopardized the entire command.

If the warriors infiltrated the timber area, immediate and decisive action would be warranted. But this so-called threat, however acutely felt, was real only in Reno's and, perhaps, Captain Moylan's minds. An ember of a rumor igniting a forest fire of innuendo was most likely responsible for unraveling the skirmish line. It should have remained intact for as long as it took Custer's five companies to strike the other end of the village. Undoubtedly, shots from hostiles across the river into the wooded area and the overall confusion bred by a lack of organized leadership from the commanding officers contributed to this impression of Indians infiltrating the timber. Hence, the perception became a reality, and as Private Newell shamefully recalled, "We went into the timber like sheep."

All Reno said about the incident was that he was near Captain Moylan, who informed him that Indians were coming up the left bank of the river and menacing the cavalry horses held in an open glade. Reno's poor understanding of the situation or bad luck caused him to rely on Captain Moylan's appraisal and judgment rather than that of the steadier and more reliable Captain Thomas French. First Sergeant John Ryan called French as brave an officer as ever rode in the 7th Cavalry, "a soldier who ascended the ranks by his own efforts and was a crack shot," but at this particular moment, he found his company abandoned on the line (see Map 7).

The earlier exodus of G Company had attenuated the skirmish line, igniting a chain reaction of withdrawal that abetted the Indians as they worked around the left of Reno's line. Along with Reno, most of G Troop withdrew into the timber. That forced A Troop to pivot to the right flank of what remained of McIntosh's G Troop, a shift-over to cover the gap left by those who withdrew. M Troop swung back to the left to create another line facing the western bluffs in reaction to the flanking attempt by the Indians. M troop turned to the left, operating like a hinged swinging door to form an L-shaped skirmish line. Within minutes, Captain Moylan's A Company suddenly abandoned its line position and retreated to the timber. That left only the cool-headed Captain French and M Company on the original skirmish line. Yet, Reno had suffered only a single fatality on the line, the unfortunate Sergeant O'Hara, with one other trooper wounded. But the way events were falling apart, it might as well have been a hundred casualties (see maps 5, 6, and 7).

At the start of the fight, the Arikara (Ree) Indian scouts managed to capture only twenty-seven mounts, and two mules before enraged, mounted Sioux cut them off and had them running for their lives as they crossed back over the ford from which they had come. The collapse of Reno's skirmish line is often blamed for the failure of the scouts' mission to stampede the horses. But there were just too many ponies and too few Rees to get the job done. Even worse, as these scouts were driven away while barely holding on to their scalps, the village warriors still had thousands of horses at their disposal to make war against their attackers. However, the warriors would be delayed in catching them.

The most significant impact of Reno's retreat into the timber was the enormous boost in morale it gave the Indians, who were misled into believing that it was their power and war-making skill, not the overreaction and misapprehension of their adversaries, that shifted the momentum of the battle. Fired up by the turn of events, hundreds more warriors from the foothills and downstream joined the fight. The arrival of Crazy Horse on the battlefield, after having completed his meticulous war preparation of painting fearsome dots of hailstone on his body and a lightning bolt across his face, also served to buoy the confidence of the warriors.

The warriors were now shifting into an offensive position. In Sitting Bull's view, it was becoming a good day to die. At first, as the Lakota Chief Crow King later noted, the Indians only wanted to delay the soldiers so the non-combatants could escape. But now, seeing the pendulum shift, they were consumed by a seething and unfettered fury to fight and kill their attackers. Reno's irresolute and poorly organized deployment effectively augmented the determination of the warriors. Two days after the battle, Lieutenant Edward McClernand arrived at the Little Big Horn, riding with the 7th Infantry of Terry's command. A snappy young officer who would later achieve the rank of general, McClernand would hear from officers of the 7th Cavalry that Reno's lack of aggressive and effective leadership had allowed the troops to slip from the offensive to the defensive and deflated the troops' morale.

Feeling trapped in the woods, a contagion of near hysteria radiated over Reno's shaky command. The scent of growing fear was unmistakable, and it pervaded the battlefield. Getting a good whiff of it from Reno's battalion, the warriors became emboldened, resolutely converging on their attackers. Meanwhile, Captain Moylan, still worried about the warriors circling into the timber from his left, imagined they would eventually seize their horses and cut off any hope of escape. These trepidations of phantom Indians running off with their horses had a debilitating effect on Major Reno, who no longer, even remotely, exercised the same command and control as when they first charged down the valley. The beleaguered soldiers, feeling surrounded in a dark wooded area, cut off from others, and increasingly bereft of leadership, felt the creeping prospect of defeat and the smothering coldness of death drawing in around them.

Whatever the general impression regarding the vulnerability of the timber's heavy grove of cottonwoods, it was a formidable defensive position for both topographical and strategic reasons, something that both Reno and Moylan failed to realize, much less exploit. The dry embankment alone offered greater protection than the valley floor. Gerard stated that he thought ten men properly placed could have stopped the Indians from coming in at that one location in the timber. Plains Indians were known to be reluctant to charge into wooded areas so as to avoid sustaining unnecessary casualties. No one knew this better than the unflappable scout George Herendeen, who had fought the Sioux two years earlier.

That the timber area was a substantial stronghold crystallized early among military personnel who closely examined its defensive potential in the immediate aftermath and post-mortem of the battle. Lieutenant Edward McClernand concluded that while the timber position was too large for a battalion of three companies to hold indefinitely, it could be held for several hours. It was a good tactical position for an advance guard since its proximity directly threatened the village. Add McClernand to the list of those who believed that Reno's lack of aggressive and effective leadership prompted the warriors to shift from a defensive to an offensive posture. He noted that the timber provided relatively good cover from enemy fire. As the soldiers fired from the wooded area, they could do little damage to the village and its inhabitants because the ground within the timber was much lower than the ground where the Indian encampment stood. But it was a strong defensive position, especially when contrasted to the bluff to which Reno's forces soon retreated. Although Reno Hill, as it would later be called, commanded higher ground, it was dangerously exposed and too far away to threaten the enemy, leaving it outside the scope of tactical operations geared to target the destruction of the village.

If Reno's battalion had utilized the outer embankment perimeter of the timber (it should be noted that timber is a questionable description since the trees were no larger than a man until you reached the side by the river), which was part of a former bed of the watercourse, the thick underbrush in conjunction with the bank that served as a natural breastwork would have made it extremely difficult to dislodge the troopers by any frontal or flank attack. Because it was next to the riverbed, the grass was too wet for the warriors to ignite a fire to burn the troops out. And unlike Reno Hill, the timber position had easy access to water. In short, had he the wherewithal and strategic sense to use all its natural advantages, the area that Reno's troops occupied in the timber was far stronger than the bluffs position he later chose. Benteen himself later acknowledged that with 120 men, the timber position could have been held for five to six hours.

With the notable exception of positioning G Company along the riverbank, Reno spent little effort deploying his troops to take advantage of the area's defensive attributes even as the fighting intensified. Moylan testified that the troops had no intention of using the timber for defense. Private Newell remembered that the sound of gunfire in the timber was one continuous roar, which caused Major Reno to see his position as a veritable death trap. Portentously, Reno later wrote that if he had chosen to stay there, he would have remained there forever. With growing desperation, he saw his present position in the timber as his graveyard and the eastern bluffs—where Custer and the greater part of the command were last seen—as his salvation. That the Indians did not forcefully drive them out underscored that the real purpose of retiring there was to reach the horses and make their escape. During the timber fight, Reno and Moylan's mutual fears ravenously fed off each other, propelling them to retreat to a less reliable, insecure defensive position.

No other explanation seems plausible since a lack of ammunition was not a pressing issue, and both Benteen and Custer would have arrived long before it was exhausted. Reno had enough men to deploy around the timber position so that both flanks could be anchored against the river. The position, fortified by eight-foot banks and thickly covered vegetation, could have been defended for several hours—not necessarily effortlessly, perhaps not definitively, maybe not even doubtlessly, but under the right leadership, adequately enough. Among those who assailed the troops in the timber, Red Feather thought that the warriors could not effectively get to the soldiers as long as they remained in the woods. With Custer arriving at the other end of the village, the tribes would have had to divide the warriors to address two different threats in two different areas, thus alleviating the pressure on Reno and likely depriving the warriors of the total victory they ultimately achieved over the 7th Cavalry.

Whatever disadvantages the timber position presented could have been overcome with resourcefulness and imagination. The soldiers deployed along the perimeter embankment could have spaced each trooper seven to eight yards apart rather than the usually prescribed five. Later, on Reno Hill, Private Charles Windolph noted how the command was spread out at twenty-foot intervals to compensate for the lack of manpower. Moylan would later testify that two hundred men were needed to defend the timber (although in a strident fit of hyperbole, Reno testified that six to seven hundred were needed—essentially the entire regiment). But Reno had self-serving reasons to say that—specifically to insulate himself from the fallout resulting from his subsequent orders to withdraw.

Undoubtedly, two hundred men in the timber would have been optimum, but Reno had a contingent of at least 140 fighting men, including the scouts supporting the battalion. Not all the scouts rode off to capture the Sioux horses. About ten Indian scouts, including Billy Jackson, Bloody Knife, Half Yellow Face, Bob Tail Bull, Goose, Little Brave, and White Swan, remained with Reno, as did Fred Gerard, George Herendeen, Charley Reynolds, and Isaiah Dorman. They all fought with Reno's command. Additionally, the horses could have been tied down, thus releasing the horse-holders to reinforce the firing line. That is exactly what Major Reno did when he later hunkered down on the bluffs.

With the troops now at the bottom of the timber, the warriors seized the high ground along its perimeter, allowing them to encircle the area, creating openings for Indian snipers firing into the woods from the south and northwest, putting the entire command in jeopardy. It was a vacuum begging to be filled, and the Indians inevitably filled it, occupying what should have been cavalry defensive positions around that same perimeter. By then, the warriors seemed to be everywhere. Gunsmoke clouded the woods, and a fusillade of bullets caused leaves and twigs to fall like hail upon the soldiers. With the warriors positioned on the high ground (the timber, at certain points in the bottom, ran as much as thirty feet lower than the ground on the surrounding prairie), the soldiers became exposed to enemy fire.

Because they were in such a vulnerable position, Sergeant John Ryan concluded that there was nothing the command could do but mount their horses and cut their way out of the woods. One soldier recalled how Captain French stood as poised in this killing field as if he attended a Sunday school picnic. With the Indians taking possession of the high ground, cutting out might have been the only viable alternative. But this was only after Moylan sent soldiers down to the glade, an open space surrounded by woods, a place where they found themselves directionless, bereft of officers, and relegated to the unprotected bottom rather than the elevated perimeter which allowed the Indians, once again, to seize the initiative by pouring heavy fire down upon them in the woods.

The soldiers' fire slackened as they retreated, encouraging more warriors to enter the timber. Still, others were encouraged by Crazy Horse to penetrate the woods since there was no firing line to retard their advance. Amid this turmoil, Reno considered sending another messenger to Custer, but the Indian scout Billy Jackson, sixteen years old, the youngest in the regiment, dissuaded him, saying that any courier would face certain death. A few soldiers were now being felled by enemy fire. Private George Lorentz was among them, shot in the back of the neck, the bullet emerging from his mouth.

During these incredibly tension-filled moments, Gerard claimed he watched Reno down his bottle of whisky. Defenses started to crumble, and what would finally break the dam of any organized resistance was the grisly killing of Bloody Knife, Custer's favorite scout, now assigned to Reno. His death would send the battalion's commanding officer, already teetering on the precipice, tumbling over the brink. Standing beside Bloody Knife, an overwrought Reno asked the scout what the warriors would do next, when a shot rang out from the side of the timber some fifty yards away and blew the back of Bloody Knife's head off, splattering Reno with gobs of blood and brain matter. The gooey and sanguinary substance bathing Reno shattered whatever semblance of composure and command he might have possessed. Fueled and exacerbated by alcohol, he displayed every sign of post-traumatic stress disorder. Vacillating wildly, Reno ordered the troops to mount and then dismount, then mount again until he finally yelled, "Those of you who wish to make your escape, follow me." He galloped out of the woods onto the prairie and toward the ford where he had first crossed into the valley.

So that there is no undue exaggeration about this pell-mell exodus, Varnum related that "I do not consider Major Reno's retreat from the woods a disorganized rout as to the head of the column, but the rear certainly was." Nonetheless, this unexpected order left many troopers behind in the timber at a loss to determine what was happening or whether an evacuation order had been given—if indeed it could be characterized as an order and not an invitation. With this order, as Godfrey lamentably noted, "the men gave up their ammunition ungraciously." While it is easy to criticize the reaction of the troopers, no trumpet calls were sounded to signal a retreat, and even worse, there was no attempt to account for the wounded. No officer directed a rear guard to cover a desperate and precipitous retreat. Many of the soldiers in the timber, including Reno's own orderly Edward Davern, heard no order. This was especially true of the troopers from G Company, who were guarding the river and were too far away to hear anything. The leading companies withdrew with some semblance of organization, but the rear, including G Company, fragmented into a helpless and muddled rabble. Trooper William Taylor recalled that as the battalion fell back, all was in confusion.

The scouts, leaving the timber alongside the soldiers, were faring no better. Lonesome Charley Reynolds's horse was shot, and the animal fell on the scouts' legs, entrapping him some 150 yards outside the timber. The frontiersman they called "Lucky Man" lucklessly died there, but not before he had reacted to the soldiers abandoning the woods by exclaiming, "What damned fool move is this?" The black scout, Isaiah Dorman, was also hit and taken down about fifty yards east of Lonesome Charley. He put up a brave fight, killing at least one warrior, but was badly wounded by a gunshot to the chest. He was bleeding out when some Indian women and youngsters (one of them may have been Moving Robe Woman, whose ten-year-old brother Deeds was shot dead at the start of the battle) ran among the fallen soldiers clubbing and butchering them to death, which included Dorman, who for good measure had his penis severed and stuffed in his mouth. During the firefight, there had scarcely been more than three fatalities in the timber, but now, some twenty personnel were left in the woods, several of them wounded. The battalion would suffer a shattering thirty-plus casualties as they dashed toward the bluffs.

What prompted Reno to make such a rash move after suffering relatively modest casualties? According to Gerard, it could be summed up like this: Reno, seeing no support from the rear, "lost his head, if he had any," and decided to run the gauntlet of the scalp-hungry Sioux. Reno's only explanation was that reinforcements were not coming to his aid, so he moved to the hill where he was visible and could deploy his men until they were reinforced. Whatever the proximate cause or causes were, the withdrawal was a total disaster, and the shock of seeing such a rout was not confined to the troopers alone. The warrior Wooden Leg was astonished to witness the soldiers fleeing the woods as he whirled his horse in pursuit of the disoriented troopers. It was pure, unadulterated anarchy among the troops as Indians, with their repeating rifles, pumped bullets into what was now a panicked mob. Lieutenant Varnum tried to rally the troops to stand and fight until, to his utter disbelief, an irate Reno suddenly appeared, remonstrating, "I'm in command here, sir." Varnum, struck dumb, said no more.

Reno's column initially sought to make their way back to Ford A, about two and a half miles up the Little Big Horn, but a swarm of Indians on their flanks forced them to the left (see Map 8). Others, like Private Thomas O'Neill, De Rudio, Gerard, and Billy Jackson, among a few more, had no other recourse but to find sanctuary in the woods, hiding there for most of the afternoon until they reunited, in two separate groups, with Reno's command, a few doing so under the safety of darkness.

For the rest, getting out was pure hell. William Taylor hauntingly remembered the warriors as "a whooping, howling mass of the best horsemen that fell upon them—the most cruel and fiercest fighters in all our country or any other. Soldiers left, and right began to be knocked off their horses." American Horse said chasing the soldiers was like chasing buffalo. A few dismounted soldiers futilely raised their hands to surrender, but the hostiles shot them down. Reno had done the opposite of what Colonel Gibbon stated about when fighting Indians—"once I dis-mount I never remount." There would be no prisoners in this battle. As is often the case on a battlefield, there were instances of extraordinary valor. Lieutenant Donald McIntosh, who was of mixed Indian heritage, approached the banks of the river that led to the bluffs and was quickly surrounded by a group of thirty warriors. A fellow soldier described McIntosh as being as calm as a deacon during church service while firing his revolver until he was killed. His courage won him no mercy from the tribal warriors who stripped him of his clothing, leaving only his neckband. They then scalped him from his forehead to the back of his neck. A nearby brush fire scorched and nearly incinerated his corpse.

Swirling around them was the terrifying sight of hostiles on the warpath, a scene vividly captured by Trooper William Slaper:

> Indians were riding about in all directions firing at us and yelling and whooping like incarnate fiends, all seemingly as naked as the day were born, and painted from head to foot in the most hideous manner possible.

Some thirteen troopers courageously tried to form some semblance of a skirmish line rear guard along the retreat route to hold off the warriors but were quickly overrun. Others, seeking safety in flight, splashed their way across a stream about forty feet wide with an eight-foot cut bank on the other side of the timber. The Cheyenne Brave Bear said that when the soldiers' mounts hit the water, the sound resembled cannon fire. The only way for the fleeing soldiers to escape was to ascend the opposite bank through a narrow buffalo path, only wide enough to permit one horse to climb at a time. It was pure chaos and discord, and the fleeing soldiers looked like a buffalo herd threading through a narrow canyon. The banks, both on the eastern side and the western side, were high-cut banks, making it extremely difficult for the troopers to negotiate their steepness and get over the top. Varnum, a superb horseman, was nearly thrown from his Kentucky thoroughbred as he attempted to cross over. Meanwhile, bunches of soldiers desperately and despairingly, their arms akimbo, thrashed about like drowning men, a disordered mass of humanity, tumultuously and pathetically clogging the stream, sometimes crawling over one another in a pitiful struggle to escape and survive.

The ferment, confusion, and clutter made the troopers easy targets for warriors positioned on the eastern bluffs just downstream. The warrior Wooden Leg remembered the Indians, including Crazy Horse, pumping bullets and shooting arrows into the fleeing soldiers. Endeavoring to slog their way up the elevated eastern side of the bank, they were shot from above by warriors who had obtained higher ground on the bluffs. Struggling to ascend the bank, Dr. DeWolf took a bullet in the chest. He was subsequently shot in the face four times with his own revolver. After receiving a bullet, Lieutenant Benny Hodgson's horse tumbled with him into the water. Somehow, while scrambling to his feet, Hodgson was shot in the lower back. In their flight to safety, mounted soldiers rushed by Hodgson, who screamed, "For God's sake, help me." Private William Morris allowed Hodgson to grab onto his stirrup, but the Lieutenant lost his grip. Hodgson then grabbed onto Trumpeter Fisher's stirrup and was finally dragged out of the river and onto the embankment before a second bullet killed him instantly. The "Jack of Clubs" had drawn a losing hand.

Climbing the bluffs, Private Morris was seriously wounded by the same enemy fire that would kill two of his comrades. One was an Englishman named Gordon, who was struck down by a bullet through the neck; the other soldier, Bill Meyer, was shot through the eye from an Indian volley. Though knocked off his horse and bleeding profusely through the breast, Morris somehow managed to grab the stirrup of his own horse that dragged him to the summit of the bluffs.

The warriors saw the whole messy matter as target practice, glorifying the slaughter as a buffalo hunt. Only the Indians' unusual warfare practices allowed any soldiers to escape. The Oglala Lakota American Horse stated that some soldiers only escaped because the warriors did not follow Reno to the top of the hill. Many warriors stopped their pursuit to count coup, scalp their victims, and further mutilate the dead and dying. Had Reno's battalion been fighting Bedford Forrest's Confederate cavalry, scarcely a man would have remained alive. This was something that Lieutenant Hare understood only too well when he later reflected, "If the Indians had followed us in force to the hill-top, they would have got us all."

Any experienced Indian fighter knew that running from Indians or even turning his back on them was inviting death. After Reno informed him earlier that he was determined to hold out in the timber,

Captain French became so incensed over Reno's rushed and disorganized retreat that he considered murdering the Major in cold blood right then and there. Varnum claimed to be mystified about what compelled Reno to shout, "every man for himself," and then depart en masse without first establishing the necessary cover fire. A disgusted Luther Hare, who called himself a "fighting son of a bitch from Texas," implored his comrades to "die like men and not to run off like a pack of whipped curs." Disregarding his entreaty, they stopped behaving like a military unit and bolted like a herd of antelope fleeing a grass fire.

As the fight in the timber concluded, Benteen was not fifteen minutes away with a force of 117 troopers, and the advance pack trains were less than a mile behind him. Although Benteen stated that he would not have entered the timber without the packs, he almost certainly meant the advance packs, which were close behind, with each mule carrying two thousand rounds of ammunition. For Benteen, it would have been completely out of character and an extraordinary dereliction of duty to passively watch the besieged Reno battalion en extremis while he passed the time of day waiting for the entire pack train to pull up before supporting his flagging comrades in arms.

Custer deserves blame for giving Reno responsibilities beyond his capability, not better informing his battalion commanders of his intentions, and over-reinforcing the pack train instead of utilizing vital manpower elsewhere. But all these criticisms are dwarfed in comparison to Reno's unforced errors. His woefully ill-advised, disorderly, and panicked retreat had far worse consequences than Benteen's dilatory actions at the watering hole and his general indifference about returning to the main trail with all due speed. When covering a withdrawal, either mounted or dismounted skirmish lines should be utilized to provide and establish an orderly retirement of forces to a more distant and secure position. Soldiers who know that their retreat will be covered will fight more spiritedly and resolutely than those plagued by doubts and uncertainties about being shot in the back.

It would be far from unjust to conclude that withdrawing in such an abrupt and disorganized manner was nothing less than criminal. The battalion suffered only a few casualties in the timber before retreating, suggesting that the troopers could have held out considerably longer. Whatever the contingencies of the moment, a commanding officer must be composed and clearheaded in times of stress. But even if Reno felt that conditions were dire enough to demand a retreat from the timber, he should have formed an elongated circle of skirmishers, with the horses protected within the circle, and methodically walked backward toward the river crossing, all the while firing upon their pursuers to keep the enemy at a distance. Once on the open plains, they could have outranged the warriors with their Springfield carbines, still not turning their backs on the hostiles, and withdrawn in an orderly fashion.

That is exactly what Custer did at the First Battle of the Yellowstone in 1873. He walked his dismounted troopers for a mile, approximately the same distance from the timber to the river crossing at the Little Big Horn, facing the enemy the entire way and suffering but a single wounded man. Perhaps Reno would not have been as fortunate, but it is a good guess that he would have evacuated his command from the timber far more effectively and with fewer losses had he adopted those same tactics. An orderly retreat would have also bought Benteen more time to reinforce Reno.

Chapter 6: Were Reno's Actions in the Valley Fight Justified?

It is beyond the scope of this book to relate all the gruesome and disastrous details of Reno's desperate, headlong flight from the timber to imagined safety. The stories of these life-and-death struggles en route to the river crossing and the bluffs, as well as the desperate and furious struggle of June 25th and 26th atop Reno Hill, have been fully and meticulously recorded in other books and articles. The final word about this appalling fiasco belongs to Lieutenant Godfrey, who renders a fitting verdict of Reno's performance while acknowledging that no love was lost between him and Reno. In short, he was the wrong man at the wrong time in the wrong place:

> Frankly, I do not believe Custer's command would have been rescued under Reno's leadership. At no time during the battle was his conduct such as to inspire confidence. His faltering advances down the valley, his halting, his falling back to the defensive position in the woods in the old riverbed before his command had suffered a single casualty [they suffered two, Private George E. Smith and Sergeant Charles White] in the ranks; his disorganized, panic retreat to the bluff with practically no resistance, his conduct up to and during the siege, and until the arrival of General Terry was not such as to inspire confidence or even respect, except for his authority.

After the battle, the talk of the horrors of that June day would not fade away. The murmuring undertones, whispers, and hushed voices were impossible to ignore. Such was the nature of the obloquy that so tenaciously attached itself to Major Reno's reputation that it would be Reno himself who demanded an official inquiry into his conduct. Some thirty months later, Reno's request was granted in a venue and climate that was as different from that hot, dusty day in southeastern Montana as one could imagine. On a frigid January morning in the windy city of Chicago, a Court of Inquiry convened at the opulent Palmer House. It commenced in a small room, but the growing crowds, consumed by the drama of it all, forced the proceedings into a more spacious room with tall windows and a gaslit chandelier powered by six mounted jets suspended from a lofty ceiling. The inquiry became a public spectacle, with newspapers feverishly covering the day-by-day proceedings. "Custer's Last Stand," as it came to be called, had already passed into legend, and the mystery surrounding it was as gripping then as it is today.

After numerous witnesses were called to the stand to give an accounting, Major Marcus A. Reno agreed to testify on his own behalf. Impeccably appareled in military regalia that included a full-dress uniform, spotless white gloves, and a sheathed sword smartly buckled to his waist, the mustached officer looked every inch the military hero in the civilian setting. Under vigorous questioning, he defended his actions, stating that he could not have done more for Custer and his command if his own brothers had been in that column. But some sensed that at least part of Reno's lengthy testimony seemed curiously amiss or downright disingenuous. The Recorder, Lieutenant Jesse Lee, picking up on the potentially misleading responses, asked whether he went into that fight feeling confident or distrustful about his commander. Reno repeated his earlier assertion that his feelings toward Custer had been friendly enough. Recorder Lee would not be put off by the response and demanded the question be answered directly. Looking at the rapt audience that packed the hearing room, Reno's response triggered a long and audible gasp when he pointedly answered, "I had known General Custer for a long time, and I had no confidence in his ability as a soldier."

It is as straightforward and crisp a condemnation as could be made. However, Reno's response may have revealed more about himself than it did about Custer. On that day and on that battlefield, there was a commanding officer in whom Reno had no confidence. But it was not Custer; it was himself. On June 25, 1876, Reno's doubts were indeed justified. While it is possible to mount a defense for Reno's actions in ordering a halt and dismount on the floor of the valley or to defend his move into the timber and perhaps find some justification for ordering a retreat toward the bluffs, it is impossible to defend his hasty and reckless withdrawal from the timber. After fleeing the woods and reaching the summit of those bluffs, the survivors, badly beaten, their command in tatters, could only watch in horror the demoralizing sight in the valley below as their helpless comrades were scalped, stripped, and gutted like hogs. Only then did it become obvious to the proud and mighty 7th how devastatingly steep a price they paid for a failure of nerve and leadership.

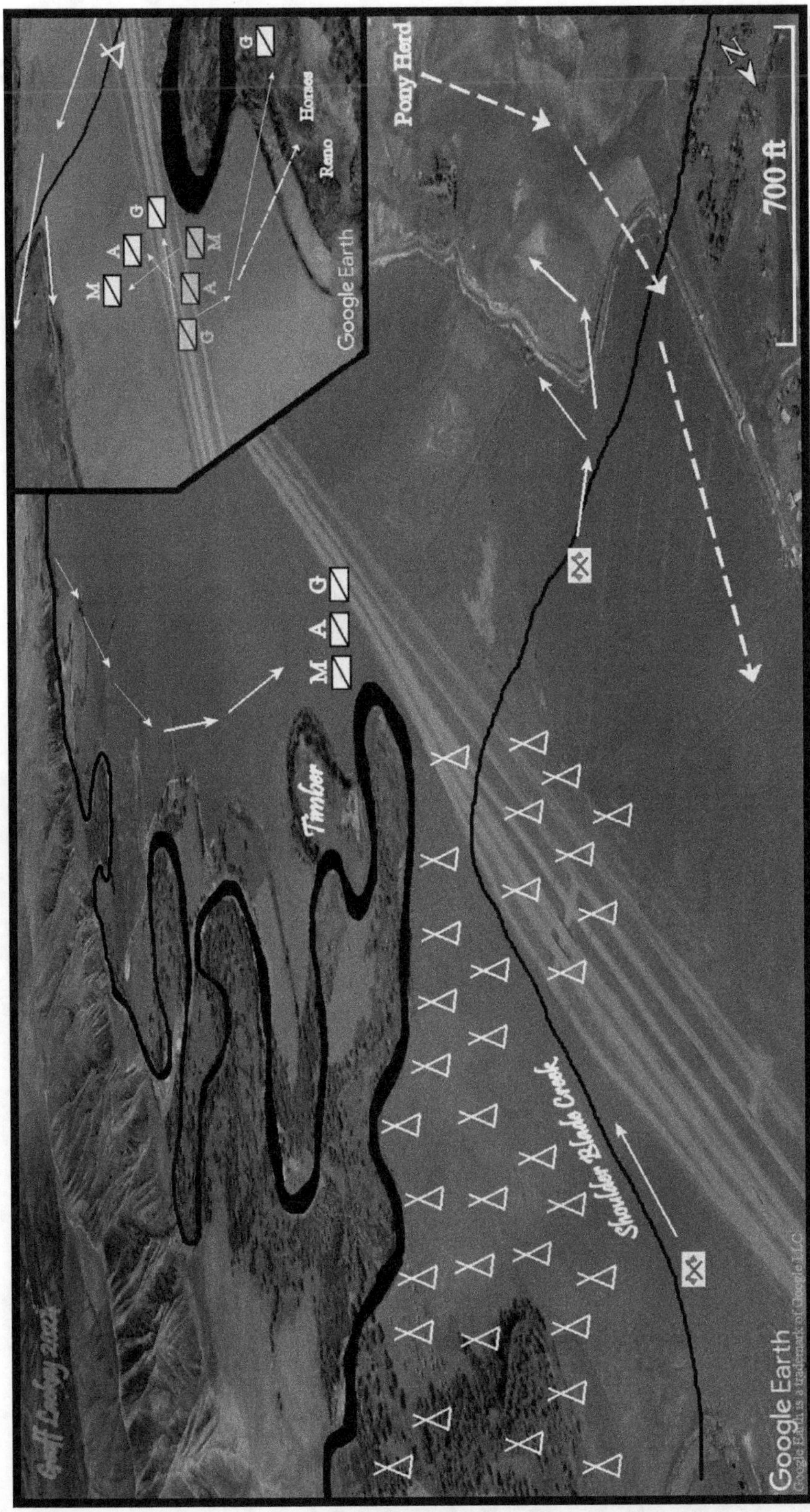

Map 4. Major Reno's battalion advances down the valley towards the Garryowen Bend of the Little Big Horn River.

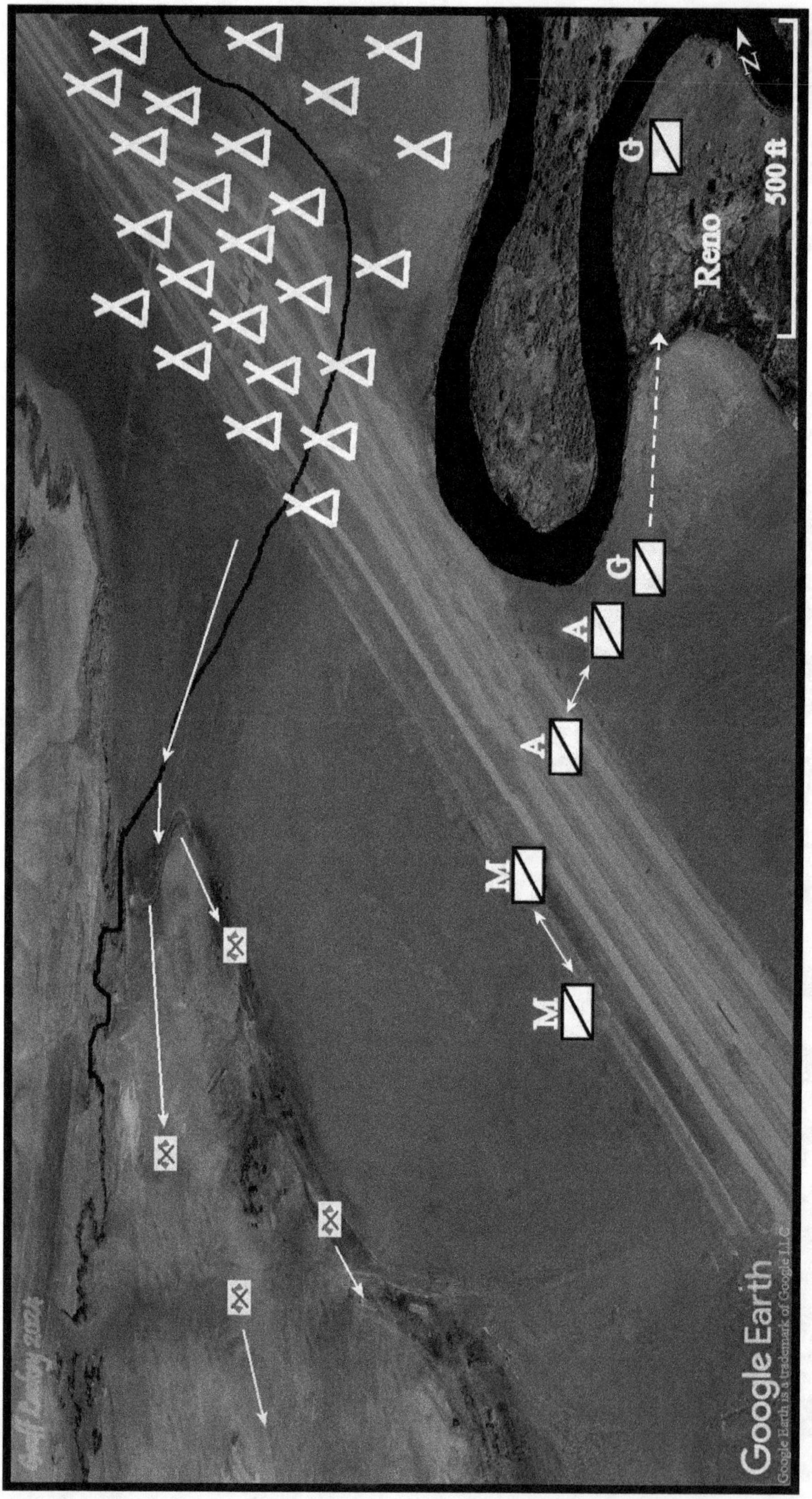

Map 5. Reno's Valley Skirmish Line as G Company withdraws to the timber, and M Company refuses the flank.

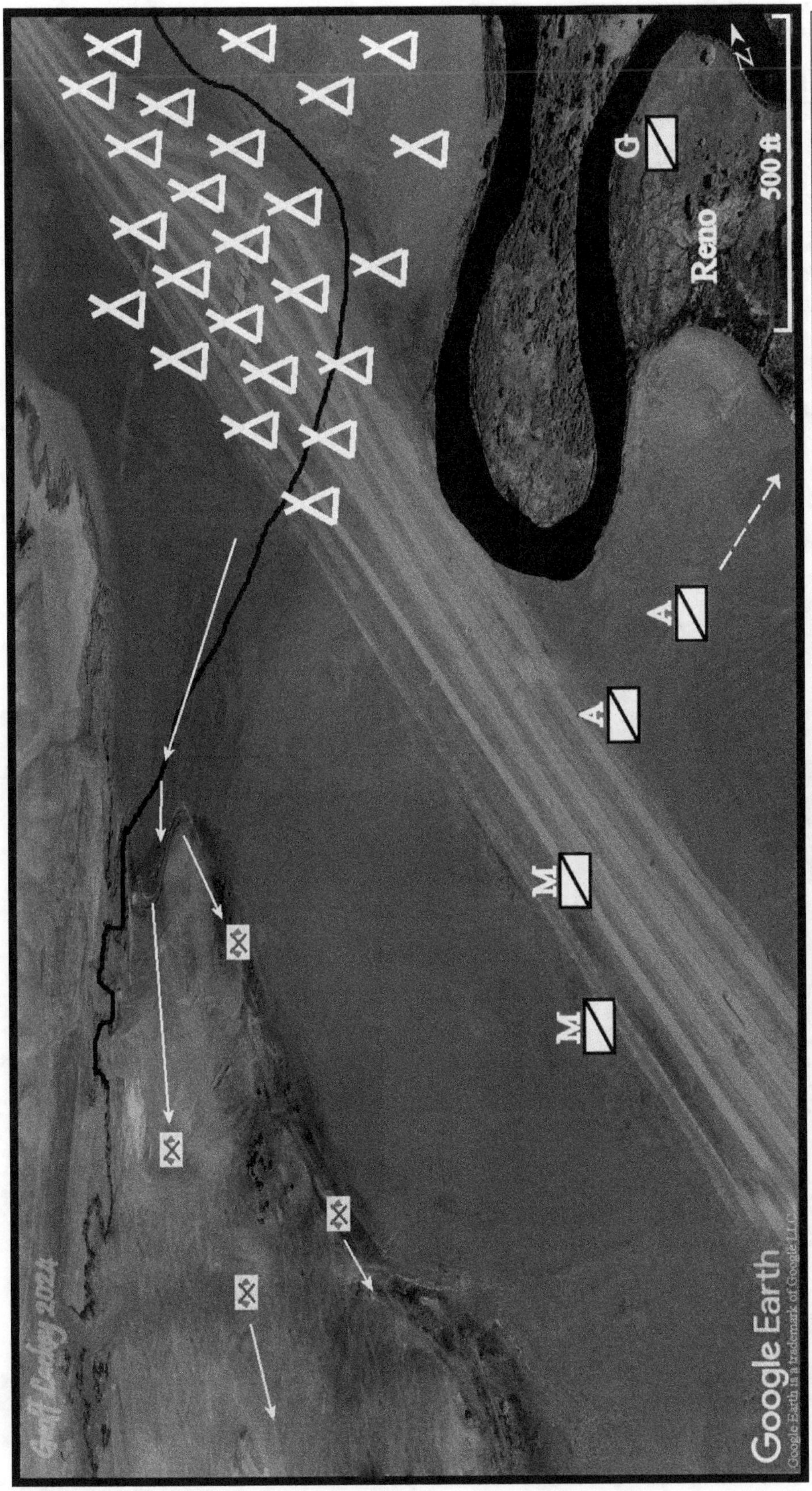

Map 6. **A Company withdraws to the timber position, leaving M Company exposed.**

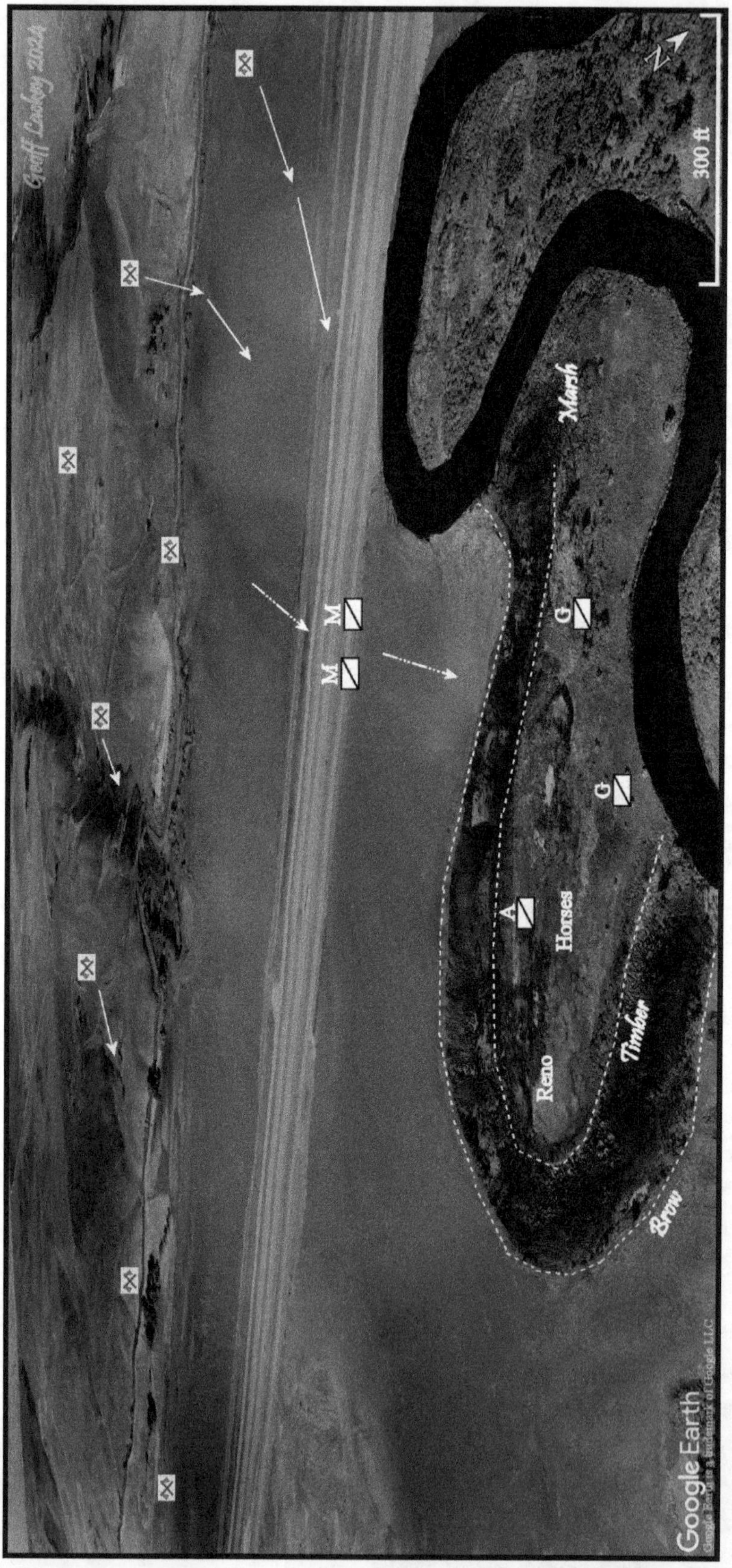

Map 7. M Company withdraws to join the others in the timber position.

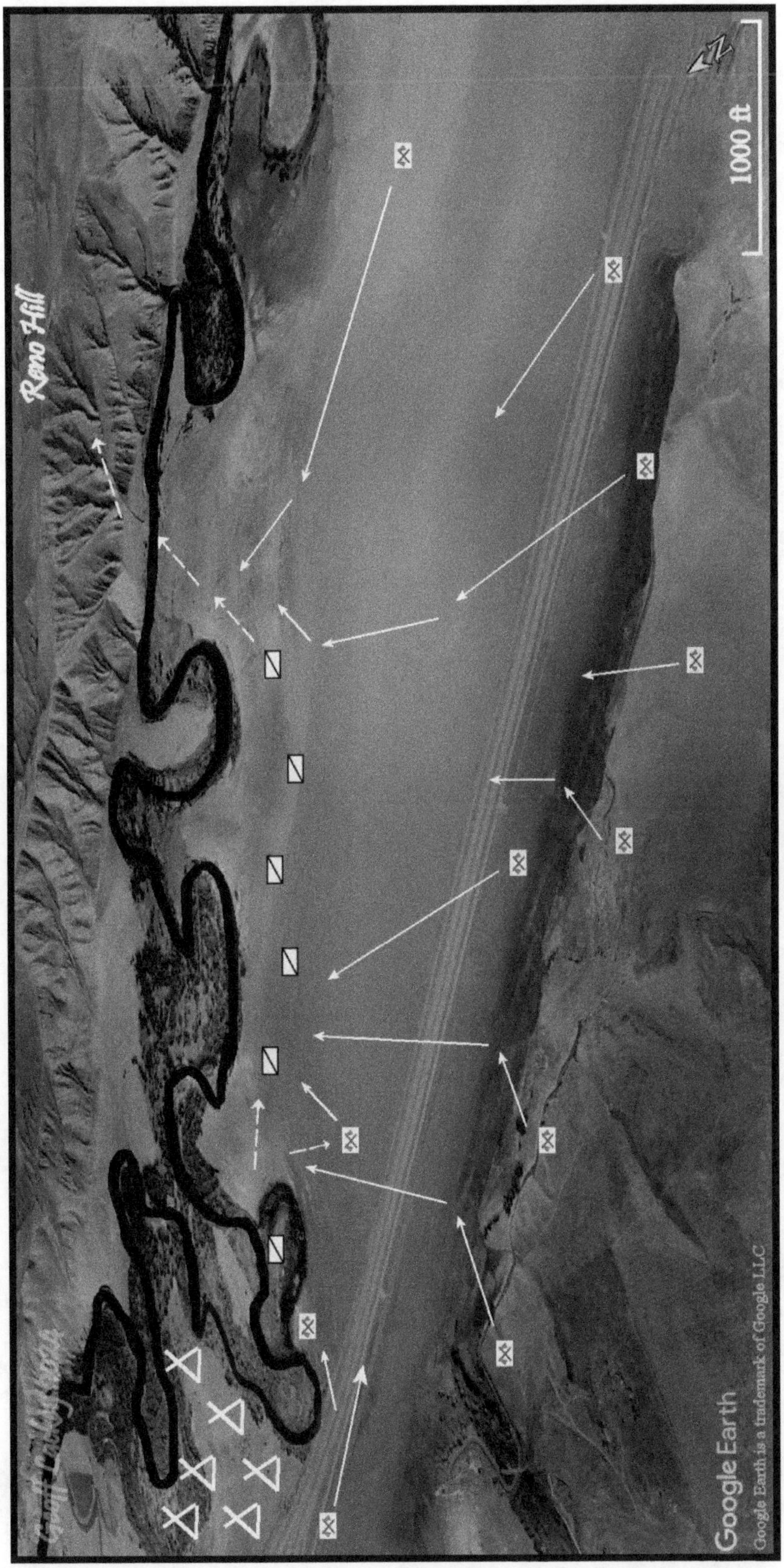

Map 8. Reno retreats from the valley and ascends the bluffs.

Chapter 7: What Route Did Custer Take to Medicine Tail Coulee?

While Reno was pursuing an Indian village, which he believed to be in flight, Custer still had not yet seen the very large encampment he had been tracking for three days. No matter, it was enough for him to know that the village was within striking distance, and timing was essential in reconsolidating the regiment he had purposely divided. The entire outfit would be needed to bring this operation to a successful conclusion, in which he intended to attack the enemy from multiple directions. This scenario was confirmed by Daniel Kanipe, who later recounted, "I have always thought that General Custer's plans were to attack the Indian camp at three different points at the same time..."

Whatever attack plan was germinating in Custer's mind, after considering conflicting intelligence reports, he concluded that the 7th had been discovered and the Indians were bolting. He would forego reconnaissance since an enemy in flight called for immediate action. This was neither impulsive nor unprecedented. Gregory Michno, a historian of Western Indian warfare, discovered that there had been 216 documented instances in which the American military had attacked an Indian village without first scouting it. The only difference at the Little Big Horn is that it was an unusually large enemy encampment, and the regiment was divided into four different battle groups, along with the slow-moving pack train protected by a disproportionate twenty percent of the regiment. The pack train had problems that somewhat delayed it, including stubborn mules carrying their much-needed 24,000 rounds of ammunition. Godfrey recalled that the pack train was a mess, and its inexperienced packers allowed supplies to slip off the mules, thereby consuming precious time as they put them back on.

Even more worrying was that Benteen's battalion had left for his scout an hour and a half earlier, and since his departure, there had been no word from him. Benteen's silence may have made Custer anxious enough to send an Arikara scout in the direction Benteen had gone. The evidence for this is sparse, but the Indian scout Stabbed claimed he had been out with a message to the south. Only Benteen's battalion would have been on a scout in that direction. Neither Benteen nor anyone else in his command ever mentioned any messenger, so the story must be relegated to one of those tantalizing dead ends of which history has no shortage.

If Custer now regretted having ordered Benteen out too early, he did not dwell on it. This is why the Crow scouts said they were happy to go with Custer; no matter the circumstances, as he never abandoned a trail. The Crow scout Curley related how Mitch Boyer once pointed to Custer and said, "This man will stop at nothing." The Crows expressed their admiration for his relentless determination once he struck an enemy trail." The Crows expressed their admiration for his relentless determination once he struck an enemy trail." It was the one thing everyone seemed to agree about: the General was relentless.

For the time being, Custer focused on freshly received intelligence from his Chief of Scouts, Lieutenant Charles Varnum, who informed Custer that he had sighted a portion of the village and it was still immobile, teeming with Indians. As far as Custer was concerned, this was welcome news, not only because of the element of surprise involved but also because the bigger the village, or the greater the odds, the greater the glory. Observing the tail end of Reno's three companies departing from the command, Varnum inquired where they were heading. "To attack the village," Custer crisply replied, stating that Varnum could ride ahead and go with them.

Although the village remained concealed behind the bluffs, its strength was evident from the large pony herd grazing on the western foothills. Whatever the warrior count, Custer harbored no doubt that he would emerge victorious. Confidence was an ever-present part of his modus operandi, and 'Custer's Luck' had yet to fail him on the battlefield. He had openly stated that the 7th could defeat all the Indians it came across. Now, fate had conveniently gathered a large force of hostile Sioux and Cheyenne around the Little Big Horn River as if the Goddess of Fortune challenged him to make good on his boast.

Custer had earlier switched mounts and now rode his sorrel Vic, short for Victory. The only thing that seemed to oppress him was the heat, causing Custer to remove his buckskin coat and tie it behind his saddle. He was getting into battle mode, and the soldiers cheered him, tied handkerchiefs around their heads, and threw their hats in the air. Many knew Custer lived by the axiom that he who hesitates is lost, and it was clear that the General had no intention of letting this golden opportunity slip through his fingers. With Reno and Benteen's battalions moving to block the hostiles' escape to the south and Terry and Gibbon blocking an escape to the north somewhere downstream, Custer's strategy of closing in and trapping the hostiles appeared to be accomplished. What mattered now was speed and how decisively Custer could bring together his divided regiment within mutually supporting distances as they shut off all avenues of flight while circling in for the kill.

As Custer advanced toward the river, perhaps a half mile east of Ford A, where Reno delayed while looking for his commanding officer's promised support, two of his most trusted officers, Cooke and Keogh, suddenly galloped back from Reno and punctured the restless silence with sudden news: The Indians were not running down the valley but were coming up in force to confront Reno. This development put an entirely different spin on things. Custer now weighed whether he should continue to the river or change course. Maybe he would not follow Reno into the valley, for if the warriors were coming out to fight, he could instead surprise them with a flanking maneuver, hitting them from the side—a more vulnerable angle and one less defended than a concentrated front.

This 'fix and flank' maneuver would involve turning right up the bluffs, thus seizing an unanticipated tactical opportunity. Once Custer decided not to follow Reno into the valley, the only logical move was to turn north and ascend the bluffs to a high enough elevation to assess the situation and decide how best to deploy his forces, now spread about the countryside.

Chapter 7: What Route Did Custer Take to Medicine Tail Coulee?

Before advancing three-quarters of a mile west along Ash Creek, Custer watered his horses at the Echleman Ford and then abruptly turned right to climb the one-hundred-foot elevation on the east side of the river. Once at the summit, Custer, with Trumpeter Martin just a few yards behind him, saw that Reno had yet to charge the village and that Benteen's column had just made its way back onto the main trail. For the first time, Custer would observe a portion of the village down in the valley, allowing him to determine whether the Indians were running, coming out in force to fight, or, as he suspected, simply forming a rear guard to delay the attackers to afford their families time to escape.

According to Sergeant Kanipe, their wing ascended the bluffs in columns of two, riding ten men across and twenty men deep. Custer's sudden and unannounced change of plans altered the trajectory of the battle and might have called for a courier to be sent to his other battalion commanders, informing them of his latest movements and plans. Only questionable and anecdotal evidence suggests Custer sent a messenger to deliver this intelligence. If he failed to inform Reno and Benteen, this oversight would be one example of the lack of timely and transparent communication between the three battalion commanders. If so, it would prove a costly mistake, especially since a notice regarding his objective and plan may have given Reno the confidence and resolve to hold the skirmish line on the valley floor or in the subsequent timber area to which he retreated, although as the commander of an advance guard battalion, that was part and parcel of his assigned mission, which theoretically did not need additional orders to emphasize.

Riding at the head of the column, Custer would resemble more of a shorn Samson than his calling card, "Long Hair," an appellation given to him by the Indians because his yellow curls fell to his shoulders. For this campaign, he promised Libbie that he would not make himself an inviting target, so his once long locks had been neatly cropped. Custer was still wearing his blue flannel shirt, white buckskin pants shoved into long-legged boots with red tops, and a bone-white, wide-brimmed hat. When he reached the top of the bluffs, there was little to impede the march north to the other side of the village. The terrain, while rugged, was not inhospitable to mounted cavalry, so all five companies were ordered to reform and march in columns of four, which is the regular marching order when traversing such terrain. Looking down, Custer obtained his first clear view of a good portion of the Indian village and the pony herds on the western foothills that had first been spotted from the elevation of the Crow's Nest. The scouts were right—this was a very large Indian encampment.

Trumpeter John Martin observed that Custer was in a big hurry, but once he reached Sharpshooter Ridge, the general paused to use Lieutenant De Rudio's Austrian field glasses to study the village's make-up and disposition more closely. He may have spent ten minutes studying the village through De Rudio's prized binoculars. The camp was easily a mile and a half in length and a half mile wide, although a hill obstructed his view to the northwest. About this time, Reno's second courier, Private John Mitchell, came galloping toward the General with another message from Reno. He conveyed nothing new, essentially repeating what the first courier, Private Archibald McIlhargey, as well as his Adjutant Cooke, had told him just a short while earlier: The warriors were not fleeing but were coming out in force to challenge Reno. As Custer's glasses scanned the area again, deliberately and methodically sweeping over the valley floor, he must have been puzzled by these messages and Reno's alarm.

In the valley below, he observed rows of pale tepees gleaming in the light of the early afternoon sun. Everything seemed to hum with tranquility, quite unexpected considering the crackling cauldron of excitement he had been expecting. No warriors were charging out of the village or mobilizing to defend themselves from an upcoming attack. The village gave every indication of an unprepared and unengaged encampment. All that could be seen, John Martin recalled, were a few ponies grazing about the tepees, children playing and swimming, a few dogs sniffing around and occasionally barking, and some squaws going about their daily duties as if this were just another hot summer afternoon. There were no visible signs of war parties forming, no indication that the village was in any way alerted to an impending attack, or that it was panicked and preparing to flee in multiple directions.

As he swiftly led the way to the other side of the village, Custer would wait for Reno's advanced guard to draw all the hostiles in the encampment toward his battalion at the south end of the village. That would provide a clear path for Custer's wing to cross the river and then charge through an empty village at the most unsuspecting moment, since the warriors would have been preoccupied with an attack on its southern end. Thus, they could successfully strike an undiscerning enemy in its rear or through its back door. But before Custer departed further north, it was important for him to determine if his divided regiments were within supporting distance of each other.

It is fascinating how many students of this battle believe Custer did not endeavor to obtain critical information about the whereabouts of the disparate parts of his regiment, which he desperately needed to unite in order to win the victory he sought. No one disputes that he knew where Reno was—but Benteen was another story. The belief that Custer did not know Benteen's approximate location before moving downriver, as some have maintained, was first perpetuated by Benteen himself, who stated that Custer could not possibly have known the location of his battalion within ten miles.

But how could that be true? Custer was positioned on the high ground; the sun had burned through the morning haze, his powerful field glasses abetted his vision, and he was being fed information by his Indian scouts, who were strategically deployed high on the bluffs and whose visual powers were even keener than his own. The day was windless, and one's vision to the south was unimpeded. They could hardly have missed Benteen's three companies, the mule pack train, and its approximate 130-man plus guard that kicked up enough dust to make his advance visible from the surrounding elevations.

More proof that Benteen's general position was known to Custer came from Private Theodore Goldin. While riding with Custer's column on the bluffs, Goldin recalled seeing the rising dust from Benteen's battalion and the pack train as he made his way toward the river, which is more proof that Benteen understood that the valley was his destination and that he was not to follow Custer over the bluffs. The point is, if Goldin saw Benteen's dust, so did Custer. This also explains why Custer sent Trumpeter Martin with the famous "Come on--Big Village—Be quick" note since he now knew from the prodigious sweep of dust rising in the east that Benteen had finally returned to the main trail. Nor does Goldin's regrettable tendency for fabrication and hyperbole, especially when featherbedding his own exploits, disqualify his observation. With reasonable assurance, Goldin could be relied on when conveying matters that had no import on his own heroics. One can safely conclude that by the time Custer arrived on Sharpshooter Ridge, he knew the general location of Benteen's battalion.

Before long, Custer's command saw Reno and his three companies charging down the valley floor. After forty days in the field, it had finally begun. Now alerted to an imminent attack, the unsuspecting and unprepared village broke into a storm of hysteria. How long Custer watched events unfold in the valley and the village is unknown. No doubt, he waited to see what course of action Reno's battalion would take as the warriors took up their traditional defenses by forming a battle screen for their fleeing families.

Confirming the sequence of events, Varnum had a glimpse of Custer's wing moving across the bluffs when Reno dismounted in the valley, verifying that Custer had not yet begun to move to the other side of the village. Custer saw Reno's charge from approximately a mile away over impassable ground between the two respective cavalry forces, Custer high on the bluffs and Reno down in the valley.

At first, it was anyone's guess whether Reno would charge into the village without his expected reinforcements. Custer would have observed that Reno's command had changed from a column to a line formation, and Custer must have then realized that Reno did not intend to plunge into the southern end of the village. Rather, his forces were likely to dismount and form a skirmish line, which they did.

Regardless, Custer was almost certainly unvexed by this development since Reno was still playing the role Custer had intended: an advance guard occupying the enemy while affording Custer time to move downstream and strike at a vantage point the Indians were not expecting. It is more than likely that Custer immediately left Sharpshooter Ridge after Reno dismounted his battalion. Custer began to move north, searching for another river crossing to attack the village from the other end. Thus, Custer did not loiter to observe the valley fight and Reno's retreat into the timber.

If anything, the village appeared startlingly vulnerable, like low-hanging fruit ripe for the taking. Custer could not have planned a more inviting scenario if he himself had set up the chess pieces on the board for the inevitable checkmate. From his vantage point on Sharpshooter Ridge, Custer saw Reno's battalion down in the valley, advancing to the attack. Looking south, he saw dust churning upward from Benteen's battalion and the pack train in his rear. Typically high-strung when on the cusp of battle, Custer exclaimed, "We caught 'em napping, boys! We will go down and make a crossing." Buoyed by the cocksure certitude of their commander, the troopers began to whoop and holler to the point where several horses became agitated. With a battle still to be won, Custer cautioned them to hold fast to their mounts, assuring them that there were "plenty of them down there for all of us."

This exuberant demonstration was another burst of confidence from Custer and the 7th, one that they would, in ways that were then unthinkable, come to rue later when the fickleness of chance worked its black magic in nebulous ways. No one expected that the whole house of cards Custer had meticulously assembled would come tumbling down in a gamble gone terribly wrong.

However, at that moment, the battalion's morale was sky-high, especially since the encampment had shown that they were unaware of the threat mounted against them. Nevertheless, the Little Big Horn was not Washita, which was a small mouthful to swallow compared to what he now confronted in this great valley—a massive feast of Epicurean proportions. Unlike Washita, which commenced in nearly pitch darkness because Indian villages were most vulnerable at pre-dawn, this attack would begin past noon, underscoring Custer's growing confidence. As of now, fortune was, as Custer always believed, favoring the bold.

After leaving Sharpshooter Ridge, Custer moved down into Cedar Coulee, heading north, before making a sharp turn to the left and moving west across the northern ridges of what would later be called the Weir Peaks, heading directly towards the river. Just shy of reaching the head of Western Coulee, Custer got a further view of the size and scope of the encampment, and knew he needed Benteen's battalion and the pack train with its ammunition to get to the valley immediately. Custer also observed that no warriors obstructed Benteen's path to reunite with the rest of the regiment. Cooke wasted no time tearing out a page from a small notebook with orders from Custer for Benteen to hurry.

If Custer brought his battalion north for a 'fix and flank', he needed Benteen to close the gap between him and Reno, if he was to reconsolidate the regiment. Custer also knew that he had to move rapidly to find a river crossing somewhere up north, from which he could attack from another direction, and determine how far downstream the village extended. This prompted him to send his trumpeter, John Martin, to hurry Benteen into the valley to support Reno, who was still the advanced guard acting to occupy the enemy and threaten the village from the south.

Even though all the parts were not yet in play, Custer had to be greatly encouraged by what he saw. His first instinct had been right; if any resistance came from the village, it would be an Indian rear-guard action allowing the non-combatants time to flee, whereupon the warriors would follow in their wake. The Lakota warrior Low Dog alluded to this tactic when he stated that the warriors would cease fighting and fall back once the women and children were safe.

Custer would seek the shortest way to the other side of the village. But what route would he ultimately take to that fateful destination? Unlike many other questions and controversies that plague the battle, this was the one subject that, historically speaking, battle students have generally agreed upon. The prevailing consensus, and by far the most popular theory of Custer's march, maintained that Cedar Coulee was the path Custer and his troops embarked upon. This long ravine, apparently named after cedar trees that once populated it, but now replaced by juniper bushes, was believed to be the route Custer would eventually take to Medicine Tail Coulee. That path would ultimately lead down to a river crossing to the other side of the village. Custer's popularly accepted route to Medicine Tail Coulee would become all but settled doctrine and remained so for a long time.

The reason the Cedar Coulee hypothesis became so widely accepted was that Walter Camp, an early historical interviewer and researcher, had lent his imprimatur to the story, effectively transforming what was speculation into a matter of sacred scripture. Camp was the editor of the *Railway and Engineering Review* who, beginning around 1890, made an Olympian effort to hunt down and interview every white and Native American who either participated or witnessed some part of the famous battle. His dogged pursuit of documenting various and numerous recollections included interviews with more than sixty survivors of Reno's command and over 150 Sioux and Cheyenne. His corpus became a gold mine for historians and scholars researching the battle, and their accounts were so thoroughly and expansively accumulated that few were willing to challenge his conclusions, which helps explain the longevity and insularity of the Cedar Coulee viewpoint despite some compelling evidence to the contrary that was always available, if not obvious.

Hank Weibert, a rancher who had spent sixty-six years in the area, scoffed at the idea that Custer had traveled through Cedar Coulee, but he was almost alone in holding that belief. Then, after much reflection, inquiry, and study of the terrain, a thirty-year-plus seasonal ranger at the Little Big Horn Battlefield challenged the orthodoxy and, in doing so, essentially changed the way students and scholars now think about Custer's march over the bluffs toward Medicine Tail Coulee. Ranger Michael Donahue, the author of several well-received and publicized books about the battle, has probably been over and across the Little Big Horn Battlefield as many times as anyone now living. Plunging deeply into the primary sources, including both written testimony and drawn battle maps, Donahue employed empirical observation, logistics, and logic to analyze the movement of Custer's wing to Medicine Tail Coulee.

This author concluded that Custer did not go all the way down Cedar Coulee, which eventually turns right, away from the river. Donahue notes that reaching a river crossing was Custer's objective and that riding all the way down Cedar Coulee would have eaten valuable time. Instead, he argues that Custer's path down Cedar Coulee was truncated for an alternate route. Donahue makes such a convincing case for his hypothesis it is now hard to fathom why so many unquestioningly subscribed to the belief that Custer traversed the entirety of Cedar Coulee rather than, as Donahue maintains, having gone down only a third of the way before cutting over toward the west to Western Coulee, thence down to the river, which is closer to how Martin better described it.

Martin's attestations regarding Custer's route had taken a back seat to Camp's insistence that Custer had gone down the whole of Cedar Coulee. Having interviewed Martin several times, Camp was well aware of the trumpeter's recollection about Custer's path but had concluded that he was mistaken. In a series of letters, a perplexed and frustrated Camp earnestly tried to persuade Martin that Custer exited Cedar Coulee at its mouth, then turned left and descended from there into Medicine Tail Coulee. But the trumpeter remained steadfast and would not budge on the question. Martin claimed that Custer's route to Medicine Tail Coulee took them within six hundred yards of the river, which would place their route immediately behind the first line of bluffs that overlook the river. And as Donahue correctly points out, this path is the closest viable route to the waterway. Getting to his destination as rapidly as possible was critical if Custer was to flank the enemy and support Reno. As stated previously, we know that Custer was moving swiftly, as John Martin made a capital point of noting that he had been trotting the whole way, as if to underscore Custer's velocity and determination. But had he decided to go down Cedar Coulee, all his hustle would have been for naught and counterproductive because he would have ended up a mile and a half from the river crossing—his targeted destination and one that gave him passage for a flank attack. By following down Western Coulee, Custer would have observed how the Reno fight was developing in the valley from the high ridges beside the river.

However, there are still more factors to consider than just the verbal testimony indicating Custer's true path to Medicine Tail Coulee, including the all-important map evidence. The maps lend credence to the fact that Custer had gone to the hill in question and thence down the ravine but did not fully traverse Cedar Coulee. After the 7th's devastating defeat, there was an early attempt to visually put the pieces together to understand Custer's last movements, hoping that it would provide a portal into Custer's thinking. One way to examine all of the pieces in the battle was to draw maps based on post-battle evidence.

On June 27, two days after the battle, an inquisitive Benteen, brimming with questions, was accompanied by Martin, and together, they followed Custer's trail on the east side of the river to reconstruct the probable route Custer took. Events were fresh in the trumpeter's mind since only forty-eight hours had passed from the time of the battle. Martin showed Benteen the spot from which he had left the Custer battalion after Cooke handed him the urgent note about a big village and the need to hasten Benteen's battalion—a location for Martin's departure point that Benteen estimated was approximately six hundred yards from the river. Armed with this information from a living eyewitness, Benteen took a pencil and paper and drew a map of what he saw while surveying the area with Martin. A careful review of that map does not show Custer going down Cedar Coulee but rather down the first ravine closest to the river.

Benteen was an intelligent and experienced field officer, and he was particular to take with him the only surviving eyewitness, John Martin, who had recently traversed this terrain with Custer, and who would, for years afterward, even under pressure from an insistent and informed student of the battle (Walter Camp), never waver or equivocate from his recollection of their route traversed. The collaboration between Benteen and Martin is perhaps the most definitive evidence available about Custer's route to Medicine Tail Coulee.

Another map that has been highly instrumental in addressing this question and ascertaining clues about the route taken was the one drawn by First Lieutenant Edward Maguire. Maguire, twenty-eight years old, was the chief engineering officer with the Department of Dakota army column commanded by General Alfred Terry. He would also be among the first to map the battlefield, and later, much to his surprise and chagrin, he would be called to testify at the Reno Court of Inquiry about his map, which would be utilized as a prime exhibit, and in fact, act as a prima facie witness. Yet, when he drew his battlefield map, Maguire had no idea he would be called to testify officially and that his hastily drawn map would be elevated to such significance. Because so much information was compiled in the immediate aftermath of the battle, Maguire drew his map more casually, extemporaneously, and certainly less punctiliously than he would have had he known it would become the subject of such a critical and highly publicized army inquiry. Nevertheless, despite its mistakes, the McGuire map remains of topological and historical interest to battle scholars. Of particular interest to our subject is the fact that the Maguire map shows Custer's route to Medicine Tail Coulee as traversing down Western, not Cedar, Coulee.

Chapter 7: What Route Did Custer Take to Medicine Tail Coulee?

In his later years, Martin, clearly caught up with his place in history and the folklore of this epic fight, would, unfortunately, though almost unconsciously, change some of his initial testimony to fit the mythological aura that had come to surround the battle and his starring role in it. However, whatever embellishments Martin undertook cannot take precedence over his earlier recollections. Going back to less than three years after the battle, while he was under oath during the Reno Court of Inquiry, Martin never voiced a doubt about the route Custer took, and it never included going over the Weir Peaks, as he insisted that only the scouts had gone up there, and that instead, Custer went to the top of a high hill—i.e., a clear reference to Sharpshooter Ridge.

Although he did not draw any maps of the battlefield (nor in his remaining forty-six years of life did he return to the Little Big Horn despite his lifelong and emotional connection to it), Martin remained unwavering about Custer's trajectory. When he said that Custer climbed the high hill, he did not mean the Weir Peaks but rather the most conspicuous promontory in the vicinity, Sharpshooter Ridge. This was also the opinion of Colonel John Gibbon of Terry's command, who carefully studied Custer's final movements and concluded that he went, not over, but to the right of the Weir Peaks. Adding to the evidence is an unpublished letter by Captain Thomas French stating that Custer's men "rode over the saddle" to their death beyond. "Over the saddle" could have been a reference to Weir Peaks, which had a curved dip that resembled a saddle, but it is also possible that this description fits another and more likely location.

While it is possible that Captain Thomas French might have been alluding to the Weir Peaks, several other contours in the landscape physically resembled a saddle, specifically on the south end of Sharpshooter Ridge toward the edge of the bluffs. Could this be the site, one that would have still camouflaged Custer from the Indians, to which French was referring? It might not be incontestably clear, but its likelihood is compelling when other accounts germane to the subject are factored in, along with the knowledge that concealing a portion of one's force from the enemy is the sine qua non of a 'fix and flank' maneuver.

There is something else to consider about the Cedar Coulee route. Had Custer gone over the Weir Peaks, again the highest point in the whole panorama of that hilly country, he could not have avoided being seen by Indians down in the valley. That is the very thing Custer tried to avoid, since the purpose of a 'fix and flank' is to remain concealed from the adversary. Going over the top of those peaks would have put the command directly in the village's line of sight.

Not surprisingly, some objections remain among respected authors and researchers regarding this battle. Larry Sklenar, who wrote a commendably researched and largely unappreciated book on the battle, *To Hell with Honor,* offers his own unique interpretation that Custer had wanted to be seen on the bluffs to create an illusion of countless soldiers waiting to pounce on the village. Such an optical stunt would likely deflate and demoralize the enemy. But in his counterintuitive surmise, we must part from Sklenar. While this theory is seductively appealing insofar as it supplies a reason for Custer wanting to be seen, it does not obviate Custer's paramount objective in necessitating a 'fix and flank', which was to avoid being seen at all.

Custer's Last Stand Demystified

The strongest evidence concerning Custer's ride to Medicine Tail Coulee is arguably from Lieutenant Winfield Scott Edgerly. On the day of the battle, Edgerly rode with Captain Thomas Weir to the Peaks, now named after him, to make contact with Custer after hearing volley fire downriver. Tortured by guilt for not coming to Custer's assistance, Weir would drink himself to death, dying some six months after the battle without committing much to paper and revealing very little about his harrowing recollections.

But Edgerly would live for many years after, and he was quite firm in saying there was no sign of any tracks or physical evidence whatsoever that Custer had gone over the top of the Weir Peaks. Five mounted companies trekking over this terrain would make an obvious and unmistakable trail, and not a sliver of evidence exists indicating horse tracks on the peaks. It is far more likely that Custer led his battalion down a portion of Cedar Coulee and then cut over the lower northern ridges of the Weir Peaks, continuing to move in a westerly direction until he dropped into Western Coulee, the route that would lead him down to the river located about six hundred yards from its junction with Medicine Tail Coulee, and destiny itself.

Chapter 8: When Did the Custer Fight Begin?

Many battle students accept that the Custer fight, commencing at Medicine Tail Coulee, began perhaps ten minutes after Reno's charge down the valley. But is that a reliable estimate, or did the Custer battle actually begin an hour later? Establishing a reasonably accurate battle timeline is key to sequencing the dizzying maze of events that day. Formulating such a timeline is indispensable to understanding how this battle unfolded and is essential to ascertaining when certain events happened.

To determine when the Custer fight began—which was separate and distinct from the valley fight and the siege on Reno Hill—we must return to Trumpeter John Martin, Custer's last messenger. What did Martin claim to see and hear en route to Benteen? With matters heating up, Martin rode to Benteen with a note from Adjutant Cooke, who was in such jittery haste that the message he wrote consisted of a few words scribbled on a piece of paper brusquely torn from his notebook.

Spurring his horse on, with hoofbeats pounding the hardened earth, Martin focused on reaching Benteen without getting shot or his scalp lifted. He briefly glanced down at the valley where Reno's command was engaging with the Indians. But while he did not see much of that battle, he certainly heard it, and in his mind's echo chamber, his auricular discernment became as disoriented as some of his later recollections when colored by the afterglow of a battle that has now passed into legend. The erroneous timing of the commencement of the Custer fight came down to gunshots from the valley fight that Martin heard, misunderstood, and misplaced.

Martin had received Cooke's note approximately six hundred yards from the river, so that was the place to determine what he saw and heard. But, as the years went by, and the aura of Custer's last stand grew in the American imagination, Martin began putting himself closer and closer to the Custer Battlefield. The question is, was Martin's 'six hundred yards from the river' a reference to the crossing at Ford B-1, which some have erroneously surmised? Martin's actual point of departure was roughly a mile and a half upriver from Ford B, undermining the argument that he witnessed the beginning of the Custer fight. Let us recall that Martin testified at the Reno Court of Inquiry that he traveled about three-fourths of a mile (he would years later estimate that distance to be closer to a mile) and arrived on the same hill (i.e., Sharpshooter Ridge) from which he had seen the village with Custer not ten minutes earlier.

At the Reno Court of Inquiry, Martin had been unsure if the shots he heard came from Custer or Reno's command. Only in later years, seemingly enthralled by the romance of being the only living soul to witness the beginning of Custer's Last Stand, did he come to believe that what he saw was the beginning of the end. To support his version of events, Martin referred to seeing Indians across the river waving blankets or robes. But while this recollection is probably accurate, Martin's sighting had nothing to do with Custer's battalion or the onset of hostilities between Custer's five companies and the tribal warriors of the Sioux and Cheyenne. Those blanket or robe wavers were likely the Indian pony herders, who used them not to signal any offensive action, but rather to drive the horses down into the camps northward and away from the threat of Reno's attack at the south end of the village.

The Cheyenne warrior Bobtail Horse corroborated this. He was half dozing when the sound of roaring thunder reached his ears. He soon recognized that thunder as the hooves of thousands of Indian ponies driven northward through the Cheyenne encampment. The Indians' herd likely caught Trumpeter Martin's attention on his ride to Benteen, feeding his imagination that a large war party was attacking Custer's wing. The dust clouds he saw down the valley toward Custer's position were created by those ponies stampeding northward, not, as Martin would later convince himself, the result of a mad rush of warriors in an all-out attack on Custer at the Ford B crossing.

Martin's ride back over the bluffs only added to his confusion when he glimpsed Custer's troops galloping northward away from the river, creating the impression that the Custer fight had begun earlier than it actually had. Custer deployed most of his battalion in North Medicine Tail Coulee—hidden from the village inhabitants behind Luce Ridge—which convinced Martin that Custer's five companies were forced from Ford B only minutes after his own departure, and that Custer's wing was in full retreat and riding away from the river! However, Martin's later interpretation of what he saw and heard so long ago could not be further from the truth.

Initially, when Martin first reached Benteen, he told a much different story. According to Martin, the soldiers were rolling up the village, and Reno was killing almost every Indian in sight. This description is the polar opposite of a Custer retreat from an onslaught of warriors at Ford B. The reality was that Custer was not being forced back anywhere. Custer's objective was to not tip his hand to the enemy and reveal his location before the warriors in the encampment had been drawn to the Reno fight at the south end of the village—lest he find himself fighting a large force at the opposite end. During the early and middle stages of the Reno fight, Custer successfully concealed his presence, waiting for the right moment to strike. Timing is critical in a 'fix and flank' maneuver. Striking too soon can be just as disastrous as attacking too late.

It was one man's misperception that launched the story that the Custer fight began almost simultaneously or, at most, ten minutes after the start of Reno's valley fight—notwithstanding Walter Camp and General Woodruff's assessment that there was at least a forty- to forty-five-minute gap between the two events. This misunderstanding still prevails today and has been the source of some confusing timelines that wreak havoc and spawn miscalculations when analyzing the sequence of battle events.

The best estimate for when the opening shots of the Reno valley fight occurred is approximately 1:35 to 1:40 p.m., and it was very likely an hour later, about 2:40 p.m., before the onset of the Custer fight near the Ford B crossing. Based on Custer's history and temperament, the notion that he would allow so much time to pass before engaging the enemy has struck many battle students, as well as at least one who fought with him that day, as downright unbelievable, if not ludicrous. Lieutenant Varnum, to cite one source, became incredulous over the idea that there was such a long gap between the start of the valley fight and the engagement at Ford B. Varnum dismissed that because he felt Custer was not constitutionally built to wait five minutes, much less nearly an hour before he jumped into the fray as quickly as if a fire was at his backside. But these protests about Custer not being psychologically oriented toward patience during a military engagement are a projection of Varnum's inclinations, not Custer's. The fact is that Custer had engaged in successful 'fix and flank' maneuvers during the American Civil War, most notably at Tom's Brook against General Thomas Rosser's Laurel Brigade.

A 'fix and flank' maneuver necessarily involves a waiting period for those doing the fixing and those doing the flanking. Because of the unusually large number of warriors in the village, Custer was forced to wait longer before striking. It is a prelude pregnant with peril, whereby the flanking force assesses from a secreted location how and where the enemy is reacting to the advance guard (the fix) so the flank attack can respond accordingly. In other words, first, you fix, and then you flank. Reno's force in the valley was the fix, and Custer's five companies were the flanking force. This was not a pincer movement where two disparate forces attacked the enemy simultaneously from different directions. The purpose of the maneuver was for Reno to draw all the warriors out of the village toward him, leaving an open back door for Custer to charge into the warriors from the opposite end.

Reno had not been sent to conquer the valley with three companies; he was the advance guard. He is to attack, ideally via charge, the south end of the Hunkpapa camp and start a fight. And he is to stay in that fight for however long it takes for the main body to strike the other end. This point is key. The main body, led by Custer, is not on some determined, inflexible schedule. There is no timetable. The purpose of the advance guard is to draw the enemy upon itself, making the enemy's flank and rear vulnerable to attack. The timing of that attack is, by definition, up to the main body. If Custer attacks before the warriors are sufficiently committed to fighting Reno, it defeats the purpose and places Custer between two bodies of warriors rather than placing the entire warrior force between him and Reno. Custer will only strike the warriors once they have sufficiently converged on Reno, however long it takes them to do so. The oft-made claim that Reno was expected to charge the entire length of the mile-and-a-half-long village is simply a misunderstanding of how the cavalry operated. If Reno was expected to charge through the entire village, then there was no reason for Custer to be moving north to strike the rear of an already-defeated village. So, while it might often be stated that it was unreasonable for Custer to expect Reno to defeat the entire village by himself, the premise is false. Custer expected no such thing. Nor, despite his later claims, did Reno ever think he did.

A 'fix and flank' had been a tried-and-true tactic employed in the American Civil War that sometimes produced spectacular results for a smaller force against a much larger one. It was most famously used against Federal forces at the Battle of Chancellorsville, where Robert E. Lee literally waited hours for Stonewall Jackson to flank General Joseph Hooker's forces, giving the Confederates their greatest victory of the war. During this famous battle, Custer was keenly aware of Lee and Jackson's tactics. He utilized the very same maneuver with great success at the Battle of Waynesboro in the

Shenandoah Valley against General Jubal Early, a renowned commander of the Confederate army. It was a brilliantly executed maneuver and is recounted here for the benefit of the reader:

> Leading the 3rd Cavalry, Custer had Colonel Pennington move around under the Confederate left under the cover of some woods near the river. Led by the division's Inspector General, lieutenant Colonel Edward Whitaker, the flanking force got into position unseen, crouching among the trees and awaiting the signal for Custer to turn them loose. Custer kept the enemy's attention riveted to the front by having Colonel Wells send forward a line of mounted skirmishers, while the 3rd Brigade was massed for a charge. When all was ready Joseph Fought, the bugler, sounded the charge and Colonel's Whitaker's flank crashed out of the thicket. At the same time, the 2nd and 3rd brigades dashed over the breastwork and Custer and his troops tore after the fleeing rebels. Colonel Stevenson of the 1st New York remarked how the Confederate infantry was totally dismayed and stampeded while also noting that Custer's gleaming saber and scarlet cravat being conspicuous among the foremost.

That would not be the last time Custer used Civil War flanking tactics. He also used a version of it during the Plains Indian wars against Lakota warriors under Sitting Bull and Gall. Because of the Indians' deficiency in command and control, this tactic against the Sioux and Cheyenne was far easier to execute than it had been against a trained conventional army. So, Custer would wait and conceal his battalion, probably longer than he had hoped or expected. But he was nonetheless compelled to do so for multiple reasons: Benteen had yet to arrive in the valley to support Reno, the enemy fighting force was substantially larger than his own, and the warriors delayed mobilizing for battle.

All these elements necessitated a longer wait because the preponderance of warriors must be drawn to Reno's position south of the village. Supporting the concept that there was an extended delay between Reno's charge down the valley and Custer's first act of aggression at Ford B was that Sitting Bull, who was closely watching the battle on the west side of the river, mistakenly believed that Custer and Reno's troops were one and the same because of the time it would take to traverse the nearly three miles from Reno Hill to Ford B.

During his trek across the bluffs to the other end of the village, Custer told Mitch Boyer to release the Crow scouts: White Man Runs Him, Hairy Moccasin, and Goes Ahead. They were not expected to fight, having done their job by helping to lead the soldiers to the village. Boyer, though also released, would remain with Custer for reasons known only to himself, only telling the other Crows that he was going to fight with the soldiers. However, a young Crow scout named Curley would later implausibly insist that he never departed from Custer but stayed on the bluffs with the command almost to the end of the battle. This appears to be another example of a participant exaggerating what he saw and did during the fight. On the tenth anniversary of the battle, when Gall heard Curley's tale of surviving the Custer fight, he sarcastically asked, "Where are his wings since nothing but a bird could have escaped the Custer field that day." Thomas LaForge,also known as White Crow, shared the sentiment, saying anyone who crossed Medicine Tail Coulee with Custer was still there.

After Martin's departure, near a location now called "Upper Boyer's Bluff" overlooking the river, Custer engaged in a "Leader's Reconnaissance" to assess developments in the valley and find a suitable crossing to attack the village. It was there that Custer finally got his long-awaited look at the entire village. He could now see how far north it extended. From Upper Boyer's Bluff, he could see a portion of the Reno fight to the south and the thousands of panicked women, children, and oldsters streaming out of the village to the north and west. It was a ripe harvest of refugees and was exactly the outcome for which Custer had hoped, affording him the opportunity to have his battalion sweep down into the village like a whirlwind of avenging angels.

At 2:00 p.m., about twenty minutes after the start of the valley fight, Custer's command remained behind the bluffs near Medicine Tail Coulee, while Reno's battalion fought in the valley. A frustrated and incensed Benteen, returning from his scout to the south, joined the main trail at about 1 p.m., where he proceeded to halt and water his horses for some twenty-five minutes at an area he called the boggy place. So far, aside from Benteen's noted absence, Custer's improvisations on the fly were coming together as well as he could have hoped under such circumstances.

By that time, Boston Custer had likely arrived at Western Coulee to meet up with his older brother. He had made a determined and courageous run from the time he left the pack train on a fresh Indian pony, to somewhere along the west side of Sharpshooter Ridge, about a mile north of Reno Hill, where he met and conversed with John Martin about the whereabouts of the General. Despite Martin's warning about the presence of Indians, Boston rode across the next ridge beyond the Weir Peaks and on into Western Coulee. Boston's itinerary was high theatre and another example of the old Custer chutzpah against daunting odds.

Riding north from Weir Peaks, it took Boston another twelve to fifteen minutes to reach Custer's command. When he arrived, the General was likely on a leader's reconnaissance near the river below Boyer's Bluff. The evidence that Custer embarked on this scout has been challenged but cannot be easily dismissed. Custer's solitary scout is based solely on the testimony of one who spotted the General, remarkably enough, riding alone. Private Peter Thompson, a soldier in C Company, part of Custer's battalion, found himself on foot after his horse had broken down. Three other stragglers, all from C Company, were left behind Custer's column. Two of the troopers probably had second thoughts about fighting an enormous Indian force and disappeared. Later, Thompson encountered Private James Watson, whose mount had fallen behind. It was shortly after the time that he ran into Watson that Thompson said he saw Custer all alone and apparently scouting the area.

While it may seem implausible that the commanding officer of a regiment would be all alone nosing around in the heart of Indian country, Custer was notorious for making these risky solo excursions away from his command, sometimes merely to hunt, other times to scout. These reports had greatly upset his wife, Elizabeth Custer, and the General found himself compelled to assure his anxious spouse that he would not stray or drift far from the regiment. However, neither could Custer stray far from who he was.

For many battle students, Thompson's sighting is an important marker because it pinpoints where Custer was and provides a clue about the time of the Custer fight. Thompson stated that firing was heard downstream approximately half an hour after he saw the General. At this ford about a mile upstream from Ford B (now designated as Thompson's Ford), the private recalled that Custer was riding at the river's edge, appearing to look for a suitable crossing, and this particular pathway apparently did not meet his expectations. There was too much timber on the opposite side of the river, and the prospect of five cavalry companies getting tangled in a thicket of heavy brush and woods would render them easy targets.

Custer motioned with hand signals for the two troopers to join him downstream, a conjunction that never came to pass and undoubtedly saved Thompson and Watson's lives. Oddly, Thompson said he also saw an Indian scout with a captured Lakota woman on the banks of the Little Big Horn, which made some battle students question Thompson's truthfulness. Instead of joining Custer, the two troopers cautiously backtracked and took part in the fight on Reno Hill, a battle in which both survived. Understandably, over the years, many have questioned the veracity of Thompson's story, including Walter Camp, who believed that the trooper, though sincere, was sunstruck, confused, or frightened to death after finding himself virtually alone in hostile Indian territory.

Many battle students believe that Thompson, a candid and straightforward character, had only imagined he had seen the General. It did not help his cause that a few years later, his companion James Watson died before Walter Camp or any other enterprising reporter could interview him to substantiate Thompson's vignette. Peculiarly, Watson did not speak to anyone about this hair-raising tale even though it involved a direct encounter with Custer just before the General's fatal fight with the Indians. Nor is there any correspondence from Watson or anyone who spoke to Watson confirming this occurrence. All that is recorded for this curious episode is Private Thompson's eyewitness account and verifiable timing factors to support his narrative. Thompson made his way back to Reno Hill, where the reactions of the enlisted men in the regiment were mixed; some supported his narrative, while others dismissed it.

In his defense, Thompson was not a blusterer or a prevaricator like Theodore Goldin (who, of all people, denounced Thompson as a fabulist), nor was Thompson given to telling tall tales like Curley and some of the Crow scouts. His fellow C Company troopers respected the private, and he was awarded the Medal of Honor for his bravery on Reno Hill. In that life-and-death drama, despite being wounded, Thompson courageously risked his life to bring water to his thirst-ravaged comrades on Reno Hill, who were beleaguered by hundreds of warriors on all sides. Even after leaving the army, he never drew attention to himself or his dramatic brush with history. He did, however, take offense when, at a dinner gathering of the 7th, some decades later, William Slaper, a former trooper from A Company, challenged the truthfulness of his story, leading to a near fistfight, although both men were by then past seventy years old.

Other than his acute sensitivity associated with being questioned about this incident, no complaint was ever registered against Thompson. Thompson's integrity was his best character witness. He became a wealthy rancher and a responsible, upstanding citizen in civilian life. Nothing in the record or his history (other than this remarkable account) ever gave anyone cause to question his sincerity and truthfulness. In light of his low-key personality and forthrightness, Thompson's enigmatic but remote encounter with his commanding officer near the river appears reliable enough to accept—with some reservations—notwithstanding a later tendency to mix up one scout's identity with another, which was more likely due to his interviewer Walter Camp's confusion than Thompson's.

After his 'leader's reconnaissance' to the river, Custer rode back and over the ridge of Upper Boyer's Bluff to rejoin his command. He moved farther north in his quest for a ford that they could cross to attack the village and support Reno. Meantime, Boyer became Custer's eyes and ears, monitoring the Reno fight in the valley and keeping a close lookout for Benteen. Custer had risked moving away from his reinforcements and ammunition, with neither of his subordinates knowing, or at least fully understanding, his intentions as he continued to search for a river crossing. Employing a successful 'fix and flank' maneuver demands the utmost cooperation from each command component. Custer depended on Reno and Benteen to fulfill their own assigned missions despite his sudden change of plans after he decided to ascend the bluffs rather than follow Reno into the valley.

Sometime during or just after this reconnaissance to the river, Boston Custer brought fresh and valuable intelligence to his brother's command. Boston's appearance had to be heartening because if Boston could make it through enemy territory unscathed, then the Indians had not yet gathered in force along the main trail. That would mean Martin stood more than a fighting chance of reaching Benteen's command to deliver Cooke's note. Custer's luck was still intact; for now, he knew the length and breadth of the entire village and was reasonably sure Martin would reach Benteen. Moreover, at the bottom of Medicine Tail Coulee, Custer had located a ford leading to the enemy encampment. Because life is random, even for celebrated Generals, this would be the last good news Custer would receive. Due to the dearth of reliable information, succeeding events involving Custer's wing necessarily become more speculative as historical events become conflated with mythical ones. No matter how deep one is immersed in this subject, personal theories are not unvarnished truths, no matter how well sourced. Despite the ever-present fog of war and the relative poverty of source material after Custer reached Medicine Tail Coulee, historians and battle reconstructionists are neither unarmed nor bereft of resources when it comes to intelligently speculating about what transpired in the Custer fight.

It took a complex set of twists and turns, along with a mesmerizing array of personal and impersonal forces, to bring about the situation that Custer found himself in when he boldly moved down the east side of the river as he contemplated descending Western Coulee. Only when viewed as a whole and judged with all other developments can some semblance of understanding be achieved when considering decisions that shaped and defined this battle. Again, Custer's sudden and unexpected detour up the bluffs, rather than following Reno into the valley, was the great axis upon which the battle turned and ultimately would be defined. Not because it was, in itself, conclusive, but because of how Custer's temporary absence from the battle, contrary to the expectations of one of his battalion commanders and the burning resentment of the other, affected everything else.

Once Custer was informed that the Indians were either coming out to fight or were making a stand, pursuit was rendered meaningless. Having learned on Sharpshooter Ridge that the Indian village was clueless that they were coming under attack, Custer decided not to follow Reno into the valley because he had never intended to charge the enemy in a frontal assault. He would only follow Reno into the valley if he believed the Indian village was in flight and on the precipice of scattering in multiple directions. One of the great difficulties in comprehending the battle is a prevailing misunderstanding that cavalry, in general, does not closely resemble infantry in tactical maneuvering. Infantry is not governed by the movements and stratagems that propel mounted soldiers. While it is true that basic tactics are essentially the same, the way each branch of the army utilizes them is significantly different. Frontal assaults are more the work of infantry, while the cavalry is tactically disposed and oriented to pincer movements or 'fix and flank' maneuvers on a wider scale.

However, each division of Custer's forces was unraveling—not only the physical regiment but its psychological unity as an indivisible force. At least two of Custer's battalion commanders were distancing themselves, perhaps inadvertently and possibly unconsciously, from their commanding officer. A cavalry regiment, regardless of any temporary fragmentation, must always act as a unified force against the enemy. But the way matters were shaping up, the 7th was devolving almost inexorably from a cohesive fighting unit into a diffused, divided, and segregated fighting force. Its disparate elements no longer acted as the responsive, precise, unitive instrument that its commander had intended.

When fragmentation—as opposed to integration—began to assert itself, Custer thought and reacted differently, no doubt compensating where he could for a ruptured and detached regiment. Although some facts remain shrouded in mystery, thanks to archeological digs and a more precise, scholarly exegesis of Native American interviews, it is now possible to determine within the general context how events unfolded before tribal warriors finally turned the weight of their numbers and fury upon Custer's five companies.

Even with the perspective of passing years, we still possess only an opaque light to penetrate the veil of mystery and complexity of this battle with all its perpetually moving parts. It is much like studying a photographic negative rather than the image itself. Its likeness, although not perfectly transparent, is still the best illumination we can hope for in this haze-filled chapter of the Plains Indian war. We are abetted in our understanding by the observations, albeit also from a distance, of the Crow scouts—especially the half-breed scout Mitch Boyer, who sat on his mount atop the bluff that is now named for him. The perch allowed him a line of sight, not only to the Indian village but to the Reno valley fight, and that helped knit together multiple loose strands to make a more seamless whole of transpiring events.

As Custer's lookout, Boyer would be in a position to signal his commander about what he saw of the Reno valley fight, which was initially encouraging and, no doubt, gave Custer a measure of confidence to drive farther north away from Benteen and Reno in search of a passable ford. But like an unexpected shift in weather patterns, encouragement rapidly spiraled downward from quiet confidence to a raging storm of disillusionment.

While serving as Custer's eyes and ears for the valley fight, Boyer witnessed Reno's mad and inexplicable dash from the valley. While he could not see the hill to which Reno's three companies were repairing, he could surely observe the rising dust in that direction and follow the audible trail of the heavy gunfire. Alarmed by this stunning turn of events, Boyer swung his mount around and dashed toward Custer, who was well down the bottom of Western Coulee, to deliver the disturbing tidings. The news of Reno's staggering retreat cannot be underestimated. The fixing force (Reno's battalion) had vanished from the valley.

With Reno's setback, it was not just eighteen lives at risk as it was with Joe Elliott at Washita, but more than 150. With no word or sign of Benteen and the trailing ammunition packs, Reno's expected support from that direction had been seriously delayed. But it was not only the unprecedented number of warriors or the absence of Benteen that was so worrisome. With Reno repulsed and in full flight, the reversal of fortunes had infused confidence and determination into what was formerly a flagging, unnerved, and astonished enemy. This Indian village was no longer unaware but fully cognizant of the threat against it. It mobilized with all the frenzy of an invaded hornet's nest.

Ultimately, seven of Custer's twelve companies were either not on the scene or were lamentably relegated to being holed up on a godforsaken hill, licking their wounds nearly four miles from his command. It must have been astounding to Custer that Reno, much less Benteen, was not in the valley but a hundred feet above, stranded on an elevation Custer had recently traversed. He now faced some difficult and terrible choices. Decisions that would mean not only the difference between victory and defeat but the life or death of his regiment. Nothing less was at stake than either transcended glory or the hell of irreparable loss and humiliation. As to which direction the pendulum would swing, only the fates could foretell. The once-so-called Boy General, always at home in the precarious milieu of war that colored his entire military career, would once again decide to defy the odds, having resolved not only to endure this crucible but to prevail through the audacity and pugnacity that had forged and honed his reputation.

Custer would enter the fray much later than is usually supposed, and it would be in direct response to Reno's retreat from the valley. Approximately ten minutes after Benteen arrived on the bluffs at 2:30 p.m., volley firing was first heard downstream, and there was no doubt that those shots came from the Springfield carbines of Custer's command. The ever-enterprising and improvising Custer, amid a vertiginous sweep of terrain, climate, and chance, had finally descended to Medicine Tail Coulee and a river ford that, because of changing circumstances, he was no longer determined to cross into the village.

Instead, he would confront the crisis head-on. With peril lurking at every turn, in what amounted to a shrewd handshake with the devil, he would attempt a feint to draw the Indians away from Reno's beleaguered command and bring the warriors onto his own, regrouping from there. Tottering on the precipice of catastrophic collapse, in a perdition of a dizzying maze of hills, ravines, and coulees, Custer unflinchingly stared down the vagaries of fortune—recently a smiling ally—but now with Reno's rollover, the two-faced Janus god swiveled its head to mock him, just as the last unwritten chapter of this immortal saga awaited history's final judgment.

Chapter 9: How Many Indians Were There?

An infatuation with numbers has long been wedded to the Little Big Horn. Many have defined the battle by this metric alone: hordes of bloodthirsty savages overcoming a tiny but intrepid band of blue. Its compelling contrast of opposing cultures and a frightening depiction of civilization overcome by sheer barbarism made for an absorbing drama. Numbers at the Little Big Horn have a numinous and panoptic quality because, through that popular and accessible prism, the battle is remembered. The numbers also have the virtue of simplicity and straightforwardness: Custer lost because he attempted, with too few soldiers, to defeat too many Indians—period—end of story.

Yet the truth is more complicated and absorbing. While numbers at the Little Big Horn were not inconsequential, they were also not determinative in deciding the battle's outcome. Whatever the impact of the numbers, the question of how many Indians were camped along the serpentine, pellucid waters of the Little Big Horn River and what percentage of them were warriors in that great valley the Indians called Greasy Grass is the cause of a host of wrangling and dispute among battle historians.

There is, however, agreement among the dissenters that this was the largest gathering of Indians anyone had ever seen. Specific numbers are difficult to establish since Indian tribes never had a method of compiling or computing a census. Even when focusing solely on the strength of one tribe, such as the Cheyenne, discrepancies between respected authorities vary significantly. Calculating the village population is a challenging task, making it impossible to determine exact numbers, and one must settle for narrowing it down to a plausible range. Indian villages consisted of highly mobile and transient populations, so even as the question begs for an answer, it is virtually impossible to ascertain the number of tribal Indians on June 25, 1876, and its percentage of warriors. By contrast, the strength of Custer's command was almost definitively answered by Army muster rolls, enlistment and discharge papers, and various other records on which to base their figures and on which they could reasonably rely.

Even an approximation of Indian strength is fraught with problems. The encampment on the Little Big Horn was migratory, with Indians coming and going, but mostly coming because of the fertility of the verdant land and hunting grounds. Accounts of the enormous aggregation of Indian people can be found in the accounts of White Bull, Foolish Elk, and Kate Big Head, among others. These hunting grounds were irresistibly inviting, especially when contrasted to the stifling and enervating government reservations, propelling the Indians to gravitate there in record numbers. According to author John Gray, the Indian population in the village nearly doubled in the week leading up to the battle. Even before the regiment's Indian scouts got their first peek at the encampment from high on the Crow's Nest, Army intelligence suspected that a staggering number of hostiles was congregating somewhere at the Rosebud or farther west toward the Little Big Horn. Because the village's numbers had swelled so quickly, the encampment feared no attack—there were just too many Indians in the valley for them to feel overtly threatened. And because of this overconfidence, the village was caught by surprise.

Traces of the Indian frontier are few and scarcely visible today, making a retrospective crunching of the numbers an uphill task. Various Indians who fought or witnessed the battle had conflicting memories—estimating the hostile's strength depended on who was interviewed. Consequently, Indian numbers are all over the ledger. Not surprisingly, in the months and years following the battle and after such a shocking defeat, the number of warriors Custer encountered around the Little Big Horn became exaggerated, enabling reporters, dramatists, and actual participants to highlight the impossible odds facing the 7th. At least two accounts from experienced army officers and knowledgeable observers who fought there had once estimated that Custer fought a preposterous seven to nine thousand warriors!

When General Terry and Colonel Gibbon's forces approached the Little Big Horn on June 27th, two days after the battle, all the Indians abandoned the encampment, most of whom had already had their fill of fighting. This en masse departure greatly complicated any accurate post-battle assessment of the Indian numbers confronted by the 7th, as the native population vanished so swiftly into the mountains and scattered with the wind. For the 340 or so soldiers on Reno Hill who witnessed the Indians leaving the valley, the staggering sight left them in amazement. Godfrey, a sober and astute observer, could not dampen his awe as he recorded his impressions of the evacuation: "It started at sunset and was in sight until darkness came. It was a straight line about three miles long and I think a half mile wide, as densely packed as can be. No less impressed was Lieutenant Edgerly, who called it, "the largest number of quadrupeds [i.e., horses] I have seen in my life—like a great brown carpet being driven over the ground." Aghast at what he saw, Reno turned to Moylan and said, "For God's sake Moylan, look at what we have been standing off."

The sweeping flow of humanity witnessed from Reno Hill is evidence enough that a massive number of Indians had been ensconced in the valley. Over time, despite observations by those who fought in the battle, the pendulum began to swing the opposite way, with more than a few battle scholars and students settling for a more modest number of Indians. For example, National Park Service personnel who today serve as guides and keepers of the tablet at the battlefield are formally instructed to tell visitors that the warriors numbered between fifteen hundred and two thousand. One ranger confided, apparently an outlier, more than once that the warrior force on June 25, 1876, numbered no more than six hundred to seven hundred, approximately the same strength as Custer's regiment. But most judge this estimate as far too low—nearly as ludicrous as estimates of nine thousand warriors.

Despite the obsession with numbers, Indian strength is one of the most misunderstood, even as it remains one of the more tantalizing aspects of this battle. While the quantitative superiority of the Indians only marginally mattered to the army in the Indian wars, it endures in the popular imagination of those bombarded by cinematic depictions and artist renditions of that epic fight. But at the outset of the campaign, warrior strength was a relatively minor consideration for those in the 7th who were to fight the battle. Whether the warriors numbered fifteen hundred, two thousand, or even three thousand was not weighed heavily in terms of the 7th's ability to defeat them. That is, until the regiment was repeatedly divided but never reconstituted its nucleus to fight as a whole.

Because of the unique nature of Plains Indian warfare, numbers were rarely an influential factor when battling the U.S. Cavalry. Dr. George Bird Grinnell's observations regarding Indian tactics are particularly insightful in understanding how Indians waged war and why sheer numbers could not overpower a regular and conventional military force. Grinnell, a naturalist and a student of Native American life, focused specifically on the Cheyenne. He had accompanied the Army during the Black Hills Expedition of 1874 and became intimately acquainted with Indian warfare. Grinnell gave a good description of the nature of Plains Indian warfare and how warriors fought against a mounted enemy, which he relates in his critically acclaimed book *The Fighting Cheyennes*:

> Anyone familiar with Indian ways, mode of thought, and war customs know very well that as a rule the Indian avoids coming to close quarters with his enemy. If his enemy charges, the Indian runs away; but as the vigor of the charge lessens, or the enemy stops, the Indian becomes encouraged, turns about and himself charges. This was characteristic of the old inter-tribal wars, which consisted largely of charges backward and forward by the two opposing forces.

Grinnell's description of Indian tactics closely reflects several of the Little Big Horn battle phases. Reno's charge down the valley, the flight from the timber to Reno's Hill, and portions of what we know of the Custer fight on the east side of the river all underscore Grinnell's observations.

History informs us that the weight of numbers was usually not the deciding factor in winning battles between tribal warriors and conventionally trained forces. A good example is the 1867 encounter on the Kansas plains. Captain Benteen's thirty-plus troopers successfully charged and drove from the field a force of two hundred Sioux warriors. Numerous other incidents yielded similar outcomes. Terrain, training, tactics, and morale carried more weight than any numerical advantage in clashes between the cavalry and hostiles, and no one, despite a couple of highly publicized omens remembered afterward, seriously believed it would be different at the Little Big Horn.

Indians were more suited to unconventional warfare and did not fight like the U.S. Army because tactical theory and conventional military organization were unknown and contrary to the Indians' temperaments and their customs of war. In light of their dwindling numbers, as Europeans overspread the continent, the Plains Indians would not swarm or overrun a military force as in the way that often occurred in those near-suicidal and celebrated late 19th century Zulu charges against British armed forces on the African continent.

One of the leading proponents of lower warrior numbers is Greg Michno, author of *Lakota Noon,* a classic work and one of the more analytical books on the battle. Although it is now somewhat dated in light of recent research, his insights remain relevant and cannot be dismissed without reservation. Michno maintained that Indian warrior numbers were significantly inflated at the Little Big Horn because Americans could not bear the thought that primitives overwhelmingly defeated an outstanding regiment led by a war hero and legendary Indian fighter. Arguing for a lower aggregate number of warriors promotes the egalitarian idea so pronounced in contemporary American culture and ethos, affirming that Indians could defeat Custer and the 7th Cavalry without the advantage of disproportionate numbers. Such presumptions, however, are often more political in nature rather than historical and will not get researchers closer to the truth.

One of the quandaries historians must face is that the concept of numbers, as understood by European Americans, was completely foreign to the American Indian. Indigenous people of the Americas interpreted numbers differently from those in societies shaped by the Enlightenment and Western Civilization. Native Americans never developed a rudimentary numerical value system, nor did they attempt to define values through integers and mathematical symbols. For example, time was measured based on natural phenomena and physical realities, such as the sun's position, when discussing the distance traveled. To represent large quantities, they often used the expression, "greater than the blades of grass." In short, arithmetical schematics and conceptual abstractions had no meaning or significance in Indian culture.

Consequently, their accounts are of little help when calculating tribal numbers, other than most of the village inhabitants agreed that they saw more Indians than they had ever seen in one place. Because of this cultural difference in understanding numbers, some enterprising individuals have sought to approximate the Indian population based on the village's resources and capacity to maintain sanitary conditions. Proponents of fewer inhabitants in the village have maintained that such a community could not sustain the needs of an Indian village of that size for any period. By the morning of June 25th, the day of the battle, the Indian population had overwhelmed the environment's capability to sustain it. They were already preparing to move downstream, seeking sufficient game and fresh grass.

Unsurprisingly, Indians left stultifying reservations in droves for open-air and verdant valleys. Government reservations were austere, depressing, and hopelessly degrading to those whose very soul was wedded with a boundless land, the cavernous sky above, whose dimensions seemed to mirror the breadth of eternity itself. Being holed up in squalor and deprived of hunting wild game left them despairing, bereft of their cultural identity. Indians confined to reservations yearned for their ancestral way of life and rebelled when reduced to vassals or wards of the state. Political corruption and greed spawned skimpy government rations distributed by racketeers. They seized upon any chance to escape this enervating existence. Even General Philip Sheridan, who harbored little warmth for the Indians, called a reservation "a worthless piece of land surrounded by scoundrels."

Meanwhile, their trek to the Little Big Horn was accompanied by a growing and real sense that their medicine was unusually strong, abetted by their perceived victory over General Crook. The mystical power of the war chief Sitting Bull symbolized, if not incarnated, the old, cherished way of life. All these auspicious factors melted in an irrepressible yearning for their ancestral homelands and, like a mighty vortex, more and more Indians were drawn to the bountiful and fertile land of the Little Big Horn. These peregrinations were fostered by a renewed lease on life as the number of "summer roamers" swelled along the crystal-clear waters of the Little Big Horn River, numbers that would steadily increase and accelerate up to the day of Custer's unexpected arrival.

The Indian tribes migrated to the Little Big Horn as the frost of a hoary winter dissipated and the welcoming rays of an awakened sun bathed the spacious land with light and warmth. Indians of every stripe assembled on this riverbank with such rapidity that the military was running a step behind ever since the order was issued that all Indians must be back on the reservation by January 31, 1876. The military authorities had neither the time nor the means to accurately track or estimate the size of this population, much less the direction in which they were heading. The Cheyenne, Oglala, Brulé, Hunkpapa, Minneconjou, Sans Arc, and Blackfeet Lakota found temporary refuge and succor in that teeming land. Despite their lagging and faulty intelligence, the military attempted to establish a reputable accounting of the strength of the hostiles before and after the Little Big Horn battle. It was a thankless task since nothing seemed more capricious, transitory, and subject to change than a migratory Indian population. Efforts to accurately document their strength often proved to be an exercise in futility.

Overreliance on these official primary sources was misleading since nothing was more common than an authoritative source inflating the strength of the hostiles. Still, others sought to base their calculations on numbers anchored to agency reservation counts and data predicated on the subsequent counts of Indians surrendering after the Little Big Horn battle. These researchers worked by counting backward—an accounting technique believed to lead to a more unerring and truthful count of village occupants and warriors on June 25. But these figures are hardly bulletproof. Thanks to dishonesty and greed, the number of Indian inhabitants was often padded to increase supplies for these agencies, thus supporting Custer's earlier testimony before Congress that there was widespread fraud in the Bureau of Indian Affairs, and this accusation of corruption was commonly believed to be not only in higher military circles but also among the ranks of civilian Indian Agents.

One sociological factor that may have augmented Indian numbers is the possibility that summer roamers contained a higher percentage of unmarried warriors who left the reservations to journey to the Little Big Horn Valley. This greater-than-usual exodus from these governmental jurisdictions may have been related to the number of wikiups (abodes that might have housed unmarried warriors) tallied in the village after the battle. But counting wikiups may be misleading—the more significant number of them were probably sweat lodges—these should not be part of the addition process in calculating the warrior population in the Indian encampment.

Sweat lodges, igloo-shaped domes that imprisoned heat and induced sweating, were primarily used for spiritual renewal and prayer, not for housing warriors. But even discounting this as a reason to abridge Indian numbers, informed sources still support a more modest number of warriors. In his article, "The Campaign Against the Sioux," published in the January 1876 issue of the *Journal of the Military Service Institution,* General Robert P. Hughes, Aide de Camp to General Terry, quotes this conversation between General Terry and Captain Benteen regarding this very subject:

> As the question of the number of warriors in the hostile camp has been raised, it may not be amiss to recall an incident that took place on the hill in which Reno took

> refuge, within a few minutes after General Terry's arrival. He and some of the officers of his staff were surrounded by a group of officers of the 7th Cavalry amongst whom were Major Reno, Colonel Benteen, Colonel Weir and Major Moylan—all officers of long and varied experience in the army. General Terry put the direct question to them: What is your estimate of the number of Indian warriors? The replies pivoted about the figure of 1,500, and I can recall Colonel Benteen's reply almost verbatim, which is as follows: I have been accustomed to seeing division of cavalry during the war and from my observations I would say there were some fifteen to eighteen hundred warriors.

That should settle the argument, since General Hughes's observations were aligned with the presumptions of the moderately lower counts often embraced by modern scholarship. Except that it does not. Notwithstanding the consensus of the officers in Hughes's reported conversation, we have not only Benteen's after-action report but also that of Reno and General Terry, all of whom agreed in the aftermath of the battle that there were about twenty-five hundred warriors. Benteen estimated fifteen hundred to eighteen hundred warriors on the morning of June 27, and then on the evening of the same day, he stated twenty-five hundred warriors. Perhaps Benteen began to weigh more heavily the exodus of the Indians from the village that he witnessed that day, and the breadth of the numbers he saw caused him to push his estimate upwards. It is hard to say for sure. Because of all the numbers being bandied about, it is possible that Robert P. Hughes misheard, misunderstood, or misstated what was said. Writing is a more exact instrument, so official written reports should carry more weight than anecdotal, gossip-spreading, unofficial representations. After a military disaster of this magnitude, everyone's senses are still recoiling, and eyewitness information is still being filtered, processed, and assessed. Battlefield numbers are fluid, and subjecting them to an inflationary bias is not uncommon.

Perhaps more time needs to pass to develop perspective and additional facts assembled by those not affected by the horrors of the battle's immediate aftermath to shed new light on the subject. Fortunately, we have the analysis of Lieutenant William Philo Clark, a brilliant U.S. Army officer during the Plains Indian wars. At the end of August 1876, Philo Clark was a trusted part of General Crook's staff and became consumed with Indian numbers only two months after the Battle of the Little Big Horn. He was with Crook when he rejoined the columns of General Alfred Terry and Colonel John Gibbon. Clark was also part of Crook's pursuit of the Lakota/Sioux during the late summer and early fall of that year when he first encountered Indian participants of the battle and closely quizzed them for information.

During the next few years, Clark spent a good deal of time with these tribal members, learning their ways and how they communicated. He discovered he possessed a genius for interpreting their language, assessing its setting and frame of reference, and the subtleties and nuances of their descriptive and peculiar phrasing. He became so proficient with their means of communication that his book, *The Indian Sign Language* (published posthumously—he died young), remains a classic in the field and is still considered the most definitive and comprehensive primary source on the sign language of the Great

Plains tribes. It is not an overstatement to say that no white man understood the language and ways of the Plains Indians as thoroughly, intimately, and comprehensively as William Philo Clark.

Despite his profound knowledge and intuition of Indian ways, Clark struggled with the conundrum of tribal numbers, sometimes changing his mind as he sifted through every piece of evidence to reach the most up-to-date count of the Indian population at the Little Big Horn. He once believed the figure to be as high as thirty-five hundred warriors. Initially, Clark did not realize that after several days of camping at the Little Big Horn, the village became more fluid, impermanent, and restive, making it problematic to determine a stable and unchanging number at any one point.

On the afternoon of June 24, the day before the great battle, the village inhabitants were preparing to move the camp north down the valley, an undertaking that had commenced early on the 25th, the day of the battle. Further compromising the accuracy of his initial computations, Clark did not fully consider the Indians' proclivity to move an encampment immediately following a battle. The Sioux and the Cheyenne harbored the belief that it was a bad omen to stay where there had been a death in a lodge or tepee. Consequently, a number of lodges were removed and relocated farther down the valley at the end of the first day's fighting. Thus, tallies naturally led to higher rather than lower numbers. These moved Indian encampments following the Battle of the Little Big Horn created more teepee circles, but not more tepees that might have housed warriors.

Determining the number of tepees was a key metric used to estimate the Indian population. Several from Custer's command sought to do just that in estimating the size of the standing village, but their numbers varied. William Cross, a half-breed Indian scout who on the day of the battle rode in pursuit of Indian ponies, had an opportunity to view the village while it was standing. After a debriefing on June 28, Cross stated that there were eight hundred or nine hundred teepees in the village. Lieutenant Hare and Lieutenant De Rudio endeavored to examine the deserted village site, with the former estimating fifteen hundred teepees and the latter estimating twelve hundred to fifteen hundred lodge sites. The additional tepee circles, however, did not, in this case, translate into more tepees and hence, more Indians. Relocating a number of their lodgings left two circles for each teepee, misleading those attempting a headcount into thinking that the Indian encampment was more extensive and domiciled more warriors than it did. General Nelson Miles picked up on this erroneous multiplier effect and attempted to correct this mistake in his account, *The Personal Recollections of General Nelson Miles:*

> The Indians further explained that after the fight was over, the Uncpapas, who, previous to the engagement, had been encamped at the upper end of the village and pitched their tents near the Cheyenne, where they remained for two days. Thus, the impression was given to the troops who afterward went over the field and examined the campground, that a larger body of Indians was encamped there than was actually the case, or in other words, there seemed to have been a larger camp than there really was.

His discussions with General Nelson Miles might have caused Philo Clark to readjust his total and his subsequent analysis of the Indian's numerical strength that day to make it more commensurate with other estimates given by eyewitnesses of the battle. Here is General Miles:

> Captain Philo Clark, who subsequently had charge of the surrendered Indians, and could readily talk the sign language, investigated this matter with great care, inquiring of the most intelligent Indians in each band of Sioux, and also among the Cheyeene [*sic*], Arapahos, etc., and he considered 2600 as the maximum number of warriors in the affair.

For the Army, targeting the number of warriors was a top priority since they still had to mobilize the resources and manpower to defeat them, which became even more imperative after the Little Big Horn battle. But the calculations were often charged with emotion, speculation, and exaggeration—a tendency that tarnished much of the historical reporting and fact-finding at the time. Even the inquest (Reno Court of Inquiry), convened nineteen months after the battle, furnished little in answering this question of Indian numbers. The succeeding decades tended to cloud rather than clarify the issue as the Little Big Horn passed into legend and was gilded with the sensationalism of storybook drama fueled by the romance of Hollywood hyperbole that nurtured myths instead of unmasking them.

In the week before the battle, the influx of roamers, who came into an already dauntingly large village, increased steadily right up to the day of the battle. Even if its full strength cannot be calculated, it had likely reached its zenith on June 25. This was also the assessment of Dr. John Gray, whose scholarly approach to the battle, replete with charts and mathematical formulas, attempted a scientific and definitive answer regarding the numerical strength of the Indian village. His methodology, consisting of parameters and postulates, effectively determined that the village's population accelerated rapidly leading up to the day of the battle. Even in the years before the battle, the annual migration of summer roamers from the agencies to the unceded territory was significant. The military patrolling the frontier was fired with resentment that these hostiles were being fed and armed by federally appointed Indian agents. Even though the year 1876 brought an arms embargo that was imposed upon Indians living on the reservations, the Army was preparing additional measures for a more forceful and decisive campaign come spring to rein the hostiles back onto tracts of land under government supervision. This more aggressive policy resulted in the Indians hoarding ponies and arms. It also served to augment the number of summer roamers to an unprecedented level.

Many roamers descended on the village between June 18 and June 25. Their zeal was heightened by the Indians' perceived success over General Crook's forces on June 17. Their success in forcing Crook's column to withdraw triggered celebrations and a convergence of teeming humanity along the banks of the Little Big Horn River. The timeline of the multiple trails leading to the village precisely segues into the immediate aftermath of the Rosebud fight. When the 7th rode above the Sundance camp on June 24, they discovered a myriad of trails crisscrossing over the valley that the scouts had determined were just a few days old. From Mark Kellogg's dispatch, we learn that Custer and his officers believed there were as many as fifteen hundred warriors with Sitting Bull. But these numbers were still subjective as there was no consensus on warrior strength from intelligence reports.

These primary sources, including those of the Army and tribal inhabitants, must be approached with the foreknowledge that amid the stress and immediacy of events, memory does not always faithfully represent what senses record. Captain Anson Mills, who served with Crook, estimated twenty thousand Indians with five thousand warriors. This was seconded by Lieutenant Hugh Scott, who was counting the sites where teepees once stood. He stopped when he reached fifteen hundred lodges. With an accepted average of two warriors to each teepee, one only has to do the math. General Terry did cite several sources from General Sheridan that were collected in camps over the Little Missouri River, reporting up to fifteen hundred Indian lodges, an estimate which agreed with Scott. This would translate into at least three thousand warriors. Sheridan dismissed those estimates by stating that no large number of Indians could keep together as a hostile body for even one week.

Still, Herendeen estimated there were three thousand warriors, while Edgerly thought four thousand fought at the Little Big Horn. In terms of executing military tactics, counting heads in an enemy village is merely an academic exercise since, as noted before, fifteen hundred or twice that number would not have intimidated Custer nor altered his plans to attack the village. As for Custer speculating that he would confront fifteen hundred warriors, some battle scholars believe he underestimated Indian forces because most of the scouts were interpreting the trails as diverging, not converging, and that this translated into a smaller village and fewer warriors as the hostiles were already believed to be going their separate ways.

But it is hardly likely that the scouts would confuse a diverging trail with a converging one. They saw the trails overlapping but traveling in the same direction as the Rosebud trail, which was headed straight for the Little Big Horn valley. Whether the evidence is interpreted as one gigantic camp or as a series of camps like Godfrey saw it, either way, the 7th would be running into a hell of a lot of Indians. But what does "a hell of a lot" mean in real numbers? Dr. Gray thought the key to finding this answer was connecting X (the number of warriors) to some constant. Gray's constant was the belief, firmly embraced by the army scouts, that the Indian lodges, on average, contained less than two warriors. This is why, at the battle of Washita, Custer reported 103 warriors killed, knowing full well that the figure was lower. But he justified this total by establishing that there were 53 lodges in the village. Two warriors to each lodge would afford him the credibility he needed to claim a death toll of more than a hundred warriors.

Gray also estimates the number of lodges in the encampment by referencing the 996 calculated by Wooden Leg, who fought in the battle, and the 949 projected by Professor Charles Eastman, an Americanized full-blooded Sioux who obtained an advanced degree at an American university. The two figures are close enough for Gray to establish a proximate baseline. Allowing for some upward elasticity, Gray rounded off this total to a thousand lodges and, with the requisite two warriors per lodge, came up with a maximum of two thousand warriors. Mathematically sound, but the problem was that Eastman was not at the Little Big Horn and Wooden Leg, though uniquely capable among tribesmen of relating large numbers in cross-culturally intelligible ways, was just eighteen years old and probably still impressionable.

The question then becomes why Eastman's conclusion would side with Gray's lower numbers rather than Benteen and Reno's higher battle count. These two experienced Civil War officers were accustomed to estimating the strength of enemy forces, each of whom calculated that the 7th faced twenty-five hundred warriors at the Little Big Horn. Gray's stated maximum number of two thousand warriors falls on the moderately low side of the equation when all the evidence is considered. Benteen and Reno had not only fought in the battle, but they had witnessed, along with more than three hundred soldiers on Reno Hill, the awe-inspiring length and breadth of the Indian caravan as it migrated away from the village and did so over several hours. That allowed ample time in the stillness of a pacific morning to reasonably assess the enemy's strength. Where do all these conflicting accounts lead us? According to the most reliable accounts, the encampment's physical size and geographic dimensions appear to be approximately one and a half miles long and half a mile wide, comprising around eleven hundred tepees or lodges.

Additionally, the ratio of warriors to non-combatants at the Little Big Horn was likely more significant than usual for an Indian encampment as rumors of war percolated pursuant to the U.S. Army's January 31 ultimatum to return to the reservation. The Indians were part of a warrior culture, with an integrated pattern of human behavior that reflected and anathematized any warrior, including those as young as fourteen, who did not live up to a code of honor necessary for the tribe's well-being. Based on the overwhelming threat posed to their survival on June 25, it can be safely assumed that every male, ranging from early adolescence to those nearly superannuated, bore arms that day, albeit at different times during the battle. The Indians even had a contingent of young warriors called the "Suicide Boys" who had sworn themselves to fight to the death.

Godfrey noted that Indian males who reached the age of fourteen were considered warriors in general combat. Fred Gerard, who also knew the Indians and their ways, later testified that any male aged twelve to fourteen who could fire a gun was considered a warrior. Once the tide of battle turned in favor of the Indians, as it did as early as the valley fight, it is unlikely that any warriors sat out the entire battle. The attack that early afternoon was an existential threat, and tribal warriors had a strong sense of honor. Any able-bodied male sitting out this fight would have invited invectives of derision and denigration.

Custer himself calculated there would be "three fighting men for every ten Indians." If the village at Little Big Horn contained about eight thousand people, the number of warriors, according to Custer's formula, would be about twenty-six hundred. Considering all of this and combining it with Lt. Philo Clark's expertise in sign language and his close involvement with tribal Indians at the Little Big Horn, the best estimate of the number of warriors fighting that day was higher than is commonly believed by many contemporaries today, likely attaining at least twenty-six hundred.

If true, this does not mean that any of Custer's battalions faced all these warriors at once or in the same place. Indian forces were staggered and spread out much as the troops of the 7th were. A few of the warriors, for example, were protecting the non-combatants in their flight from the village to where many of them sought refuge along Squaw Creek at the north end of the valley. Yet the tide of battle turned heavily in favor of the tribes once Reno's forces were routed in the valley and Custer was isolated on his hill. The effect was nothing less than catalytic as warriors, galvanized by the smell of blood, would not miss out on a final triumph over the bluecoats and all the booty that would come with it.

In the final analysis, one hastens to the most salient point by restating that numbers might have been the least of Custer's problems compared to a lack of communication and cooperation among the disparate parts of his regiment. The more significant point is that the 7th unquestionably faced a determined and resolute enemy that, whatever its precise numbers, was fighting for their lives and those of their families. The Indians fought bravely, ferociously, with unity and a sense of purpose, buoyed by the warriors perceived success over Crook at the Battle of the Rosebud, Sitting Bull's mystical sun dance, and, finally, their unflagging perception during the battle that the soldiers were retreating from them because of the strong medicine that pervaded the village.

Chief Gall proudly remembered that the "Great Spirit was present and riding over the fields that day." This fueled their belligerency and gave the warriors momentum and courage as they skillfully utilized the terrain, which they knew intimately, to their best advantage. While it is not entirely accurate to say the Indians won the battle more than the regimental command lost it, a plethora of mistakes by the 7th had nonetheless been fully and skillfully taken advantage of by tribal warriors. In another context and against another enemy, no one said it better than Confederate General George Pickett. When asked why the South lost the war, Pickett, pausing ever so briefly for effect, laconically replied, "I always thought the Yankees had something to do with it." Similarly, so did the warriors at the Little Big Horn on June 25, 1876, when General George Armstrong Custer and his 7th Cavalry met their match.

Chapter 10: Did Reno Deny Weir Permission to Assist Custer?

Riding his bay mount Dick, with Adjutant Cooke's urgent message securely tucked away, Benteen briskly moved from a walk to a trot as he and his battalion followed the main trail back to the regiment. Fresh horse droppings and shod hoofprints indicated that the column he was following split into two trails, one taken by Reno and the other by Custer. Somehow, Trumpeter Martin had neglected to mention that Custer had again divided the regiment. Just as curious, Benteen failed to ask Martin about the meaning of the diverging trails. Benteen would change his story after the Reno Court, saying that the two diverging trails he encountered created "a two-horn dilemma," sowing doubt about which path he should take. But in his correspondence following the battle, Benteen indicated no confusion as to his belief that the Custer and Reno columns were down in the valley fighting; therefore, the picture Benteen later painted of being perplexed when he arrived in the valley appears not to have been entirely, if at all, accurate.

As the column marched on, the crescendo of gunfire grew louder by the minute, spurring Benteen's column to a gallop with Captain Weir and D troop following the trail to the left and Lieutenant Godfrey with his troop to the right. Benteen rode between them and had his orderly give the order to draw their pistols in anticipation of meeting hostiles.

Soon, the rugged, broken landscape gave way to the sweeping valley of the Little Big Horn. Despite its breathtaking vista, he saw curling, black smoke rising into the cloudless sky. They had entered a war zone, a prairie all aflame. Contrary to the encouraging reports of Kanipe and Martin about the Indians being routed and shot down, hundreds of mounted warriors (Benteen estimated nine hundred) could be seen circling Reno's troops. Many troops were in a desperate headlong flight, trying to make their way toward the bluffs and away from the pursuing hostiles.

Along the retreat route in the valley, Benteen also saw a skirmish line of twelve or thirteen dismounted troopers trying to organize some rear-guard action parallel to the river. There is disagreement over where and when Benteen saw these troopers, with some battle students believing he was witnessing a platoon of M Company falling back to the timber position from the valley skirmish line. But this cannot be true because it was too great a distance—some two and a half miles—making the sighting of M Company by Benteen's troop impossible. From Benteen's observation point near the river, these men were clearly dismounted stragglers from the timber position seen along the retreat route, attempting to provide a makeshift rear-guard action for Reno's retreating battalion. Their skirmish line was parallel to the river, in a general east-west direction. These men would comprise nearly half of Reno's casualties in his mad dash from the valley.

Whoever they were, these valiant few did not stand a chance against the mounted warriors who swept over the soldiers like a wave over a sandbar. At first glance, it seemed to Benteen that two separate engagements were taking place, one on the bluffs and the other in the valley. Things looked bleak, and with only 117 men in his command, Benteen did not see much sense in charging into nine hundred fired-up, mounted warriors. Before long, several Crow scouts appeared, gesticulating and pointing to the north. Following the scout's direction, Benteen turned to the right and made his way up the same slope that Custer had ridden over an hour ago.

Atop a hundred-foot-high hill, Benteen found Reno's gasping and wrecked command. Benteen appeared unfazed and, according to Private Morris, he moved toward Reno's position as slowly as if he were going to a funeral. Stopping to survey the spectacle, Benteen spotted a hatless and agitated Marcus Reno. The bright red handkerchief tied around his neck at the river crossing was now tied around his forehead. Alarmed that this much-needed support would ride past him, Reno pleaded, "For God's sake, Benteen, halt your command and help me—I've lost half my command." Reno was hardly exaggerating. Counting those left behind in the timber, he had lost 40 percent of his men, including thirty-seven who lay dead in the valley and timber.

Maintaining an impassive posture, Benteen matter-of-factly asked, "Where's Custer?" Reno replied that he did not know and had not seen or heard from him since he went off downstream. After breathing in the import of these words, Benteen counseled that "we should make a junction with him," demonstrating at least early on an intention to reunite with the forces of the 7th. That the fractured regiment never came together is arguably the most controversial aspect of this battle and, in all probability, the principal reason for its overwhelming defeat. As desperate as things looked, one bit of good news on Reno Hill was that Doctor Henry Porter had survived, reaching the summit of the hill unscathed to treat and give succor to the wounded. Earlier, the doctor had related to Reno how demoralized the men were about their retreat from the valley, to which Reno bizarrely replied, "No, that was a charge sir."

Private Charles Windolph, of H Company, remembered that it was "no use pretending that the men on the hill from Reno down were not discouraged and downright frightened." While the Arikara scout Red Bear may have exaggerated when he said that the Major's eyes were rolling and that he was foaming at the mouth, there was no doubt that Reno was profoundly agitated and unsettled. Had Benteen's battalion not arrived when it did, and had Custer not initiated an attack when he did at Ford B, Reno's command, in its disorganized and demoralized state, would have surely perished. But now reinforced, skirmish lines were hurriedly formed, instilling some semblance of military order and readiness in the discombobulated command. Benteen shared with Reno the contents of Adjutant Cooke's scribbled note urging him to be quick and bring packs, igniting another debate among battle students about whether Custer's absence invested Reno with the authority to override Cooke's directive and stay on the hill, a decision to which Benteen readily acquiesced.

Chapter 10: Did Reno Refuse Weir Permission to Assist Custer?

Those who believe that Benteen disobeyed written orders by stopping and not going directly to Custer miss the point of the order. Again, Custer never intended for Benteen to go over the bluffs and join his command but rather to support Reno in the valley as he (Custer) moved to the other side of the village, intending to strike from the north. Benteen had now joined Reno, as Custer had intended, though not in the valley as he anticipated, but rather high on the bluffs, the same hill where Reno's charging soldiers last saw Custer's wing. This entirely unexpected development, where Reno and Benteen ended up on an exposed hilltop, set everything topsy turvy.

Once Custer concluded that the village was not in flight, his plan had evolved from chasing down fleeing Indians to stealthily striking the other side of the village. At the same time, his two battalion commanders would keep the Indians occupied in the valley. But seven of his twelve companies were now marooned on top of a hill that he himself had recently traversed. The consequences of this dramatic repositioning of forces cannot be overstated because they neither threatened the village nor supported Custer. These troops were now more or less inactive, neither making offensive nor defensive preparations.

While technically still in command, Reno showed no inclination to move off the hill and form a junction with Custer. Instead, he attended to the wounded and waited for the slow-moving pack train and rear guard to arrive before going to the sound of the firing downstream. Private William Taylor later moaned, "for over two hours after the arrival of Benteen's command, we remained there on the bluffs unmolested in any manner."

Reno and even Benteen seemed to be under the impression that Custer should find and support them rather than the other way around. Questions percolated about where Custer had gone since neither Benteen nor Reno had been officially informed of Custer's intention to move farther north. On the other hand, several men in Reno's command saw Custer's column on the bluffs. One must ask where else could he be heading other than to attack the other side of the village? How much Benteen knew of Custer's sighting on the bluffs at this time is unknown, but it stands to reason that someone had informed him where Custer was last seen.

Benteen continued to think the worst. He was sure that Custer had abandoned them, much as he had abandoned Major Joel Elliott at Washita. Benteen was not the only one, as it turned out. George Wallace vividly remembered hearing a great deal of swearing about General Custer "running off and leaving us." Reno sarcastically quipped, "Where is the Murat of the American army?"—a pointed reference to Custer often being compared to Napoleon's great cavalry commander.

Roughly ten minutes after Benteen joined with Reno on the bluffs, the heavy turbulence of warriors besieging Reno's position on the bluffs suddenly subsided to a zephyr. Hundreds of hostiles turned from the soldiers and were seen racing downstream, heading north. With their unforeseen and inexplicable departure, Reno, in a pique of acute stress, senselessly began discharging his revolver at the receding enemy despite their having ridden hopelessly out of range. Varnum, exhausted, sleep deprived, heartsick over the battalion being whipped so severely and his comrades brutalized and mutilated, also broke down. He cried and cursed as he fired his rifle at the fleeing warriors in what Lieutenant Edgerly described as a sheer act of defiance.

Astonished by the bewildering exodus of an enemy who seemed to have had them at their mercy, the command waited with mounting anxiety, wondering what they should do now that they were no longer under attack. For several interminable minutes, all was eerily quiet. But then, reports of gunfire downstream smashed the uneasy silence. Like rolling waves of thunder, volleys echoed throughout the landscape and reverberated off the sprawling hills.

Lt. Charles Varnum, his equilibrium restored, turned to Lieutenant George Wallace, exclaiming, "Jesus Christ, hear that?" Wallace sure did—and so did nearly every soldier on that hill. "How's that for abandonment?" Varnum lashed out at those who believed that Custer had scooted away to join General Terry or was somewhere downstream grazing his horses. These volleys could mean only one thing: Somewhere out there and not far away, Custer and his five companies were heavily engaged with tribal warriors, and scarcely a man on that hill doubted it. Everyone knew what discharging Springfield carbines sounded like. Herendeen, still holed up and concealed in the timber with a contingent of men, was startled by the same report of rifles going off from the north. Herendeen later testified that he heard the volleys about a half hour after Reno retreated from the woods. "There were a great many volleys," he remembered, "with a great many scattered shots between the volleys."

More signs emerged that a desperate fight was taking place. Spotting a large contingent of warriors churning up a dust cloud down at the horizon, Reno's orderly, Davern, remarked to Captain Weir that it must be Custer fighting down at the bottom. After seeing the turbulent, powdery cloud for himself, Weir could only concur. Despite these portentous developments, the command remained on the hill. Why? Certainly, exhaustion, anger, fear, and a lack of direction from Custer all played a part in what was an abject failure to do something—anything. But there was also something else at play that held them in place. Although there was a clear sense of urgency over what they were hearing but not fully seeing, there was also an unspoken confidence that whatever fighting was going on downstream, Custer, with five picked companies, could take care of himself. He always had.

But not everyone was so sanguine. A now-composed Captain Myles Moylan (a disgusted Benteen said he had found him blubbering like a whipped urchin when he first arrived on the hill) peered over the edge of the bluff and solemnly stated to those around him, "In my opinion, gentlemen, General Custer has made the biggest mistake of his life by not taking the whole regiment at once in the first attack." This despairing note came as they helplessly gazed down at the tribal elders, squaws, and youngsters who had come in the wake of the warriors to finish the job of plundering and mutilating the dead and dying soldiers by the river crossing below—some wielded stone hammers to crush the skulls of their helpless victims.

Gunfire downstream continued intermittently—even the hard-of-hearing Lieutenant Edward Godfrey heard it, although, strangely enough, Reno and Benteen insisted they heard no heavy firing. Yet two days later, in his after-action report, Reno wrote, "We heard firing in that direction [downstream], and we knew it could only be Custer." Yet, some nineteen months later, at the Reno Court of Inquiry, the

Major reversed his statement again, saying he did not hear any gunfire. Others not only heard it but saw physical signs of a mounting fight. Davern and Weir saw evidence of a battle being fought several miles away. Pointing to a dust cloud to the north, First Sergeant Michael Martin approached Lieutenant Edgerly on the hill, saying, "There must be General Custer. I guess he is getting away with them." Edgerly agreed. Martin then dared to say what many were thinking, "I think we ought to go there." Hearing this, Edgerly, apparently lost in thought, stared blankly out into the empty distance and remained troublingly silent. Since time immemorial, the code of the military was to ride to the sound of the guns.

Only a few warriors remained around the hill. Certainly, nothing could stop three or four resolute companies from saddling up and moving north to ascertain Custer's condition and endeavor to reunite the command or at least position the respective battalions within supporting distance of each other. The uncomfortable truth was that while Custer was fighting for his and the lives of his command, most of his regiment remained on a hill some three miles away, the result of fear, indecisiveness, and a not-so-disguised animosity that had been festering toward their commander. And so men delayed and rationalized. Reno did not even send a reconnaissance detachment to assess the source of the gunfire.

Finally, Reno dispatched Lieutenant Luther Hare to hurry along Lieutenant Mathey and the advance pack train with the ammunition packs. Hare, having discovered that his horse's jaw had been shattered from a gunshot wound, promptly borrowed a mount from Godfrey, his company commander, and, in a flash, galloped off. Hare met the advance pack mules about a mile south of their hilltop position and was back in twenty minutes flat with two of the lead ammunition mules in tow.

Meanwhile, much to Edgerly's disbelief, atop the hill, Reno seemed only to have enough energy to muster a dozen or so troopers to descend the bluffs to locate his Adjutant Benny Hodgson, who had been left mortally wounded and was probably now lying dead by the river. Even Lieutenant Wallace, a reliable Reno apologist, thought it odd that Reno would leave the troops who needed him to find his adjutant and friend, thinking he should have sent someone else. Leading his contingent down the bluffs, Reno would not be seen for close to thirty minutes—an unconscionably long time to be absent from a command that had been pummeled, demoralized, and in desperate need of leadership.

Bruce Liddic noted that Reno's single-mindedness to retrieve Hodgson's body is inextricably connected to what psychologists call 'idee fixe,' a state of mind that allows those who have been traumatized to retain whatever sanity they have left by holding tightly to one piece of reality. Liddic perceptively wrote: "Errors of judgment and possible mental breakdowns may be used to explain Reno's decision... but it does not excuse them." Whatever psychodynamics were involved in Reno's attempt to retrieve Hodgson, his efforts were as fruitless as they were pointless. Though they managed to locate the lieutenant's body, Reno and his party soon returned to the hill, with Reno managing only to retrieve Hodgson's ring and a few other personal effects.

Shortly after Benteen joined with Reno on the bluffs, and prior to Reno's descent to the river to look for the body of his adjutant, gunfire was again heard from the north. Having observed that the hostiles had nearly deserted the bottom, Edgerly turned to an anxious and pacing Weir and said, "We ought to go down there." Yet, for the next twenty minutes, they remained. Why, then, stay rooted on the hill? Neither indifference nor callousness thoroughly explains their inertia. Realizing this, Captain Weir asked Lieutenant Edgerly if he would be willing to go to Custer with D Company if the rest of the command did not. Edgerly replied that he would, adding, "we certainly should go." Weir then said he would ask permission of Reno and Benteen, and moved off in their direction."

Frustrated by the command's paralysis, Weir rode over to Reno and reportedly asked permission to go toward the sound of the guns. When permission was reputedly denied, an epic quarrel erupted between the two, including the waving of arms, recrimination, and even personal threats. In direct contravention of Reno's orders, this clash would supposedly lead an exasperated Weir to ride north to the peaks with a company of soldiers to determine the situation for himself. After his fiery exchange with Weir, Reno went missing for nearly a half-hour in an attempt to retrieve Hodgson's body.

If there is to be an honest retelling of this story, it is important to understand that once the full scope and breadth of the catastrophe at Little Big Horn were realized—one that ultimately resulted in the utter destruction of five companies of the 7th and the defeat, quartering, and battering of the remaining seven—consciously and unconsciously, post battle lines were drawn around the Reno, Benteen, and Custer camps. It would be an endless cycle of finger-pointing on what and who was to blame. The fusillade of bickering between partisans of these camps has been historic and unrelenting. A century-and-a-half later, it shows no sign of diminishing, much less ending. This ongoing dispute between feuding partisans, arguably more divided than the regiment that day, has made it difficult to answer these vexing questions. The only way to unearth the truth, or as close as we can get to it, is to reexamine the historical record using context and logic where gaps in the story cannot be filled.

The Reno–Weir confrontation maintains prominence in the story because it touches on the core of Reno's fitness for command. It also invites critical questions: Was victory on the east side of the river, where Custer was fighting, still possible? Could at least some of Custer's battalion have been saved if Reno and Benteen had acted promptly and with dispatch to support him in his hour of trial? As it would be of any competent cavalry commander, Custer's ultimate objective was

to secure the cooperation of all his battalions in defeating the warriors and destroying the hostile village. But as minutes, then hours, ticked by, this prospect looked more and more remote as neither Reno nor Benteen, notwithstanding the latter's initial inclination to make a junction with Custer, showed little if any enterprise to do so, even in the face of discernible and continual gunfire downstream and during a period when the troopers on Reno Hill were under minimal duress.

Of course, there were the wounded to think about, along with the arrival of the ammunition packs. On the other hand, skirmish lines had been deployed quickly upon the arrival of Lieutenant Mathey. Most of the ammunition packs had arrived within thirty minutes of Benteen, and managing the wounded at that point was not as onerous as purported, since there were only seven wounded on the hill when firing downstream was first heard and never exceeded ten by the time they finally set out for the Weir Peaks. Most importantly, the command was no longer being harassed by the enemy with any credible force that threatened their survival or seriously impeded their march to their besieged comrades-in-arms.

Chapter 10: Did Reno Refuse Weir Permission to Assist Custer?

Again, Custer surely deserves blame on multiple fronts: Not communicating better with his battalion commanders—assigning too many soldiers to protect the pack train—not fully appreciating the challenges and risks of dividing the regiment several times—a misplaced faith in his subordinates' willingness or resolve to comprehend and cooperate with a 'fix and flank' maneuver.

However, none of these factors alter the fact that Custer could and should have had more support from the rest of the regiment. From the time of the valley fight, everything in the 7th's divided command seemed a tempo behind when speed and timeliness were utterly critical for success. Once Reno retreated to the bluffs, it freed the warriors to concentrate on Custer, who was threatening the encampment from the opposite end of the village by the river, roughly where the non-combatants were headed. Reno's retreat afforded the tribes the opportunity to direct the entire weight of their forces upon Custer, now confined to poor defensive ground. Minutes flew by while Reno insisted on waiting for the main pack train and rear guard after the advanced ammunition mules had already arrived on the hill, with each ammunition pack mule carrying two thousand rounds of ammunition. This was more than sufficient to order an advance, especially since sounds of battle continued to be heard. The civilian packer Benjamin Churchill, who drove one of those advance pack mules, later told the Reno Court that he heard four or five not very plain volleys, but volleys just the same.

Weir was not the only officer alarmed by a command that was not disposed to take action. Captain McDougall, escorting the main pack train with Company B, arrived on the bluffs about one hour after Churchill and the advance ammunition mules, and was quietly chaffing about the lack of response to what was indisputably an armed conflict involving their own comrades. The rear guard had been assigned to protect the main pack train, and upon encountering Reno, McDougall reported that he heard clear sounds of gunfire downstream. Reno seemed unfazed by this intelligence, preoccupied only with Hodgson lying dead in the valley. Benteen was also there to hear McDougall's report but strangely walked away with an air of indifference. In short, after multiple attestations of hearing gunfire and with the entire pack train now present, Reno had no excuse for not moving his wing to Custer's last known position.

As for the Weir–Reno quarrel, a few dissenters claim there was no disputatious demonstration between these senior officers. The weight of evidence, however, makes it almost impossible not to conclude that there was a loud outburst between the two. There are too many primary sources to discount the story. It also fits with the timing of Weir's obvious distress that the regiment was not reacting to the distant gunfire. One who maintained that no argument took place was Lieutenant Winfield Edgerly.

Edgerly, a West Point graduate and an executive officer in D Company under Weir's command, stated he saw the Captain ride in the direction of Reno, only to stop and do an about-face—ostensibly never going to Major Reno. Edgerly later reported that he did not see or hear an argument. Additionally, when Weir returned, he did not mention to Edgerly any altercation with Reno. Seeing Captain Weir and his orderly return, and then watching him ride north along the crest of the bluffs, Edgerly assumed Weir received permission from Reno to leave on a reconnaissance towards Custer. Edgerly and D Company dutifully followed Weir and made the one-and-a-quarter-mile journey to the Peaks, heading down the ravine to the west of and parallel to Sharpshooter Ridge—bringing them much closer to the gunfire. But other than moving to a nearby location, nothing changed. They remained immobile and unresponsive for the next two hours.

At least that was Edgerly's story, and he never wavered from it. In a private letter dated January 17, 1886, nearly ten years after the battle, Edgerly was emphatic that Reno never denied Weir permission and that the two officers never conversed, much less argued over the matter. Yet, this explanation is problematic as Edgerly stated that Weir "left me, started toward Reno and Benteen, and then instead concluded he would go out to the high point and take a look before asking permission to go down." If Edgerly did not witness Weir speaking to Reno, then he cannot, on his own cognizance, jump to the conclusion that Weir did not go to secure permission to advance, which Reno then promptly denied. Since Weir said nothing to him, Edgerly could not know what he did not see or hear.

From Edgerly's perspective, D Company's advance to the Peaks was not an act of disobedience or insubordination but rather the result of a misunderstanding or a mistaken assumption on Weir's part. Weir felt he did not have to wait for permission to do what was obviously necessary—go to the sound of the guns. Weir and Edgerly had already advanced without authorization when they thought Benteen, their commanding officer, was taking too much time watering his horses at the morass. The reason for their advance was gunfire heard from the valley, and now it was being heard downstream. When guns beckon, you ride to the commotion. According to Edgerly, Weir received no direction from Reno until Lieutenant Hare came out with orders from Reno directing him to "open communications with Custer, if possible."

Be that as it may, Edgerly remains the only primary source to deny that a Weir–Reno altercation occurred. Even Weir, according to Edgerly, denied it, telling the lieutenant afterward, "I did not have permission, nor did I ask for any." Others, however, saw and heard differently. Private John Fox said he overheard Reno denying Weir permission, stating, "You cannot go. For if you try to do it, you will get killed and your company with you." He also remembered that Benteen and Moylan were present at the time of this encounter, and neither showed any support for Weir.

Private Charles Windolph essentially told the same story, stating that it was common knowledge among the outfit that Reno and Weir had argued fiercely. Fred Dustin also recounted how he was told by another soldier that there was a back-and-forth exchange of uncomplimentary language between the two officers, Weir and Reno, with even threats being exchanged. Private Theodore Goldin stated that Lieutenant Wallace told him that Weir and Reno exchanged hot words. Trumpeter John Martin was even more emphatic that a serious argument between the two occurred and stated so:

> It sounded like a big fight was going on, and the men thought it was about General Custer, and that he was whipping the Indians, and we all wanted to hurry on and join him, but they wouldn't let us go. Captain Weir had some heated words with Reno, and I could tell by the way he (Weir) was acting that he was excited and angry. He waved his arms and gestured and pointed down the river. Then we heard some volleys and Captain Weir

jumped on his horse and started down the river all alone. But his troop followed him right away.

Not just enlisted men said there was an argument between Reno and Weir. Lieutenant Francis Gibson, usually a steadfast defender of Reno and Benteen, also confirmed the story. Gibson had no axe to grind or score to settle against Reno. Even after forty years had passed, Gibson took the trouble to correct an article critical of Benteen for joining in with Reno and denying Weir permission to go to Custer. Gibson wrote to Richard Roberts, Captain George Yates's brother-in-law, stating, "Colonel Weir did ask Reno, not Benteen, for permission to go forward and was refused." Decades later, Gibson thought it important to correct the historical record, going so far as naming names, which goes a long way in establishing his credibility that he had seen and heard the heated exchange between Weir and Reno.

There is more. Years later, Colonel Edward Godfrey wrote a letter stating that Weir had asked Reno for permission to go but Reno refused to give it. Godfrey also claimed that later he (Reno) tried to make it seem as if Weir had left without his permission—meaning that his later order via Hare to open communications with Custer was nothing more than a deliberate attempt to conceal his reluctance in investigating the gunfire they had now all been hearing for some time. Private William O. Taylor from A Company would later reflect on those two hours spent on Reno Hill as the sounds of fighting clearly could be detected downstream:

> We heard firing off in the general direction Custer was supposed to have gone. "Why don't we move was a question asked by more than one." The 3 troops that were engaged in the valley were, it is true, somewhat demoralized, but that was no excuse for the whole command to remain inactive. A few of our men had been wounded, but none so seriously that they could not ride with the pack train. All of the officers must have known that Custer was engaged with the Indians and quite nearby for he had not time to go a great way. The sudden withdrawal of that strong force of Indians who had driven us from the close vicinity of their camp could indicate but one thing, and that was another attack on their camp, real or threatened by a force from another direction.

And what about Frederick Benteen? When Captain Weir sallied out toward Custer, the Captain called it "a fit of bravado." Would the cantankerous and often headstrong Benteen obsequiously obey Reno's orders to stay put simply because he was not in command and Reno was? If he had strongly objected to Reno's orders, it would be utterly out of character for Benteen not to challenge it. Military rank and protocol never stopped Benteen from challenging Custer, which he had volubly done more than once, even on this campaign. It was not in Benteen's makeup to be unquestioningly subservient to anyone. Still, perhaps his animosity toward Custer made him curiously apathetic to Reno's lack of initiative—at least for a time until he was propelled, as he was, by Weir's actions at the first morass, to follow Weir's example by leading Companies H and K out to the Peaks.

Witnessing the departure of several companies heading downstream without his orders, a near-frantic Reno ordered his trumpeter to sound recall "continuously and assiduously." But when Benteen ignored it, Reno knew he had no choice but to acquiesce. Only by rank alone could Reno be considered in command. As for the Weir–Reno contretemps, any mention of it by Benteen only serves to bolster the evidence that a confrontation did occur. In a letter Benteen wrote to Goldin, he acknowledged that Weir, in leaving the hill, had sallied out without orders and, therefore, was insubordinate—although he does not say to whom. But who else could Weir have been insubordinate to but Reno? And what else could he have been insubordinate about except that he left the hill to march toward Custer against the orders of his commanding officer?

With more than two hundred able-bodied soldiers present, there was little excuse to remain on Reno Hill. The ammunition pack mules had arrived, and the wounded were no longer a pressing consideration. Indeed, the absence of any hostiles finally resulted in Herendeen and eleven or so others leaving the wooded area where they had remained behind after Reno's sudden departure whereupon they made their way unscathed to the command atop the hill. Herendeen remembered seeing just five Indians, only one of whom shot at them.

All told, a compelling case remains that Reno and Weir had a heated, gesticulating, finger-pointing argument about going to Custer. It is likely that Lieutenant Edgerly's denial of disobedience on Weir's part in moving out to the Peaks had more to do with an effort to protect both Weir and himself from potential charges of insubordination should the truth come out. In short, the Reno Court was mistaken when it concluded that there was nothing in Major Reno's conduct that day requiring animadversion from the Court, neither for the manner of his retreat from the timber nor his dilatory actions in moving to the sound of the firing.

Aside from the Weir–Reno quarrel, perhaps the biggest question is why there were not more reports and accounts regarding other questionable actions taken on that hill. Undoubtedly, it dovetails back to Custer's ultimate destruction and the need, officially and psychologically, to insulate the regiment from castigation, if not incrimination. As a result, rationalization abounded, including from Varnum, who, looking back, said he saw no circumstance where Reno and Benteen could have rescued Custer's command. That point, for the honor of the regiment, needed to be made loud and clear. It was repeated not only by Varnum but also by Benteen, Reno, and others.

We also must not discount the ongoing bitterness that Reno and Benteen felt toward Custer, a mindset that seemed to say, he got us into this fix; let him get us out of it. Benteen was still fuming about being removed from leading the regiment and sent over terrain so rough and unwieldy that he knew no Indian would ever traverse. Then there was Reno, who had been chastised by both Terry and Custer over disobeying his orders by scouting to the Rosebud, that caused him to be consigned to the rear of the regiment until he was plucked out and ordered to charge the village, only, at least in Reno's mind, to be abandoned by Custer. This only made the already toxic relations between Custer and his battalion commanders worse.

Chapter 10: Did Reno Refuse Weir Permission to Assist Custer?

When pressed, Benteen conceded that he felt no trepidation or uneasiness for Custer, believing that with five companies, he could take care of himself. Reno felt similarly, and, in this case, there is no reason to disbelieve them. At the Reno Court, when Recorder Lee asked Reno, "Considering the number of Indians you saw around him, did it not occur to you that Custer might need someone to 'be quick' as the order indeed said,"—in other words, to come to his aid with all due dispatch. Reno tersely replied, "It never occurred to me at all 225 men could not hold off quite a number of Indians if they are properly disposed." Additionally, Reno later said that he believed the 7th Cavalry could have whipped them if they were properly handled, since he (Reno) had stood them off with half a regiment on Reno Hill. What Reno stated about five cavalry companies being able to thwart any number of tribal warriors is more than defensible. No one, however, including Reno, brought up the lack of ammunition in Custer's command—just 124 rounds per trooper, not enough to hold out during a sustained siege.

Accusations were made, the whispering grew louder, the press picked up on nasty rumors—the grapevine was long and even Reno felt compelled to call for an investigation. When the Army finally convened the Reno Court of Inquiry in January 1879 to examine Reno's conduct at the battle, the Army justly feared that reputations could suffer, and not just Reno's. The proceedings lasted twenty-eight days, and twenty-four witnesses were called to the stand, most of them survivors of the Little Big Horn battle. A noticeable difference, however, soon emerged between what witnesses were willing to say during questioning before the hearing and what they said under oath. Many saw the subsequent testimony by surviving members of the 7th as amounting to a collective exercise in reticence, collusion, disingenuousness, and subterfuge.

Recorder Jesse Lee, appointed by the Court to determine Reno's guilt and responsibility, was among the first to sense that something was amiss among many who testified. He saw a pattern emerging among the witnesses who testified to the events at Little Big Horn as a subtle, deliberate, and concerted attempt to insulate the regiment from a more extensive and harsher truth. Mainly, what Lee saw was a failure of nerve and decisiveness among the leadership. Lieutenant Carlo De Rudio would later admit as much about a conscious dissembling of the facts—but nearly everyone else, including Lieutenant Charles Varnum, emphatically asserted that there was no collusion among the officers to shield Reno.

Sitting in the gallery during the Reno Court proceedings was an intensely interested spectator seen furiously scribbling down notes, someone who had become convinced that something more sinister was going on behind the scenes at this so-called inquest. He came to believe there was much more to tell than those testifying were willing to say or admit. This included what the official after-action reports of the battle had indicated or maybe, even more importantly, did not indicate. Frederick Whittaker was a most attentive and interested spectator. If anyone was responsible for the Reno Court hearings coming to fruition, it was this prolific writer who was versatile enough to write a thorough and analytic treatise on cavalry tactics but, curiously enough, also turned out a series of adventure dime novels, a remarkable eighty in total, with stories based on the colonial frontier, the Wild West and Napoleonic Europe.

As if this literary output was not prolific enough, Whittaker also wrote a two-volume biography of George Armstrong Custer. Overlooking the work's partisanship, unvarnished sentimentality, and the author's almost unctuous, undisguised adulation of its subject, the book was profitable and entertaining. But while his version of Custer's life titled *Complete Life*, was more hagiography than biography—a mythological slant worthy of Parson Weems—essentially a treatise of a star-struck author with an unflagging tendency toward hyperbole and romanticism, the novelist-biographer-investigator had nonetheless contributed valuable insights about the battle, and in doing so, hit a raw nerve. It is not surprising that Whittaker's efforts to redeem Custer and castigate his critics caught the attention and won the plaudits of Custer's adoring widow, Elizabeth, or Libbie, who was livid over what she saw as the withering and calumnious criticism directed at her martyred husband, not only by Reno and Benteen but by the higher echelon of the military brass.

Strong, beautiful, educated, and unafraid, Elizabeth Bacon Custer was not going to stand by complacently and see the husband she idolized being used as a scapegoat to screen the incompetence and culpability of others. She knew that some of those now condemning her husband were once his most fervent and unstinting admirers. But now a gamut of military luminaries, beginning with his two battalion commanders but ascending to General Terry, Sheridan, and even President Grant, were, she believed, complicit in orchestrating a bald-faced lie wholly attributing the disaster to her husband's glory-hunting, unsated ambition, and sheer recklessness.

There was some truth to the grief-stricken widow's complaints. Many of the country's pro-Republican newspapers blamed Custer for being rashly imprudent, as did his once supportive and admiring colleagues. Only General William Tecumseh Sherman mildly demurred, expressing shock, saying it was just too terrible to be true, that the news of what befell Custer must be exaggerated since he believed there were not enough Indians on the Great Plains to destroy his command. The truth, as always, was more complicated and tangled—beginning with when his long-suffering wife first learned of her husband's terrible fate.

Elizabeth Bacon Custer had received news of her husband's death and the defeat of his regiment while waiting for his return with other wives at Fort Abraham Lincoln. After being informed early that morning by Captain William McCaskey and two others of the catastrophe, she wept inconsolably, and then, being a true soldier's wife, she courageously put on a shoulder wrap and, with admirable stoicism, walked outside to the quarters of the other women, and bravely delivered the hard-to-believe tragic news. There were now thirty-eight other widows at Fort Lincoln, and she felt it was her obligation, even amid her own convulsive sorrow, to console them in their bereavement. She would quietly hide behind the black veil of sorrow until the dime novelist of romantic pulp gave her a reason not only to hope but to go on living: The vindication of her maligned, heroic husband demanded she not only publicly defend him but keep the flame of his proud memory burning.

A year younger than Custer, Whittaker met the "Boy General" only once, not long before his untimely death. He was immediately drawn in by Custer's celebrity, charisma, and fabled reputation. Wounded at the Battle of the Wilderness, Whittaker served as a Second Lieutenant in the Civil War, where he first learned about the young, dashing commander and grew to esteem Custer's unrelentingly bold leadership and hair-raising exploits on the battlefield. The "Boy General's" swashbuckling appeal made Custer a man after his own heart, and Whittaker made no secret or pretense about his admiration.

Back in Monroe, Michigan, Libbie, still depressed by the death of her beloved Autie, as she affectionately called him, and infuriated by the posthumous damage done to his reputation, thought, who better than Whittaker to set the record straight and rehabilitate the fortunes of her late husband. She was convinced that Reno and Benteen had railroaded her beloved to cover up their own bitter failings and that Whittaker was the right man to unsheathe the avenging blade and cross swords with his detractors. The dime novelist did not disappoint as he intrepidly jumped headfirst into the roiling waters of controversy.

With his customary zeal and rhetorical artillery, Whittaker embarked upon a one-man crusade to rescue Custer from the obloquy attached to him and see justice done to his sacred memory. To those who scarred Custer's good name, Whittaker would spare nothing and mince no words in his thundering rebuke: "The nation demands that such a court vindicate the name of a dead hero from the pitiless malignity which first slew him and then pursued him beyond the grave." Whittaker became the voice for the voiceless Custer, his unflagging defense attorney, against what he perceived as self-serving and deceitful accusations. He petitioned hard for a Congressional investigation, believing that with Custer dead, the U.S. Army had characteristically circled its wagons and that only under the auspices of duly elected public officials would there be a fair and independent hearing and a transparent adjudication of the facts. It was Whittaker's incessant pleading to the War Department that compelled Reno, whose fortunes had steeply declined after the battle, to head Whittaker off at the pass by requesting an investigation from the Army, a less censorious body than a Congressional committee, for his own vindication.

Reno had been reeling from a *New York Herald* 1877 interview with the famous war chief Sitting Bull, who, when recalling the bloody afternoon of June 25, 1876, chided Major Reno's efforts on the hill, stating that "there were only squaws and papooses" opposing their march to Custer. It was a stunning rebuke from the famous Indian Chief, whose warriors humiliated Reno in the valley fight and annihilated Custer's command. A deeply mortified Reno, in a last-ditch effort to save his honor and reputation, wrote both to the House Committee of Military Affairs and then to President Rutherford B. Hayes, imploring that a military court of inquiry be convened—the equivalent of a grand jury in civil matters—to remove this stain from his escutcheon and restore some luster to his tarnished standing as a military officer who served with honor in the United States Army.

Meantime, Whittaker reached out to Captain Weir, who was rapidly spiraling downward into his own personal hell. Whittaker saw Weir as the principal antagonist against the official narrative of the battle. But Weir was not responding to Whittaker's letters that exhorted him to tell his side of the story. Holed up in an apartment building on Hudson Street on the west side of New York City, Weir was about as far as he could get from the vale of death he had survived in southern Montana. Fragmenting and paranoid, Weir struggled under the shadow of melancholia, besotted by drink. His heart had plunged into darkness.

Having battled alcoholism for years, the trauma of the Little Big Horn greatly exacerbated his drinking and dissipation. By the time Whittaker tracked Weir down (like Custer, he met Weir just once), he found a desperately sick and troubled soul. All we know from that meeting between the two is Whittaker's conversation with Weir, in which Weir supposedly exonerated Custer and pilloried Reno.

Whittaker claimed that Weir stated that he had not responded to his importunities for an interview because he believed his testimony would be seen as biased and, therefore, would do more to undermine the truth than restore it, though he more than hinted at the dishonorable behavior of certain personages of the 7th. He refused to sign an affidavit stating his version of events (though Whittaker indicated that Weir had done just that—possibly referencing a written statement Weir may have given to the journalist James O'Kelly shortly after the battle) until, finally, after a primarily profitless interview in which Weir refused to commit anything to writing, there was little Whittaker could do but take his leave.

Still, wishful thinking or not, Custer's first biographer continued to hope for enlisting Weir in the cause, writing to assure an anxious Libbie that once the fight had begun, Weir would take his place front and center among their ranks. Unfortunately, two weeks after Whittaker's visit, on December 9, 1876, less than six months after the battle, Weir was dead, the cause later attributed to "congestion of the brain" and "intemperance"—but in truth, he became just another casualty, another statistic, in the most famous battle ever fought on the Western frontier.

But Whittaker's failure to enlist Weir in the cause did not end the probes of inquisitive newspapermen determined to scratch out the facts and burn through the thick fog clinging to the battle. James O'Kelly, a prominent journalist from the *New York Herald*, who had covered the recent Sioux war, picked up the same scent of deceit as Whittaker, having independently reached the same conclusion that the Army was doing more concealing than revealing about the battle. While his early articles about Little Big Horn gushed with praise over the actions of Benteen and Reno, O'Kelly became more skeptical of the official account with every conversation he had with survivors of the 7th.

O'Kelly had joined General Terry's command in July of 1876, affording him a unique opportunity to interview many of the surviving officers and enlisted men of Custer's regiment. At first, he noted, they clammed up whenever the conversation gravitated toward the battle. But O'Kelly's persistence, frankness, and natural bonhomie won over many who served in the 7th, and as their inhibitions eroded, they eventually spoke more openly and freely. Slowly, a different story emerged from the one transcribed in the official record. O'Kelly began to correspond with Custer's widow and with Whittaker. The more he dove into the matter, the more convinced he became that Benteen's tardiness and Reno's pusillanimity were responsible for the staggering dimension of the defeat. He, too, pushed for a full-scale investigation into the Little Big Horn fiasco.

O'Kelly insisted that some of the after-action reports "were full of inaccuracies and had later been read with something approaching astonishment and disbelief by those who fought in the battle." He, too, was convinced that surviving members of the regiment were merely echoing the "official language" and that the truth about the Custer fight remained unwritten and could only be brought to light through a government investigation. O'Kelly also believed that Weir's D Company went to the Peaks only on their own accord and without orders from Reno. "They did so," said O'Kelly, "only when they tired of the inaction of Reno, while rifle volleys were telling them that Custer's command was being done to death." O'Kelly vented on, "What were seven companies of cavalry doing gathered upon a hill [Reno Hill] when four miles away their comrades were fighting desperately for their lives?"

These critical stories had a cumulative effect in awakening a public outcry for the truth. Just prior to the opening of the Reno Court of Inquiry, the *New York Herald* republished O'Kelly's probing articles under the headline "The Long-Delayed Investigation." The Herald left no doubt about where they stood on the matter:

> The correspondent's statements—which we republish for the information of the court—were based on confidential communications made to him during the campaign by officers and men who had been present for the fight. These charges made against Major Reno by his brother officers were first brought to public notice by the Herald—Had the government voted promptly on the Herald's full and impartial statement of the charges made against Major Reno, the evidence of the gallant Colonel Weir would have settled forever the question of responsibility.

What can the interested historian or searching student conclude about the putative Reno–Weir encounter? Nothing less than the weight of the evidence showed there was an explosive exchange between the two officers over what should be done about the roar of gunfire downstream. There are five primary, credible sources directly attesting that there was a Weir–Reno argument: Trumpeter John Martin, Private John Fox (who said some of the officers wanted Weir arrested for disobeying orders), Private Charles Windolph, Lieutenant Edward Godfrey, and Lieutenant Francis Gibson all supported this version of events, although as a rule, the officers, as opposed to enlisted men, were much less likely to raise their voices against those of their own rank.

Men of higher rank tended to keep their own counsel because no one would blame them for words unspoken. But like a blade of grass irrepressibly protruding through the cracks of a concrete jungle, serious questions about what really happened began to sprout. The revealing epistolary of the Custer-hating Captain Benteen, now retired from the Army, disclosed some heretofore unknown facts. In response to Theodore Goldin's written inquiry, some fifteen years after the battle, regarding the Reno–Weir incident, Benteen did not deny it, even labeling Weir as insubordinate when referring to his unauthorized move to the Peaks.

Both journalists, especially O'Kelly, would be fed bits and pieces from survivors of the regiment that strongly suggested that much more could have been done to help Custer. The wounded on the hill were being cared for, even with only one doctor on hand. It would be inconceivable that the seven, later ten, wounded on the hill that day would have stopped either the Union or Confederate army from obtaining a military objective. That is not how battles are won, or successful wars are waged. While some of the horses from Reno's battalion were spent, the mounts of Benteen's three companies, while not fresh after their grueling scout, also were far from depleted. Neither were the horses from McDougall's company guarding the pack train. The distance from Reno Hill to the Weir Peaks was only one-and-a-quarter miles, hardly an insurmountable distance for mounted soldiers.

It is hard to accept that forty-five minutes to an hour after Benteen arrived on the hill, three or four companies could not be mustered to contact Custer. Later, in a candid moment, Benteen admitted as much when he said, "We could have gotten there [meaning to Custer], and we would have been there yet." Benteen's trenchant point was unmistakable: Going to the Custer battlefield would not have saved Custer's command but sacrificed the whole regiment to their commanding officer's foolhardiness and recklessness. Edgerly agreed with Benteen, asserting, "There was no chance to have saved Custer's command or any considerable part of it from destruction if Reno had advanced at once upon Benteen's junction with him and without waiting for the ammunition, and I believe that such a move would have only resulted in the practical annihilation of the regiment."

Playing war games of "what ifs" is risky. There are enough snares and pitfalls involved in counterfactual history to deter anybody with sense from pronouncing definitively the outcome of any one scenario. But even if the east side of the river heralded Armageddon, there were other tactical options open to Benteen and Reno than just a headlong charge into a couple of thousand warriors assailing Custer. The two battalion commanders could have taken advantage of their long-range Springfield carbines by planting themselves on the Luce/Blummer Ridge Complex opposite Calhoun Hill and Finley Ridge, commanding the low ground of upper Deep Coulee. Positioning themselves thus might have taken the pressure off Keogh's troops by targeting the warriors with their longer-range weapons and creating havoc since, as every officer in the army knew, lack of command and control made Indians unsuited to fight a two-front battle. If not obtaining victory, this strategy may have been enough to enable the survival of at least a portion of Custer's command.

In retrospect, it seems a better and more feasible option than essentially embarking upon a do-nothing strategy for more than two hours. But Benteen would never entertain any possibility other than Custer's extreme recklessness for foolishly outdistancing his reinforcements as the principal cause of the overwhelming calamity. Any other narrative would detract from his newly won sobriquet "Savior of the 7th," an honorary title he jealously guarded.

Major Marcus Reno is a more complicated case. Charges of cowardice were something the widow Custer, smoldering with bitterness, had no compunction leveling at him. It was a description that found favor with both Whittaker and O'Kelly. While it does look damning that Weir's D Company made it to the Peaks without resistance from the enemy, at the same time, Reno, wallowing in a state of inertia, stayed put on his hill, making no effort to respond to the firing downstream, the authors still believe that the incendiary accusation of cowardice appears to be overly judgmental and emotional.

What must be acknowledged is that Reno obviously had some sort of mental breakdown in the timber and ceased functioning with anything that might resemble military leadership. That he haphazardly and without apparent forethought bolted during the furious firefight from the timber and toward the bluffs in a blind rush for safety and survival tells us as much. His performance on the hill as a commanding officer did not glorify him. Only when Lieutenant Hare arrived on the Peaks with Reno's belated order to contact Custer did those on Reno Hill respond to the firing downstream. Weir possibly may have had a second unpleasant encounter with Reno since he was again seen on the hill. The point, however, is that Reno remained idle for an unjustifiable length of time, and it is almost impossible to conclude that he was not culpable of failing to go to Custer's aid sooner, regardless of whether such an effort would have succeeded.

David Evans, in his deeply sourced and superb book *Custer's Last Fight,* speculates that, in the close fighting phase of the battle, both in the timber and the flight to the bluffs, where soldiers were being picked off one by one, everything happened too quickly for the officers first to comprehend and later react to what was happening away from them downstream. He finds no evidence that Reno, though having ceased to command, refused to allow the companies to go to the aid of Custer. Evans surmises that even if Reno had issued an order preventing his wing from leaving for Custer, based on the previous refractory actions of Weir, French, and Benteen, it is doubtful it would have been obeyed if they were genuinely determined to make a junction with Custer.

The weakness of this hypothesis is that it cannot be squared with the measured and persuasive accounts of Lieutenant Edgerly about the Weir-Reno encounter, and it simply gives Reno too much benefit of the doubt. Edgerly forthrightly notes that when Benteen arrived on the bluffs, or shortly thereafter, "We then stood to horses"—meaning that the battalion, or companies thereof, were holding their horses ready to mount, waiting only for the order to advance, which never came—at least not for a long time. Edgerly also does not indicate in his accounts that, according to Evans' view of events, "things happened too quickly for the officers to first comprehend and then react."

Granted, leaving the hill with a portion of the command would have meant yet another division of the regiment, a tactic that Captain Myles Moylan had already said was the greatest mistake of Custer's life. Yet advancing toward the clamor of a battle was still the right thing to do, in what was an extraordinarily challenging and trying situation. Even Benteen, boiling over with animosity toward Custer and believing him responsible for their present predicament, realized there was no other honorable choice, so even he ultimately decided to move to the Peaks without orders.

By the time an unraveled Reno had retreated from the valley to the top of the bluffs, he had already seen enough to last a lifetime. The best we can say for him is that based on his previous military record, his largely deplorable and inexcusable actions from when he decided to leave the timber mainly were an anomaly and out of character. His only previous blemish as a combat officer came in the last year of the Civil War when John Mosby's marauders ambushed him. It was no shame, however, to be caught unexpectedly by such an elite guerilla force. There was a reason Mosby was called the "Gray Ghost." Who can say for certain how past experiences affect and influence future actions? It is not implausible that Reno, fighting an unfamiliar enemy obscured by dust and smoke, could have possibly awakened memories of the Mosby episode, heightening his fear of being similarly ambushed.

Such speculation, however, is highly subjective. The prospect of Reno being ambushed at Little Big Horn was next to nil. While it is true that the village could not be seen from Ford A or until passing the timber and getting within a third to a half mile of it, it is unlikely that Reno thought, with his experienced scouts out front and on his flanks, that he would be surprised by the enemy when charging down the valley or when he was on the bluffs around Reno Hill. Reviewing the matter, Captain McDougall had the best explanation for Reno's foot-dragging and unwillingness to support Custer, opining, "Men are different. Some are dashing, and others have a quiet way of going through. I think Reno did as well as anyone could do." Perhaps that is the most charitable explanation that could be given that day on the performance of this troubled and contentious personality.

One thing is certain. If he thought him incompetent and a poltroon, Custer would not have allowed Reno to lead an advance guard battalion into battle, regardless of military protocol. It behooves the student of this battle to be extremely cautious when guessing the motives of others since one is more likely to project his own motivations rather than Reno's or anyone else's, for that matter. But it is worth saying that neither Reno nor Benteen nor the other soldiers on Reno Hill believed their hesitancy would lead to the annihilation of Custer's command. Otherwise, they would not have been shocked by the news that Custer's battalion had been wiped out, thinking that, at worst, Custer had been repulsed and was moving north, trying to make a junction with General Terry. But there is no gainsaying that on June 27, two days after they last saw Custer ride off, every soldier in the Reno/Benteen contingent was astonished by General Terry breaking the news of Custer's stunning fate.

Until he finally took charge on Reno Hill, there was little to credit Captain Benteen's leadership, although he was not culpable for all that transpired. While Benteen understood those parts of the battle where he was engaged, he seemed unable to connect them in relation to the whole. His weakness was an inability to connect the dots, to understand what a particular occurrence on one part of the battlefield might have disclosed about what was happening elsewhere. For example, he had prematurely accepted Sergeant Kanipe and Trumpeter Martin's report that the Indians were already defeated. And despite some incoherence in Cooke's written order, he did not fully appreciate the need to 'be quick and bring the advance ammunition packs.'

He later failed to comprehend the full implication of the sudden withdrawal of the warriors from the bluffs when he first arrived. Benteen's bitterness caused him to falsely compare his present situation with that of the late Major Joel Elliott at the Washita, believing that Custer had also abandoned him. He also misunderstood the significance of the volleys of gunfire downstream, erroneously believing (as did Reno) that, with five companies, Custer could handle anything he encountered. It was a fatal assumption, but one that served to lessen his own obligations to the regiment. Lastly, Benteen then wrongly concluded that engaging the warriors on the east side of the river, where Custer was fighting, no matter the tactical approach, was a suicide mission. In sum, Benteen's perceptions and prejudices created blind spots for this experienced and thoroughly competent officer who focused only on what he saw as Custer's sins and shortcomings, not his own.

We have now come full circle back to Captain Thomas Weir. His descent into his own purgatory was swift and pitiful, striking him nearly dumb about what he knew and what he had seen that afternoon. Correspondent James O'Kelly claimed that Weir told him that he took D Company to try to help Custer—

but what, Weir moaned, could one company do? This, however, was another documented conversation for which we only have O'Kelly's word, casting a convenient pall of indissoluble doubt in the minds of Benteen's and Reno's defenders. Fueled by alcohol and depressed in his New York City apartment, the bottle served as his only refuge from the dark days he had spent in a Confederate Prisoner-of-War (POW) camp during the American Civil War, along with the recurring nightmares of the Little Big Horn battle. The bloodbath in the valley and around the hills of southeastern Montana left a lingering sense of personal guilt for his role there, deepening his mental decline. Like Hamlet, he became his own termite—slowly gnawing away at his marrow, self-respect, and peace of mind.

No knife cuts deeper than remorse, and Weir felt its cold blade mercilessly twisting inside him. Long before General Terry officially informed the surviving members of the 7th about the demise of Custer's wing, Weir knew the awful truth as he almost certainly witnessed some of the final action on Calhoun Hill and Finley Ridge. From where he stood, helpless and impotent on the Peaks, the whole valley was visible to him. For Weir, it was a tragedy of Aeschylean proportions, a reality too dreadful to contemplate, much less internalize, so these feelings were either unconsciously buried or compartmentalized until, over time, they bored into Weir's soul, giving him no rest or solace.

We know that sometime in the late afternoon or early evening of June 25, 1876, scanning the ground to the north from the elevation now named for him, Weir and a few others could vaguely see hundreds of warriors on horseback riding around and firing at objects on the ground. Several guidons could be spotted, and at first, Weir thought it was Custer's troops. But Sergeant James Flannagan saw things more clearly and told Weir to look through the field glasses. Gazing hard through the binoculars, Weir discovered the bitter truth pitilessly staring back at him.

Custer took great pride in never losing his Regimental colors in the Civil War or during the Plains Indian wars. But not far from where Weir now stood, two of the five company guidons of Custer's wing were in the hands of Sioux and Cheyenne warriors. Meanwhile, Private Edward A. Pigford, who was at the northernmost part of the Peaks and gazing in the same direction as Weir, described what he saw:

> I looked over toward Custer's ridge and the Indians were firing from a big circle until they seemed to converge into a huge black mass on the side hill toward the river and along the ridge.

The side hill, toward the river and along the ridge, was where soldiers were making a stand until they were overwhelmed by that same huge black mass. Afterward, far away from the dust-filled and blood-soaked battlefield, Weir tried to put distance and time between that day and the new life he tried to adapt to in New York. But no matter how far he ran, he could not run from himself even as he hunkered down and tried to anesthetize his demons with liquor.

Drawn inexorably back to those sights and sounds, his grief and guilt intermittently leaked out in self-accusatory correspondence. He would write to the family of the late Lieutenant William W. Cooke, an officer with whom he had served nine years and remembered with deep fondness. The remorsefulness Weir expressed to Cooke's family had the sound of the lash cutting deep into his flesh, an exercise of self-flagellation and condemnation: "With our cherished ones' deliverance within our grasp we waited a breathless two hours," he forlornly wrote, "for the order that never came." With little to do but think and remember, long-hand script became his way of bearing his cross and unburdening himself. He wrote tantalizing, semi-confessional letters to Libbie Custer, for whom he also felt much tenderness and affection, stating that he had so much to tell her in person that he was, apparently, unwilling to commit to paper.

Years earlier, there were unfounded rumors that the relationship had grown to something more than a friendship—gossip that Libbie did not actively discourage, perhaps as a measure of revenge over her husband's long absences, loose talk, and suspicions about his roving eye. Rumors of dalliances, even if never proved, must have hurt. It had all come to a head in mid-July of 1867 when Custer impulsively bolted from his command at Fort Wallace (for which he was later court-martialed with additional charges of shooting deserters, withholding medical treatment from prisoners, and unauthorized use of government property), and with an escort of seventy-six troopers made a mad dash for Fort Riley where his suffering and neglected wife had been anxiously awaiting his return.

Some fifty-seven hours later, after a blistering trek in a sauna of unrelenting heat, Custer arrived at his heart's desire. For Libbie, her enduring object of affection had returned at last. She found her beloved's daring jaunt amid so many dangers hopelessly romantic and, in her memoirs, not too subtly hinted at a furious night of passion. Custer would justify his flight from his command by improbably claiming to his superiors an outbreak of cholera in the Fort Wallace area. Colonel Andrew Smith did not buy it and was livid that Custer took leave of his command for reasons he saw as self-serving and fraudulent. There is little doubt that cholera was a smokescreen and that the more likely explanation for his being AWOL (Absent Without Leave) was that Custer got word that the handsome Captain Weir, a charmer and pleasing conversationalist, was spending more time with his wife than he felt comfortable with. Captain Benteen told an interviewer that Lieutenant Charles Brewster had written to Custer anonymously, stating that he "ought to look after his wife a little closer." Lieutenant Edward Mathey essentially said the same thing. These suspicions caused a rift between George Armstrong Custer and Thomas Benton Weir that might never have been fully bridged at the time of the General's death.

So, Weir and Libbie knew each other. There may indeed have been a strong, subliminal attraction, as the beautiful, intelligent, and cultivated Elizabeth Bacon Custer was sadly childless (clearly not for lack of trying, but very possibly because "bad boy George," unbeknownst to her, had contracted a case of gonorrhea, perhaps from patronizing New York City brothels when he was a plebe at West Point, and thus rendering him sterile). Insulated and cosseted in her youth by her father, an esteemed judge and the most prominent citizen in Monroe, Michigan, Libbie now found herself a dutiful soldier's wife, often alone in the vast nowhere of the American frontier, while her dashing husband, first on the battlefields of the Civil War, and now in pursuit of Indians, galivanted across the endless rolling Great Plains.

Tragically, her deepest fears were realized when he was killed by the Indians, leaving her to prematurely embrace a martyred widowhood, surviving him by fifty-seven years, during which she remained faithfully and unrepentantly betrothed to his ever-present, divisive ghost. Whatever history's feelings were toward Custer, Libbie had become an object of enormous sympathy and affection for the public. In Weir's sensitive, empathetic, post-battle letters to the bereaved widow, he promised to unmask the hidden truth about the Little Big Horn that had been left unspoken in the official records. His correspondence exuded a profound respect for the sanctity of her sorrow, and that she would find solace in his still unspoken truth. He was obviously haunted, and what he says is haunting, though the fullness of what he knew never found a voice.

In September, not three months after the battle, Weir told Libbie about the black memory of June 25, 1876: "You know I can't tell you now but will sometime tell it to you." Then, only weeks before his own death, he wrote once again, almost cryptically, the aching words still unspoken but silently coursing like the plaintive moan of an ebbing tide from the isolated and besieged shore of his wounded conscience. Pausing, as if he were about to finally unveil the hidden truth to his most intimate and trusted friend—as if somehow the revelation itself would be cathartic, an expiation, an atonement, he wrote:

> It is my life business to vindicate my friends of that day .... I know if we were all of us alone in the parlor, at night, the curtains all down and everybody else asleep, one or the other of you would make me tell you everything I know.

Reno was unmoved, but Weir knew, in his hour of need, he had failed to provide the support his commanding officer looked for, as he stayed planted on the peaks as the Custer death fight slowly bled out to its bitter end. In Captain Weir's own words, "With our cherished ones deliverance within our grasp, we waited breathless two hours, for the order that never came." He had let down his friends, comrades, and Custer, leaving only his inexpungable, unsated guilt to relive over and over. But that tortured, unrelenting, accusatory voice inside his head, perhaps mercifully, was stilled by his own not-unwelcomed death. Whatever secrets he carried with him were suddenly and forever swept behind a shroud of impenetrable silence, inaccessible, forever mute before the parade of history and lost in the suffocating mist of what remains an irrepressible and gripping mystery.

This final, haunting sentiment came from a previously unpublished letter by Captain Thomas Weir on October 22, 1876, currently residing in the George Kush Collection: *"With our cherished ones deliverance within our grasp, we waited breathless two hours, for the order that never came."*

Chapter 11: Did Custer Cross the River?

Turning westward from Cedar Coulee toward Western Coulee, Custer and his five mounted companies headed toward the river at the north end of the village, moving as quickly as possible without burning out the horses. He must have expected Reno to hold out in the timber for at least an hour, which would draw all the warriors out of the village. That would leave Custer an open back door to enter the empty encampment for a timely 'fix and flank' maneuver. But before Custer moved farther north from Sharpshooter Ridge, he made sure he had a reasonably good idea of Benteen and the pack train's proximity. Providing his orders were followed, he thought the entire regiment would soon be within mutual supporting distance.

Knowing that Captain Benteen's return to the main trail could only mean that there was no significant warrior force to the south, and no satellite villages planted downstream would assure that the command would not be surprised by any hostile bands as they were at the Battle of the Washita. All the Indian camps at Little Big Horn were to the north, concentrated in the expansive valley. The good news was that the preponderance of hostiles was gathered in a contained area. As Custer rode from Sharpshooter Ridge to the opposite end of the village, he took no unnecessary chances. He positioned his scouts on the high ridges to watch the warriors and the whereabouts of his respective battalion commanders. Through their eyes, he could monitor the river crossing where he was headed and the bluffs from which he had just come, all the while taking every precaution to keep his column out of sight below the ridge.

Moving down Western Coulee toward a wide ravine called Medicine Tail Coulee, Custer finally found what he was looking for: a break in the bluffs leading to a river crossing (see Map 10). This crossing was later labeled "Ford B" by the Reno Court of Inquiry. While atop Boyer's Bluff (the western bluff closest to the river south of Ford B), Custer could finally see the terminus of the village—the biggest he had ever seen—situated about a half mile north of Ford B where the Cheyenne camp circle was located. He could also see an exodus of non-combatants fleeing the village, a good many of whom were heading toward the Cheyenne camp and beyond. Their departure was more of a stampede as they fled down the valley from Reno's charge. The Indian Chief Gall recalled that these refugees were told to move downstream where many congregated by Squaw Creek, a small tributary well north of the village.

Multiple Indian accounts support this terror-stricken crush of humanity dashing from the village seeking safety. Black Elk, a 13-year-old Lakota warrior, vividly remembered the women and children running downstream. "…I looked back and saw them all running and scattering." Iron Hawk, a young Hunkpapa warrior, also recalled the women and children running wildly down the valley. Soldier Wolf, a Cheyenne, saw those same women and children racing to his Cheyenne camp tribal circle, which was north of the other tribal gatherings and, for the moment, deemed the safest refuge. It was evident to Custer that the village was breaking up and that Reno's charge, though ultimately aborted, was having the desired effect.

From a ridge overlooking the river, Custer watched the non-combatants flee toward the Cheyenne camp. From the time they realized Reno was charging down the valley, war parties left the village to catch their horses grazing on the western foothills. Having captured their mounts, they rode directly out to confront Reno's soldiers who attacked them in the valley as Custer's troops remained out of sight behind the bluffs. This took time, but Reno's aborted charge gave the Indians an opportunity to catch their breath and more mounts.

While on Boyer's Bluff, it is possible that Custer saw Crazy Horse's war party (though he would be unaware that it was the legendary Lakota warrior) leave the village—one of the last war parties to have mobilized for battle. Their departure allowed Custer to strike the village, now mostly empty of warriors, from behind those who had opposed Reno's advance. This maneuver would open a second front against the warriors, who were vulnerable to a flanking or envelopment maneuver. With non-combatants and warriors in his crosshairs, it became a point of contention among battle students whether Custer intended to charge into the village and destroy it or to bring the hostiles back onto government reservations by seizing the noncombatants as hostages to compel the warriors to surrender and return to their government supervised status.

That Custer's paramount objective was purely military is beyond debate. He intended to destroy the village and utterly crush the warriors. If anything, the non-combatants were merely a secondary consideration. The shortest and surest path to victory is to defeat the warriors. Rounding up the non-combatants happens only after the battle is won. Because Custer had apprehended fifty-three hostages at Washita, some battle students also insist he was employing the same strategy at the Little Big Horn. The circumstances were significantly different at Washita than at Little Big Horn. At Washita, the non-combatants were still in the village, and the attack took place at the crack of dawn. Custer could not help but run straight into them. Seizing captives at Washita was more incidental than intentional. At no time in Custer's military career did he single out capturing non-combatants to use as a bargaining chip to get the warriors to capitulate. Custer's objective at the Little Big Horn was to achieve an uncompromising military victory by capturing and destroying the hostile village and horse-herds. Capturing non-combatants would always be a secondary concern.

At Washita, there was no opportunity for the Indians to form a rear guard and scarcely any time for the non-combatants to flee since Custer never hesitated to barrel into their village, which was a far smaller encampment than the one around the diaphanous and limpid waters of the Little Big Horn River. Destroying the village and defeating the warrior force would compel those who fled to return to the reservations or face the impossibly harsh consequences of the elements and the prospect of exposure and starvation. Being without food and shelter, usually plentiful in the locus of the village but meager on the frontier, made survival in winter climates, whether faced individually or tribally, extremely remote.

Chapter 11: Did Custer Cross the River?

Up to this point, things for the 7th could not have gone better, from locating the village so quickly, to unprecedently surprising the hostiles in broad daylight, to finding the Indians all clustered in one place, to Reno's charge and the subsequent skirmish line that distracted the enemy and opened a rear door for Custer's wing to strike the village, thus initiating a potentially devastating envelopment. Custer must have been highly encouraged by where things stood. But the vicissitudes of war are often unpredictable, and fortune is a capricious and fickle mistress. Expecting the unexpected is a reliable rule in war, and the Little Big Horn would be no exception.

The half-breed scout Mitch Boyer, peering from a long ridge later named for him, was the first to pick up signs that something had gone terribly wrong down in the valley where Reno was fighting. Indian scouts had an almost preternatural ability to detect sounds and follow dust trails. Soon, it became obvious to Boyer that the direction of the rising dust and the trailing sound of gunfire presented an ominous turn of events. Reno was not advancing, not even holding the line, but was retreating from the valley, probably toward Ford A (only later was it discovered that Reno went up the bluffs), where he had first crossed to assail the Indian village directly.

Although Custer knew Reno halted the charge, dismounted, and organized his forces into a skirmish line, he remained confident that his battalion commander was in a strong enough defensive position to distract and engage the enemy, thereby still functioning as an advance guard. But when the spine of Reno's defensive posture was broken, his command came under heavy attack and was routed. Upon receiving this news, Custer undoubtedly recalled how he had been pilloried for not doing enough to rescue the highly regarded and popular Major Joel Elliott and seventeen of his troopers at Washita. Elliott had been pursuing warriors fleeing the camp when other warriors from unknown satellite villages surged in and made short work of Elliott's small detachment.

The stakes here were now much higher. Reno's panicked retreat meant not just eighteen lives were imperiled but more than 150 troopers and scouts under his command. Reno's debacle in the valley meant that any hope for a 'fix and flank' maneuver faded away. The fix had vanished, and the flank was rendered impotent, seriously jeopardizing not only Reno's command but the success of this entire campaign against the hostile Sioux and Cheyenne. Scarcely an hour had passed since Custer bullishly told his command on Sharpshooters Ridge, "Courage, boys, we've got them, we will finish them up and go back to our station." The men cheered lustily, but now that swagger and audacity seemed a lifetime ago. Fast action and new tactics were needed to address this emerging calamity, as battles often hinge on tempo, and Custer had to believe that retreating over the bluffs back to Reno's position would destroy their momentum, leaving the campaign against the hostiles in tatters.

But when faced with adversity, Custer would battle on tenaciously. The thunder of rolling volleys, the reek of gunpowder, and the deafening, ever-swelling crescendo of war drove him to dance to the music faster and faster. He reveled in its fearful pageantry where even death, glorious death, was part of the libretto of this grand and terrible opera.

Sheer recklessness, his critics cried. But Custer had always been lucky in war, which was due in no small part to his unrelenting, aggressive temperament—his willingness to challenge the fates and disdain the odds. For Custer, it was about audacity—always audacity. Yet this very zeal was what many of his obituarists claimed led to his doom. Edward Curtis, the celebrated Indian photographer who compiled *The American Indian*, a massive 20-volume work, wrote in the *New York Herald* on November 10, 1907, "When the true story of Custer's fight is told, it is probable that the telling will add no luster to the Custer name unless a willingness to gamble with the lives of others to accomplish a personal end is considered laudable."

Much of that same criticism simmers today among scholars who view Custer that day as a poker player, a daredevil chasing his losses, refusing to fold, and digging an even deeper hole for himself and his command. An incorrigible crapshooter who refused to come to terms with the idea that the risks he was taking were no longer commensurate with the potential rewards, adopting the premise of 'the greater the odds, the greater the glory.' Their view of Custer would have him rather be damned with the devil and his fallen angels than surrender the initiative to the enemy.

But war is an imprecise instrument. The fog of battle is real; clearly, Custer could not gauge how rapidly his campaign was disintegrating at the Little Big Horn nor how desperately determined the warriors were to defend their women and children. He failed to appreciate how quickly the best-laid plans could go awry with a divided regiment in the face of a much larger enemy force. As one battle student insightfully observed, "When you disperse power among your command, be you a Roman Legionnaire in Gaul, a Napoleonic officer in Eastern Europe, or an American officer in Afghanistan, it entails serious risk."

Nonetheless, Custer was lightning-fast at reading battle situations and just as quick to act on what he saw. He responded to Reno's fiasco by moving toward the Ford B river crossing to do one of three things: a) cross the river and attack the village, b) engage in reconnaissance that might be necessitated by the lack of visibility from the bluffs, or c) create a demonstration to draw the Indians off Reno by threatening the fleeing non-combatants (see Map 9). Custer appears to have opted to create a diversion to draw the Indians away from Reno Hill. That would give Reno time to regroup and meet with Benteen's reinforcement, which Custer knew was coming up along the main trail. As a result, that would give the reunited 7th another opportunity to act in concert to defeat the Indians.

While some battle scholars still insist there was no feint or demonstration at the river, there is no other rational explanation for Custer moving down Medicine Tail Coulee toward the Ford B river crossing except to draw attention to himself and away from Reno. To cross the river and strike what was now an abandoned village with one battalion (Reno) already in a headlong retreat and the other (Benteen) still not on the battlefield makes no tactical sense. One account from a rather obscure source states that an inconvenient beaver dam at Ford B prevented Custer from reaching the Indian village on the other side of the river. But not much credence can be afforded to that explanation when the beaver dam did not impede the Indians from crossing in heavy numbers to attack Custer.

Despite the feint, Indian testimony indicates that Custer's troops did not get down to the river crossing of Ford B. The Lakota warrior Kill Eagle said the soldiers did not come any closer than six hundred yards from the river, and the overwhelming testimony of Indian accounts is that none of the troopers in Custer's command crossed over to the side occupied by the village. Though arriving at Ford B late, after the troops had withdrawn from the area, Gall was told by other warriors that the troops never came within a half mile of the ford.

Additional factors strengthen the conclusion that Custer's troopers approached the river tentatively—an indication that this was indeed a feint, not a charge. Foolish Elk remembered that the soldiers sat on their horses and fired across the river into the village without getting into the water. He further stated that the Indians were always between Custer and the river, coming up and getting around to the east of him, passing around both his front and rear. He Dog remembered that there were fifteen to twenty Sioux on the east side and some soldiers down near the river, but not much shooting, which would have occurred if there had been an attempted charge across the ford.

After the battle, several soldiers and scouts examined the bank on the village side for wear and tear typical of horses crossing over a banked river. Frederic Gerard noted that the bank on the village side was eighteen inches higher than the water, stating that he found no evidence of horses or other animals crossing there. Because of the elevation and the miry footing, there would have been physical evidence if mounted soldiers had trampled over it. Godfrey agreed that Custer's command never got close to the river. Two Moons said the soldiers never got to the river and Respects Nothing also said Custer never came directly to the river.

Often overlooked is that there are two ford crossings at that juncture of the river, known as B-1 and B-2, respectively. B-1 is at the base of Boyer's Bluff at the mouth of Medicine Tail Coulee and is hampered by heavy timber on its western side. That ford had enough cover for hostiles to conceal themselves in firing positions to try thwarting an attempted crossing. The B-2 crossing, however, is several hundred yards downstream at the outlet of Deep Coulee, and those natural impediments do not obstruct it. So, B-2 would make for a far easier crossing. While a few troopers might have come close to the river at Ford B-1 to scout it out, none came within several hundred yards of the second or lower B-2 ford, despite the more accessible crossing it provided.

A host of reasons, both hypothetical and practical, argue against the proposition that Custer sought to cross the river at either of these two fords to attack the village:

1. With Reno's battalion having fled the valley, and Benteen's battalion having yet to arrive, crossing the river would have isolated Custer's five companies in the valley with no visible means of support from any of his remaining combat battalions. His presence there would draw the entire hostile force toward his isolated command, which would be forced to engage them unsupported on poor defensive ground. Custer's better option would be to seek a junction with his separated battalions and the pack train east of the river on better ground.

2. If Custer were serious about attempting a crossing at Ford B-1, he would have sent a platoon or company up Boyer's Bluff, the higher ground commanding the crossing, to suppress any counterfire from across the river. This is not only standard operating procedure but a tactic that a cavalry officer of Custer's experience would have instantly recognized and employed. If he had been determined to cross

at Ford B-2, then a detachment of sharpshooters would have been deployed on Greasy Grass Hill overlooking the approach to that ford. Custer would not attempt to cross either ford without first occupying the high ground. Military tactics call for defended fords to be supported first by dismounted fire. Since Custer sent no skirmishers to Boyer's Bluff to provide cover for a charging contingent of soldiers bears witness that he had no intention of crossing Ford B-1. But Custer did leave open the possibility of crossing Ford B-2, as he occupied the high ground of Greasy Grass Hill with a small detachment, suggesting an anticipated junction with Benteen and Reno, reforming their ranks and resuming their offensive posture. Although this never materialized, it would have put in motion the pieces needed to cross the ford and sandwich the enemy between two forces from opposite directions.

3. After the battle, Benteen began composing a letter to his wife stating that Custer had crossed the river into the village. But later, Benteen returned to his unfinished letter with a changed opinion. About whether Custer crossed the river to the west side, Benteen had this to say: "The latest and probably correct account is that none of Custer's command got into the village." We can reasonably surmise that Benteen had acquired fresh information, causing him to change his mind about Custer charging the village and now understood that his earlier position could no longer be supported.

4. Reno, unlike Benteen, thought that Custer's C and I Companies and possibly a portion of E Company crossed the river only to be driven back by staggering fire. But this cannot be squared with the scout Frederic Gerard's observation that the bank on the west side (Custer would be attacking from the east side of the river) was nearly two feet higher in elevation than the river. If cavalry horses had gone that route, they would have torn off chunks of the embankment that Gerard said were still intact. A properly organized cavalry charge would have caused a significant physical disturbance around the eastern bank which was not even faintly observable to the scouts who closely examined the terrain following the carnage.

5. Based on plausible timelines, there was insufficient time for Indians to assemble in any significant numbers at Ford B to stop a full-scale mounted cavalry attack since the fighting was upstream at Reno Hill. Perhaps twenty or more warriors had assembled at Ford B, though a hardy few did come over to the east side of the crossing to challenge and fire at the troops. Such a paucity of warriors was not going to stop five determined companies of cavalry under a highly aggressive and skilled commander like Custer from crossing the river if that is what he intended. Custer had charged into Confederate battle lines, facing not only gunfire but a barrage of artillery to cross into enemy territory. Two dozen or so warriors with limited marksmanship would not stop him, much less force him to retreat, especially with his skirmishers being smartly positioned to cover the charge and pick off the enemy.

6. If Custer wanted to cross the river and attack the village, he would have brought down all five companies of his command and not leave, as he did, more than 40 percent of his strength three-quarters of a mile to the rear. There is not a chance that an experienced battle commander like Custer would have so markedly reduced his strength for a mounted charge into enemy territory. There is no precedent for him doing this during the Civil War or anywhere else among the many charges he made.

7. Walter Mason Camp interviewed scores of Indians from the Battle of the Little Big Horn, and not one of the accounts he transcribed stated that any of Custer's troops crossed the river. Since then, additional Indian accounts have produced only one or two questionable testimonies from Indians who said that Custer crossed the river as opposed to some 225 extant Indian sources that make no mention of it. A repulsed cavalry charge would be big news in an Indian village and most likely exaggerated for effect. Yet if there was a charge across the ford, nearly every Indian account, aside from the generally discredited White Cow Bull story, stands peculiarly mute about it. These factors overwhelmingly argue that Custer made no real effort to cross the river into the Indian village.

8. Not much credence can be given to the theory, popular in some circles, that Custer moved down Medicine Tail Coulee merely for reconnaissance purposes. Custer could see what he wanted to see well enough from the bluffs. He could see more from the heights than he could up close. If reconnaissance was Custer's primary motivation in his descent to the river, then he was either ignorant of Reno's retreat from the timber, indifferent about the fate of Major Reno's battalion, or had no compunction about telegraphing the whereabouts of his five companies to the enemy.

Some maintain that an attempt was made to cross the river but that it was abruptly arrested because Custer was seriously wounded or even killed, leading the charge. A commanding officer being shot would be prima facie evidence that a charge across the ford was indeed attempted. One Indian account did state an officer in a buckskin jacket with binoculars, a wooden box, and a compass was shot near the river. Custer was wearing a buckskin jacket that afternoon (but, according to Trumpeter Martin, he removed it as the temperature rose and tied it to the rear of his saddle), and so were several officers in his command.

If, for argument's sake, Custer had been shot at the river, the idea that his battalion would have dragged a wounded, much less dead Custer, one and a half miles north where his body was ultimately discovered on Last Stand Hill is beyond the realm of reason or the norms of credulity. If Custer had been shot, military protocol would have called for Captain Myles Keogh to have taken command of the right wing, and Cooke, as Regimental Adjutant, would have then accompanied Keogh throughout the battle. But that did not happen. Cooke's body was also found on Last Stand Hill, maybe twenty feet from Custer and about a half mile from Keogh's remains, so the body locations alone discount that theory, although it continues to live on. The appeal of Custer being an early casualty helps explain the disintegration and one-sided defeat of the 7th Cavalry by a native warrior force. It fits snugly into the old maxim that once the shepherd is struck down, the sheep scatter.

The rumor that Custer was shot and disabled at Ford B is not in accord with the battalion's future movements. The evidence plainly points to just the opposite—that Custer remained a vital and vigorous presence almost to the very end. If Custer had been killed or even wounded at Ford B, it would have made no sense for the battalion to move farther north, away from Benteen/Reno and the ammunition packs they were so critically dependent on. If he had been incapacitated, either Adjutant William Cooke or Captain Tom Custer would have immediately fallen back to the position on Luce/Blummer Ridge

Complex, and the decisive battle would have been fought in that vicinity. Instead, Custer was a presence throughout the battle, and the fact that cartridges from his own Remington Rolling Block Sporting rifle were found underneath and around his body on Last Stand Hill is all but irrefutable evidence that he was alive and fighting until the end.

The only extant first-hand account of a buckskin-clad officer being shot at Ford B came from White Cow Bull. Having lived a remarkable sixty-six years after the battle afforded him numerous opportunities to tell his story, which he did with relish, even though his account was published only once. White Cow Bull went so far as to say that he not only witnessed Custer being shot but claimed that he had shot him. This is an extraordinary claim, and it is dutifully noted that scholarship has determined that the accounts of White Cow Bull are frequently tarnished with hyperbole, if not characterized by outright grandeur and delusion. Like the charge across the river, the problem lies with corroboration. As Greg Michno rightly points out in his book *Lakota Noon*, there is not a single Indian account that attests to White Cow Bull's claims. An officer being shot at Ford B, especially Custer, would be a huge event, a memorable story to be told around the campfire for generations. And yet none of the other accounts mention a word about White Cow Bull taking down the great white chief.

Custer is not the only candidate for an officer in buckskin being shot at Ford B. A young Second Lieutenant named James "Jack" Sturgis was also reputed to be wearing a buckskin jacket that day. The twenty-two-year-old was fresh out of West Point—having graduated from the academy a year before the battle. His father, Colonel Samuel Sturgis, was the official regimental commander of the 7th but was not on active field duty. A veteran of the Mexican War, the elder Sturgis spent several years fighting Indians before entering the Civil War. Because of his advancing age and inactivity, Sturgis was never seriously considered to lead the regiment on this campaign. Nevertheless, young Sturgis was assigned to the 7th, undoubtedly with the help of his proud and influential father, and he served in Lieutenant Algernon Smith's E Company. So, James Sturgis rode with Custer, and he wore a buckskin jacket. But it is highly unlikely that the son of the regimental commander, with just twelve months of experience in the field, would have been singled out to lead such a daring charge across a river and into the teeth of enemy gunfire.

Even more unbelievable is that having been shot, the young lieutenant, personally popular with the regiment, would be left near the river at the mercy of the hostiles. If Sturgis had been shot off his horse near Ford B, every effort would have been made to either rescue the son of the regimental commander or retrieve his dead body to spare his remains from being mutilated. There is, however, little to no documentation in any of the Indian accounts, aside from White Cow Bull, of an officer being shot and rescued.

No evidence exists that any of the men from Sturgis's E Company was shot around Ford B. Only one dead cavalry horse was found there. A cavalry assault is likely to produce many more dead mounts since a river crossing would slow the momentum of a charge, making these large animals easier targets for the Indian skirmishers positioned nearby. A sure way to repulse a mounted charge is to kill the soldiers' horses. Even Sturgis's father, based on his own interviews with Indian captives after the battle, eventually came to believe that his son was killed just downriver by the mouth of Deep Ravine, near Ford C, close to the end of the fight. Some early battle researchers later mistook that area for Ford B as the location of young Sturgis's death.

Some believe that the officer who was shot at Ford B was Captain Myles Keogh. The same age as Custer, he also fought in the Civil War and had the distinction of serving on the staff of General George B. McClellan, commander of the Army of the Potomac. Being a combat-experienced and colorful officer minted in the same firebrand mold as his counterpart, he commanded the lead battalion of his wing. Could Keogh then have been mistaken for Custer, and was he the officer shot near the crossing? There is some circumstantial evidence that the officer in question could have been Captain Keogh. According to Private Gustave Korn, Keogh was leading troops down to the river, which would make him an inviting target for Indian sharpshooters. Moreover, one of Keogh's bloody canvas leggings and shoe was found near the riverbank (a posthumous examination of his body disclosed a wound above the knee), causing some to conclude that this was indeed the officer in a buckskin jacket who was shot near the river crossing. Keogh's horse, the most famous mount in 7th Cavalry lore, had a matching bullet hole in his flank where Keogh's knee would have been. But even if Keogh were considered a likely candidate, we still would face similar puzzling questions about the command's subsequent movements in reacting to such a high-ranking officer's death or incapacitation.

While it is possible that Custer still would have marched north even with his battalion commander (Keogh) seriously wounded, a debilitated Keogh certainly would have been brought to Calhoun Hill, the strongest defensive position in the area. This location would have then been secured for the protection of the wounded rather than a low-lying stretch of land a quarter mile farther north, now known as Keogh's Swale, where the dapper Irishman's body was later found. Moreover, rather than Lieutenant James Calhoun, who logically would have inherited the command upon Keogh's incapacitation, Custer would have likely left Captain Tom Custer or his Adjutant William Cooke to direct operations around what would later be known as Calhoun Hill while he subsequently rode off north with two companies and Headquarters Staff. Instead, both of those officers accompanied Custer on his northern route and died with him on Last Stand Hill.

It should be noted that Keogh's bloody legging and shoe being found at the river is not hard evidence that he was the officer cited in that Indian account. After the battle, the bodies of the soldiers were stripped by the Indians, who then proceeded to wear the soldier's clothing or bring it into the village as battle souvenirs. With warriors riding back and forth over the ford, it would not be unusual for the belongings and apparel of soldiers, amid all the excitement and chaos, to be unintentionally dropped there. Clothes and other belongings of the soldiers traveled even if their bodies did not accompany them.

Additionally, the leggings that Keogh was wearing were not easily removed. Since Keogh was wounded above the legging, it is highly improbable that any time or effort would be expended in removing them in the middle of a firefight with bullets flying everywhere. Military protocol and common sense dictated that the wounded be removed from the area of conflict and treated in a more secure place than in the middle of a battlefield. When factoring in all the evidence, it is doubtful that any officers were killed near Ford B. However, the possibility exists that a Non-Commissioned Officer (NCO) may have been hit there. Perhaps the best source for determining this was John Stands In Timber, a grandson of the Northern Cheyenne warrior Wolf Tooth, born just eight years after Little Big Horn and who would walk the battlefield with his grandfather. After listening to the accounts of all the warriors who fought Custer that day, he concluded that the firing did not last very long and that few, if any, soldiers were killed at Ford B.

There is general agreement among most battle scholars that Custer had the right wing split into two squadrons, one commanded by Keogh and the other by Captain George Yates, each initially with two companies. Yates was assigned F and L Companies, and Keogh, the senior Captain, was assigned C and I, although that would change during the battle. Lieutenant De Rudio and Lieutenant McClernand (the latter served on Brigadier General Gibbon's staff and arrived on the battlefield two days later) observed a double set of fours approaching Ford B with the trail cutting up to Butler Ridge, an indication that at least two companies of Custer's wing had approached the ford area but did not cross the river. Most likely, E Company was serving as Custer's Headquarters escort that day, using it in a manner to support whichever of his two squadrons needed reinforcement. This would also give him added support to cover greater distances than he might otherwise attempt with only Headquarters Staff.

A better understanding of what was an evolving situation at Medicine Tail Coulee can be better illustrated by examining the strange adventure of Private Korn of I Company, who was one of the last soldiers to see Custer and his comrades alive. The Korn affair, much like the earlier discussed Peter Thompson incident, is beneficial, provided it can be believed. It furnishes information at critical moments where facts are painfully hard to come by, including where Custer was just before he engaged with the hostiles. Private Korn, shortly after arriving with his company at Medicine Tail Coulee, asked and was granted permission to halt and adjust his saddle before advancing toward the river at Ford B. But, as Korn would tell it when he attempted to remount his horse, it spooked and took off with its astonished rider past the squadron to the river, through a gauntlet of Indians who fired at him from all sides until, almost miraculously, he arrived safely at Reno Hill. Luckily for Korn, the warriors were generally poor shots. His horse was less fortunate—taking five hits and succumbing to its wounds.

Korn's hell-bent-for-leather ride to safety raised questions about what spooked his horse and how he made it back to Reno Hill. Korn's tale was met with skepticism by more than a few in the regiment, and his credibility was done no favors when various individuals recounted several different versions of his ride. Many thought Korn's nerves got the better of him and that he took off on his own. Others, however, maintained that there was no denying Korn's reputation as a stalwart and honest soldier (one of his comrades called him the bravest man he knew) and believed his story.

The matter was later investigated, with Korn cleared of wrongdoing or dereliction of duty. But the idea that he rode through the valley past all those warriors without receiving so much as a scratch does appear to be colored by exaggeration. A more likely scenario was that he crossed at Ford B-1 and then recrossed the river at Thompson's Ford (named for where Peter Thompson said he sighted Custer riding alone) about a mile upriver from Ford B. Korn himself never did say he rode through the Indian village. But others said they had heard Korn describe his ride to safety in those terms.

Based on what we know of Korn, the subsequent inquiry into his conduct, and character references about his veracity, he appears credible enough to be given the benefit of the doubt, even if his story contained an embellishment or two. War can be a matter of inches, hinging on trifles of happenstance and often fraught with caprice and strange occurrences. But while this is a titillating story, we learn a few additional details from Korn. Because he was from Keogh's I Company, Korn's story firmly places that company in the vicinity of Ford B and not on Luce Ridge, which is likely where Yates's squadron was deployed, contrary to popular belief.

Returning to the demonstration at Ford B, it is now 2:25 p.m. Custer, alerted by Boyer about Reno's retreat, sent his escort, E Company (which may have been split into platoons—a platoon being roughly half a company, in this case numbering eighteen to twenty men) down to Ford B-1 to threaten the village and incite fear among the hostiles that the non-combatants were in mortal danger (see Map 9). While troop movements at this stage are speculative, Custer likely rode into North Medicine Tail Coulee behind Luce Ridge to confer with his squadron commanders Keogh and Yates, ostensibly on the explosive turn of events triggered by Reno's retreat. E Company was ordered to approach the vicinity of Ford B-1 at the mouth of Medicine Tail Coulee. Custer brought the Keogh and Yates squadrons up onto the Luce/Blummer Ridge Complex, which was in full view of the Indian camps below.

On this campaign, Custer did not have his regimental band to inspire his command with the rollicking strains of Garryowen. Still, he did his best to draw attention to his battalion as it moved down Medicine Tail Coulee, with the trumpets blaring and the soldiers shooting off their guns as they approached the ford's crossing. Indian testimony strongly supports that this occurred. Yellow Nose heard what he called "music" at Medicine Tail Coulee, and Custer's Crow scout Curley also said he heard bugles sounding off in that area. Mrs. Spotted Horn Bull recollected the cacophony near the ford: "From across the river, I could hear the music of the bugle and see the column of soldiers turn to the left, to march down to the river where the attack was to be made." The overpowering effect of this demonstration was not only auricular but visual, for Custer had orchestrated the distinctive Gray Horse Company to make its way down Medicine Tail Coulee. That had a decided and lasting impact—an image indelibly burned into the Indians' collective memory. The Cheyenne White Shield would long remember the Gray Horses because these animals were "pretty white" and could be seen a long way off.

Meanwhile, Short Bull was astonished at the cavalry's sudden appearance, thinking that the only blue coats were coming from the south. He remembered that it looked as if a million soldiers were up in the hills. Runs the Enemy said that as he looked over the line of the ridge, the soldiers seemed to fill the hill, and when he looked into the eyes of his fellow Sioux, he saw fear. Red Horse remembered that when the warriors saw the women and children running into the jaws of mortal danger, they abandoned Reno to save the non-combatants. Crow King remembered that the only purpose for confronting Reno was to give the women and children time to reach safety, and now, to his horror, it looked as if their escape led them right into the clutches of their attackers.

Custer's tactic worked. Nearly all the Indians harassing and pressing a beleaguered Reno now turned around and, with an agitation bordering on panic, headed to the sound of the firing and the new and unexpected threat against the women and children downriver. The roles of Custer and Reno were now reversed. When Reno fled the valley, Custer endeavored for his wing to become the new fixing force, drawing the Indians toward him, while Reno and Benteen, if tactics fell into place, became the new flanking force. Whether this reversal of roles would have worked will never be known because while Custer fixed, Reno and Benteen never flanked. Reno's command remained immobile on the bluff for more than two hours. Reno was either unable or unwilling to do anything to shift his command from an exclusively defensive position that was unnatural for cavalry that relied on mobility and swiftness for success.

During the demonstration, the trumpeters of all four troops, as well as Headquarters Staff, assembled there to accompany Keogh's squadron's advance down to the river by making as much noise and fanfare as possible to draw attention to themselves and away from Reno's shattered command on the bluff, perhaps even playing "Garryowen" for the last time. It was at Ford B that the Custer fight finally began. E Company, led by First Lieutenant Algernon Smith, usually assigned to Company A but now commanding Company E, was initially ordered down to the crossing at Ford B. Supported by Keogh's squadron consisting of C and I companies, E Company, as it neared the river crossing at Ford B-1, had commenced firing across the river toward the village. Meanwhile, the Yates squadron (F and L companies) was now deploying on Luce and Blummer Ridges to facilitate the expected junction with Benteen and Reno (see Map 11).

The opening gunfire was heard by any number of officers and troopers on Reno Hill and by those few still ensconced in the timber. The demonstration drew nearly all the warriors away from Reno Hill to confront Custer's command. The Yates squadron on Luce and Blummer Ridgeengaged warrior forces that unexpectedly appeared from the east (this was Wolf Tooth's band) and to the south against those warriors arriving from the Reno fight over the bluffs. Custer's Headquarters Staff and E Company were deployed on Butler Ridge halfway between the Keogh and Yates squadron to protect Keogh's left flank. Again, it is important to emphasize that Custer's hand is not forced in moving back up the coulee. He is in control, confident that his own wing is not endangered in any serious way. Custer has Keogh's squadron fall back up Deep Coulee toward Finley Ridge, with a squad of flankers moving up parallel with them on the high ground from Greasy Grass Hill to join them there. E Company was falling back from Butler Ridge to Luce Ridge to join the Yates squadron, which would soon move from Luce Ridge to Blummer Ridge (see Map 12).

Both sides did fire across the river, with nearly all the Indian participants agreeing that the fight at Ford B was sharp but brief, lasting perhaps a little longer than twenty minutes. Again, it was primarily a demonstration to draw the Indians off Reno, not an attempt to charge across the river. Nor were the soldiers, as Reno later stated, driven back by staggering fire. This misperception was a mixture of wishful thinking and the Indians' misunderstanding of cavalry tactics. As they did when interpreting Reno's actions in the valley, they believed the soldiers' retreat up Medicine Tail Coulee from the river was the direct result of their medicine, fighting skills, and bravery, which, though undeniable, had nothing to do with Custer's squadron retreating.

What the warriors witnessed was the cover fire of one company to mask the withdrawal of another. Lieutenant Oscar Long, who had interviewed Indian combatants after the battle, realized this when they related how one company started to run when they were near the river. This covering fire mystified the warriors, who thought that the soldiers were shooting at their own troops when, in reality, they were shooting over their heads. The warriors made these judgments by appearances alone since they were ignorant of this tactical maneuver. The retreat of Custer's forces up Deep Coulee was a planned withdrawal, not a forced one.

If a flood of warriors had driven back the columns, the battle would have ended much earlier than it did. Instead, Custer retreated of his own volition and in good order, knowing that his best chance for victory was to reconsolidate his command somewhere east of the river and then resume the offensive. Despite the Reno setback, Custer was still thinking offensively and aggressively and still fighting to win. He conducted a retrograde redeployment on the eastern ridges, firing to keep the enemy at a distance while the command waited for reinforcements and then find another location to 'fix and flank' (See Map 12).

As Dale Schoenberger stated in his book *The End of Custer,* it was highly doubtful that Custer, who invariably thought in terms of offensive tactics, would have turned his squadron away in the face of such weak warrior opposition without another offensive battle plan in mind. If Custer had been thinking defensively, he might have seized Luce/Blummer Ridge Complex since he would have been closer to his support and obtained excellent visibility all around. But that would mean Custer was settling for a stalemate, an indeterminate conclusion not only to the hostilities around the Ford B crossing but to the outcome of the entire battle. But that was not the case—Custer was very much in an offensive mode. What he might have had in mind entailed moving downstream, forcing the warriors to follow him, and trapping them between his own column and his expected reinforcements led by Benteen and Reno.

It is probable that during this retreat up to the high ground, Custer first realized that Reno did not retreat in the direction from which he came—Ford A—but rather up onto the bluffs. There is a line of sight from East Ridge to the bluffs and, as Gregory Michno astutely points out, while the Indians could see his flankers on this ridge, Custer's flankers (a small detachment of five or six troopers from F Company) could also see the Indians on the bluffs—and possibly even Benteen approaching Reno Hill giving Custer encouragement that both battalions would unite and soon be on their way to join the battle downstream.

This demonstration by Ford B might also have had a secondary purpose besides drawing the Indians off Reno—to concentrate the Indian force in the Ford B area. It is difficult and ineffective to fire upon warriors scattered all over the battlefield. But if you can compel the enemy to concentrate their numbers in one place, they are susceptible to being flanked. Getting behind them lures them into a crossfire—providing your reinforcements arrive in a timely fashion—an eventuality which, sadly for Custer's command, would not come to pass.

As Custer moved farther north, there was a brief, lively, and sudden encounter with a band of some fifty warriors returning to the village, a party of Indians commonly referred to as Wolf Tooth's band. Heated gunfire was exchanged between the warriors and the soldiers at Luce and Blummer Ridge, first with Yates's squadron, soon to be joined by others in the battalion. The short engagement was of no consequence for either side—just another unwelcome reminder that the 7th was deep in hostile territory. In the meantime, Wolf Tooth's band, outnumbered and outgunned, was careful to keep out of range of those army Springfields that easily outdistanced their weapons. Custer remained unfazed by the encounter, most likely seeing it as a distraction. After all, what are the bites of a few horseflies to a spirited horse? Once the gunfire subsided, Custer's path was open to journey farther north, and Wolf Tooth resigned himself to following Custer's battalion at a guarded distance—across Battle Ridge—down to the northern flats—and finally back to Last Stand Hill, where he would ultimately position himself and his warriors to join in the kill. But that would be later—much later.

It was roughly 3:00 p.m. Headquarters time. The temperature had ascended to its meridian, and soldiers and warriors were fighting and maneuvering in the stifling heat. Custer's plan to bring the warriors to the ford and away from Reno had worked only too well, as they were now mainly directed and arrayed against Custer's wing. Despite the geographical shift in the action, from Reno Hill to Medicine Tail Coulee, the engagement was pretty much at a standstill—resembling a pregnant pause more than a ferocious firefight. Thus far, the start-stop-start episodes characterized the battle—it was not defined by continuous firing. According to the Cheyenne woman Kate Big Head, there would be an interval of long-range firing between Custer's forces and the Indians as events gradually progressed. With his keen eye for topography, Custer saw that the east side of the river he occupied was poor defensive ground. Even more concerning, despite his earlier expectations that there would be an imminent conjunction with his reinforcements, there was still no sign of an oncoming Reno and Benteen, which was what the feint at the river crossing was supposed to facilitate. It all boiled down to one thing: Custer was alone.

Even now, events surrounding Custer's battalion were alarming but not yet critical. Custer gave every sign of regrouping and carrying on offensively. He still was optimistic that the battle, despite setbacks, could be won. But recent developments had not been encouraging, and now, having retreated from the river crossing, there was still no sign of reinforcements. But, the warriors were convinced that they had not only thwarted the bluecoats from crossing the river but were emboldened by the notion that they themselves had been the proximate cause of the retreat. When Reno charged the village, rather than the expected scattering, there was a concentration of Indian forces, while the 7th Horse remained fragmented. Notwithstanding the unanticipated circumstances the men of the 7th found themselves in, there was still an opportunity for the regiment to come together and win the battle. But perhaps faster than any in Custer's command realized, time was running out.

Map 9. Custer's wing moves north along the bluffs toward Medicine Tail Coulee.

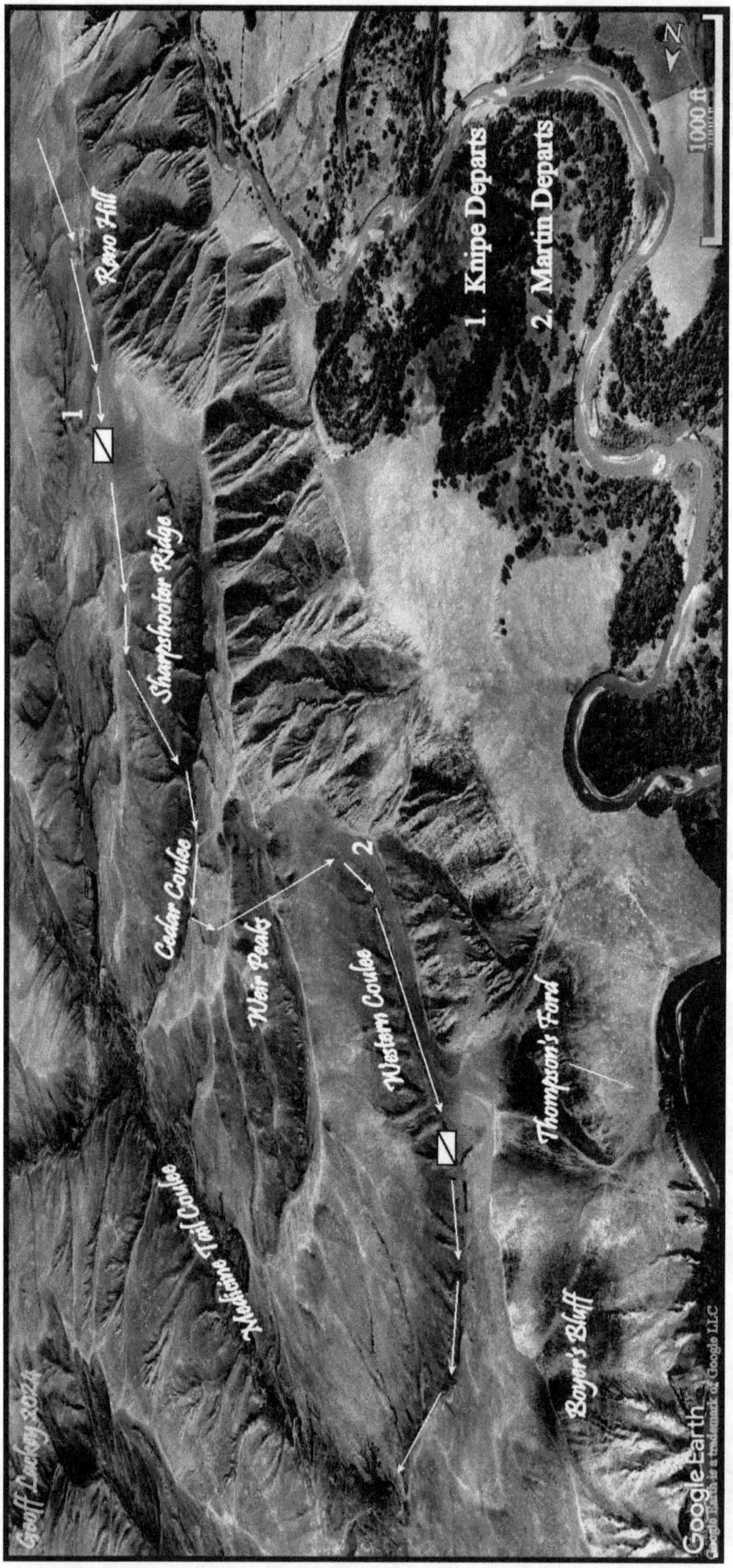

Map 10. Custer's advance toward Ford B and Luce Ridge.

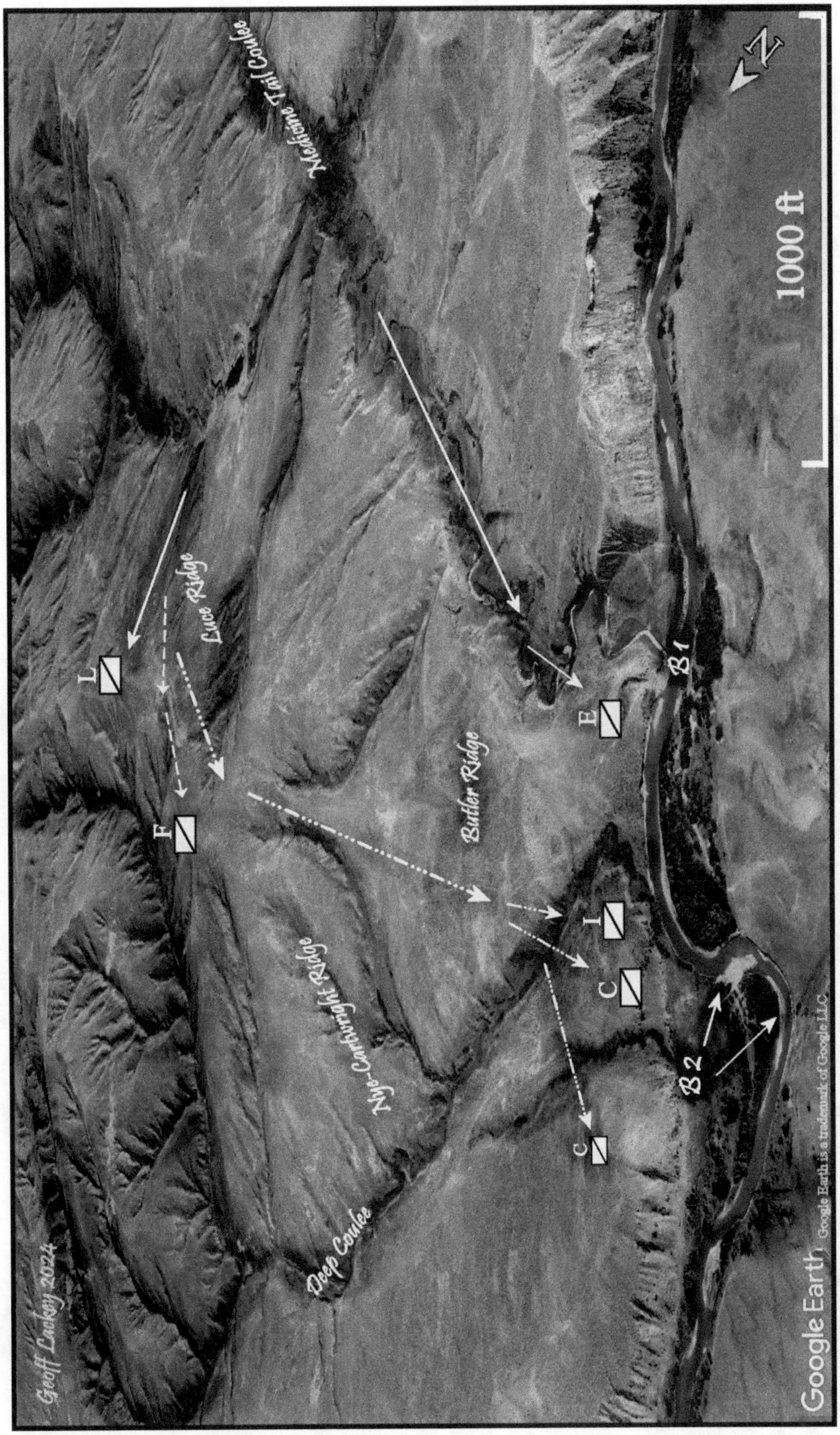

Map 11. Keogh's squadron demonstrates at Ford B while Yates's squadron holds Luce Ridge.

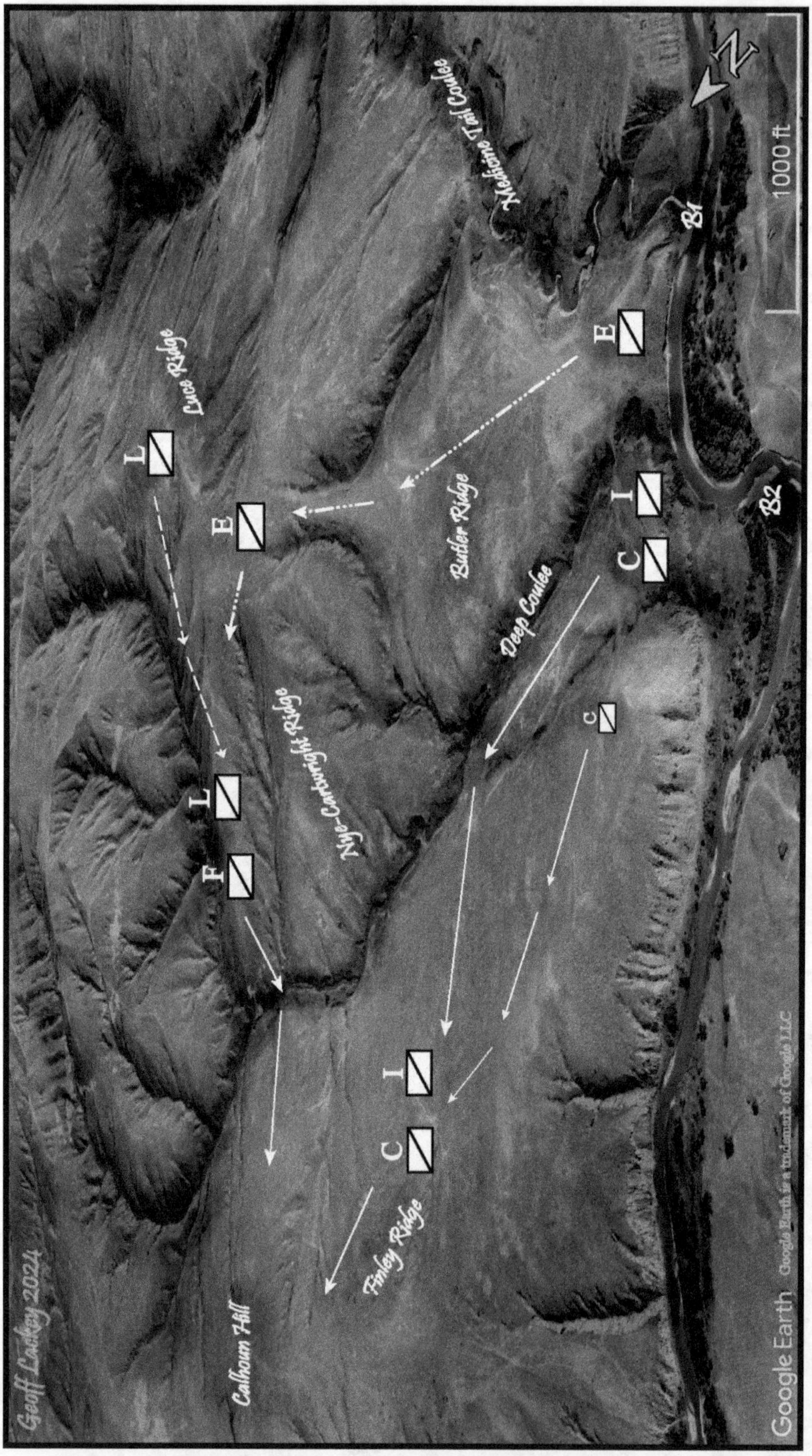

Map 12. Custer withdraws from Ford B. Keogh's squadron moves to Finley Ridge, while Yates's squadron moves along Nye-Cartwright (Blummer) Ridge, with all units heading toward Calhoun Hill.

Chapter 12: Why Was Keogh's Command Destroyed on Battle Ridge?

Since the time of the Sumerians, few military maxims have been more venerated than "seize the high ground." Custer's command attempted to do just that, even as they retreated from Medicine Tail Coulee. They would regroup about three-quarters of a mile northeast to assess the situation at Ford B, check the progress of Benteen's reinforcement, and decide how best to take back the momentum, considering the element of surprise had long vanished. This regrouping would lead to a crucial struggle on and around Calhoun Hill that would result in the implosion of Keogh's command on the east side of the river.

It is critical to understand the Keogh engagement and how it relates to the whole battle. This chapter attempts to present a plausible and reliable model of why and how Keogh's command was destroyed, and it will differ from what many battle students and scholars now believe. The riddle can be unraveled only by adhering to military logic, the law of necessity, source documentation, body locations, and how Captain Myles Keogh deployed his three troops on Battle Ridge that afternoon.

It was around 3:00 p.m. near Medicine Tail Coulee when the firing grew heavier, so the Yates squadron moved from Luce to Blummer Ridge. The probable intention of this withdrawal was to relocate farther north, across Deep Coulee, and on to Calhoun Hill. Keogh's C and I companies were safely falling back along Deep Coulee near Ford B-2 to Finley Ridge, protected by E Company deployed along Butler Ridge and Yates's squadron on the Luce/Blummer Ridge Complex. A line of flankers trotted toward the edge of the bluffs north of Deep Coulee, a tactic designed to protect the horses of Keogh's battalion as they moved to Calhoun Hill's higher elevation.

This was a classic retrograde—using military tactics. Custer observed and directed all these movements from Butler Ridge even as he endeavored to stymie warriors approaching from the south. He kept one wary eye on the hostiles, the other watching for signs of Benteen. Custer's Headquarters Staff and E Company moved back from Butler Ridge toward Luce Ridge, acting as a rear guard to Yates's squadron as Custer moved northward off Luce Ridge and away from Medicine Tail Coulee. Many warriors had gathered at Ford B and the gullies from the south and east. More warriors had retrieved their horses and merged onto the eastern hemisphere of the battlefield. They congregated slowly but inexorably below the ridge, all the while being sure to keep their distance from the soldiers' Springfield carbines.

With no sign of reinforcement, it is possible that Custer sent another messenger to Benteen, although this would have to be weighed against the odds of a courier getting through. Given the number of warriors beginning to encircle them, Custer knew he needed every man. Trumpeter Henry Dose, whose body was found halfway between Custer and Reno with multiple arrows protruding from his back and sides, might have been that luckless courier. It is also possible that as the battle was going to hell, Dose died while desperately attempting to reach the troops seen on the Weir Peaks as the Indians overwhelmed the command. The final moments of many would be forever lost in the seething mayhem that soon swept over the Custer battlefield.

Nevertheless, some evidence suggests that a courier was sent. A lone account documented by several Philadelphia newspapers reported that a Sergeant Thomas Harrison of D Company—serving as an orderly for Custer's Headquarters detachment—claimed to have been sent with a message to Captain Benteen to move his battalion up in support. Harrison recounted that he left about twenty minutes after the start of the Custer fight and rode a mile and a half before delivering the communication for Benteen. That would make Harrison, rather than Korn, the last to see Custer's command alive. If true, Harrison's likely point of departure from Custer's wing would have been at Luce Ridge, one and a half miles from the Weir Peaks, where Harrison would have met D Company shortly after its arrival there. Harrison claimed that after Benteen's juncture with Major Reno, the senior captain was preoccupied with dealing with the chaos on the bluffs and could not comply with Custer's order for support.

Was Harrison telling the truth? It is hard to know with anything approaching certainty since no other individual, including Benteen, ever publicly mentioned a courier named Harrison with a message from Custer. On the surface, Harrison appears to be a reliable character. He was an experienced Non-Commissioned Officer, having served in both the Crimean War and the Civil War. He was also part of Custer's Headquarters Staff—a position where straightforwardness and reliability are valued attributes. He can be afforded a degree of credibility, even though there is no other known source attributing Harrison as a messenger. But, lending weight to Harrison's claim, an unpublished source indicates that on the back of Sergeant Harrison's discharge papers, is noted that he did serve as a messenger for Custer that fatal day. This was attested to by Captain Thomas Weir, his company commander, who might have accompanied Harrison back to Reno Hill to deliver Custer's request for support to Captain Benteen. Several Philadelphia newspapers considered Harrison's account believable enough to print, and his account does fit into the known time and distance factors. Together with the personal note by Weir on his discharge papers, the authors believe that this event likely did occur, although not mentioned by certain officers for obvious reasons.

Returning to Custer's departure from Medicine Tail Coulee, many maintain he waited for all five companies to rendezvous at Calhoun Hill. Though not an optimum defensive position, the small hill nonetheless commanded the surrounding area, which served as an adequate base to repel an attack and provide cover to shelter the wounded. It was also an area adequate to conceal the horses from enemy fire and capture. A dependable horse was the mounted soldier's most valuable asset on the open plains. Protecting them was critical since there was nothing warriors desired more than to latch onto a good cavalry horse. It has been said that a cavalryman's first responsibility is to his horse, and this would certainly be true at the Little Big Horn, but in the maelstrom of violence and killing, preserving their mounts soon became impossible for many troopers.

Chapter 12: Why Was Keogh's Command Destroyed on Battle Ridge?

When Custer positioned himself on Calhoun Hill, using it as a temporary holding position, it is doubtful that all five companies were with him. Such an assembly would be extremely unlikely since some of his unmounted troopers led their horses, requiring more time to traverse the terrain and reach Calhoun Hill afoot. Keogh's C and I Companies dismounted after their demonstration down by Ford B and were withdrawing under fire. If Benteen arrived, Custer must have considered how he could provide those troops' safe passage to Calhoun Hill while the warriors from the Reno fight were gathering in strength at Deep Coulee. What if Benteen did not arrive? Charging into a fully aroused village without the support of two-thirds of the regiment would likely lead to a crushing and ignominious defeat. Retreating over the bluffs to Reno would consolidate the regiment but would also sacrifice momentum and position, likely turning earlier prospects for victory into inevitable defeat.

If reinforcements failed to materialize, Custer would be forced to fight on the east side of the river on poor defensive ground with limited ammunition. There was, however, no good reason for Custer to believe that his senior captain would not do his duty and come to his assistance. Despite a querulous personality, Benteen was a good officer and a first-class fighter and would not be expected to neglect to follow written orders to "come" and "be quick.". The distance between Custer's battalion and those of Benteen and Reno was certainly bridgeable—the inexperienced Boston Custer made it to his brother's command over the bluffs on his own initiative without escort or support.

Surely, a top-shelf combat cavalry officer like Benteen, with an entire battalion, could reach the valley to support Reno and then, with their combined forces, move toward the gunfire to secure a junction with Custer to defeat the warriors and win the day. When that did not occur in a punctilious manner, Custer must have attributed Benteen's delay to the time it took him to meet up with Reno on the bluffs, secure whatever wounded there were, and make sure the ammunition train was safeguarded before advancing. No matter how proficient a fighting force he had, Custer must have realized that relying on just five companies occupying substandard defensive ground against more than two thousand aroused warriors defending their homes and families would be tempting even his good luck. Nevertheless, fighting alone was becoming a distinct and unwelcome possibility. No matter how the options were weighed, once Custer withdrew to Calhoun Hill, as author Frederic Wagner pointed out, there was no turning back.

Deep in enemy territory, Custer waited impatiently, and he could not discount the need to improvise, to maneuver in whichever way he could survive if not prevail. In his rapid-fire way of speaking when excited, he barked out orders as the seeds of an alternate plan germinated in his mind. He informed Lieutenant James Calhoun, Commander of L Company, that he would use the hill as a bulwark and wait for Keogh's C and I Companies to arrive. These three companies would hold the high point, first to serve as a signpost and support for the reinforcement and, secondly, to keep his rear secure against the hundreds of warriors attempting to circle his position.

In a risky move, Custer would take Yates's squadrons E and F, Headquarters Staff, and about eighty-five to ninety troopers on a reconnaissance to ride north along the ridge and west toward the river. Historians still debate over which of several possible goals he might have had in mind. If this battle was to turn out to be a drawn-out siege, the first objective was to find good defensive ground, likely identified by his most experienced scout, Mitch Boyer—a stretch between The Flats known as Ford D-1 and Ford D-2, sites with access to ample water. If a junction with Benteen and Reno did not materialize, he could establish a defensive formation there, dig in, and send for Keogh's command on Calhoun Hill to join him while waiting for General Terry to arrive from the north.

Likely, Custer's primary objective for holding Calhoun Hill was the expectation that the arrival of his reinforcements would allow him to catch the enemy in a deadly crossfire. Again, because of their general lack of command and control, warriors, like most irregular forces, did poorly when facing military attacks from different directions, converging from several locations on the battlefield, with opposing forces working against them in tandem. It appears Custer was attempting to set up another 'fix and flank' maneuver, with Keogh's battalion now acting as the fixing force and Reno and Benteen acting as the maneuver element. Another possibility has gained significant traction over the last few decades—a hypothesis based on some documented history and what some might today call a "Hail Mary" pass—a move borne out of desperation to seize the women and children rushing down the river, use them as shields and compel the warriors to surrender.

Whatever scenario Custer was contemplating, he was taking the biggest chance of his life. An inveterate gambler, he would risk everything on a single roll of the dice. But that characterization would be true only in the most general sense. Riding off with his two companies and Headquarters Staff to an area more distant from his reinforcements is not as extreme a tactic as it might sound. Trained mounted troopers could adequately defend themselves against large numbers of warriors using the terrain and the river as Custer had done in the past. When executed under skilled and experienced leadership, Custer's move north away from Keogh was a risky, though not reckless, maneuver.

But was victory worth the risk? Was it wise to weaken and expose the regiment by dividing it again? In leaving Keogh's command at Calhoun Hill, with roughly 120 to 125 men, the General must have believed that Keogh was under no real duress and that he was on good enough ground to form a nearly impregnable defensive perimeter. At some point ahead, Custer figured that he would make a junction with Keogh, who would then be hopefully reinforced. If not reinforced, Custer would unite with Keogh on better defensive ground somewhere to the north and wait for General Terry.

Custer's recent history demonstrated that embracing these assumptions, although perilous, was viable. If, at the first Yellowstone battle on August 4, 1873, Captain Thomas Custer, deploying about fifteen to eighteen troopers, successfully repelled a charge of nearly three hundred warriors with their volley fire, then approximately 125 troopers Keogh had on and around Calhoun Hill could hold off warriors many times their numbers. As long as their ammunition held out, their long-range Springfield carbines would likely prevent them from being overrun by tribal warriors. Although Indians were highly skilled fighters, tactical deployments of organized military units using disciplined fire trumped superior numbers, "whether it is the Roman Legions versus the Germanic tribes, Alexander's Macedonians against the Persian hordes," or, in this case, the 7th Cavalry versus the Sioux and Cheyenne warriors of the Plains.

Chapter 12: Why Was Keogh's Command Destroyed on Battle Ridge?

Satisfied that Keogh's defensive position was in no immediate danger, Custer struck out with a detachment of two companies and his Headquarters Staff. Custer took only two companies because he needed to leave Keogh enough manpower to deal with whatever threat emerged from the hostiles, but he also needed to furnish a detail strong enough to bargain with whatever his contingent might run into, since a threat to their survival could happen anywhere and at any time.

Moving north, Custer would conduct a reconnaissance in force, not in a search for Indians, but for good, solid defensive ground in the event of being tied down and belabored by hundreds, if not a couple of thousand, warriors. The soldiers still maintained an advantage because, despite the warriors retrieving some of their mounts, they still lacked mobility on a large scale. Many remained on foot and could not keep pace with mounted cavalry, especially as they trekked upward toward higher and more difficult ground. Even as the troops moved back from the ford, many Indians, according to the Oglala Foolish Elk, were still busy catching their horses. But by then, enough warriors had secured their mounts and were coming up in large numbers.

Young Two Moons recalled that when the soldiers attacked, the Indians came rushing out like ants to defend their loved ones, but they were still unhorsed and had to struggle up the gullies northeast of the soldiers, which expended time and energy. The tempo of the battle, which drastically subsided after Reno Hill, was slowly building up steam east of the river, now focusing on Custer's wing. Perhaps no one has better explained Custer's intention at this stage of the battle and why he pivoted and formed a base at Calhoun Hill than the late battle scholar Bruce Liddic:

> Custer could see the Indians around Ford B and their movements toward the troops. By establishing a position on the crest of Calhoun Hill, support could be given to Benteen, especially if warriors tried to interdict him and the pack train as they crossed over the ground. From Calhoun Hill, Custer could also deny the use of Deep Coulee to the warriors, thus facilitating Reno's and Benteen's hammer. In addition, occupation of this hill would serve two additional purposes, one would be to act as a blocking force to cut off any Indians in pursuit of his troops. The second purpose would be to act as an anvil to Benteen's hammer and catch those warriors who had been bold enough to follow between two fires.

Waiting for Benteen, however, increasingly became like waiting for Godot. In Samuel Beckett's famous play, Godot never showed up, and neither, as it turned out, did Benteen or Reno. The stunning failure of the 7th to gather and unite what had been scattered is undoubtedly the paramount reason for their infamous defeat. In the meantime, back around Calhoun Hill, in Custer's absence, Keogh deployed his forces (see Map 13). More than a few battle students maintain that this deployment of forces was either defective or inadequate, leading to its catastrophic collapse. The popular and prevailing belief is that Captain Myles Keogh chose to keep I Company, one of the three under his command, relatively inactive in a reserve role east of Battle Ridge—an area known today as "Keogh's Swale"—where they were ultimately overpowered and destroyed after a relatively abbreviated but robust resistance.

The proposed model for this scenario is that C Company was the first of Keogh's forces to be overrun, which in turn caused L Company to collapse under the pressure. Having never been deployed for action, Keogh's I Company remained essentially idle in Keogh's Swale until they were suddenly engulfed by a Lakota tsunami. Some battle students have suggested that I Company was destroyed while marching in column, with no flankers or security posted, towards either Calhoun Hill or Last Stand Hill. Still, others postulate that Keogh and his subaltern Lieutenant James Porter were already dead, leaving their entire battalion either leaderless or poorly led by a subordinate officer.

However, none of these theories, still widely in circulation, accurately depict the deployment that would have been utilized by an officer of Keogh's experience and wherewithal. Based on the location where the bodies of Keogh's troops were ultimately found, it is an oversimplification to construct a hypothesis predicated on any of these scenarios. To enhance the narrative's explanatory power, it is necessary to develop a more realistic model and a more likely disposition of Keogh's three troops on Battle Ridge. A model, as stated earlier, is based on the demands of military logic, Indian accounts, and body locations. The best guess is that Custer, anticipating Benteen's imminent arrival, ordered Keogh to defend and hold Calhoun Hill while he headed to the northern flats looking for a better defensive position. After Custer's departure, the number of warriors on the battlefield facing Keogh steadily increased. Satisfied that the soldiers they had left on the bluffs (Reno Hill) were making no serious effort to reunite with the soldiers on the east side of the river (Custer's forces), the warriors became more belligerent and aggressive toward Keogh's forces.

Back on the bluffs some four miles away, neither Reno nor Benteen seemed motivated to make a juncture with Custer. We know that Lieutenant Hare had hurried the advance pack train along, and as soon as Lieutenant Mathey understood the urgency for bullets, even repeating an order for Benjamin Churchill and another packer to cut out two ammunition mules carrying in total some four thousand rounds of ammunition, to move with all dispatch to Reno Hill. This was followed within minutes by the rest of the ammunition mules in the advance pack train. For whatever reason, this urgency slowly dissipated when the ammunition finally arrived, to the point where it was never even distributed among Reno's soldiers.

An almost interminable delay of two cavalry battalions to move off Reno Hill all worked to the advantage of the tribal warriors and greatly endangered Custer's beleaguered wing. With Major Reno's forces no longer an immediate threat to the village and with no visible effort being made to reach Custer, the hostiles not only had numbers but time on their side. This proved deadly because the collapse of Captain Myles Keogh's unreinforced battalion on Battle Ridge would break the back of Custer's defenses, directly leading to the unraveling of Custer's command and its final destruction.

Though this was a relatively small battle in terms of numbers and weaponry, the Keogh fight around Calhoun Hill is the most complex part of what is a very complicated battle. The narrative is as difficult to trace and comprehend as any battle dating back to antiquity. Understanding how Keogh's defense caved remains a challenge even for those who have methodically dissected and examined the battle down to its very essentials. Despite admirable efforts by others to map out what went so terribly wrong, something has always been lacking in analyzing the Keogh fight. To not adequately account for an action so integral as Keogh's defense of Calhoun Hill and relate it to the logic of the whole battle makes any understanding of the Custer fight fundamentally incomplete. To assist the dedicated student in looking deeper into this phase of the battle, the entire toolbox is utilized to make the reasons for the implosion of Keogh's command understandable to the reader. This includes illustrated maps and diagrams to make this model as clear and transparent as possible.

It is 3:15 p.m. Department Headquarters watch time. Captain Keogh has deployed his forces on the high ground around Calhoun Hill. His troops are armed with the longer-range Springfields, which the enemy does not possess. He recognizes an attempt by the warriors to encircle him (see Maps 13 and 14). On Calhoun Hill, L Company is deployed to cover the southern approach from Deep Coulee. One platoon (half a company) of L Company faces south, and the other faces east. C Company has a ten-man detachment facing east and a twenty-man detachment facing west on either side of Horse-Holder's Ravine, where the horses of C and L Companies are held.

At the same time, I Company's first platoon is divided into two detachments of roughly ten men each, under the dual command of First Sergeant Varden and Sergeant Bustard. They dismount along the military crest of Battle Ridge with one detachment providing cover fire toward the hostile position behind the eastern ridge—henceforth called Crazy Horse Ridge—and the second detachment covering a position on the west side of Battle Ridge just above Deep Ravine—a position henceforward referred to as Keogh's Crest which is roughly midway between Custer Hill and Calhoun Hill.

Providing further support, a platoon from E Company, not positioned by Keogh but placed there by Custer, is deployed farther north on Battle Ridge Extension and just north of what would later become Last Stand Hill in order to protect Keogh's position from warriors attempting to come up the swale from the north. Lieutenant Porter's platoon from I Company is held in reserve in the swale to protect C and L Company horses in Horse-Holder Ravine. Keogh had deployed a formidable defensive perimeter on the highest ground there. It is a classic, mostly circular, hedgehog defense. As long as the ammunition holds out, it appears nearly impregnable, tightly protecting the horses and all the flanks, leaving no avenue open and no sector vulnerable to attack (see Map 14).

Evidence supporting Keogh deploying his battalion in this manner is provided by John Stands In Timber, his account published in Margot Liberty's book *A Cheyenne Voice*, wherein he relays the testimony of his grandfather Wolf Tooth, who witnessed these events from behind Crazy Horse Ridge as he looked directly down into Keogh's Swale. Wolf Tooth supports the belief that Keogh deployed a detachment of I and C Companies on the eastern side of Battle Ridge to suppress enemy fire from behind Crazy Horse Ridge. John Stands In Timber recalls that Wolf Tooth said the hottest part of the fight occurred within the compass of Keogh's Swale to the top of Last Stand Hill. Here is his account, in which bracketed annotations are included for purposes of explication and clarity:

> I know for sure [that] he [i.e., Wolf Tooth] killed one horse [down in Keogh's Swale or Horse-Holder Ravine east of Battle Ridge]. They were shooting [at the soldiers and their horses] whenever they got a chance to peep over [from the eastern Crazy Horse

> Ridge down into Keogh's Swale], but when the soldiers [of I Company] saw them, they would shoot, too. Several Sioux were killed right on top [of the eastern Crazy Horse Ridge] trying to shoot [at] them; the soldiers [under Keogh] were already [deployed and] aimed [their carbines] waiting for them [to show themselves] ... The soldiers [of Keogh's battalion] running up here [on Battle Ridge] were lined [up] in different places [by detachment or platoon], and when the hand-to-hand [fighting started in Keogh's Swale], up here on top [of Last Stand Hill]. The hottest fight was over here.

Significantly outnumbered by the warriors, the cavalry wanted to avoid close-quarter, hand-to-hand combat. Their objective was to stand off and firefight, using their longer-range Springfield carbines that enabled them to avoid fire from the shorter-range weaponry of the enemy. If Keogh had maintained his initial deployment on Battle Ridge, where he covered all potential areas of attack, it is doubtful that his battalion would have been overrun the way it was. Why did Keogh, an experienced combat commander who had fought over thirty battles in the American Civil War and was a ten-year veteran of the Plains Indian Wars, alter that deployment?

That answer can be derived from Little Big Horn historian Charles Kuhlman in his classic *Legend Into History,* in which he backhandedly and serendipitously stumbled upon the answer when he proposed a fascinating theory on what motivated the command's inexplicable move around Calhoun Hill. Ironically enough, Kuhlman says it was when the much-awaited reinforcements, Captain Weir and his Company D, appeared on Weir Peaks, engendering a misunderstanding on Keogh's part about what the sighting of those reinforcements promised. With a clear line of vision from Calhoun Hill, Keogh clearly saw Weir's reinforcements but misconstrued or wrongly guessed their future actions. This misperception ignited a chain of events that led to Custer's downfall. While Kuhlman's theory is generally correct conceptually, his details regarding the battle location of "South Skirmish Line," once thought to have been employed by Custer south of Cemetery Ridge. But over the years, the accumulated evidence has eroded the belief that there was a South Skirmish line.

To rectify Kuhlman's mistake and understand what most likely occurred, we must supplant Kuhlman's proposed position at the so-called "South Skirmish Line." Kuhlman maintained that Custer had deployed a portion of his command in the "Basin" below Last Stand Hill but would mistake that position for C Company's actual deployment on what would later be called Finley Ridge. Once that substitution is made, the pieces of the puzzle regarding the Calhoun Hill fight begin to fall into place. When Keogh spotted Captain Weir's D Company on Weir Point, he almost certainly thought that Benteen's entire battalion was making its way to reinforce him at what is identified today as the Custer Battlefield. As a result, Keogh redeployed a portion of his battalion to accommodate their anticipated arrival, fatally weakening his defensive perimeter by throwing out a skirmish line to the southwest on Finley Ridge to support their expected advance to the Luce/Blummer Ridge Complex, which was on a straight path to Calhoun Hill.

Chapter 12: Why Was Keogh's Command Destroyed on Battle Ridge?

There is a direct line of fire from Finley Ridge toward Upper Deep Coulee at the northern extremity of Blummer Ridge. This deployment is supported by the archeological digs that have uncovered shell casings and cartridges in this area. This deployment of C Company by Keogh would be based on a misunderstanding of the readiness and resolve of the advance elements of Benteen's battalion, led by Captain Weir, to join them on Calhoun Hill, which would ultimately result in Keogh's command being so devastatingly overrun. Thus, what had been an almost airtight defense around Calhoun Hill was abandoned to give cover and support to the expected reinforcements that never left the Peaks from which they were seen, thereby turning the battle into a series of ambuscades leading to the destruction of Keogh's battalion.

Other misunderstandings and misrepresentations continue to distort and plague the Keogh fight. Almost from the aftermath of the battle, it was believed that C Company was deployed, not on Finley Ridge, but at the bottom of Calhoun Coulee to relieve pressure on L Company from the west as the Indians were firing upon them from Greasy Grass Ridge. But the notion that Keogh sent C Company charging into the bottom of Calhoun Coulee to disperse the warriors makes no tactical sense. No competent cavalry officer—and Keogh certainly was that—would ever order a senseless dismount at the bottom of a coulee with both flanks exposed, surrendering the high ground to the enemy on either side to fire down upon them. Instead, by providing fire from Finley Ridge toward an area of upper Deep Coulee at the base of Calhoun Hill, Keogh's intent was to facilitate and support the progress of the reinforcements he saw on the Weir Peaks whom he believed were advancing toward the Luce/Blummer Ridge Complex high ground the reinforcements would have to traverse.

But because these expected reinforcements did not advance toward Luce Ridge, it would be the deployment of C Company on Finley Ridge that exposed Keogh's position and led to the subsequent decimation of his battalion, allowing the warriors to focus entirely on C and L Company's held horses without having to worry about Benteen. This deployment resulted in C Company's horses being sheltered in Upper Calhoun Coulee behind Finley Ridge, necessitating Keogh to deploy a detachment of about ten men from C Company to protect the backdoor leading to the horses from Harrington Ridge (named after Second Lieutenant Henry Harrington, Commander of C Company) just north of Calhoun Coulee. Upper Calhoun Coulee would be the logical choice to shelter C Company's horses and serves as an important clue as to most of C Company's deployment on higher ground adjacent to this coulee—Finley Ridge to the south (see Map 15). Unbeknownst to Keogh, only D and M Companies of Benteen and Reno's combined seven companies had thus far managed to reach the Weir Peaks, but they had insufficient manpower to advance past Medicine Tail Coulee, much less onto the Luce/Blummer Ridge Complex toward where the Keogh fight was heating up.

With pressure growing on both sides and cavalry reinforcements sputtering to an unexpected halt on Weir Point, the dam was about to break, releasing a flood of warriors first upon Keogh and then upon Custer. Once it was seen that the soldiers were making no effort on the Weir Peaks to advance to Keogh's battalion and with no threat materializing from their south where the Reno and Benteen battalions were, the warriors could concentrate on dismantling Keogh's new defensive alignment on Battle Ridge. By now, hordes of warriors were lined up along a ridge called Greasy Grass parallel to the river, and that area would serve as a launching pad for the warriors' assault. At the Cheyenne camp about one mile north of where she had been hiding, the Cheyenne Kate Big Head had witnessed the battle's progress from a place not too far away from the action. She remembered that after the attack or demonstration at Ford B by Custer, there would be an extended but relatively subdued period of long-range fighting with little damage done on either side.

Fraught with anxiety over the looming fight, Kate Big Head attempted to distract herself by singing songs she believed strengthened the heart of her young nephew, Noisy Walking, who was fighting the bluecoats that day. She remembered that warriors approached the soldiers on and around Calhoun Hill with mostly quasi-guerrilla-like tactics using gullies and jumping from knoll to knoll. Some crawled to closer ridges to fire upon the soldiers; other Indians who were mounted rode up Greasy Grass Ridge. Meanwhile, Crazy Horse likely crossed the river at Deep Ravine, following the gully leading toward the soldiers. Like a creeping, crawling, unrelenting force of nature, the Indians were moving steadily to the east, trying to circle the soldiers from behind Crazy Horse Ridge.

Thus, to summarize this new deployment by Keogh, sometime between 3:50 and 4:00 p.m. Headquarters watch time, the situation became precariously tenuous. C Company advanced west in mounted formation from Battle Ridge in two separate detachments, the first to occupy Harrington Ridge just north of Calhoun Coulee and the second designated to occupy Finley Ridge to establish a field of fire directed at the southern base of Calhoun Hill. C Company's horses would be led down to the shelter of upper Calhoun Coulee between these two dismounted detachments of C Company that were deployed on either side of the coulee. D and M Companies on the Weir Peaks remained immobile and out of the fight.

The Battle of Little Big Horn is popularly called Custer's Last Stand, but in reality, it was a series of last stands as the soldiers' separate strongholds were sequentially overwhelmed by the plains warriors. Tension slowly mounted in this standoff until a fusion was finally reached, triggered when a young Cheyenne-adopted Ute warrior, Yellow Nose, burst forth with a chilling, visceral cry, "Let us charge," unfurling a pent-up outburst of raw energy, whereupon a rush of warriors torpedoed toward Keogh's command. Although both charges he led toward the soldiers were driven back, his brave example emboldened and inspired the warriors poised to descend on Keogh's command.

Watching Yellow Nose's attack from the tall grass near Calhoun Coulee and Deep Ravine was a thirty-seven-year-old Southern Cheyenne war chief named Lame White Man. Seen as brave and wise by his people, he entered the fight wearing nothing but a blanket wrapped around his waist. He had been completely undressed, entering his sweat lodge when word arrived that soldiers were attacking the Indian encampment. He now rose and turned to the warriors around him and roared, "Young men, come with me. Show yourselves to be brave. Come, we can kill them all." The effect was combustible, igniting an explosive mix of excitement and bloodlust that swelled into a torrent of young warriors unhesitatingly following Lame White Man, come victory or death.

Chapter 12: Why Was Keogh's Command Destroyed on Battle Ridge?

Until now, the Indians were cautiously moving on either side of the soldiers, taking advantage of the terrain whenever possible. But now, Yellow Nose and his followers were closing in. Throwing caution to the wind, and in a wild upswing of sheer determination, warriors poured out of the coulees on the west side of Battle Ridge. Two Indian forces drove the soldiers to the top of the ridge—a geographical dagger that pointed directly at the heart of Keogh's command, thus exposing C Company's held horses in Calhoun Coulee. Those startled horses were stampeded by the Indians, causing the immediate redeployment of I, C, and L Companies. Keogh's defense did not immediately collapse, but he was now bleeding from multiple wounds (see Map 16).

At this point, Keogh has successfully repelled the charges of Lame White Man and Yellow Nose despite the stampeding of C Company's horses from upper Calhoun Coulee into Deep Ravine. Keogh's detachments on Harrington Ridge have fallen back to the top of Keogh's Crest. At the same time, L Company's second platoon redeploys from the east to the western side of Calhoun Hill to help defend against this unexpected assault. More critical is C Company's detachment on Finley Ridge, primarily dismounted and severed from the remaining elements of Keogh's battalion, resembling a lone cork bobbing on waves as an ocean of hostiles surrounded them (see Map 17).

The assault had markedly and fatally compromised Keogh's defenses, but his troopers were not immediately driven down the ridge into the swale—at least not yet. As it is so often characterized, this surge would not be the devastating and overwhelming Indian charge that overran Keogh's position on Battle Ridge. His deployment was able to successfully repel this charge, leading to the death of the Cheyenne war chief, Lame White Man, and Kate Big Head's nephew, Noisy Walking, as well as an unnamed Sioux warrior. It also led to the wounding of many other warriors. One young Lakota warrior had his lower jaw shot away, the bullet violently tunneling through both sides of his mouth; his teeth grotesquely resembled a necklace around his neck, and his jaw limply and pathetically hung down on his breast. The carnage was unremitting, and Lame White Man, in a tragic twist of fate, was not only shot dead but, in war's ever-present fog, was scalped by another warrior who mistook him for an enemy scout.

What is surprisingly overlooked in this melee is that both Lame White Man and Noisy Walking were killed on the west side of Battle Ridge, well short of its summit. These warriors were most likely shot from a detachment of dismounted I Company troopers deployed directly above on Keogh's Crest. This was evidence that not all of I Company, as many may believe, were left idle in reserve on the east side of Battle Ridge. The death sites of both Lame White Man and Noisy Walking were too far away to have been caused by firing from Calhoun Hill or C Company.

The attack that would finally uproot Keogh's defensive perimeter came shortly thereafter when the warriors behind Crazy Horse Ridge noticed Keogh had stripped his eastern defenses, leaving the backdoor of his defensive perimeter open after he bolstered his response to Lame White Man and Yellow Nose's charges on the west side of Battle Ridge. Consequently, Keogh exposed L Company's held horses in Horse-Holder Ravine to warrior infiltration led by Crazy Horse, White Bull, and Red Horse. More than anything else, this incursion led to the battalion's virtual annihilation. Keogh faced the impossible task of holding a position too large for a small force to defend from sustained multiple attacks and from different directions (see Map 17).

Sensing opportunity, the Indians were a storm unleashed. Crazy Horse was a terrifying sight as he rode over the Keogh battlefield, his unbraided black hair wildly whipping around his head, a lightning bolt painted across his countenance, and his near-naked body covered in white painted hailstones. Paul Noisy, a full-blooded Sioux whose father fought that day, recounted to his son how Crazy Horse was the indisputable leader in the battle. It was, said Paul Noisy, the custom of Indians in battle to have some bold soul ride out in front of the warriors, and this is the role that the intrepid Crazy Horse took upon himself.

Racing back from the Reno fight, Crazy Horse and his warriors moved up Deep Ravine across Battle Ridge north of Keogh's position and eastward behind the ridge bearing his name. His freakish ability to elude enemy bullets did not desert him at the Little Big Horn. Red Hawk marveled how a soldier fired his revolver at Crazy Horse from close range but miraculously missed him. Legends are born from such quirks of fate. Without hesitation, Crazy Horse shot the soldier's horse, dismounting him, whereupon Red Hawk struck the soldier on the side of his head with his Spencer carbine, disabling him before the coup de grâce was administered.

With L Company's horses stampeded by Crazy Horse and White Bull's charges, a catastrophic occurrence for any cavalry detachment, Keogh's weakened eastern defenses allowed warriors to penetrate their perimeter, resulting in the stampede of L Company's horses from Horse-Holder Ravine (see Map 18). Desperately trying to salvage what was a rapidly deteriorating situation by engaging warriors now pouring into Keogh's Swale, Lieutenant James Porter and his reserve platoon attempted to intercept and counter this attack. At the same time, Keogh endeavored to rally his dismounted troopers atop Battle Ridge to support Porter's engagement in the swale.

The warriors now advanced in strength into Keogh's Swale. One group directly attacked Lieutenant Porter's reserve platoon, still fighting over L Company's stampeded horses. It turned into a messy, close-quarter death fight with hundreds of mounted warriors following Crazy Horse, White Bull, and Red Horse as they hurtled through the soldiers in the swale. Having rallied his dismounted skirmishers from C and I companies atop the western side of Battle Ridge, this position was now abandoned when Keogh's troops charged into the swale area to support Porter's platoon in saving the led horses near Horse-Holder Ravine.

Keogh led a dismounted charge into the swale to attack this war party from the rear while other warriors, led by Yellow Nose, advanced up the western slope of Battle Ridge, now stripped of its defensive firing positions. Keogh was forced to fight on the low ground of the swale in a desperate attempt to prevent the capture of their horses, now their only hope for deliverance. The dearth of cartridge cases in the swale indicates little or no time to reload their Springfield carbines, with most troopers using their revolvers. Amid the pungent gun smoke, war cries, and death shrieks, the fight was bloody, fierce, and, despite its brevity, appallingly savage (see Map 19).

Chapter 12: Why Was Keogh's Command Destroyed on Battle Ridge?

As Keogh's battle raged in the swale, E Company's reserve platoon was now deployed in a skirmish line just east of Last Stand Hill, about one half to two-thirds of a mile north of Keogh, positioned where they could support his engagement while still maintaining contact with the Yates squadron. The final stages of the defense on Battle Ridge are shown on Map 20. Red Horse witnessed the final destruction of Keogh's battalion in the swale:

> The party we killed [i.e., Custer's wing] made 5 different stands [before they were completely overwhelmed]. Once we charged right in [note: likely a reference to the Indian charge into Keogh's Swale to attack the horse-holders and Lieutenant Porter's reserve platoon] until we scattered the whole of them, fighting among them hand to hand. One band of soldiers was right in the rear of us [a likely reference to the dismounted detachments of I and C Company deployed on the top of Battle Ridge above them]; when they [i.e. Keogh and these dismounted troopers] charged [down into Keogh's Swale in support of Lieutenant Porter's platoon] we fell back and stood for one moment facing each other. Then the Indians got courage and started for them in a solid body. We went but a little distance, when we spread out and encircled them [down in Keogh's Swale]. All the time I could see their officers [i.e., Keogh, Porter, Harrington, Varden, Bobo, and Bustard], and hear them shouting to their men. It was in this charge that most of the Indians were killed. We lost 136 killed [note: this may be a transcribed error from 36, which tallies closer to other Indian accounts of their deaths at Little Big Horn] and 160 wounded. We finished up this party right there in the ravine [i.e., Keogh's Swale].

With the battle around Calhoun Hill lost, it appears that those with horses attempted to get to Last Stand Hill, maybe two-thirds of a mile away, where a platoon of E Company under Lieutenant Smith lent support and cover fire. The evidence is unclear, but the possibility remains that Porter and Harrington were among the few survivors who reached Last Stand Hill—scarcely more than a dozen from as many as 125 under Keogh's command made it to what proved but a temporary sanctuary until the inevitable curtain of death closed upon the gory and blood-stained scene. As a testament to the ferocity of the fight, Indian casualties around Calhoun Hill and Keogh Swale were among the highest on the battlefield. Some young bucks, in their excitement and limited visibility because of the smoke and dust, killed each other with arrows and war clubs. Fifty or so troopers died fighting in Keogh's Swale, with roughly half that number found in a tight circumference around Keogh's position in a slight depression near Horseholder Ravine. In a picturesque frame of mind, Two Moons remembered warriors circling these clustered soldiers "like water around a stone." The remaining troopers sought refuge at Last Stand Hill.

As for L Company, their fate was quickly sealed when they abandoned their strong defensive position to support Keogh's defense of their horses in the swale. As L Company fell back to the low-lying, depressed ground, they were quickly overpowered by mounted warriors charging from the south and east into their earlier position. The first platoon, along with Lieutenants Calhoun and Crittenden, fell at the top of the north end of Calhoun Hill after abandoning their skirmish line position at the south end of the hill. Most of their second platoon fell between Calhoun Hill and Keogh's Swale around the vicinity of Horse-Holder Ravine, with a few reaching Keogh in the Swale and several others who were still mounted reaching Last Stand Hill.

With C, L, and I Companies essentially destroyed, the connective tissue between E Company's platoon that served as a bridge from Keogh's command and Custer's farther north was irrevocably ruptured. Harassed and pummeled by warriors, the few survivors fled along Battle Ridge to Lieutenant Smith's position. He could do little other than absorb these refugees. Meanwhile, Sergeants Finley and Finkle and about twenty men of C Company found themselves stranded on Finley Ridge, most unmounted and completely isolated—an island amid an archipelago of tribal warriors. Soon, their own last stand would also take place. As for Calhoun's men, they died in a vain attempt to support Keogh as he rallied his company. Both platoons were killed in a bunch, and other troopers who fled under unbearable pressure were shot down or mauled to death (see Map 20).

Most of the dead of L Company were found on the south end of Battle Ridge, which runs parallel to the Little Big Horn River. Among them were Lieutenant James Calhoun, Custer's brother-in-law, and Lieutenant Crittenden, who lay dead not far from him. The handsome Calhoun was scalped, a trophy no doubt to be worn on the belt of some warrior, and so mutilated he had to be identified by the fillings in his teeth, while Crittenden was shot full of arrows, including one that shattered his glass eye.

Down in the swale, on the east side of Battle Ridge, was Captain Myles Keogh's body. Like Custer, he had been a bit of a popinjay, a true soldier, and a survivor. His sporting goatee and mustache, the way he cocked his hat, and the devil-may-care glint in his eye giving more than a hint of Irish swagger were all part of his colorful persona. Keogh had survived the great potato famine in his native Ireland, survived fighting in the Saint Patrick (Irish) Battalion in the Papal Wars, survived the arduous journey over the Atlantic to America, and fought and survived more than thirty battles in the American Civil War. But his luck, like Custer's, ran dry on those bloody banks along the Little Big Horn River.

His distinctive voice that lustily sang along to "Annie Laurie," "For He's a Jolly Good Fellow," and, of course, "Garryowen" was forever silenced in a cacophony of warrior shrieks and whoops. His Irish crooning interlaced with streams of colorful profanity was no less endearing to his brothers-in-arms, especially when cursing a blue streak at those infernally stubborn mules pushing along the long, dusty trail. But now he lay lifeless and strangely silent in the swale that was now his grave, his body stripped of everything but a pair of socks. Whatever he once was, he was now reduced to a pale-complexioned, unclad, bloody corpse. A portion of his battalion lay united in one great cluster of death about him: First Sergeant Edwin Bobo, First Sergeant Frank Varden, Corporal John Wild, Trumpeter John Patton, Willis Wright (at age seventeen, the youngest soldier in the regiment), and roughly thirty others died as they apparently surrounded and shielded a wounded Keogh in what proved to be yet another "Last Stand."

Forced into this low-lying, unhallowed ground just below the rough-hewn ridge, Keogh had apparently made one last attempt to cover the retreat of his men toward the safety of the platoon on Last

Stand Hill. Mounted on his tough, claybank gelding, ironically with the Indian name Comanche, Keogh fought to the last. Though the horse was punctured and bleeding from several bullet holes, it remained upright along with its rider. But then another shot rifled through Keogh's left knee, shattering his joint and then burrowing into Comanche's left flank to inflict another wound on the still unbowed little horse. Comanche would survive the battle and live another fifteen years in legend and lore, while in death, his body was stuffed, mounted, and memorialized in Kansas University's Natural History Museum in Lawrence, Kansas. But its intrepid rider breathed his last.

Though neither a fatalist nor a sentimentalist, journeying into the valley of the Little Big Horn, Myles Keogh sensed death's cold shadow hovering over him. He had premonitions, maybe all soldiers do. In war, death is an occupational hazard, a leering, somber, ever-present specter. But now, just before the Little Big Horn campaign and anticipating what fate might have in store for him, he prepared his last wishes. As he boarded the steamer *Far West* on the Yellowstone River, he sought out Lieutenant John Carland, a 6th Cavalry Infantry officer and a lawyer. He asked Carland to compose his Last Will and Testament. He confided to Carland, "I don't know what may happen to me, and I have not disposed of some things I have. I want to make a will. If anything should befall me, open the envelope and send it to my sister."

This time, Keogh's deepest fears were realized. If his sister found any consolation from the news of his violent and untimely death, it was that her brother was among the few not mutilated. Keogh, though stripped to his naked skin, was unscathed save for his gunshot wounds while many of his comrades lay strewn like so many butchered hogs. Was this apparent act of leniency just another unexplained curiosity in battle, a mere caprice of war over a widespread field of mayhem? Perhaps—or maybe there was some other influence at work. More pious than others might suspect, Keogh wore around his neck a Catholic medal and scapular with the inscription *Agnus Dei*, which means Lamb of God. The Indians were wary of spiritual talismans, choosing for that reason perhaps not to desecrate his body

Prostrate and lifeless, Keogh remained an unshorn victim; his flesh spared the blade of the knife and tomahawk. For hundreds of years, Jews brought lambs to the temple, offering them up as a sacrifice for their sins. Christianity brought out that imagery to its fullest, making Christ, the anointed and sinless one, the eternal spotless lamb, Agnus Dei. Christians believe that Christ became the perfect and eternal sacrifice for the sins of the world. Yes, Keogh was dead, but unlike so many others on this battlefield, he lay unflayed and uncrucified.

While Keogh's command had been mercilessly slaughtered, the battalion's few survivors made one last dash to Last Stand Hill. The battle was not yet over. Custer's final fight would generate more books, paintings, reenactments, and cinematic creations than any other battle on the North American continent. Its pathos would soon play out its final drama on a little hill, an American Golgotha, a site just north of Keogh's decimated battalion where a remaining, resolute few huddled closely together to make the most famous last stand in American history.

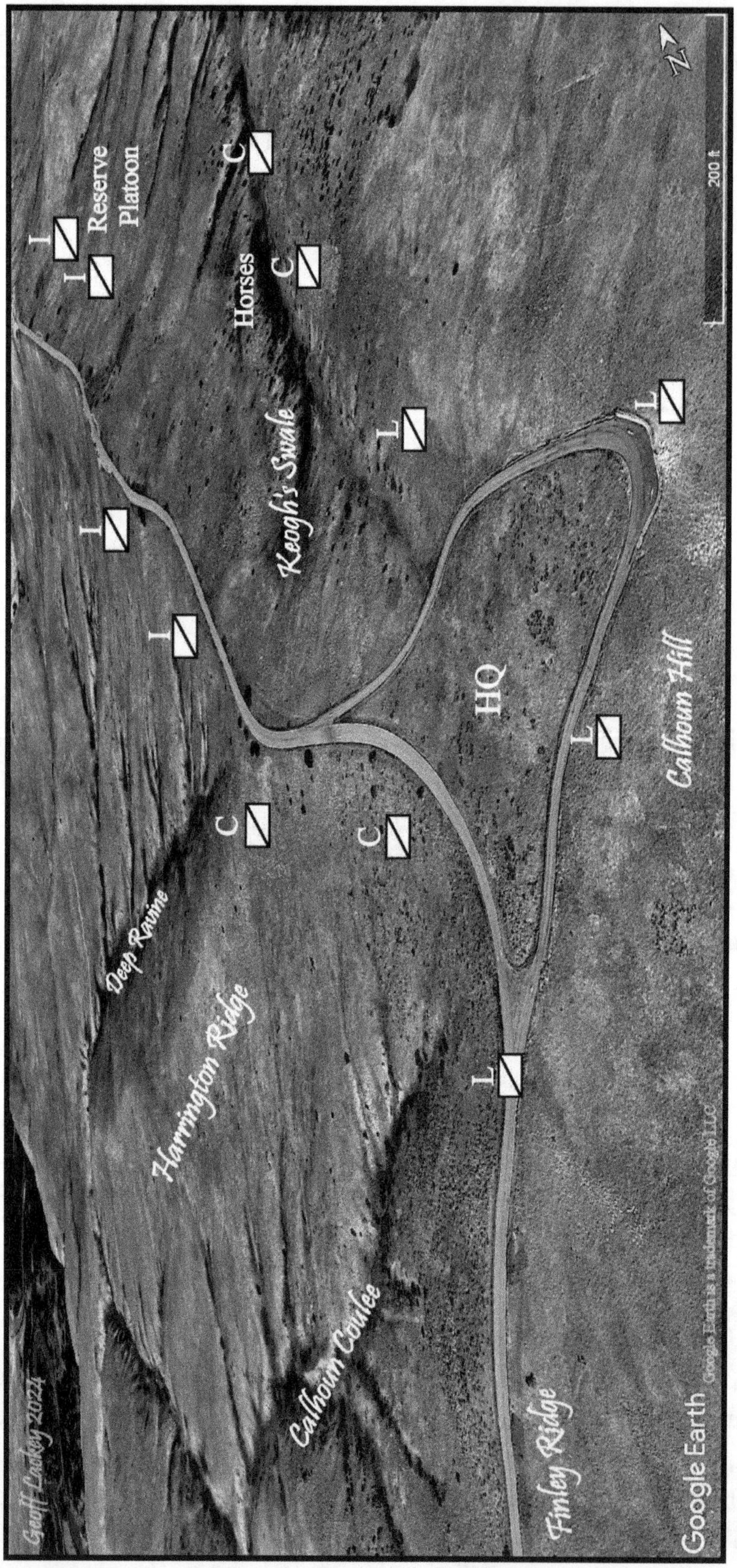

Map 13. Troop and warrior positions and major terrain features along Battle Ridge.

Chapter 12: Why Was Keogh's Command Destroyed on Battle Ridge?

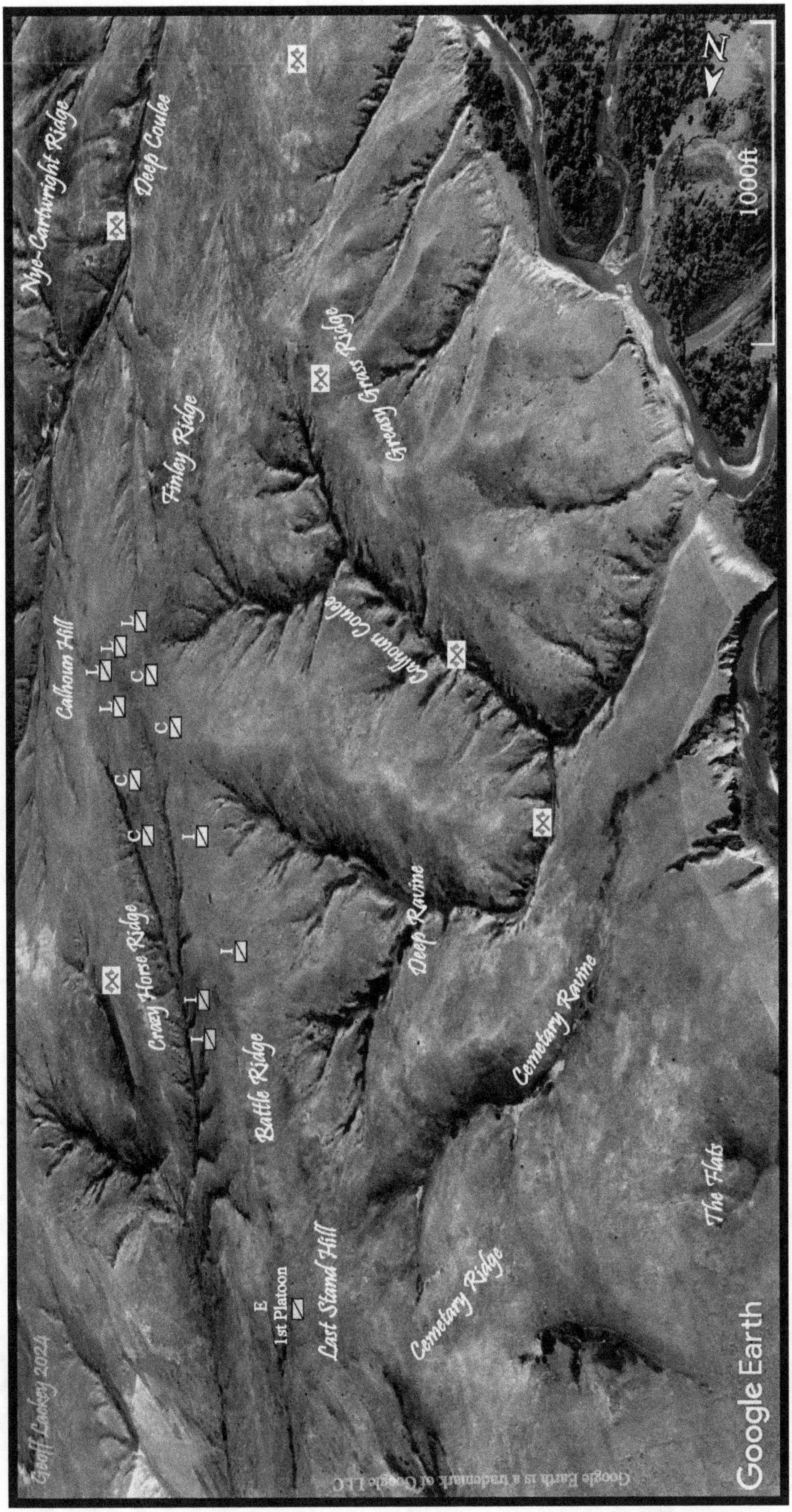

Map 14. Keogh's platoons deployed on Calhoun Hill at the south end of Battle Ridge.

Map 15. Keogh deploys C Company to Finley Ridge to support the arrival of troops seen on the Weir Peaks.

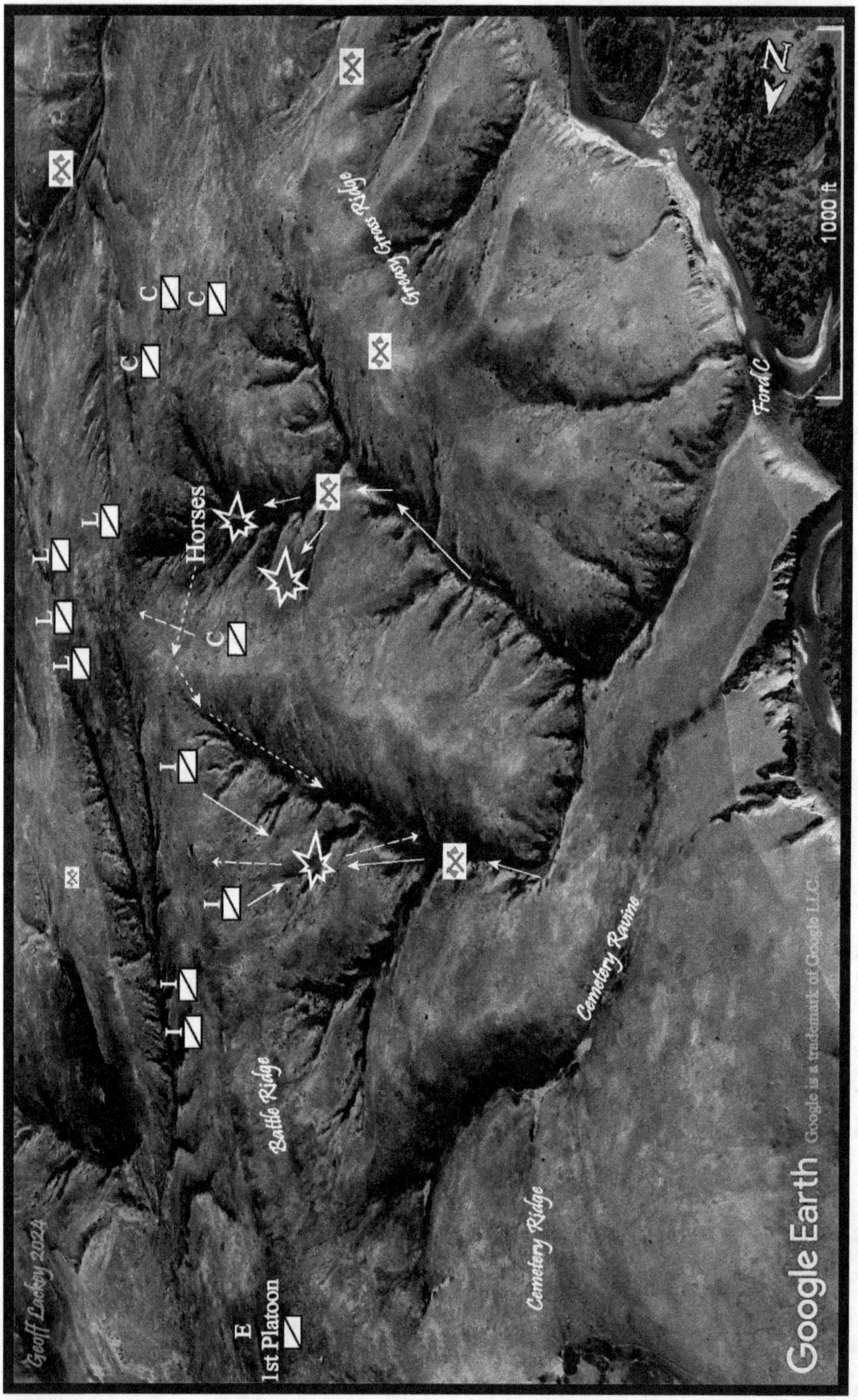

Map 16. Attacks led by the Cheyenne Chief Lame White Man and Yellow Nose via upper Calhoun Coulee against C Company's held horses.

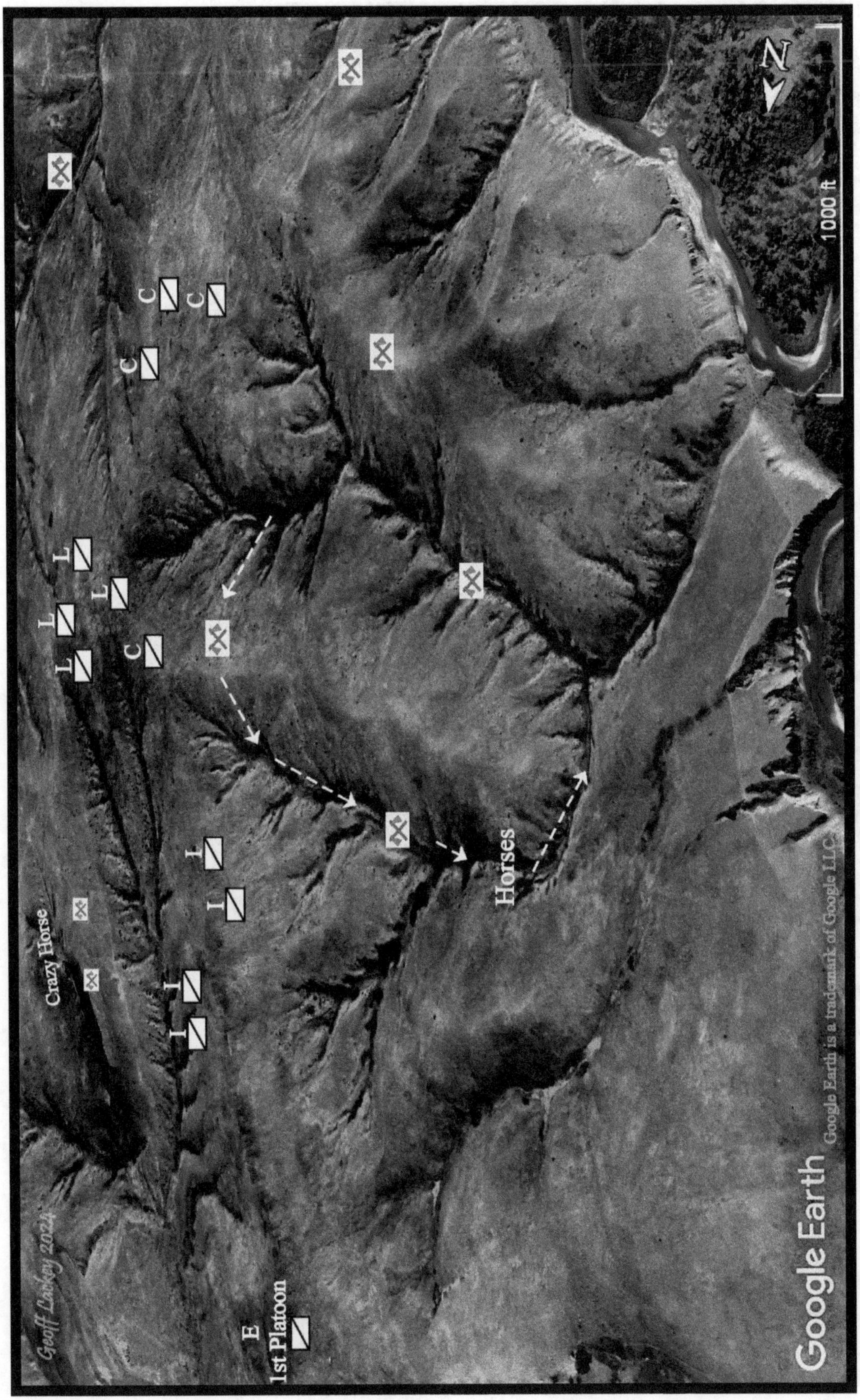

Map 17. C Company's horses held in upper Calhoun Coulee are stampeded to the river via Deep Ravine. C Company detachment on Harrington Ridge regroups on Battle Ridge.

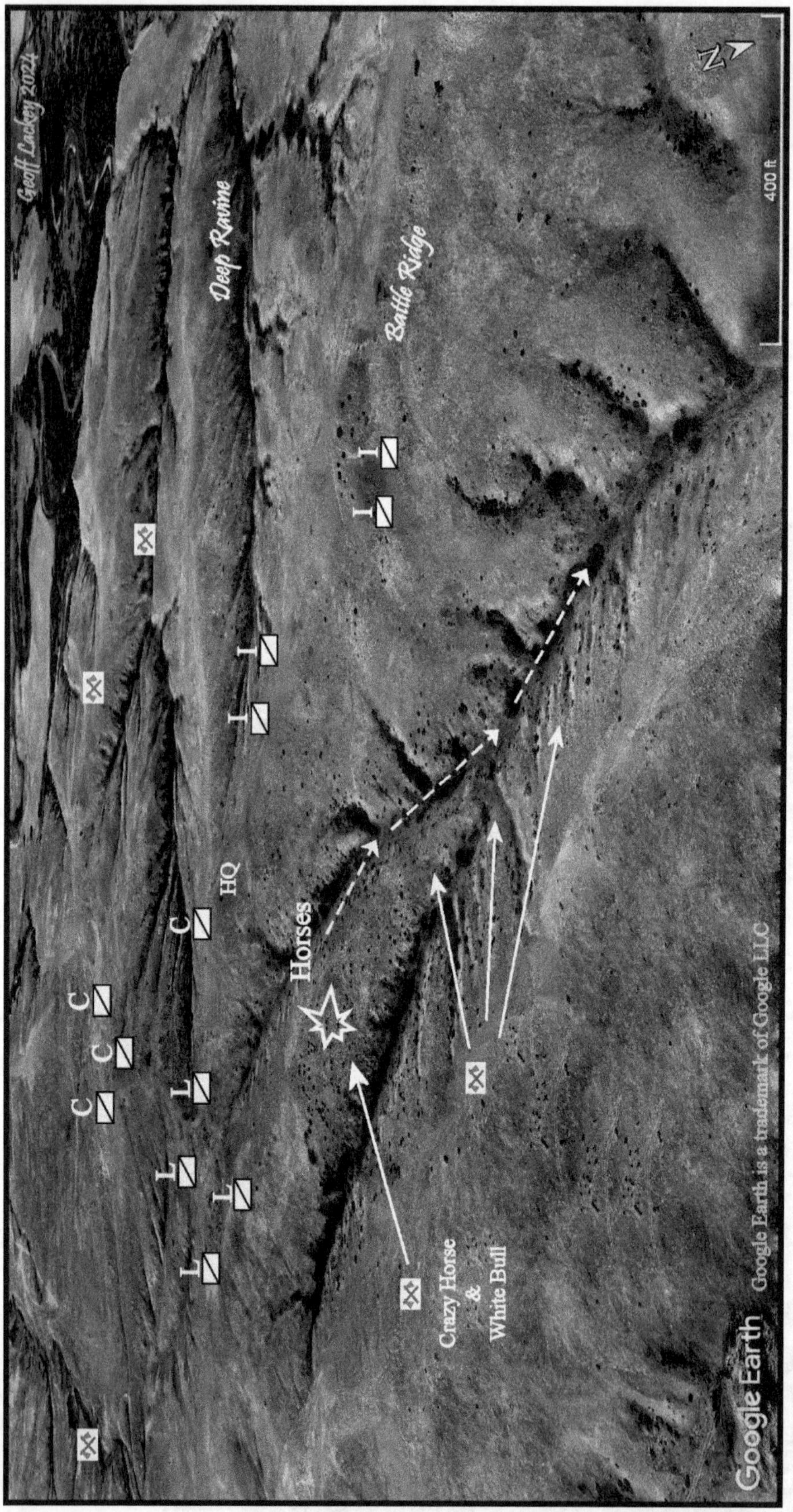

Map 18. Crazy Horse and White Bull lead attacks on Keogh's horses in Horse-holder Ravine after Keogh weakened his eastern defenses while repelling the attacks of Lame White Man and Yellow Nose on the west side of Battle Ridge.

Map 19. The Keogh battalion falls back to the swale to fight for their horses while being attacked from both east and west of Battle Ridge.

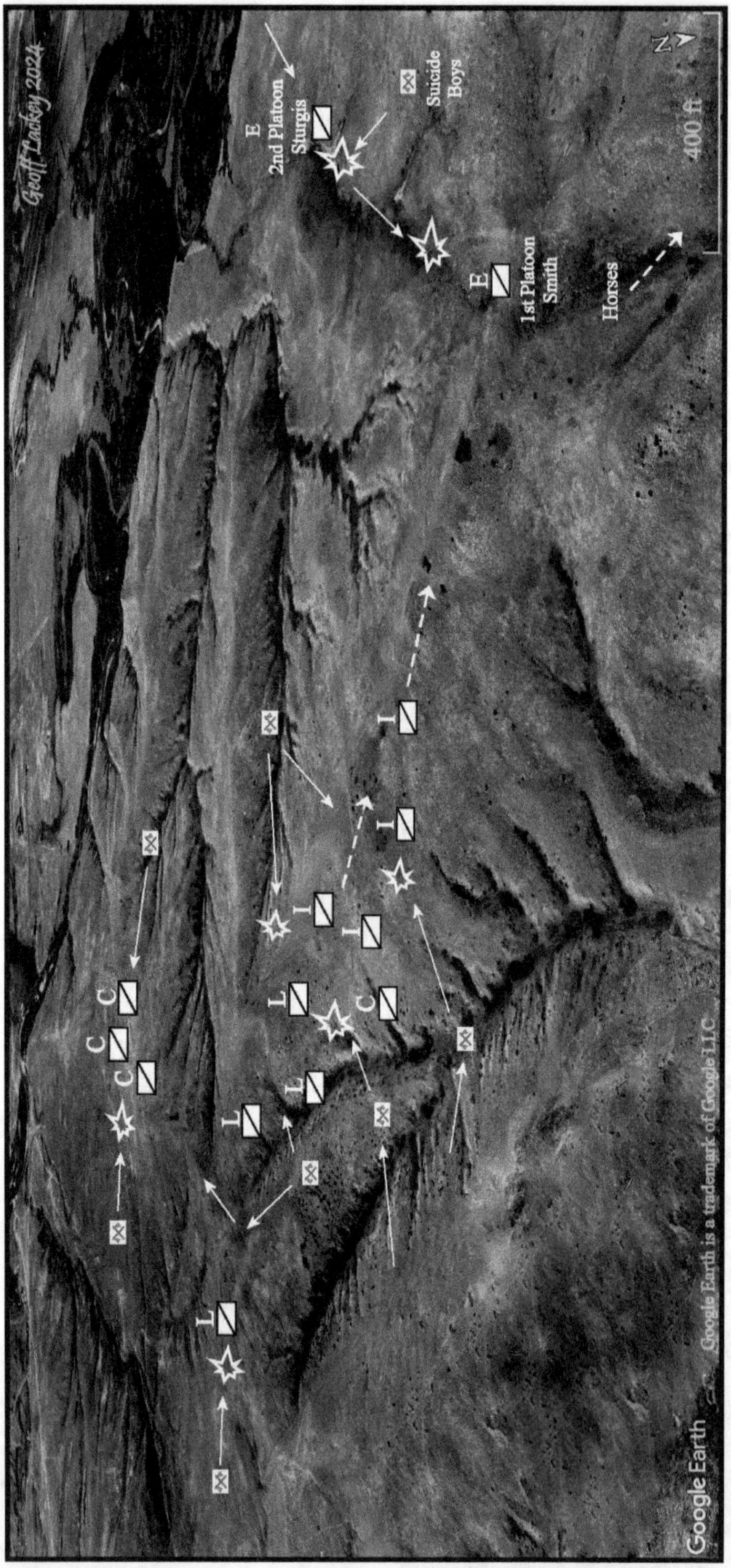

Map 20. Keogh's battalion fights for its life in the swale while survivors attempt to reach E Company's platoon on Last Stand Hill.

Chapter 13: Did Custer Move North to Seize Hostages?

As Keogh deployed his troops on and around Calhoun Hill, Custer was moving north. Trampling over the grass and sagebrush, he rode along the east side of Battle Ridge with his regimental staff, his two younger brothers Tom and Boston, nephew Autie Reed, and Yates's two companies all in tow. Left behind with Keogh's command was his brother-in-law, Lieutenant James Calhoun, who once told Custer that when the time of testing came, he would not find him wanting. Riding ahead of Custer, probably several hundred yards to his right front, the distance depending on the contour of the terrain, were several F Company flankers and a good guess would be another small detachment on his left front flank accompanied by the half-breed Indian scout Mitch Boyer who had been over this terrain before.

Speculation that Custer was running for his life at this point is unwarranted and vastly overstated. If Custer's command had been in such mortal straits, he would never have separated from Keogh and divided the squadrons of his regimental wing. Some battle students maintain that in his journey north, Custer took all five companies down to one of the lower fords, later designated as Ford D. A few Indian accounts state that the soldiers were all together the entire time and never operated as separate units, casting doubt on whether Custer had separated from Keogh. But these Indians were most likely describing only Keogh's command, as the Yates squadron, attached to Custer, had already moved out of their line of sight. However, the evidence suggests that there were two separate and distinct squadrons, one of them moving north on the high ridge known as Nye-Cartright or Blummer Ridge and another traversing from Ford B-2 to Calhoun Hill.

Most of the warriors were focused on Keogh, unaware that Custer had separated from Keogh and was advancing stealthily north with another portion of his command, his movements hidden behind the natural landscape. Although many Indians told John Stands in Timber that Custer was heading toward the Little Big Horn River to capture non-combatants, his more likely objective was securing the flats near the later-named Willy Bends place.

Custer left Calhoun Hill with a squadron (two companies) of his wing and his Headquarters Staff. Old-fashioned detective work and the power of deduction can spell out the reasons. If Custer had taken all five companies northward to Ford D, the remains of his command would not have been so physically segregated from one another when their bodies were subsequently discovered on different parts of the battlefield, i.e., Calhoun Hill, Finley Ridge, Battle Ridge, Last Stand Hill, and Deep Ravine. Parsing out who died where on the battlefield allowed for a general determination about which companies were where, and which soldiers belonged to what companies. The post-mortem evidence strongly suggests that a small command under Custer rode farther north to a different part of the battlefield. Most bodies from C Company in Keogh's battalion, for example, were found at Battle Ridge and Finley Ridge—

those bodies found in Deep Ravine, not far from Last Stand Hill, were primarily from E and F Companies of Yates's command.

Another scenario is that Custer took five companies to Ford D, only to be driven back by a swarm of hostiles before they died in separate locations on the battlefield. That Custer journeyed north with all five companies originates from what many battle students see as the unlikelihood that Custer would again divide his men and move on with just a fraction of his force. But that view has little support in the post-battle body locations nor in most Indian accounts. Custer had deployed a platoon of E Company, some twenty men, on Battle Ridge Extension, just north of Last Stand Hill, to create a bridge between his command and that of Keogh's. Custer's strength was reduced to no more than seventy troopers traveling to what appeared to be the northern flats between Fords D-1 and D-2. But what was the purpose of heading north? It certainly was not to capture thousands of non-combatants with so few troopers.

General Philip Sheridan's post-battle analysis shows he pored over the details of the battle to find a convincing explanation for the disaster. Grieved and shaken, Sheridan was determined to learn how Custer could suffer such a total and one-sided defeat. After scrupulously reading all the after-battle reports and speaking with many of the survivors of the 7th, he concluded that:

> The Indians were actually surprised and in the confusion arising from the surprise and the attempt of the women and children to get out of the way, Colonel Custer was led to believe that the Indians were retreating and would escape him ... I do not attribute Colonel Custer's action to either recklessness or want of judgment but to a misapprehension of the situation and a superabundance of courage.

Could Custer's misapprehension of what was happening on the battlefield, as Sheridan believed, be the catalyst in dividing his right wing? Anthropologist Richard Fox, in his book *Archeology, History and Custer's Last Battle,* first popularized the notion that Custer, along with the Headquarters Staff and Companies E and F, moved north, leaving Keogh with three companies on Calhoun Hill to wait for Benteen's anticipated arrival and to protect the back door of his command against an attack from the rear. If true, this invites another question: If Custer remained aggressive, which was his defining characteristic throughout his entire military career, then what could his objective have been by riding north away from his expected reinforcements and with so few men other than, as Fox maintained, to seize captives to compel the warriors to surrender?

One explanation is that circumstances dramatically changed when Custer retreated from Medicine Tail Coulee and regrouped on Calhoun Hill. Since the Indians had been driven out of the village, it would make no sense to attack an empty encampment. Moreover, Indians typically reacted to a cavalry charge not by meeting it head-on but by eluding it and then reforming for another attack on their own terms. If Custer had charged across the river toward the village, whether from Ford B or farther north from Ford D, any warriors in his path would have side-stepped the initial thrust while other warriors would have converged on Custer's flanks.

Chapter 13: Did Custer Move North to Seize Hostages?

On the south side of Ford D, Custer would have had no higher ground or timber position to rely on, while taking his command even farther away from his reinforcements. Closing the gap between himself and the hostiles would place his soldiers in an untenable position, reducing their advantage of long-range firepower and enhancing the warriors' most effective weapon—fighting at close quarters.

While all this is undoubtedly true, it was not in Custer's makeup to wait passively for reinforcements. From his vantage point on Calhoun Hill, he probably realized that the longer he waited, the more time it allowed the warriors to get their horses, mobilize, and attack his command from every direction. The warriors could stay put on Greasy Grass Ridge, a good thousand yards away from Calhoun Hill, and shoot at the troops all afternoon with their limited supply of ammunition. Likewise, Custer could keep them at a distance from now until kingdom come, providing the ammunition held out, since the warriors had no intention of sacrificing scores of braves to shorten the distance between them and Custer's disciplined lines of fire.

Custer was determined to turn the tide of battle. While on Calhoun Hill, there were still no signs of reinforcements coming over the Weir Peaks onto the Custer battlefield—no Benteen—no Reno—no pack trains—not even the dust from their respective commands could be seen. Custer would have to control rather than react to events to take back the initiative. Again, retreating to Reno would end any chance of victory. So, instead, despite the risks, he would seize fate by the short hairs and take his chances. He would leave Keogh's battalion behind while he moved north, searching for offensive and defensive locations. After thirty years of Indian warfare on the Plains, there was never a known case of warriors breaking through solidly formed skirmish lines, which Keogh would deploy. Once a classic hedgehog defense was formed and the higher ground on all sides around Calhoun Hill was secured, it did not matter how many warriors came at Keogh as long as his supply of ammunition held out. The warriors could not overwhelm the tightly knitted military defense, which is why Benteen needed to bring the ammunition pack mules to their fighting position.

But it is one thing to believe Keogh's battalion was secure, quite another to divide his forces for the third time in the face of large numbers of the enemy. Questions abound on what compelled Custer to counterintuitively execute such a risky strategy. One school of thought is that the Indians forced Custer, suggesting tribal warriors had succeeded in splitting the soldiers into smaller groups to overwhelm and kill them. This scenario fits neatly with the notion that whatever organized resistance materialized on the east side of the river, it would come in pockets as if all the units were forced into separate clusters and then, one by one, summarily overpowered by the warrior forces.

But this theory does not square with the known facts of how the battle was fought. The Custer fight did not unfold with a sudden and unremitting onslaught by the warriors. Rather, it progressed with long-term sniping until it evolved into a slow, almost methodical encroachment by the warriors against the soldiers' defensive positions. Simply believing that it was a Zulu-like frontal assault by the warriors that split Custer's command on Calhoun Hill relieves the historian or researcher of the burden of determining what compelled Custer to further fracture his command after already dividing it into four separate columns. For the sake of simplicity, it is easy enough to say that the whole matter came down to warrior aggression and overwhelming numbers.

However, such tactics were not a staple of Plains Indian warfare. If, for argument's sake, they altered typical forms of combat, they lacked the command-and-control capability to bring off such an assault successfully. It is a mistake to ascribe to Plains Indians the same tactics that the U.S. military might have utilized in a similar situation. For any of these questions to be relevant, they must be approached and understood in the context and through the prism of traditional tribal warfare. Warriors did not matriculate at West Point; they did not study military science; they did not drill; they did not establish sub-units in the form of military companies or platoons; they had no officers or noncommissioned officers to direct the fighting and the combatants. They knew nothing of the military tactics of Alexander the Great, Julius Caesar, or Napoleon Bonaparte, nor did they ever analyze their great battles.

While they were known to be skilled at luring an enemy into a trap, there was neither time nor opportunity for that at the Little Big Horn. The village had been taken by surprise, and they did not have the luxury of setting an ambush or designing snares to trick their attackers. Additionally, if the warriors had sought to split Custer's command by sheer force of numbers, then the terrain would have been littered with dead warriors and ponies. Yet, in the area of conflict, there was little carnage.

Other battle students have hypothesized that Custer would not have attempted this objective with just two companies. Instead, Custer moved north to open another front. While there is some logic in Custer utilizing the Yates squadron of his battalion as a reconnaissance in force to determine the feasibility of crossing that far down at Ford D, such a maneuver would have squandered precious minutes waiting for Keogh to join him. Moreover, that kind of painstaking, deliberate reconnaissance was long past as events had rapidly overtaken them. While any number of unnamed fords were at the Little Big Horn, including one just below Cemetery Ridge, Custer was not interested in crossing at any of them. The move north only makes sense as a reconnaissance and not an offensive strike against the enemy. It is unlikely that it was to determine the feasibility of a river crossing because no crossing was attempted at Ford B with five companies, so why would Custer endeavor to cross, after leaving a platoon on Battle Ridge Extension with less than two companies at Ford D?

Another theory suggests that Custer's departure from Calhoun Hill allowed Keogh to keep the warriors defending Ford B engaged while Custer rode to the lower fords to cross the river and then returned to attack the enemy from the rear. This would relieve the pressure of the hostiles besieging Keogh's command, drawing the warriors away from that portion of the battlefield and opening the way for Benteen's battalion and the pack train to arrive. Attacking the rear of the warriors at Ford B, however, would have been an overcomplicated an overburdensome maneuver—one based on leading the warriors to think he intended to strike the women and children (and if this was his intention, Custer succeeded—since more than a few Indian accounts mention it) to draw the majority of hostiles from their position south of Calhoun Hill, behind the ridge at Greasy Grass, where warriors were growing in alarming numbers.

Chapter 13: Did Custer Move North to Seize Hostages?

While this theory sounds plausible, it fails on several fronts. First, Custer would have had to travel much too far a distance only to circle back to Ford B to accomplish this objective. Secondly, it would have taken Custer back to the west side of the river, where there is a preponderance of Indian strength, with scarcely seventy men against an enemy not only acutely alert to the danger the soldiers presented but mobilized to combat it. The small command, with no recourse to good defensive ground, would have been quickly spotted and surrounded by a mass of warriors.

As alluded to earlier, when Custer left Calhoun Hill, Keogh's command was not under heavy pressure from the enemy. There was no reason for Custer to believe that Keogh needed support, much less rescuing. Finally, Custer could have executed this feint more smartly and without expending so much time by approaching Ford C (See map), sometimes called Cheyenne Ford, with the same effect of posing a serious threat to non-combatants and drawing the warriors toward himself—if that was indeed his intention. Circling back to Ford B pays no strategic dividends and dangerously invites an inordinate number of risks to Yates's under-strength squadron in attempting to accomplish it.

Another reason given for Custer going north is that he had observed the warriors heading in the same direction, believing they were attempting to flank and destroy his command. To prevent that, Custer angled Yates's column to the right by moving downstream to foil the warriors' designs. But this hypothesis is flawed for the same reasons as before: The enemy's lack of command and control. The warriors would not flank Custer's command from such a distance since they did not tactically operate as a modern army and were likely to confront Custer from the shortest possible route. The cavalry artifacts found near Ford D are unlikely to prove that Custer was heading there to intercept the enemy from flanking his command. A percentage of these items were left after, not during, the battle, dropped haphazardly by the hundreds of warriors who took them off dead soldiers as they rode across the grounds and fired their guns in the air or at the deceased.

So many of these artifacts proved red herrings, leading battle students not closer but further away from the truth. They were of no value in reconstructing how the battle was fought. In the following days, Major Reno and Colonel Gibbon crossed Ford D, and neither reported any physical evidence of an attempted crossing by Custer's command there.

The best surmise is that upon leaving Keogh, Custer moved along the eastern slope of Battle Ridge, keeping well hidden from most of the warriors, since most failed to report this split in command. One notable exception came from the Cheyenne tribal historian John Stands In Timber, grandson of Wolf Tooth, who heard from his grandfather and others that his band fired at Custer's troops from a distance and continued to do so until Custer's forces passed out of sight, heading toward the river and the flats below. While Wolf Tooth's band did no more than pester Custer, their recollections provide valuable information about the trajectory of Custer's forces that clearly shows they were moving in a northwest direction (see Map 28).

Because there were one or more bodies from Custer's command found within a hundred yards of Ford D, and a few Indian accounts that state there was some fighting lower downriver, many have taken the unsupported leap that this is evidence that Custer attempted to cross that ford. But, again, it is difficult to draw such an inference since searching for the lower fords inevitably increased the distance from Custer's operating base and his only available support. The archaeologists Doug Scott and Peter Bleed got it right in their insightful and revealing 1997 study, *A Good Walk Around the Boundary*, published as a booklet by the Nebraska Association of Professional Archeologists and the Nebraska State Historical Society. In this study, a convincing argument was made for the course Custer ultimately took:

> It appears reasonable to conclude that ... Custer did move north through Last Stand Hill and northwesterly along Custer Ridge Extension at least 1/4 of a mile. The presence of scattered Indian caliber bullets on the ridge in the adjacent ravines suggests the soldiers' movement was under fire by a few warriors ... it is possible that some of Custer's men returned fire and may have deposited a few of the army caliber cases found on Custer Ridge Extension.

Scott also voiced doubt about an earlier Richard Fox assumption that the Yates squadron got as far as the river but an attempted crossing at Ford D was repulsed, with Scott explaining, "Rich [Fox] thinks they [the cavalry] almost got to the river. I do not think they did. They got away, but certainly not clear down the river. That is where Rich and I would differ, but that's how I see the physical evidence."

One of the more disconcerting aspects of this battle is not, as many believe, the lack of viable scenarios from which to choose but rather selecting the most feasible ones. Over the greater part of thirty-five years or more, the most popular theory about why Custer moved north—a theory accepted among some very prestigious and erudite battle students—is that Custer, seeing the battle as unwinnable by conventional means, made a desperate attempt to seize non-combatants as hostages, targeting the women and children, whom the warriors valued most, as a means to get them to capitulate.

Educated to the harsh realities of plains warfare, or so the argument goes, Custer saw the utility of taking Indian hostages based on his experience in previous battles and his encounters with native belligerents. As it turned out, these hostages had come in handy, or so the proponents of this theory maintained, not only for negotiating favorably with the hostile tribes but also as a means to defeat and subdue the warriors. Using hostages as negotiation chips is now championed as an all but obvious component of the 7th's thinking that day and the paramount reason that Custer moved north after finding himself without reinforcements. In some circles, the question has become a closed case.

Refuting this hypothesis does not mean that Custer did not consider hostage-taking. In war, the unexpected must be expected. It was simply a matter of how crucial this would be to a military campaign. One prominent authority of this battle states that Custer had taken Indian hostages on five different occasions while fighting the warriors only four times. This theory asserts that hostage-taking is key to understanding Custer's actions at the Little Big Horn and that too many students of the battle have either ignored or neglected this most likely and fetching possibility. Professor Michael Donahue, a seasonal park ranger at the Little Big Horn, champions this thesis in his well-received and informative book, *Where the Rivers Ran Red*:

Chapter 13: Did Custer Move North to Seize Hostages?

> Many contemporary writers and historians have avoided scrutinizing Custer's hostage taking strategy. Rather than looking at it as a functioning and fascinating stratagem in dealing with the southern tribes, it seems to receive only cursory notice. Few authorities search for clues into his initial mindset, and how it may have affected his actions throughout the Indian wars. Custer's own words, and more importantly, his actions, are extremely revealing to anyone who reads the lines he wrote as well as in between them. Custer defenders would rather not talk about this aspect of his manipulation of Indians because of the stigma mentioned, while his detractors emphasize it as the pinnacle of defeat, dishonesty, treachery and ultimate character flaw. This historian takes it for what it was—a strategy that was effective and cost few soldiers' lives. It occurred, no matter how one wishes to interpret, ignore, rationalize, or relabel it.

While the highly respected Donahue is an accomplished author, artist, and historian of the battle, it is untrue that contemporary writers and historians have avoided scrutinizing Custer's hostage-taking. If anything, they have overemphasized it as the raison d'être for why Custer moved north from Calhoun Hill. Since the publication of Dr. Richard Fox's *Archeology, History and Custer's Last Battle* in 1993, arguing in favor of this possibility, battle students have and continue to strongly embrace this scenario. Because there were prior incidents of hostages being taken, specifically at the Battle of the Washita, it provided a credible and satisfying answer to the question of why Custer moved toward the lower flats of his own volition rather than being driven there by the hostiles. The distinction many have failed to grasp is that there is a considerable difference between taking hostages as a military strategy and having hostages fall into your hands during the battle. There was not a single instance during his military career that Custer sent units into the field—including at the Washita—to expressly hunt down and round up hostages to use as human shields to insulate his troops—certainly not during an ongoing battle when every man was needed.

In late November 1868, Custer and his 7th Cavalry surprised and attacked an Indian village along the Washita River, where, during the battle, Indian women and children were seized as hostages. But at the Washita, Custer, with far more soldiers at his disposal than he had at Little Big Horn, the 7th Horse began taking prisoners before most had time to flee. And ever since, historians have consistently placed the captive cart ahead of the battle's horse. Captives resulted from the victory at the Battle of Washita, not the objective. So much the more at Little Big Horn.

As he left Keogh at Calhoun Hill, Custer could see the thousands of noncombatants gathered across the river north of Ford C. They would have been hard to miss. That very size is one of the primary reasons his goal was not to take captives. There were too many, and the logistics would not work. Custer and his command would have to travel a mile and a half north and west to Ford D to prevent them from fleeing downriver. The families would not wait to be caught; they would scatter when he moved their way.

Much of this misunderstanding has resulted from the accounts of the Cheyenne warrior Magpie, an Indian survivor at the Battle of the Washita. Unfortunately, over the years, many battle historians have extracted more than Magpie actually said to prove that hostage-taking became a staple of Custer's battle operations against the Indians. When more than a thousand warriors from an encampment downstream surprised Custer at the Washita, Magpie states the following:

> The Indians were very angry [regarding Custer's assault at dawn upon the unsuspecting Cheyenne village of Black Kettle] and they would have attacked the soldiers had they not been afraid that such an attack would bring death to the women and children that the soldiers had captured and were holding in the village. Shortly after dark, the soldiers suddenly fell into line with the women and children in the direction from which they had come—to the northwest. Some of the Indians started to follow them but were called back for fear harm might come to the prisoners.

The Cheyenne Moving Behind Woman, another Washita survivor, had a similar recollection of not shooting at the soldiers for fear of inadvertently "hitting the women and children." First Sergeant John Ryan remembered that "General Custer took the squaw prisoners and put them on the side of the camp the firing came from. When the bullets would come close to these prisoners, they would send up an awful yell, and it caused the Indians to cease shooting from that direction." No doubt, the warriors harbored fears of bringing harm to their loved ones, but this bears little weight when considering that never once did the warriors surrender to the army because they were holding non-combatants as hostages. Often overlooked in this entire narrative of hostage-taking is that Custer himself weighs in about this occurrence at the Washita, in which he draws no firm conclusions about Indian motivations for not shooting at the soldiers:

> Whether the fact that they could not fire upon our advance without endangering the lives of their own people, who were prisoners in our own hands, or some other reason prevailed with them,_they never offered to fire a shot or retard our movements in any manner, but instead assembled their outlying detachments as rapidly as possible, and began a precipitate movement down the valley in advance of us, fully impressed with the idea no doubt that our purpose was to overtake their fleeing people and herds and administer the same treatment to them that the occupants of the upper village had received. This is exactly the effect I desired.

It is, perhaps, more accurate to conclude from Custer's statement that he referred to the capture and destruction of the remaining hostile camps and horse herds rather than the capture of additional non-combatants. Also, when considering hostage-taking at the Washita and assuming that this same objective was in play at the Little Big Horn, the logistics, population, and size of the respective villages must be factored in. At the Washita, Custer had some eight hundred soldiers and ended up capturing fifty-three non-combatants who were still in the village where the troops were attacking. At the Little Big Horn, he had nearly two hundred fewer soldiers, and there were at least six thousand non-combatants, a number that dwarfs the Washita. Moreover, these non-combatants were no longer in the village; dust clouds to the north and west attested to their flight from the attack.

Discounting the platoon of some twenty troopers left on Battle Ridge Extension, the notion that the other sixty-five to seventy men in Custer's command were sufficient to round up this multitude of refugees constitutes a failure of the imagination to consider the odds in subjugating and containing such a moving human tide. He would then have to herd them back across the river and upcountry to rejoin Keogh, all the time being harassed by warriors. He had minimal time, no means, and too little manpower for the task. Even if Custer somehow managed the task of apprehending hostages at Little Big Horn, he would have to push them across the river to rejoin Keogh on Calhoun Hill or Battle Ridge. Moreover, to capture enough of these non-combatants to repel an attack, much less force a wholesale surrender of the warriors, would have required a pursuit that would have sent the soldiers chasing these refugees well past Ford D.

Researcher Lorne Langley, in an article for the *Little Big Horn Associates Research Review*, persuasively writes how great the odds were, much less the risks entailed, of having small detachments chasing down Indian non-combatants. As Langley states, "It has never made any sense for soldiers to ride so far away to chase fleeing Indians." At the Washita, Major Joel Elliott and seventeen other troopers lost their lives when they tried to rescue captive Clara Blinn and her two-year-old son Willie. Mrs. Blinn and her child had been captured from a wagon train. In that bloody encounter, all the men were killed along the Arkansas River in what is present-day Colorado. During the battle, Elliott allegedly espied from a distance some of the Indians escaping with Clara and her young son. Fully cognizant of the risk he was taking, Elliott shouted out the old army dare, "Here's to a brevet or a coffin," before galloping downstream to the east to chase the fleeing hostiles. In this case, the price was tragically costly as two weeks later, not only were the two captives found dead and frozen, but so were Elliott and his entire force when they were surprised by hundreds of armed warriors from other villages.

The effectiveness of "hostage-taking," as would be true of any military tactic, must be weighed by the results it achieved. At the Little Big Horn, these non-combatants were not immobile or confined to the village but were running for their lives. So few soldiers could not easily corral this large herd of humanity. In the face of mounted cavalry, they would be as elusive as a covey of quail in the earshot of gunfire, quickly taking off in multiple directions, making it impossible to target any number of them with any reasonable return of success. The survival instincts of Indians were keenly honed in the perilous world in which they lived. Any misstep could lead not only to their own death but also to the extinction of their tribal band.

Furthermore, taking hostages is more of a civilian mindset than a military one. Why would General George Armstrong Custer, whose appetite for war was matched by few others, consider fleeing non-combatants a more worthy military objective than the defeat of the warrior force? His overriding objective was always to kill the enemy and destroy the village that sustained their war-making abilities.

Custer's Last Stand Demystified

Custer was a West Point graduate, trained to defeat and destroy the enemy, not capture mostly unarmed non-combatants and use them as bargaining chips to coerce a warrior force to capitulate. The United States Cavalry, with its proud heritage and strong sense of honor, would have perceived such a tactic as contrary to the military code and even beneath contempt. Where was the glory, so close to Custer's heart, in bagging old men, women, and children? Was such an act worthy of Grant, Sherman, and Sheridan's exploits in the late war, much less his own? In reflecting upon this, Michael Donahue was right when he stated that there was a stigma attached when a military unit resorted to the capture of hostages.

One reason Custer seized hostages at the Washita was that he was ordered to do so by General Sheridan, his commanding officer. But Sheridan only issued this order to avoid more bloodshed than necessary. The last thing Sheridan wanted was another Sand Creek Massacre, an atrocity that occurred just four years earlier under Colonel John Chivington, commander of a Colorado militia. The Sand Creek incident was the *My Lai* of its day (At *My Lai*, American soldiers shot down unarmed civilians during the Vietnam War), as Indian women and children were mowed down in cold blood, leading to condemnatory and censorious headlines in the eastern press tarnishing the army's reputation, labeling them as bloodthirsty, genocidal aggressors. Additionally, there had never been any discussion or plan prior to the battle at Little Big Horn about capturing hostages, nor was it ever mentioned at any officer's call prior to discovering the village or when orders to attack were finally given. Nor can we find a single reference in the numerous letters, diaries, newspaper accounts, and military inquests by participants, including the Reno Court of Inquiry, that any such strategy was ever contemplated, much less implemented.

At no time did General Terry give any orders, written or verbal, intimating such an objective as seizing noncombatant hostages. When Custer was dispatched with twelve companies of cavalry to find the Indians, Terry only emphasized that Custer hold on to his wounded—a cautionary reference to a battle fought at the Powder River some months earlier, where wounded military personnel were left in enemy territory. Undoubtedly, with his wide breadth of battle experience in the Civil War as well as on the western frontier, Custer was quite aware of the reality of tribal warfare and all its facets. He certainly considered the vulnerability of native women and children and the solicitous and protective attitude the warriors showed toward them. He even wrote about it in his biography *My Life on the Plains*, where he recounts his meeting with Cheyenne warriors several months after the Battle of the Washita:

> Indians contemplating a battle, either offensive or defensive, are always anxious to have their women and children removed from all danger thereof. By our watchfulness we intended to let the Indians see that there would be no opportunity for them to take us by surprise, but that if fighting was intended, it should not be all on one side. For this reason, I decided to locate our camp as close as convenient to the village, knowing that the close proximity of their women and children, and their necessary exposure in case of conflict, would operate as a powerful argument in favor of peace when the question of peace and war came to be discussed.

But this passage cannot be considered a template or blueprint for hostage-taking. In this case, the Cheyenne, cognizant of the danger posed to the most vulnerable and valued members of their population, had decided to surrender without firing a shot because their chances of survival in what was a veritable dustbowl were dire and remote. Additionally, the Cheyenne warriors did not number anywhere near what was present at the Little Big Horn, nor were they in any way sustained by the abundance of wildlife that was found in the valley of the Little Big Horn's fruitful environs.

The Cheyenne had fled to a starkly barren area in Texas known as the "Staked Plains," where food and supplies were scarce, and the threat of mass starvation was always present. The almost hopelessly arid and depleted landscape was the reason that Custer only brought a single company with him, for fear that was all he could sustain and feed. It was only by the weaponizing of food that Custer, with a mere handful of men, eventually forced the Cheyenne, who outnumbered his own troops, to surrender. Such a strategy could work in the naked and sterile landscape of West Texas, but it would not work along the fertile banks of the Little Big Horn River.

In short, employing tactics to capture non-combatants to compel the surrender of warriors during the Plains Indian Wars was never an effective strategy and was never seriously employed by the U.S. army. While it may be possible to temporarily thwart aggression from a superior number of warriors in the same manner that terrorist groups in the Middle East have used women and children as human shields to insulate themselves from retaliation, but you are not going to compel the opposing side to surrender.

This is not what happened at the Washita, and there is no other known case, not only in the Plains Indian wars but in any other in the annals of military history, with the sole exception of the ancient Battle of Gabiene, where the surrender of an army or warrior force took place because their non-combatants were captured.

The appeal of hostage-taking is that it builds a useful and explanatory bridge over the yawning abyss of hard-to-come-by answers. Expediency, however, is no substitute for truth. The hostage-taking explanation has nonetheless proved saleable, accruing enormous currency amid the length and breadth of contemporary scholarship. It has now anchored itself in the historical consciousness as a compelling reason why Custer moved north. The need to know is deep, and amid so many unanswered questions, the temptation to fill those gaps in the story with credible answers can be overwhelmingly seductive, even to the point where mere conjecture passes into received wisdom.

It is more likely that when Custer found himself on the east side of the river saddled with limited defensive terrain and far fewer troops than he now needed, he was playing high stakes chess (a board game he reveled in) on the battlefield, looking for opportunities to capitalize upon and win, by trying to force the enemy into making mistakes. He was seeking a pin, even a checkmate while knowing that he was operating at a keen positional disadvantage—and a deficit of pieces on the board—due to the absence of seven of his twelve companies. Those seven were neither on his battlefield nor within supporting distance for the concerted action needed to avert the full brunt of tribal warriors bent on his destruction. While still thinking about victory, Custer was also compelled to prepare for survival if forced to fight alone without the help of Benteen and Reno. With every passing moment, this gloomy prospect grew more likely, and that meant looking for a better fighting ground to dig in and hold out.

Another important element not infrequently overlooked is the weighty and often decisive importance of the terrain at Little Big Horn. It is one of the keys to understanding this battle. Reading and using the terrain was fundamental to how Custer strategized and fought. He had a genius for topography and knew how to utilize ravines, rivers, and the contours of the land to his best advantage. In 1873, in the first Yellowstone fight on another river, he fought for three and a half hours after ordering a dismount and walking back more than a mile from where he was first attacked because the surrounding area was unfavorable. He would not fight on the grounds of the enemy's choosing or on grounds that disadvantaged his troops.

For Custer, the situation always dictated the tactics one should use, whether that meant fighting mounted or dismounted, charging the enemy, or forming a hedgehog defense while waiting for reinforcements. Contrary to popular belief, Custer was not a prisoner of one-dimensional thinking. He was not simply offensively oriented and impulsive, he could also respond to a situation by acting defensively, resourcefully, and with deliberation when the occasion demanded it.

Despite being a great risk-taker, he also prided himself on adapting to the moment and whatever challenge presented itself. Having been previously stung by criticism that he behaved rashly and without restraint or calculation, he had, in a rare moment of introspection, sharply responded to his critics by defensively penning these words:

> I am not impetuous or impulsive. I resent that. Everything that I have ever done has been the result of the study that I have made of imaginary military situations that might arise. When I become engaged in campaign or battle and a great emergency rose, everything that I have ever read or studied focused in my mind as if the situation were under a magnifying glass and my decision was the instantaneous result.

If it was not the hostages Custer was after, then the only logical argument for moving north was to find defensible ground beside the lower fords, near water, where he could wait for Benteen to launch another attack or, absent that, though less desirable, wait for General Terry to come from the north. As would be the case for their thirst-ravaged troops on Reno Hill, access to water was critical. Yellow Nose recalled that it was very hot on the day of the battle and that many of the warriors suffered badly from lack of water. The river by the lower fords provided an avenue to this life-sustaining force. Combined with its superior defensive character, especially when contrasted to Calhoun Hill, Battle Ridge, and most especially Last Stand Hill, establishing a defensive perimeter there with Keogh, with whom Custer would ultimately affect a junction, whether Benteen or Reno arrived on the battlefield or not, made more sense than chasing and securing hostages amid several thousand enemy warriors.

Chapter 13: Did Custer Move North to Seize Hostages?

Lack of ammunition would still be a serious problem without access to the advanced pack mules, but that would be the case no matter where they positioned themselves on the east side of the river. Three years earlier, at the first battle of the Yellowstone, Custer had skillfully used the river to fight off a larger, well-armed Indian force, masterfully utilizing the terrain to ensure the survival of his command. General David Stanley arrived at the first Yellowstone fight in the proverbial "nick of time," just when Custer's ammunition was running dangerously low. Custer recalled looking far to the right and over the crest of hills "... all eyes were turned to the bluffs and coming almost with the speed of the wind, four separate squadrons of Uncle Sam's best cavalry..." 'Custer's Luck' was smiling again.

There is no reason to think that Custer would not do the same at the Little Big Horn. As things turned out, there would be no General Stanley—no General Terry—and, unfortunately, no Benteen or Reno. Confronted by Reno's retreat and the absence of support from seven companies of his regiment, Custer would temporarily seize holding ground and concentrate on uniting his command with that of Keogh in an area accessible to water. But he was ready to go on the offense immediately if his much-anticipated reinforcements arrived.

If Custer's intention had been to seize non-combatants as hostages, he could have ridden directly to Ford C, the Cheyenne Ford, which was closer to his target area. In his battle classic, *Centennial Campaign*, Dr. John Gray rhetorically asked, "But why had Custer attempted so much with so little?" The answer is simple—he didn't. The primary objective of war is to defeat and destroy the enemy. This was drilled into cadet Custer at West Point and became second nature by virtue of his extensive experience on North American battlefields. That was how Grant, Sherman, and Sheridan waged war, and Custer unhesitatingly reflected the same pugnacious mindset throughout his military career.

By a careful process of elimination, considering all the alternatives and comparing them closely with Custer's history, seeking better defensive ground while not surrendering the opportunity to snatch victory from the jaws of defeat, is the explanation that carries the fewest assumptions and the most plausible reason why Custer, riding with just forty percent of his five-company wing, made the fateful decision to venture north with less than two companies of soldiers.

Chapter 14: Did Weir Peaks or Reno Hill Have the Better Defensive Position?

Was Reno Hill a better defensive position than the Weir Peaks? Did Benteen and Reno, having retreated from the Peaks to the Hill, inadvertently put the command in greater peril than it already was? Questions abound about most aspects of this battle, and the two-day struggle on Reno Hill is no exception. While this part of the fight seems anticlimactic, it was a bitter, life-and-death struggle as seven companies moved back from the Weir Peaks to make a heroic defense on Reno Hill.

Though separate from the Custer fight, the decisions made on Reno Hill are inextricably connected to what unfolded on Battle Ridge. Those decisions also provide context and understanding of the larger picture since so much of what went wrong at the Little Big Horn resulted from command decisions. Why, for example, did it take an excruciating two hours or more for four companies to move one and a quarter miles from their position on Reno Hill to the Weir Peaks that overlooked the Custer battlefield, despite the persistent and audible volley firing coming from that direction?

Writing to his wife nine days after the battle, Lieutenant Gibson defended the decision to stay on Reno Hill. "We could neither abandon our wounded men nor the packs of the whole command." However, in light of a fuller examination of the evidence, that assessment must be challenged since most of the warriors had moved away from Reno Hill north toward Custer's wing, which at the time appeared to be threatening the women and children. Plains warriors characteristically reacted to a target that presented the greatest threat to their security. They generally did not attack on multiple fronts that would split their forces unless those targets were a direct threat to their village, horse herds, or noncombatants. From the onset of the Custer fight until the destruction of the Custer wing, Reno Hill was neither under attack nor surrounded by hostiles. Most warriors instinctively moved toward the more significant menace—Custer's forces to the north. Because that diverted most of the warriors from Reno Hill to the Ford B crossing, Weir and his one company could move to the Peaks unmolested. No good reason appears to justify why three or four companies could not have been mustered to directly support Custer within forty-five minutes of the juncture of Reno and Benteen on the bluffs (see Map 21). The argument that a shortage of ammunition prevented an attempted juncture fails notably because most of the advanced packs of ammunition had already been hustled to Reno Hill just thirty minutes after Benteen's arrival.

Major Reno and Captain Benteen's delay and failure to make any serious attempt to link up with General Custer is the most contentious aspect of this battle. It is a subject so loaded with dynamite that the reverberations can still be heard today. Of the questions covered by the Court of Inquiry, Dr. Kuhlman correctly noted that this was the most implicating not only for Reno and Benteen but all the officers called to testify. Efforts to reach Custer were so anemic and damning that, over the next fifty years, Lieutenant Edgerly made it a point to insist that there was no chance that Custer's command could have been saved: "If Reno had advanced at once upon Benteen's junction with him and without waiting for the ammunition, such a move would have resulted only in the practical annihilation of the regiment."

But was Edgerly correct in stating that Custer's command was beyond the reach of any assistance? After reviewing every angle as to why it took so long to go from Reno Hill to Weir Peaks, the answer boils down to this: command decision (see Map 22). Shortly after arriving on the Hill, Major Reno had decided that he would not ride to the sound of the guns downstream until the entire, slow-moving pack train arrived. He decided that even though the advance ammunition packs, as Lieutenant Mathey testified, had arrived forty-five minutes earlier than the rear guard and main pack train. When the Reno Court questioned Mathey about why everyone on the hill was waiting, he replied, "I don't know exactly." Neither did Captain Weir, who was so exasperated by the command's immobility that he left without orders, confident that the source of the discernible gunfire was coming from Custer's beleaguered wing, a point the senior commander on Reno Hill seemed either strangely indifferent about or reluctant to act upon (see Map 21).

As for Captain Benteen, much like when he was watering horses at the morass, he seemed unconcerned about firing in the distance. He was only shaken from his lethargy after Captain Weir, fed up with the inaction, took the lead in an attempt to contact Custer. Reno and Benteen regarded Weir's martial display as rank insubordination.

There were actually three separate, belated troop movements to the Weir Peaks (see Maps 21-24). When the last group finally moved north, it would be for naught since Custer's command may already have been obliterated by the warriors, who were now busy finishing off whoever remained alive and claiming their bloody trophies of war (see Map 24). Excuses of wounded on the hill delaying the march are questionable at best. Sergeant John Ryan of M Company, arriving with the second troop movement, supports the timing of the destruction of Keogh's forces. Despite clouds of hovering gun smoke, Ryan still could see the carnage taking place on the east side of the battlefield:

> We saw at a distance from a mile and a half to two miles parties, whom we supposed were Indians, riding back and forth and firing scattered shots. We thought they were disposing of Custer's wounded men, which afterwards proved to be true.

The Weir Peaks not only provided an excellent observation point of the Custer battlefield but, as the aerial photograph (*See* Exhibit A) clearly shows, it served as a natural fortress against an attacking force. Professor Kuhlman thought it was by far the best defensive position in the battlefield, and for good reasons. The Peaks had a far higher elevation than Reno Hill, and its steep incline included three promontories resembling an "L" on its side. Its points can be traced along the river's bluffs, two located west of the present road. East of the road is another large peak and long ridge extending northward.

Chapter 14: Did Weir Peaks or Reno Hill Have the Better Defensive Position?

Today, a century-and-a-half later, its appearance has changed since the eastern and western projections were connected by sloping sides, which have since been graded down to construct a roadbed. But with a bit of imagination, one can visualize how it essentially appeared in June of 1876, and it scarcely takes a military frame of mind to see how daunting the Weir Peaks were, being the highest point on the battlefield..

Captain Benteen arrived on the Peaks with Companies H and K at about 5 p.m. to assess the situation. Lieutenant Edgerly had previously deployed D Company at a small hill off the easternmost ridgeline (see Map 24); while Captain French deployed M Company to the west and slightly behind D Company, while First Lieutenant Godfrey ordered K Company to deploy a little to the rear of M Company's position.

The summit at the Weir Peaks gave Benteen his first full view of the sprawling Indian village and the surrounding terrain. Even with binoculars, it was difficult to get a clear sighting of the situation past the Peaks. Clouding their view was a storm of gun smoke and dust, making it nearly impossible to determine whether a battle still raged or if they were witnessing a mop-up operation in the aftermath. If it was indeed the latter, it did not appear that the U.S. Cavalry was doing any of the mopping.

Whatever was happening, Benteen continued to be oddly detached, focusing primarily on his battalion rather than the entire command. Still, he was aware enough to take a sweeping look over the battlefield and his impressions he later memorialized:

> I saw enough to cause me to think that perhaps this time we had bitten off quite as much as we would be able to chew. Then I got the guidon of my own troop and jammed it down in a pile of stones which were on the high point, thinking perhaps the fluttering of the same might attract attention from Custer's command if any were in close proximity.

One thing Benteen clearly discerned was that the battlefield was thick with hostiles. Once the soldiers were seen on the Peaks, it triggered a furious response as a large body of warriors (Lieutenant Hare estimated fifteen hundred) began steamrolling in their direction. Lieutenant Godfrey essentially observed the same thing:

> The air was full of dust. We could see stationary groups of horsemen, an individual horsemen moving about. From the manner of grouping and the manner in which they sat their horses we knew they were Indians ... Looking toward Custer's field, on a hill two miles away [Calhoun Hill] we saw a large assembly. At first our command did not appear to attract their attention, although there was some commotion observable among those near to our position. We heard occasional shots, most of which seemed to be a great distance off, beyond the large groups on the hill. While watching this group,

> the conclusion was arrived that Custer had been repulsed, and the firing we heard was the parting shots of the rear guard.

While the Indians charged toward the troops seen on the Weir Peaks, Major Reno and the still slow-moving pack train and rear guard were one-half to three-quarters of a mile short of reaching Weir and Benteen's position on the Peaks. Reno was about to be caught in the unenviable position between Reno Hill and the Weir Peaks, having no place to take cover. Seeing from the height of the Peaks, the oncoming warriors, and Reno's stark vulnerability, an alarmed Benteen turned to Lieutenant Gibson, exclaiming, "This is a hell of a place to fight Indians. I am going to see Reno and propose that we go back to where we lay before starting out here."

If Reno had not tarried so long on the hill and reached the Peaks in a timely manner, it is doubtful that Benteen, now supported by Reno's troops, would have fled back to Reno Hill rather than ordering the troops—Indians be damned—to dig in where they were and make a fight of it. Instead, Benteen rode off the Peaks, abandoning his strong position, to convince Reno to retreat back to a lower elevation on the bluffs, an area far more vulnerable to enemy fire from higher elevations. Yet, many battle students still believe Benteen made the right decision to leave the Peaks for his former position at Reno Hill. Four companies on the Peaks would have been sufficient to cover Reno's command and likely divert the onrushing warriors' attack toward Benteen rather than Reno, who was dangerously exposed (see Map 25).

In his book, *The Last Stand,* Nathaniel Philbrick supports Benteen and Reno's decision to retreat to a location on the bluffs near their former position, stating that "the battalion needed to find a place where the steepness of the bluff facing the Little Big Horn protected at least one side of their entrenchment from attack. Reno Hill, the location Benteen eventually chose, argued Philbrick, was certainly not perfect, but it was as good as they would find under the circumstances. What Philbrick fails to appreciate, however, is that the steepness of the bluffs on the east and west sides of the Weir Peaks established a natural defensive barrier that Reno Hill lacked. Benteen later contended, albeit mistakenly, that if the command had remained on the Peaks, the Indians could have passed around them to the east and by the river flat at the west, putting the command in the undesirable position of finding hostiles in their rear. What Benteen did not see, or perhaps more truthfully, did not want to later acknowledge, was that the Indians could just as easily pass around their rear at Reno Hill, which, as it turned out, is exactly what happened.

By declaring, as he did, "this is a hell of a place to fight Indians," Benteen was almost certainly alluding more to the open ground that Reno's rear guard and pack train were traversing, not the Peaks itself. Also, unlike Reno Hill, there was no ground in that area higher than the Weir Peaks, which commanded all areas of the surrounding ground. Unfortunately, for Reno's men, the defensive position taken on Reno Hill allowed the greater elevation of Sharpshooter Ridge, along with positions to the east for Indian sharpshooters to fire down at the soldiers, resulting in over fifty casualties in the next twenty-four hours. The natural folds within the northern arms of the Weir Peaks would have provided more protection for the wounded and the horses and mules compared to the treacherous, low, open ground at Reno Hill. The only real weakness associated with the Weir Peaks was its lack of easy access to water, a strategic disadvantage that Reno Hill shared.

When called to testify at the Reno Court of Inquiry, General Gibbon, who examined the terrain around Reno Hill just days after the battle, was asked what his examination revealed about Major Reno's position on the hill and if he would describe that position by showing its advantages or disadvantages as a defensible position. General Gibbon's testimony is revealing:

> I spent probably 2 hours up on the hill conversing with the officers and attending to official business. I did not go into a very extended examination of the position. I don't even know where all the troops were posted. I saw some rifle pits and examined the position generally. My conclusion in regard to it is that it is an exceedingly weak position for defense. Naturally, it was commanded at tolerably long range by the hill [he is referring to Sharpshooter Ridge] which bordered it on the downstream side. The country was quite broken by a succession of little rolling hills and valleys, behind which attacking forces could conceal themselves. Then the manner in which the animals were exposed was very bad for the command. I think I counted 48 dead horses in one little valley; and then the fact of them being cut off from water, not entirely so, as I understood, but very much cut off from an ample supply, to render the point weak for a prolonged defense.

Dr. Charles Kuhlman supported the crux of Gibbon's analysis, observing that in contrast to the soldiers, the warriors were able to pour gunfire onto Reno Hill, making the men "a perfect target" for the warriors positioned to the north and northwest. Kuhlman also believed "it was impossible to cover the whole position" [on Reno Hill] without exposing large sections of the line to long-distance crossfire. Studying both positions, Kuhlman would unequivocally state that the Weir Ridges formed an almost perfect defensive position in preventing a successful attack from an aggressor.

It is near inconceivable to believe that Benteen was oblivious to the obvious defensive advantages of the Peaks over Reno Hill, despite what he later officially said. As noted before, although there is no doubt about his attributes as a combat officer, Benteen had a formidable and self-justifying personality. He loathed being second-guessed or having his decisions questioned, much less admitting error. The proud and haughty Virginian was hypersensitive to criticism, which he would not tolerate from others or himself. It would be a mistake for historians and scholars to rely entirely on what he said about anything regarding this battle unless it is independently supported.

When testifying before the Reno Court about what he observed from the high point, Benteen said, in part, "Some of the officers say that the battlefield was in sight, but I know positively that it was not, having gone over it two or three times since." This is an astounding statement, as any fair-minded observer familiar with that observation point would only conclude that this is yet another example of Benteen either playing fast and loose with the truth or retrospectively deluding himself. His sworn statement clearly runs counter to what many others saw and recorded. Lieutenant Varnum, for example, had little trouble seeing the Custer battlefield and describing what he saw:

> We moved to a point that overlooked where the Custer fight took place, but it was covered with Indians riding in every direction. A considerable firing was heard, but there was no body of troops in sight. I saw some white objects that I thought were rocks but found they were the naked bodies of men.

With the retreat to Reno Hill in full swing, Lieutenant Godfrey ordered Second Lieutenant Luther Hare to take ten men and occupy the southern terminus of Sharpshooter Ridge to prevent the hostiles from flanking their position (see Map 25). Neither the Weir Peaks nor the southern terminus of Sharpshooter Ridge was later vacated due to positional weakness but rather because their commanding officers inexplicably decided to abandon the stronghold. The warriors did not drive Hare and the soldiers off the Ridge any more than they had forced Custer away from Ford B and up Deep Coulee. Rather, the soldiers left of their own accord pursuant to orders given by their superior officers.

The general assumption by those who first manned the Peaks was that there was no intent other than to hold their ground; otherwise, why deploy there in the first place? In his narrative of the battle, Lieutenant Godfrey states that he had every expectation of holding his position on the Peaks and believed Major Reno would support him. It was almost unimaginable for the soldiers to abandon Sharpshooter Ridge, since the elevation commanded Reno's position at the corral he had formed on the lower bluffs of Reno Hill. If the high ground was abandoned to the enemy, the troops would be in a gravely precarious position. As a result, the command gravitated from the highly defensible Weir Peaks to the nearly indefensible Reno Hill. The consequences of vacating the Peaks for Reno Hill would be severe and must raise questions not only about Benteen's judgment, but why others so blindly followed.

Lieutenant Francis Gibson thought it was the flood of warriors heading toward them that finally broke Weir's resolve to advance, recalling the captain hyperbolically saying, "there were enough Indians over there to eat up his company a hundred times over." Other battle students believe that, although Weir was the first to look beyond the Peaks, he could scarcely see farther than Calhoun Hill and Finley Ridge. But he perhaps saw enough to sense that Custer had been badly repulsed. Weir admitted as much to Lieutenant Godfrey, concluding that his commanding officer, whose martial ardor he admired, was, at best, now riding north to link up with General Terry.

Regardless of what Weir thought about during those two harrowing hours on the Peaks promontory, he saw more than he ever wanted to see or know—perhaps even intuitively sensing the worst but not fully admitting it to himself. Weir left his company to Lieutenant Edgerly, who found himself with only Captain French's M Company and Lieutenant Godfrey's K Company as the ominous hoofbeats of Sitting Bull's warriors thundered closer to the troops' position.

Chapter 14: Did Weir Peaks or Reno Hill Have the Better Defensive Position?

Foreground: The Weir Peaks, where several of Reno's companies abandoned their attempt to join with Keogh's battalion, who was then on Calhoun Hill, just over two miles to the north.

Soon, Godfrey, French, and Edgerly would be ordered off the Peaks. In subsequent years, Benteen never wavered in his belief that the Peaks provided little cover or protection against the warriors amassing to their north and northwest. Reno must have concurred, though he never arrived at the Peaks to assess their defensive capabilities for himself. Instead, Reno obligingly followed Benteen's counsel, retreating to that same dubious sanctuary they had occupied after leaving the timber.

Delayed by an attempt to bury Lieutenant Hodgson, Lieutenant Varnum may have been the last officer to arrive on Weir Peaks shortly after 5:00 p.m. Headquarters time. Varnum trailed Lieutenant George Wallace's call to advance, whereupon he belatedly reached the Peaks, which he called the "point of the ridge downstream." There, he found Weir's D Company firing at scattered groups of Indians some eight hundred yards away. Simultaneous events seemed jumbled, but Varnum specifically recalled that at that time, the assault against their Weir Peaks position was not very heavy. Indeed, the distant plain below was filled with warriors. Blood-lusting and barbarism ran amok as the Indians killed, overkilled, or dispensed the coup de grâce to whatever troopers remained alive on the Custer battlefield. Based on the observations of Varnum and others, the overriding question remained: How imminent was the threat of onrushing warriors to the soldiers on the Weir Peaks?

Hare's estimate of some fifteen hundred warriors heading toward the Peaks is probably sound—but how fast and resolutely were they coming? Many warriors were three miles away at the Custer site when they realized more soldiers had arrived on a promontory to their south. Most of the Indian accounts state that by the time the soldiers were noticed on the Peaks, Custer's five-company wing had been wiped out. D Company and M Company remained on the Peaks for some two hours or more, so they would likely have witnessed the last gasps of the soldiers posted on Finley Ridge. However, events unfolding on Last Stand Hill were clearly beyond the range of human visual observation.

As to the intensity of the fight on and around the Weir Peaks, Varnum remembered that for a little while, the firing was "quite long range…a slow fire—a shot now and then at quite a little distance"—which would be typical of what occurred early on with the firing at Calhoun Hill when little more than back and forth skirmishing took place. From Varnum's testimony at the Reno Court and scraps of information from the recollections of others, this rather inconsequential shooting back and forth from the peaks may have lasted for well over an hour. Varnum also remembered that the Custer battlefield was covered with Indians. But now they turned toward the Peaks, coming as fast as they could, with their numbers swelling rapidly.

However, the warriors did not arrive at the Peaks all at once. Riding three miles chewed up time, and the warriors did not want to play out their horses. If the soldiers panicked and fled as had occurred during the valley and timber fight, the warriors would want their horses fresh to savor "another buffalo hunt" with the soldiers on the run.

Visions of warriors storming hedgehog defenses and established skirmish lines, as depicted in cinematic classics such as *They Died With Their Boots On* and *Fort Apache*, is a triumph of imagination over reality. As the warriors closed in on the Peaks, they decelerated, much like a locomotive puttering out of steam, grinding slowly to a halt. In a *Century Magazine* article published in 1892, a puzzled Godfrey recalled seeing the command firmly rooted in a strong defensive position followed by their inexplicable and abrupt retreat from the Weir Peaks:

> Weir and French's troops were posted on the high bluffs and to the front of them; my own troop along the crest of the bluffs next to the river; the rest of the command moved to the rear, as I supposed to occupy other points in the vicinity, to make this our defensive position. Busying myself with posting my men, giving direction about the use of ammunition etc., I was a little startled by the remark that the command was out of sight. At this time Weir's and French's troops were being attacked. Orders were soon brought to me by Lieutenant Hare, Acting-Adjutant, to join the main command. I had gone some distance in the execution of this order when, looking back, I saw French's troop come tearing over the bluffs, and soon after Weir's troop followed in hot haste. Edgerly was on top of the bluff trying to mount his frantic horse, and it did seem he would not succeed, but he vaulted in his saddle and then joined the troop.

Chapter 14: Did Weir Peaks or Reno Hill Have the Better Defensive Position?

The initial deployment of troops on the Peaks had "fight" written all over it, not retreat. However, as the warriors approached, their resolve weakened and collapsed. Yet, the onslaught was not an unrelenting, irresistible avalanche but one that had slowed to a tentative advance, incremental and sporadic rather than explosive and volcanic. Like Varnum, Edgerly described the fighting on the Peaks as 'general' rather than heavy and overwhelming. The Indians did not drive the command from their position, but the troops withdrew only after being ordered to do so.

The fallback began after the withdrawal of Godfrey's K Company, followed by French telling Edgerly to withdraw from their position. Edgerly reluctantly complied, but because of the confusion among the troops over whether an order had been given, the subsequent retreat from the Peaks did not occur in an orderly manner. In addition, the command to withdraw was understated and delivered unhurriedly. It cannot be stated strongly enough that, with Reno's A, B, and G Companies and the ammo pack train nearby, the four companies on the Peaks were well-positioned to meet an attack. The only missing element was leadership, without which discipline and determination labor in vain.

Benteen showed little, if any, resolve to hold a sturdy position even when it was not seriously challenged. Incredibly, not a single round was fired by any of Benteen's troops on the Peaks. Yet, he blamed French as the weak link when he precipitously broke from the position. As for Major Reno, he must have been hallucinating when he wrote in his official report that the Indians nearly overwhelmed the troops at Weir Point, and only a heavy fire kept them at bay.

The truth is that the warriors were reluctant to attack more aggressively because the Peaks were an imposing position. The warriors were biding their time, hoping the soldiers would abandon the elevations much as they had vacated a redoubtable timber position in the valley. What then explains the ensuing staggered but hasty departure, other than the battalion commanders' fears, when they began to perceive, almost certainly erroneously, that they were about to be overrun? And so, under the illusion of an unstoppable wave of warriors, flight prevailed over fight.

The sudden evacuation of the Weir Peaks can also be attributed to the direct result of irresolution on the behalf of the two battalion commanders. By contrast, Custer's natural self-assuredness and fighting spirit inspired others and was an indispensable part of his leadership prowess. His coolness in the face of danger was contagious and inspiring. While it cannot be denied that the magnitude of Custer's defeat can be found in the Biblical admonition, "Pride cometh before destruction and a haughty spirit before a fall," the collapse at the Weir Peaks was just the opposite, a result of irresolution and overreaction. Like the valley and timber fight, the Weir Peaks sudden withdrawal exposed a lack of determination and discipline, thus the abandonment of a formidable position.

The sequence of events can be traced back to Benteen leaving the Peaks to seek out Reno, followed by the decision to have the pack train, wounded and rear guard elements halt and return, and subsequent retreat back to the environs of Reno Hill. Neither Lieutenant Godfrey nor his company realized Reno and Benteen were heading back to their former position until they were far enough south to be hidden behind the hills. But after Lieutenant Hare prodded Godfrey with the retreat order, Godfrey directed the company to mount and begin their march south to Reno Hill (see Map 25).

Soon thereafter, Captain French received orders to withdraw his command, whereupon he yelled the retreat order to Edgerly. Not having heard the order, Edgerly and his orderly were unwilling to leave their post. Perhaps they did not realize others had already left or were about to leave. Or maybe they were in a pugnacious mood, itching to take a couple of shots at the oncoming warriors. Whatever the reason, one warrior had fired at Edgerly, causing his horse to panic. The terrified animal was wheeling, whinnying, and stamping up dust as Edgerly sought to steady and mount him. Drawing their pistols, the troopers opted to shoot their way out of the jam, barely escaping with their lives. Private Charles Sanders later remembered a hail of Indian bullets spattering around them, but providentially, they remained unscathed. Even as they fired their Colt 45s, the soldiers remained composed. In a burst of breezy bravado, Sergeant Thomas Harrison appeared more amused at the Indians' poor marksmanship than he feared being shot.

As Edgerly, Harrison, and color bearer George Washington Wylie dashed away, they encountered a pitiful sight. A redheaded troop farrier, now in his second enlistment, had been unhorsed by a bullet that had torn through his flesh and coursed through his hips. The badly wounded soldier was twenty-seven-year-old Swiss-born Vincent Charley. Crawling on his knees, blood oozed freely from his head while almost helplessly trying to balance himself upright with one hand. Preoccupied with their own flight to safety, other troopers had almost insensibly galloped past the private. Every trooper serving on the frontier knew the warrior culture had no compunction about slowly and sadistically killing captives or the wounded—cruelty and depraved indifference in war were an intrinsic part of how warriors dealt with their enemies. To be at their mercy was no mercy at all.

Edgerly and Harrison had halted by the wounded soldier, and knowing that this was his last chance, Vincent Charley begged not to be left behind. The Indians were on the heels of the fleeing soldiers, and having no rear guard for cover, Edgerly and Harrison could not dismount under a fusillade of bullets and then lift the grievously wounded trooper onto a saddle. That would probably have resulted not in saving his life but in losing their own.

No code is more venerated or sacred to the military than not to abandon the wounded. These were the very words of caution that General Terry had strictly confided to Custer: "Whatever you do, Custer, hold on to your wounded." No doubt this weighed on Edgerly and Harrison as they were in this morally agonizing dilemma. Those not on the field of fire must be careful about making armchair judgments concerning the actions of those on the battlefield. Without dismounting, Edgerly and Harrison might possibly have grabbed one arm each and dragged Charley with them, no matter how agonizing or deadly an ordeal it might have been for the farrier. But events were happening too fast to think clearly about anything. In those few moments of indecision, a bullet had already pierced color bearer Wylie's canteen, and another split the wooden pole of his guidon. With no time to lose, and despite Charley's beseeching importunities, they told him to find a ravine to hide in, swearing they would rescue him as soon as they could bring reinforcements.

Chapter 14: Did Weir Peaks or Reno Hill Have the Better Defensive Position?

Vincent, exhausted and barely able to move, must have realized he was facing a slow, agonizing death. Many years later, in an interview with the journalist Walter Camp, Sergeant Harrison admitted that after they rode off, he and Edgerly looked behind them and were sickened at the sight of the Indians lamentably and savagely disposing of their ill-fated farrier. Charley was later found with a stick jammed down his throat, probably the portion of the stick that was shot off from Wylie's guidon, along with other atrocities promiscuously committed upon his person. The memory of leaving Charley behind would weigh heavily on Edgerly's conscience for the rest of his life. Writing to his wife just nine days after the battle, Edgerly remorsefully and heartbrokenly penned that it was the thing he "regretted more than any other thing that happened to me, for I had promised that wounded man I would get him out and wasn't able to raise a finger for him."

That was an entirely understandable and tormenting plaint. Less forgivable were the narratives Edgerly began writing in the early 1880s, perhaps a subterfuge for the pangs of a guilty conscience, in which Edgerly would dishonorably blame Captain Weir for Charley's abandonment. Weir died six months after the battle, and dead men neither tell nor refute tales. So, Edgerly concocted a story as early as the Reno Court hearings, perhaps even believing his own distortion to be the truth. He said that he petitioned Weir to return and save Charley, but the captain refused, saying it was too late.

Sergeant Harrison, on the other hand, remained silent. He received his military discharge in 1876 and chose not to look back at that painful and regrettable incident. But Edgerly, an ambitious career army officer who later obtained the rank of Brigadier General, may have feared that the tragic and painful story of Vincent Charley could tarnish his reputation and permanently damage his rapid ascent up the military ladder of command. Many years later, before the truth about that whole piteous and regrettable episode became known, Harrison, probably struggling under the onus of his own remorse and dark memories, finally unburdened himself, recounting that both Edgerly and he had known Charley was already dead before Edgerly ever reached Captain Weir.

Many more Vincent Charley episodes could have occurred because the stampede from Weir's Point quickly might have imploded into catastrophe if not for the bravery and quick thinking of Lieutenant Edward Godfrey. Startled to see the troops galloping down from the Peaks, Godfrey feared another disastrous retreat with the Indians running down panicked soldiers. Even though they managed to reach Reno Hill safely, many, now routed for the second time, would likely have found themselves too demoralized and enervated to put up much of a defense. Godfrey, seeing heavy warrior fire coming from the rear, made a split-second decision to take K Company and deploy them dismounted in a skirmish line some five hundred yards north of Reno Hill. This rear guard covered the regiment's retreating soldiers and gave them time to deploy to where they could organize some kind of defense.

Hastily arranging a skirmish line under those conditions was a formidable task. K Company, incredibly, had not yet fired a shot, but a livid Godfrey was up to the challenge. He harangued the troops in the bluest language and threatened to shoot anyone who turned their back on the enemy and ran. This had a beneficial effect of sobering the men who decided that it was safer to fight the Indians than to be shot in the back by an enraged Godfrey. While mobilizing his company and giving the necessary covering fire, Godfrey expected the rear guard he singlehandedly organized to be supported. Instead, he received a mind-blowing order delivered by Trumpeter George Penwell to retire from the field. Rather than receiving reinforcements to repel and hold off the Indians, Reno ordered another retreat.

Even making every allowance, during the key moments of this battle, Reno made the wrong decision at the wrong time. Hare would later try to cover for Reno, saying that the Major ordered the cover fire through Weir and French. But this is untrue. More than thirty years later, Hare would confess to Walter Camp and admit that he lied to the Reno Court about the Major ordering a cover fire and that Godfrey was the true hero. This was not the first time that Godfrey demonstrated such cool-headedness and tactical skill under enormous stress. Eight years earlier, at the Battle of the Washita, he ordered a dismount with every fourth man assigned as a horse-holder as his company was about to be overtaken by hostile warriors. He then divided his platoon and deployed it into two alternating skirmish lines, and as one line withdrew, the other provided covering fire. It worked at the Washita, and it would work again at the Little Big Horn as those soldiers retreated to Reno Hill.

Nonetheless, his heroics were ignored. Godfrey received no plaudits from his commanding officers. Captain Benteen patted only himself on the back with a bald-faced lie, claiming he ordered Godfrey to check the Indians' advance. Major Reno was no better, not even bothering in his after-action report to mention Godfrey's dauntless and laudable actions, which saved the command. Godfrey also made a belated attempt to save Vincent Charley after learning the farrier had been left behind. Even more impressive was that K Company suffered not a single loss in what can only be described as an extraordinary performance of soldiering and leadership by Godfrey.

The only other redeeming outcome in retiring back to the hill was that Benteen would now subtly take the reins of command from Major Reno and, in an astonishing turnaround marked with a remarkable blend of bravery and skill, save what was left of the regiment. Captain McDougall, after observing Reno's inability or refusal to take command, briskly rode up to Benteen and said matter-of-factly, "Say, old man, what is going to be the outcome of this if we don't have a commanding officer pretty damn soon." McDougall alluded to the Fetterman massacree ten years earlier, where a force of eighty-one soldiers and civilians was annihilated because of poor command decisions. Neither knew at the time that several miles over the ridge to the north, five companies of their regiment were lying dead, stripped, and mutilated.

Only smiling at McDougall's invitation, Benteen kept his own counsel. But McDougall had planted a seed that rooted, earning Captain Benteen his much-deserved sobriquet, "Savior of the Seventh," a title he prized and guarded to his dying day. There had been one misadventure after another: Reno's battalion abruptly retreating from the valley skirmish line, the rout from the timber, the inordinate delay on the bluffs, and then the snail-like march to the Peaks, followed by its abrupt abandonment and return to a weaker position. But whatever criticism was leveled at Benteen's earlier performance, his cool and intrepid actions saved the command from Custer's fate. Standing firmly at the helm, Benteen would stop the bleeding from a thousand cuts. The wearied and bedraggled troops on Reno Hill needed all the encouragement they could get, and Benteen provided it with a swashbuckling combination of strategic wherewithal, unwavering confidence, and a wholesale display of courage in the face of a redoubtable and determined enemy that greatly outnumbered his own forces.

The purpose of this book is not to describe the valiant defense of Reno Hill, which was nothing less than an intense, two-day struggle for survival. Stirring and lively accounts of what transpired there can be found elsewhere. However, that part of the battle has always played a subordinate and anticlimactic role in the story, given the gore, mystery, and drama surrounding Custer's last moments. The point here is that the defensive ground on Reno Hill, with its limited and open ground, was inferior to the Weir Peaks and, for that matter, the timber area in the valley. It was a strategic oversight to consider Reno Hill a better defensive position than the Weir Peaks. Reno Hill has been described as a large saucer with part of the rim broken away. Others have depicted it as a horseshoe with one elongated prong, the lower portion of it being a swale (see Map 26). Either way, it was not suitable ground to make a stand.

A long and terrible day soon faded into darkness around the beleaguered soldiers on Reno Hill. Exhausted soldiers listened to the strident sounds of Indian victory blanketing the prairie. Like terrestrial stars, countless Indian campfires illuminated the blood-drenched earth around the dead and dying. Black Elk remembered that all over their large campsite, big fires and kill dances continued all night long. A gloomy Godfrey recorded what the surrounded and hunkered down soldiers heard what could only be described as outpourings of celebratory exuberance for their victory and heart-wrenching bereavement for their dead:

> The long shadows of the hills and the refracted light gave a supernatural aspect to the surrounding country, which may account for the illusions of those who imagined they could see columns of troops … All night long the Indians continued their frantic revels: beating tom toms, dancing, whooping, yelling with demoniacal screams, and discharging firearms. We knew they were having a scalp dance.

Tomorrow, the soldiers would face another crucible as both the troopers and the animals of the regiment were left grievously exposed on Reno Hill. The casualties to both man and beast were lamentable. They had paid a heavy price for not taking advantage of the higher elevation and other defensive attributes of the Weir Peaks. In roughly twenty-four hours, Reno's command lost eighteen more dead, more than thirty wounded, including at least forty-eight horses and mules. Even three-quarters of a century later, a now superannuated Private Charles Windolph darkly remembered, "We were a million miles from nowhere. And death was all around us."

As the men dug in on Reno Hill, the Indians were not the only enemy. Thirst and isolation preyed upon them, bereft as they were of their leader and five of their best companies. Suffering exhaustion and mind-numbing fear, as well as the uncertainty that comes from separation, they remained in the direst straits until, two days later, a teary-eyed General Terry finally arrived to rescue the besieged regiment, bringing the battered battalions much-needed reinforcements and relief. But he also brought news about a devastating and incomprehensible defeat, several miles downstream from where they themselves were fighting for their lives.

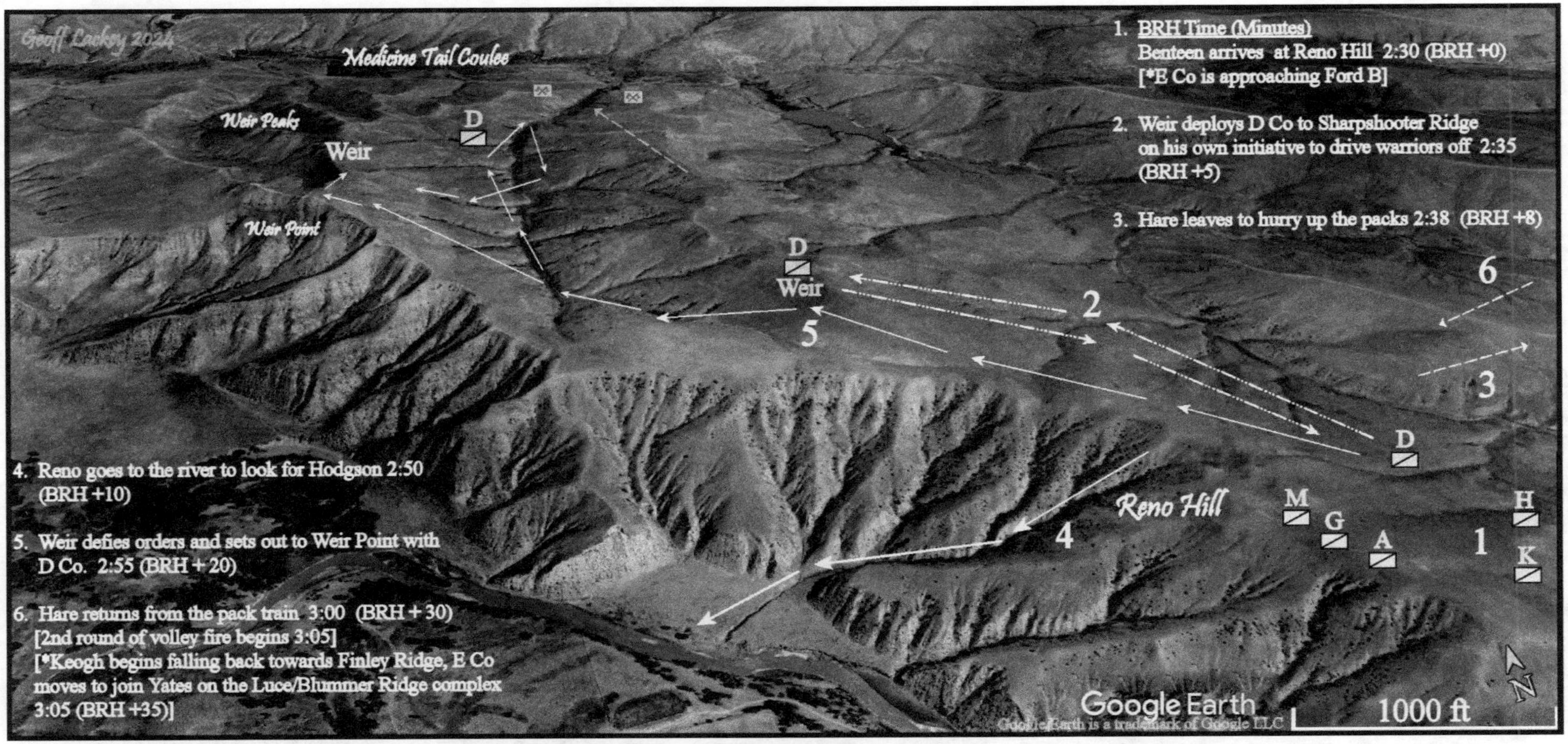

Map 21. Troop and courier movements on the bluffs, including D Company's advance to the Weir Peaks.

Map 22. Troop and courier movements on the bluffs, including M Company's advance to the Weir Peaks.

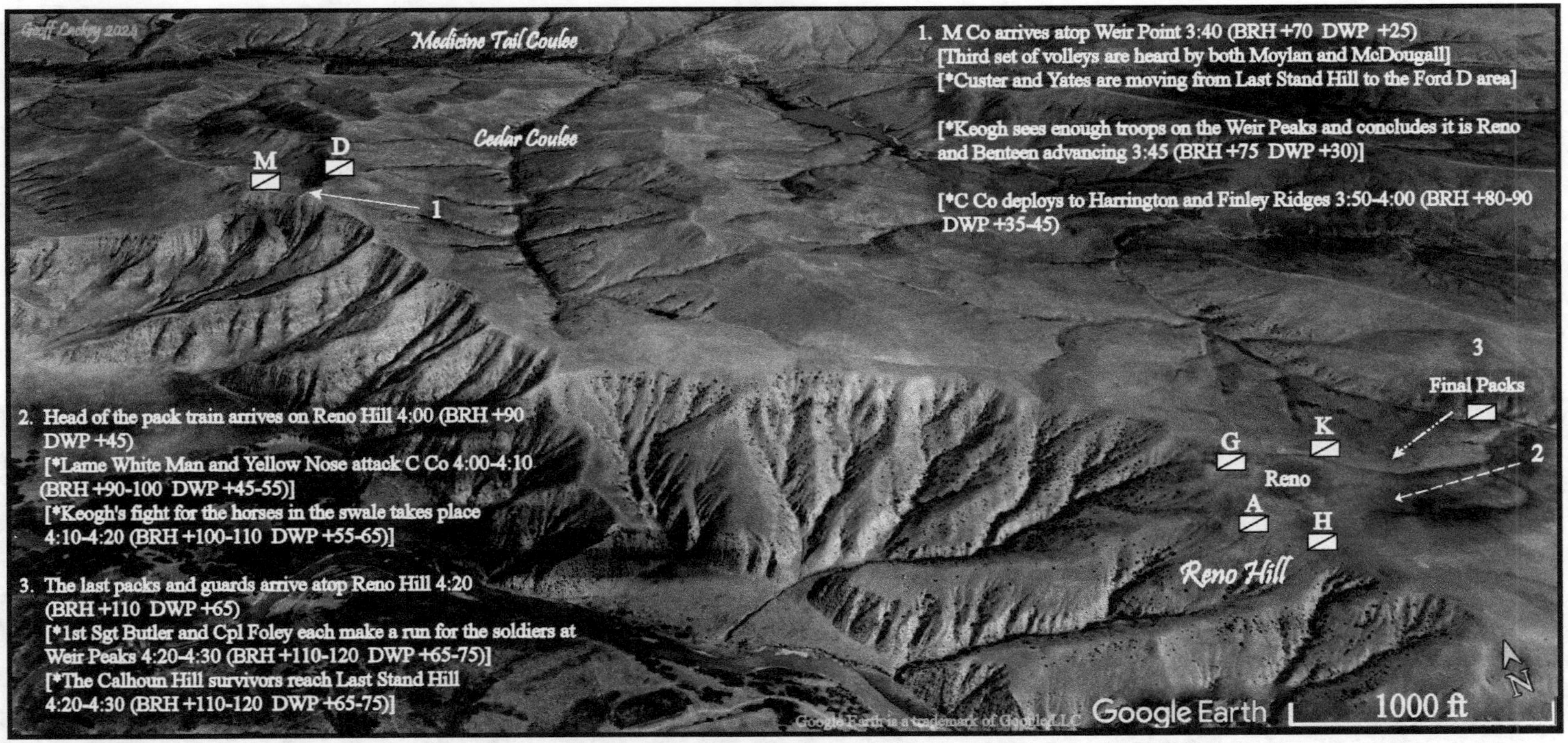

Map 23. Troop movements on the bluffs, including the arrival of the main pack train.

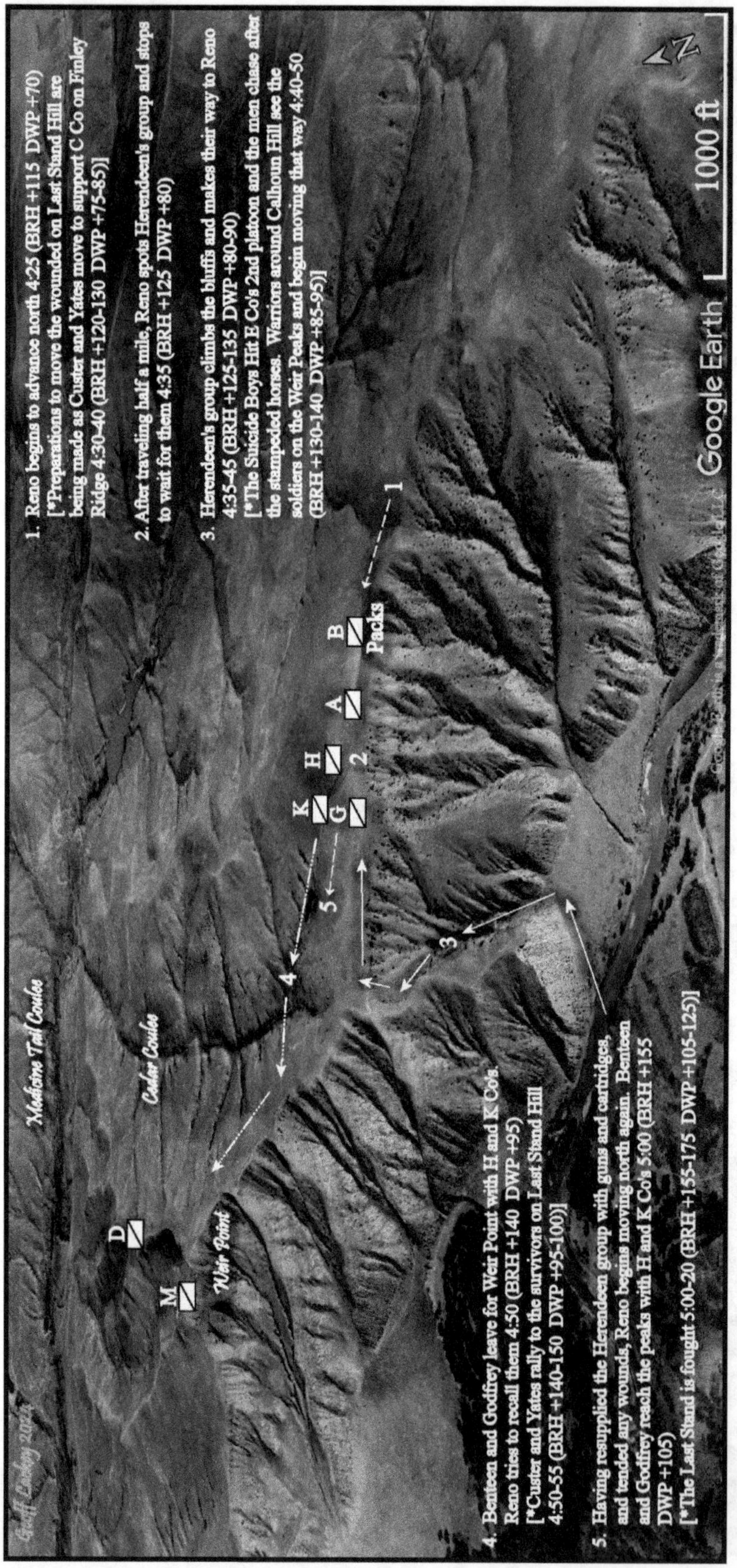

Map 24. Reno's troops advance to the Weir Peaks while Herendeen's valley survivors join with Reno on the bluffs.

Map 25. Reno's troops retreat from the Weir Peaks back to Reno Hill.

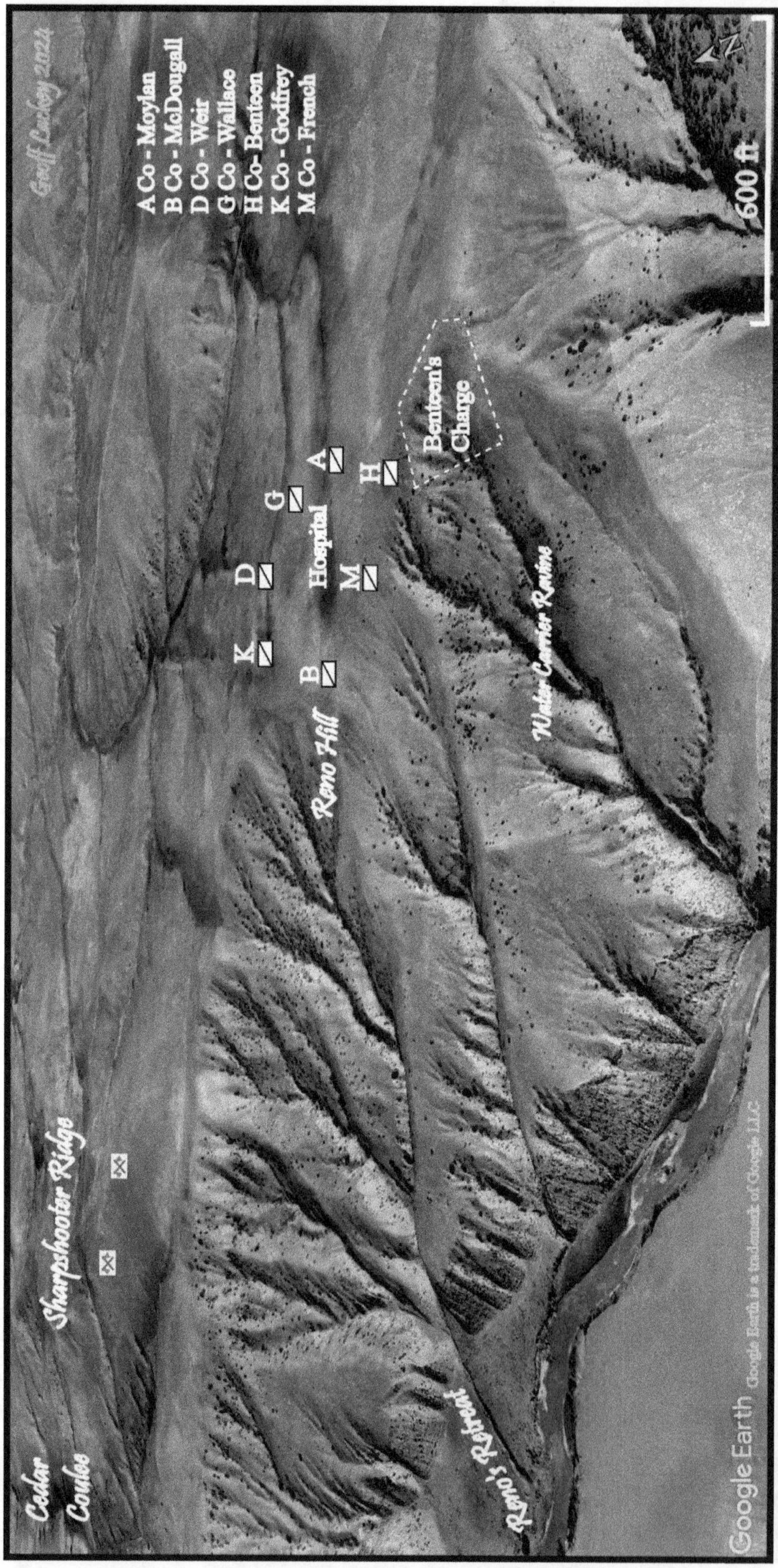

Map 26. Reno's troop deployments on Reno Hill on June 26th.

Chapter 15: What was the Last Stand?

The first time Custer gazed down into the valley at the huge encampment of Indians strung out along the banks of the Little Big Horn River, dreams of glory for himself and his beloved 7th seemed all but assured. Perhaps he recalled his dear friend, the stage actor Lawrence Barrett, gently chiding him, "Well old fellow, hard at work making history are you?" A ringing triumph here would gild his already laureled reputation, silence his critics, stop the chronic sniping at him by senior officers of his own command, and put an end to the attacks and mudslinging of his enemies who found him such an inviting target. This battle would be the final vindication in his long, drama-filled, roller-coaster military career, known for its dazzling peaks but also pockmarked with precipitous valleys.

Considering the massive size of the village he was about to attack, Custer was all too willing to accept the risks and dangers he had boldly courted all his life. Fortune might well favor the bold, but even for the boldest and most fortunate, a battle can seethe in a myriad of microscopic fluctuations where a minor overlooked detail, a defect in the command structure, unwarranted expectations, or the vagaries and vicissitudes of a rapidly changing situation could all conspire to upset the best-laid plans.

Negativity was a state of mind inhospitable to Custer's buoyant temperament. Right down to his incendiary core, Custer was an optimist, a winner, and an unwavering believer in his destiny. The General did not bring along reporter Mark Kellogg of the *Bismarck Tribune* because he needed the company of a fellow scribe. Kellogg was to archive for the world and unborn generations the triumph of the 7th Cavalry over the ferocious, hostile northern tribes that had caught the imagination of the civilized world. This was intoxicating wine for a spirit as expansive and adventurous as Custer's. No doubt he envisioned the loud, celebratory headlines of the nation's newspapers trumpeting his mythmaking victory over the warring Indian nations that, as chance would have it, would fortuitously come just in time for America's Centennial gala. Believing he was fated for great things, perhaps the timing of this victory would be no coincidence, but something providentially ordained. Looking down from Sharpshooter Ridge, smartly mounted on his horse Vic, Custer saw that all the glory he ever dreamt, desired, or hoped for was there for the taking in a sleepy village that, for the moment, appeared strikingly vulnerable and ready for annihilation. With three columns of cavalry converging on an unsuspecting encampment, what could possibly go wrong?

Two days later, astride his black horse, Appaloosa, Major Marcus Reno crossed the bluffs toward where eyewitnesses reported that Custer's five companies were heading. Arriving at the battlefield to witness what had once promised to be the biggest trophy of Custer's whirlwind, storied military career, Reno, instead, found an appalling scene of ruin and devastation. Profoundly unnerved by the carnage,

Reno would remember and record the gruesome images that haunted and overshadowed the rest of his life:

> The scene at Custer's battlefield was beyond description, and it filled us with horror and anguish. The dead had been mutilated in the most savage way, and they lay as they had fallen, scattered in the wildest confusion over the ground, in groups of two or three, or piled up in an indiscriminate mass of men and horses … Many of their skulls had been crushed in, eyes had been torn from their sockets, and hands, feet, arms, legs and noses had been wretched off. Many had their flesh cut in strips the entire length of their bodies, and there were others whose limbs were closely perforated with bullets, showing that torture had been inflicted while the wretched victims were still alive.

Bodies littered the field in contorted and grotesquely twisted mounds of white flesh. Some were now blackened. After being subjected to a pitiless blast of tribal rage, they had been stripped naked to lie blistered and bloated after more than two days under a scorching sun. More than a few of the troopers and burial crews were overcome, gagging at the overpowering stench. Some, like Lieutenant Varnum, wept uncontrollably. "Oh, how white they look," lamented a shocked and heartsick Captain Weir. One officer thought the sentiment poetic. But there was no poetry in the nauseating effluvia of death emanating from every pore of rotting human and horse flesh on the battlefield. At Calhoun Hill and Finley Ridge, slain soldiers lay prostrate in a heterogeneity of death, even as they were uniformly positioned in a skirmish line formation.

Searching among the human wreckage of their fallen brothers in arms, officers recognized at the southern end of the long high ridge, now known as Calhoun Hill, the bodies of Lieutenant James Calhoun and Second Lieutenant John Crittenden. Crittenden's father, General Thomas Crittenden, had used his rank and prestige to pull strings to obtain for his son a commission in Custer's command. Now, his naked and broken body was riddled with too many arrows to count, including one sticking out of his glass eye, the result of a hunting accident in his youth. Sunlight shimmered off the glass. Lying near him was the handsome Lieutenant Calhoun, Custer's brother-in-law and husband of his beloved sister Maggie, cleanly scalped, exposing the top of his skull, his features so marred that he had to be identified by the fillings in his teeth.

Two-thirds of a mile north of Calhoun Hill, on a small hill encompassing a seventy-five square-yard circle, lay the decaying bodies of forty-two soldiers and thirty-nine horses. Five officers, including Headquarters Staff, were bunched together, resembling a swept-up pile of debris hit by hurricane-force winds. Adjutant William Cooke, the tall, good-natured Canadian, was found among them with half his face skinned and just one of his much-talked-about dundrearies remaining. One scout noted that the hideous, gaping wounds were sickeningly filled with a host of "buzzing, quarreling blue flies" as if engaged in a noisy dispute over the booty of death. Descriptions of gutted corpses with detached, half-flattened heads, smashed by primitive stone hammers, filled the postmortem descriptions. Others recorded how private parts were stuffed into the soldiers' mouths and how the skin of the dead peeled off as if they were trimmings of overripe fruit. One soldier recalled the first body he buried in a shallow grave:

> All the bodies we heaped mounds of dirt over were mutilated more or less, mostly more. Will give a description of the first we buried: he was scalped; the skull was bare to the ears, the crown of the head was chopped out, his cap put into the cavity … there were 12 or 15 places where no doubt a spear or knife had been thrust to the hollow. Blunt arrows were driven in and left in the wounds.

Conspicuous among all the unredeemed carnage was Captain Tom Custer, who appeared a particular target of the warriors' venom. He lay face forward; a stone hammer had flattened his head to the point where its thickness was no greater than the width of the palm of one's hand. The metal point of an arrow buried in his head was bent like a fishhook from which no amount of effort could extricate. The muscles of his arms and legs were horribly slashed. A plethora of arrows protruded from his grossly disfigured body. He was castrated. His stomach was sliced open like a gutted antelope, with a deep cut that rode sharply up the torso, and another slash across the abdomen. Disemboweled, his entrails and intestines spilled out. A tattoo on one arm depicting the Goddess of Liberty and the American flag next to the initials TWC—standing for Thomas Ward Custer—enabled his identification. In contrast, there was the strange, incongruous sight of Private Weston Harrington, who was unmutilated and largely unmolested but wrapped in a blanket as a sign of respect.

In this circle of mayhem, within a thirty-foot diameter ring of dead sorrel horses, near a dead white horse was the now-forever-still figure of General George Armstrong Custer. His body lay just ten feet from his brother Tom, his shoulders resting on top of two other soldiers with only the small of his back touching the ground. Around him in this intimate circle of death were Sergeant Jack Vickory (aka Groesbeck), Sergeant Robert Hughes, Lieutenant William Cooke, Custer's Adjutant, and Lieutenant Algernon Smith from E Company. Custer's tilted head lay on his right arm as if he had been napping. He was less mutilated than others, although there was talk that some of his wounds were hushed up to spare the sensibilities of his grieving widow. Like most of the command, he was stripped bare, and perhaps because his hair was cut short before the campaign, he was not scalped. There were two bullet wounds. One had pierced his left breast near the heart, exiting through the ribs, possibly a post-mortem wound since there was little evidence of blood, while the probable killing shot plowed into his left temple.

Whatever else one might say about Custer's last moments, he did not leave this vale of tears quietly or peacefully. Scattered around and beneath his body were expended cartridges from his Remington rifle, giving ample testimony to that fact. Sergeant Ryan counted five or six shells, but closely inspecting the ground underneath him, another soldier counted seventeen cartridges from his hunting rifle. Blood had oozed down his face and congealed in the hairs of his mustache, but the sanguinary fluid had long dried in the torrid heat. A portion of one finger had been sliced off, perhaps to seize a ring, and one of his thighs had a deep slash, assuring that in the next life, he would be incapable of mounting a horse to wage war. An arrow had been jammed into his penis. A report that a trace of a smile creased his lips sounded like cheap romanticism, but at least one witness swore to it.

Tales of the storied dead never run in short supply. The Cheyenne woman Kate Big Head said she and other squaws recognized Custer and intervened to spare his body from further desecration. They told the others that he was a relative—the Indian camps had been full of gossip that Custer had fathered a child from Monahseetah, a young and comely Cheyenne—a blood tie with the tribal people he so unrelentingly fought against. Allegedly, a woman then jabbed a sewing awl deep into his ears, puncturing his eardrums. The Southern Cheyenne had once warned Custer that he would die if he ever attacked them. They now hoped that he would be able to hear them better in the next world.

In short, the sheer and utter devastation of Custer's five companies was beyond belief. So much for glory and honor to be won this day for the regiment.

Insatiably curious and famished for details, Captain Benteen embarked on his own journey to inspect the battlefield, still struggling to come to grips with the recent tragic news delivered by General Terry. He rode downstream with Captain Weir, Lieutenant De Rudio, Lieutenant Henry Nowlan, and Lieutenants Edgerly and Gibson. Following Custer's trail, he soon arrived at the harrowing scene of destruction. They first came across the body of First Sergeant James Butler lying near his dead horse. He was either trying to escape from the Indians or had been sent as another messenger to Benteen, pleading for help as their world was rapidly and irrevocably crashing down around them. Whatever the truth of Butler's demise, the expended shells from his rifle indicated that he, like much of the command, put up a stubborn and brave fight.

Now galloping northeast, the little party came upon the detritus of what was once Company I. Here, Lieutenant Nowlan gazed down at his good friend and drinking buddy Myles Keogh. His body lay amid a small apocalypse. They stared stonily at the cadaverous spectacle of their butchered comrades, wondering how things could have gone so dreadfully wrong. Moving northward, they espied an animated trooper by a flat crest on a hillock frantically waving Benteen in and pointing to a circle of dead soldiers and horses occupying a small perimeter around the top of the hill. Benteen trotted over into another scene of extermination. Focusing on one of the dead, he needed no one to point out or identify him. There are different accounts about precisely what Benteen said when he first viewed the body of Custer, against whom he had, for the better part of a decade, waged his own personal war—the most dramatic of which had Benteen saying, "There he is, damn him, well he won't fight anymore."

If these were not Benteen's exact words, they closely represented his sentiments. The acid-tongued Captain would never waver from his conclusion that the Custer fight "was a rout, a panic, till the last man was killed, there was no line formed...you can take a handful of corn and scatter it over the floor and make such lines, and there were none," he said derisively dismissing that there was any organized defense except around Last Stand Hill when the battle was all but over.

Benteen's hasty conclusion was his first attempt at a preemptive strike, putting his own spin on events before they could be defined and characterized by others or before any official investigation. As usual, there was some truth in what he said. On that portion of the field, from the middle of Keogh's Swale stretching to Last Stand Hill, the command looked in shocking disarray, lacking any credible defense. As Benteen saw it, the rout started at the river and spread to Finley Ridge, where the first markers on the battlefield locating the dead would be later erected. Over the next two decades, he would painstakingly paint for his rapt listeners and readers the unmistakable impression that the Custer fight was over in thirty minutes and that there was nothing he, nor anyone else, could have done to avoid this tragic denouement once Custer had crossed what would be his Rubicon, Medicine Tail Coulee, inexplicably and senselessly moving away from his reinforcements.

But Benteen was not telling the whole story. However, it was a quick-witted ploy, an anticipatory feint, to establish an irrefutable narrative that Custer's annihilation was entirely brought on by Custer's impulsiveness and recklessness. Endorsing Benteen's interpretation, Reno claimed that Custer's command had already been destroyed by the time Reno's command bolted from the timber. Further, Benteen and Reno agreed that there was only one author of this epic and monumental defeat: George Armstrong Custer.

When pushed on the timeline, however, Benteen would later concede that he could have reached Custer but quickly added that had he done so, "they would have been there yet"—a view that Lieutenant Edgerly readily seconded. In other words, Benteen going to Custer would have destroyed the entire regiment, not just Custer's five companies. While such an eventuality could not be categorically ruled out, considering the number of hostiles and their unprecedented aggression, there is still no avoiding the fact that Benteen and Reno could have ridden toward the sound of the guns, perhaps an hour or more earlier than they did. Had they acted as Custer typically would, who could say with any authority what the battle's outcome would have been? There are no rules in counterfactual history.

Still, the provocative question remained. Even without receiving reinforcements, how could Custer have been defeated and wiped out to the last man by an undisciplined, essentially leaderless fighting force lacking military science, untutored in the conventional tactical art of maneuvering or maximizing the employment of forces in combat? Custer, it was agreed, should have at least survived and held out, even if he had just three companies, much less five. He had been largely heralded as the finest U.S. cavalry commander of arguably the finest cavalry regiment in the army.

But with no survivors to tell the tale, theories and speculations filled gaps where facts were few. No matter how conscientious and impartial the researcher is, any reconstruction of the events occurring east of the river must be approached with the utmost caution and humility. This was as true then as it is today, a century and a half later. As the author James Donovan said in his critically acclaimed *A Terrible Glory*, perhaps the best overall treatment of the saga of the Little Big Horn, what happened to Custer's battalion after trumpeter John Martin galloped away with a message from his commander must always remain an open question. "We will never know," says Donovan, "without a reasonable doubt, why Custer … and his 210 men were so totally and one-sidedly defeated."

It is a sober conclusion and a fair one. But it must also be acknowledged that any historian, scholar, or interested student does not enter this thorny thicket blind and unarmed. They are tutored by the tactics employed by mounted cavalry in the 19th century and the strategy Custer utilized in previous battles on the Plains. The archeological and forensic evidence amassed from the landscape after the dramatic and sweeping 1983 fire denuded the battlefield and revealed in its wake an arsenal of artifacts such as shells and arrowheads that have enabled at least a tentative and crude reconstruction of events.

Now unmuzzled, history began to speak and was abetted by a careful and scholarly examination of the copious Indian accounts, many recorded more than half a century after the fight by participants who fought or witnessed the battle. Over time, some of the long-segregated dots have been connected, and loose ends have been tied in with what has most likely been, and almost certainly remains, the single greatest mystery in United States military warfare. Though the answers to some puzzling questions will forever remain hidden, much has been learned about the events and circumstances that eventually led to Custer's Last Stand.

After having saved Reno by a feint at Ford B, Custer was faced with snatching victory from defeat. The ploy down Medicine Tail Coulee near the river's edge had the desired effect of drawing almost the entire weight of the enemy off Reno and toward Custer, presumably giving Reno time to regroup and join up with Benteen, who had been given a direct order to move to the valley with all dispatch: "Benteen, Come on. Big village. Be quick. Bring packs" is still a cry that reverberates loudly off the cavernous walls of history's echo chamber.

Custer's demonstration down toward the river concentrated almost all the Sioux and Cheyenne warriors on his five companies. Despite the ominous events, Custer's forces were still relatively secure and unthreatened. Stalemate, however, was not the General's objective—victory was. Reno's headlong retreat forced Custer to fold the winning hand he recently held. With circumstances radically changed, Custer had to plan anew. With support from his two other battalions still not on the field with him, his regiment could prevail. But neither time nor tide was on Custer's side. After regrouping on Calhoun Hill, he could see the warriors accumulating to his south in a depressed area called Deep Coulee that leads northeast from the river. That had to be addressed before he decided what to do next.

Retreating to Benteen and Reno was out of the question. A withdrawal would be an admission of the campaign's failure and unrecoverable loss of initiative, allowing the enemy to escape again and fight another day. Undoubtedly, a more cautious commander would have opted for that safer alternative. But Custer was a cavalier whose very name invoked audacity, always audacity. Retreating rarely wins battles, and with the element of surprise gone, Custer had to rely on tactics and join with his anticipated reinforcements to reverse the momentum that had shifted against him. Still, he had reason to hope. Despite the frequent flare-ups, Custer knew that Benteen was a first-class combat officer. Had Captain Benteen not come through for him at the Battle of Washita? There was no reason to suppose he would not respond to the repeated volley firing, much less ignore orders to come quick and bring the ammunition packs.

Of course, Custer still had much to consider. What if Benteen did not arrive? He could not just plant himself on Calhoun Hill and wait for General Terry. Physical land features had to be considered, and for five companies of cavalry with limited ammunition to linger indefinitely on Calhoun Hill made no sense. The terrain was inextricably wedded to battle tactics and had been since the time of the Assyrians. All the great generals, from Alexander to Julius Caesar to Hannibal to Napoleon, understood this and wisely used the ground to their best advantage in winning battles. Custer knew that mounted cavalry does not optimally operate from a fixed position. Speed, deft maneuvering, and being on the offensive make the cavalry a formidable force in war. Calhoun Hill was too small for three companies to man a defensive position, much less five. Most critically, Calhoun Hill had no access to water. Given the rising mercury, dust, and frenetic pace of battle, thirst could become more of a mortal enemy than bullets.

Mitch Boyer, who had been over this area before, may have informed Custer about the defensive advantages of the Northern Flats nearly two miles northwest of Calhoun Hill. That would be the most practical place if reinforcements failed to arrive for them to dig in and hold out. Even if such assurances were given, Custer never entirely relied on others, even those of a scout with instincts and knowledge like Boyer. As Regimental Commander, Custer would have wanted to personally assess the defensive possibilities without forfeiting the chance of a new offensive, and, in concert with his other battalions, enveloping and defeating the enemy.

Leaving Keogh at Calhoun Hill to sight and support Benteen's expected arrival serves both defensive and offensive purposes (see Map 27). Had Custer been able to foretell that he would only have his immediate command to rely on, he almost certainly would not have separated from Keogh and deployed his limited manpower over such a vast expanse of terrain against a foe of such strength fighting with uncharacteristic aggression. He would have then been forced to either fight his way back over the bluffs to join forces with Benteen and Reno or move to the Northern Flats to await General Terry's expected arrival. Custer was trying to serve two masters: one being offensive in waiting for Benteen's arrival to strike the enemy, and the other defensive in scouting out the Northern Flats to survive without reinforcements (see Map 28). Serving this double mindset would be one of the primary reasons for Custer's disastrous defeat.

There is general agreement that Custer rode north from Calhoun Hill along Battle Ridge with Yates and Companies E and F. His reason for doing so has been the source of much controversy and theorizing. As discussed previously, the most common explanation is that Custer was attempting to seize a large number of captives in the valley near Ford D. The following examination of his movements and actions will illustrate why such would not have been Custer's goal and unveil his true objective. Custer and his men did approach the families near Ford D, but it had nothing to do with taking prisoners.

Instead, Custer would move north for better terrain, access to water, tactical advantage, or to make a junction with Terry. After Custer ascended Calhoun Hill, a number of possibilities confronted him. The warriors below in Deep Coulee were in a perfect location to be crushed between the men on the hill and the other half of the regiment when they arrived from the south. But what if Benteen and Reno did not come, or the warriors moved out from the bottom of that deep gully? If either happened, the command would have to move. Calhoun Hill was a rock-solid defensive position, towering up to two hundred feet over the vast stretches of open ground it commanded. But for all its virtues, as stated earlier, it lacked water and was not large enough to hold the entire regiment. Whether Reno and Benteen joined them or not, Custer had to move.

For the second time that afternoon, Mitch Boyer may have assumed a critical role. While Custer and the 7th did not know this valley, Boyer did. He may have advised Custer that if the regiment had to dig in, he knew of good ground that could serve the entire 7th. This position was about two miles away and had access to water. If push came to shove, they could hole up safely as they awaited Terry's arrival.

Custer had not given up on victory, and his actions indicate that he was still very much on the offensive. Calhoun Hill was a secure waystation for the command to reunite, hopeful that the rest of the 7th would advance from the south. Most warriors were afoot in Deep Coulee, between the margins of Greasy Grass Ridge and Crazy Horse Ridge. No braves charged about on war ponies and none crept up on the soldiers—another popular myth—through half a mile or more of grass. Many warriors carried short-range weapons, suitable for around 150 to 200 yards in capable hands, and many of the rifles lacked the mechanical elevation in their sights to allow aimed shots beyond that. The cavalry carbines had an effective range for about three hundred yards or more, and aimed volley-fire made it incredibly dangerous for a warrior to linger in the open.

The situation on Calhoun Hill was apparently secure as the participants reported hearing at this time only intermittent long-range firing. The soldiers stayed on the hill because Custer left with E and F Companies, something he would never have done if the position on Calhoun Hill was in jeopardy, or if he had known that Reno and Benteen were not coming. In that case, then all five companies would have rejoined the rest of the 7th somewhere south of Weir Point or ridden northward to select a spot to hole up or wait for Terry.

Regardless of what Reno and Benteen might do, Custer's five companies eventually had to leave Calhoun Hill. Whether to mount a new attack with the reunited regiment or find a place to weather the storm, Custer needed to see what was to the northwest. So, he set out along Battle Ridge with two companies as a reconnaissance in force. He left C, I, and L Companies under the command of Captain Keogh on Calhoun Hill, where they could see and cover the long-awaited advance of Reno and Benteen from the south. Should that support be too long in coming or the situation become untenable, at his own discretion, Keogh would be expected to remount his companies and rejoin Custer to the north.

As Custer advanced, he deployed a platoon from E Company on Battle Ridge Extension. They would later move back a distance to a high knoll at the Battle Ridge terminus, now known as Last Stand Hill. This deployment provided a clear line of sight between Keogh and Last Stand Hill, as well as between Last Stand Hill and Custer. Though physically spread out, no portion of the command was visually isolated. The hillock allowed the first platoon of E Company under Lieutenant Algernon Smith to guard the northern approach to Horse-Holder Ravine and Keogh's Swale below Calhoun Hill and cover the high ground east of Cemetery Ridge. An easily observable location, Custer could fall back there should the need arise. But it was never meant to be a serious defensive redoubt; it was to provide a physical and visual link between the command's two squadrons and to facilitate the unobstructed movement of one to the other.

Leaving Last Stand Hill, Custer shifted west into a broad coulee and followed it down to the river and Ford D-1, then continued farther on to the Northern Flats, which lay between Ford D-1 and D-2 along a timbered stretch of the river, which later was called the Willy Bends place. According to John Stands In Timber, as related to him by his grandfather Wolf Tooth, his grandfather led a small band of warriors who trailed and sniped at them from small ridges to the north for the rest of the journey. It appeared to the Indians that the soldiers being led by Custer were heading to the river bottom by Ford D, near the women and children on the other side.

This reputed trajectory proposed by some battle students does not prove that Custer intended to cross the river at Ford D to seize hostages. While it is true that Indian prisoners told Lieutenant Philo Clark about Custer's route, even drawing a map on the floor of Clark's room to show soldiers heading north beyond Battle Ridge, this still does not demonstrate that the river crossing was their destination. It was another case of the warriors misinterpreting what was happening on the battlefield, as they had at several important junctures of the fight.

Arriving at or around 3:45 p.m. Headquarters time, Custer evaluated the ground for defense and then turned back east, swinging southeast of Ford D-1 and making his way to flat ground seated on the lower portion of Cemetery Ridge. This probably appeared to Custer as the best defensive ground near Last Stand Hill. Then, Custer moved to the southern edge of Cemetery Ridge, placing himself closer to Keogh's position on Calhoun Hill. Perhaps he had concluded that this second area was the better tactical choice. The Cemetery Flats was only four-fifths of a mile from Calhoun Hill, straight across Harrington Ridge and Deep Ravine, less than half the distance Keogh would have to travel to reach the Northern Flats between the two D fords. The warriors continued to gather in strength, and Smith's single platoon on Last Stand Hill could only hold that critical junction for so long.

The Flats below Cemetery Ridge might have been a little less advantageous than the Northern Flats, but the location had its advantages. The most obvious was that it was closer to Keogh, which would behoove the Reno and Benteen columns if they successfully joined with Keogh's forces on Calhoun Hill. It also allowed Custer to support the platoon on Last Stand Hill, if needed, while holding that position open for the passage of the rest of the command. Whether to dig in for defense or use it as a rallying point for a renewed offensive, its location was more readily accessible to the rest of the command.

The topography of the Cemetery Flats made it a sound choice. It was not the highest ground, but the cavalry carbines far outranged the weapons carried by the great majority of the warriors, giving the soldiers a significant advantage over any would-be snipers seeking to fire at them from the top of the ridge. Most guns possessed by the warriors lacked the range to reach the Cemetery Flats from any direction. And while it does not command the surrounding country by towering over it like Calhoun Hill, any warriors seeking to charge the position would have to do so by crossing many hundreds of yards into the aimed and disciplined fire of nearly half the 7th without themselves being able to engage, an extremely costly proposition that the always casualty-averse bands were extremely unlikely to try. Again, massed charges against ready defenders were not part of the Plains Indian fighting style. Further, the Cemetery Flats' elevation dominated the enemy from across the low, wide basin to its south.

To the west, bluffs along the river prevented any attack from that direction and sheltered the Cemetery Flats from sight and fire from the valley. A low ravine passes through the center of the position, deep and broad enough to provide shelter for the horses, any wounded, and any soldiers not guarding the line. No other ravines or coulees in the area afforded shelter to warriors seeking to close in on that position. Additionally, the ravine by the lower Cemetery Flats that led to the river gave Custer's command an avenue, a lifeline, to water. While largely unimportant when fighting a conventional military force, these topographical features become critical when opposing an enemy using irregular and guerrilla-like tactics, as they would facilitate those tactics. The warriors would crawl and crouch in them, using every fold of ground for protection as they closed in on their target.

Having found his ground, Custer deployed the remaining E Company platoon under Lieutenant Sturgis to the southern edge of Cemetery Ridge to oversee The Basin area in that direction. F Company was divided into its two platoons, with one platoon staying near Custer and the Headquarters detachment, while the second platoon deployed along the northern edge of Cemetery Ridge. Here, Custer would wait for Keogh and, hopefully, Reno and Benteen to arrive (see map 29).

It is probable that by this time, Keogh had sent a messenger to inform Custer that the long-awaited reinforcements were finally in sight on the Weir Peaks (i.e., D and M Companies, *see* Chapter 11). Being so informed, Custer likely sent a message to Keogh instructing him to advance to the Cemetery Flats position as soon as the reinforcements and ammunition packs arrived. And if, for some reason, that was not forthcoming, he should disengage from Calhoun Hill and move to the Cemetery Flats with his three companies if practicable. Supporting this scenario are the observations of Lieutenant Edgerly, who, after studying the location of the bodies in Keogh's command, concluded that the men had been in the process of just such a withdrawal before disaster struck.

If Benteen and Reno had failed to arrive on Battle Ridge, Keogh would rejoin with Custer on the Cemetery Flats, whereupon the entire command would then likely have moved down to the Northern Flats to dig in and hole up near what was later termed the Willy Bends place. There, they would await the arrival of General Terry, who was scheduled to arrive the next morning at the junction of the Big and Little Horn Rivers just fifteen miles to the north.

It was a sound plan. It was also going to fail.

A secondary reason for the theory's persistence that Custer moved north to take the families captive is that the Indians insisted such was the case. And from their point of view, it made sense. While European armies focused on destroying other armies, the slash-and-burn raids that made up so much of Plains Indian warfare were very different. Women and children were every bit as legitimate a target as were warriors. For just one example, until Custer learned and put a halt to it, the Osage scouts at Washita had set about killing the women and children in revenge for what the Cheyenne had done to them many times in the past. Plundering an enemy village and taking captives, or worse, was a regular part of tribal warfare.

Aside from the small group of warriors with Wolf Tooth, Custer's reconnaissance to the Northern Flats had gone largely unnoticed by the warriors. A minor clash occurred near Ford D-1, where warriors fired at the soldiers by the brush near the river, whereupon the soldiers moved back to where the Little Bighorn Battlefield National Monument cemetery now is. The body of newspaperman Mark Kellogg, who had been riding a mule, was an unfortunate casualty. His body was found later about a hundred yards from the river. Custer's arrival so far north astonished the warriors, most of whom had been wholly focused on the soldiers atop Calhoun Hill. For the third time that day, a heretofore unknown body of cavalry had materialized where none had been expected—the initial attack, Custer's appearance at Medicine Tail Coulee Ford, and now here at Ford D. And to these warriors, the natural conclusion was that an attack was being made on the women and children, which fueled their ferocity. Charles Corn, an Oglala, said it succinctly, "You tried to get our children and wives, so I was willing to die for them that day." Roughly a couple of thousand other warriors felt the same way. When threatened that day by U.S. Cavalry forces, the warriors typically fell back to establish a covering force for the non-combatants to escape; however, following Reno's precipitate rout from the valley, the warriors' confidence and combativeness grew in confronting Custer's troops on Battle Ridge.

Custer found himself waiting on the Cemetery Flats for far longer than he ever intended, perhaps up to thirty minutes, and within view of the perplexed Indians in the valley below. Years later, John Stands In Timber would recall some elderly warriors joking that Custer was surveying the women across the river, wondering which to pick to keep him warm that night. But perplexed or not, the number of warriors continued to grow in the surrounding areas of the Cemetery Flats. Their presence could not have gone unnoticed by Custer, but for the time being, he had no choice but to hold this ground. It made no sense to surrender this stronger position to the warriors and fall back to the weaker, waterless position held by Keogh on the far end of Battle Ridge. So, Custer deployed his remaining squadron to their best advantage, using the long-range punch of the carbines to keep the warriors at bay, who, at this time, were not pressing the attack. But not far from the Cemetery Flats, the situation on Calhoun Hill and Battle Ridge was a different story, as it was under heavy siege that would undermine their defenses. Soon, the relatively secure positions of both Keogh and Custer were about to implode.

At the sight of D and M Companies on the Weir Peaks, Keogh concluded that the long-awaited reinforcements were finally coming. (*See* Chapter 12.) A platoon from C Company was deployed out onto Finley Ridge to provide cover for the anticipated advance, not to drive encroaching warriors from the death trap of naked ground that was Calhoun Coulee, as so many propose. It would also be expected that this forward position would result in the warriors in Deep Coulee being crushed between the soldiers on Finley Ridge and Calhoun Hill with the companies arriving from the south, forcing, if not their destruction, at least a rapid retreat to the river, perhaps even chased by riders from Keogh sweeping down upon them from Calhoun Hill.

But the reinforcements did not advance, and the platoon on Finley Ridge found itself overextended and isolated with warriors to its front in Deep Coulee and the rear in the area of Greasy Grass Ridge. In what was probably a desperate attempt to reach the soldiers on Weir Peaks, it seems Sergeant Butler was dispatched from Calhoun Hill, perhaps as a messenger to the platoon on Finley Ridge, and thence onwards to the soldiers on the Peaks. His body would be discovered near Blummer Ridge on the 27th by Benteen, Weir, and Nowlan as they rode to inspect the Custer battlefield.

Around this time, Corporal John Foley of C Company, with the C Company guidon tucked in his shirt, and perhaps riding with Sergeant Butler, attempted to break through and reach the Peaks, possibly as a messenger. Spurring his horse across Deep Coulee, he ascended Butler Ridge and had ridden clear of the warriors giving chase. There, in what might be among the saddest and most inexplicable events witnessed by the battle's survivors, Foley shot himself in the head with his revolver. The revolver may have been held in the same hand with which he was whipping his horse with the reins, and it will forever be an open question of whether it was a case of panicked suicide or a terrible accident.

Meanwhile, the C Company platoon on Finley Ridge remained isolated as the disaster on Calhoun Hill played out, and the soldiers on Weir Peaks could only look on as Keogh's last stand in the swale concluded. The surviving troopers of Keogh's battalion on Battle Ridge now fought and ran its length to its last island of possible refuge, Last Stand Hill.

The sudden eruption of sustained firing in the Keogh sector did not go unnoticed on the Cemetery Flats. Though Custer could not see what was happening at Calhoun Hill, something had gone very wrong. The twenty-plus C Company troopers on Finley Ridge had been visible to Custer for some time. There is a straight line of sight between the Cemetery Flats and Finley Ridge, and it is only about three-quarters of a mile away. Their presence would have been expected as a covering force when Reno and Benteen advanced up from the south, but now they were suddenly being attacked from at least two directions. With Yates and the First Platoon of F Company, Custer moved south into The Basin and began to make his way to support the men on Finley Ridge and hopefully reunite his disjointed command. A small detachment from the Second Platoon of F Company would dismount and deploy on the low Basin Ridge that overlooked the lower reaches of Calhoun Coulee and Deep Ravine, protecting E Company's held mounts from threats from that direction (see Map 30).

No other reason exists for Custer to leave the defensibility of the Cemetery Flats and descend into The Basin unless it was to go to the aid of the soldiers now seen isolated on Finley Ridge. Indian battle veterans related to John Stands In Timber that such a move was made. The Cemetery Flats was under no serious pressure when Custer entered The Basin, just like Calhoun Hill was under no threat when he moved north. Clearly, he would not have abandoned either position if it were under attack. While Custer moved south, preparations were underway on Last Stand Hill for Lieutenant Smith's first platoon of E Company, along with the arriving refugees from Keogh's Swale, to move to the safety of the Cemetery Flats on the lower western flank of Cemetery Ridge to rejoin the Yates squadron there. Their horses and all the spare ammunition were being held in a sheltered low spot on the western side of the hill, below the crest, where they could be best guarded from attack.

Chapter 15: What was the Last Stand?

If anything was certain in Plains Indian warfare, the warriors would go for the horses—they always went for the horses. Knowing this, Custer took appropriate precautions. The dismounted F Company detachment (about ten men) held their horses individually, not with horse-holders in sets of fours, on the Basin Ridge as the rest of the F Company moved onto Harrington Ridge on their way south toward Finley Ridge. With no intention of abandoning the good ground on the lower Cemetery Flats, Custer had Lieutenant Sturgis's second platoon of E Company deployed on the southern edge of the ridge, with their mounts sheltered just below in The Basin. The situation was far from ideal in the Custer sector, but barring some bolt from the blue, it was relatively under control. The E Company troopers on the southern end of Cemetery Ridge held a strong position, and their carbines kept the gathering warriors out of engagement range and would do so as long as the ammunition held out. As illustrated on Map 30, attempting to cross those long stretches of open ground to attack the soldiers deployed on Cemetery Ridge would be tantamount to a death wish. Unfortunately for Custer and his men, a group of young warriors was intent on making just such a sacrifice.

Having secured the area and the mounts, Custer crossed The Basin between Cemetery and Deep Ravines and ascended Harrington Ridge, heading to the beleaguered C Company soldiers still holding out on Finley Ridge. He had not gone far when all hell broke loose to his rear, and events took a stunning turn for the worse. According to John Stands In Timber, in the aftermath of the battle with Crook at the Rosebud a week earlier, a group of young Sioux and Cheyenne took an oath to die fighting the soldiers when they were next encountered. Such oaths were far from the norm but not unheard of, and among tribal members, such a commitment came with a major boost in personal prestige.

Their communal pledge stiffened their spines to make a charge at Custer's soldiers. This took great courage, since much of the soldiers' tactical formations would have been difficult or impossible to observe from their location in a coulee off to the north. The disposition of the soldiers' mounts in The Basin below Cemetery Ridge certainly was hidden from their sight. Nevertheless, they were about to deliver the death blow to these soldiers.

Whooping and howling, these Suicide Boys charged out of the coulee Custer had earlier taken on his reconnaissance to the north and swept over the upper neck of Cemetery Ridge. Turning this way and that way from the fire they were surely taking from the troopers on Last Stand Hill, the Suicide Boys veered to the right, charging directly into the path of the held horses of E Company secured in The Basin, driving them directly into the held horses of the F Company detachment deployed on the Basin Ridge, causing confusion and pandemonium among the troopers holding them. Some of the young warriors drove the stampeded horses of E and F Companies down to Deep Ravine and Cemetery Ravine. The rest of the Suicide Boys swung about and charged directly toward Last Stand Hill and the position of E Company's first platoon under the command of Lieutenant Algernon Smith and the survivors of Keogh's battalion. This unexpected thrust caught the second platoon of E Company, deployed dismounted on Cemetery Ridge, flat-footed as they helplessly watched their horses and ammunition being driven toward the river. Consequently, F Company's dismounted detachment on The Basin Ridge, along with E Company's platoon under Lieutenant Sturgis, would then turn and race down into The Basin following their mounts toward Deep Ravine in a desperate attempt to recover their horses carrying their spare ammunition (see Map 31).

Based on the body locations of these troopers descending from higher ground, the dead were mainly from E Company, along with about ten or so troopers from F Company and a few troopers from the Keogh battalion, who may have previously moved from Last Stand Hill down to Cemetery Ridge. It was probably here that Mitch Boyer, Custer's most able scout and guide, was killed. Seeing the ensuing chaos enveloping Last Stand Hill, Custer and Yates arrested their movement toward Finley Ridge, and although still not under heavy attack, Yates's first platoon of F Company and Custer's Headquarters staff immediately rode toward the beleaguered soldiers on Last Stand Hill being besieged by the Suicide Boys who made a valorous charge but paid a steep price, as most of them were shot down by heavy gunfire coming from Last Stand Hill. The charge of the Suicide Boys broke the backbone of the soldiers' defense and stampeded their horses carrying their spare ammunition and with it any hopes of survival. The Suicide Boys' brazen attack and the gaping hole it left in the already battered battalion's defensive line left no doubt that Custer's command was rapidly spiraling to its doom. Custer was prophetic in once stating, "I would rather fight a veteran Indian than boys of fourteen years because the latter are fearless and are without the caution of old braves; consequently, they are dangerous."

After many of the horses were run off, the soldiers gravitated to the best defensive position around, finding themselves marooned on what would become the most storied hill in American military history, where they joined E Company's platoon for a last-ditch defense. Other than attempting to escape with Yates to the north, Custer had no choice but to have Yates's first platoon and Headquarters staff gallop back to Last Stand Hill to help hold that position and rally whatever support they could find. Custer's final movement from The Basin area back to Last Stand Hill was rapid, with little resistance against him and with few, if any, casualties. That is the best that can be said for what was a hopelessly disintegrating command and battle situation.

Despite its grave defensive liabilities, Last Stand Hill became a final rallying point to the leftover remnants of Keogh's shattered command as they haphazardly floated over a sea of grass like driftwood from a shipwreck. Barely more than a dozen survivors from what was once three solid companies struggled through a melee of hand-to-hand combat across Battle Ridge to Last Stand Hill. Only covering fire from the hill permitted these troopers to survive a swarming Indian attack that pummeled them every step of the way. Unfortunately for the soldiers who sought refuge there, this hill was neither steep nor high enough to offer any real protection. Still, there was no other recourse since the cavalry had lost its most valuable asset, mobility.

On Last Stand Hill, with one glance, Custer must have read how hopelessly desperate matters had become. Like a doctor checking his own vital signs, he saw his wing, or what was left of it, flatlining into a lethal descent. Defensive, disciplined lines of fire and offensive thrusts that could hold off large numbers of Indians were no longer possible. After ten years of pursuing Plains warriors, Custer, the hunter, was now the hunted. With Benteen and Reno's reinforcements nowhere in sight and General Terry's command still too far north to be of any help, the man the Crow called Son of the Morning Star had run out of options, and that quixotic "Custer's Luck" that had charmed his wartime career was now gone with the wind.

In this terminal phase, Two Moons recalled that a lone trumpeter would forlornly blow out commands. No one knows what those notes signaled or what orders the trumpeter was trying to communicate, but as the resounding tones rose over the deafening din of battle, it might as well have been a dirge—or the sounding of "Taps." What had been a proud military outfit on the precipice of its greatest victory had turned into a tragic spectacle of a sputtering and dying command.

The warriors had breech-loading and repeating rifles at their disposal, and many of these weapons, along with their bows and arrows, were directed at the beleaguered but still fighting command. A Hunkpapa woman, Moving Robe, remembered how the soldiers were trapped on Last Stand Hill, with the Cheyenne attacking from the north and the Sioux from the south. Bravely making her way to the closing scene, Kate Big Head said there were hundreds of warriors for every white soldier left alive—an exaggeration for sure, but it underscores the hopeless odds.

With every passing minute, more warriors were being drawn into the fatal vortex. They crossed the stream at various points, getting between Custer and the river, encroaching upon his rear. The terrain around Custer was amenable to guerrilla warfare, giving warriors cover to take potshots at their hopelessly surrounded prey. According toOglala Chief Lone Bear, the soldiers made a stand on top of the hill, no more than a small knoll, where the Little Bighorn Battlefield National Monument memorial—the very centerpiece that now commemorates the battle—stands today. As one author plausibly surmised, Custer probably placed his best marksmen at the top of that knoll to make every shot count. With the bell tolling the final knell, little else could be done but to die as a soldier.

With the small command sheltered alongside the horses on the west side of the hill, one might imagine an overwhelmed George Lord, the regimental surgeon, doing his best to attend to the wounded or, at this point, maybe even shouldering a gun. Wolf Tooth's band and others made their way behind the ridge line to the east, firing up at Custer's troops. There was heavy fire by the warriors on both sides of Last Stand Hill. Custer ordered the horses shot, six sorrels from the survivors of C Company. A breastwork of horseflesh was constructed behind which Headquarters Staff fired at the enemy. In *My Life on the Plains,* Custer wrote, "There is nothing Indians dislike more in warfare than to attack a foe, however weak, behind a breastwork of any kind."

But more than breastwork was needed to stem the angry red tide irrepressibly encroaching upon them from every angle. Limiting casualties was always paramount among tribal warriors, and they were not about to forego that caution now. They took their time, methodically positioning themselves to better advantage, crawling in gullies, firing from behind sagebrush and hillocks in a deadly, ever-tightening circle. The grass was not as thick and high as it is today, so the warriors, as some battle students mistakenly believe, could not use it as cover. Because of today's ash-infused soil, the grass grows more robustly than it did in 1876. Besides, the black powder emitted from their guns would have immediately disclosed the warriors' locations to the troopers. Instead, when they got close enough to the hill, about one hundred yards away, they shot arrows into the air, traveling in high, long arcs. They fell upon the soldiers and thudded into the flesh of frightened, screeching horses. Picking off the bluecoats one by one, they waited for the shooting to die down before exposing themselves for the final killing rush.

Most of the mounted warriors circled at a distance, beyond the range of the soldiers' carbines. That only a few Indian ponies were later found dead on the battlefield is evidence that most of the warriors around Last Stand Hill fought on foot. The number of warriors circling the hill rapidly swelled to 1,000 … 1,500 … perhaps even 2,000. No one was counting. Though the soldiers managed to hit some hostiles, including foiling a charge on the east side, it was not nearly enough to stop the inevitable rush of warriors.

At 5 p.m., whatever fragment of hope remained alive on a hill hardly worthy of its name soon shriveled up like an unwatered flower. In the claustrophobic madness of a death fight, the soldiers defending this modest elevation uttered shouts of alarm and curses merged with the fearsome rapports of firearms. But none of it could drown out the hideous, savage cries of the painted warriors implacably converging upon them. Swirling around Last Stand Hill, as one warrior remembered, was a violent sweep of dust and smoke, making everything as black as evening. As remorseless as this extermination scene was, in some tortured but poignant sense, as is true of all last stands, there was also something noble and inspiring about it. The overwhelming testimony from the Indian accounts is that the soldiers fought bravely and obstinately. In the engulfing cacophony of a darkening day seemingly gone amok in a melee of unremitting violence, a prolonged and pregnant pause lingered, as if both sides took one last breath before the final, closing, bloody act.

Wooden Leg remembered that the soldiers' firing had ceased. Perhaps they were all dead; perhaps they were making a last effort to reload; perhaps they were out of ammunition. The Indians could only guess. Unbearable tension. Then all at once, everything unfurled into an explosive climax. The warriors propelled themselves forward in one mighty surge, racing desperately toward the hill before the soldiers could fire upon them. Custer's defenses fell like the spume of a wave. All Red Hawk could recall was "an Indian yell, bone whistles, and a stampede …" The troopers on Last Stand Hill had been last seen focusing on repelling an attack from the west. Instead, they were hit from behind by mounted warriors east of Last Stand Hill, and then, like storm winds whipping up on every side, they were deluged by a massive crush of warriors as the fighting degenerated into a dogfight of tangled and twisted bodies, which ended in the death of every trooper as the enraged warriors vented upon the dead and the dying.

Finding himself in the middle of the uproar, Standing Bear thought it was like being in a bad dream as the last of the fighting plunged into a deeply personal business with knives, tomahawks, hatchets, and stone hammers doing the killing and overkilling. Tall Bull said that once the Indians got to the top of the hill, the soldiers were quickly killed. How long did the fight on Last Stand Hill last? Gall said it took about as long as it takes a hungry man to devour his lunch.

The hemorrhaging did not stop at the end of the battle. Following in the wake of the warriors were thousands of non-combatants that swarmed over the soldiers like locusts in an orgy of destruction and barbarity. The women and children thirsted to avenge their dead in a mayhem of retribution and degradation. Several soldiers, more dead than alive, were apparently dragged into the village and tortured. The scout George Herendeen said that four decapitated heads from the regiment were found in the Indian encampment. One, according to Lieutenant Godfrey, was the head of young Lieutenant James Sturgis.

At or just before this warrior avalanche upon Last Stand Hill, Custer was mortally wounded in the left temple. Every indication suggests that before the fatal shot, he was conscious and fought to the end, as seventeen cartridges from his gun were found nearby. Some later speculated that Custer had saved the last bullet for himself and that the fatal shot rifled into his left temple came from his own hand. Such a fate was often considered preferable to falling into the hands of the Plains warriors. But from what is known about Custer's personal history and the physical evidence surrounding his demise, death by suicide is very unlikely. A second bullet lodged in his breast may have been inflicted post-mortem.

In 1869, while serving in the Texas panhandle during a campaign to bring Kiowa and Cheyennes back onto the reservation, Custer rode into a Cheyenne village to negotiate with Chief Stone Forehead for the release of two white women. It was a daring and courageous thing to do, and his life hung in the balance. Custer was offered a derringer by Benteen (of all people) so that if something went awry, he could end his own life rather than suffer the alternative. But he refused, presumably because his inner code of conduct would not permit it, perhaps thinking that dying by one's own hand was either dishonorable or unmanly, whatever the circumstances.

Physical evidence for suicide is also lacking. Lieutenant Godfrey and Private Jacob Adams examined the General's body, and both emphatically stated that there were no powder burns, which would have been present if Custer had fired a gun point-blank at his head. In the 1870s, the army used black powder loads in their guns. A self-inflicted gunshot wound would have left his face smudged with black powder burns, but there was no evidence of this on Custer's unmarred countenance. Nor was typical physical damage found to indicate the General had administered the final bullet at close range. Besides, Custer was right-handed, and he was shot through the left temple, which would have been an unlikely entry point if the wound were self-inflicted.

We can also discount the story of Tom Custer seeing his idolized older brother mortally wounded and rushing in to end his suffering, which seems more than a little melodramatic. Powder burns and physical damage from a bullet fired by his brother at close range would have been just as evident as if the shot had come from Custer's own hand. Finally, nothing is odd or peculiar about a soldier being shot in the head during a battle, especially with the rain of bullets fired into the confined circle where Custer ultimately died.

Another pernicious myth was the mass suicide when soldiers realized the end was near and killed themselves out of desperation, instead of enduring the slow wrath of Indians torturing them to death. Private Charles Windolph had recalled the officers telling the soldiers that if they were ever wounded or in danger of being captured in an Indian fight, to save the last cartridge for themselves. The message was clear: it was better they end their own lives than the Indians finish them off. As the battle drew to a close, it had become too stressful for six or seven troopers who bolted from the hill. They did not get far. The Indians quickly brought them down.

Custer's younger brother, Boston, and nephew Autie Reed may have been among them. Both were found shot dead about a hundred yards from the hill. But none of these deaths indicate suicide. While felo-de-se may have possibly been the fate of a trooper here and there, the mass suicide stories would later be recanted by the very warrior who first perpetuated them. Wooden Leg disavowed his initial testimony, laughingly suggesting he had been misquoted about the mass suicides on Last Stand Hill. Most of the warrior testimonies attested that Custer and his band died fighting.

Because of anecdotal rumors about panic and rout, some claimed that Custer's final fight was misnamed. There was no last stand—it was only another fictitious byproduct of romance wedded and gilded by the legend of the storied old West. This was Dr. Richard Fox's thesis in his *Archeology, History and Custer's Last Battle.* But just as historians may not make the best archeologists, archeologists do not always make the best historians. Although a scholarly work rich in research, Fox draws the wrong supposition, focusing disproportionately on archeological finds to make some dubious conclusions based on the paucity of cartridges found on Custer Hill, whose remnants were greatly reduced to the point of scarcity from decades of souvenir hunting, rather than the more reliable eyewitness accounts.

Whatever else happened at Little Big Horn, this was indeed a last stand. Three of them occurred almost simultaneously on different parts of the battlefield, of which Last Stand Hill is the most famous. As Greg Michno, author of *Lakota Noon,* rightly noted, most Indian casualties on Custer's field were inflicted by soldiers defending the northern portion of the battleground. Time spent on the fight and the result of their shooting are all the proof we need to show that they defended their ground tenaciously. "… I call this a Last Stand," said Michno. "So did the Indians." And so do we.

While life ended for Custer on a little hill in southeast Montana, he was almost immediately resurrected in a fever-pitched blend of mythology and history that in the cultural wars that would later follow eventually morphed into a cauldron of controversy and mystery. Whatever he had been in life, hero or villain, dashing cavalier or reckless wild man, a fearless leader or an unrepentant glory hunter, Custer's body was nothing more than a partially mutilated, naked corpse broiling under the relentless Montana sun. Many saw his squadron commander, Captain Yates, as being prophetic when three years earlier, at the second Yellowstone Battle, he said, "One day Custer would get us in a hole that he would not be able to get us out."

Custer may have had an outstanding supporting cast in making Yates's prediction come true. Still, as commanding officer, Custer would be held predominantly responsible for what happened to his command on that afternoon of June 25th. Custer would never be able to speak for himself or talk about the actions he took that afternoon. However, his restless ghost lurks in those ever-present shades of history, both dark and bright, to haunt and mesmerize those who encounter and wrestle with the chronicle of the "Little Big Horn." This has never been truer than in the political and cultural turmoil of the 21st century. Though he would never give us his version of what had gone so terribly wrong at the Little Big Horn, others have been only too willing to fill the silent void with their own voices, many self-serving and censorious against Custer.

Chapter 15: What was the Last Stand?

The battle would soon be popularly known as Custer's Last Stand—a tag with a highly charged and dramatic ring of finality. But neat, definitive endings have never been the language of this battle, and the debate of who and what was to blame for the disaster goes on as heatedly as ever. The Indians did not know Custer was in the battle—Gall, among many others, later admitted that he had no idea who he was fighting. And yet Custer remains the embodiment of the battle, and the emotions evoked by his legacy seem immune to apathy or obscurity. All the heroic qualities and human imperfections are presented as caricatures of his person, and he remains a political lightning rod for today's political wars. Layers of obscuring legend have so accreted around this story that describing his improbable life and the dizzying circumstances that led to his untimely demise defies, at least in part, discovery.

In the midst of the great Sioux War, perhaps America's greatest Plains Indian War, the Lakota and Northern Cheyenne tribes overwhelmingly and definitively, amid the geographical labyrinths of dissected uplands, rugged towering bluffs, and along a 138-mile-long tributary, defeated the best equipped and supplied regiment in the United States Army. Although the Sioux took these lands from their ancient enemies, the Crows, the famous battle called the Battle of the Little Big Horn and known by the Indians as the Battle of Greasy Grass was the culmination of what had been a centuries-long incursion by white settlement, the railroads, cattle drives, the Pony Express, gold prospectors, surveyors, and the ideological wake of Manifest Destiny that held it was ordained by Providence to overspread the continent with those values, customs, and ideals which defined and propelled the civilization of the white race. The bodies of blue soldiers scattered across the valley, the timber, the hills and hillocks was but a temporary repudiation of the inevitable.

Custer, however, was not the only one who suffered a last stand, For, as soldiers fought and died on a nondescript hill in Montana, a sweeping drama would be indelibly painted on an outsized canvas of history. Two smaller military contingents on that battlefield were, roughly at the same time, making their own desperate last stands—a subject which, along with some post-mortem controversies, will be examined in the next chapter.

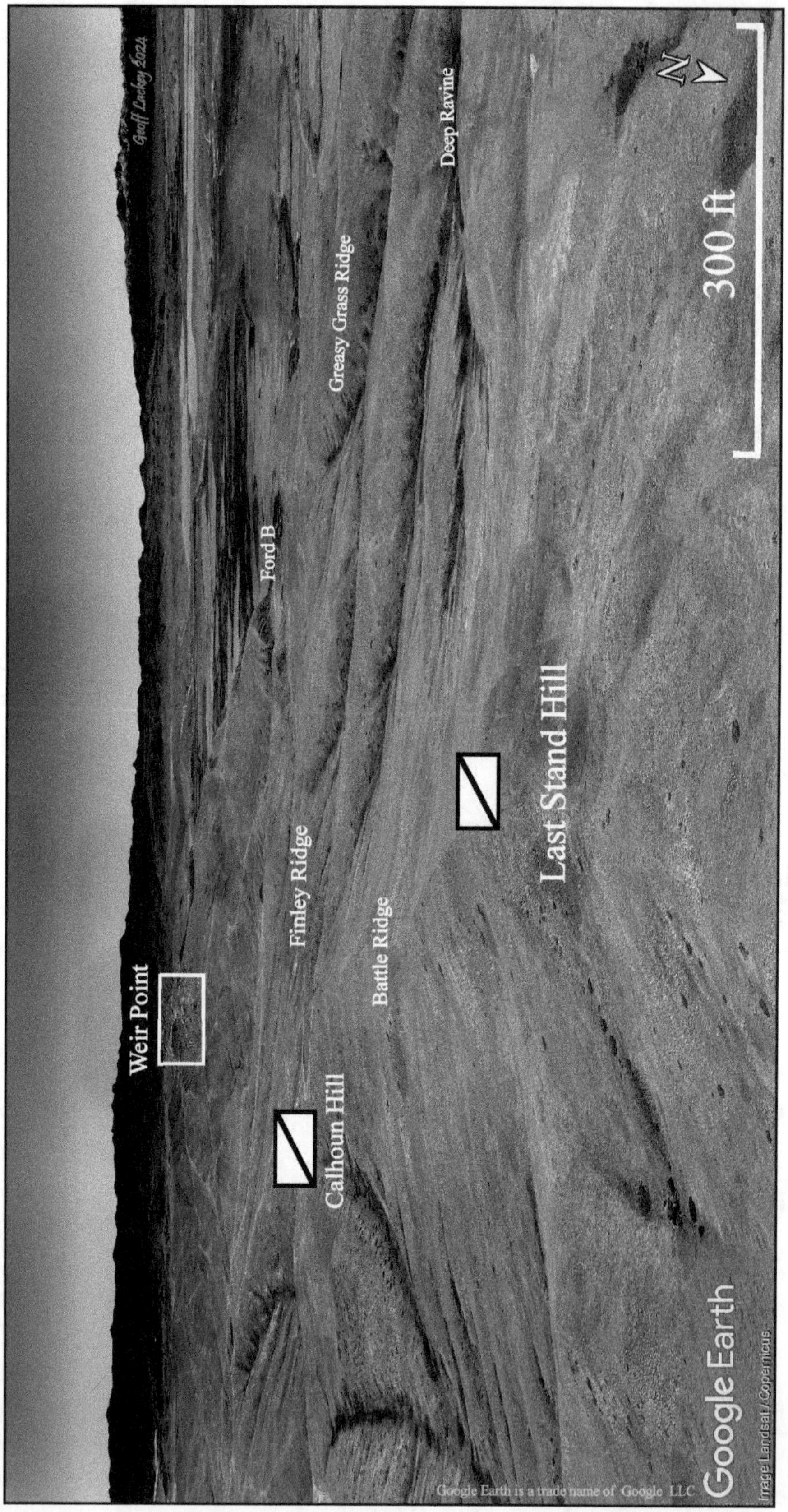

Map 27. The viewshed from Last Stand Hill and Calhoun Hill toward the Weir Peaks.

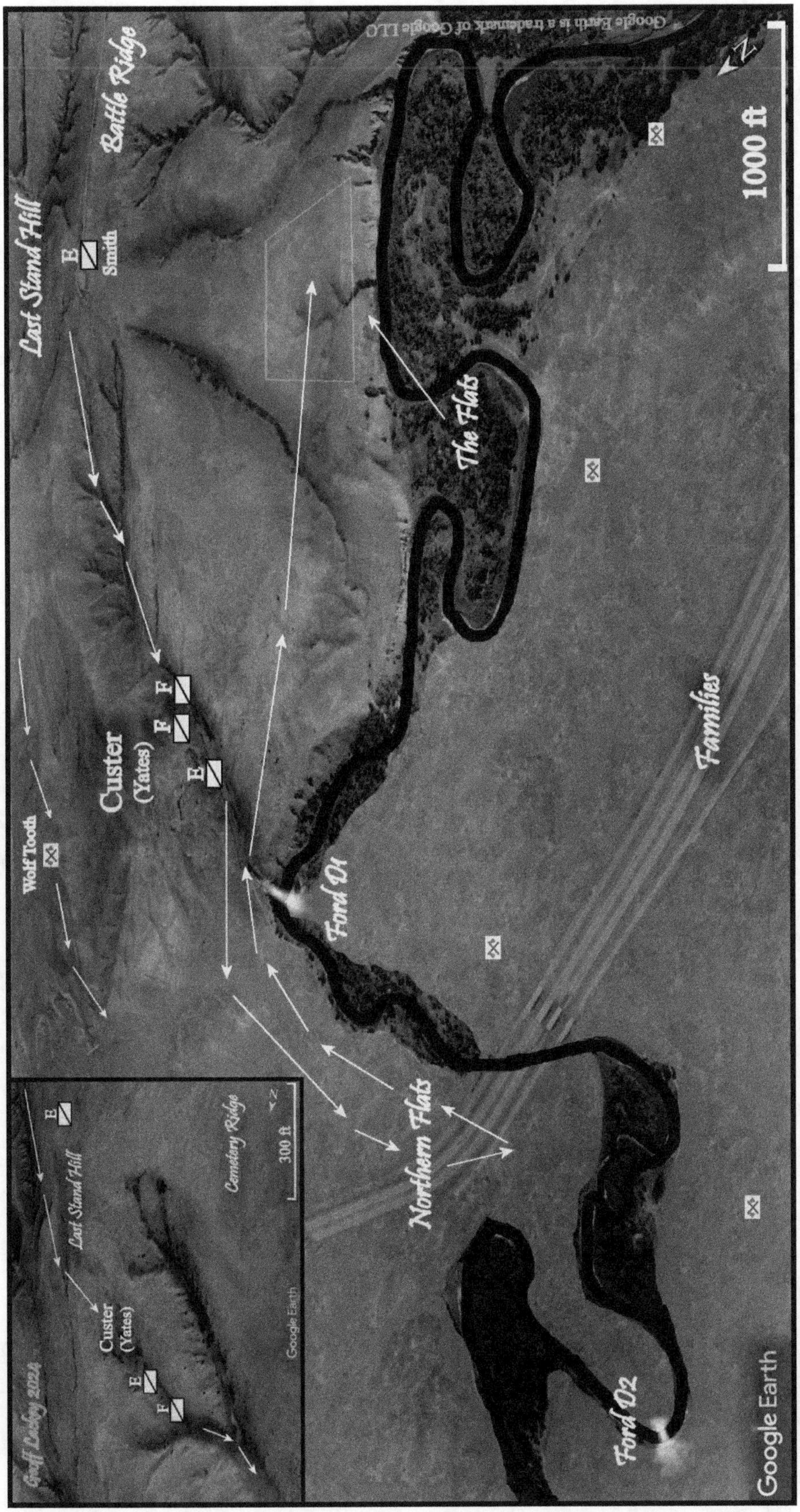

Map 28. The Custer/Yates squadron moves to the flats by Ford D-2 and returns to the lower Cemetery Flats.

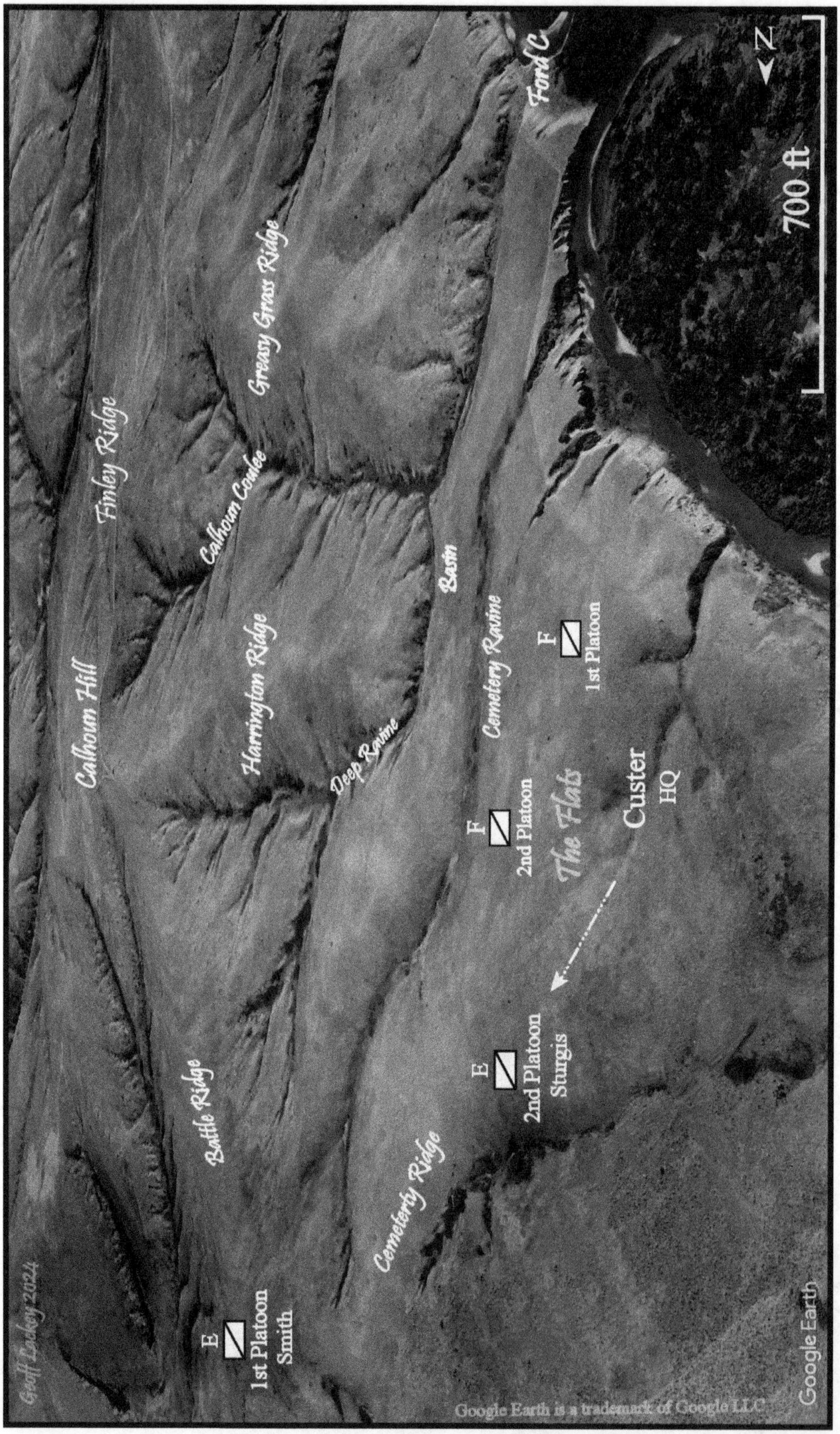

Map 29. Custer deploys the Yates squadron on the lower Cemetery Flats in preparation for the move north.

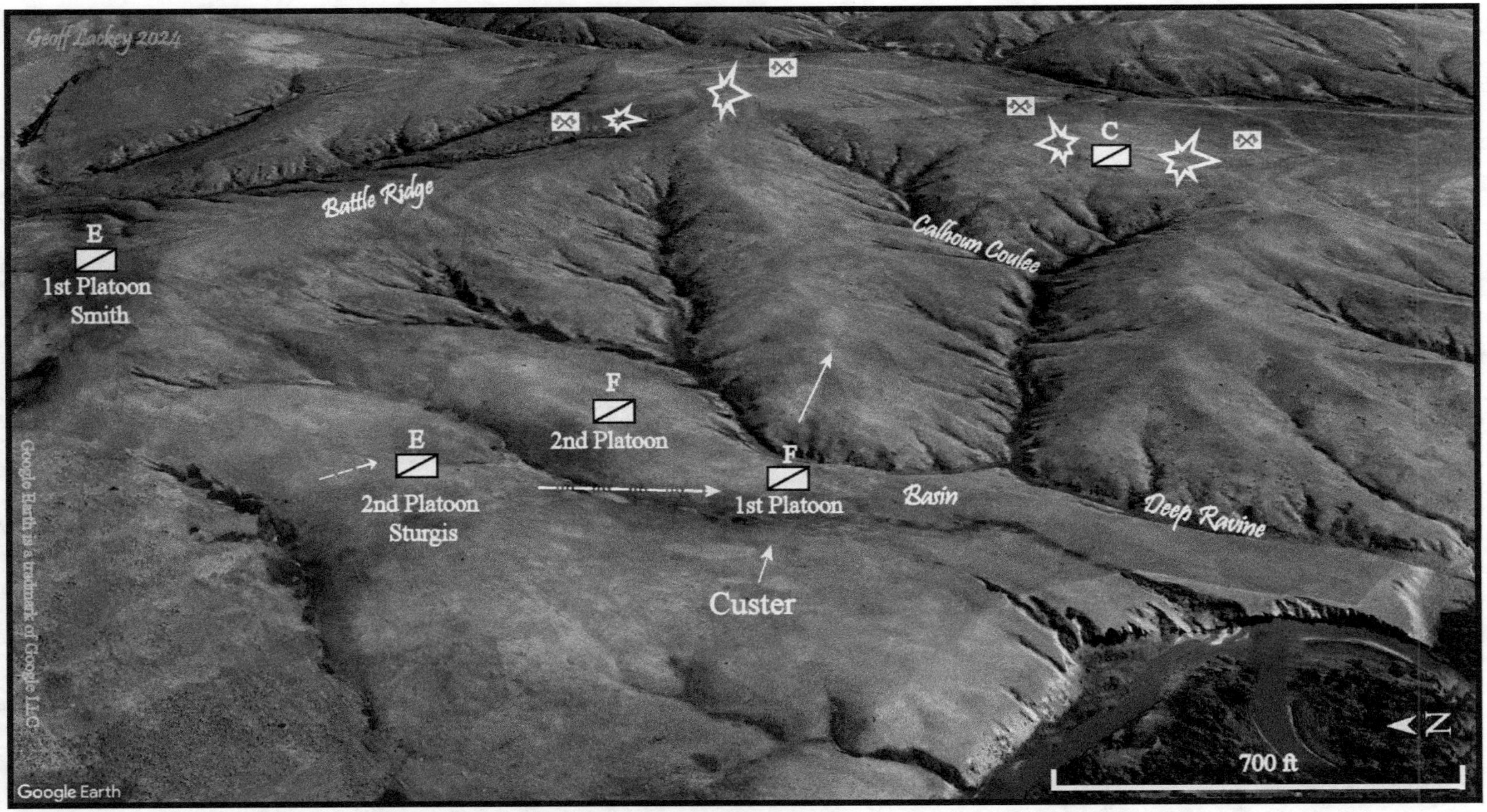

Map 30. Custer moves F Company into the Basin area to support the C Company platoon cut off on Finley Ridge. E Company deploys with platoons on Last Stand Hill and Cemetery Ridge.

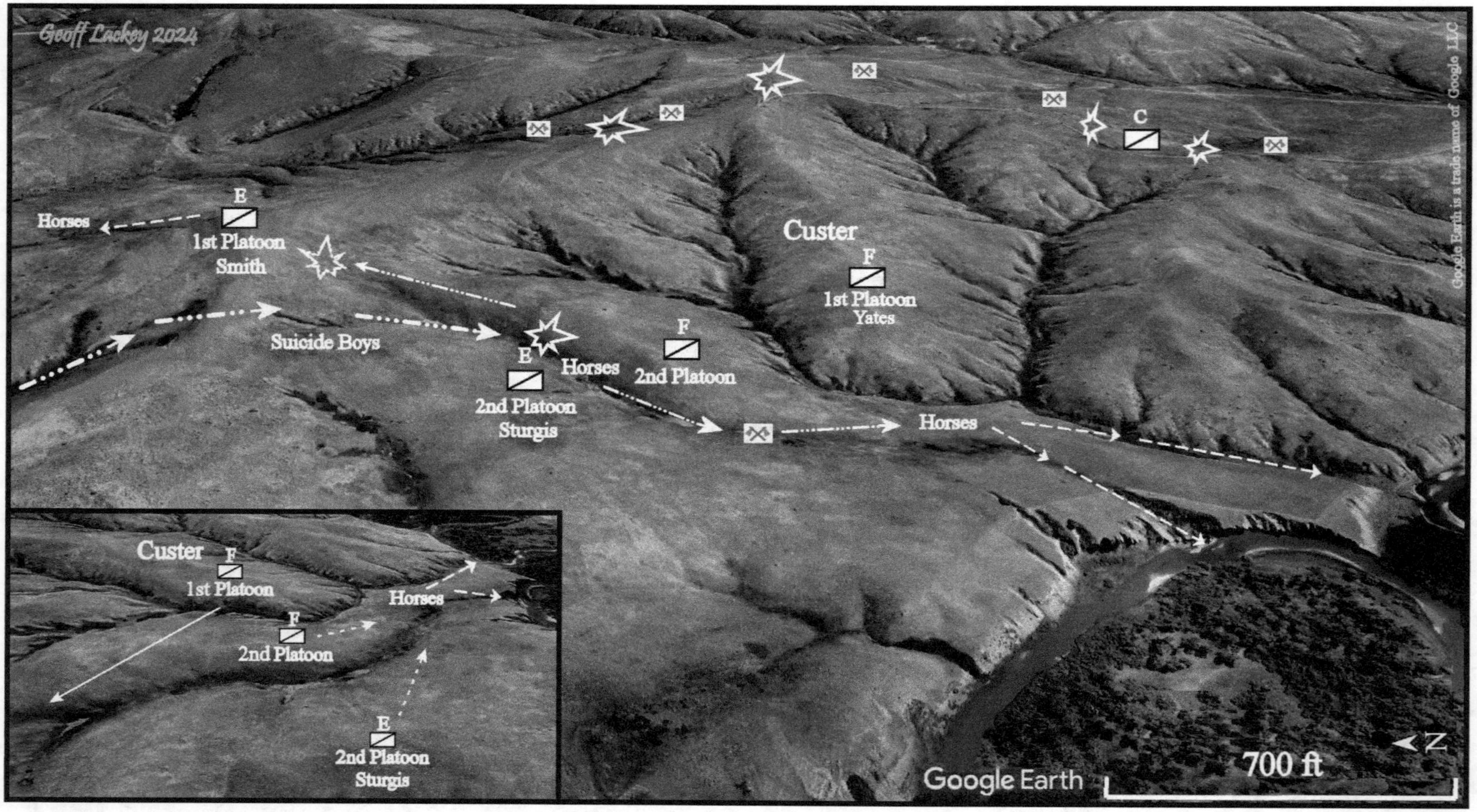

Map 31. The final phase of the battle. The Suicide Boys attack the horse-holders in The Basin and Last Stand Hill while Custer attempts to reach Finley Ridge. Stampeded horses are driven to the river via Deep and Cemetery Ravines. Surviving troopers fall back.

Chapter 16: Where are the Bodies Buried?

To know where the bodies are buried is to claim knowledge of a closely guarded secret. Somewhere a skeleton lurks in the closet. Though the bodies of soldiers from the 7th Cavalry were grotesquely scattered all over the Little Big Horn Battlefield, the mystery of the dead still looms large in fact and legend. As many as twenty-eight soldiers from E and F Companies may still lie below tons of mud and earth in Deep Ravine, though not so much as a fingernail, bone chip, or a strand of hair has ever been found. Numerous primary sources have stated that near the battle's end, with warriors in furious pursuit, over a score of troopers fled to Deep Ravine where they were slain. Three days later, a burial party arrived to properly dispose of their remains of those soldiers who died at Little Big Horn, but they purportedly had left the bodies to rot in a large gash in the earth.

According to one source, at the battle's climax, these soldiers bolted from Last Stand Hill to Deep Ravine, where they were trapped and killed. But this scenario is not entirely accurate. The soldiers who ended up in Deep Ravine were not fleeing from Last Stand Hill, a distance too far for anyone to run. Rather, these troopers almost certainly came from Cemetery Ridge, pursuing their stampeded horses carrying saddlebags of spare ammunition. The Cheyenne Yellow Nose, whose battle honors include capturing one of Custer's battle flags, was seen down by the river chasing a cavalry mount. Horses were prized booty for the Plains tribes, and horses took priority over almost everything else. Many of these same horses were captured by Indian women who had found them floundering aimlessly about in the river.

Reams of words have been written about the mystery surrounding the troopers from E Company, who were said to be strewn in and around Deep Ravine. This was odd because why would any remnant of their presence be found when they were not deployed there as an organized military force? They were not, as some still maintain, decimated at the supposed south skirmish line in The Basin. A platoon of E Company was deployed at the south end of Cemetery Ridge, northwest of Last Stand Hill to protect their held horses in The Basin area below, and to deny Cemetery Ridge to the warriors who would have hastened to occupy the high ground.

True, only a few dead horses were found by Deep Ravine, and Captain Tom McDougall noted that "there were few or no cavalry horses between [the] top of [Battle] Ridge [at Last Stand Hill] and [the] deep ravine where [many] E Company men were found." But this objection of no dead horses—for where dead cavalry troopers are found, logically there also will be their dead mounts—does not carry the weight its proponents maintain since most of these soldiers were dismounted when their horses were stampeded by the Suicide Boys—and nothing is improbable about a man on foot failing to overtake a spooked runaway horse.

Standing atop Cemetery Ridge and taking in the length and breadth of the distance involved draws a clear distinction between an overwhelming likelihood from where the troopers had fled and a remote one. The first markers placed on the battlefield signified where the bodies of soldiers were discovered a short distance south of Cemetery Ridge. The Cheyenne warrior Wooden Leg reported seeing six or seven troopers fleeing from Custer Hill toward the river at the end of the fight. In his communication to Dr. Thomas Marquis, Wooden Leg signed that "The soldier's fire ceased: the warriors [then] rushed forward. All the soldiers were dead but seven, who ran toward the river. But they were surrounded…this ended the fighting on Battle Ridge."

Wooden Leg obviously was referring to the northern terminus of Battle Ridge (Last Stand Hill) and not Cemetery Ridge. It is hardly surprising that in the dusty, smoke-filled swirl of battle, the witnesses to these two separate events conflated the exodus of troopers from Cemetery Ridge with the troopers in flight from Last Stand Hill. Two Moons, observing the battle from farther south, also mistook that portion of Cemetery Ridge for Last Stand Hill, adding to the confusion about where the fleeing soldiers originated. Muddying the waters further, historian Charles Kuhlman did not think this was an honest mistake, calling Two Moons "the Indians' prized liar." Early historians of the battle, such as Charles Kuhlman and Colonel William Graham, had little or no use for the Indian accounts, seeing them as fraught with hyperbole and hopelessly contradictory. Graham became so frustrated with their accounts that he dismissed them entirely, while Kuhlman omitted anything that varied from his model of how the battle was fought.

Still others, while acknowledging the flight of thirty to forty troopers from Cemetery Ridge in the direction of Deep Ravine, have argued that these troopers were not in pursuit of their runaway horses but instead had been ordered to attack the warriors assailing Keogh's position on and around Calhoun Hill. This line of thinking maintains that it would have been an instinctive measure of war for E Company, supported by fire from F Company on Last Stand Hill, to have swung around to push the warriors off Keogh's position by charging the west side of the ridge. As these soldiers drove forward toward Calhoun Hill, or so the theory goes, they plunged into an unexpectedly huge ditch known today as Deep Ravine, where they were trapped and ultimately killed.

But that scenario has major problems. How would the command on Cemetery Ridge somehow divine Keogh's desperate plight when Battle Ridge obstructed their line of sight to Keogh's Swale? Moreover, even conceding that they could discern it, the distance on foot between these two points would have been almost impossible to traverse. Nearly every account of the soldiers charging off Cemetery Ridge has them unmounted, except for a few who were likely either officers or horseholders. Trying to reach Battle Ridge two-thirds of a mile away from Cemetery Ridge on foot in the middle of a raging battle and not covered by a skirmish line would have been extremely foolhardy, if not suicidal. The documentation that these soldiers were dismounted is supported not only by Indian accounts but also by Captain McDougall, who stated there were few if any, dead cavalry horses between the top of the Ridge and where E Company was found in Deep Ravine. No dead horses meant no mounted charge, and that meant no charge at all. One can only reasonably conclude that no trooper sallied forth from Cemetery Ridge to come to the aid of Keogh's besieged command.

Of course, when a battle stimulates so much speculation, the quaesitum of why soldiers moved toward Deep Ravine may never be fully explained to anyone's satisfaction. After the Suicide Boys barreled into them and stampeded their horses, soldiers on Cemetery Ridge had little choice but to make a desperate attempt to retrieve their lost mounts, and, more importantly, their lost ammunition. It was a situation of "do or die."

When no evidence surfaced over the next few decades to indicate bodies were buried in Deep Ravine, questions arose about whether any troopers, regardless of from whence they came, were buried there. When, over the next century and a half, no evidence of human remains materialized in Deep Ravine, some battle students concluded that the oft-quoted twenty-eight-body figure bandied about in the immediate aftermath of the battle must have been erroneous. Even today, doubt hovers over whether anyone lies beneath Deep Ravine's soil.

The question of who was buried and where was most notably taken to task by the author Greg Michno in his now-controversial and much-debated book, *The Mystery of E Company*. Michno later published a highly regarded book on the battle, *Lakota Noon*, wherein he successfully sought to make sense of the vertiginous and contradictory Indian accounts revolving around the battle. However, in the earlier E Company book, he employed his superb investigative skills to explore what happened to E Company and whether nearly thirty unaccounted troopers were entombed in Deep Ravine. Michno concluded that this story was more an exercise in mythmaking than fact. He noted that no bones, clothing, a button, or other physical evidence of human remains had been found there. In an article in *Wild West Magazine,* Michno expressed his grave doubts that the remains of as many as twenty-eight soldiers lay in Deep Ravine:

> Interpretation should be based on historical and physical evidence whenever possible. Battle relics and bones have been found on virtually every part of the Little Big Horn Battlefield. Where they have not been found is in the trench of the Deep Ravine. When the archeological records show no signs of bodies it ought to be matched with the appropriate historical record—that there were few, if any, bodies in the Deep Ravine. It is incredible that diametrically opposed historical and archeological interpretations are presented as facts.

However, this analysis too easily dovetails into the cautionary adage that "absence of evidence does not mean evidence of absence." Deep Ravine is a long outlet south of Cemetery Ridge that travels several hundred yards to the river. It was deep—some eight to ten feet deeper at the time of the battle than it is today. It had perpendicular banks filled with runoff mud, making any major dig in that area maddeningly problematic. From its physical characteristics alone, it was a death trap for soldiers attempting to mount a defense there, which naturally led to questions as to why troopers would foolishly jump into a deep gully, making themselves easy targets for an enemy to shoot down at them from high on the perimeter of the ravine.

Some eyewitnesses in the immediate aftermath of the battle, as well as those who arrived a bit later to investigate the scene of carnage, cast doubt that dead soldiers had been piled and buried in Deep Ravine. Lieutenant James Bradley, an efficient and conscientious officer serving under the command of General John Gibbon, was emphatic that no bodies were buried at Deep Ravine. But his eyewitness account was contradicted by that of Private Dennis Lynch, who first rode along with the pack train but was later assigned to the burial detail, where he recalled that bodies were indeed in Deep Ravine, although several of them, perhaps seven or so, were removed and buried just outside the ravine, subtracting from the original sum of twenty-eight bodies.

Two years after the battle, Lieutenant Charles Roe, who oversaw digging up and reburying soldiers in shallow graves, stated that he did not see or find any bodies in or by Deep Ravine. He even went to great lengths to investigate a report of twenty men escaping from the battlefield and being killed near the Rosebud River. He marshaled a cavalry troop, deployed them at suitable intervals, and marched to the mountains but discovered not a scintilla of a single soldier's remains. Recollections, however, especially during times of trauma and stress, can be erratic and fickle. Others who were at the scene and had examined the battlefield, including Deep Ravine, did recall seeing bodies there. As it happened, corroborating testimony on both sides of the vexing question was available, but it was difficult to draw a single conclusion because they were at variance.

These troubling, contradictory accounts can be explained by recalling the awful circumstances under which the burial parties were working just three days after the battle. Bodies of naked troopers, a reeking and tangled gore of blood-spattered corpses and intestinal material, baked under a sweltering Montana sun. The soldiers assigned these grisly tasks reeled at the shocking state of the men they knew, served with, and befriended. Inconsolable that their comrades in arms had been so gruesomely slaughtered, some of the more peripheral details may have become repressed as a means of coping with this nightmare. The sight on the battlefield was revolting, their losses unbelievable, and the nauseating stench from the decaying corpses of men and horses lingered in their nostrils. An overriding sense of urgency drove them to get the loathsome job of burying the dead finished as quickly as possible. With burial details thrusting their spades into the dirt, Arthur Wellesley, 1st Duke of Wellington's description of war rings loudly: "The only thing on earth worse than a battle won, is a battle lost."

Hurrying to get the dirty business over with, imprecision crept into documenting the burials. This fostered doubts and raised questions when a new generation arose to consider and explore old mysteries. The last thing on these burial crews' minds was to consciously remember these disturbing details for posterity and history. The hurried entombments neglected the protocols and rituals of a decent burial as the dead were quickly and indecorously interred. "We only had a few tools," Sergeant John Ryan remembered, "we simply dug a little dirt aside from the bodies and threw it over them. In a great many instances, their arms and legs protruded out."

Another nettlesome problem was burying bodies with too few shovels at their disposal. In most cases, they made do with chopping down sagebrush and arranging it over the bodies—an easy feast for hungry scavengers. Years later, even Private Dennis Lynch, a member of the burial detail, questioned his own memories of that blurry day of turmoil—second-guessing himself on whether they did go into Deep Ravine and remove bodies. We have only this extract to go on:

> Lynch saw [the] 7th infantry men carrying the dead out of [the] deep gully [i.e. Deep Ravine]. They got "7 men on the bank" and he remarked: "There are a whole bunch of them there yet." Lynch says that these [seven] men in [the] gully were carried up and buried on the ridge [just south of Deep Ravine]}. [He] says that Sergeant Dave Heaton was one of these 7th infantry men [who is] now dead. One of [the] men in [the] deep gully [i.e., Deep Ravine] was Timothy Donnelly of F Troop. Lynch identified Briody by [the] sailor's mark on his arm.

There is no official documentation of an effort to retrieve the bodies in Deep Ravine. But if a conscious effort had been made to retrieve these corpses and bring them out of the ravine, these reburial parties could not have removed every bone, hair, piece of equipment, and every last stitch of clothing. Geologists who excavated the site have noted that the ravine soil is alkaline, which is conducive to preserving organisms and inhibiting their erosion and disintegration. The chances of finding no trace of human remains in Deep Ravine must be considered unlikely. However, natural decomposition processes, soil erosion, animal activity, and human disturbances cause remains to become inaccessible and unrecoverable.

Over the years, a theory took hold that heavy rains flushed down the physical evidence into the river. However, this hypothesis must be considered extremely remote since the geomorphology of this ravine is characterized more by the accumulation of sediments than acute erosion. In the *Mystery of E Company*, Greg Michno stated that all but three bodies were interred in the nearby Cemetery Ravine that was mistaken for Deep Ravine. But the notion that the bodies of E Company troopers along with those of F Company were buried in Cemetery Ravine is a nonstarter. In his analysis, Michno makes a rare mistake in failing to recognize that Cemetery Ravine has neither the dimensions nor the depth to have trapped these soldiers who could have easily escaped from those confines.

Regardless of the number, the description of where the bodies of E and F companies were buried is clearly that of Deep Ravine and not Cemetery Ravine. Neither now nor in 1876 could Cemetery Ravine be described as the deep gully with a headcut. Additionally, archeologist Douglas Scott, who did a masterful job of collecting and excavating raw material and has provided invaluable interpretive references regarding Little Big Horn, has stated that Michno's argument that Cemetery Ravine is the true Deep Ravine is a misinterpretation of the documentary and of the archeological evidence. Dr. Scott pointed out that Michno had ignored the absence of archeological materials in Cemetery Ravine and overlooked that the ravine has not been an erosional feature in that area for several thousand years.

That still leaves those accounts purporting that Deep Ravine had no bodies. Recalling that Lieutenant Charles Roe claimed that he saw no bodies in Deep Ravine does not seal the case that none were buried there. In 1878, two years after the battle, Roe returned to the battlefield to oversee reburying bodies but was not actively or conscientiously recording information. Instead, he was preoccupied and pressured by his limited time to get the job done while working under physically onerous conditions. Roe did the best he could, mostly reburying bones that had been interred in skin-deep graves, which had become exposed because of wind, winter frost, and edacious animals digging for food. As he attempted to pick up all the loose ends and whatever physical remnants he could find, he forthrightly stated that none of the remains of the enlisted men were buried in the trenches around the monument until after it was erected in 1881. As thorough as Roe was in his work, considering the ravine's depth and length, and the fact that it became infilled with dirt and mud, he would not have been able to detect any evidence of bodies buried therein.

By 1878, large sectors of Deep Ravine had closed in over these bodies. Questions then arose about whether Roe had ever dug in that location. In other words, Roe likely never dug or searched for any remains in Deep Ravine. Roe always maintained that there were never twenty-eight dead men in the Ravine. Perhaps. It is also possible that the burial detail might have inadvertently contributed to the cave-in of a section of the ravine, pushing these bodies deeper into the earth and thereby permanently entombing them. Nevertheless, despite the ongoing insistence of some dissenters that few, if any, bodies are buried in Deep Ravine, considerable evidence can be based on formal statements and eyewitness accounts that there may well have been a little under thirty bodies seen there. But, because of the scramble, troublesome geography, the absence of time, insufficient resources to do the job competently, noisome conditions, and the ongoing threat of a victorious and dangerous enemy not far away, the bodies had been mostly left where they were found. In the oft-expressed phrase, the dead were left to bury the dead.

Despite the careless documentation, including that of bodies buried at Deep Ravine, six maps drawn by the army and three Indian maps indicate dead soldiers were found there. One of these maps was drawn by Sergeant Kanipe, who stated that he counted twenty-eight bodies in Deep Ravine. Kanipe went to the battlefield in 1908 with researcher and archivist Walter Camp. He painstakingly drew a map and marked where he saw those bodies. That same map can be found today in the Brigham Young University archives with Kanipe's "X" marking the location of the buried soldiers—a site measuring some 2,100 feet away from Last Stand Hill. In other words—squarely in Deep Ravine.

Captain Thomas McDougall commanded the rear guard of the pack train and was ordered by Major Reno to oversee operations of the burial detail that included Deep Ravine. He was best suited for the job since he had commanded E Company for over five years, was intimately familiar with the troopers in that company, and was fully capable of identifying the dead from the outfit. McDougall first stated that about half of the twenty-eight men from E Company were found in the ravine, and the other half were laid out on a line outside. He further noted that all the troopers outside the ravine were lying face down and appeared to have been shot mostly in their sides. This observation is particularly relevant since it was a well-known Cheyenne practice to turn their victims face-to-the-ground since they deemed it bad luck to allow a dead enemy to face the sky. Decades later, on May 18, 1909, in a letter addressed to General Godfrey, McDougall went into more detail about what he witnessed at Deep Ravine, stating that the preponderance of bodies there belonged to E troop:

> "In the ravine I found most of the troop who had used the upper sides of the ravine for a kind of breastwork, falling to the bottom as they were shot down … this was not very far from the village … Only a few of the men could be recognized. I knew Sergeant Hohmeyer at once; he had one sock left on his foot with his name on it." [Note that McDougall could only recognize "a few of the men" which he assumed were from E Company.]

So, apparently, McDougall believed the troopers used the top of the ravine as sort of a defense, meager though it was, making a final stand before they were shot down and sent tumbling back into its cavernous depths. As noted earlier, doubts emerged about troopers racing toward Deep Ravine because it was a massive ditch, not only inescapable for those plunging in but utterly indefensible ground. But what is too often overlooked is that the soldiers on that blistering afternoon, unlike the historians and researchers of today, were completely unfamiliar with the contours of the terrain, having encountered it for the first time and doing so under extreme duress—indeed, what amounted to a life-or-death struggle. The troopers had no idea that the ravine was tantamount to a terrestrial black hole and that entering it to escape a determined enemy surrendered all hope of survival.

Another eyewitness who saw dead bodies in the ravine was Captain Benteen, who remembered that the bodies were located somewhere in the vicinity of fifty to seventy-five yards from the river. Benteen counted twenty-two bodies and, because these soldiers could not shoot out of the ravine, concluded they never went into it in the first place. Essentially, they found themselves exposed on open ground. Deep Ravine was perceived as their only chance for survival, not realizing that the terrain offered no defensive possibilities, and by casting themselves into the abyss, they had inadvertently thrust themselves into an abattoir.

Once in the ravine, some soldiers desperately tried to claw their way upward out of the trench in a hopeless attempt to climb to the top of those perpendicular walls. They made pitifully slow-moving targets, and Indians around the rim of the ravine quickly shot them down. Some half dozen Indian cartridge casings from Winchester and Henry rifles uncovered near and around the deep end of the ravine confirms that the Indians pumped bullets one after the other at whatever was alive and still moving down there.

That essentially accounts for the dead soldiers there, but what about the soldiers found just outside the ravine's steep walls? It is possible that the June 28th burial detail may well have dragged some dead soldiers, at least those who were not too far down, back up to the top of the rim to bury them with other soldiers. They may have been the troopers' bodies McDougall said were laid out on a line outside the ravine's perimeter.

However sensible this explanation might sound, more accounts support the conviction that the majority of the bodies were buried in Deep Ravine than were laid out at the top of the rim. That is what Lieutenant Charles De Rudio claimed to have seen. He recalled that no enlisted men were identified in the ravine, nor was he sure about sighting E Company troopers there. Still, he did see "a heap of men in a gully" (Deep Ravine) and said that the dead horses nearest the river were gray ones belonging to E Company. Private George Glenn of H Company also remembered seeing something similar. Glenn said that the bodies lay thick in there (Deep Ravine) and that at first glance, some of the troopers were so taken by the slaughter they exclaimed, "Here lies the whole company." With one body piled upon another, it must have looked as if an entire company was now sharing a mass grave. Lieutenant Luther Hare not only saw these dead but counted them—he noted twenty-eight bodies in the Ravine.

There is more testimony regarding bodies buried in Deep Ravine. In his journal, Captain Henry Freeman, later an Army General but then an Infantry Company Commander under Colonel Gibbon, wrote that after slogging into the valley of the Little Big Horn at about 3:00 p.m. on June 27th—"the hardest marching I ever made," he had counted twenty bodies in Deep Ravine. Lieutenant Edward Maguire, the Chief Engineering Officer for the column led by General Alfred Terry, was directed to draw a survey and map of the battlefield. Although dated July 10, 1876, some two weeks after the battle, McGuire had been surveying the battlefield since shortly after his arrival on June 27, just two days after Custer's command was annihilated. McGuire was among the first to map the battlefield and was assigned the requisite men and equipment to complete the job. He would also be the first to report twenty-eight dead soldiers in Deep Ravine.

On his map, McGuire marks the spot where the bodies were found with the letter "H"—and that spot was indeed in Deep Ravine. Captain McDougall referred to the "H" marking on the McGuire map when he testified that was where the bodies of E Company were found: "To the best of my recollections," McDougall said, "about half were in the ravine and the other half on a line outside." McGuire's maps are littered with mistakes, erroneously drawing the course of the river, omitting the Garryowen Bend of the river, and incorrectly drawing not only the bluffs but also the Weir Peaks and Sharpshooter Ridge. In addition, the terrain he had sketched around Ford B, Luce and Nye-Cartright Ridge hardly qualified as a paradigm of accuracy. However, while McGuire's depiction of the surrounding landscape was sometimes grossly subpar, his map also has features that can be accepted with confidence, mainly since his "H" marking corresponds perfectly with the testimony of other reliable witnesses.

Also, Captain Myles Moylan of Reno's Battalion, testifying at the Reno Court, spoke of the twenty-odd bodies of E Company found on a map marked with the letter "H." In what has to be considered a most depressing and harrowing piece of testimony, Moylan went on to attribute the marks dug into the perpendicular walls of Deep Ravine as those of the trapped soldiers trying to scratch their way back to the top but unable to find a firm grip—painting a stark image of their terrifying final moments. Moylan specifically pointed out that those scratch marks only extended halfway up the bank, most likely indicating a frenzied attempt to avoid being shot down at the bottom of the ravine where they would drown in a pool of their own blood.

Walter Mason Camp, in his unfettered zeal to document everything about Little Big Horn, provided a letter written to General Edward Godfrey, who, on June 25, 1876, held the rank of lieutenant and was the commander of K Company, part of Benteen's battalion. In that missive, Camp questions

Godfrey about the marker placement for the dead found on the Custer battlefield. His letter, though lengthy, sheds some light on the mystery of Deep Ravine:

> Dear Sir: In order that the subject may not grow cold, I will bring up a question that I imagine will be of some interest to you, at least in a general way, as well as to me for particular reasons. I refer to the absence of markers for the twenty-odd-bodies that we have found in the deep gully or ravine about half-way between the point where the General's body was found and the river. In your [Century] magazine article published in 1892, you refer to the bodies of 28 men that were found in this [Deep] ravine area together. There seems to be abundant evidence that the bodies were found there, but I will ask if this is a fact that came within your knowledge, and whether … you know the bodies were buried in the [Deep] ravine or anywhere near it. I once overheard a conversation at this spot between the custodian [the then Superintendent of the battlefield] and some man [never identified] who professed to have been present when the dead were buried on June 28, 1876. This party told the custodian that he saw bodies lying in the [Deep] ravine and had counted 20 in a heap, and he could not understand why no markers had been placed there. Now let me digress to say that between this gully and the top of the ridge where the monument is [Last Stand Hill] there are some 40 or more markers pretty much in line [note that early researchers of the battle would later refer to these markers as a the "South Skirmish line," a fictitious formation that was accepted for years as gospel by many early 20th Century battle scholars] not counting 10 or a dozen scattered at the same distance from this imaginary line, and not counting any of the 50 or more markers in the group where that of the General stands, on the side of the [Last Stand] Hill about 15 or 20 yards below the monument. Returning to the conversation of the veteran referred to above, this party said, in substance, something like this: It looks to me that though the markers should have been placed for the men found in that [Deep] ravine strung out between here and the monument giving the [false] impression of men fighting in a [skirmish] line. I am sure that I saw only a few dead bodies lying between [Deep] ravine and the group where the General lay at the top of the ridge, whereas there are more than 40 markers distributed along with considerable regularity over this interval of 600 to 700 yards.

The essence of Camp's letter confirms that there was no mythical South Skirmish Line deployed in The Basin area—not surprising since forming a skirmish line in such a depressed area makes no tactical sense—and that the bodies found in and around Deep Ravine were soldiers who had come down from Cemetery Ridge, not Last Stand Hill. Finally, Camp's letter discloses that placing the markers on the battlefield was an exercise marked by carelessness and no small quantity of negligence.

Those who were later put in charge of recording the burials of bodies did what was more practical, or at least convenient, rather than what was best and proper for the deceased. Accounting for all the dead bodies was an almost impossibly formidable task, as they lay in eight distinct battlefield regions. In 1881, the Army sent Captain Owen Jay Sweet to oversee marker placement on graves, which were mistakenly thought to be indicated by depressions left in the ground where the bodies had once lay. While many of these placements did represent where soldiers died, some were misplaced, and too many were erected on the Custer battlefield, some markers representing men from Reno's battalion who were killed several miles away. The National Park Service, though aware of erroneous marker placements, decided not to address it for reasons of their own. Nor is it altogether clear whether some of the markers placed just outside of Deep Ravine, in reality, represent soldiers who died in it.

The testimonies of participants and eyewitnesses in the battle are the best markers. Tall Bull spoke of the last soldiers being killed in Deep Ravine. He stated, "I was near [point] H [Note: at Deep Ravine] and heard a big war whoop that [said some] soldiers were coming [toward Deep Ravine]. [These] soldiers came [running] on foot and ran right through us [and] into [the] deep gully [i.e., Deep Ravine], and this was the last of the fight; and the men were killed in this gully." He was not the only warrior who spoke of the soldier's flight and fate in Deep Ravine. Standing Bear stated that the soldiers killed (between Custer and the river) were men on foot who were trying to make it to the river, and they were killed in Deep Ravine. Good Voiced Elk added that there was a deep gully without any water in it: "I saw many jump over the deep bank into this gully in their effort to escape, but these were all killed. There were probably 25 or 30 of them."

The accounts of these Indians, along with those of the soldiers, strongly indicate that the last of the battle occurred at Deep Ravine, either just after the fall of Last Stand Hill or, perhaps, simultaneous with it. It is hard to say for sure. But the overriding point is that at Little Big Horn there were multiple Last Stands. One of these stands ended around Keogh's Swale with the virtual destruction of Keogh's command. Three others, Last Stand Hill, Finley Ridge, and Deep Ravine, occurred nearly concurrently, so close in time, that it is hardly worth the trouble to assess the difference in sequence.

Rumors and attestations of bodies buried and unaccounted for were so intriguing that the world of science and technology was brought in to weigh in on the mystery of the bodies in Deep Ravine. Sonar equipment was first brought to the battlefield in 1984–1985 to comb over Deep Ravine to answer this question. The technology then used was not able to detect anything underneath the earth. If bodies were indeed there, they were buried deeper than the then-available sonar equipment could reveal. There are eight major geologic strata identified at Deep Ravine, which shows the impressive density of the earth in that area. Its per-unit volume proved to be a logarithm of opaqueness, seriously hampering the ability to disclose physical objects. The anthropologist Dr. Richard Fox has accurately described Deep Ravine as a ravine over a ravine.

However, in 1994, the National Park Service decided to peer deeper by employing more sensitive and sophisticated ground penetrating technology, a sonar device dubbed the "sled." With its sonic and supersonic waves, this apparatus had the capacity to locate deeply submerged objects. After painstakingly dragging the sled back and forth several times over the area, a procedure known as "mowing the grass," the technicians found an anomaly, something that was not part of the ravine's natural terrain. The significance of this find lies in the fact that this undefined object was situated in the very spot Sergeant Kanipe marked as the place where the soldiers were buried. This finding is deeply telling, as it is the only location at Deep Ravine where an irregularity was found. Reviewing the data, Dr. Douglas Scott opined, "Geophysical investigations in Cemetery Ravine revealed a few small near-surface magnetically susceptible soils. But no anomalies that could be even remotely construed as burial sites were identified. The only anomaly consistent with such an interpretation is the one in Deep Ravine proper."

So compelling was the primary and scientific evidence about bodies still buried in Deep Ravine that, after an exhaustive investigation of the matter, Douglas Scott, Richard Fox, Melissa A. Connor, and Dick Harmon concluded that the one thing that could be agreed upon was that there were likely soldiers buried in Deep Ravine. In fact, there is not a single convincing account to the contrary that explains this anomaly. This consensus corroborated what Fox had already established in his previous work, that bodies were still in Deep Ravine, where the 1876 surface is more than six feet below today's gulch floor. The whole question is much like solving a giant jigsaw puzzle. Once the pieces are assembled and fitted together, they form a fairly convincing picture.

The mystery of E troop eluded a number of historical detectives because they failed to realize that E Company, like other companies, was operating on a platoon system rather than a troop basis. That is why they were found on different parts of the battlefield rather than in one place together. That is why the best evidence indicates that troopers from E Company were found both at Deep Ravine and Last Stand Hill.

Based on the scientific findings, the knowledge that Custer's wing was operating on a platoon level, and primary source testimony, there is more than sufficient reason to conclude that the only mystery about Deep Ravine and the bodies buried there is why there is any mystery at all. The bodies are undoubtedly interred there, right where they were left back in June of 1876. The only question that remains is whether to leave them undisturbed or embark on a search and recovery mission.

But exactly what to do with the dead in Deep Ravine is a problematic and momentous question. Despite the violence associated with their deaths, these men have peacefully slumbered in their common grave for 150 years. The question of whether they should be left alone or exhumed and interred with their comrades around the monument on Last Stand Hill, or in the adjacent cemetery, is controversial and fraught with disputation. Obviously, a great stretch of time has passed since the battle was fought, long enough for the turning of one century and then another. Until now, no one has had the time, inclination, or, most importantly, the resources to exhume the dead of Deep Ravine. It is quite possible that the National Park Service, along with the various organizations dedicated to commemorating and preserving the memory of the Battle of the Little Big Horn, may well favor bringing those so long lost to a more appropriate resting place.

Whatever the practicality and moral probity of such an endeavor, all the plans suffer from one common liability: They cost too much. If two score years earlier, science was able to locate the remains of the Titanic, thirteen miles down in the North Atlantic, in what is among the roughest areas of ocean in the world, to successfully bring haunting pictures and artifacts of the tragically famed ship to the surface, then recovering the remains of E and F Company soldiers in the relatively fixed and stable terrain of southeast Montana should not be considered beyond the means of modern resources and technological know-how.

Nor would excavating these bodies be a question of grave moral dimensions for those who might now be wrestling with the issue. No doubt such an operation would be undertaken with all the reverence, sensitivity, and delicacy such an effort would entail. History is a seductive and tantalizing lure that has the power to bridge a warring and cultural divide into a peaceful and ritual remembrance of a shared and historically consequential event. The reinterring of the remains of those who died at the battle could be observed and overseen by those in solemn communion with what befell the 7th Cavalry that day while also profoundly acknowledging the shattering and enduring consequences the battle had for Native Americans who had roamed and ruled the Great Plains for centuries.

The driving motives for westward expansion that had been sullied by greed and commercial exploitation would be consigned to a lamentable past as both sides of the cultural divide forge ahead in a symbolic gesture of unity and reverence for a shared history and our common future. It would be a last best hope of reconciliation where piety rather than profit, understanding rather than ignorance, and commemoration rather than partisanship.

It is not cheap romanticism or tasteless sentiment to envision a memorial service that solemnly acknowledges not only what was lost but also what should be remembered and preserved of a fabled frontier. If a successful effort can be mounted to bring the long lost into the clear, disinfecting sunlight, then, perhaps, we can hear one more time the reverential Indian death songs and be uplifted by the spirited strains of "Garryowen"—anthems honoring the dead on both sides and paying tribute to an enduring saga of a national memory that continues to define us.

Conclusions

With a battle so complex and fraught with so many variables, for purposes of summation, the authors enumerate contributing reasons—some before the outset of the campaign—for Custer's monumental defeat. While much of the following has already been discussed in the text, it might prove beneficial for the reader to review and consider:

1. In the Plains Indian Wars, the United States Army failed to come to terms with irregular or unconventional warfare, a problem that would persist to perplex the U.S. military well into the modern era. Elsewhere, in the Philippines, Vietnam, and the post-war insurgency in Iraq, the enemies of the United States were neither impressed nor intimidated by America's industrial and military complex. All those conflicts underscore the difficulties the United States military had when confronting an enemy that employed guerrilla warfare tactics, regardless of setting, although by every metric, they faced an enormous military, logistic, and economic disadvantage. It was no different in 1876 when nomadic tribes used the land with its wide-open spaces as an effective strategy against the military superiority that confronted them. It is ironic that these very tactics were highly instrumental in the American Revolution, where local, colonial militia battled the world's greatest empire and endured against the full might of Great Britain until France's involvement in the war resulted in larger, more conventional victories that secured independence.

2. Overconfidence in the 1876 campaign to subdue and contain the native tribes was partially responsible for the military's inability to adopt new strategies to meet the challenges of unconventional warfare. This was the same U.S. army, though much attenuated, that vanquished a gigantic rebellion waged by a modern Confederate army, abetted by heavy artillery and led by West Point-trained generals, many of whom already earned their plaudits during the Mexican War a half a generation earlier. Compared to those, what hazards could a technologically impoverished native people, often hampered by internecine tribal conflict, present to the U.S. military? The Centennial Campaign, as the conclusion of the Plains Wars has been sometimes called, was a classic case of grossly underestimating one's enemy—an old but very real story in human warfare.

3. The objectives first outlined in the 1876 campaign were eventually brought to a successful conclusion in reaction to the epic defeat of the 7th Cavalry by tribal warriors. It was because of the staggering failure at the Little Big Horn that military operations were removed from the department commander's and senior staff's personal control. Humiliated, shocked, and chastened by what occurred at the Little Big Horn, the army was more willing to put the

matter of subduing the Indian populations in the hands of younger, dynamic officers such as Nelson Miles and Ranald Mackenzie who were not committed to old, tactical assumptions and were willing to think anew. These young commanders were ready to eschew the military orthodoxies of the past to employ whatever was necessary to meet the guerrilla hit-and-run tactics that plagued the military over the last decade rather than seeing the present conflict through the prism of fighting the last war.

4. The predominant reason often given for Custer's defeat, and the one that still feeds the public imagination, is that Custer imprudently ignored the danger associated with being deep in enemy territory and hopelessly outnumbered. Consequently, it was his chutzpah and recklessness that ultimately led to the slaughter of his entire command. But this seriously oversimplifies why the battle was so infamously lost. While it is true that the numerical superiority of tribal warriors at the Little Big Horn was real, perhaps 2,500 to 3,000 warriors, they were not all concentrated at any one time or point against Custer's forces. In truth, the numerical disparity was the least of the 7th Cavalry's problems that afternoon. The situation must be viewed as it then appeared to the military rather than retrospectively or through the eyes of modern presumptions. Even though Custer's Indian scouts, in memorable and colorful language, expressed alarm about the size of the village, no one in the army officer corps had expressed anxiety about warrior numbers and the inability to defeat them, much less harbor fear of being destroyed in detail. Indian scouts had been known to exaggerate enemy numbers. The most critical and gnawing concern was finding the Indians and preventing their escape rather than fighting and defeating them. Dominating strategic thinking was how to prevent or minimize the tribal warriors' oft-conspicuous defensive tactic: disperse and scatter. The most exasperating element of Indian warfare was their unwillingness to stand up and fight against the U.S. Army. Although efforts were indeed made to ascertain the strength of the hostile Indian population and warrior force, it was a secondary consideration and not a matter for an entire regiment of cavalry to preoccupy itself. When Custer's wing moved north on the high bluffs with five companies, the conventional wisdom was that three solid and properly led companies of cavalry were sufficient to defend themselves against any number of Indians, either by tactical mobility or when constituted in a defensive skirmish formation, always providing that their ammunition held out and that they maintained a distance where the advantage of their Springfield carbines could be maximized. That Custer was defeated in detail with five of the picked companies of the 7th was a result that no one in the command even remotely anticipated.

5. There was no expectation that a warrior force, especially when taken by surprise, as indeed they were on the afternoon of June 25, would move aggressively against their attackers. Custer's primary objective was preventing the village from escaping without giving enough consideration to the unexpected, where he would fight a standing, pitched battle against these warriors, one that would ultimately be directed against him by nearly the full weight

of the enemy against his divided forces. It was only after Custer's total defeat that the overwhelming number of Indian warriors became a convenient rationale for the military to explain to the uninitiated and uninformed why the defeat was so one-sided. This explanation segued nicely with Custer's rejection of Terry's offer of four companies from the 2nd Cavalry. That supported Custer's well-known overconfidence and even perceived rashness, leading others to conclude that Custer was hungering only for glory and unwilling to share the plaudits of victory. Moreover, the explanation that it was solely the sheer force of numbers and Custer's unwillingness to accept assistance was uncomplicated and straightforward, something the average citizen could readily understand and accept. It also served to conveniently insulate senior military leadership from explaining their shortsightedness or the lack of effort by Custer's subordinates, who were within earshot of gunfire that could only have been directed against Custer's forces. It became a neat and handy excuse to say that the reason he was so disastrously defeated was Custer himself. In other words, it would be easy for the public to conclude that the entire explanation for the catastrophe could be chalked up as Custer using too few soldiers to fight too many Indians.

6. News of Custer having divided his command to attack the village only reinforced the perception of his metabolic impulsiveness to attack, his failure to gauge the odds, and his overreliance on what he saw as 'Custer's luck.' Overlooked was that dividing his forces in the face of an irregular enemy force, regardless of numbers, was considered standard operating procedure, which had, especially in small wars, been used effectively by Custer and others. The key to success when dividing forces was to confuse and surprise the enemy with a multi-directional attack and then reunite those forces in a timely fashion, whereby they could act in concert with each other for maximum effect. At the Little Big Horn, it was not dividing the regiment that proved fatal but rather the inability or unwillingness to reunite it. This failure was due in part to Custer's decision to send Benteen off on a reconnaissance in force while still fifteen miles from the village, along with the failure of subordinate battalion commanders to rejoin the fight after reaching the bluffs.

7. Although Custer had reasons to act as he did, we must remember that he was not the only officer in this campaign who operated in the field without firm intelligence. Generals Crook, Terry, and Gibbon made no detailed plans for the campaign as it was deemed not needed and even impractical as a means of conducting military operations against hostiles. Gathering intelligence took time, and the Indians were a primarily mobile and elusive population, so speed was of the essence against perpetually migratory tribal communities, moving to where grass and game were plentiful. So, the military was mostly in the dark about what the Dean of Western historians, Robert Utley, so astutely capsuled in a well-crafted synopsis of why Custer lost:

> Never before or after were the Northern Plains tribes better prepared for war, they were numerous, united, confident, superbly led, emotionally charged to defend their freedom, and able through design or good fortune, to

> catch their adversary in unfavorable tactical situations.... In large part the generals lost the war because the Indians won it.

8. Even prior to Custer leaving to pursue the Indians on June 22, others of high rank and positions were inadvertently setting the stage for his defeat. They were guilty of the very mistakes they would attribute to Custer. General Terry had divided his force in the presence of a large enemy and did so without the benefit of proper reconnaissance when he sent Custer in pursuit of the Indians. Reno's scout on the Tongue River was supposed to provide information on the Indian's whereabouts. Although he discovered their trail, he failed to follow it, leaving the entire command uninformed about their present location and the growing size of the warrior population. It was common knowledge that younger warriors would leave the reservations during the warmer months and head to the unceded territories. Terry could have kept his command together until the enemy's strength and location were determined. He then could have assessed the resources needed to defeat the enemy and destroy their village and resources. Additionally, from the time he sent Reno on his scout, Terry's express wishes were more directed at determining where the Indians were not (the Tongue River) rather than where they were, that being west of the Rosebud River.

9. From the outset, Terry's plan was flawed because it foolishly presupposed that a large body of Indians would remain a stationary target as converging military forces zeroed in on them from different directions. General Miles aptly stated, "It is folly to suppose that either a small or large band of Indians would remain stationary and allow one body of troops to come up one side of it while another body came up on the other side and engage it in battle." In other words, Miles maintained that Terry's belief that he could establish a definitive timeline for cooperative action between Custer and Gibbon's column was nothing more than a fictitious supposition that showed Terry's inexperience as a field commander. Indeed, Terry did not mention a word about cooperative action between the two respective columns in his famous and controversial letter of instructions to Custer on June 22.

10. While Terry believed that the Indian village was somewhere in the Little Big Horn Valley, he was unsure of their precise location. Had the village been at either the source or the mouth of the Little Big Horn River, Gibbon or Custer would have needed an extra day to come within supporting distance of that location. This created another problem since Gibbon's forces were a mix of infantry and cavalry, which made catching the Indians almost impossible. Additionally, Terry's plan sealed off only one of the Indians' escape routes without securing other avenues. Because the military was demobilized after the Civil War, Terry and Custer only had about eleven hundred men between them, insufficient to cover such a wide swath of territory. If, for example, Custer had accepted the four companies of cavalry Terry offered, Gibbon's trapping force, from whom these companies

would be taken, would have rendered them too weak to effectively confront such a large body of hostiles.

11. Meanwhile, General Gibbon's performance at the start of the campaign, dating back to May 16 when he first discovered the Indians only to let them out of his sight, was conspicuously tentative and inadequate. Later, on June 14, Gibbon declined to ford the Yellowstone River despite repeated provocations from the Indians. He justified his lack of initiative with what turned out to be his implausible and overliteral interpretation of Terry's orders to remain on its north bank. Dr. Holmes Paulding, accompanying the command, saw it differently and took time to write his mother about the slothful and inexcusable lethargy of Gibbon, which, for the benefit of the reader, is quoted a second time here:

> A large camp was found up the Rosebud about 18 miles off, but our genial commanding officer [Gibbon] did not deem it advisable to attack it, a chance any other commander would give any price for, and after laying there for 10 days with the Indians showing themselves every day in plain sight, as though they knew what a harmless concern they were dealing with, he at last began to do something. Not to cross the command in boats and attack—but to go away.

Just as curious was that in a dispatch, Gibbon, an experienced Civil War officer, only off-handedly mentioned this sighting to Terry as if it had no import for the campaign. Inexplicably, Gibbon used none of his Crow scouts to keep a watchful eye on the moving village and later, as if to cover his failure, denied the village was ever sighted. Not surprisingly, the Indians vanished from the army's sight, which ultimately resulted in Custer, much like Robert E. Lee at Gettysburg, being deprived of Jeb Stuart's eyes and ears, going into the great valley nearly blind and merely guessing at the enemy's strength and location.

12. Leading the Wyoming Column, separate from Terry and Custer, General George Crook moved steadily northward, spoiling for a fight with the Indians. He got more than he bargained for when over seven hundred warriors, showing unprecedented brazenness, surprised him by attacking his twelve-hundred-strong force after he ventured toward their village. This encounter became known as the Battle of the Rosebud. Communicating knowledge of this attack and the Indians' newfound aggressiveness would have been valuable intelligence for Terry. At the Rosebud, the warriors were outnumbered close to two to one. Still, they fought a professionally trained army led by a renowned Indian fighter, resulting in a draw that forced both to withdraw from the field. Even then, he failed to send a single courier to either Terry or Gibbon informing them of the outcome. He did report the engagement to General Sheridan, but Sheridan was not in the field tracking the Indians—Terry and Gibbon were. Typically, warriors fled from the cavalry to avoid combat whenever possible unless their village was directly threatened. The Rosebud fight strengthened the

understanding that if the warriors believed their village and families were in danger, they would fight ferociously no matter the odds. The Indians' aggressive, yet indecisive engagement at the Rosebud also strengthened the warriors' confidence when facing the U.S. Army.

13. Five days later, oblivious of the Rosebud fight, Custer was sent to pursue these same Indians. Three days after that, he would face a warrior force more than three times the strength that Crook faced with half the number of men, divided into four separate columns. Had Custer received word of the Rosebud fight, he would have known that the Indians were unusually confident and combat-ready—far more likely to fight than scatter when their village appeared threatened.

14. One of the more galvanizing points about how ill-prepared the Lakota/Cheyenne village was for an attack is that while the hostiles were always alert for an approach through the valley, they did not expect soldiers to approach across the Wolf Mountain divide from the east. They had known about Gibbon's command for weeks and would have expected him to come up the valley from the north. Likewise, they would expect Crook, following his defeat, to come down the valley from the south. Most Indians were unaware of a third column in the field. Moreover, the natural corridors of travel were from the river valleys, which were north and south, like the Little Big Horn, Rosebud, Powder, and the Tongue. Hence, the Indian guards watched for the soldiers where they were not. So, not expecting an attack in full daylight, they were surprised when Custer attacked them from an unanticipated direction.

15. Yet, Custer's attack failed, attributable in no small degree to battalion commanders who could not come together at the critical time. More than any other single factor, Custer's aplomb, daring, and unfettered aggression were responsible for his meteoric ascent. A national hero after Gettysburg, Custer's persona, his attention-getting flamboyance both in attire and action, his showy brand of leadership, unflagging self-confidence, thirst for glory, death-defying charges, and extraordinary battle successes all factored into what happened at the Little Big Horn. Custer's many lauded triumphs proved an object of admiration but also envy and resentment among more than a few of his peers, a couple of whom he would later call upon and find wanting on that hot, dusty day at the Little Big Horn.

16. From their first meeting a decade earlier, Frederick Benteen intensely disliked Custer, and that animosity grew. The aura of the vainglorious hovered about Custer. And Benteen resented Custer's preening pretentiousness, coming to view his success more as luck than valor and skill. At the Washita, any hope of reconciliation was obliterated when, despite their victory, the 7th lost the popular Major Joel Elliott, who was found frozen stiff and mutilated with seventeen other troopers. For Benteen and some others in the 7th, anger festered over Custer leaving the field without determining Elliott's whereabouts or recovering their dead. Benteen considered that an unforgivable affront to the decency and

honor of the regiment. Benteen would state that Custer had abandoned Elliott to a cruel fate. Neither Benteen nor Custer's other critics seemed troubled that Elliott had bolted from the command without orders or that more than a thousand warriors materializing from downstream might have influenced Custer's actions that day. The deaths of Elliott and his small contingent of men darkened the hearts of some officers in the 7th, and the bitterness lingered with Benteen at the Little Big Horn. Benteen would anonymously use the press to snipe at Custer. Perhaps alluding to Benteen's involvement, Custer threatened to horsewhip the perpetrator of such slander to within an inch of his life, causing Benteen to reach for his pistol in an overheated moment that could have easily spun out of control. Benteen later mocked Custer's autobiography, *My Life on the Plains* as 'My Lie on the Plains.' As far as Benteen was concerned, Custer's autobiography was another example of commercial aggrandizement and undeserved glory-mongering. Later, in retaliation for Benteen's barbs and occasional recalcitrance, Custer accused the pasty-faced Captain of cold-bloodedly shooting down an Indian boy at the Washita, in what Benteen angrily insisted was an act of self-defense.

17. This toxicity between the two did not foster esprit de corps, and the narrative of a not-so-disguised hostility on Benteen's part would continue to the fateful day of the Little Big Horn battle. Other issues also plagued the regiment. Custer was known as a martinet. Rigid and severe, in his early years out west, he issued scores of orders and directions to straighten out and toughen up his regiment. One officer complained about how vexing these iron directives were, describing them as a swarm of evils from Pandora's box. Eleven of Custer's subordinates complained about his harsh methods, which involved a series of charges and a court-martial. In one notorious incident, Custer ordered sick men out of ambulances and into wagons to make room for his hounds whose paws had been wounded by cactus pricks. Benteen, of course, witnessed some of this and seethed over the man he held responsible. Also infuriating to Benteen were Custer's relatives within the regiment, including two brothers, a nephew, and a brother-in-law, Lieutenant James Calhoun, who was married to Custer's sister Maggie. An indignant Benteen resented the nepotism which may have fueled his animosity toward his Regimental Commander on that fateful June day.

18. Custer and Benteen had flare-ups even on the day of the battle. It is easy to conclude that Benteen's visceral hatred of Custer influenced his actions at the Little Big Horn. Custer may have sent Benteen on that scout to the left, over challenging terrain, as punishment for manipulating company inspection rules to covertly secure the honor of leading the regiment. Benteen certainly saw Custer's order as an act of spite, robbing him of his just due and reducing him to an even lower subordinate role that again deprived him of the glory that he had hoped to garner for himself in leading the advance.

19. Meanwhile, riding off to the south and ignorant of subsequent alarms that the village was fleeing, Benteen smoldered over his assignment. He did not send a report to Custer about the findings of his scout but instead took to carping about valley scouting for Indians "ad

infinitum"—Indians he was sure were nowhere near his assigned path. He undoubtedly malingered after he returned to the trail, especially at the first morass where the horses and mules were watered, in apparent spitefulness before rejoining the regiment as ordered. Nevertheless, Benteen's tardiness has been overstated as a reason for Custer's defeat, even if Benteen did spend too much time nonchalantly watering the animals despite the shooting he heard in the distance. Consequential or not, the sound of gunfire did not spur Benteen forward. That was the same gunfire prompting Captain Weir to advance with his company ahead of Benteen, confirming that he was in no hurry. On the other hand, two messages from Custer, the first purportedly delivered by Sergeant Kanipe and then from Trumpeter Martin, reached Benteen, despite the content of Martin's note to be quick and bring packs, both giving the unmistakable impression that the battle with the Indians was virtually won, with Martin mistakenly claiming that Custer was essentially cleaning up whatever little resistance remained. Based on reports from Kanipe and Martin, haste was unnecessary since the battle was over, and, in his mind, Custer had again won glory for himself while relegating Benteen to anonymity. In fairness to Benteen, after receiving Martin's note from Cooke, he did pick up the pace toward Custer's trail, despite Private William E. Morris's description of Benteen's command being as slow as if he were going to a funeral.

20. Also in Benteen's defense, by the time Private Morris saw him—Morris was in Reno's battalion—Benteen had already arrived on the bluffs. More important is what Benteen did after he arrived on the hill, where the command's actions are disturbingly questionable. Benteen realized that Kanipe and Martin could not be more wrong. Reno's command was defeated and demoralized, and Custer was nowhere in sight. Benteen understood then that Custer and Reno had separated——important intelligence Kanipe and Martin had never reported. In fact, "Where's Custer?" was Benteen's first question to Reno. Until that point, Benteen was more or less following Custer's orders until he was diverted from the main trail by the sound of firing and seeing the three Crow scouts. That sent him up on the bluffs instead of down to the valley where Custer had intended for him and the pack trains to go and where he had been heading. The village was there, and that is where Reno was supposed to be. But Reno had abandoned the valley and the subsequent timber position in a disorganized and panicked retreat to the higher ground where Custer was last seen. Plenty of legalities have been vetted and agonized over what Benteen should have done about who had precedence of command, Custer or Reno, in response to the note Martin brought from Adjutant Cooke. But from the time these orders were first issued, matters had taken a dramatic turn for the worse. Instead of supporting Reno's intact command as originally contemplated by Custer, Benteen was rescuing his shattered one.

21. This puts an entirely different slant on whether Benteen disobeyed orders to "be quick and bring packs." Many battle students have pointed out that Benteen knew he was disobeying orders. They contend that his later testimony at the Reno Court of Inquiry stated that Custer was dead (thereby releasing him of the obligation to obey Custer's orders) by the time he

reached Reno, a sentiment Reno also echoed when he testified that Custer's command was destroyed by the time he had left the timber. But Benteen stating that Custer was already dead, even if he did not believe so when he was on Reno Hill, was not meant to shield himself from temporarily stopping to shore up Reno's shattered command. In retrospect, who would be so pedantically legalistic and rigid to court-martial him about not explicitly following Custer's orders when the situation had gone terribly wrong from the time, at Custer's behest, Cooke had first sent Martin with orders for Benteen? By claiming Custer was already dead, Benteen cleverly sought to insulate himself from criticism that his command took more than two hours to saddle up to go to the Weir Peaks to investigate the sound of heavy gunfire downstream that he had been claiming not to hear.

If the relationship between Benteen and Reno had been something more than feeding off each other's mutual loathing of their commanding officer, they might have moved with more purpose toward the distant gunfire. If it had been Major Joel Elliott's detachment, whose carbines had been urgently barking downstream, might Benteen and Reno have reacted with more haste? No matter how it is diagnosed, Benteen owns his belatedness, and no rationalization on his part to twist the facts can justify his almost unconscionable tardiness in responding to the loud gunfire downstream.

22. Reno's culpability for the catastrophe is even more palpable than Benteen's. It is not a question of what Reno did wrong that day but, if anything, what he did right. Others have gone as far as to accuse Reno of cowardice, but the authors stop short of calling him little more than a no-good poltroon. How can anyone calumniously smear someone who fought at Antietam, the bloodiest day in American warfare, and at Mechanicsville, where he conducted himself commendably, and then slander his entire military career with such one-sided absoluteness? Reno may have turned out, in the pungent assessment of the historian Robert Utley, to be a "besotted mediocrity;" however, before the Little Big Horn, he had shown no outward signs of cowardice or incompetency during his nearly twenty-year military career. It is entirely fair to say that he experienced an acute psychological collapse, a loss of emotional equilibrium, and a critical absence of nerve at the Little Big Horn. No doubt, June 25, 1876, would be his day of infamy, by far the worst day of a military career that was otherwise colorless and relatively undistinguished but unmarked by disgrace or condemnation.

23. Those who knew Reno had to notice how morose he had become after being widowed two years before the battle, his only child now being raised by his late wife's relatives. Facing the loneliness of living on a frontier outpost, he began to rely heavily on the bottle for consolation. Moreover, Reno had little experience fighting Indians and possessed none of the inborn self-confidence that so motivated and directed Custer's primitive lust for battle. This seemingly innate reluctance did not stop him from trying to wrest command of the regiment from Custer when political troubles in Washington belabored him. Status in the military was important. Reno, despite his problems, was not indifferent to that. But while

Custer's fuse might have burned fast, he was not one to hold a grudge for long, and the rift between them had been relatively mended by the time of the Little Big Horn campaign. However, their relationship took a downturn after Reno botched a scout. Ordered to scout as far as the Tongue River, he disobeyed Terry's detailed and explicit orders. Instead, Reno went to the Rosebud, despite Terry's prohibition, and discovered an Indian trail but failed to follow it to ascertain its destination, which left the regiment no more enlightened than before. Reno's contretemps led to an outburst of emotions, especially from Custer and Keogh, who were scalding in their criticism and concerned that Reno, for no worthwhile reason, may have alerted the hostiles to their presence.

After a series of sharp questions by Custer and curt responses from Reno, Terry interceded to diffuse the situation. The next day, Custer, still fuming, wrote an anonymous letter to the *New York Herald* stating that a Reno court-martial was more than justified. It must have been particularly galling and humiliating for Reno when Terry removed him from an independent command of the 7th and returned it to Custer. Amazingly, despite all of Custer's political troubles, including with President Grant (the President himself initially forbade Custer from even participating in the 1876 campaign), he now found himself leading the 7th and commanding the strike force Reno desired. This was a tonic to Custer, as the one-time Boy General reveled that his old "Custer's Luck" was alive and well. But for Reno, it had to be mortifying to be returned to serve under the command of someone he now saw as an enemy, an officer whose military leadership he would later profess at the Reno Court of Inquiry to have no confidence in.

24. During the march from the Yellowstone, Reno was relegated to the rear until Custer ordered Benteen to scout to his left in search of Indian encampments leading to the Little Big Horn Valley. That opened the curtain for Reno to play a leading role in the campaign. He had been persona non grata since the regiment left the Yellowstone River, whereupon Custer ignored him, not even deigning to give him any orders or command responsibilities. Reno was so far out of the loop that even on the day of the morning march, he was stunned when he was given a command. He rode a distance back along the column when Custer summoned the officers. When Reno did not appear, Cooke went to fetch him, advising the Major, "The General directs you to take command of Companies A, G, and M," led respectively by Moylan, McIntosh, and French. These three companies amounted to one battalion of the regimental wing. When Reno noticed that Benteen was taking another battalion in a different direction, he looked back at his three companies and asked Cooke, "Is that all?" Reno was none too pleased to see Benteen, a formidable cavalry officer, ride off with three companies of the wing whose presence would have been profoundly reassuring to himself as they penetrated deep into enemy country. Though never one to engage in much conversation with Benteen, when he turned off the trail, Reno nonetheless almost solicitously asked where he was going. Benteen could only dryly reply, "To drive everything before me."

25. Being thrust back into a position of command may have been a psychological adjustment for Reno, who had been the picture of detachment and anonymity for more than a week. But now, he found himself leading a battalion intent on engaging an Indian village of unprecedented size. For days, Reno, like much of the command, had been hearing about the swelling number of hostiles in the valley and warnings from their Indian scouts since his return from his misbegotten scout on the Rosebud. His whiskey flask provided some comfort and courage, but some evidence suggests that he remained uncertain and insecure about his new assignment. Because no overall plan had yet to materialize, Reno betrayed his feelings several times before he was ordered to make his famous charge into the valley. Even though Cooke advised him that the entire outfit would support him, Reno continued to look for additional reassurance by seeking his commanding officer out and telling him, "Let us keep together," to which Custer, as Reno remembered it, tipped his wide-brimmed hat toward him as if to say, "I hear you."

26. It is possible that Custer sensed Reno's apprehension and decided to send Cooke and Keogh to tag along, stiffen Reno's spine, and ensure that his orders were being carried out. In fact, Custer may have initially intended for Keogh's battalion to enter the valley directly behind Reno's advance before being notified that the warriors were coming out in force to confront Reno. Keogh commanded Custer's lead battalion, which explains why he, rather than Captain Yates, accompanied Cooke to Ford A. When Reno departed with the advanced guard battalion, he believed that he was chasing a running village and that Custer's five companies would dutifully follow him into the jaws of battle. When Reno got to the river's ford leading into the valley, he appeared visibly nervous when he realized, contrary to his expectations, that Custer was not close behind. He was seen taking another swig from his flask of whiskey. When De Rudio's mount splashed water on him as he made his way across the ford, Reno reacted sharply, "What are you trying to do, drown me before I'm killed?" Reno was in no hurry to charge because he had hesitated long enough to send two messengers back to Custer, most likely seeking additional assurance he would be supported. Apprehensive about the diminished strength of his wing, rumors over the village's size, Benteen's absence, and the need for a sign of his commanding officer's promised support, he appears to milk every minute in a conscious attempt to buy time. With Keogh and Cooke close by and shouts of alarm about Indians coming up the valley, Reno finally ordered the advance. Reforming their ranks, Reno's battalion moved straight toward the village.

27. With a smarting Benteen five miles away, though reluctant, Reno had taken the 7th's first aggressive action against hostiles of still unknown strength and disposition. Custer, in receipt of contradictory reports about whether the Indians were running or making a stand in the valley, had not decided whether he should follow Reno into the valley, thus he ascends the high bluffs to see the situation for himself. Speed and decisiveness in a battle can make the difference between victory and defeat, so this was not an auspicious

beginning for an operation that demanded rapid and concerted action. Still, as he rode down the valley, Reno's military training and experience kicked in, and, to his credit, he led an orderly advance with his flankers doing their jobs and his scouts off like clockwork to attempt to stampede the enormous Indian pony herd on the western foothills. Reno was in command and control as they advanced down the valley. But for various reasons, including trepidations that he was inadvertently heading straight into an ambush, fears accentuated by the growing dust ahead, and visions of Indians pouring out of an imaginary ditch that existed nowhere in front of him, Reno ordered a dismount to form a skirmish line. That move sparked a heated debate that continues today about whether he should have charged headlong into a surprised village or whether his discretion in dismounting saved his command from being swallowed up into the belly of the beast.

To this day, a vocal contingent of battle students remains fervently supportive of some of Reno's actions that day, believing him to be prudently cautious in approaching an enemy of unknown strength and disposition. Many also credit him for adequately protecting his horses and placing safety over risk. Without unduly sidling up to the orthodoxies on either side of the question, this book affords Reno the benefit of the doubt—at least in the early stage of the battle. He was, after all, in charge, and the responsibility for his command rested solely upon his shoulders. If Reno continued to act as an advance guard to keep the enemy occupied by threatening the village, he would fulfill the objective Custer had intended for him. Whatever the should-haves, could-haves, might-haves—as far as Reno's actions up to this point—so far, so good.

28. On the skirmish line, however, questions began to multiply about the judiciousness of command decisions made by Reno and Captain Myles Moylan. History has too easily allowed Moylan to escape responsibility for the valley skirmish line collapse, the first in a series of calamities that would plague the 7th that afternoon. As Reno entered the timber to investigate rumors that the Indians had entered the woods and threatened the horses being held there, Moylan, an experienced combat officer, started to unravel on the skirmish line as he felt the pressure from the hostiles. He led A Company off the line and into the woods, either because he feared that Indians in his front were flanking him or they would stampede cavalry horses held in the timber. It is anybody's guess as to precisely what precipitated this premature move. But, by removing A Company from the skirmish line where that company had sustained no casualties, Moylan weakened the line and helped initiate a concatenation of events from which Reno's battalion never recovered. The Indians were doing little more than demonstrating before the soldiers with a few bravery charges to test the young braves' mettle while providing time for the non-combatants to escape from the village. But soon, much like a chain reaction, the remaining troopers, seeing their line weaken and contract, began to abandon it, following A Company into the timber. This retreat into the battlefield's wooded area caused the spirits of the tribal warriors to soar,

believing their prowess, rather than an officer's shaken nerves, had arrested the charge and driven the soldiers' withdrawal to a more secure position.

29. Although Reno's troopers had sought refuge in the timber, they occupied a strong defensive position, and its proximity still threatened the village. Moylan, however, continued to fracture psychologically, jeopardizing the battalion. Instead of deploying the troops defensively around the perimeter of the timber, he permitted them to move to the bottom where the horses were. That allowed the Indians to penetrate the outskirts of the woods and shoot downward at the highly vulnerable troopers. It was around this time that Bloody Knife, Custer's favorite scout, and someone Reno relied on, had part of his skull blown off by an Indian sniper while standing beside the Major. Covered with blood and brain matter, Reno's brittle confidence disintegrated at the gruesome death of Bloody Knife. Reno, under the mistaken impression, largely due to Moylan's poor deployment of the troopers in the timber, believed that the Indians were infiltrating the woods and threatening to steal and stampede the horses. His succeeding actions became a case study of a commander utterly untethered and teetering on the brink of derangement.

30. A traumatized Reno rushed from the timber with no effort through the sounding of trumpet calls to notify his troops that the battalion was preparing to withdraw from their defensive position. Nor did he attempt to secure and transport his wounded in this anticipated retreat. That meant almost half of G Company was left behind during his withdrawal. Bizarrely, several of his officers and men would later support his decision at the Reno Court of Inquiry, inviting speculation of collusion among them to save Reno's reputation and preserve the honor of the regiment, even as other officers, scouts, and civilian personnel questioned Reno's actions. Later accounts from the Indians show that even they were puzzled by the sudden exodus.

31. One after another, the mistakes continued to compound. During the withdrawal, Reno failed to take the most elementary steps to cover their retreat as his command crossed the Little Big Horn River. Mounted warriors compared the debacle to a buffalo hunt, riding right up to the soldiers and, in some cases, pulling them off their horses. Reno had broken a cardinal rule—an army outfit never turns its back on mounted warriors while attempting to outrun them to another defensive position. It was an egregious error and dereliction of tactical soundness for which his command paid a terrible price. Reno failed to establish a rallying point at his intended destination, only exclaiming for the few who heard him, "Any of you men who wish to make your escape, follow me." More than once, on his retreat and ascent up the bluffs, he had an opportunity to establish a screen of fire to cover the retreat but failed to do so. Reno not only lost a third of his outfit, but his actions further strengthened the resolve and confidence of the warriors, who had been afforded the time and opportunity to recover from their initial shock of being attacked in broad daylight. For the first time in the battle, the Indians seized the all-important initiative and, from then on, would never surrender it.

32. With this sudden turn of events, the warriors could fully concentrate their manpower on Custer's wing, which was now spotted at the north end of the village on terrain not suitable for defense. General Nelson Miles later seized on this fact and was insistent that the principal cause of Custer's defeat was Reno and Benteen's passivity and inflexible defensive posture on the bluffs. His words thundered with condemnation:

> "No commanding officer can win victories with seven-twelfths of his command remaining out of the engagement within the sound of his rifle shots."

General Miles succinctly summed up why the Indians would not be vanquished that day and would instead overwhelmingly destroy one-third of the command and almost mortally wound the remainder of the regiment before Terry's reinforcements rescued it.

33. With Benteen's arrival on the summit of the hill, no serious attempt was made to move toward the sound of the gunshots, which were heard by most of the officers, save Benteen, Reno, and Wallace. Although in his after-action report, Reno was compelled to admit that "we heard firing in that direction and knew that it could only be Custer." Yet there was no shortage of excuses about why a junction with the general's command could not be made: The wounded could not be moved off the hill—they were still waiting for the packs to arrive—although most of the advanced ammunition packs had arrived early enough. Better still was the supposed belief, later expressed by Captain Benteen, that Custer was not fighting but was somewhere down the Little Big Horn grazing his horses or making a junction with General Terry. Benteen arrived on the battlefield when no more than ten men of Reno's battalion were wounded, most of whom could have ridden a mule, even if difficult. But even supposing they couldn't be moved, enough able-bodied soldiers could remain on the hill to protect the wounded since nearly all the warriors had departed the vicinity of Reno Hill. Sitting Bull derisively said, "The squaws could deal with them. There were none but squaws and papooses in front of them that afternoon." Surely, three or four companies could have been mustered and sent just one-and-a-quarter miles to the Weir Peaks to make visual contact with Custer. Benteen's three companies were still mostly fresh and unscathed. The more likely reason for the inaction had nothing to do with the wounded but instead resided in Reno's fear that the Indians had set a trap and that he would be ambushed if he left the hill. As for ammunition, Moylan's company may have exhausted most of theirs, but the Reno battalion still had plenty of reserve ammunition. Battle survivors of the 7th were tight-lipped over the fact that the packs had arrived within thirty minutes or thereabouts of Benteen's arrival on Reno Hill.

34. The inertia was a direct result of the ineffectiveness, if not the outright failure, of the command to take the initiative. Military organizations are top-down driven. But, in this case, no one was driving until Benteen removed the reins from Reno's loose, slippery grip

and took control. This book argues that while fighting in the valley, Reno was not fully impaired by alcohol. But as time went on, there is good evidence that he became increasingly intoxicated, no doubt abetted by the trauma he experienced in the timber, Custer's absence, and the disastrous retreat to the bluffs.

35. The issues of distance and wounded soldiers were not prohibitively disqualifying in having at least three or four companies on Reno Hill sent to Custer. Of course, there is no guarantee that had Custer been reinforced earlier (say within an hour of Benteen's arrival on the hill), the 7th would have triumphed. As Benteen later insisted, it is possible that had he gone to Custer, he would have been there still—meaning that the entire 7th would have met Custer's fate that afternoon. It is also possible that boldly led reinforcements, in concert with Custer's forces, could have turned the battle in their favor or at least saved a good portion of Custer's command. Too many variables prohibit speculation on such a scenario, so hypothesizing must be left to the wargamers and theorists. Still, there is no denying that the adage to "ride to the sound of the guns" still echoes loudly in this battle and throughout history. This sorry episode of delay remains an inexpungable blemish, and most officers and soldiers on Reno Hill either knew it then or, at some level of consciousness, came to believe it later.

36. Custer certainly deserves his share in assessing blame for this catastrophic defeat. It was once said that Custer's actions were wrong for the right reasons. That might sound like an oxymoronic observation, but it bears a good measure of truth. In retrospect, Custer made some obvious errors, including neglecting to change the command structure of the 7th, partly because of his deference to military protocol. His command would have been much stronger had he addressed the fragmented relationships among the leaders or relegated to a lesser role any who disagreed with his thinking or were less able to cooperate. The ongoing bitterness that festered among some in the 7th Horse should have been excised before tackling a village of that size. Cooperation and trust among battalion commanders are key ingredients for battle readiness and success.

37. As a regimental commander, Custer may have been too reticent, too close-mouthed with his subordinates, unwilling to explain his intentions in detail, apparently believing that they could discern his plans by his actions. Reno was an officer clearly in need of direction and reassurance that day—Custer probably realized this too late, which may be why he sent Cooke and Keogh to accompany Reno when he was ordered to pursue a running village. But once alarms were raised that the Indians were coming out in force, Cooke and Keogh left to advise Custer of this new development, leaving Reno to his own resources. Meanwhile, upon hearing the news from multiple sources that the warriors were not scattering but coming out in force, Custer, apparently still skeptical, ascended to the bluffs to assess the situation for himself. Observing that the village was largely unaware of an imminent attack, he decided to attempt a 'fix and flank' maneuver against the enemy.

It is probable that Custer sent Private Theodore Goldin with a message to Reno about the change of plans. If he did, it was another error in judgment. Goldin, an inexperienced seventeen-year-old, likely failed in his mission to deliver it. Because of Goldin's unfortunate tendency to muddy his credibility by repeatedly lying about events concerning himself, it is impossible to draw a definitive conclusion on whether he was sent as a courier. On the other hand, there were occasions when Goldin's testimony could be verified. One feasible scenario is that Goldin was dispatched with a message for Reno, and when he crossed Ford A, he witnessed Reno's retreat with howling warriors in close pursuit. Young Goldin thus abandoned his mission whereupon he was later discovered after Reno's retreat at the base of the bluffs. Custer had also wrongly assumed that his battalion commanders would think similarly to himself when in the same situation. This proved an enormously erroneous assumption, and, in the end, Custer had mistakenly extrapolated his own motivations and dispositions of placing mission first to those of his subordinates with fatal consequences.

38. Custer sending Benteen on a scout twelve miles from the valley was also a highly questionable move. It is possible that Custer mistook the satellite camp at the mouth of the South Fork of Ash Creek, as one can easily do when sighting from the Crow's Nest, as part of the Little Big Horn Valley, and sent Benteen to the valley of the South Fork in an effort to flank the enemy from the south. But it is also possible that having enough of Benteen's backtalk and manipulative behavior, the scouting excursion was a punitive measure that had Benteen chasing ghosts. Or it could have been a combination of both. Custer discerning a dozen or so tepees from the Crow's Nest, even if he suspected that they were indicative of many more, is not sufficient justification to order nearly one-fourth of his attack force several miles away from the rest of the regiment. Of course, it is always prudent to verify the existence and disposition of enemy forces, as Washita had taught him, but a small reconnaissance party rather than a battalion of an entire regimental wing would have been sufficient to determine the strength of any satellite camps without depriving Custer of the majority of Benteen's command.

39. As soon as Custer was alerted by Fred Gerard, who was heard yelling, "There go your Indians, General, running like devils," shortly after being informed by Half-Yellow Face of the rising dust seen in the distance, indicating large numbers of Indians fleeing, he was compelled to give the order to attack the village. And while this is speculation, there is a good chance Custer must have regretted sending Benteen with an entire battalion to the left to act as a reconnaissance in force. But such regrets came too late, and there was too little time, based on the information he was receiving, to wait for Benteen. He had to act immediately and do his best with what he had or have the villagers scatter in multiple directions, resulting in the campaign ending in an abject failure. The dust cloud pointed out to Custer came not from a fleeing village but from Indian herders who got word of the oncoming bluecoats and drove their ponies off the western foothills to avoid their capture.

But the shattering news of the arrival of the Long Knives, as the warriors sometimes called the soldiers—although the 7th carried no sabers that day—was information that had yet to radiate through the sleepy, nearly two-mile-long village. Custer's most salient problem now was bringing together what he had divided.

However, analyzing what Custer knew and when he knew it makes it hard to figure out what he could have done differently. He overestimated the Arikara scouts' ability to disperse such a large pony herd, which allowed the warriors to mobilize by seizing horses that the scouts were supposed to have stampeded away, however, the horse herds were primarily located on the opposite side of the village, making their successful capture questionable under any circumstances. Additionally, because Custer had almost lost his pack train to war parties back at the Washita, he overcompensated for that mistake at the Little Big Horn by assigning too many troops as its guard. There was no need for the pack train guard to be so large, especially since there was no sign of an enemy in his rear, and any attack from the front would have to first go through Benteen. Many of these guards could have been better utilized by being assigned to himself or Reno—though it almost certainly would not have changed the result.

40. Then there is the criticism that Custer pounced upon such a huge village without a full reconnaissance, and in a manner that would have given other cavalry commanders pause. However, doing so was actually standard operating procedure during the Plains Indian Wars. The author Greg Michno put to rest the idea that cavalry attacking an Indian village without prior reconnaissance was a rare or seldom attempted maneuver. After some solid and incisive research, Michno made a capital point when he stated that this has never been the case:

> There were 206 incidents in which the military followed the Indians into the camps or the villages and attacked, regardless of the numbers. That was the whole point of the pursuit: Find the Indians and attack. No commander from the 2nd lieutenant to the general, would expend time and energy to track Indians only to call it off at the crisis point, even with unfavorable odds … Of the attacks that followed a pursuit, in which the record is complete enough to tally, we find only 60 of them were planned out; 119 were spontaneous. With this in mind it is easy to see why Lieutenant Colonel George Armstrong Custer made his attack on Little Big Horn. He had found a village; it was his duty to attack. Those who say Custer attacked too hastily, without thorough reconnaissance, can see from the data that his actions were well within the norm. Those who condemn him do not understand the basic tactics of the Western Indian wars.

41. After Reno reached the river crossing, today known as Ford A, Custer was informed that the Indians were making a stand and were coming out in strength. Custer acted on that intelligence, breaking off from following Reno and ascending the bluffs instead. He may have unsuccessfully attempted to inform Reno of this change in plans. Arriving high on Sharpshooter Ridge, he saw with his own eyes the location of the village and that the Indians were not only remaining but were completely unaware that they were about to be attacked. Custer galloped to the north away from Reno, though he purposely made himself visible to Reno's troopers down in the valley, whereby he signaled to them by waving his hat. This move toward the other end of the village has all the earmarks of a 'fix and flank'. Reno's advance guard battalion would serve as the fixing force, and Custer would act as the flanking force. When Benteen arrived in the valley, he was expected to support Reno when Custer struck the enemy's rear. It is not unreasonable to surmise that an experienced battalion officer like Major Reno, who had fought in the Civil War, would have understood Custer's intentions when he and others saw the General's battalion riding downstream. While it is true that a commander more cautious and conservative-minded than Custer might not have acted so aggressively or undertaken greater risk, especially without consulting his battalion commanders beforehand. But Custer lived and, as it turned out, died by the maxim that fortune favored the bold. That was his creed.

42. Through no fault of his own, Custer acted on information that he had good reason to believe was accurate. He had been advised, wrongly as it turned out, that the village was running, causing him to dispose of a fuller reconnaissance and commit to a hasty attack that day, rather than the next morning as planned, and doing so without the immediate assistance of Benteen's battalion. But the Indians were not running—nor were they anticipating an attack. Thus, he orders Reno's battalion, as an advanced guard ahead of his own, to run down the fleeing village and, in concert with the rest of the regiment, destroy whatever remnants of the camps or pony herds that remained behind, kill whatever warriors who resisted, and destroy their ability to survive the coming winter, pursuant to the January 31, 1876, ultimatum.

43. Meanwhile, even before Benteen had been summoned by Cooke's note of "Come on. Big village, Be quick, ..." he understood that his battalion, along with all other forces, was to go to the valley. Private Davern heard Cooke tell Reno, which he later testified to before the Reno Court of Inquiry, that Benteen would be given the same orders as himself, which was to pursue and attack the hostile village in the valley and to arrive on Reno's left. Once the village was destroyed, the Indian population, especially the non-combatants, no matter where they escaped, would be forced to give themselves up and return to the reservation. Even when Custer changed his plans and rode up the bluffs for more accurate surveillance of the enemy, his objective was still to move down into the valley and strike the village, albeit from another direction.

44. Somewhere near Medicine Tail Coulee, Custer's plans for a 'fix and flank' maneuver changed when he learned that Reno had abandoned his position in the valley. The Indians were no longer fleeing or slumbering as they were when he first sighted the village from Sharpshooter Ridge but were now fully aroused and fighting fiercely. Reno's retreat freed several thousand warriors to concentrate on Custer, who, after making a demonstration at the river crossing at Ford B to draw the Indians off Reno, retreated back up Deep Coulee to Calhoun Hill. There, he divided his command for a third time. This tactical maneuver shows Custer was not about to curl up in a defensive posture and call it a day. He was still counting on Reno to reform his battalion and join with Benteen, who had been ordered to make haste and rejoin the command. With all his forces acting as one, he still could reverse the tide and win the battle. But once Custer retreated from Ford B to Calhoun Hill and Battle Ridge, reuniting the regiment would be his only hope to turn things around and seize victory. It was risky for sure, but Custer thrived on risk.

    Had Custer been less driven, he might have taken the safer course and prioritized survival, not victory. If he had resigned himself to the likelihood that reinforcements would not arrive soon enough, he likely would have deployed his forces much differently. He would have avoided a divided wing spread out against an enemy that outnumbered them by more than ten to one, especially considering the ground's notable defensive liabilities. He would have also chosen to fight on ground with access to water and an unobstructed field of fire, which Finley Ridge, Calhoun Hill, and Last Stand Hill did not afford him. If he had moved his forces to Cemetery Ridge earlier or to the northern Flats, he would have surrendered any possibility of winning the battle. However, with all five companies under his direct command, he would have almost certainly resisted the sweeping and total destruction that sealed his fate, and the battle would have been called something less ringing than "Custer's Last Stand." Custer was not always and inalterably offensively minded. While it is true that offense was his natural reaction in battle, he had proven that he could fight effectively and skillfully on the defensive at both Yellowstone fights—August 4th and 11th—in 1873.

45. Many battle students believe that Custer lost the battle because he was wedded to his own presuppositions, mostly borne out of personal experience and certain supposedly immutable facts connected to Plains Indian warfare: (1) Indians would not aggressively attack a large body of U.S. Cavalry; (2) they dispersed in the face of a resolute Cavalry charge; (3) that a disciplined skirmish line, with organized fire, could thwart any number of warriors; and (4) that Indians were loath to suffer casualties even for the price of victory.

    Many of these factors held true at the Little Big Horn, but they were not all hard and fast rules. The warriors did attack a body of cavalry when they destroyed Fetterman in 1866. They attacked General Crook at the Rosebud despite being outnumbered. And, true to form, they would attack the 7th Cavalry aggressively at the Little Big Horn when their non-combatants were threatened. The warriors did not, however, overrun disciplined skirmish lines at the Little Big Horn, whereas they did assault weak, fractured, and retreating ones.

Some warriors, like the Suicide Boys, were willing to suffer casualties to achieve an objective, but most were not. While the warriors often acted according to the generalizations that were believed about them, they sometimes reacted unconventionally and counterintuitively.

46. Regarding Custer's deployment on Calhoun Hill and Finley Ridge, some have derided the notion that Custer formed the line to facilitate Benteen either by holding back the warriors or by closing in on their rear if they left to thwart Benteen's crossing over Medicine Tail Coulee toward Luce Ridge and beyond. This was Professor Charles Kuhlman's thesis (See *Legend Into History*, p. 118, 183). The troopers who were intended to facilitate Benteen's passage were those sent by Keogh to deploy on Finley Ridge after ostensibly spotting troops on the Weir Peaks. Kuhlman, however, mistook the deployment of C Company on Finley Ridge for a skirmish line deployed in the Basin area. Under this new model, the scenario becomes quite plausible that Benteen would proceed forward to Luce Ridge under cover fire from the troops on Finley Ridge.

47. Instead of scattering and diffusing their strength by fleeing from the cavalry, the Indians gathered and became more united in action. At the same time, Custer's command was, pursuant to his own orders, fragmenting. In other words, the warriors were slowly coming together in strength on terrain favorable to them while Custer's wing was being weakened and fragmented, and they occupied less favorable defensive terrain. This gathering of warriors in one specific location was probably welcomed by Custer when first sighting the village since such an aggregation signaled the possibility of any satellite encampments downstream was not likely. With the warriors congregating on the river's east side, Custer saw an opportunity to set a trap for another fix and flank that would conclude the battle in his favor. But this could only work if Benteen and Reno arrived on the Custer battlefield, which never happened. In the end, Custer realized all too late that there would be no reinforcements or cooperation for a new fix and flank, no directed and organized crossfire between commands, and that his five companies, deployed on relatively poor defensive ground while anticipating badly needed support, could no longer sustain themselves without support. That resulted from Custer trying to keep both offensive and defensive possibilities open. He could not forestall the flood of warriors rapidly closing in on him and ultimately deluging his entire command.

48. The Custer battle was a moving fight in the way battles involving cavalry often are. Dismounted skirmish lines were established at Calhoun Hill, Finley Ridge, and Last Stand Hill, but only as a temporary holding action against enemy forces until all of Custer's attack battalions could be reunited for further action. When that failed, Custer's squadrons were separated in an attempt to find and hold better defensive ground. When the enemy attacked in close-quarter combat, they were unable to provide a unified defense and lacked sufficient ammunition for a sustained siege. Given the situation, it became only a matter of time before events would be brought to a bloody close. But despite these setbacks, it would be

wrong to assume that the Custer battlefield was a snapshot in chaos and disorder—until the final few minutes, it was not. For most of the fight, the units formed an organized and resolute defense. While some soldiers fled for their lives toward the end, the command did not immediately implode. Unlike what Benteen would later describe (aside from the final few minutes of the battle), the troops were not scattered about like corn. There were pockets of defense, points of stubborn and even valiant resistance. To reach such a conclusion as Benteen did would be a gross historical misinterpretation. Custer's Last Stand, whatever mystifying elements still surround it, was aptly named.

49. In the end, Custer's wing was utterly destroyed. There were 268 men of the 7th Horse who either died during the battle or soon thereafter from their wounds. This is an established fact. But how many died with Custer is still debated in some circles, and the only certainty is that no one in his command escaped or survived. From the beginning, a vigorous effort was made to determine what the figure was—with each remaining company sweeping over the battlefield to locate their dead so that no cavalryman, scout, or civilian who rode with Custer would be overlooked.

"In war," Winston Churchill said, "truth is often the first casualty." And the question of fatalities is a topic often subject to distortion. For quite some time, the National Park Service claimed that Custer and all 225 in his immediate command died in the battle. Long before that, other numbers were bandied about, including those there during the battle or who arrived in its immediate aftermath. General Terry stated there were 204 who died with Custer, which echoed what Major Reno had told him. Captain Benteen thought the body count totaled 203, but he omitted more than a handful of soldiers that later would be found some distance north of the battlefield. Lieutenant Bradley posited a figure of 206. Sergeant Ryan said 207 were killed, and Lieutenant Godfrey, including dead troopers he found in the village, stated the true number was 212. Over the years, historians have retrospectively weighed in on the question. Robert Utley, long considered the dean of Western historians, thought the figure was around 215. As for the authors, while recognizing additions and subtractions and that the total will continue, we have settled on a figure of 210 for no better reason than Little Big Horn scholar Frederic Wagner, III—who was acutely precise in compiling numbers and who did an exhaustive study of the 7th Cavalry's muster rolls—believed that was the correct figure. Indian casualties are much harder to come by. Regardless of the number, good and brave men died on both sides, and that is what really matters and should be remembered.

Afterword

Years ago, if you drove along one lonely, open highway in Montana, an enchanting endless expanse of land undulating under an immense azure sky, your reverie would be suddenly broken by a large billboard sign: "Custer Died for Your Sins." The phrase came from the title of a book written by Vine DeLoria, Jr., published in 1969 when the repercussions from the countercultural movement and the intense controversy over the Vietnam War had come home to roost when America's moral foundations were being seriously challenged. While it is true that the most memorable thing about that book was the title, seeing that inscription on a looming billboard was emotionally arresting, even a little shocking—like a giant accusatory finger pointing directly at one's guilty conscience. Simmering with political turmoil and controversy, one realizes how morally allegorical the great, mythological West had become. The stark and primitive beauty, with breathtaking vistas so magical that eyes feast on a seemingly never-ending flattened surface, is enough to make one question the earth's curvature. Yet the real story is not the majesty of its physical dimensions or its scenic panoramas but the flesh and blood people who settled, lived, fought, bled, and died upon the North American continent's ancient crust. If there ever was a place on this earth made for storied epics, this was it.

Making sense of what happened at the Little Big Horn often leads to the all-too-familiar tendency to jump to unsupported and even wild conclusions that had begun even before the bodies were buried in their shallow, makeshift graves. No one had all the answers—and no one ever will. It is a subject loaded with dynamite. A figure as controversial and strife-ridden as George Armstrong Custer lies at the center of what ended these Indigenous people's way of life—where a violent clash of cultures concluded the longstanding "Indian problem" that had thrived on these rolling plains for centuries. Custer's role has immensely complicated and suffused the story with multiple layers of controversy. Remembering our resolve not to delve into all the intricate, moral questions of the rights of man, Western imperialism, and the ineluctability of historical determinism, we still are left with the battle itself. In the immediate months and years following Custer's defeat, as much as the army wanted to sweep away this military catastrophe, it could not avoid an accounting. "I don't want millions of rubles," protested Alyosha, one of the protagonists in Dostoyevsky's *Brothers Karamazov*, "I want answers to my questions." Ever since General Terry broke the news of Custer's death, the questions have never stopped. The battle, having aroused such public interest, cried out for a full hearing to unearth the facts wherever they might lie.

After nineteen long months of finger-pointing, guesswork, and mutual recrimination, time enough for the army to absorb the initial shock and put things into some semblance of order, they reluctantly agreed to make an official effort to come to terms with and determine the causes of the 7th Cavalry's ignominious downfall on that torrid afternoon of June 25, 1876. With so many reputations at stake, including that of the U.S. Army, this proved easier said than done. The journalist H. L. Menken once pointedly noted that the world makes a big show about the truth, yet most do not want to tell or hear it.

Menken's observation pretty much encapsulates the Army's attitude when it finally got down to the business of holding an inquest in the windy city of Chicago, Illinois, which commenced on Monday, January 13, 1879, in Room 14, on the mezzanine of the eight-story Palmer House. Three somber judges sat at the crossbar of tables arranged in the letter "T," their objective was to review the actions and commands of Major Marcus Reno and get to the bottom of a mystery that, for more than thirty months, had mesmerized an entire nation. This was a deeply disconcerting task for the army, knowing nothing about the 1876 campaign was redeeming. The high command had no desire to relive the circumstances surrounding that tragedy and was more than anxious to wash their hands of the matter. In the end, they realized that it was better for the investigation to be under their control than that of a wider-ranging and more searching congressional committee.

The first reports filtering out from the military after the battle, forming the first tenuous draft of history, were as subtle as a brass band trumpeting where the blame for the entire mess lay. Of course, what happened and its magnitude were shocking. General Sherman thought the news was almost too terrible to be entirely true—that it must be exaggerated: "I cannot believe that Custer and his command would be swept away," the old General mused disconsolately, "I don't think there were enough Indians there to do it like that." But many reactions flooding in gave voice to Custer being preponderantly at fault and the primary cause of the disaster—and all of it happening right at the cusp of the nation's much-awaited Centennial celebration.

General Sheridan wrote that Terry's column was sufficiently strong to have handled the Indians if Custer had waited at the junction. Sheridan's message was unmistakable—rash impudence brought this catastrophic and stunning defeat. "Had the 7th Cavalry been kept together it is my belief that it would have been able to handle the Indians, and under any circumstances, it could have at least defended itself." Sheridan wrote to General Sherman in the same vein, "I deeply deplore the loss of Custer and his men. I feel it was an unnecessary sacrifice due to misapprehension and a superabundance of courage—the latter extraordinarily developed in Custer." Meanwhile, General Terry, alluding to Custer, said, "Every soldier here who is suffering with wounds is the victim of a terrible blunder, a sad and terrible blunder." Later, writing to General Sheridan, Terry fleshed out his criticism of Custer, claiming he had consciously diverted from the plan that had been carefully laid out: "For whatever errors he [Custer] made he has paid the penalty and you cannot regret his loss more than I do, but I feel that our plan must have been successful had it been carried out, and I desire you to know the facts."

General Terry also had gone as far as to say that if Custer had survived, he would likely have been court-martialed for disobeying orders. Major Marcus Reno joined in with this indictment, making sure to list in his official report what he saw as Custer's fatal mistakes, ultimately concluding that the defeat resulted from his impulsiveness. Samuel Sturgis, Commanding Colonel of the 7th, having sat out this campaign, was scathing nonetheless in his criticism of its commanding officer. His son, Lieutenant

James Sturgis, had died with Custer, and the Colonel, convulsed with grief and embittered to the core, vented heavily: "He [Custer] was guilty of disobedience and of sacrificing good men's lives to win notoriety for himself." Believing history should condemn Custer, he added for good measure, "If a monument is to be erected for General Custer, for God's sake let them hide it in some dark valley, or veil it, or put it anywhere where the bleeding hearts of widows, orphans, fathers and mothers of the men so uselessly sacrificed to Custer's ambition can never be wrung by the sight of it." President Grant, who so admired Custer during the late War Between the States, sizzled with condemnation: "I regard Custer's massacre as a sacrifice of troops that was brought on by Custer himself, that was wholly unnecessary—wholly unnecessary."

Thomas Gray's poem "Elegy Written in a Country Churchyard" warned, "The paths of glory lead but to the grave." Perhaps a fitting obituary for Little Big Horn, considering Custer's known and unrelenting appetite for fame and notoriety. But it has never been easy to proffer one solitary, cautionary moral in this engrossing tale of many loose ends and unanswered questions. For the gravediggers who were left with the morbid task of burying their dead and mutilated comrades, it was as if a fire from heaven, or maybe hell, had scorched the landscape that lay mute, withered, gaunt, and bleak. Their dauntless leader slain, they went about lowering him just beneath the earth's surface. Sergeant Ryan captured the mood as he succinctly and plainly described Custer's and his brother's interment:

> At the front of the knoll we dug a grave 18 inches deep and then laid the General in it. We then took the body of Tom Custer and then laid him beside the General. We wrapped the bodies in canvass and blankets and then set them side by side in the shallow grave and then covered them with dirt. We took a blanket from an Indian Travois turned it upside down and laid a row of stones around the edge to keep the wolves from digging them up.

Requiescat in pace? That would be wishful thinking. With his martyrdom, Custer drifted into not only being more of a myth than a man but also an unenviable symbol of defeat, a sobering caution against hubris and overconfidence. A cautionary tale for those with the panache of a conquistador—an austere reminder that "Pride goeth before destruction, and a haughty spirit before a fall." The truth, however, is that the real story behind this staggering defeat was more complicated and complicitly entangled among the command than anyone at the time dared to admit. Like so many sentries, a steward of lies guarded the so-called official version of what happened. But enough of the facts surrounding the battle had begun to seep out, with a different version of events pointing to others who had shared the blame with Custer. Most of the allegations targeted Major Marcus Reno regarding what he did and failed to do that day. All the innuendos and carping began to make Major Reno very uncomfortable, even as the army did its best to ignore the matter in the vain hope that it would ignore them. Reno was stung to the core by the backbiting and criticism and was determined to clear his name from the incessant prattle and scandalmongering that was blackening his reputation. With an insatiably curious public also demanding answers, the Army had little choice but to open old wounds and conduct an official investigation into Reno's conduct during the battle.

Or so they at least publicly maintained. However, conducting a fair and transparent inquiry into the facts relating to Reno's conduct in the campaign proved problematic, as witnesses called to testify often appeared to be less than candid and forthcoming. The collective testimony given at the Reno Court can be characterized more by what was left out of the record than what was put in. There had been talk of Reno's largesse and hospitality at his hotel suite before the inquest, which may have worked to influence and soften up what might be said under oath. Lieutenant Charles Varnum always emphatically denied this, but there is enough anecdotal evidence to invite suspicion that the deck had been partially stacked. Godfrey suggested as much, and so did De Rudio, whose statements would border on suggesting that such cooperation was wrought to the point of collusion—although other officers vehemently denied any such machinations behind the scenes.

Yet, the hearing was characterized mainly by the winks and nods of more than a few of the twenty-three witnesses called to testify under oath. Attempts at cross-examination were mostly ineffective. It became all too clear that there was more intent to conceal than to reveal in what was possibly a concerted effort to uphold the regiment's honor. Lieutenant Gibson, who rode with Benteen on his left flank march, was not even called to testify, nor was Captain French, who fought in the valley with Reno. Those who did testify remained mute about what they knew—or did over the length of the battle—or about what was said or believed while on Reno Hill.

The Arikara interpreter Frederic Gerard caught onto these tight-lipped tactics, saying, for example, that Captain McDougall was a good soldier but would not dare tell the truth. Many, no doubt, justified their studied reticence to tell all they knew by concluding that Custer had made a fatal mistake in dividing the regiment before a vast, numerically superior enemy and had done so without proper reconnaissance. To attribute the downfall of the 7th to anything else, no matter its relevance to the truth, was merely tangential and counterproductive. After all, the controversial, enigmatic, irrepressible, flamboyant lightning rod known as George Armstrong Custer was dead. Nothing that would be said or done, including besmirching the high reputation of the 7th at this hearing, would bring him back to life.

Bruce Liddic, the outstanding Little Big Horn scholar, had it right when he said there was enough whitewash at the Reno Court to cover half of Chicago and that Benteen was not the only one who dipped his brush into the whitewash bucket. Ultimately, and with little fanfare, Reno was exonerated of charges of cowardice and incompetence against him but in a somewhat backhanded way, with General Wesley Merritt memorably saying that "the officers of the [7th Cavalry] gave us nothing, and so we damned him [Reno] with faint praise." As much as the Army hoped that memories of this tragic event would disappear over time, both the battle and the figure at the barometric center of it all refused to fade away.

History came to see the events that transpired that day differently than the Reno Court, but not with unanimity. General Merritt was right that the verdict of the Court was largely exculpatory regarding Major Reno, and the only comment that was at all censorious came in the form of a mild admonition reading:

> The conduct of the officers throughout was excellent, and while subordinates in some instances did more for the safety of the command by brilliant displays of courage than Major Reno, there was nothing in his conduct that requires animadversion from this court.

Almost immediately after the conclusion of the Court of Inquiry, ugly rumors circulated about collusion among the officers who stacked the proceedings to insulate the regiment. The Arikara interpreter, Frederic Gerard, and General Wesley Merritt, a judge on the Reno Court, who had served on General Phillip Sheridan's staff for two years and was given command of the 5th Cavalry, both referred to collusion among the officers of the 7th Cavalry. They were hardly alone. Lieutenant Carlo De Rudio saw the proceedings as a circus, a sham, a whitewash. He told Walter Camp that there was a private understanding among some officers to do all they could to protect Reno. Others, however, vehemently denied that there was any such surreptitious agreement between the officers.

But there were just too many things that did not seem honest, and many came to believe, as the saying goes about corruption, that things were not just 'rotten in Denmark' but also in the windy city of Chicago. Accusations of manipulation and cajoling of the witnesses continued to whistle like a noisy teapot. Frederic Gerard remembered how Reno sent for him in the most hospitable manner. There was plenty of dining and wining (imbibing) where the witnesses were present during the trial. Gerard further stated, "Reno had me talk to his lawyer." Readers can make up their minds regarding this interference, but to the authors, this sounds like witness tampering. At the close of the hearing, Gerard remarked that it was amazing to see how badly some memories failed in the space of three years. Just as curious was the transformation of the morose, humorless Reno. The Arikara called him "the man with the dark face" and, we might add, a personality to match his physiognomy. In Chicago, however, he was a picture of conviviality, and his bonhomie was on full display with his fellow officers, entertaining them in his hotel suite where liquor flowed freely. Author and Custer biographer Frederick Whittaker even made the unsupported claim that "Ladies of Pleasure" were brought into Reno's suite.

Although Reno had escaped censure, his career took a major and irrevocable nosedive after the Court of Inquiry. His drinking grew worse, and he was caught peeping through a window of the quarters belonging to Colonel Sturgis's daughter, a young woman admired for her comeliness. An outraged Sturgis pressed charges, and Reno was dishonorably discharged from the army, where he was already alienated from many of his fellow officers. He married again, but it did not last, and he drifted about in a lowly government job until tongue cancer took his life at the age of 54. Despite the criticism he received from many regarding his actions at the Little Big Horn, he still has a group of vociferous supporters who defend most of his actions and point out that his military career before the Little Big Horn debacle, while unspectacular, was marked mainly with competency for which he was recognized.

Benteen fared better than Reno, but not by much. He held his ground and, unlike Reno, at least after the Court of Inquiry, was aggressive in defending his actions and damning Custer at the Little Big Horn. How much truth was shaded or unspoken by others at the inquest could only be guessed. Benteen, himself, was forthcoming in his comments and openly admitted to shielding Reno in his response to a letter from Theodore Goldin, with whom he regularly corresponded: "As to the query before the Court of Inquiry, these I would answer as I did then and shield Reno quite as much as I did then." Protecting Reno would be an understatement if Benteen is to be believed. Five years after the battle, Benteen told Godfrey quite explicitly, and later on would hint as much to Captain Moylan, though without any corroboration, that on the night of the 25th, when the command was isolated and surrounded on the hill, Reno had proposed abandoning the wounded and making a forced march to safety. Even if some of these claims were exaggerated or shaded, enough post-inquiry evidence and documentation exist in letters and newspaper interviews to say that the truth, the whole truth, and nothing but the truth, did not prevail.

Why so many were so reticent to speak out can only be surmised. They all knew there was enough blame to go around, notwithstanding Reno's shortcomings as a commanding cavalry officer that day. Examples abound: failure to provide covering fire for soldiers withdrawing from the timber, failure to provide cover fire a second time when they fell back from Weir Point to Reno Hill—leaving behind the wounded, including Farrier Vincent Charley, to an enemy known to torture its captives, and more than three hundred soldiers lingering on Reno Hill for some two hours as they heard gunfire downstream. There was an overwhelming sense of not only personal failure but also a failure of regimental proportions, even if Custer had repeatedly and questionably divided the command and overplayed his hand.

Referencing Napoleon's invasion of Russia, the diplomat Talleyrand said that it was worse than a crime; it was a blunder. This was precisely how the Army chose to view the Little Big Horn campaign—as a blunder. As in Shakespeare's Hamlet: "The rest is silence." And I suppose one could add that some things are better left unsaid and unwritten. And yet, the one officer in the outfit who never bowed to silence and never wavered about being unrepentant was Captain Frederick Benteen. Blustery and irascible, he not only refused to quail before the judgment of history but also attempted to brazenly inform and shape it. How else to explain, shortly before his death, his vitriolic blast to Theodore Goldin about there being a kind of divine judgment against Custer and his epigones:

> The Lord in his own good time had at last rounded the scoundrels up [no doubt referring to the Custer Clan as he called them], taking, however, many good and innocent men with them. You know enough of me to know I'd have gone through to him had it been possible to do so. At the same time, I'm only too proud to say I despised him [Custer] as a murderer, thief and a liar—all of which I can prove.

Benteen again wrote Goldin, seemingly tired of the whole thing, stating, "I'm out of the whirlpool now, it's a dead, dead issue." But no sentiment could have been further from the truth. While it is hard not to believe that there was some kind of schadenfreude at work in Benteen's uncompromising attitude after the battle, and while he expressed regret about the death of the undeserving, he clearly was delighted that his nemesis got his just rewards, a chastisement long overdue. In this, he never disguised his feelings. Custer had paid the price for his arrogance and self-absorbed importance: "Watch ye therefore for ye know not when the master of the house cometh." And so, for Benteen, the righteous hand of the Lord could be stayed no longer and the decimation of Custer and his legion of soldiers was a divine judgment from which Benteen gained no small measure of satisfaction in contemplating.

But Benteen's role in the result continued to tread water as he sank deeper into the whirlpool he claimed he was no longer in. Indeed, the "dead, dead" issue was never more alive. No matter how much distance Benteen put between himself and those memories of the fatal June 25, the sound of the hoofbeats of the 7th galloping into the valley of death was never far behind. The story and his role in it rotated with dizzying and febrile excitement, where he would be pulled down into a tumult of confusion and intrigue. He would find succor from its vertiginous, spiraling violence and even redemption on "Reno's hill," a mound where he became transfigured before the world as the "Savior of the Seventh." Still, accusing voices would beckon him back in time to the wreckage of five cavalry companies on the east side of the river "Styx." Not even the slumber of death could insulate him from the endless questions, speculation, and inferences of suspicion about what he did and what he could have done on that battlefield. A century and a half later, the Court of Inquiry remains unadjourned, and the bitter quarrel rages on.

As for Custer, he would live on both famously and infamously. For all the glory and plaudits he had won in his military career, he would be forever tainted, almost solely remembered for this one overwhelming defeat. Moreover, with the flavor and appetite of contemporary American politics to loath its own history and being awash in national guilt over its past transgressions, Custer, alongside Christopher Columbus and others, has been frequently depicted as a remorseless, genocidal killer. Yet, closer to the truth would be saying that despite fighting the Indians for the better part of a decade after the Civil War and the brutality and pitilessness that war brings, he was surprisingly empathetic about his foe, understanding more than most, the tragedy of what would be their inevitable future. In some ways, he even aligned himself with the native population and the mounted warriors of the plains, writing with admiration about these majestic, defiant horsemen and their ferocious, determined insistence not to bow to a world they little knew or understood:

> If I were an Indian, I often thought I would prefer to cast my lot among those of my people attached to the free open plains rather than submit to the confined limits of a reservation, there to be the recipient of the blessed benefits of civilization with its vices thrown in without stint or measure.

As for the native tribes, they had gloried in their fight with Crook at the Rosebud and especially Custer at the Little Big Horn, but they must have known that the sun had come to set on their way of life. What would popularly be called "Custer's Last Stand" paradoxically, as noted countless times, became their own. Just fifteen months after the Battle of Little Big Horn, Chief Joseph of the Nez Perce, seeing the inevitable onslaught of Western expansion and Manifest Destiny, offered a desperate plea to the warring tribes, a plaintive epitaph fitting not only for his people but for all Indian nations on the continent:

Custer's Last Stand Demystified

> I am tired of fighting. It is cold and we have no blankets, the little children are freezing to death. No one knows where they are. I want to have time to look for my children and see how many of them I can find. Maybe I shall find them among the dead. Hear me my chiefs. I am tired; my heart is sick and sad. From where the sun now stands I will fight no more forever.

So, we have told the tale again, in our own way, by our own lights, and in our own words. No doubt we have done so imperfectly but never dishonestly. After all, the Little Big Horn is a risky subject with more pitfalls than safe landings. Our approach harbored no illusions; we fully understood that there is no scythe sharp enough to slice through all the layers of complexity, mysteries, and unknowns to account for the battle fully. We are conscious that all histories are provisional insofar as there can be no final word, and this work will inevitably serve as a prologue for future discussions and debates. Nevertheless, we believe that new and important ground has been explored with this book and new tangible theories offered. It is not our boast we dare to say, but our sincere conviction that the scenario presented on how the battle unfolded and the reasons why it unfolded as it did is the best explanation yet offered in untangling the Gordian knot of many of its nettlesome and unrelieved mysteries.

Today's silent guns and tranquil skies over the great valley of the Little Big Horn are another timeless illusion. For as long as this battle is remembered and talked about, the din of thunderous debate and the cannonading of controversy will rumble ceaselessly over its outlying bluffs, muted hills, and hidden ravines, pushing its way forward with the inevitability of nature's laws and decrees, only to be fought and struggled over a thousand times anew amid the labyrinthine, boundless arc of history. Whatever new information is uncovered or discovered, the subject remains as vast, deep, and mysterious as the cosmos, mesmerizing those who want to know and understand. FINI

Appendix A: Participant Timeline of the Battle of the Little Big Horn

The authors would like to acknowledge our debt to Dr. John Gray, who, among other works, authored *Centennial Campaign: The Sioux War of 1876 & Custer's Last Campaign: Mitch Bouyer and the Little Big Horn Reconstructed.* Dr. Gray pioneered the time-motion study of the battle. Although Dr. Gray's data has been challenged, including by the authors of this book, his method to achieve a better understanding of the battle is a landmark in Little Big Horn scholarship.

The Participant Timeline of the Battle of the Little Big Horn took a period of over seven years of study to develop. It was created as a joint product between the late Little Big Horn scholar Frederic C. Wagner III, and I (Bill Rini) using every available primary source material on the battle, including over 250 witnesses and battle participants. Their input forms the backbone of this timeline. The rest comes from an extensive first-person examination of the ground, topographical, aerial, and hand-drawn maps, photographs, archeological and artifactual findings, measured distances, and average horse speeds in formation deduced from numerous mounted reenactments on the battlefield.

A proper understanding of a timeline is an essential criterion for any serious student of the battle to comprehend what happened, and when, on June 25, 1876, on that storied battlefield in southeast Montana. Without this resource, the Battle of Little Big Horn cannot be completely understood and will remain an enigma for most. Together with my good friend and colleague Frederic Wagner, author of *Strategy of Defeat and Participants in the Battle of the Little Big Horn*, we began working on this timeline back in 2007.

Of course, Fred and I didn't always draw the same conclusions from the available evidence. As Edgar Stewart, author of the classic *Custer's Luck,* famously said, "No two experts agree on the Battle of Little Big Horn—it's against the rules." We were no exception, but through it all, our friendship endured even when Fred and I pursued alternative paths in developing the Participant Timeline. Fred had this to say about the necessity of using a timeline to enhance a more advanced study of the battle, and I believe it is right on the money:

"… without an understanding of the time that various events took place, the Battle of the Little Big Horn cannot be understood properly … For the historical event of the Little Big Horn, time rules all. And there are too many impressions extant, formed not by fact but by the opinions of others."

The participant timeline presented here seeks to correct many false impressions that have contributed to the mystery and confusion surrounding this epic event in the American Plains Indian Wars. A time-motion study is useful regarding questions about when the battle began, how long it lasted, and how different events at different times interacted with each other. Still, the authors readily acknowledge that no system is definitive and absolute, no matter how meticulously and scientifically organized. No time-motion study can conclusively prove what happened during every minute of the fight at the Little Big Horn. Such a paradigm lacks flexibility and will quickly place battle students and researchers into a conundrum from which they cannot extricate themselves. Instead, the authors employ a range of approximate times to establish a chronology of events, some of which occur simultaneously. By using a general range of time with supporting documentation—not an implausible minute-by-minute account—we believe this time-motion analysis can greatly enhance the understanding of the trajectory of this complex battle.

Bill Rini: As a student of 19th-century mounted military tactics and, most importantly, a reenactor who has ridden over the battlefield more times than even I would care to remember over the last 35 years, I have developed a deep appreciation of the importance of terrain in this battle.

I have learned that it is a different battle on horseback than when seen walking or driving the battlefield. The Participant Timeline unfolds and is amplified in seeing the battle as it entirely evolved, at least for the cavalry, on horseback. Conversely, for most of the battle, the majority of warriors were not on horseback since many of their mounts were grazing outside the village and had to be retrieved, which hampered and slowed down their movements in confronting the cavalry and these timing elements have been factored into the Participant Timeline.

Finally, readers should note that references to sun times are local to southeastern Montana and are 57 minutes (or roughly one hour) prior to Department Headquarters watch times. The latter times are based on the Military Department of the Dakota Headquarters time in Saint Paul, Minnesota. Please note that all the listed timeline accounts of Sunday, June 25, 1876, with an accompanying asterisk (*) are speculative due to a lack of specific participant witness accounts:

RCOI refers to testimony given at the Reno Court of Inquiry.

Please note: Timeline accounts with an asterisk (*) are speculative due to the lack of specific participant witness accounts. The times listed below are based on Regimental Headquarters St. Paul Time, or roughly an hour (i.e., 57 minutes) after local sun time.

8:10 a.m.: General Custer left for the Crows Nest.

8:30 and 8:45 a.m.: The 7th Cavalry departs from Halt 1 on the east side of the divide. The lead elements leave about 8:25, while the pack train and rear elements leave about 8:45 a.m.

9:45 a.m.: The lead elements of the Regiment arrive at Halt 2.

10:07 a.m.: The rear elements begin to arrive at Halt 2.

10:00 to 10:05 a.m.: Custer returns from the Crows Nest, sounds Officer's Call.

10:15 a.m.: The slow-moving pack train and Rear Guard arrive at Halt 2, about the time Custer began his last Officer's Call.

10:15 to 10:30 a.m.: Officer's Call occurs.

10:45 a.m.: The 7th Cavalry leaves Halt 2 at the base of the Crows Nest.

10:55 a.m.: The head of the 7th Cavalry crosses the divide.

11:12 a.m.: Benteen's battalion leaves on his flank and marches to the left.

11:15 a.m.: The main pack train departs from Halt 2.

12:35 p.m.: Boston Custer leaves the main command just west of the first (eastern) lone tepee.

12:45 p.m.: Reno's battalion departs from Custer's command at the Echleman Ford 1.2 miles east of Ford A with orders to attack the hostiles. (Note: Reno may have actually gotten his initial attack orders as early as 12:30 p.m. while in the vicinity of the first lone tepee.) Boston Custer would begin his ride back to the pack train to acquire a fresh mount.

12:50 p.m.: Boston Custer reaches the advance pack mules about 2.75 to three miles farther east or six miles east of Ford A.

12:55 p.m.: Reno reached Ford A.

1:05 p.m.: Benteen's battalion returns to the main trail about five miles east of Ford A.

1:05 p.m.: The advance pack mules are on Ash/Reno Creek, about six miles east of Ford A.

1:05 p.m.: The head of the main pack train and Rear Guard are on Ash/Reno Creek about seven miles east of Ford A.

1:05 p.m.: Reno's battalion crosses the Little Big Horn River at Ford A.

1:07 p.m.: Benteen stops to water at the eastern morass.

1:10 p.m.: Reno's battalion reaches the open prairie west of Ford A, where they dismount to prepare for their advance to combat.

1:10 p.m.: Boston Custer passes Benteen's battalion at the eastern morass, 4.75 miles east of Ford A.

1:20 p.m.: General Custer ascends the bluffs at Sharpshooter Ridge for his first glimpse of the Indian village.

1:20 p.m.: Sergeant Kanipe leaves Custer's column at the south end of Sharpshooter Ridge.

1:20 p.m.: Reno begins his advance (at a fast walk) for about a third of a mile before reforming his command into line for his charge down the valley.

1:25 p.m.: Reno begins his charge down the valley.

1:25 p.m.: Boston Custer reaches the vicinity of the second lone tepee about 1.7 miles east of Ford A.

1:30 p.m.: Benteen leaves the first morass just as the advance packs arrive there.

1:30 p.m.: Reno halts his battalion's advance and dismounts northwest of the timber position about 100 to 200 yards from the Garryowen Bend of the River.

1:30 p.m.: Custer departs Sharpshooter Ridge and heads for Western Coulee.

1:35 p.m.: Trumpeter Martin leaves Custer's command about three-quarters of a mile to a mile north of Sharpshooter Ridge (from Western Coulee).

1:40 p.m.: Reno's valley fight begins after the skirmish line deploys and advances to the Garryowen Bend of the river.

1:40 p.m.: Benteen's battalion has just passed the first (eastern) lone tepee and is now 3.75 miles east of Ford A.

1:40 p.m.: The advance ammunition pack mules have just left the eastern morass 4.75 miles east of Ford A.

1:40 p.m.: Custer's command is now in Western Coulee.

1:40 p.m.: Boston Custer reaches the top of the bluffs at Sharpshooter Ridge.

1:43 p.m.: Boston meets Trumpeter Martin in Cedar Coulee near the northwest slope of Sharpshooter Ridge.

1:45 p.m.: The Pack Train and Rear Guard reach the eastern morass 4.75 miles east of Ford A.

1:45 p.m.: Martin reaches the top of the bluffs and views the Reno fight as he crosses the bluffs.

Appendix A: Participant Timeline of the Battle of the Little Big Horn

1:45 p.m.: Most of Custer's command is now out of sight in North Medicine Tail Coulee, and Custer has begun his leader's reconnaissance with E Company in Western Coulee and has now reached upper Boyer's Bluff.

1:50 p.m.: Martin crosses over the top of the bluffs and begins his descent to Ash/Reno Creek.

1:50 p.m.: Benteen is watering D Company's horses at the Hartung Morass and meets up with Sergeant Kanipe about 2.75 miles east of Ford A.

1:50 p.m.: The Main Pack train and Rear Guard reach the eastern morass 4.75 miles east of Ford A.

1:50 p.m.: G Company is taken off the line and into the timber.

1:55 p.m.: Boston reaches Custer's command in North Medicine Tail Coulee.

1:55 p.m.: Custer is soon spotted at a river crossing (i.e., Thompson's Ford) by Private Peter Thompson.

1:58 p.m.: A Company falls back to the brow of the timber.

2:00 p.m.: Benteen reunites his battalion, finishes watering at the Hartung morass, and begins his advance.

2:00 p.m.: Sergeant Kanipe reaches the advance pack mules at the first (eastern) lone tepee about 3.75 miles east of Ford A.

2:00 p.m.: M Company falls back to the timber position. Custer has returned to the top of upper Boyer's Bluff and realizes Reno's command has fallen back to the timber position with no sign of Benteen's battalion having yet arrived in the valley to provide him with support. Benteen's battalion has just left the Hartung morass 2.5 miles from Ford A.

2:02 p.m.: Sergeant Kanipe reaches Captain McDougall just west of the eastern morass about 4.25 miles east of Ford A.

2:05 p.m.: Sergeant Kanipe reaches the main Pack Train at the eastern morass 4.75 miles east of Ford A.

2:05 p.m.: Reno's troops abandon their defense and race for their horses in the timber bottom.

* **2:05 p.m.:** Custer has left upper Boyer's Bluff and rejoined E Company, and they are making their way back along Western Coulee to the rest of his command in North Medicine Tail Coulee. Mitch Boyer arrives at upper Boyer's Bluff.

2:05 p.m.: Benteen's battalion has reached the second (western) lone tepee location 1.7 miles from Ford A.

2:10 p.m.: Benteen meets up with Trumpeter Martin about one-half mile west of the second (western) lone tepee with Custer's last orders.

2:10 p.m.: The advance pack mules are now 150 yards behind Benteen's battalion at this point.

2:10 p.m.: Reno's battalion abandons the timber position and retreats from the valley.

* **2:10 p.m.:** Custer has reached the junction of Western Coulee, Medicine Tail Coulee, and North Medicine Tail Coulee. Boyer witnesses the beginning of Reno's retreat from the timber.

2:10 p.m.: The three Crow scouts have left elevation 3,411 and are now riding south of Reno Hill toward Ford A. (note: These three scouts may be the individuals spotted on the bluffs by Lieutenant De Rudio at about this time that he mistook for Custer, Cooke and another.)

2:15 p.m.: Benteen reaches the vicinity of Ford A to view Reno's retreat from the valley.

2:15 p.m.: The head of Reno's battalion has now reached the retreat crossing of the Little Big Horn River and is in the process of crossing to the east side.

* **2:15 p.m.:** Mitch Boyer has left upper Boyer's Bluff and is riding down Western Coulee to alert Custer regarding Reno's retreat from the valley. Curley leaves Boyer and rides back up the bluffs away from the river.

2:20 p.m.: The head of Reno's command reaches the top of the bluffs.

2:20 p.m.: Benteen leaves the vicinity of Ford A, meets the three Crow scouts, then begins to ascend the bluffs to Reno Hill.

2:25 p.m.: The main Pack Train and Rear Guard leave the eastern lone tepee 3.75 miles east of Ford A.

* **2:25 p.m.:** Custer has now been alerted by Boyer as to Reno's plight and sends his escort E Company down to Ford B to demonstrate against the village, while Custer himself rides into North Medicine Tail Coulee to confer with his squadron commanders Keogh and Yates on this new development leading to an immediate change of plans.

2:30 p.m.: Benteen's battalion arrives on the bluffs at Reno Hill.

2:30 p.m.: The advance ammunition mules under Lieutenant Mathey arrive on a hillock about a mile south of Reno Hill.

* **2:30 p.m.:** E Company is approaching the vicinity of Ford B at the mouth of Medicine Tail Coulee while Custer has rejoined his command in North Medicine Tail Coulee and is bringing both Keogh and Yates's squadrons up onto the Luce/Blummer Ridge complex into full view of the Indian camps below. The trumpeters of all four troops and that of Headquarters Staff were assembled here at this time to prepare for the Keogh's squadron's advance down to the river with as much noise and fanfare as possible.

2:30 p.m.: The Pack Train and Rear Guard have passed the first (eastern) lone tepee and are now about 3.5 miles east of Ford A.

Appendix A: Participant Timeline of the Battle of the Little Big Horn

2:38 p.m.: Lieutenant Hare leaves Reno Hill and rides back to the advance pack mules about a mile southeast of Reno Hill.

2:40 p.m.: The rear elements of Reno's scattered command arrive on the bluffs.

*** 2:40 p.m.:** Custer's fight in the vicinity of Ford B begins. E Company is now at the mouth of Medicine Tail Ford (i.e., Ford B-1) and has begun firing across the river while Keogh's squadron of I and C troops are rapidly advancing in support from Luce Ridge down to Butler Ridge with all their trumpeters making as much noise as possible on this advance, the purpose being to attract as much attention to his force converging in the area of Ford B in order to draw as many warriors off of the beleaguered Reno battalion and allow them to rejoin with either Benteen's battalion or the pack train rapidly approaching his position. Yates's squadron (F and L Companies) is now deploying on Luce or Blummer Ridge. The initial volleys from Custer's command are heard by a number of officers and troopers both on the bluffs as well as those left behind in the timber position in the valley. Gerard indicates that these volleys were fired by one company near the river at Ford B. The initial volleys from Custer's command are heard by a number of officers and troopers both on the bluffs as well as those left behind in the timber position in the valley. Gerard indicates that these volleys were fired by one company near the river at Ford B. The result of this volley of fire was the immediate departure of nearly all the warriors from Reno and Benteen's position on the bluffs, who are now riding hard to confront Custer's command demonstrating on the east bank of Ford B.

2:40 p.m.: The main pack train and Rear Guard are about three miles east of Ford A.

2:45 p.m.: Nearly all the warriors have withdrawn from Reno's position on the bluffs to concentrate on Custer's five companies in the north.

2:45 p.m.: Lieutenant Hare reaches the advance ammunition mules.

2:50 p.m.: Major Reno and Sergeant Culbertson leave the bluffs to return to the River to find Lieutenant Hodgson's body immediately following Major Reno's refusal to grant Captain Weir permission to advance to the sounds of the firing.

2:50 p.m.: Captain Weir is mounting his horse and heading back towards his own company, now deployed on and west of the south end of Sharpshooter Ridge.

2:50 p.m.: Lieutenant Mathey has sent a messenger back to Captain McDougall informing him of the situation on Reno Hill after receiving Lieutenant Hare's message and request for ammunition. Lieutenant Hare is now starting back on his return ride back to Reno Hill, about a mile north of Mathey's hillock.

2:50 p.m.: The main pack train is now over a mile east of the second (eastern) lone tepee, about 2.5 miles east of Ford A.

3:00 p.m.: Lieutenant Hare returns to Reno Hill after delivering his message for the ammunition mules. He sees D Company advancing toward the peaks some distance out. Reno is still down at the river, checking on Lieutenant Hodgson's body.

3:00 p.m.: Lieutenant Mathey's advance pack mules are now moving towards Reno Hill and will reach that position within the next ten or fifteen minutes, although the lead two ammunition mules would arrive within five minutes of Hare's arrival on the Hill.

3:00 p.m.: Lieutenant Mathey's messenger has now reached Captain McDougall at the main pack train, approaching the second lone tepee.

* **3:00 p.m.:** The Ford B demonstration has ended, and Keogh's squadron is falling back up Deep Coulee towards Finley Ridge, with a squad of flankers moving up parallel with them on the high ground from Greasy Grass Hill to join them there. E Company is falling back from Butler Ridge to Luce or Blummer Ridge to join the Yates squadron there.

* **3:00 p.m.:** The Yates squadron moves from Luce Ridge to Blummer Ridge.

3:05 p.m.: The first of the ammunition mules reaches Reno Hill from the advance pack mules.

3:05 p.m.: The Pack Train and Rear Guard reach and halt to regroup at the second lone tepee, 1.7 miles east of Ford A. Captain McDougall organizes the pack train in a compact group with his rear guard protecting it from the front and rear with his two platoons of Company B.

* **3:05 p.m.:** The second set of volleys are heard fired from Custer's troops, likely from the Luce/Blummer Ridge complex farther back from the river, and likely fired at warriors entering Medicine Tail Coulee from the bluffs on the north side of the Weir Peaks, or at Wolf Tooth's band of warriors east of Blummer Ridge.

* **3:05 to 3:30 p.m.:** The firing during this phase is described as heavy and likely takes place on Luce and Blummer (Nye-Cartwright) Ridge. It ended at about 3:30 p.m., likely indicating the withdrawal of this squadron of Custer's command farther north across Deep Coulee on their way toward Calhoun Hill.

* **3:10 p.m.:** Keogh's squadron has reached Finley Ridge, while the Yates squadron has moved down to Nye-Cartwright or Blummer Ridge, while Custer's Headquarters staff and E Company has moved back from Butler Ridge towards Luce Ridge acting as a rear guard on Yates's move northward off the Luce Ridge complex and away from Medicine Tail Coulee.

3:15 p.m.: The last of the advance pack mules and ammunition reaches Reno Hill.

3:15 p.m.: Reno and Culbertson return to the bluffs from their foray down to the river to check on Hodgson's body.

3:15 p.m.: Captain Weir and D Company have arrived on the Weir Peaks.

3:15 p.m.: Reno sends Lieutenant Hare to Weir with orders to make contact with Custer if possible.

Appendix A: Participant Timeline of the Battle of the Little Big Horn

3:15 p.m.: The Pack Train and Rear Guard leave the second lone tepee and head to the North Fork of Ash/Reno Creek, heading directly towards the troops seen on Reno Hill.

* **3:20 p.m.:** Custer's Headquarters staff, the Yates squadron, and E Company (acting as the rear guard) arrive on Calhoun Hill.

* **3:25 p.m.:** Custer's Headquarters staff and the Yates squadron depart Calhoun Hill and begin their reconnaissance in force to locate good ground to defend in the vicinity of the north fords. L Company is left behind to hold Calhoun Hill for Keogh's squadron, now making its way up Finley Ridge.

3:25 p.m.: Captain French leads M Troop from Reno Hill and begins to advance to the Weir Peaks.

3:25 p.m.: Lieutenant Hare arrives on the Peaks and delivers his orders to Captain Weir.

3:25 to 3:45 p.m.: The Pack Train and Rear Guard cross over the North Fork of Ash/Reno Creek.

* **3:30 p.m.:** Keogh arrives and deploys on Calhoun Hill and Battle Ridge with his own squadron reinforced by L Company to await the expected advance of Benteen and the ammunition mules.

3:30 to 3:45 p.m.: The third set of volleys fired from Custer's command are reportedly heard by Captain McDougall as he was ascending the bluffs to Reno Hill with the Pack Train and Rear Guard.

3:40 p.m.: M Troop arrives on the Weir Peaks, passing Lieutenant Hare on his way back to Reno Hill.

* **3:40 p.m.:** Custer and the Yates squadron, after dropping off Lieutenant Smith with a platoon of E Company on Last Stand Hill, are on their way towards the flats between the northern fords.

3:40 p.m.: Captain McDougall hears these volleys as he approaches the southeast side of Reno Hill. Captain Moylan reports hearing them as well.

3:45 p.m.: McDougall and the first platoon of Rear Guard arrive on Reno Hill.

* **3:45 p.m.:** Keogh holds Calhoun Hill and Battle Ridge. Keogh notices the arrival of troops on the Weir Peaks and sends a messenger to Custer to inform Custer of the expected arrival of Benteen's battalion. Keogh has L Company fire several volleys to attract the attention of the troops on the Peaks.

* **3:50 p.m.:** Custer and the Yates squadron have reached the flats between the northern fords and have begun to reconnoiter the area for good defensive ground. They focus their attention on the flats by the area of Ford D-2 (at the Willy Bends Place).

* **3:50 to 4:00 p.m.:** Keogh deploys his battalion to support the advance of troops now clearly seen on the Weir Peaks. Keogh orders C Company on a mounted advance west from Battle Ridge in two separate detachments to occupy Harrington Ridge just north of Calhoun Coulee while the second detachment to occupy Finley Ridge for the purpose of establishing a field of fire from a squad under Sergeant Finkle directed at the southern base of Calhoun Hill in upper Deep Coulee. A secondary field of fire from a squad under Sergeant Finley would be directed west towards Greasy Grass Hill. C

Company's horses would be led down to the shelter of upper Calhoun Coulee between these two dismounted detachments of C Company deployed on either side of the coulee.

*** 3:50 – 4:00 p.m.:** Custer and the Yates squadron are now heading back from their reconnaissance to the flats around the northern fords, stopping at the lower Cemetery flats to examine the ground there for an adequate defensive position. I speculate that Custer sent an additional messenger to Keogh telling him to move ahead to his position on the Cemetery flats once a successful junction has been made with Benteen's battalion and the ammunition mules. If it appears that no advance is forthcoming, Keogh should use his discretion to disengage his battalion and move north immediately to reconnect with Custer and the Yates squadron on the lower Cemetery flats above the river.

4:00 p.m.: The head of the main pack train arrives on Reno Hill.

4:00 to 4:10 p.m.: the fourth set of volleys fired by Custer's command are reportedly heard by Captain Moylan, McDougall and Godfrey, and are reported to Major Reno at this time.

*** 4:00 to 4:10 p.m.:** Lame White Man and Yellow Nose lead a hostile charge up Calhoun Coulee into the horse-holders of C Company in the upper end of the Coulee. C Company's first platoon under Lieutenant Harrington and First Sergeant Bobo fall back up Battle Ridge to rejoin with a portion of I Company deployed on the top of the ridge. The second platoon of C Company remains dismounted on Finley Ridge as the Indian charge by-passes their position as they charge up Calhoun Coulee and circle around to the north along the west side of Battle Ridge as they drive the stampeded horses of C Company down to Greasy Grass Ridge and Deep Ravine. Keogh throws every available trooper on Battle Ridge to the west side of the ridge—with the exception of Lieutenant Porter's reserve platoon, which remains east of the swale by his own led horses—to repel this charge against C Company's horses. Keogh's troopers, deployed on the west side of Battle Ridge near the top, shoot down Lame White Man and Noisy Walking as they circle around from Calhoun Coulee, moving to the north on the west side of Battle Ridge. Many more hostiles are wounded in this charge, which succeeds in stampeding most of C Company's horses but is repelled with heavy loss by Keogh's troopers now deployed on the western military crest of Battle Ridge.

*** 4:00 to 4:10 p.m.:** L Company remains in position on Calhoun Hill during the charge of Lame White Man and Yellow Nose, as does C Company's detachment deployed on Finley Ridge.

*** 4:00 to 4:10 p.m.:** Custer hears the sounds of heavy firing from Keogh's position and realizes that Keogh's battalion has come under heavy attack. He mounts up Captain Yates F Company and his own Headquarters staff which begins to move south across Cemetery Ridge from the lower flats. E Company's second platoon under Lieutenant Sturgis remains dismounted and deploys on Cemetery Ridge at its southern edge to provide cover fire for F Company's mounted advance back south into the Basin area below Last Stand Hill.

4:20 p.m.: The last of the pack mules and Rear Guard arrive on Reno Hill. Reno sends Lieutenant Varnum and a small detail to the river to bury Lt. Hodgson.

Appendix A: Participant Timeline of the Battle of the Little Big Horn

4:25 p.m.: Reno's remaining five companies on Reno Hill begin their advance towards the Weir Peaks with the intent to join with Custer's command in the north.

4:30 p.m.: Herendeen's group was first seen by elements of Reno's command as Herendeen's group was crossing the river north of where Reno retreated across the river. Varnum's detail is recalled to the bluffs.

4:35 p.m.: Reno's command, having moved about a half to a quarter mile downriver along the bluffs, halts their advance upon seeing Herendeen's group of survivors making their way out of the valley towards their position on the bluffs.

4:35 to 4:40 p.m.: Herendeen's group passes Billy Cross and his party of Rees near the top of the bluffs as the latter group was descending to the river to water their horses.

4:45 p.m.: Herendeen's group ascends the bluffs and joins Reno's command north of the ridge where Dr. DeWolfe was killed. They are then outfitted with more cartridges for their weapons, and medical attention is given to Sergeant White while they quickly debrief and report the situation in the timber to Major Reno.

4:50 p.m.: Benteen and Godfrey move out with H and K Troops to the Weir Peaks while Reno orders his trumpeter to sound the Halt to prevent their departure from the rest of his command.

5:00 p.m.: Godfrey and Benteen arrive on the Weir Peaks with K and H Companies to join with D and M Troops already deployed on the Peaks.

5:00 p.m.: Colonel Reno, after having resupplied the survivors from the valley and inspected and tended to Sergeant White's wounds, has now begun his advance to the Peaks with the remaining Companies under his command (G, A, the second platoon of B, and the pack train).

5:05 p.m.: Captain McDougall and the first platoon of B would likely have moved out to the Weir Peaks at a trot at this time, leaving Reno's slow-moving command behind and arriving at the Peaks a few minutes after Benteen.

5:15 p.m.: H Company leaves the Weir Peaks; the Indians begin to move on the troops seen on the Weir Peaks. D Company engages in long-range fire with a group of these Indians on the lower ridges of the northern Peaks.

5:20 p.m.: Reno begins his retreat back to his defensive position on the bluffs. This would involve companies A, G, B, and H along with the pack train, while Troops D, M, and K remained out on the Weir Peaks a while longer to engage the hostiles and give time for Reno and Benteen to locate a defensive position on the bluffs for the command.

5:35 p.m.: B, A, G, and H Troops, as well as the Pack Train, reach Reno Hill and begin to deploy there.

5:35 p.m.: K Company leaves the Weir Peaks after being recalled by Lieutenant Hare acting on orders from Major Reno, with M and D Troops acting as a Rear Guard.

5:40 p.m.: M Company leaves the Weir Peaks. Captain French advises Lieutenant Edgerly to do likewise.

5:45 p.m.: D Company leaves the Weir Peaks.

5:50 p.m.: Trooper Vincent Charley is shot off his horse about 650 yards south of the Weir Peaks.

5:50 p.m.: Lieutenant Godfrey dismounts K Troop about 500 yards north of Reno Hill to provide cover fire for M and D Company's withdrawal from the Peaks.

5:55 p.m.: M Troop arrives at Reno Hill. D Troop passes K Troop on their retreat from the Peaks.

6:00 p.m.: D Troop arrives at Reno Hill.

6:00 to 6:40 p.m.: K Troop engages in a dismounted Rear Guard skirmish action until repeatedly recalled to fall back to Reno Hill.

6:45 p.m.: First Sergeant Winney of K Troop is killed at Reno Hill, marking the general start of heavy skirmish firing directed at Reno Hill.

9:00 p.m.: The heavy firing at Reno Hill ends, concluding the first day's fighting on June 25th, 1876.

For those who prefer a timeline not based on a clock time:

H hour (baseline): Regiment leaves Halt 1 east of the divide.

H = 1 hour and 15 minutes: The Regiment arrives at Halt 2.

H = 2 hours and 15 minutes: The Regiment leaves Halt 2.

H = 2 hours and 25 minutes: The Regiment crosses the divide.

H = 2 hours and 42 minutes: Benteen's battalion leaves on his flank and march to the left.

H = 2 hours and 45 minutes: The Pack Train leaves Halt 2.

H = 4 hours and 5 minutes: Boston Custer leaves the main command just west of the first lone tepee.

H + 4 hours and 15 minutes: Reno's battalion separates from Custer and begins their advance to the attack.

H = 4 hours and 20 minutes: Boston reaches the advance pack mules 2.75 to three miles further east or 6 miles east of Ford A.

H = 4 hours and 25 minutes: Reno reaches Ford A.

Appendix A: Participant Timeline of the Battle of the Little Big Horn

H = 4 hours and 35 minutes: Benteen's battalion returns to the main trail.

H = 4 hours and 35 minutes: Reno's battalion crosses the Little Big Horn River at Ford A.

H = 4 hours and 37 minutes: Benteen stops to water at the eastern morass.

H = 4 hours and 40 minutes: Reno's battalion reaches the open prairie west of Ford A where they dismount his to prepare for their advance to combat.

H = 4 hours and 40 minutes: Boston Custer passes Benteen's battalion at the eastern morass, 4.75 miles east of Ford A.

H = 4 hours and 50 minutes: Custer ascends the bluffs at Sharpshooter Ridge for his first glimpse of the Indian village.

H = 4 hours and 50 minutes: Sergeant Kanipe leaves Custer's column.

H = 4 hours and 50 minutes: Reno begins his advance (at a fast walk) for about a third of a mile before reforming his command into line for his charge down the valley.

H = 4 hours and 55 minutes: Reno begins his charge down the valley.

H = 4 hours and 55 minutes: Boston reaches the vicinity of the second (western) lone tepee about 1.7 miles east of Ford A.

H = 5 hours: Benteen leaves the eastern morass just as the advance packs arrive there.

H = 5 hours: Reno halts his battalion's advance and dismounts northwest of the timber position about 100 to 200 yards from the Garryowen Bend of the River.

H = 5 hours: Custer leaves Sharpshooter Ridge and heads for Western Coulee.

H = 5 hours and 8 minutes: Trumpeter Martin leaves Custer's command about three-quarters of a mile northwest of Sharpshooter Ridge.

H = 5 hours and 10 minutes: Reno's valley fight begins after the skirmish line deploys and advances to the Garryowen Bend.

H = 5 hours and 10 minutes: Boston reaches the top of the bluffs at Sharpshooter Ridge.

H = 5 hours and 13 minutes: Boston Custer meets Trumpeter Martin in Cedar Coulee on the northwest slope of Sharpshooter Ridge.

H = 5 hours and 15 minutes: Pack Train and Rear Guard reach the eastern morass.

H = 5 hours and 15 minutes: Martin reaches the top of the bluffs and views the Reno fight as he crosses the bluffs.

H = 5 hours and 20 minutes: Martin reaches the end of the bluffs and begins his descent to Ash/Reno Creek.

H = 5 hours and 20 minutes: Benteen meets up with Sergeant Kanipe 2.75 miles east of Ford A before stopping with D Company to water again at the Hartung morass.

H = 5 hours and 22 minutes: G Company is taken off the valley skirmish line and into the timber.

*** H = 5 hours and 22 minutes:** Custer's command is out of sight in North Medicine Tail Coulee, and Custer has begun his leader's reconnaissance with E Company in Western Coulee and has now reached upper Boyer's Bluff.

H = 5 hours and 25 minutes: Boston reaches Custer's command in North Medicine Tail Coulee.

H = 5 hours and 25 minutes: A Company falls back to the brow of the timber.

H = 5 hours and 25 minutes: Custer is soon spotted at a river crossing below by Private Peter Thompson.

H = 5 hours and 30 minutes: Benteen finishes watering at the Hartung morass after reuniting with his battalion 2.5 miles east of Ford A.

H = 5 hours and 30 minutes: Sergeant Kanipe reaches the advance pack mules at the first lone tepee 3.75 miles east of Ford A.

H = 5 hours and 30 minutes: M Company falls back to the timber position.

*** H = 5 hours and 30 minutes:** Custer has returned to the top of upper Boyer's Bluff and realizes Reno's command has fallen back to the timber position with no sign of Benteen's battalion having yet arrived in the valley to provide him with support.

H = 5 hours and 35 minutes: Sergeant Kanipe reaches Captain McDougall about 4.25 miles east of Ford A.

H = 5 hours and 35 minutes: Sergeant Kanipe reaches the main Pack Train at the eastern morass 4.75 miles east of Ford A.

H = 5 hours and 35 minutes: Reno's troops abandon their defense and race for their horses at the bottom of the timber position.

*** H = 5 hours and 35 minutes:** Custer has left upper Boyer's Bluff and rejoined E Company and they are making their way back along Western Coulee to the rest of his command in North Medicine Tail Coulee. Mitch Boyer arrives at upper Boyer's Bluff.

H = 5 hours and 35 minutes: Benteen's battalion has reached the second (western) lone tepee location 1.7 miles from Ford A.

H = 5 hours and 40 minutes: Benteen meets with Trumpeter Martin 1.5 miles east of Ford A to deliver Custer's last orders.

Appendix A: Participant Timeline of the Battle of the Little Big Horn

H = 5 hours and 40 minutes: The advance pack mules are 150 yards behind Benteen's battalion at this point.

H = 5 hours and 40 minutes: Reno's battalion abandons the timber position and retreats from the valley.

*** H = 5 hours and 40 minutes:** Custer has reached the junction of Western Coulee, Medicine Tail Coulee, and North Medicine Tail Coulee. Boyer witnesses the beginning of Reno's retreat from the timber.

H = 5 hours and 40 minutes: The three Crow scouts have left elevation 3,411 and are now riding south of Reno Hill towards Ford A. (Note: These three scouts may be the individuals spotted on the bluffs by Lieutenant De Rudio at about this time that he mistook for Custer, Cooke, and another.)

H = 5 hours and 40 minutes: The head of Reno's battalion has now reached the retreat crossing of the Little Big Horn River and is crossing to the east side.

*** H = 5 hours and 40 minutes:** Mitch Boyer has left upper Boyer's Bluff and is riding down Western Coulee to alert Custer regarding Reno's retreat from the valley. Curley leaves Boyer and rides back up the bluffs away from the river.

H = 5 hours and 45 minutes: Benteen reaches the vicinity of Ford A to view Reno's retreat from the valley.

H = 5 hours and 50 minutes: The head of Reno's command reaches the top of the bluffs.

*** H = 5 hours and 50 minutes:** Custer has now been alerted by Boyer as to Reno's plight and sends his escort E Company down to Ford B to demonstrate against the village, while Custer himself rides into North Medicine Tail Coulee to confer with his squadron commanders Keogh and Yates on this new development leading to an immediate change of plans.

H = 5 hours and 50 minutes: Benteen leaves the vicinity of Ford A, meets the three Crow scouts, then begins to ascend the bluffs to Reno Hill.

H = 6 hours: Benteen's battalion arrives on the bluffs at Reno Hill.

H = 6 hours: The advance ammunition mules under Lieutenant Mathey arrive on a hillock about a mile south of Reno Hill.

*** H = 6 hours:** E Company is approaching the vicinity of Ford B at the mouth of Medicine Tail Coulee while Custer has rejoined his command in North Medicine Tail Coulee and is bringing both Keogh and Yates squadrons up onto the Luce or Blummer Ridge complex into full view of the Indian camps below. The trumpeters of all four troops and that of Headquarters Staff were assembled here at this time to prepare for the Keogh's squadron's advance down to the river with as much noise and fanfare as possible.

H = 6 hours: Pack Train and Rear Guard have passed the first (eastern) lone tepee and are now 3.5 miles east of Ford A.

H = 6 hours and 8 minutes: Lieutenant Hare leaves Reno Hill and rides back to the advance pack mules about a mile southeast of Reno Hill. He delivers his message to Lieutenant Mathey, who in turn sends a messenger back to the main pack train several miles to his rear and down on the flats between Ash/Reno Creek and the North Fork with instructions to hurry on to the bluffs.

*** H = 6 hours and 8 minutes:** E Company is now at the mouth of Medicine Tail Ford, while Keogh's squadron of I and C troops are rapidly advancing from Luce Ridge down to Butler Ridge with all their trumpeters making as much noise as possible on this advance, the purpose being to attract as much attention to his force converging in the area of Ford B to draw as many warriors off of the beleaguered Reno battalion and allow them to rejoin with either Benteen's battalion or the advance pack train rapidly approaching his position.

H = 6 hours and 10 minutes: The main Pack Train and Rear Guard are about three miles east of Ford A.

*** H = 6 hours and 10 minutes:** Custer's Fight in the vicinity of Ford B begins. E Company has begun firing across the river by Ford B-1, while the Keogh squadron (I and C Company) is rapidly approaching the Ford area to support them there. Yates's squadron (F and L Companies.) is now deploying on Luce or Blummer Ridge. The initial volleys from Custer's command are heard by a number of officers and troopers both on the bluffs as well as those left behind in the timber position in the valley. Gerard indicates that these volleys were fired by one company near the river at Ford B. The result of this volley fire was the immediate departure of nearly all the warriors from Reno and Benteen's command who were now riding hard to confront Custer's command demonstrating on the east bank of Ford B.

H = 6 hours and 15 minutes: Nearly all the warriors had withdrawn from Reno's position on the bluffs to concentrate on Custer's five companies in the north.

H = 6 hours and 15 minutes: Lieutenant Hare delivers his message to Lieutenant Mathey ordering the advance ammunition mules ahead to Reno Hill.

H = 6 hours and 20 minutes: Reno and Culbertson and several others leave the bluffs to return to the river to find Lieutenant Hodgson's body immediately following Reno's refusal to grant Captain Weir permission to advance to the sounds of the firing in Custer's direction.

H = 6 hours and 20 minutes: Captain Weir is mounting his horse and heading back towards his own company, now deployed on the south end of Sharpshooter Ridge west towards the end of the bluffs.

*** H = 6 hours and 20 minutes:** Custer's command is fully engaged against the hostile force in and around Medicine Tail Coulee. Keogh's squadron is actively demonstrating at Ford B-2, while the Yates squadron is deployed on Luce and Blummer Ridge, covering this force by engaging in hostile forces appearing from the east and the south. Custer's Headquarters staff, along with E Company is deployed on Butler Ridge midway between the Keogh and Yates' squadrons.

H = 6 hours and 20 minutes: Lieutenant Mathey has sent a messenger back to Captain McDougall informing him of the situation on Reno Hill after receiving Lieutenant Hare's message and request for ammunition. Lieutenant Hare is now starting back on his return ride back to Reno Hill, about a mile north of Mathey's Hill.

H = 6 hours and 20 minutes: The main pack train is now just under a mile east of the second (western) lone tepee, about 2.5 miles east of Ford A.

H = 6 hours and 30 minutes: Lieutenant Hare returns to Reno Hill after delivering his message for the ammunition mules. He sees D Company advancing toward the peaks some distance out. Reno is still down at the river, checking on Lieutenant Hodgson's body.

H = 6 hours and 30 minutes: Lieutenant Mathey's advance pack mules are now moving towards Reno Hill and will reach that position within the next ten or fifteen minutes, although the lead two ammunition mules would arrive within five minutes of Hare's arrival on the Hill.

H = 6 hours and 30 minutes: Lieutenant Mathey's messenger has now reached Captain McDougall, who is now organizing the main pack train near the second (western) lone tepee in a compact group with his rear guard protecting it from the front and rear with his two platoons of Company B.

*** H = 6 hours and 30 minutes:** The Ford B demonstration has ended, and Keogh's squadron is falling back up Deep Coulee towards Finley Ridge, with a squad of flankers moving up parallel with them on the high ground from Greasy Grass Hill to join them there. E Company is falling back from Butler Ridge to Luce or Blummer Ridge to join the Yates squadron there.

*** H = 6 hours and 35 minutes:** The second set of volleys are heard firing from Custer's troops, likely from the Luce/Blummer Ridge Complex further back from the river, and likely fired at warriors entering Medicine Tail Coulee from the bluffs on the north side of the Weir Peaks, or from Wolf Tooth's band of warriors east of Blummer Ridge.

*** H = 6 hours and 35 min to 7 hours:** The firing during this phase is described as heavy and likely takes place on Luce and Blummer (Nye-Cartwright) Ridge. Its ending about H = 7 hours likely indicates the withdrawal of this squadron of Custer's command further north across Deep Coulee on their way towards Calhoun Hill.

*** H = 6 hours and 40 minutes:** Keogh's squadron has reached Finley Ridge, while the Yates squadron has moved down to Nye-Cartright or Blummer Ridge, while Custer's Headquarters Staff and E Company have moved back from Butler Ridge towards Luce Ridge, acting as a rear guard on Yates's move northward off the Luce Ridge complex and away from Medicine Tail Coulee.

H = 6 hours and 45 minutes: Reno and Culbertson return to the bluffs from their foray down to the River.

H = 6 hours and 45 minutes: Captain Weir and D Company have arrived on the Weir Peaks.

H = 6 hours and 45 minutes: Pack Train and Rear Guard leave the second (western) lone tepee and head to the North Fork of Ash/Reno Creek, moving towards the troops seen on Reno Hill.

* **H = 6 hours and 50 minutes:** Custer's Headquarters staff, the Yates squadron and E Company (acting as a rear guard) arrive on Calhoun Hill.

* **H = 6 hours and 55 minutes:** Custer's Headquarters staff and the Yates squadron depart Calhoun Hill and begin their reconnaissance in force to locate good ground to defend in the vicinity of the north fords.

H = 6 hours and 55 minutes: Captain French leads M Troop on its advance to the Weir Peaks.

H = 6 hours and 55 min to 7 hours and 15 minutes: The main Pack Train and Rear Guard cross over the North Fork of Ash/Reno Creek.

* **H = 7 hours:** Keogh arrives and deploys on Calhoun Hill and Battle Ridge with his own squadron reinforced by L Company to await the expected advance of Benteen and the ammunition mules.

H = 7 hours to 7 hours and 15 minutes: The third set of volleys fired from Custer's command are reportedly heard by Captain McDougall as he was ascending the bluffs to Reno Hill with the pack train and Rear Guard.

H = 7 hours and 10 minutes: M Troop arrives on the Weir Peaks, passing Lieutenant Hare on his way back to Reno Hill.

* **H = 7 hours and 10 minutes:** Custer and the Yates squadron, after dropping off Lieutenant Smith with a platoon of E Company on Last Stand Hill, are on their way for a reconnaissance on the flats between the northern fords.

* **H = 7 hours and 10 minutes:** Keogh holds Calhoun Hill and Battle Ridge. Keogh notices the arrival of troops on the Weir Peaks and sends a messenger to Custer to inform him of the expected arrival of Benteen's battalion. Keogh has L Company fire several volleys to attract the attention of the troops on the Peaks.

H = 7 hours and 10 minutes: Captain McDougall hears these volleys as he approaches the southeast side of Reno Hill. Captain Moylan reports hearing them as well.

H = 7 hours and 15 minutes: McDougall and the first platoon of Rear Guard arrive on Reno Hill.

* **H = 7 hours and 20 minutes:** Custer and the Yates squadron have reached the flats between the northern fords and begun reconnoiter the area for good defensive ground. They focus on the flats near the Willy Bends Place (near Ford B-2).

* **H = 7 hours and 20 min to 7 hours and 30 minutes:** Keogh deploys his battalion to support the advance of the troops now clearly seen on the Weir Peaks. Keogh orders a mounted advance for C

Company to move west from Battle Ridge in two separate detachments—one under Lieutenant Harrington and First Sergeant Bobo to occupy Harrington Ridge just north of Calhoun Coulee while the second platoon, under Sgts. Finley and Finkle to occupy Finley Ridge for the purpose of establishing a field of fire from a squad under Sergeant Finkle directed at the southern base of Calhoun Hill in upper Deep Coulee. A secondary field of fire from a squad under Sergeant Finley would be directed west towards Greasy Grass Hill. C Company's horses would be led down to the shelter of upper Calhoun Coulee between these two dismounted platoons of C Company deployed on either side of the coulee.

*** H = 7 hours and 20 minutes to 7 hours and 30 minutes:** Custer and the Yates squadron are now heading back from their reconnaissance to the flats around the northern fords, stopping at the lower Cemetery flats to examine the ground there for an adequate defensive position. We speculate that Custer sent an additional messenger to Keogh telling him to move ahead to his position on the Cemetery flats once a successful junction has been made with Benteen's battalion and the ammunition mules. If it appears that no advance is forthcoming, Keogh should use his discretion to disengage his battalion and move north immediately to reconnect with Custer and the Yates squadron on the lower Cemetery flats above the river.

H = 7 hours and 30 minutes: The head of the main pack train arrives on Reno Hill. The rest of the pack train has just finished crossing the North Fork of Ash/Reno Creek just southeast of Reno Hill.

H = 7 hours and 30 minutes to 7 hours and 40 minutes: The fourth set of volleys fired by Custer's command are reportedly heard by Captain Moylan, McDougall, and Godfrey and are reported to Major

*** H = 7 hours and 30 minutes to 7 hours and 40 minutes:** Lame White Man and Yellow Nose lead a hostile charge up Calhoun Coulee directly into the horse-holders of C Company in the upper end of the Coulee. C Company's first platoon under Lieutenant Harrington and First Sergeant Bobo fall back up Battle Ridge to rejoin with a portion of I Company deployed on the top of the ridge. The second platoon of C Company remains dismounted on Finley Ridge as the Indian charge by-passes their position as they charge up Calhoun Coulee and circle around to the north along the west side of Battle Ridge as they drive the stampeded horses of C Company down to Greasy Grass Ridge and Deep Ravine. Keogh throws every available trooper on Battle Ridge to the west side of the ridge—with the exception of Lieutenant Porter's reserve platoon, which remains east of the swale by his own led horses—to repel this charge against C Company's horses. Keogh's troopers, deployed on the west side of Battle Ridge near the top, shoot down Lame White Man and Noisy Walking as they circle around from Calhoun Coulee to the north on the west side of Battle Ridge. Many more hostiles are wounded in this charge, which succeeds in stampeding most of C Company's horses but is repelled with heavy loss by Keogh's troopers now deployed on the western military crest of Battle Ridge. L Company remains in position on Calhoun Hill during the charge of Lame White Man and Yellow Nose, as does C Company's second platoon deployed on Finley Ridge.

*** H = 7 hours and 30 minutes to 7 hours and 40 minutes:** Custer hears the sounds of heavy firing from Keogh's position and realizes that Keogh's battalion has come under heavy attack. He mounts up

Captain Yates's F Company and his own Headquarters staff which begins to move south across Cemetery Ridge from the lower flats. E Company's second platoon under Lieutenant Sturgis remains dismounted and deploys on Cemetery Ridge at its southern edge to provide cover fire for F Company's mounted advance back south into the Basin area below Last Stand Hill.

H = 7 hours and 50 minutes: The last pack mules and Rear Guard arrive on Reno Hill. Reno sends Lieutenant Varnum and a small detail to the river to bury Lieutenant Hodgson.

H = 7 hours and 55 minutes: Reno's remaining five companies on Reno Hill begin their advance towards the Weir Peaks with the intent to join with Custer's command in the north.

H = 8 hours: Herendeen's group was first seen by elements of Reno's command as Herendeen's group were crossing the river north of where Reno retreated across the river. Varnum's detail is recalled to the bluffs.

H = 8 hours and 5 minutes: Reno's command, having moved about a half to a quarter mile downriver along the bluffs, halts their advance upon seeing Herendeen's group of survivors making their way out of the valley towards their position on the bluffs.

H = 8 hours and 5 to 10 minutes: Herendeen's group passes Billy Cross and his party of Rees near the top of the bluffs as the latter group is descending to the river to water their horses.

H = 8 hours and 15 minutes: Herendeen's group ascends the bluffs and joins Reno's command north of the ridge where Dr. DeWolf was killed. They are then outfitted with more cartridges for their weapons, and medical attention is given to Sergeant White while they quickly debrief and report the situation in the timber to Major Reno.

H = 8 hours and 20 minutes: Benteen and Godfrey move out with H and K Troops to the Weir Peaks while Reno orders his trumpeter to sound the Halt to prevent their departure from the rest of his command.

H = 8 hours and 30 minutes: Godfrey and Benteen arrive on the Weir Peaks with K and H Companies to join with D and M Troops already deployed on the Peaks.

H = 8 hours and 30 minutes: Col. Reno, after having re-supplied the survivors from the valley and inspected and tended to Sergeant White's wounds, has now begun his advance to the Peaks with the remaining Companies under his command (G, A, the second platoon of B, and the pack train).

H = 8 hours and 35 minutes: Captain McDougall and the first platoon of B Company would likely have moved out to the Weir Peaks at a trot at this time, leaving Reno's slow-moving command behind and arriving at the Peaks a few minutes after Benteen.

H = 8 hours and 45 minutes: H Company leaves the Weir Peaks; the Indians begin to move on the troops seen on the Weir Peaks. D Company engages in long-range fire with a group of these Indians on the lower ridges of the northern Peaks.

Appendix A: Participant Timeline of the Battle of the Little Big Horn

H = 8 hours and 50 minutes: Reno begins his retreat back to his corral on the bluffs. This would involve companies A, G, B, and H along with the pack train while Troops D, M, and K remained out on the Weir Peaks a while longer to engage the hostiles and give time for Reno and Benteen to locate a defensive position on the bluffs for the command.

H = 9 hours and 5 minutes: B, A, G, and H Troops, as well as the Pack Train, reach Reno Hill and begin to deploy there.

H = 9 hours and 5 minutes: K Company leaves the Weir Peaks after being recalled by Lieutenant Hare acting on orders from Major Reno, with M and D Troops acting as a Rear Guard.

H = 9 hours and 10 minutes: M Company leaves the Weir Peaks. Captain French advises Lieutenant Edgerly to do likewise.

H = 9 hours and 15 minutes: D Company leaves the Weir Peaks.

H = 9 hours and 20 minutes: Trooper Vincent Charley is shot off his horse about 650 yards south of the Weir Peaks.

H = 9 hours and 20 minutes: Lieutenant Godfrey dismounts K Troop about 500 yards north of Reno Hill to provide cover fire for M and D Co's. withdrawal from the Peaks.

H = 9 hours and 25 minutes: M Troop arrives at Reno Hill. D Troop passes K Troop on their retreat from the Peaks.

H = 9 hours and 30 minutes: D Troop arrives at Reno Hill.

H = 9 hours and 30 min to 9 hours and 40 minutes: K Troop engages in a dismounted Rear Guard skirmish action until repeatedly recalled to fall back to Reno Hill.

H = 9 hours and 45 minutes: First Sergeant Winney of K Troop is killed at Reno Hill, marking the general start of heavy skirmish firing directed at Reno Hill.

H = 12 hours: The heavy firing at Reno Hill ends, concluding the first day's fighting on June 25th, 1876.

CONCLUSION

Please note that the action and movements of Custer's five companies during this time are based in large measure on speculation together with my interpretation of Indian accounts and body locations. and archaeological evidence.

2:40 p.m.: The Custer Fight begins with action at Ford B about ten minutes after Benteen has joined Reno's battalion on the bluffs. Lieutenant Hare has just left Reno Hill on his ride back to the advance ammunition pack mules one mile to the rear. Several of Reno's officers and men—including Gerard,

Herendeen, Private O'Neill, and Lieutenant De Rudio in the valley timber position—hear volleys fired in the vicinity of Ford B, marking the opening of the Custer Fight.

* 3:00 to 3:10 p.m.: One squadron moves from the flats near Ford B up to Finley Ridge, while the second squadron moves from Luce Ridge to Blummer Ridge to Calhoun Hill. Captain Weir and D Troop have left Reno Hill and are approaching the Peaks. Lieutenant Hare has just returned from his ride to the advance pack mules, who are just now arriving on Reno Hill. Private Davern, Lieutenant Godfrey, and Lieutenant Hare hear several volleys fired by Custer's troops on Luce Ridge.

* 3:10 to 3:20 p.m.: Custer's five troops reunite on Calhoun Hill. Captain Weir and D Troop have just reached and deployed on the Weir Peaks. Colonel Reno has just returned from his trip to the river to find Lieutenant Hodson's body. He sends out Lieutenant Hare to the Peaks with instructions for Captain Weir to attempt to make contact with Custer's command.

* 3:20 to 3:30 p.m.: Custer's Headquarters staff and the Yates squadron depart Calhoun Hill and begin their reconnaissance in force to locate good ground to defend the flats in the vicinity of the north fords. Keogh remains on Calhoun Hill and Battle Ridge with his own squadron reinforced by L Company to await the expected advance of Benteen and the ammunition mules. M Troop leaves Reno Hill and advances to the Weir Peaks.

* 3:30 to 3:40 p.m.: M Troop arrives and deploys on the Weir Peaks. Custer and the Yates squadron, after dropping off Lieutenant Smith with a platoon of E Company on Last Stand Hill, are well on their way towards the flats between the northern fords. Keogh holds Calhoun Hill and Battle Ridge. Keogh notices the arrival of troops on the Weir Peaks and sends a messenger to Custer to inform him of the expected arrival of Benteen's battalion. Keogh has L Company fire several volleys to attract the attention of the troops on the Peaks. Captain McDougall hears these volleys as he approaches the southeast side of Reno Hill. Captain Moylan reports hearing them as well.

* 3:40 to 3:50 p.m.: Captain McDougall and a platoon of the Rear Guard reach Reno Hill in advance of the main pack train. He reports hearing volleys fired from troops in the north. Keogh holds his position on Calhoun Hill and has L Company fire several more volleys in an effort to attract attention to his location to the troops now visible to him on the Weir Peaks. These volleys are heard by a number of officers and men on Reno Hill, among them Lieutenant Godfrey and McDougall. Custer and the Yates squadron have reached the flats between the northern fords and have begun to reconnoiter the area for good defensive ground. They focus on the flats between Ford D-2 (at the Willy Bends Place) and Ford D-1 (at the northwest corner of Cemetery Ridge). Custer receives Keogh's message about troops spotted on the Weir Peaks and understands that the expected arrival of Benteen's battalion is finally occurring. A messenger is likely sent back to Keogh at this time to provide all the support necessary to facilitate the successful arrival of Benteen and the ammunition mules to the Luce/Blummer Ridge complex.

* 3:45 to 3:55 p.m.: Keogh, having previously received instructions from Custer to deploy his battalion to support the advance of Benteen's battalion, now clearly seen on the Weir Peaks, begins to do so.

Keogh orders C Company to advance west mounted from Battle Ridge in two separate detachments—a small detachment of about eight troopers under First Sergeant Bobo to occupy the east-west ridge (i.e., Harrington Ridge) just north of Calhoun Coulee, while a larger detachment of about twenty-two troopers, under Sergeants Finley and Finkle to occupy Finley Ridge for the purpose of establishing a field of fire from a squad under Sergeant Finkle directed at the southern base of Calhoun Hill (in upper Deep Coulee). A secondary field of fire from a smaller squad under Sergeant Finley would be directed west towards Greasy Grass Hill and Ridge. C Company's horses—likely held by four troopers, each holding up to eight horses were led down to the shelter of upper Calhoun Coulee between these two dismounted detachments of C Company, deployed on either side of the coulee. At the same time, Keogh deploys a small detachment of about eight to twelve men of I Company to deploy dismounted on the west side of Battle Ridge a short distance north of the eight-man detachment of C Company on Harrington Ridge. Custer and the Yates squadron are now heading back from their reconnaissance to the flats around the northern fords, stopping at the lower Cemetery flats to examine the ground there for an adequate rally point and defensive position. We speculate that Custer sent an additional messenger to Keogh telling him to move ahead to his position on the Cemetery flats once a successful junction has been made with Benteen's battalion and the ammunition mules. If it appears that no advance is forthcoming, Keogh should use his discretion to disengage his battalion and move north immediately to reconnect with Custer and the Yates squadron on the lower Cemetery flats above the river.

* 3:55 to 4:15 p.m.: Lame White Man and Yellow Nose lead a hostile charge up Deep and Calhoun Coulees directly into the horse-holders of C Company in the upper end of the Coulee. Once they penetrated far enough up the coulee to reach and stampede these horses, the dismounted troopers of C Company were helpless to repel the assault without hitting their own horses or horse-holders. Many of C Company's horses are driven down towards Greasy Grass Ridge, then along that ridge towards Deep Ravine. C Company's small detachment on Harrington Ridge, under Lieutenant Harrington and First Sergeant Bobo, fell back up Battle Ridge to rejoin with a portion of I Company deployed on the top of Battle Ridge. The larger detachment of C Company remains dismounted on Finley Ridge as the Indian charge bypasses their position as they charge up Calhoun Coulee and circle around to the north along the west side of Battle Ridge as they drive the stampeded horses of C Company down to Greasy Grass Ridge and Deep Ravine. Keogh throws every available trooper on Battle Ridge to the west side of the ridge—with the exception of Lieutenant Porter's reserve platoon, which remains east of the swale by his own led horses—to repel this charge against C Company's horses. Keogh's troopers—with eight to twelve having previously been deployed on the west side of Battle Ridge near the top, shoot down Lame White Man and Noisy Walking as they circle around from Calhoun Coulee to the north on the west side of Battle Ridge. Many more hostiles are wounded in this charge, which succeeds in stampeding most of C Company's horses but is repelled with heavy loss by Keogh's troopers now deployed on the western military crest of Battle Ridge. L Company remains in position on Calhoun Hill during the charge of Lame White Man and Yellow Nose, as does C Company's detachment deployed on Finley Ridge. Custer hears the sounds of heavy firing from Keogh's position and realizes that Keogh's battalion has come under heavy attack. He mounts up the Yates squadron and his own

Headquarters staff and gallops south off Cemetery Ridge and down into the low ground known as 'the Basin' west of Last Stand Hill. D and M Troops are still deployed on the Weir Peaks at this time. The main pack train is now arriving on Reno Hill.

* 4:00 to 4:10 p.m.: As most of Keogh's redeployed troopers focus their attention on repelling the charge of Lame White Man and Yellow Nose west of Battle Ridge, Crazy Horse and White Bull lead a mounted charge from the low ground east of Calhoun Hill directly into the led horses of L Co in Keogh's swale on the now exposed east side of Battle Ridge. These horses of L Company are stampeded northwards along the swale directly into the led horses of I Company and are met by I Company's second platoon mounted in reserve under Lieutenant Porter. Colonel Keogh rallies the dismounted detachments of C and I Companies deployed on the western military crest of Battle Ridge and leads a charge into the basin of the swale in an effort to save his horses and ammunition carried in the saddlebags. As Keogh's force in the swale—numbering about forty-six troopers of I and C Company—engage in melee combat with the warriors of Crazy Horse and White Bull, Lieutenant Calhoun, recognizing the fatality of their position if their horses are stampeded, orders L Company's skirmish line to fall back and move to support Keogh's desperate struggle for their horses down in the swale just north of their position on Calhoun Hill. It is at this point that First Sergeant Butler is sent by Calhoun down to Finley Ridge in an attempt to get a messenger through to the Weir Peaks for immediate support from the troops now seen clearly from those heights.

* 4:10 to 4:20 p.m.: Troopers from L Company begin arriving at the swale to support Keogh's last stand in the fight for their horses. Lieutenants Calhoun and Crittenden, along with the rearmost platoon of L Company, are killed during the retreat to Keogh's swale. First Sergeant Butler has reached C Company's main detachment on Finley Ridge and informed Sergeants Finley and Finckle of events transpiring on Battle Ridge and the desperate need to get a messenger through to the troops seen on the Weir Peaks. Colonel Keogh has been shot in the leg. He rallies the remaining dismounted troopers from I and C Company and the platoon from L Company, who have managed to reach his position in the swale—about sixty-five men—for a last stand. First platoon of E Company under Lieutenant Smith moves dismounted southward from Battle Ridge Extension to Last Stand Hill to provide cover fire and support for Keogh's battalion battling for their horses in the swale. Lieutenant Porter and I Company's second platoon mount their horses and ride north to Last Stand Hill in an effort to bring support for Keogh's troopers making their stand in the swale. Lieutenant Harrington leaves Keogh's swale and rides south, east of Calhoun Hill and across upper Deep Coulee and Medicine Tail Coulee, thence into South Medicine Tail Coulee (east of Sharpshooter Ridge) in an effort to reach Reno or Benteen on the bluffs there. Meanwhile, Custer's Headquarters Staff and the Yates squadron have arrived at the low ground in the Basin area south of Cemetery Ridge and west of Last Stand Hill. The second platoon of E Company under Lieutenant Sturgis is dismounted there with their horses held in the low ground of the small Basin just south of Cemetery Ridge, while the dismounted troopers of E Company's platoon move back to the upper edge of Cemetery Ridge and deploy a skirmish line there. Captain Yates's F Company separates into two platoons, with the second platoon under Lieutenant Riley dismounted and

deployed along the next small ridge, i.e., Basin Ridge—south of Cemetery Ridge (overlooking Deep Ravine).

* 4:20 to 4:30 p.m.: First Sergeant Butler and Corporal Foley ride off southwest from Finley Ridge in a desperate effort to reach the troops seen on the Weir Peaks. Butler's horse is shot down in the vicinity of Deep Coulee, where he is unhorsed. Corporal Foley crosses over Deep Coulee and ascends what is today called Butler Ridge, where he shoots himself in the head. Keogh's Last Stand in the swale is now over, and the surviving troopers of I, C, and L fight their way—or simply run—dismounted towards Last Stand Hill. Most do not make it. Lieutenant Porter and between ten to fifteen survivors of Keogh's battalion do manage to reach Last Stand Hill. Lieutenant Harrington is shot and killed by pursuing warriors in South Medicine Tail Coulee east of Sharpshooter Ridge. The survivors of Keogh's battalion rally with the first platoon of E Company under Lieutenant Smith on Last Stand Hill. E Company's second platoon under Lieutenant Sturgis maintains its skirmish line on the upper southern edge of Cemetery Ridge, protecting the led horses in the small Basin area beneath them and holding open the route down to the lower Cemetery flats. The main detachment of C Company—just over twenty men—remains cut off from support on Finley Ridge. D and M Troops remain on the Weir Peaks. Meanwhile, Reno has just given orders to prepare to advance the remaining elements of the Regiment, Pack Train, and Rear Guard to the north to seek a junction with Custer.

* 4:30 to 4:40 p.m.: Reno begins his advance with the remaining seven companies, halting a short distance north of Reno Hill to await the arrival out of the valley of the Herendeen group of twelve survivors from the valley fight and the Varnum burial party. On Last Stand Hill, preparations are being made to prepare the wounded—which might include Lieutenants Smith and Reilly—for transport from Last Stand Hill to Cemetery Ridge by the first platoon of E Company along with the few survivors of the Keogh battalion. The main detachment of C Company remains cut off from support on Finley Ridge. Meanwhile, F Company's first platoon under Captain Yates, along with Custer's Headquarters staff, continue moving southwards into the next area of low ground known today as the 'Big Basin.' They continue on to deploy on the high ground just beyond (and overlooking the lower reaches of Calhoun Coulee) in an effort to support the withdrawal to their position of the C Company detachment now cut off and trapped on Finley Ridge.

* 4:40 to 4:50 p.m.: The led horses and spare ammunition of E Company's second platoon are held on the low ground just below their position in the small Basin. The horses of F Company's second platoon are kept in the same area. The horses of Yates's first platoon are in the 'Big Basin' area; F Company's second platoon under Lieutenant Reilly are deployed on the small Basin Ridge between these two depressions, while their first platoon has deployed on the next ridge line to the south end of the 'Big Basin'—i.e., Harrington Ridge—to prepare to cover the withdrawal of C Company's detachment on Finley Ridge to their position. Reno continues to delay his advance with the remaining seven companies of the Regiment while resupplying the arrival of the survivors from the valley fight. With the destruction of the Keogh battalion, a number of warriors in and around Deep Coulee, Calhoun Hill, and Finley Ridge notice additional troops on the Weir Peaks and begin to move that way. Godfrey and

Benteen, tired of waiting, begin to advance to the Weir Peaks once again. Reno attempts to halt this advance through his trumpeter but his orders are ignored by Benteen and Godfrey.

At this critical juncture, the suicide boys mentioned by John Stands In Timber's account launch their mounted charge over the top of Cemetery Ridge and down into the northernmost smaller Basin just below Cemetery Ridge, stampeding the led horses of E and F Company being held there. They are supported in this charge by additional warriors coming up from Deep Ravine to join them in the small Basin. Upon reaching this Basin, the suicide boys split up, with some driving the led horses of E and F Company down to Deep and Cemetery Ravines and the rest swinging back up towards the remaining troopers on Last Stand Hill. The second platoon of F Company deployed on the Basin Ridge (just south of the small Basin), and the second platoon of E Company, under Lieutenant Sturgis, dismounted on the southern edge of Cemetery Ridge, turned, and raced down into the small Basin where their horses are being stampeded by the suicide warriors. This move is noticed by warriors who spoke of seeing about forty troopers running dismounted from higher ground down into the small Basin towards Deep Ravine. Seeing the loss of their horses and spare ammunition contained in the saddlebags, the platoons of E and F Company charge down into the small Basin in an attempt to recover their mounts and, more importantly, their ammunition. This charge into the small Basin of roughly forty troopers was accompanied by Mitch Boyer. Yates's first platoon and Custer's Headquarters Staff, now deploying around the big Basin area further to the south, immediately moved to ride to the Basin Ridge—just north of the big Basin (where the second platoon of F Company had been deployed) in an effort to provide support for this attack.

* 4:50 to 5:00 p.m.: On reaching the Basin Ridge, Yates's first platoon and Custer's Headquarters staff see two things occurring at once. In the small Basin below them, they see the horses of E and F's second platoons being stampeded down into Deep and Cemetery Ravines, with both dismounted platoons following them down into this ravine in an effort to retrieve their mounts. At the same time, they notice the other half of the suicide boys charging uphill directly toward the position of the first platoon of E Company (under Lieutenant Smith) on Last Stand Hill. Custer makes a quick decision to gallop Yates's first platoon and his Headquarters Staff back to Last Stand Hill in an attempt to hold that position and rally the survivors of Keogh's battalion and provide support there for his remaining platoons now fighting for their mounts in Deep Ravine. Nearly all of the suicide warriors who made this charge are killed by the troopers.

* 5:00 to 5:10 p.m.: Custer is able to stabilize the situation on Last Stand Hill briefly, and he orders whatever remaining horses there to be shot down as breastworks at the top of the hill. Six horses of the survivors of C Company are thus dispatched in this manner, providing breastworks for Custer's Headquarters detachment. Captain Benteen and Lieutenant Godfrey arrive and deploy on the Weir Peaks with H and K Company, and Reno has begun his advance northwards with the remaining troops and Pack Train under his command. Some warriors begin to converge on the high ground south of Medicine Tail Coulee in reaction to the troops on the Weir Peaks. D Company moves to the forward slopes of the Peaks and deploys there to confront them.

* 5:10 to 5:20 p.m.: As the second platoons of E and F Company's—most dismounted except for the officers and a few Non-Commissioned Officers (NCOs)—continue their dismounted charge from the small Basin into Deep Ravine, following their stampeded horses directly towards the river and Ford C, hundreds of mounted warriors follow them off Cemetery Ridge down into the Basin. Some of these warriors assault the remaining troopers on Last Stand Hill, who direct their fire westward against these warriors in the Basin and on Cemetery Ridge. While the troopers on Last Stand Hill were focused on repelling the attack west of Last Stand Hill, they were hit from behind by mounted warriors east of Last Stand Hill, wherein the fighting on Last Stand Hill devolved into melee combat, ending in the death of the remaining troopers there. At roughly the same time, the C Company detachment isolated on Finley Ridge come under heavy attack. About seven to nine C Company troopers from Finley Ridge mount and race alongside Calhoun Coulee towards the Yates troopers seen in the Basin area, but most are killed before reaching it. The remainder of C Company's detachment soon after go down fighting on Finley Ridge. During the final minutes of the fighting, the remaining troopers under Porter and Sturgis who had made it into Deep Ravine—Boyer had likely been shot down a short distance before reaching the ravine—were killed in the attempt to recover their horses by warriors positioned on the high ground above and on either side, shooting down into the ravine. Roughly twenty-eight men were killed in this fashion, with several making it to the river and at least one trooper—Private William Brown of F Company—making it onto the west side of the river before being shot down. Both Porter and Sturgis were killed in or near Deep Ravine or at the river, with Sturgis possibly being captured alive and burned later in the evening. The troopers in Deep Ravine were killed at roughly the same time as the final troopers were killed on Last Stand Hill, ending the final action on Custer's field at about 5:20 p.m. Groups of warriors now begin engaging in long-range skirmishing with D Company out on the forward slopes of the Weir Peaks.

* 5:10 to 5:20 p.m.: The warriors from the Custer Fight begin to head south to the bluffs to confront the troops now seen skirmishing on the Weir Peaks with warriors who had reached this position earlier.

Participant Timeline For Little Big Horn Battle: Continuum

1. THE REGIMENT LEAVES HALT 1 ON THE DIVIDE

The actual participants of the Battle of the Little Big Horn tell us a great deal about the time that a particular event of the battle took place. As we explore and analyze what these participants had to say in the following sections, we will develop what may well become the most accurate of the battle timelines ever seen before. Note: All watch times are based on Department Headquarters Time in St. Paul, MN. These times will be about 57 minutes later than local sun time in southeastern Montana.

We begin our timeline with an examination of when the Regiment left what has been called Halt 1 (H) about 4.5 miles east of the divide on the morning of June 25, 1876. Custer himself had previously left with a few of his scouts up to the Crows Nest for his first view into the valley. Here is what our participants had to say about this first movement of troops that morning:

1) Private Gallenne—M Company (letter to Father Genin, July 5, 1876, with my annotations in brackets): "...[we marched] all night; [M Company] halted at 3 a.m. ... Rested 2 hours [till 5 a.m.] and marched out till 9 a.m. ..."

ANALYSIS

Private Gallenne's account tells us that M Company was traveling nearer the end of the Regiment, which was strung out along the east side of the divide between the Rosebud and Little Big Horn rivers. The lead elements of the command had covered about eight miles during the night, reaching Halt 1 early that morning. M Company apparently did not do so, having stopped at some location east of Halt 1 at 3 a.m., where they rested for about two hours before resuming their march eastward again at about 5 a.m. and continuing on their march until they stopped at some location, likely Halt 2, again at 9 a.m. What we learn from Gallenne's account from his letter above is that elements of the Regiment were scattered along the trail east of the divide and not all conveniently camped in one location. Apparently, not all the companies made it as far as Halt 1 by the morning of June 25.

2) Private Pigford—M Company (*Washington Observer* Interview, October 5, 1932), with our annotations in brackets. "The march was resumed [from Halt 1] at 8 o'clock ..."

ANALYSIS

Private Pigford, traveling with M Company, tells us that they resumed their march at 8 a.m., presumably from Halt 1, which appears to contradict Private Gallenne's implications.

3) Private Donahue—K Company (*Bismarck Tribune* Interview, January 17, 1888): "about 4 o'clock in the morning on the 25th, General Custer made camp. ... About 6 o'clock in the morning we were again in the saddle leading our pack animals. ..."

ANALYSIS

It would appear from Private Donahue's account above that K Company—or more likely the Pack Train—was traveling near or at the rear of the Regiment, not stopping during the night march until about 4 a.m., whereupon they rested for about two hours before continuing on again at 6 a.m. His comment about "leading our pack animals" may indicate that he was traveling with the pack train as opposed to his own company at this particular time.

4) Private Taylor—A Company (*With Custer on the Little Big Horn*): After daylight ... General Custer, who had gone ahead to the point on the divide from whence the scouts had seen the smoke rising from the Indian village and the pony herds grazing in the valley near it, some twelve or fifteen miles away, had returned, and a little before 8 o'clock came riding bareback, and we think also bare-headed, around to several troops giving the officers the information that the Indians had been located, and saying that the command would move at once. The men began to saddle up, and we were soon in motion up the divide. ...

ANALYSIS

Private Taylor has a number of errors in his account of the battle, and one of them can be seen in his account above. Other participants recall that Custer left for the Crows Nest before the command began their advance to Halt 2 Gerard, who accompanied Custer to the Crows Nest that morning, indicated that Custer gave orders for the command to remain in camp at Halt 1 and did not go with them to advance at all. There are other accounts that also recall Custer riding bareback and giving instructions to his troops, yet these accounts place this event just before Custer leaves his command for his first trip to the Crows Nest. This latter interpretation is likely correct, as Custer's horse had not yet been saddled, nor was he fully attired (being bare-headed). It is my belief that Private Taylor mistook General Custer with his brother, Colonel Tom Custer, who very likely made the decision to order the Regiment to advance onwards to Halt 2 at the base of the Crows Nest shortly after the General had departed for the Crows Nest with his scouts. Private Taylor does indicate that the command was soon moving out towards Halt 2 just after 8 a.m.

5a) Lieutenant Godfrey (*Godfrey's Narrative*): they started marching at 8 a.m. on June 25 ...

ANALYSIS

Lieutenant Godfrey agrees with other participant accounts, which place the march from Halt 1 at about 8 a.m.

5b) Lieutenant Godfrey (*Field Notes*): about 8 a.m. on June 25 the command moved from Halt 1 on the divide to Halt 2. Prior to this move, Custer came around to all the companies and informed them that the Indian village was in view. They then continued the march.

ANALYSIS

Godfrey again says the command left Halt 1 at about 8 a.m., shortly after receiving word from Custer that the scouts had spotted the village.

6) Capt. Moylan (Reno Court of Inquiry [RCOI]): On the morning of June 25, the command began to move up toward the divide at about 8 a.m.

ANALYSIS

Captain Moylan agrees with Godfrey, Taylor, and Pigford that the command left Halt 1 at about 8 a.m.

7) Lieutenant Hare (Camp Interview, *Custer in '76*): says the 7th began to move again up the divide between 8 and 9 a.m. Custer had already left earlier for the Crows Nest.

ANALYSIS

Lieutenant Hare gives us what I believe may be the most accurate assessment yet concerning this move from Halt 1. He tells us that the command moved out sometime between 8 and 9 a.m., which makes more sense to us than the 8 a.m. departure time usually given. For reasons that will be developed later, I believe the command left Halt 1 at a time closer to 8:30 or 8:45 a.m. instead of 8 a.m., which would be closer to the time that Custer himself left for the Crows Nest. Lieutenant Hare also makes it clear that Custer had already left for the Crows Nest before the command moved out towards Halt 2. This directly contradicts Private Taylor's assertion that Custer had returned from an earlier trip to the Crows Nest and led the command out from Halt 1.

8) Lieutenant Wallace (RCOI): The march commenced at 8:45 in the morning.

ANALYSIS

Lieutenant Wallace supports Lieutenant Hare's recollection that the command left Halt 1 at some time between 8 and 9 a.m. My own speculation—based on the distance the command would travel to Halt 2 (about 4.5 miles)—is that Wallace's time for the commencement of the march at 8:45 a.m. was very likely correct for at least part of the command. I do not believe the evidence will support the notion that the entire command rested at Halt 1 before moving out again towards Halt 2. The command was likely scattered for a distance along the trail east of the divide.

9) Packer Ben Churchill (RCOI – *Army and Navy Journal*): the pack train moved forward on the divide at 7 a.m., although it might have been an hour later.

ANALYSIS

Packer Churchill indicates that the pack train moved out from Halt 1 at 7 a.m. This would be sun time or 8 a.m. Headquarters watch time. Then he adds that it might have been closer to an hour later, which is probably correct, indicating that the Pack Train and Rear Guard left Halt 1 at a time closer to 9 a.m. watch time—or more likely at 8:45 a.m. as Lieutenant Wallace suggests in his itinerary above.

10) Private Peter Thompson (*Peter Thomson Narrative*): At half past six, the cook was awakened to prepare breakfast for the men. ... It was half past eight o'clock when we moved out of camp.

ANALYSIS

Private Thompson, in his Narrative of the battle, tells us that the cooks were ordered to prepare breakfast for the men at 6:30 a.m., and the command was ordered to move on up the divide toward Halt 2 at 8:30 a.m.

11 a) Lieutenant Varnum (via General Nelson Miles' unpublished article in 1896, published by Tom O'Neil, Arrow and Trooper Publishing, 1993):

... namely, the statement of Lieutenant Varnum ... proves, in Lieutenant Varnum's own language, that neither he, General Custer, nor any officer knew the location of the Indian village until 8 o'clock on the morning of the 25th. Up to that hour, General Custer was in ignorant about whether the Indian village was in the valley of the Little Big Horn, Tullock's Fork, or Rosebud, or wherever it might be, and there is not a particle of evidence up to that time that he intended to fight on the 25th.

ANALYSIS

General Miles relates that Lieutenant Varnum indicated above that it was not before 8:00 a.m. on June 25th that Custer discovered the general whereabouts of the hostile village.

11 b) Lieutenant Varnum (From Col. Robert P. Hughes, in his article "The Campaign Against the Sioux in 1876", published as an appendix in Graham's *The Story of the Little Big Horn* wherein he quotes Captain Charles Varnum describing his trip to the Crows Nest):

"I got to the Crows Nest about 2:30 A.M. ... I threw myself down and fell asleep; but in about three-quarters of an hour [i.e., or about 3:15 A.M.], or an hour [i.e., about 3:30 A.M. or 4:30 A.M. HQ watch time], I was waked up. It was then just daylight. ... I sat down and wrote a dispatch to Custer and sent it off at about 4:45 [local sun time, or about 5:45 A.M. HQ watch time.]. ... Custer got my message at about

8 o'clock [HQ watch time—or about 7 a.m. local sun time] and started soon after and the dust of his column could be plainly seen as soon as he did so, though not his troops."

ANALYSIS

Again, Varnum indicates that Custer is informed of the general whereabouts of the hostile village at about 8 a.m. HQ watch time, whereupon the regiment moved out "soon after."

OVERALL ANALYSIS

Of the 11 accounts listed above, most, if not all of the accounts, indicate that the command left Halt 1 at some time between 8 a.m. and 8:45 a.m., just as Lieutenant Hare and Varnum indicated. I believe the lead elements of the command likely left Halt 1 at about 8:30 a.m., about 25 or 30 minutes after General Custer and his scouts left the command on his ride to the Crows Nest. The remaining elements of the command, who had lagged several miles behind during the night march, likely departed the area of Halt 1 at about 8:45 a.m., as Wallace indicated.

CONCLUSION

Between 8:05 and 8:10 a.m.: General Custer left for the Crows Nest.

Between 8:30 and 8:45 a.m. (H = 0, baseline): The Regiment departs Halt 1.

For those not using clock times: **H = Regiment leaves Halt 1 (baseline).**

2. THE REGIMENT ARRIVES AT HALT 2

Here are the statements and analyses of our participants regarding the time that the Regiment arrived at Halt 2 on the divide. Halt 2 was about a quarter mile east of the divide, not far from the base of the Crows Nest. It was about 4.5 miles west of Halt 1. This would take the command about an hour and 15 minutes to cover that distance at a walk.

1) Sergeant Kanipe (Camp Interview, *Custer in 76*, with my annotations in brackets): Custer returned from his view from [the Crows Nest] and rejoined the command sometime between 10 and 11 a.m.

ANALYSIS

Custer returns from his trip out to the Crows Nest to rejoin his command at Halt 2 sometime between 10 and 11 a.m. Based on other accounts, we believe this time would be much closer to 10 a.m. than 11 a.m.; I would suggest about 10:10 a.m.

2) Lieutenant Hare (Camp interview., *Custer in 76*, with our annotations in brackets): says the 7th Cavalry began to move again up the divide [from Halt 1] between 8 and 9 a.m. Custer had already left earlier for the Crows Nest. They halted [at Halt 2] near the divide ... less than a half mile from the top of the divide. They stopped and lay concealed for over an hour. ... This was [sometime] between 10 a.m. and noon. During this time [likely after the last Officer's Call], Custer again goes to the Crows Nest to look at the Indians.

ANALYSIS

It is important to note that it would take the command, moving at a walk, about one hour and fifteen minutes to cover the 4.5 miles between Halt 1 and Halt 2 on the divide. If the command left Halt Number1 at 8:30 a.m., as I believe it did (see the discussion and analysis for The Regiment Leaves Halt 1), the lead elements of the command would reach Halt 2 at about 9:45 a.m. with the trailing elements arriving at or just after 10:00 a.m. During this halt, Officer's Call would be sounded as soon as Custer returned from the Crows Nest, a short time later, and the command would rest for just over an hour before moving out again to cross the divide. According to Lieutenant Hare's recollection, the time frame for this event would be between 10 a.m. and 11 a.m. We should make note of the discrepancy between Hare's comment that the command rested at Halt 2 for one hour, yet then gives a timeframe of two hours, saying it was sometime between 10 a.m. and 12 noon. This makes little sense, especially when we factor in the realization that the lead elements of the command likely arrived at Halt 2 at about 9:45 a.m., covering just 4.5 miles in an hour and fifteen minutes. A rest of a little over an hour would begin at about 9:45 a.m. and end at about 10:55 a.m.

3) Capt. Moylan (RCOI): on the morning of June 25, the command began to move up toward the divide at about 8 a.m. He says they halted at 10:30 or 11; he is not sure.

ANALYSIS:

Here, we encounter a serious problem with Captain Moylan's account. He first gives us a time of 8 a.m. as the time when the Regiment left Halt 1, then tells us that they arrived, presumably at Halt 2, at 10:30 or 11 a.m. As the total distance marched was only 4.5 miles, which should take about one hour and fifteen minutes to cover at the pace of a walk, Moylan suggests that it took the Regiment an improbable 2.5 to three hours to cover this distance, moving at the pace of a sloth at about two miles per hour! Not very likely. It would appear more logical that the Regiment either left a bit later than Moylan tells us or that it arrived a good deal sooner, or very likely, both.

4a) Lieutenant Godfrey (Field Notes): About 8 a.m. on June 25 the command moved from Halt 1 on the divide to Halt 2 (about four miles). Prior to this move, Custer came around to all the companies (before he left for the Crows Nest) and informed them that the Indian village was in view. Then they continued the march. At Halt 2, the command hid in a ravine near the divide; soon came a report of Sergeant Curtiss finding the box of lost hardtack, etc. Crow scouts also reported the command seen by hostiles. Tom Custer rode out to the Crows Nest to report this news to the General and the General returned to their bivouac and had Officer's Call sounded. Godfrey goes to Headquarters and hears Bloody Knife say they would find enough Sioux to keep them fighting for two or three days. Custer laughed and said he thought they would get through them in one day. Custer announces they will move to attack immediately—officers to prepare their companies for battle. They cross the divide. ...

ANALYSIS

Lieutenant Godfrey gives us a very accurate picture of events in the passage above. First, he says the command moved out about 8 a.m. from Halt 1. I have this event occurring at about 8:30 a.m. This would place the arrival of the lead elements of the command at Halt 2 at about 9:45 a.m. Captain Tom Custer and Lieutenant Calhoun ride out towards the Crow's Nest and meet the General sometime between 9:45 and 10:00 a.m. to report on the box of lost hardtack being discovered by the hostiles. The General returns to his command at Halt 2 between 10:00 and 10:10 a.m., whereupon he has his last Officer's Call, likely between 10:20 and 10:40 a.m. The officers then went to prepare their companies for battle between 10:40 and 10:50 a.m., during which time Custer goes up to the divide for his second look into the valley, this time with De Rudio's field glasses. The Regiment moves out to cross the divide shortly thereafter, about 10:55 a.m., fulfilling Custer's desires that they "move to attack immediately." Godfrey does not indicate an unnecessary and illogical delay of an hour and a half, as Lieutenant Wallace would have us believe.

4b) Lieutenant Godfrey (Godfrey's narrative from *Custer's Last Battle*): they started marching at 8 a.m. on June 25 and marched until 10:30 a.m. when the command halted in a ravine for concealment.

ANALYSIS

Here, we have the same dilemma discussed above with Captain Moylan's account. Godfrey indicates an 8 a.m. departure of the command from Halt 1 and a 10:30 a.m. arrival at Halt 2, or an improbable 2.5 hours to cover 4.5 miles, which amounts to an absurd pace of two miles per hour! Something is clearly wrong here. Either the command left Halt 1 later than 8 a.m. or arrived at Halt 2 well before 10:30 a.m., or both. As we can expect the command to cover this distance of 4.5 miles in about an hour and fifteen minutes moving at a regulation cavalry walk, there is no way this event should have taken anywhere near this long to negotiate. Thus, the timing estimates for this account, like Moylan's above, must be regarded as suspect.

5) Private Taylor (*With Custer on the Little Big Horn*): ... General Custer, who had gone ahead to the point on the divide from whence the scouts had seen the village and the pony herds grazing in the valley near it, some 12 to 15 miles away, had returned, and a little before 8 o'clock came riding bareback, and we think also bare-headed, around to several troops giving the officers the information that the Indians had been located, and saying that the command would move at once. The men began to saddle up and

we were soon in motion traveling up the divide. ... We did not stop until about 10:30 a.m. when we came to a halt in a ravine, some four or five miles from the summit of the divide. Here we were ordered to keep concealed and to preserve quiet ... But in the meantime, our presence having been discovered ... "it was necessary to act at once," and we started off again, crossing the divide a little before noon.

ANALYSIS

There are a number of errors in Private Taylor's reminiscence given above. His first error was in stating that Custer returned from his first trip to the Crows Nest back to Halt 1 before leading his command out again at about 8 a.m. towards Halt 2. This contradicts other witnesses, namely Fred Gerard, who accompanied Custer out on his first trip to the Crows Nest that morning. Gerard tells a very different story in that Custer did not want the command moved ahead from Halt 1 but that an officer, presumably Tom Custer, took it upon himself to move the command without authorization to do so, meeting Custer a short distance from Halt 2 as they were descending from the Crows Nest. Private Taylor appears to have mixed up this event, perhaps mistaking General Custer with his brother Tom, who was apparently dressed similarly. Taylor's second mistake is describing the location of Halt 2 as being four or five miles from the summit of the divide. That would be the correct distance of Halt 1 from the divide. According to Lieutenant Edgerly's recollection, halt 2 was about a quarter mile east of the divide. Lieutenant Hare would support that when he described its location as being less than a half mile from the divide. The third problem Taylor confronts us with is the same one given by Moylan and Godfrey in their accounts, this being an inexplicable 2.5-hour span of time for the Regiment to move just 4.5 miles from Halt 1 to Halt 2. Something is clearly wrong here. Either the command left Halt 1 well after 8:00 a.m., or the command arrived at Halt 2 well before 10:30 a.m., or both. Consequently, Taylor's time estimate of crossing the divide a little before noon must also be suspect, especially when we consider his statement that "it was necessary to act at once." This would preclude an unnecessary delay of an hour and a half before advancing on the hostiles, as a noon crossing of the divide would suggest.

6) Private Donahue—K Company (*Bismarck Tribune* interview, 1/17/1888): About 10 o'clock a.m. a halt was ordered.

ANALYSIS

Private Donahue tells us that the command reached Halt 2 on the divide at about 10 a.m. The rear elements of the command would be expected to have arrived at this location at about 10 a.m. or a few minutes later.

7) Private Kennedy—I Company (Olin Wheeler account, 1900, our annotations in brackets): After Officer's Call [at Halt 2 just east of the divide], we marched. ... The officers at this time [at Halt 2] got orders to bunch the mules together, with the exception of the ammunition mules, which were taken on ahead.

ANALYSIS

Private Kennedy gives us two important pieces of information above. First, the command marched out again after the Officer's Call at Halt 2. Kennedy does not indicate any lengthy delay of an hour or more before moving out, as Lieutenant Wallace's account would suggest. Secondly, Kennedy tells us that the ammunition mules were separated from the main pack train and were "taken on ahead," which explains why Lieutenant Mathey and these advance packs were several miles ahead of the main pack train when Lieutenant Hare rode back to them from Reno Hill.

8a) Lieutenant Edgerly (*The Custer Myth*, 1881 account, our annotations in brackets): At about 10 o'clock in the morning we were ... 15 miles from the hostile camp. Our force was then all together [at Halt 2]. We halted while Custer [earlier that morning] went on a hill with the Crow and Ree scouts to take a look at the Indian camp, which was in sight. When General Custer came down from the hill, Officer's Call was sounded. The officers all went to where he was and he told us that ... we would press on as quickly as we could and attack them in the village if possible. ... He then ordered troop commanders to mount their troops and report when they were in readiness to move on. In about a minute, every troop commander had reported.

ANALYSIS

Lieutenant Edgerly tells us that the command had all reached Halt 2 at about 10 a.m. When Custer returned from the Crows Nest a bit later, presumably about 10:10 a.m., Officer's Call was sounded, whereupon the officers were told that they "would press on as quickly as they could" and attack the hostiles "in the village if possible," thus indicating the need for speed at this point. When Officer's Call ended, presumably about 10:30 or 10:40 a.m., the troop commanders were told to "mount their troops and report when they were in readiness to move on. In about a minute, every troop commander had reported." Lieutenant Edgerly tells us here that there was no hour-and-a-half delay before moving out to cross the divide, as Lieutenant Wallace would infer from his deceptive itinerary. That extended delay makes no sense whatsoever.

8b) Lieutenant Edgerly (RCOI): Officers Call was at 10 a.m. on June 25, about a quarter mile east of the divide.

ANALYSIS

Edgerly agrees with most other sources that Officer's Call was about 10 a.m. and Halt 2 was about a quarter mile east of the divide.

8c) Lieutenant Edgerly (letter to wife, July 4, 1876): we ... moved on again till about 10 a.m. when Officer's Call was sounded again. After nearly an hour's rest and a reconnaissance by Gen'l. Custer and some scouts, we were told that the Indians had undoubtedly discovered our presence as several had been seen on the bluffs, and the only thing for us to do was to move on them at once. ... We then mounted and moved forward about half a mile, halted and he (Custer) reassigned the companies into battalions ...

ANALYSIS

Edgerly says that the command reached Halt 2 about 10 a.m., followed shortly thereafter by Officer's Call (after the return of Custer and his scouts from their reconnaissance at the Crows Nest). He says that the command rested at Halt 2 for "nearly an hour's rest," which indicates that the command rested from about 10 a.m. until just before 11 a.m. when they moved out again to cross the divide.

9) Captain Benteen (RCOI): Officer's Call was held at 10 a.m.

ANALYSIS

Captain Benteen agrees with the majority of our participants that the last Officer's Call, held shortly after the command reached Halt 2, was held at about 10 a.m.

10) Packer Ben Churchill (RCOI and *Army and Navy Journal*): the pack train moved forward on the divide at 7 a.m., although it might have been an hour later.

ANALYSIS

Packer Churchill indicates that the pack train moved out from Halt 1 at 7 a.m. This would be sun time or 8 a.m. Headquarters watch time. Then he adds that it might have been closer to an hour later, which is probably correct, indicating that the pack train and Rear Guard left Halt 1 at a time closer to 9 a.m. watch time—or more likely at 8:45 a.m. as Lieutenant Wallace suggests in his itinerary above. Moving at a walk of about three miles per hour, the pack train and Rear Guard would likely have arrived at Halt 2 about 10:15 a.m., just about the time Custer signaled his last Officer's Call.

11) Lieutenant Wallace (RCOI, our annotations in brackets): The march [from Halt 1] commenced at 8:45 in the morning. The first halt [at Halt 2] was at 10:07.

ANALYSIS

Lieutenant Wallace gives us his itinerary for the movements of the command above. He indicates a departure of the command at 8:45 a.m. or a full forty-five minutes later than nearly all other time estimates given by our participants. It is very possible that the rear elements of the command actually left Halt 1 at this time. Had it been the head of the command leaving at that time, most of our participants would likely have referenced a 9 a.m. departure time from Halt 1 instead of the 8 a.m. time they gave. Eight-thirty a.m. would be a more likely time of departure for the head of the command. The head of the command would thus be expected to reach Halt 2 at about 9:45 a.m., covering the 4.5 miles in about an hour and fifteen minutes. The rear of the command likely arrived at this location by 10:07 a.m., as Wallace indicated in his itinerary above.

OVERALL ANALYSIS

All of the participant accounts above indicate an arrival time at Halt 2 as occurring sometime between 10 and 11 a.m. Three accounts (those of Moylan, Taylor and Godfrey) indicate an arrival time of 10:30 a.m. Five accounts, (those of Hare, Donahue, Edgerly, Benteen and Wallace) indicate an arrival of 10:00 a.m. or a few minutes later. Timing and distance factors from Halt 1 indicate that the earlier time estimates of 10 a.m. would be more accurate.

CONCLUSION

Based on the participant accounts listed above, I believe the lead elements of the Regiment left Halt 1 on the divide at about 8:30 a.m. and arrived at Halt 2 at about 9:45 a.m. with the rear elements leaving Halt 1 at about 8:45 a.m. and arriving at Halt 2 at about 10:07 a.m. The slow-moving pack train and Rear Guard likely did not arrive until 10:15, about the time Custer sounded his last Officer's Call. Custer returned from the Crows Nest sometime between 10:05 and 10:10 a.m. and then came Officer's Call. Officer's Call occurred between 10:20 and 10:35 a.m. The officers are then told to prepare and mount up their companies for an immediate advance on the hostiles, which occurred between 10:35 and 10:50 a.m. with the command ready to move again, leaving Halt 2 at 10:50 a.m. and moving up to cross the divide at 10:55 a.m.

8:30 a.m. (H = 0, baseline): The Regiment leaves Halt 1.

10:07 a.m. (H + 1 hour and 15 minutes): The Regiment arrives at Halt 2.

3. THE REGIMENT CROSSES THE DIVIDE

Many students of this battle believe the divide crossing is about noon, placing their full faith upon the unsubstantiated itinerary of Lieutenant Wallace. I believe this faith to be both misguided and unfounded and has led to a number of confusions and misunderstandings that have prevented to this day the publication of an accurate timeline for this battle. Lieutenant Wallace, being the official itinerary archivist of the Regiment, was in a position to provide accurate times for most events of the battle. Many battle students today place their faith in his statements without bothering to examine the full scope of his RCOI testimony, which was riddled with errors or misstatements of facts. It is my belief that Lieutenant Wallace intentionally misled the RCOI by attempting to add at least one hour of time to the events that afternoon in an effort to protect his colleagues—and the reputation of his Regiment—from public criticism that they delayed their advance to support Custer during the height of the battle. An entire article could be written regarding the problems endemic in Lieutenant Wallace's RCOI testimony, but this is not the time nor place for such an analysis. (Note: just such an article was written about Wallace's inconsistent

testimony in the 2000 Winter edition of the Little Big Horn Associates Research Review by Larry Sklenar, author of the well-researched book on the Little Big Horn battle, *To Hell With Honor*.) Instead, let us examine what our battle participants had to say about when the command likely crossed the divide between the Rosebud and Little Big Horn Rivers:

1) Major Marcus Reno (RCOI): claims he received command of his battalion at 9 a.m.

ANALYSIS

This time estimate of Major Reno is simply not supported by other participants in the battle. According to our other battle participants, at 9 a.m., the Regiment had left Halt 1 and was making its way up the divide 4.5 miles to Halt 2. Custer was still at the Crows Nest at that time and had not yet returned to the Regiment to give Reno command of his battalion. We can safely dismiss this time estimate from Reno as inaccurate.

2) Lieutenant George Wallace (RCOI, with our annotations in brackets): The first halt [i.e., at Halt 2] was at 10:07. We started out [to cross the divide] at 11:45. At 12:05 it halted, and the divisions into battalions was made, and it moved on at 12:12....

Question: Is that memorandum you refer to the original itinerary?

Wallace: No, sir; that is a copy. I did not compare it, but Lieutenant Maguire said it was an exact copy.

ANALYSIS

Lieutenant Wallace gives us his itinerary for the movements of the command above. The rear of the command likely arrived at this location by 10:07 a.m., as Wallace indicated in his itinerary above. The problem with Wallace's itinerary is that he states that the command delayed its advance from Halt 2 for an hour and thirty-eight minutes—at a time when Custer realized his command had been discovered and the need for a rapid advance was at hand if Custer had any hopes at all of catching the hostiles before they were warned of his approach. It is truly amazing how many battle students fail to pick up on this one obvious 'faux pas' in Wallace's itinerary. It makes no sense. It appears that Lieutenant Wallace, for reasons of his own, decided to add roughly one hour to his time estimates from this time forward. Had the Regiment arrived at Halt 2 at 10:07 a.m., shortly before Custer's return from the Crows Nest, there would be no reason why the command could not have commenced their advance to the valley of the Little Big Horn by at least 10:50 a.m.

There are other red flags in Wallace's itinerary. He claims the command began their advance to cross the divide at 11:45 a.m. and moved all of a half mile before halting once again to divide the regiment into battalions. He described Halt 2 as being a quarter mile east of the divide and the stop where they divided into battalions as a quarter mile west of the divide. Wallace then claims that it was 12:05 p.m. when this stop occurred, indicating that it took twenty minutes for the command to cover a half mile, which equates to a crawl of an absurd 1.5 miles per hour! And this at a time when the command had been discovered! There are too many red flags in Wallace's itinerary—red flags consistently ignored by too many battle students today. I find the following question put to Lieutenant Wallace at the RCOI to be of interest:

> Question: Is that memorandum you refer to the original itinerary?
> Wallace: No, sir; that is a copy. I did not compare it, but Lieutenant Maguire said it was an exact copy.

It is a pity that the Recorder did not demand that Lieutenant Maguire produce the original itinerary of Lieutenant Wallace so that the court could decide for itself if it was truly an exact copy, as Lieutenant Wallace so alleged. Unfortunately, Lieutenant Maguire was not asked to do so, and to this day, no one has ever seen that original itinerary. One wonders what could have happened to it. Instead, we are left entirely with Lieutenant Wallace's word for it. The choice is yours. You can 'Believe It–Or Not!' Needless to say, I am in the 'Not' category on this one.

3) Sergeant Daniel Kanipe – C Company (Camp interview, *Custer in 76*, with my annotations in brackets): Custer returned from his view [from the Crows Nest] and rejoined the command sometime between 10 and 11 a.m.

ANALYSIS

Custer returns from his first trip out to the Crows Nest to rejoin his command at Halt 2 sometime between 10 and 11 a.m. Based on other accounts, I believe this time would be much closer to 10 a.m. than 11 a.m. I would suggest about 10:10 a.m. Although Kanipe does not give us a time estimate for the crossing of the divide, his statement above suggests a time closer to 11 a.m. as opposed to noon, as other sources tell us that Custer had returned and called for his last Officer's Call at a time close to 10 a.m.

4) Lieutenant Luther Hare (Camp interview, *Custer in 76*, with my annotations in brackets): the 7th Cavalry began to move again up the divide [from Halt 1] between 8 and 9 a.m. Custer had already left earlier for the Crows Nest. They halted [at Halt 2] near the divide .... less than a half-mile from the top of the divide. They stopped and lay concealed for over an hour.... This was [some time] between 10 a.m. and noon. During this time [note: likely after the last Officer's Call] Custer again goes to the Crows Nest to look at the Indians.

ANALYSIS

It is important to note that it would take the command, moving at a walk, about one hour and fifteen minutes to cover the 4.5 miles between Halt 1 and Halt 2 on the divide. If the command left Halt 1 at 8:30 a.m., as I believe it did, the lead elements of the command would reach Halt 2 at about 9:45 a.m., with the trailing elements arriving at or just after 10:00 a.m. During this halt, Officer's Call would be sounded as soon as Custer returned from the Crows Nest a short time later, and the command would rest for just over an hour before moving out again to cross the divide. The timeframe for this event, according to Lieutenant Hare's recollection, would be between 10 a.m. and 11 a.m. We should note the discrepancy between Hare's comment that the command rested at Halt 2 for one hour yet gives a two-hour timeframe, saying it was sometime between 10 a.m. and 12 noon. This makes little sense, especially when we realize that the command's lead elements likely arrived at Halt 2 at about 9:45 a.m., covering just 4.5 miles in an hour and fifteen minutes. After a rest of a little over an hour, the command would move out again at about 9:45 a.m. and end at about 10:55 a.m.

5) Private William Taylor – A Company (*With Custer on the Little Big Horn*): We did not stop until about 10:30 a.m. when we came to a halt in a ravine, some four or five miles from the summit of the divide. Here we were ordered to keep concealed and to preserve quiet.... But in the meantime, our presence having been discovered ... "it was necessary to act at once," and we started off again, crossing the divide a little before noon.

ANALYSIS

There are a number of errors in the reminiscence of Private Taylor given above. One obvious mistake Private Taylor makes is describing the location of Halt 2 as being four or five miles from the summit of the divide. That would be the correct distance of Halt 1 from the divide. According to Lieutenant Wallace and Lieutenant Edgerly's recollection, Halt 2 was about a quarter mile east of the divide. Lieutenant Hare would support that when he described its location as being less than a half mile from the divide. Another problem Taylor confronts us with is the same one Moylan and Godfrey gave in their accounts, this being an inexplicable 2.5-hour span of time for the Regiment to move just 4.5 miles from Halt 1 to Halt 2. Something is clearly wrong here. Either the command left Halt 1 well after 8:00 a.m., arrived at Halt 2 well before 10:30 a.m., or both. Consequently, Taylor's time estimate of crossing the divide a little before noon must also be suspect, especially considering his statement that "it was necessary to act at once." This would preclude an unnecessary delay of an hour and a half before advancing on the hostiles, as a noon crossing of the divide would suggest. Taylor was likely basing these time estimates mentioned above on what he had previously read from Lieutenant Godfrey's *Century Magazine* article on the battle as opposed to his own personal recollections. Of course, Taylor's reference to "a little before noon" might actually reference a time closer to 11 a.m., as indicated by other participant accounts.

6) Lieutenant Winfield Edgerly (*The Custer Myth*, 1881 account, with my annotations in brackets): At about 10 o'clock in the morning we were ... 15 miles from the hostile camp. Our force was then all together [at Halt 2]. We halted while Custer [earlier that morning] went on a hill with the Crow and Ree scouts to take a look at the Indian camp, which was in sight. When General Custer came down from the hill, Officer's Call was sounded. The officers all went to where he was and he told us that ... we would press on as quickly as we could and attack them in the village if possible.... He then ordered troop commanders to mount their troops and report when they were in readiness to move on. In about a minute, every troop commander had reported.

ANALYSIS

Lieutenant Edgerly tells us that the command had all reached Halt 2 at about 10 a.m. When Custer returned from the Crows Nest a bit later, presumably about 10:10 a.m., Officer's Call was sounded, whereupon the officers were told that they "would press on as quickly as they could and attack" the hostiles "in the village if possible," thus indicating the need for speed at this point. When Officer's Call ended, presumably about 10:30 or 10:35 a.m., the troop commanders were told to "mount their troops and report when they were in readiness to move on. In about a minute , every troop commander had reported." Lieutenant Edgerly tells us here that there was no hour and a half delay before moving out to cross the divide, as Lieutenant Wallace would infer from his deceptive itinerary. A delay of such duration makes no sense whatsoever.

6b) Lieutenant Winfield Edgerly (RCOI): Officer's Call was at 10 a.m. on June 25, about a quarter mile east of the divide.

ANALYSIS

Edgerly agrees with most other sources that Officer's Call was about 10 a.m. and Halt 2 was about a quarter mile east of the divide.

6c) Lieutenant Winfield Edgerly (letter to wife, July 4, 1876): we ... moved on again till about 10 a.m. when Officer's Call was sounded again. After nearly an hour's rest and a reconnaissance by Gen'l. Custer and some scouts, we were told that the Indians had undoubtedly discovered our presence as several had been seen on the bluffs, and the only thing for us to do was to move on them at once.... We then mounted and moved forward about half a mile, halted and he (Custer) reassigned the companies into battalions....

ANALYSIS

Edgerly says that the command reached Halt 2 about 10 a.m., followed shortly thereafter by Officer's Call (after the return of Custer and his scouts from their reconnaissance at the Crows Nest). He says that the command rested at Halt 2 for "nearly an hour's rest." This indicates that the command rested from about 10 a.m. until sometime just before 11 a.m. when they moved out again to cross the divide.

7a) Lieutenant Edward Godfrey (RCOI): crossed the divide "at about 12 o'clock."

7b) Lieutenant Edward Godfrey ("Custer's Last Battle," *Century Magazine*): They crossed the divide a little before noon.

ANALYSIS

Godfrey's time estimate of a divide crossing at or a little before noon would suggest that Custer delayed his advance for about an hour and a half after his last Officer's Call when Custer announced his plans for an immediate advance on a soon-to-be-expected fleeing village. This is exactly what Lieutenant Wallace suggested in his account. We believe both officers are wrong about this. No other sources indicate such a lengthy and inexplicable delay occurred at that time.

8) Lieutenant Carlo De Rudio (RCOI, with my annotations in brackets): the command was broken into battalions [after crossing the divide] around 11 or 12 o'clock.

ANALYSIS

De Rudio gives us an accurate range here of sometime between 11 a.m. and 12 noon that the command stopped to divide into battalions after traveling only a half mile from Halt 2 on the divide. I believe the correct time would be shortly after 11 a.m.

9) Captain Myles Moylan (RCOI): claims the organization into battalions was made about one and a half miles west of the divide about 12:30 p.m., but added, "I don't know definitely about the time."

ANALYSIS

Captain Moylan's account of this event contains some serious errors. His most glaring error was that the division into battalions was made one and a half miles west of the divide. More reliable accounts from Edgerly and Wallace place this halt to reorganize into battalions at only a quarter mile west of the divide, the command having only traveled a half mile from Halt 2 east of the divide. It would appear that Moylan was attempting to support Wallace's claim of a noon crossing of the divide, yet then claims it took thirty more minutes to move a half mile to the point where the Regiment divided into battalions. This would put their rate of advance at an absurd one mile per hour! Perhaps this is why Moylan chose to relocate this halt a further 1.25 miles west of where it actually took place. In our view, Moylan's account above is entirely unreliable and not supported by our other battle participants. At least he attempted to qualify these inaccuracies by saying: "I don't know definitely about the time."

10) Private J.B. Gallenne – M Company (letter to Father Genin, July 5, 1876): at about 11 a.m. we perceive a wigwam of war about 10 miles ahead of us.

ANALYSIS

Private Jean Gallenne tells us that shortly after crossing the divide at about 11 a.m., they could see a lone tepee about ten miles ahead (or west) of the command. It is interesting to note that the second lone tepee was almost exactly ten miles west of the divide crossing.

11) Lieutenant Edward Mathey (RCOI): "they started from 11 or 12 from where the long halt was made" on the divide.

ANALYSIS

Lieutenant Mathey indicates that the command started out to cross the divide at some time between 11 a.m. and noon. I believe this time range is essentially correct, with the command crossing the divide closer to 11 a.m. than noon.

12) Captain Thomas McDougall (RCOI): got his orders to act as Rear Guard about 11 a.m. on the divide. They started out twenty minutes after the command.

ANALYSIS

Captain McDougall's account above suggests a divide crossing closer to 11 a.m., with the main Pack train starting out about twenty minutes later.

13) Lieutenant Charles Varnum (*I, Varnum*): The column arrived at the trail-crossing of the divide about 10 a.m. ....

ANALYSIS

Lieutenant Varnum is likely using a local sun time reference of 10 a.m. for the command's crossing of the divide. The Headquarters watch time would actually be 10:57 a.m. or a few minutes before 11.

14) Private Charles Windolph – H Company (*I Fought With Custer*): It was around noon on this fatal Sunday of June 25 when we crossed the divide between the Rosebud and the Little Big Horn.

ANALYSIS

Private Windolph's memoirs were written by Frazier & Robert Hunt and published in 1947. We strongly suspect that the times he gave for the events were heavily influenced by Col. Graham's timelines that were previously published in his book *The Story of the Little Big Horn* in 1926, as Windolph's times follow Graham's unique and popular times—based almost exclusively on those given by Lieutenant Wallace at the RCOI and thus contradicting the evidence provided by the great majority of the battle participants, both red and white. The Hunts reproduce and insert sections from Graham's book to supplement Windolph's own memoirs. Thus, our conclusion is that Windolph's times are not to be relied on.

15) Dr. Henry Porter (*St. Paul Pioneer Press*, May 3, 1878 interview): "It was about ten o'clock when the command was divided."

ANALYSIS

Dr. Porter indicates that Custer divided the regiment into separate battalions at "about ten o'clock." This would likely be a reference to local sun time, with a of being closer to 11 a.m. shortly after crossing the divide.

16) Edward Curtis interview with three Crow Scouts in *The North American Indian*: "In the early forenoon the command [crossed the divide and] moved down the western slope of Wolf mountains and out on the plain...."

ANALYSIS

The Crow scouts confirmed to Curtis that the regiment crossed the divide in the forenoon, not at noon, which supports the Participant Timeline crossing at about 11:00 a.m. on June 25th.

CONCLUSION

Based on the participant accounts listed above, the lead elements of the Regiment arrived at Halt 2 at about 9:45 a.m., with the rear elements arriving at Halt 2 at about 10:07 a.m. Custer returned from the Crows Nest between 10:00 and 10:05 a.m. and sounded Officer's Call. Officer's Call occurred between 10:15 and 10:30 a.m. The officers were then told to prepare and mount up their companies for an immediate advance on the hostiles, which occurred between 10:30 and 10:45 a.m. with the command ready to move again, leaving Halt 2 at 10:45 a.m. and crossing the divide at 10:55 a.m. as supported by the accounts of Dr. Porter, Lieutenant Varnum, Captain McDougall, Lieutenant De Rudio, Lieutenant Edgerly, Lieutenant Mathey, Lieutenant Hare and Private Gallenne. In doing so, we must reject the later time estimates given us by Lieutenant Wallace, Captain Moylan and Private Taylor.

10:45 a.m. (H + 2 hours and 15 minutes.): The Regiment left Halt 2.

10:55 a.m. (H + 2 hours and 25 minutes.): The Regiment crossed the divide.

4. BENTEEN'S MARCH TO RENO HILL

We have very few indicators from the participants regarding the length of time Benteen's battalion took to cover the time between his departure from Custer, roughly one-quarter mile west of the divide, and his return to Ash/Reno Creek, about seven miles west of the divide. On a straight line ('as the crow flies'), this distance would be about 7.7 miles, but we will use a figure of about eight

miles as being more accurate. The participant accounts below can help us determine the times of Benteen's flank march to the left and his return:

1a) Captain Benteen (RCOI): claimed it took him three hours from when he left Custer near the divide to when he reached Reno on the bluffs.

ANALYSIS

Benteen gives us a rough estimate of time—about three hours—for his battalion to cover the eight or so miles of his flank march to the left, plus the remaining 6.5 miles to reach Reno Hill on the bluffs. The total mileage for this would come to about 14.5 miles covered in about three hours (although I actually believe the time was closer to three hours and twenty minutes). That gives us the rate of a trot of about five miles per hour using Benteen's own estimate and a rate of about 4.5 using my own time estimate).

1b) Captain Benteen: (Benteen's Narrative, with my annotations in brackets): my command getting to it [i.e., the main Indian trail on Ash/Reno Creek], just ahead of the train of packs ... this being along about 1 o'clock p.m. ...

ANALYSIS

Benteen indicates that his battalion, after riding about eight miles on their flank march to the left, returned to the main trail about 1 p.m. As Benteen's battalion was likely sent off of his independent mission at about 11:12 p.m. (using Lieutenant Wallace's corrected times, subtracting exactly one hour from the times he gave at the RCOI), his 1 p.m. arrival back on the main trail indicates that Benteen's battalion had traveled these eight miles in about one hour and forty-five minutes, or a rate of speed about 4.5 miles per hour. He indicates that he arrived on the main trail just ahead of the pack train, which was seen approaching about a mile east of where he arrived on Ash Creek. Thus, the head of the pack train would have traveled about six miles in an hour and twenty-five minutes (recall that the pack train did not start out until twenty minutes after the Regiment did—about 11:15 a.m.) or about a three miles per hour walk, which is exactly what would be expected of them under the circumstances. All of this ties in very well together and supports Benteen's recollection of a 1 p.m. return to Ash/Reno Creek at the junction of No Name Creek (about five miles east of Ford A) as well as an 11 a.m. crossing of the divide.

It is interesting to note that, at the time Benteen reached Ash/Reno Creek, he was about 6.5 miles away from Reno Hill. It would take Benteen about an hour and a half to cover this distance, due mainly to his decision to take a lengthy water break at one or both watering holes along the way. His rate of speed once he returns to the main trail now drops down to about a 4.4 miles-per-hour walk, which includes the time taken for watering his horses.

1c) Captain Benteen (Benteen's Narrative): The whole time occupied in this march was about an hour and a half .... I then decided to rejoin the main trail ... About three miles from the point where Reno crossed the ford, I met a sergeant bringing orders to the commanding officer of the rear guard, Captain McDougall, to hurry up the pack trains. A mile further, I was met by my trumpeter .... A mile or a mile and a half further on, I first came in sight of the valley and Little Big Horn.

ANALYSIS

In the above account, Benteen tells us that he aborted his flank march to the left about an hour and a half after he left Custer near the divide and decided to rejoin the main trail. It would take about fifteen or twenty minutes for his command to reach Ash/Reno Creek via the valley of No Name Creek. He would return to the main trail at about 1:00 or 1:05 p.m. He then says that about three miles east of Ford A—or just past the first, or eastern lone tepee, and west of the first watering place or morass, he met up with Sergeant Kanipe. (The actual distance would be closer to 2.5 miles.) About a mile beyond that, he met up with Trumpeter Martin with Custer's last message. Then, about a mile or mile and a half—the actual distance here would be about a mile and a quarter—he came to the vicinity of Ford A, where he could see into the valley and got his first glimpse of the Little Big Horn River.

2a) Lieutenant Godfrey (RCOI): they stopped at 2 p.m. to water their horses. ... They stopped for 20 to 30 minutes.

ANALYSIS

Although Godfrey claims they stopped to water their horses at 2 p.m. and then says they took twenty to thirty minutes more to water their horses before moving on, Godfrey claims elsewhere that Benteen's battalion joined with Reno on the bluffs at 2:30 p.m. Clearly Godfrey is wrong about one of these times. I believe it more accurate to say that by 2 p.m., Benteen's battalion had just finished watering their horses. Furthermore, I believe that this watering took place at the Hartung morass, about 2.5 miles east of Ford A, as opposed to the eastern morass, about two miles farther east. Had it been the eastern morass, Benteen's battalion would not have been able to reach Reno Hill by about 2:30 p.m., as many independent participant accounts so indicate. The evidence suggests that Benteen's battalion actually watered their horses twice, once at the eastern morass, for about thirty minutes, as Godfrey suggests in his account, and a second and shorter pause for watering—about ten minutes or so—at the Hartung morass, as Edgerly recalled, and which ties in nicely with Sergeant Kanipe's account of when he met up with Benteen's battalion on his way back to the pack train.

2b) Lieutenant Godfrey (Godfrey's narrative, Custer's Last Battle, *Century Magazine*): claims they watered their horses about 2 p.m. After watering, they continued their march very leisurely, and not long after watering, a trumpeter brought a note from Colonel Cooke that the village was in front and to bring up the packs. We then increased our gait. ... some time after getting on the [main Indian] trail, we came to a waterhole, or morass, at which a stream of running water had its source, Benteen halted the battalion. While watering, we heard some firing in advance .... we passed a burning tepee.

ANALYSIS

As indicated above, Godfrey's 2 p.m. stop for a water break does not fit with the evidence and contradicts Benteen's 1 p.m. return to the main trail. It is far more likely that Godfrey meant that they finished watering about 2 p.m. Godfrey then claims that the battalion continued marching west at a very leisurely pace—indicating a walk—when not long after watering, Trumpeter Martin delivered his note to Benteen. This comment by Godfrey indicates that this watering place was at the Hartung morass, about 2.5 miles east of Ford A, and not the eastern morass, as Martin described the point where he delivered his note to Benteen as being only 400 yards east of where Custer watered his command at what I believe to be the Echleman Ford about 1.2 miles east of Ford A. Only after receiving Martin's note did Benteen increase his gait to a lively trot to the river for the final mile or so. Godfrey's comment about hearing firing in advance would likely refer to the Ree scouts in the valley ahead as they attempted to capture a group of Sioux ponies before Reno's attack in the valley. Had this been the eastern morass (over 4.5 miles from Ford A), Godfrey, who was partially deaf, would likely never have heard this firing. The Hartung morass, being two miles closer to the river, was the likely morass referenced by Godfrey in this account.

3) Lieutenant Edgerly (RCOI): Benteen watered his horses at the morass after returning to the valley (of Ash Creek) about seven to eight miles from the divide and halfway to Reno's position on the bluffs. This morass was a half mile from where they struck the trail on Reno Creek. He claims they watered their horses for eight to ten minutes, and the packs arrived just as they pulled out. The packs stretched back about two miles.

ANALYSIS

Lieutenant Edgerly indicates that Benteen's battalion first stopped at the eastern morass, located about a quarter mile west of the point they returned to Ash/Reno Creek via No Name Creek. This morass would be about four and three-quarters miles east of Ford A and about seven and a quarter miles west of the divide, roughly halfway to Reno's position on the bluffs. He indicates that the head of the pack train—likely the advance ammunition mules—arrived at this location just as they pulled out. As it would appear that Benteen was moving at a fast walk or trot, it would have taken them about five minutes or so to reach this morass after their arrival back on Ash/Reno Creek at 1 p.m. However, Edgerly's recollection of a ten-minute watering stop does not fit with the overall evidence from other sources unless he refers only to the time his own company watered. Godfrey's recollection of a twenty to thirty minute stop would fit better here, whereas Edgerly's ten-minute water break would fit better with Benteen's second stop at the Hartung morass about two miles further east.

4) Lieutenant Wallace (RCOI), with my annotations in brackets: Benteen got his order and moved to the left almost immediately after the separation [of the Regiment into battalions]—about two hours before the order to Reno [to attack the hostiles].

ANALYSIS

The Participant Timeline indicates that Reno received his attack orders shortly before 1 p.m. Based on that time, Lieutenant Wallace's comment above would suggest that Benteen was given his flank march order about two hours earlier, or close to 11:00 a.m.

5) Private Charles Windolph, (*I Fought With Custer*): I recall we saw the lead pack mules breaking for the morass just as we were pulling out. Two or three of them got out of control, and were so greedy for water they got stuck in the bog. It showed how slow the pack train was, for here we'd been on this miserable scout over in the hills to the south for a couple of hours and yet we were still ahead of the packs when we reached this spot on the Custer-Reno trail.

ANALYSIS

Private Windolph, traveling with Benteen's H Co., indicates that their scout to the south took "a couple of hours," which is very close to Benteen's estimated "hour and a half." It is likely that the entire time taken for Benteen's flank march fell within this range.

OVERALL ANALYSIS

After crossing the divide about 10:55 to 11:00 a.m., Benteen's battalion leaves Custer's command about a quarter mile west of the divide at about 11:10 a.m. and returns back to the main Indian trail on Ash/Reno Creek via No Name Creek at a point some five miles east of Ford A at about 1 p.m. I would suggest a likely time of about 1:05 p.m. after having traversed a distance of about eight miles. They arrived at the main trail about a mile west of the advance pack mules, which had covered about six miles in the same span of time—about one hour and forty-five minutes (or one hour and twenty-five minutes for the pack train). Indications are that Benteen watered his horses at the eastern morass—arriving there about 1:07 p.m. before moving further west again at about 1:30 p.m.; he continued on at a trot (of about six miles per hour), where he stopped to water again at the Hartung morass about 2.5 miles east of Ford A, arriving there about 1:50 p.m. whereupon he met Sergeant Kanipe. Afterward, he stops for a ten-minute watering, waiting for his two trailing troops to catch up, ending this water break at about 2 p.m. He then met up with Trumpeter Martin about a mile west of the Hartung morass (about 1.5 miles east of Ford A), then rode the remaining mile or so at a trot until reaching the area of Ford A. at about 2:15 p.m.

CONCLUSION

11:12 a.m. (H = 2 hours and 42 minutes): Benteen's battalion leaves on his flank march to the left.

1:05 p.m. (H = 4 hours and 35 minutes): Benteen's battalion returns to the main trail on Ash/Reno Creek via No Name Creek five miles east of Ford A.

1:10 p.m. (H = 4 hours and 40 minutes): Benteen stops to water at the eastern morass.

1:30 p.m. (H = 5 hours): Benteen leaves the first morass.

1:50 p.m. (H = 5 hours and 20 minutes): Benteen meets up with Sergeant Kanipe as he approaches the Hartung morass.

1:50 p.m. (H = 5 hours and 20 minutes): Benteen stops to water again at the Hartung morass.

2 p.m. (H = 5 hours and 30 minutes): Benteen finishes watering at the Hartung morass.

2:10 p.m. (H = 5 hours and 40 minutes): Benteen meets up with Trumpeter Martin.

2:15 p.m. (H = 5 hours and 45 minutes): Benteen reaches the vicinity of Ford A to view Reno's retreat from the valley.

2:20 p.m. (H = 5 hours and 50 minutes): Benteen leaves the vicinity of Ford A and begins to ascend the bluffs to Reno Hill.

2:30 p.m. (H = 6 hours): Benteen reaches the top of the bluffs at Reno Hill.

5. THE PACK TRAIN: FROM THE DIVIDE TO RENO HILL

1) Colonel Benteen (Benteen's Narrative, with my annotations in brackets): my command getting to it [i.e., the main Indian trail on Ash/Reno Creek], just ahead of the train of packs ... this being along about 1 o'clock p.m. ...

ANALYSIS

Benteen indicates that his battalion, after riding several miles on their flank march to the left, returned to the main trail about 1 p.m. As Benteen's battalion was likely sent off of his independent mission at about 11:12 p.m. (using Lieutenant Wallace's corrected times, subtracting exactly one hour from the times he gave at the RCOI), his 1 p.m. arrival back on the main trail indicates that Benteen's battalion had traveled these eight miles in about one hour and forty-five minutes, or a rate of speed about 4.5 miles per hour. He indicates that he arrived on the main trail just ahead of the pack train, which was seen approaching about a mile east of where he arrived on Ash Creek. Thus, the head of the pack train would have traveled about six miles in an hour and twenty-five minutes (recall that the pack train did not start out until twenty minutes after the Regiment did—about 11:15 a.m.) or about a three miles per hour walk, which is exactly what would be expected of them under the circumstances. All of this ties in very well together and supports Benteen's recollection of a 1 p.m. return to Ash/Reno Creek at the junction of No Name Creek (about five miles east of Ford A) as well as an 11 a.m. crossing of the divide.

2) Captain McDougall (RCOI): He got his orders to act as Rear Guard about 11 a.m. back on the divide. They started out twenty minutes after the command. They were delayed about twenty minutes at the morass due to several mules getting stuck. He guessed the distance from the morass to Reno Hill to be about eight miles. Four miles later, they came to a tepee. He told Mathey to halt nearby to allow the train to close up. About a mile farther on, Mathey told him a fight was going on. He told Mathey to hurry the mules as fast as possible. About two miles later, he saw figures on Reno Hill; he divided the Rear Guard into two platoons to cover the front and rear of the main packs and advanced. About fifteen minutes before reaching Reno Hill, he heard firing off to his right. He said he reached Reno Hill at 3:45 or 4 p.m. and heard two volleys, which he reported to Reno. He thought the volleys were four to four and a half miles away. He spoke to Benteen on the bluffs after he arrived there and then threw out a skirmish line. He said they moved downstream about thirty minutes later. His B Company was the second from the last on its move towards Weir Point. They only went a quarter mile before turning back. The fight ended at 9 p.m.

ANALYSIS

Captain McDougall's Rear Guard started out to cross the divide at about 11:15 a.m. (twenty minutes after the command started out); according to Benteen's statement, the advance pack mules had covered about six miles by the time Benteen returned to the main trail on Ash/Reno Creek at about 1 p.m., with the main pack train and rear guard about a mile farther back, which indicated an advance of about a three miles per hour walk. It would have taken the head of the pack train another twenty to twenty-five minutes or so until it reached the eastern morass, placing the arrival of the advance pack mules there at about 1:30 p.m. We know this is the eastern morass, as McDougall tells us that its distance to Reno Hill was about eight miles. The actual distance was closer to six miles, but clearly McDougall is not referring to the Hartung morass, several miles farther west. The main pack train and rear guard would not arrive at the eastern morass until about 1:45 p.m. McDougall then estimates a twenty-minute delay there, moving out again about 2:05 p.m. He then claims to have come to a burning tepee about four miles farther west. Again, he overestimates his distances as the western lone tepee would actually be about three miles farther west. When McDougall reaches the lone tepee, he sends a messenger to Lieutenant Mathey telling him to halt and wait for the pack train to close up. Mathey receives this order and halts his advance group of mules about a mile south of Reno Hill at about 2:40 p.m. Lieutenant Hare reached Lieutenant Mathey—the time was about 2:45 or 2:50 p.m. and brought back two of the ammunition packs to Reno Hill. Mathey now continues to Reno Hill after sending back a messenger to McDougall, informing him that the battle was going on. The message reaches McDougall about a mile west of the western lone tepee—although it was closer to a half mile west of the lone tepee, as McDougall seems to overestimate his distances consistently; this was likely at a point near the junction of Ash/Reno Creek and the North Fork of Ash/Reno Creek. McDougall sends a message back to Mathey telling him to hurry the mules on as fast as possible, but Mathey is already on Reno Hill before this message can reach him. McDougall claims to see figures on Reno Hill about two miles later, but this distance is also wrong. At the time of this sighting, McDougall would have just crossed the North Fork and was about a mile and a half from Reno Hill. This is where he divided the Rear Guard into two platoons to cover the front and rear of the main pack train as they advanced up the final mile and a half to Reno Hill. The time would be about 3:30 p.m. at this point when McDougall heard firing off to his right—likely indicating the firing from Luce and Blummer Ridge as Custer began withdrawing his force to Calhoun Hill. McDougall gets to Reno Hill about fifteen minutes later—at 3:45 p.m. with the remaining packs arriving there about 4 p.m. when he heard two more volleys fired about four to four and a half miles away, which he reported to Reno after first speaking with Benteen. McDougall then says that Reno began his advance to the Weir Peaks about thirty minutes later or about 4:30 p.m. B Company was now the second company from the rear. The rearmost company now being A Troop bringing along the seven to ten wounded men.

3) Packer John Frett (RCOI, with my annotations in brackets): got the first reports of a battle while the [advance] pack train was "at the watering place near the tepee—the last tepee that was there before we got to the battlefield—the one with the dead Indian in it"—when word reached them through a sergeant from some company who said they should hurry up; that Custer was attacking the Indians. This occurred between 11 a.m. and 1 p.m. [local sun time, or between 12 and 2 p.m. Headquarters watch time.] The pack train then moved ahead two and a half to three miles, stopped to get them all together, then moved ahead again. The train walked this distance.

ANALYSIS

Packer John Frett, who was traveling with Lieutenant Mathey and the advance packs, tells us that the advance packs had just reached the eastern morass four and three-quarter miles east of Ford A, just east of the first lone tepee. This places the advance packs at the eastern morass when word reached them through a sergeant—likely Sergeant Kanipe—who told them to hurry up as Custer attacked the Indians. Frett puts this time as occurring between 11 a.m. and 1 p.m. sun time or between 12 and 2 p.m. Headquarters watch time.

4) Lieutenant Edgerly (RCOI): Benteen watered his horses at the morass after returning to the valley (of Ash Creek) about seven to eight miles from the divide and halfway to Reno's position on the bluffs. This morass was a half mile from where they struck the trail on Reno Creek. He claims they watered their horses for eight to ten minutes, and the packs arrived just as they pulled out. The packs stretched back about two miles.

ANALYSIS

Lieutenant Edgerly indicates that Benteen's battalion first stops at the eastern morass, located about a quarter mile west of the point they returned to Ash/Reno Creek via No Name Creek. This morass would be about 4.75 miles east of Ford A, about seven and a quarter miles west of the divide, and roughly halfway to Reno's position on the bluffs. He indicates that the advance ammunition pack mules arrived at this location just as they pulled out. As it would appear that Benteen was moving at a fast walk or slow trot, it would have taken them about five minutes or less to reach this morass after their arrival back on Ash/Reno Creek at 1 p.m. Edgerly's reference to a ten-minute watering here is contradicted by Godfrey and others that indicate a much longer watering at this morass, closer to twenty-five minutes. This would suggest that Benteen's battalion left the eastern morass at about 1:30 p.m. or roughly the same time the advance packs arrived. Edgerly's estimate of a ten-minute water break likely refers only to D Company's length of watering at the eastern morass or Benteen's second stop of a much shorter duration—about ten minutes—at the Hartung morass, where he met up with Kanipe and waited for his two remaining companies to catch up with Weir's advance along the trail.

CONCLUSION

After starting out from Halt 2 to cross the divide at about 11:15 a.m., the main pack train was moving at about three miles per hour, with the advance packs moving a bit faster—closer to three and a half miles per hour. The advance packs were about fifteen minutes behind Benteen's battalion when the latter returned to Ash/Reno Creek at 1 p.m., about five miles east of Ford A. When Benteen's battalion pulled out at 1:30 p.m., the advance pack mules were just arriving. After a short delay, they move on, with the main pack train reaching a point about two and a half miles east of Ford A at about 2:45 p.m. The advance packs have halted on a hill about a mile south of Reno Hill a short time earlier, where Lieutenant Hare met them about 2:45 p.m. Mathey reaches Reno Hill with the advance packs between 3:05 and 3:15 p.m., with the main pack train and rear guard reaching the vicinity of the western lone tepee where they halted to regroup. After a short break, McDougall and the rest of the packs crossed the North Fork and reached Reno Hill about 4:00 p.m., with the first platoon of B Company arriving about fifteen minutes earlier at 3:45 p.m.

11:10 a.m. (H = 2 hours and 40 minutes): The main pack train begins its advance from Halt 2 on the divide.

12:50 p.m. (H = 4 hours & 20 minutes): Boston Custer reaches the advance pack mules about 2.25 to 2.5 miles further east -- about 6 miles east of Ford A.

1:30 p.m. (H = 5 hours): Advance Pack mules reach the eastern morass

1:52 p.m. (H = 5 hours & 22 minutes): Sgt. Kanipe reaches the advance pack mules near the eastern lone tepee about 3.75 miles from the Little Big Horn River, where he is seen by packer John Frett.

2:02 p.m. (H = 5 hours and 32 minutes): Sgt. Kanipe reaches Capt. McDougall about 4.25 miles east of Ford A.

2:05 p.m. (H = 5 hours and 35 minutes): Sgt. Kanipe reaches the Pack Train at the eastern morass.

2:45 p.m. (H = 6 hours and 15 minutes): The advance pack mules with Lieutenant Mathey halt on a small rise about a mile south of Reno Hill.

3:00 p.m. (H = 6 hours and 30 minutes): Two of the ammunition mules race ahead and reach Reno Hill.

3:05 to 3:15 p.m. (6 hours and 35-45 minutes): The rest of the advance pack mules and ammunition arrive on Reno Hill with Lieutenant Mathey.

3:45 p.m. (7 hours and 15 minutes): Captain McDougall arrives on Reno Hill with the 1st platoon of B Company.

4:00 p.m. (7 hours and 30 minutes): The head of the main pack train arrives on Reno Hill.

4:20 p.m. (7 hours and 50 minutes): The rear elements of the main pack train arrive on Reno Hill.

6. RENO GETS HIS ATTACK ORDERS

Where and when did Major Reno receive his orders to attack the hostiles, and where on Ash/Reno Creek did he separate from Custer? Here's what the participants say about this critically important event:

1a) Captain Benteen (RCOI): claims that it would take Custer about one and three-quarters hours to two hours for his command to reach the lone tepee after Benteen left him.

1b) Captain Benteen (Benteen's Narrative): The whole time occupied in this march was about an hour and a half .... I then decided to rejoin the main trail...

ANALYSIS

From the point where Benteen left Custer, about a quarter mile west of the divide (according to both Wallace and Edgerly), to the second (western) lone tepee is about ten miles. Benteen tells us that Custer's command would take about an hour and forty-five minutes to two hours to travel this ten-mile distance, which puts Custer's average pace down Ash/Reno Creek at a slow trot of between five to 5.7 miles per hour. I would favor the faster of these times (one hour and forty-five minutes and a pace of 5.7 miles per hour). This is significant because it was said that Reno first got his orders to attack the hostiles at or near the lone tepee. It is also significant to realize that Benteen estimated roughly the same time of about an hour and forty-five minutes for the duration of his own flank march to the left, describing above how his march took an hour and a half before he decided to return to the main trail—which would take him about fifteen to twenty minutes to return—or the same hour and forty-five minutes it took Custer to reach the second (western) lone tepee. If I determined his route correctly the distance came out to about eight miles, whereas Custer traveled about two miles more to reach the second (western) lone tepee in the same span of time. This tells us that Benteen's command returned to the main trail at about the same time that Custer and Reno reached the westernmost lone tepee and that Custer was traveling about 5.7 miles per hour while Benteen was traveling at a pace of about four and a half miles per hour during this hour and three-quarters span of time. I marked the time of Benteen's departure from Custer at about 11:12 a.m. (by simply subtracting one hour from Lieutenant Wallace's time) after having crossed the divide at some time close to 11 a.m. Benteen himself estimated his return to the main trail to be about 1:00 p.m., which fits very well into this timeframe, with Benteen returning to the main trail at the mouth of No Name Creek and Custer in the vicinity of the second (western) lone tepee at some time approaching 1:00 p.m. I estimate that Benteen's battalion was about five miles behind Reno's battalion at this time, and about four miles behind Custer.

1c) Captain Benteen (letter to wife, July 25, 1876): 12:30 pm: "Just one month ago today—at just about this time of day, General Custer and his command commenced the attack on the Indian village."

ANALYSIS

Captain Benteen, in this private letter to his wife, confides to her that Custer's command (i.e., Reno's battalion) commenced their attack on the Indian village at about 12:30 p.m. This would be the time that Reno received his attack orders near the first, or easternmost lone tepee, about three and three-quarters miles east of Ford A. This statement supports the time recollection of Colonel Reno and is within range of those given by Gerard, Porter, and Varnum.

2a) Lieutenant Wallace (RCOI): said Reno got his attack orders at 2:15 p.m., but he is not sure of that time...we made the mile and a half or two miles in about fifteen minutes.

2b) Lieutenant Wallace (RCOI), with my annotations in brackets: Benteen got his order and moved to the left almost immediately after the separation [of the Regiment into battalions]—about two hours before the order to Reno [to attack the hostiles].

ANALYSIS

2a) Lieutenant Wallace's time estimations are not supported by any of the other participants. As Wallace indicated a time of 12:12 p.m. when Custer began his advance down Ash/Reno Creek a quarter mile west of the divide, he is suggesting that Custer advanced a full two hours before Reno was given his attack orders, implying a slower pace of about five miles per hour down Ash/Reno Creek. Wallace's times conflict with those given by Benteen above and most other battle participants—including the Indians themselves; thus, I reject them as being highly unreliable. I find it revealing that he freely admits that he is not sure of his estimated time. I do, however, accept Wallace's distance estimate of Reno moving about a mile and a half to two miles down to Ford A. That distance would be supported by Gerard, Porter, White Man Runs Him, and Varnum.

2b) Lieutenant Wallace indicates that Major Reno got his attack orders about two hours after Benteen got his orders to begin his flank march to the left. If the Participant Timeline is correct in concluding that

Benteen got his flank march orders shortly after 11:00 a.m., then Wallace is telling us that Reno would have received his attack orders in the vicinity of 1 p.m.

3) Colonel Reno (RCOI - *Army and Navy Journal*): said he received his attack orders at 12:30 p.m.

ANALYSIS

Although Reno's time for crossing the divide (about 8 a.m.) and his time for reaching the easternmost lone tepee (about 11 a.m.) are not supported by the other participants nor even basic logic, his time of 12:30 p.m. for receiving his attack orders comes much closer to the truth, as it falls within the range of estimates given by other independent sources and also suggests that Reno likely received his orders to take the advance in the vicinity of the first or easternmost lone tepee about three and three-quarters miles east of Ford A. His command would likely separate from Custer's command about fifteen to twenty minutes later or closer to 12:45 or 12:50 p.m.

4a) Dr. Porter (RCOI): heard Lieutenant Cooke, about 1 o'clock, and near the lone tepee, give Reno his attack orders to "charge the Indians." Porter guessed at the time but did not check his watch. This order was given between one and one and a half miles from Ford A.

4b) Dr. Porter (*memoirs*): We rode over to where Dr. Lord was, and Custer spoke to him about the contemplated arrangement. The Doctor replied: Not much. I am going with you. The poor fellow, in those few words, saved my life and sealed his own doom. I went with Reno. We had proceeded but a short distance when Captain Cook, Custer's adjutant, came up and said: "The Indians are right ahead of you, and you are ordered to charge them as fast as possible."

ANALYSIS

Dr. Porter tells us that he heard Lieutenant Cooke give Reno his attack orders about 1 p.m. near the second (western) lone tepee (about 1.7 miles east of Ford A) but then says the order was given between one and one and a half miles east of Ford A, which is a short distance west of the lone tepee and close to the Echleman Ford, 1.2 miles east of Ford A. This is very likely the point where Reno actually separated from Custer's command, about a half mile west of the second (western) lone tepee. Porter's recollections agree with Gerard's.

5) Lieutenant Varnum (RCOI, with my annotations in brackets): Reno left Custer's command about 2 p.m., but bases that time on Wallace's time of a noon crossing of the divide ... he reported to General Custer about a mile from where Reno afterward crossed the Little Big Horn. When I left General Custer he was ... moving at a walk. Major Reno pulled out at a trot ... I left the head of General Custer's column to overtake the battalion that was passing us to get ahead of the troops themselves to scout again, and the head of the column soon after crossed a little tributary of the Little Big Horn [note: this would be a

reference to the Echleman Ford on Ash/Reno Creek about 1.2 miles east of Ford A]. I was about the middle of the column as it passed [over Ash/Reno Creek] there, and I was forced off the trail ... and didn't join the command until it was just crossing that ford. One company had got across the ford at the time I got across myself.... As soon as the column passed [the ford] I was joined by Lieutenant Hare, who had been detailed to assist me in scouting. There had evidently been a village upon that tributary [i.e., Ash/Reno Creek] of the Little Big Horn that we came down in going to the fight. There was an old tepee there, and a piece of another tepee, and from the signs around there, I should judge there had been a village there .... I should judge that tepee was about a mile [note: actually 1.7 miles] from where Colonel Reno crossed the river on his advance.

ANALYSIS

Varnum's 2 p.m. time was qualified by himself as based entirely on the acceptance of Wallace's disputed noon crossing of the divide. If the command actually crossed the divide at 11 a.m., then Varnum's time for Reno leaving Custer's command would change to 1 p.m., which is supported by Gerard and Porter's recollections. Varnum, however, does indicate that Reno separated from Custer about a mile from the Little Big Horn River and describes the crossing of Reno's battalion at the Echleman Ford, where he was subsequently forced off the trail while crossing there. He does indicate that Reno's battalion was moving at a trot and that they separated from Custer's command in the vicinity of the second, or western, lone tepee.

6) Private Taylor (*With Custer on the Little Big Horn*): "the two commands, Custer's and Reno's, continued until about 12:30 p.m., when they were within about two miles of the river. Reno was then ordered to move forward at as rapid a gait as prudent and charge afterwards."

ANALYSIS

Private Taylor agrees with Benteen and Reno that the attack orders were given about 12:30 p.m., about two miles from the river. His location would suggest the second (western) lone tepee which was located just under two miles east of the river, but his time of 12:30 p.m. was very likely taken from Reno's own after-battle report.

7a) Trumpeter Martin (*Trenton Evening Times* interview, 1904): claimed they saw several hundred tepees across the river at 11:30 a.m.

ANALYSIS

All of the times given by Martin were local sun times, so the correct Headquarters watch time for the first sighting of the village should be at about 12:30 p.m., which would put Custer's command at or near the first or easternmost lone tepee. It is doubtful if any of Custer's troops saw any portion of the village from this location, yet Mitch Boyer and the Crow scouts were very likely to have seen part of the village from the white bluffs just behind and above this lone tepee location. Word likely reached Custer's command of this sighting just as the command arrived at this lone tepee site, which is probably what Martin is referring to in his account above.

7b) Trumpeter Martin (RCOI, with my annotations in brackets): "About 4 or 5 miles [this would be the first, or eastern lone tepee, actually closer to 3.75 miles] from the Little Big Horn River, there was a tepee on the right [or north] bank of the river [actually Ash/Reno Creek]. Reno's column went ahead of Custer at that tepee ... we went on a jump [i.e., a good trot] all the way." After Reno had gone ahead, "we struck a little creek [note: likely the Echleman Ford over Ash/Reno Creek] where we watered our horses." The halt was for about 10 minutes.

ANALYSIS

Martin here tells us that Reno's command first crossed over to the north side of Ash/Reno Creek in the vicinity of the first or easternmost lone tepee. Martin's distance guess of four miles would be closer to the actual distance of three and three-quarters miles east of Ford A. It would be at this point that Reno's command would take the advance, but both columns rode in close proximity to each other until they reached the vicinity of the second (western) lone tepee—about two miles farther west—where Reno was sent on ahead to begin his advance into the valley. He would not actually separate from Custer until he reached the Echleman Ford, about 1.2 miles east of Ford A.

8) Lieutenant De Rudio (RCOI, with my annotations in brackets): "Major Reno's command marched ahead of General Custer after we passed that [abandoned] village [note: containing the first, or eastern lone tepee] ... four or five miles [note: the actual distance was three and three-quarters miles] from the ford [A] ... We were going at a trot over broken country .... It took about half an hour to reach Ford A" [from the time they left the first, or eastern, lone tepee].

ANALYSIS

De Rudio supports Martin's account that Reno's battalion took the advance shortly after passing the first, or eastern, lone tepee about three and three-quarters miles east of Ford A. De Rudio's estimate of four miles is only a quarter mile off. De Rudio gives us an important clue that Reno's battalion was moving along at a trot over the broken country.

9a) White Man Runs Him (Scott/McCoy interview, *The Custer Myth*, 1919): "About nine miles down the Upper Fork of Ash Creek, we found a lodge with a dead Indian inside. He had probably died from wounds received in the battle with General Crook on the Rosebud on June 17th. Custer halted his command on a small flat about a mile and a quarter from the mouth of Ash Creek and ordered Major Reno to swing out to the left, cross the Little Horn, and attack the upper end of the Sioux village."

Appendix A: Participant Timeline of the Battle of the Little Big Horn

ANALYSIS

White Man Runs Him refers to a lodge with a dead Indian inside about nine miles from the head of the Upper Fork of Ash Creek (which the main Indian trail followed down from the divide). This distance identifies the second (western) lone tepee, which was located about ten miles west of the divide and about nine miles west of the head of Ash/Reno Creek. This lodge is about 1.7 miles east of Ford A. White Man Runs Him tells us that Custer halted his command on a small flat about a mile and a quarter east of the mouth of Ash Creek. This location would refer to the Echleman Ford, about a half mile west of the lone tepee he mentioned above. According to White Man Runs Him, Major Reno swung out to the left at this point, crossing the Echleman Ford, and left Custer's command to attack the Sioux. Reno's separation point is 1.2 miles east of Ford A.

9b) White Man Runs Him (Frank Shively, 1913 interview): White Man Runs Him recalled the meeting of Custer and Reno at the Lone Tepee "when it was only about 11 o'clock."

ANALYSIS

White Man Runs Him's reference to "about 11 o'clock" when Custer and Reno met near the first or eastern lone tepee would convert to a time of about noon Headquarters watch time. Major Reno puts this time more accurately at about 12:30 pm.

10) Private Charles Windolph (*I Fought With Custer*): It was about 2 o'clock when the two columns reached a burning tepee .... One of the white scouts riding ahead shouted back to Custer that a number of hostiles were just ahead of him. Custer now motioned Major Reno to cross the little creek they were riding down, and the General instructed Adjutant Cooke to order Major Reno to cross the Little Big Horn two miles farther on the west and then swinging to his right, charge the Indian village and, that the whole outfit would support him.

ANALYSIS

Windolph, nearly eighty years of age at the time he related to Robert Hunt in his memoirs of the battle, recalls Reno getting his attack orders at the second (western) lone tepee, located just under two miles from the river, as well as Gerard's pointing out a group of Indians seen in the distance. He claims that Reno crossed Ash/Reno Creek at this point to get his orders, but other accounts indicate that Reno had already crossed over the creek about two miles farther east near the eastern lone tepee. After sixty years, it would be easy to confuse these two events and merge them together. Private Windolph's memoirs were written by Robert and Frazier and published in 1947. I strongly suspect that the times he gave for the events were heavily influenced by Colonel Graham's timelines that were previously published in his book *The Story of the Little Big Horn* in 1926, as Windolph's times follow Graham's unique and popular times—based almost exclusively on those given by Lieutenant Wallace at the RCOI and thus contradicting the evidence provided by the great majority of the battle participants, both red and white.

The Hunts reproduce and insert sections from Graham's book to supplement Windolph's own memoirs. Thus, I conclude that Windolph's times are not to be relied on.

11) Fred Gerard (RCOI): said that Reno got his orders to attack at noon. He changed this time the next day to 1 p.m.

ANALYSIS

Fred Gerard's first estimated time of noon was likely based on local sun time, which he corrected the following day to 1 p.m. Headquarters watch time.

12) Private Davern (RCOI): Davern: I heard Adjutant Cook give him [i.e., Major Reno] an order. ... The order was, "Gerard comes back and reports the Indian village three miles ahead and moving. The General directs you to take your three companies and drive everything before you. Those I believe were the exact words."

Question: Was there anything else said?

Davern: Yes sir. "Colonel Benteen will be on your left and will have the same instructions."

Question: Are you sure about those orders?

Davern: Yes sir.

Question: Do you remember passing a tepee with Major Reno's column, before crossing the river?

Davern: Yes sir. We halted a short distance from that tepee.

Question: How far was that from where the command crossed the river?

Davern: A mile and a half to two miles. [Note: The second, or western, lone tepee is located 1.7 miles from Little Big Horn River.]

Question: Was that the place where the orders were given that you heard?

Davern: Yes sir.

ANALYSIS

Private Davern indicates several things of note here. The hostile village (their main objective) has been spotted and is three miles ahead and moving. Reno's advance guard battalion is thus given a mission to pursue the fleeing village and bring them to battle. This was an advance to contact. Davern also mentions that Colonel Benteen will have received the same instructions (i.e., a pursuit and attack order) and is to be expected to enter the Little Big Horn valley on Reno's left. Davern indicates that Reno received these orders at a halt beside the second (i.e., western) lone tepee about 1.7 miles from the river.

OVERALL ANALYSIS

Of the twelve participant accounts listed above, four of them indicate a time of 12:30 p.m. for Reno receiving his attack orders, while three participants place the time closer to 1 p.m. Only one account (from the Crow scout White Man Runs Him) indicates an earlier time of noon. Martin references seeing the village across the river for the first time—likely by the Crow scouts up on the white bluffs—which sighting likely caused Custer to call Reno over to his side of Ash/Reno Creek at that time. The evidence suggests that Reno was first called to cross over Ash/Reno Creek with his command in the vicinity of the first, or easternmost, lone tepee about three and three-quarters miles east of Ford A for the purpose of giving him his orders to advance to the attack. I believe this incident would occur at about 12:30 p.m. At this time, both commands rode side by side on the north side of Ash/Reno Creek until reaching the vicinity of the second, or westernmost lone tepee about 1.7 miles east of Ford A, whereupon Reno's command took the advance at about 12:45 p.m., giving the impression to some that the attack orders were actually given closer to 1 p.m., or, as Porter said, between one to one and a half miles east of Ford A, which indicated the location of the Echleman Ford, crossed by Reno to return to the south side of Ash/Reno Creek after departing from Custer. White Man Runs Him confirms this separation point when he mentions the correct distance of a mile and a quarter from the mouth of Ash/Reno Creek.

CONCLUSION

Reno's battalion departs from Custer's command at the Echleman Ford 1.2 miles east of Ford A with orders to attack the hostiles at about 12:45 p.m. Reno may have actually gotten his initial orders to take the advance as early as 12:30 p.m. while in the vicinity of the first (eastern) lone tepee. At the time of Reno's separation from Custer, Benteen's battalion had aborted their flank march and was returning to the main trail on Ash/Reno Creek via No Name Creek. The head of the pack train would be about five miles west of the divide at this time. Boston Custer would begin his ride back to the pack train to acquire a fresh mount.

12:30 p.m. (H + 4 hours): Reno's battalion takes the advance in the vicinity of the eastern lone tepee.

12:45 p.m. (H + 4 hours and 15 minutes): Reno's battalion separates from Custer and begins their advance to the attack:

7. RENO'S BATTALION REACHES FORD A

The key to determining when Reno reached Ford A—and from what point—lies in first establishing his likely pace along Ash/Reno Creek once his battalion separated from Custer's. The participants answer this question for us:

1) Lieutenant Charles Varnum (RCOI): Major Reno pulled out at a trot ... about a mile from where Reno crossed the river on his advance.

ANALYSIS

Varnum suggests Reno left Custer's command about a mile east of Ford A (or actually 1.2 miles east of Ford A) near the Echleman Ford, and advanced at a trot. This would project into a time span of between ten to fifteen minutes going at a trot of six to eight miles per hour.

2) Lieutenant Edward Godfrey (Godfrey's Narrative, *Custer's Last Battle*): although not a participant of this event, Godfrey would have intimate access to many participants who did witness this event. He said that Reno's battalion moved at a trot to the river.

ANALYSIS

At a distance of 1.2 miles from the Echleman Ford to Ford A, this should take about ten to fifteen minutes moving at a trot of six to eight miles per hour.

3) Captain Myles Moylan (RCOI): after receiving his attack orders, Reno moved at a trot for three to three and a half miles to the river. He claimed Reno got there in twenty to twenty-five minutes.

ANALYSIS

Moylan references the first or eastern lone tepee—three and three-quarters miles east of Ford A—as his starting point for when Reno first was given his attack orders, then claims it took Reno about twenty to twenty-five minutes to reach Ford A while moving at a trot. This would indicate a trot of about eight miles per hour.

4a) Lieutenant Carlo De Rudio (RCOI, with my annotations in brackets): Major Reno's command marched ahead of Gen. Custer after we passed that [abandoned] village [note: containing the first or eastern lone tepee] .... 4 or 5 miles from the ford .... We were going at a trot over broken country .... It took about half an hour to reach Ford A ....

ANALYSIS

The first or eastern lone tepee was actually closer to three and three-quarters miles east of Ford A. De Rudio's time estimate of about thirty minutes to reach Ford A from that location would indicate a trot of about seven miles per hour.

4b) Lieutenant De Rudio (RCOI as recorded by the *Chicago Times*):

Question, Mr. Gilbert: Where was Mr. Girard, the Indian interpreter, on the night of the 25th, and when did he become separated from you?

Answer: He and a half-breed named Jackson were mounted and were with me. We went out from the woods together. I was holding on to the tail of Girard's horse. The plain was full of Indians going backward and forward. We were hunting for the ford that Major Reno's command had crossed at noon.

ANALYSIS

Lieutenant De Rudio's reference to Reno's battalion crossing the river at Ford A at noon was a likely reference to local sun time, which was about an hour earlier than Headquarters watch time of about 1:00 p.m.

5) Lieutenant George Wallace (RCOI): moving at a gallop, I think we made the mile and a half or 2 miles in about 15 minutes.

ANALYSIS

Wallace is very likely wrong about Reno's pace being a gallop. All other participants put the pace at a trot. Wallace's distance estimate of a mile and a half to the river is quite close to the actual distance of 1.2 miles from the Echleman Ford, where Reno likely departed from Custer. Wallace indicates it took Reno's command about fifteen minutes to cover this distance, indicating a trot of about six miles per hour. If Wallace was referencing the second (western) lone tepee as his starting point—1.7 miles east of Ford A—Reno's pace would be closer to seven miles per hour.

6) Lieutenant Luther Hare (RCOI): Reno covered a distance of about 5 miles at a fast trot in about 20 to 25 minutes.

ANALYSIS

It is important to realize that Lieutenant Hare—who was out scouting with the Crows—may not have been actually present when Reno got his orders to take the advance upon the command reaching the vicinity of the first (eastern) lone tepee, about three and three-quarters miles east of Ford A, not the five miles Hare mistakenly estimated. Despite Hare's error in distance, he correctly estimates the time it would take Reno, moving at a fast trot, to reach Ford A from the eastern lone tepee, agreeing with Captain Moylan that it would take between twenty to twenty-five minutes, indicating a fast trot of about eight miles per hour.

9) Dr. Porter (RCOI): said it took 10 to 15 minutes for Reno to reach Ford A.

ANALYSIS

Dr. Porter's estimate of ten to fifteen minutes for Reno to reach Ford A while moving at a trot indicates that Reno would have covered the 1.2 miles from the Echleman Ford to Ford A at a pace of between six and eight miles per hour.

10) George Herendeen (RCOI): said it took Reno 5 to 6 minutes to reach Ford A (but he also added, "I can't tell") going at a slow lope, but estimated the distance at only three-quarters of a mile.

ANALYSIS

Of course, none of our other participants support Herendeen's time and distance estimates here, so we must look for a different interpretation. I would suggest that Herendeen estimated the time and distance he and the other scouts covered in reaching Ford A. According to Lieutenant Hare, the scouts got to Ford A about a half mile in advance of Reno's battalion, so it is probably significant that Herendeen places his distance to Ford A at about three-quarters of a mile west of Reno's separation point from Custer. Reno likely left Custer's command at the Echleman Ford—1.2 miles east of Ford A—while Herendeen references a distance of three-quarters of a mile east of Ford A, which would be about where many of the scouts would be in advance of the command. It would take these scouts about five or six minutes to cover this distance to Ford A.

11) One Bull (Hunkpapa Sioux): a party of Oglala ... headed back toward the Agency was turned around by Custer's approaching column in Ash Creek. One of them had a watch from one of Crook's soldiers, and it read 1 o'clock when they ran into Custer's men.

ANALYSIS

One Bull's account, based on the time from a watch captured from one of Crook's soldiers six days earlier, supports our timeline estimate that Reno's battalion, with Custer about a mile to his rear, was approaching Ford A at about 1 p.m.

12) Private Davern (RCOI): Question: What time do you suppose Major Reno's command crossed the river to go to the timber? Davern: About 1 o'clock; maybe after 1.

ANALYSIS

Private Davern supports One Bull's watch time of 1 p.m. when he states that it was about 1 pm when Reno reached the river, or perhaps a bit after 1 pm by the time he actually crossed it.

13) Blacksmith Henry Mechling's Letter To James Braddock, July 16, 1921, published in Gordon Harper's *Fights on the Little Horn Companion: Full Appendices and Bibliography*: ... about 1 o'clock, as near as I can recollect, the fighting began June 25, 1876.

ANALYSIS

Blacksmith Mechling tells us that the fighting began at about 1 p.m. which references the start of the fighting after Reno crossed the river and entered the Little Big Horn valley at this time.

14) Fred Gerard: An interview given by Fred Gerard published in the *St. Paul Pioneer Press* on Jan. 31, 1879, with my annotations in brackets:

> We then remained in the timber until about 9 o'clock in the evening, when we started; Jackson, the Scout, and myself being mounted. ... we could see the Indians passing [us] seventy-five or a hundred yards off. At this time we supposed the troops had gone in to camp when [*sic* where] Reno had crossed [the river at Ford A to advance into the valley] about 1 p.m. ....

ANALYSIS

Fred Gerard referenced his time estimates based on his pocket watch and indicated that Reno's battalion crossed the river "about 1 p.m."

CONCLUSION

Reno's battalion, after separating from Custer at the Echleman Ford 1.2 miles east of Ford A, moves off at a fast trot of about eight miles per hour and reaches Ford A in about ten to twelve minutes, or between twenty to twenty-five minutes if you choose to use the first or eastern lone tepee, three and three-quarters miles east of Ford A as Reno's point of departure, as several participants above have chosen to do. As the timeline has taken us to a separation point at the Echleman Ford at about 12:45 p.m., we can conclude that Reno's battalion reaches Ford A at about 12:55 p.m. If we choose to use the eastern lone tepee as the starting point (at about 12:30 p.m.), then Reno's battalion would reach the river in about twenty-five minutes, or about the same time of 12:55 p.m. These times, of course, are all approximate, give or take about five minutes either way and would be supported by Fred Gerard and One Bull's watch time of 1 p.m.

12:55 p.m. (H = 4 hours and 25 minutes.): Reno's battalion reaches Ford A.

8. RENO'S BATTALION CROSSES THE RIVER AT FORD A

Here are the participant accounts dealing with Reno's crossing the Little Big Horn River at Ford A:

1) Dr. Henry Porter (RCOI): it took Reno 5 to 10 minutes to cross Ford A and form up.

ANALYSIS

I would suggest that a more correct interpretation of Dr. Porter's account would be that it took Reno's battalion between five to ten minutes to water and cross the river and then an additional five to ten minutes to form his command up on the other side of the river.

2) Sergeant Ferdinand Culbertson (RCOI): It took 5 to 8 minutes for Reno to cross the Little Big Horn River.

ANALYSIS

Sergeant Ferdinand Culbertson's time estimate supports that of Dr. Porter's above.

3) Lieutenant Charles Varnum (RCOI): Reno crosses Ford A in 10 to 15 minutes after leaving Custer but admits to some delay in crossing.

ANALYSIS

Varnum suggests there was some delay, probably due to the watering of the horses, while crossing at Ford A, which Varnum put at some time between ten to fifteen minutes.

4) Lieutenant George Wallace (RCOI): It was probably 20 minutes after 2 when we crossed the creek first.

ANALYSIS

As usual, Wallace's watch time estimate of a 2:20 p.m. crossing of the divide is not supported by any other participant accounts. As Wallace indicated earlier that Reno received his attack orders at 2:15 p.m., he claims that Reno covered the 1.2 miles to Ford A in only five minutes, indicating an all-out full run at about twenty-five miles per hour, which is simply ridiculous. Of course, Wallace also contradicted himself in his own testimony when he previously estimated that it took Reno about fifteen minutes to cover the distance to the river crossing after leaving Custer.

5) Lieutenant Luther Hare (Camp interview, *Custer in 76*): The scouts got to Ford A a half mile ahead of Reno and watered their horses. Then Reno came up and stopped there, taking plenty of time to water. He says it is not true that Reno did not give his men time to water. He says Reno remained at the river for ten to fifteen minutes. While Reno was watering, Hare went down the valley with the scouts about a mile when some of the Rees took off after a herd of Sioux ponies and ran them off.

ANALYSIS

Lieutenant Hare agrees with Varnum's estimate of a ten to fifteen minute crossing at Ford A, with Reno taking plenty of time to water his horses. During this time, Hare accompanies his scouts down into the valley for about a mile whereupon a group of Ree scouts took off after a herd of Sioux ponies. The early firing heard by Martin and others likely came from these Ree scouts capturing the Sioux ponies in the valley while Reno's command was still at Ford A.

6) Young Hawk and Strikes Two—Ree scouts (Camp interview, *Camp and Custer at the Little Bighorn*): say the Rees captured the Sioux ponies quite a while before Reno's battle started.

ANALYSIS

This statement by two of the Ree scouts supports the recollection of Lieutenant Hare that the Rees pursued and captured a herd of Sioux ponies while Reno's battalion was still in the vicinity of Ford A, and "quite a while before Reno's battle started."

7) Sergeant Daniel Kanipe (Camp Interview, *Custer in 76*): says that Reno forded the river at Ford A about noon.

ANALYSIS

All of Kanipe's time estimates referenced local sun time. Thus, Kanipe indicates that Reno forded the river at about 1 p.m. Headquarters watch time.

8) Private Edward Davern (RCOI): Reno crossed the river at Ford A at 1 p.m.

ANALYSIS

Private Davern gives us the correct Headquarters watch time of 1 p.m. and supports Kanipe's similar local sun time recollection of this event.

9) Lieutenant Edward Godfrey (Godfrey's Narrative, *Custer's Last Battle*): Reno's battalion moved at a trot to the river, where he delayed about 10 to 15 minutes watering the horses .... Both ... Keogh and ...

Cook were at this crossing for a short time. Reno now sent a message to Custer that he had everything in front of him and that the enemy was strong.

ANALYSIS

Lieutenant Godfrey—who was not present for this event but had extensive contact with those who were—agrees with both Varnum and Hare that the watering and crossing at Ford A took between ten and fifteen minutes. This was also the likely location where Reno sent back his first messenger to Custer (Private McIlhargy of I Company).

10 a) Private William O. Taylor (*With Custer on the Little Big Horn*): The horses having been watered, we rode out of the river and through the underbrush and then out a few yards onto the prairie, where we dismounted and tightened our saddle girths....

ANALYSIS

Private Taylor does not give a time estimate for Reno's crossing at Ford A, but he does support other participants' claims that the horses were watered there. There does seem to be evidence—via Private Morris' account—however, that M Troop, the first to arrive at the ford, did not stop to water their horses.

10 b) Private William O. Taylor (An excerpt from a letter written by William O. Taylor to General Edward Godfrey, Feb. 21, 1921, from the *Francis R. Hagnar Collection*, Box 1, Manuscript Dept., New York Public Library): "... I did not see any evidence of drunkenness or anything that was suggestive of it, from the mouth of the Rosebud until after we had crossed the Little Big Horn, about 1 p.m., June 25...."

ANALYSIS

Private Taylor places Reno's crossing at Ford A "about 1 p.m.," thus supporting the times given by both Davern and Kanipe.

11) Sergeant Thomas Harrison (1905 interview by the *Arkansas Gazette*): Custer soon saw that he had caught up with a large body of warriors, and the fight between them began about noon.

ANALYSIS

Sergeant Harrison's reference to the start of the fight as being "about noon" would refer to local sun time. Headquarters watch time would place the start of the fight at about 1 p.m., thus supporting the Taylor, Kanipe, and Davern estimates placing Reno in the Little Big Horn valley at about 1 p.m. or shortly thereafter.

Appendix A: Participant Timeline of the Battle of the Little Big Horn

OVERALL ANALYSIS

Three of the participant accounts (Sergeant Kanipe, Private Davern, and Private Taylor) agree that Reno crossed the river at about 1 p.m. Headquarters watch time. This fits very well into our previous timeline analysis that Reno arrived at Ford A at about 12:55 p.m. This time estimate is supported by Sergeant Harrison's start of the battle occurring about noon (local sun time or 1 pm Headquarters watch time). Lieutenant Wallace's time, as usual, is not supported by any other and can be safely dismissed. Two of the participants (Dr. Porter and Sergeant Culbertson) estimate Reno's crossing to have taken between five and ten minutes, whereas three accounts (Lieutenant Varnum, Hare, and Godfrey) claim a slightly longer time of ten to fifteen minutes. All told, this would span a range of between five to fifteen minutes. To try to gain a bit more accuracy, I would suggest that the shorter times likely indicated the time that this individual—and the company he was traveling with—took to water and cross the river. In contrast, the longer times likely indicated the total time for the battalion to water and cross the river. Thus, I favor the slightly longer time estimates of Varnum, Hare, and Godfrey for Reno's crossing at Ford A.

At the time Reno is crossing Ford A, the Custer column, having finished watering their horses at the Echleman Ford on Ash/Reno Creek, is rapidly approaching the vicinity of Ford A itself, being about a half mile or so behind Reno. This was why Adjutant Cooke and Captain Keogh could accompany Major Reno down to Ford A. Custer was about a half mile behind and, I believe, intent upon following Reno across the river at Ford A.

Benteen's battalion at this time had stopped at the first, or eastern morass, for his first water break. He would reach this morass about four and three-quarters miles east of Ford A, a few minutes after 1 p.m., as Reno is crossing the Little Big Horn River.

CONCLUSION

Reno's battalion crosses the Little Big Horn River at Ford A after two of his three companies water their mounts in about ten to fifteen minutes. The time for this event would be between 1:05 and 1:10 p.m.

1:05 p.m. (H = 4 hours and 35 minutes.): Benteen reached the first (eastern) morass.

1:08 p.m. (H = 4 hours and 38 minutes.): Reno's battalion crosses the River at Ford A.

9. RENO'S BATTALION BEGINS THEIR ADVANCE-TO-CONTACT

Reno's battalion has now reached the west bank of the Little Big Horn River across Ford A at about 1:08 p.m. and will reform just beyond the far edge of the timber about fifty yards west of the river before initiating his advance towards the village. Here's what the participants tell us about this time:

1) Captain Myles Moylan (RCOI): after crossing the ford, they paused to allow the companies to close up, then advanced about a third of a mile in a column of fours when they then deployed from column into line formation.

ANALYSIS

Captain Moylan's account tells us that Reno stopped to allow the companies to close up before ordering an advance towards the hostile position. During this interval, Reno likely sent back his two messengers to Custer (Privates. McIlhargy and Mitchell, both of I Company). Moylan does not tell us how long this stop lasted. He also indicates that Reno reformed his command from column into line about a third of a mile west of Ford A before beginning his advance.

2) Lieutenant Edward Godfrey (Godfrey's Narrative, *Custer's Last Battle*): Reno's battalion moved at a trot to the river, where he delayed about ten to fifteen minutes ... reforming the column on the left bank of the stream. ... Reno now sent a messenger to Custer that he had everything in front of him and that the enemy was strong. Reno moved forward in a column of fours for about a half-mile, then formed the battalion across the valley with the scouts on the left.

ANALYSIS

Godfrey, who was not present at this event, but was in close communication with those who were, indicates that Reno paused to reform his column on the left (west) bank of the Little Big Horn River just beyond the belt of timber about fifty yards west of Ford A, which, he suggests, may have taken between ten to fifteen minutes. He also indicates that Reno moved forward about half a mile before reforming his command from column into line, with his remaining scouts on his left flank.

3) Private William Taylor (*With Custer on the Little Big Horn*): the horses having been watered, we rode out a few yards onto the prairie where we dismounted and tightened our saddle girths, and in about ten minutes were heading down a long but rather narrow valley .... We were then at a fast walk.

ANALYSIS

Private Taylor tells us that Reno's battalion, after moving through the fifty yards of timber on the west side of Ford A, reaches the open prairie where they dismount, tighten their saddle girths and prepare for contact with the enemy. This took about ten minutes before they moved on down the valley at a fast walk. This fast walk pace (about five miles per hour) likely lasted for about a third of a mile (see Moylan's account above) when Reno changed his formation from column to line. It is actually unclear whether Reno halts the command a second time during this maneuver or conducted it while moving at a fast walk. My guess is that the battalion did so while halted.

4) Private Thomas O'Neill (Camp interview, *Custer in 76*, with my annotations in brackets): Custer came within 300 or 400 yards of the river before he turned up to the right. ... On [the] other side of [the] river [at Ford A], there were timber and fallen logs, and [it] took some time to get through.

ANALYSIS

Although Private O'Neill does not give us any time estimates, he tells us that it took Reno's battalion "some time to get through" the timber and fallen logs in the fifty yards of wooded terrain on the west side of Ford A. We would need to factor in five to ten minutes to do this.

5) Corporal Stanislas Roy (Camp interview, *Custer in 76*, with my annotations in brackets): [There was] about fifty yards of timber on [the] west bank [of] Ford A .... After passing the ford , we formed in line ....

ANALYSIS

Corporal. Roy tells us that there were fifty yards of timber on the west bank of Ford A. He also mentions Reno reforming his battalion into line formation, but gives us no time estimates.

6) Richard Roberts (from his monograph, *Reminiscences of General Custer*, p. 34): "He [i.e., Reno] recrossed the stream [i.e., Ash/Reno Creek], moved down it, crossed the Little Big Horn, halted his columns (Robert's note: no one knows how long, some soldiers say twenty minutes, some say a half hour. Fatal halt, he should have formed in line without halting)...."

ANALYSIS

Richard Roberts (Custer's personal secretary) was not present during this event; however, he relates what troopers who were there told him. Although this is a secondary account, it is worth noting as an addition to the other primary source accounts of this event. According to Robert's sources, Reno, after crossing the Little Big Horn River, halted his command for twenty to thirty minutes before beginning his advance down the valley to attack the hostiles.

OVERALL ANALYSIS

Based on the participant accounts above, the evidence suggests that it took Reno's battalion about eight minutes or so to traverse the fifty yards of timber and fallen logs just after crossing the Little Big Horn at Ford A. Upon reaching the open prairie, Reno dismounted his command to tighten their girths and prepare for the advance down the valley and his imminent contact with the hostile force. This delay lasted about ten minutes before they moved down the valley again for about a third of a mile (or about six hundred yards) at a fast walk (about five miles per hour), taking about another four minutes before Reno reformed his command from column into line formation, bringing G Company up into line on the left and keeping his remaining scouts out on his far-left flank. From this position, Reno would begin his charge on the village, roughly a third of a mile west of Ford A. After crossing the river at Ford A, it would take Reno about twenty-two minutes before he was in position to charge the hostiles.

CONCLUSION

1:05 p.m. (H = 4 hours 35 minutes): Reno's battalion had crossed the river at Ford A.

1:10 p.m. (H = 4 hours 40 minutes): Reno's battalion reached the open prairie west of Ford A at 1:10 p.m., then dismounted his command to prepare for their advance.

1:20 p.m. (H = 4 hours 50 minutes): Custer has already ascended the bluffs at Sharpshooter Ridge and had his first glimpse of the Indian village well before Reno had made his appearance to the hostiles in the valley. His command remained in the vicinity of Sharpshooter Ridge for at least another ten minutes. Reno's battalion mounts and moves out at a fast walk in column formation for about a third of a mile before halting again to reform his command into line formation for his advance down the valley. Custer is spotted on the bluffs at this point by a number of troopers and officers of Reno's battalion.

1:25 p.m. (H = 4 hours and 55 minutes): Reno begins his advance (i.e., charge) down the valley, first at a trot, then at a gallop. Benteen's battalion was just leaving the eastern morass and was about four and three-quarter miles east of Ford A at this time. The advance packs and ammunition mules under Lieutenant Mathey had likely just reached the eastern morass and was about four and three-quarter miles east of Ford A. The main pack train and Rear Guard would be about five and a quarter miles east of Ford A at this point.

10. CUSTER IS SEEN ON THE BLUFFS

When and where Custer ascended the bluffs has been a source of controversy among battle students for many years. Here is what the participants had to say of this event:

1) Trumpeter John Martin (RCOI, with my annotations in brackets): Custer stopped to water his horses [note: likely at the Echleman Ford on Ash/Reno Creek about 1.2 miles east of the Little Big Horn] for about 10 minutes. There appeared to be a lodge pole trail there; then they moved straight ahead for about 300 yards, following the lodge pole trail. They moved very fast. Custer then moved a little to the right going another 4 to 500 yards. [Note: Custer's battalion has now traveled just under 1/2 mile west of the Echleman Ford, or about 3/4 of a mile east of the Little Big Horn.] He then turned up to the bluffs by a "big bend on the hill". He turned on this bend and rode on to the top of the [Sharpshooter] ridge. He says he got to the top of the bluffs about 12 noon.

ANALYSIS

Martin gives us a very good account of Custer's route from the time he left the Echleman Ford to when he ascended the bluffs "about 12 noon." Martin's reference to time is based on local sun time. His 12 noon is actually closer to 1 p.m. Headquarters watch time. He indicates that Custer's battalion moved "very fast," likely indicating anything from a fast trot to a lope (eight to ten miles per hour) on this route. The timeline would indicate that Custer likely finished watering his command at about 12:55 p.m. or about the same time that Reno had reached Ford A. From this point on, Custer moves at a fast clip of between eight to ten miles per hour, which would likely put him on Sharpshooter Ridge between twenty to twenty-five minutes later (a distance of just under three miles from the Echleman Ford). We can project from Martin's account that Custer reached the south end of Sharpshooter Ridge at sometime between 1:15 and 1:20 p.m.

2) Private Edward Pigford (*Washington Observer* Interview, October 5, 1932): Outlined against a blue sky on a far distant hill, back over the Little Big Horn, they beheld the figure of their commander. He sat on his horse, waving his hat to them for an instant, and then disappeared over the ridge.

ANALYSIS

Private Pigford does not identify the whereabouts of this sighting and hat waving of Custer. It is difficult to place this sighting with any degree of accuracy; however, Pigford suggests that the sighting took place before Reno began his advance down the valley, likely at the time Reno was redeploying his battalion from column to line across the valley floor about a third of a mile west of Ford A.

3) Private Henry Petring (Camp Interview, *Custer in 76*): while in the bottom going toward the skirmish line, I saw Custer over across the river on the bluffs waving his hat. Some of the men said, "There goes Custer! He is up to something, for he is waving his hat."

ANALYSIS

This sighting of Custer, along with his hat waving, described by Private Petring, likely took place during Reno's reformation of his battalion from column into line on reaching the open prairie about a third of a mile west of Ford A.

4) Corporal Stanislas Roy (Camp Interview, *Custer in 76*): After passing the ford, we formed in line and while forming up, I heard some of the men say, "There goes Custer!" He could be seen on hills to our right and across the river.

ANALYSIS

Corporal Roy supports the accounts of both Petring and Pigford, stating that Custer was seen on the bluffs at the time when Reno was forming his battalion into line. This sighting would take place about one third mile west of Ford A, according to Captain Moylan's account.

5) Private Thomas O'Neill (Camp Interview, *Custer in 76*, with my annotations in brackets): Custer came within three hundred or four hundred yards of the river before turning to the right.... On [the] other side of [the] river [i.e., on the west side of Ford A], there were timber and fallen logs, and [it] took some time to get through. When [they] were about half way down to where the [valley] skirmish line was formed, he saw Custer and his whole command on the bluffs across the river, over to the east, at a point which he would think was about where Reno afterward fortified, or perhaps a little south of this. Custer's command was then going at a trot.

ANALYSIS

Private O'Neill agrees with the accounts of Corporal Roy and Privates Pigford and Petring with his sighting of Custer's command moving at a trot as they ascended the bluffs south of Reno Hill, likely at the time when Reno deployed his command from column into line across the valley floor.

6) Lieutenant Carlo De Rudio (RCOI): claims to see Custer, Cooke and another person on the edge of the bluffs just north of Reno Hill about five minutes before Reno retreated from the timber position.

ANALYSIS

There are major problems with De Rudio's alleged sighting of Custer and Cooke on the edge of the bluffs. The first problem is that Trumpeter Martin never mentioned Custer riding to the edge of the bluffs here to look into the valley. He does, however, describe a hill set back from the edge of the bluffs, which matches up very well with the south end of Sharpshooter Ridge, a few hundred yards to the east of elevation 3411.

Secondly, De Rudio estimated the distance of this sighting on the bluffs to be about a thousand yards from where he was standing. However, the actual distance was closer to eighteen hundred to nineteen hundred yards from De Rudio's position in the timber to the edge of the bluffs at elevation 3411. This distance is over a mile away, and it is extremely doubtful that De Rudio—or anyone else—could identify someone on the bluffs with the naked eye (and De Rudio specifically stated that he was not using field glasses for this sighing, as he had already loaned them to Custer).

The third problem is the time. De Rudio claimed that he saw Custer and Cooke on the bluffs about five minutes before Reno retreated from the timber position—or about 2:05 p.m. HQ watch time. This time does not fit with the evidence at hand. For this to be true, Custer would have had to remain idle on top of the bluffs for nearly the entire duration of Reno's valley fight—a fact known from the accounts of Trumpeter Martin, Private Peter Thompson, Private Pickard, and Sergeant Kanipe to be not true.

So, how does one explain De Rudio's sighting? It is not difficult to interpret. De Rudio very likely did see three figures on the edge of the bluffs—and very likely at or near elevation 3411—about five minutes or so before Reno retreated from the valley. These figures, dressed in buckskin clothing and blue shirts, with one of them having what appeared to De Rudio to be long side-whiskers on either side of his face, were likely either the three Crow scouts (Goes Ahead, Hairy Moccasin, and White Man Runs Him) or three of the Ree scouts. De Rudio, in his mind, imagined them to be Custer and Cooke. The long, dundreary side-whiskers he thought were Cooke's were very likely the long hair of one of the Indian scouts. It is interesting to note that in several of De Rudio's other accounts of the battle, given to Walter Camp, he neglects to mention this sighting, indicating that perhaps he recognized the improbability that this was Custer or Cooke. Lieutenant Godfrey, in his narrative of the battle, makes mention of De Rudio's sighting of Custer on the bluffs but changes the time of the sighting to when Reno was deploying his troops on the skirmish line at the start of the battle—a more logical interpretation.

7) <u>Lieutenant Edward Godfrey</u> (Godfrey's Narrative, *Custer's Last Battle*) says that Captain Moylan admitted seeing "Custer's party" on the bluffs about the position of Reno Hill. This party was heard to cheer and seen to wave their hats as if to give encouragement, and then they disappeared behind the hills. ... De Rudio thinks he saw Custer on the ridge about opposite from where Dr. DeWolfe was killed. He says Custer and his brother Tom were dismounted, looking at them through field glasses. Reno was then developing into a line of skirmishers.

<u>ANALYSIS</u>

Lieutenant Godfrey, who was not actually present during these events but was in close contact and later in communication with those who were, relates two accounts told to him by Captain Moylan and Lieutenant De Rudio. Moylan indicates that Custer's command was seen on the bluffs near Reno Hill and was heard to cheer and wave their hats before disappearing behind the hills. This sighting likely took place during the time that Reno's command was deploying from column into line about a third of a mile west of Ford A. At this time, Reno's command would be spotted by some of Custer's troops on the bluffs just before riding out of sight behind the south end of Reno Hill.

Godfrey also mentions the sighting of Custer on the bluffs by Lieutenant De Rudio. It is interesting to note that De Rudio's account—at least the one given to Godfrey—has now changed from the flawed one he gave at the RCOI. Now, there is no mention of seeing Adjutant Cooke at all. He is replaced by Captain Tom Custer. De Rudio now places this sighting on a point opposite from where Dr. DeWolfe was killed, which places the sighting near or at elevation 3411 on the edge of the bluffs. He says they were dismounted and looking at them through field glasses at the time Reno was first forming up his valley skirmish line. Again, we would suggest that this sighting—even with the timing discrepancy fixed—would still indicate a likely sighting of the Crow scouts, this time, more likely Mitch Boyer and Curley than the other three Crows.

8) Lieutenant Charles Varnum (RCOI): Varnum sees the Gray Horse troop moving north across the top of the bluffs as Reno begins to form his valley skirmish line.

ANALYSIS

Varnum spotted the Gray Horse Troop (E Company) moving north across the highest bluff just north of Reno Hill—which almost certainly identifies it as Sharpshooter Ridge—just as Reno halted his charge and began to dismount his battalion. Varnum would have been ahead of Reno's command in the vicinity of the Garryowen Bend of the river when he made this brief sighting, indicating that Custer's column was departing from the area of Sharpshooter Ridge just as Reno dismounted his command.

9a) Fred Gerard (RCOI): he claimed to see Custer's command on the bluffs about 10-15 minutes before Reno left the timber position. He saw them south of Reno Hill. Gerard would later clarify and correct the time he saw Custer's command riding north at a trot along the bluffs. He now says (p. 108, Graham) he saw them about 1 1/2 to 2 miles north of Ford A. He now claims he saw Custer's command there about 10–12 minutes after he crossed at Ford A. Gerard (p. 109, Graham) also indicates that he saw Custer himself on the bluffs while the men were still on the skirmish line (or several minutes after Varnum's sighting of E Company). This may indicate Custer remaining on elevation 3411 for a few minutes to observe the deployment of Reno's battalion on their skirmish line.

ANALYSIS

Gerard's comments on seeing Custer and his command are the source of much confusion among serious battle students. First, he claimed to have seen Custer's command on the bluffs south of Reno Hill about ten to fifteen minutes before Reno left the timber position. This would place the time of the sighting between 1:55 and 2:00 p.m., which does not fit in with either supporting evidence or logic. We would suggest that Gerard actually may have seen the party of Ree warriors who had captured some Sioux ponies who were now some distance south of Reno Hill at this time, as Gerard stated.

In any event, Gerard would later modify and clarify his remarks on seeing Custer's command on the bluffs, eventually stating that he observed them riding north at a trot about one and a half to two miles north of Ford A. He claims he made this sighting ten to twelve minutes after he crossed the river at Ford A. The distance Gerard references would cover the area between Reno Hill and Sharpshooter Ridge on the bluffs. Gerard appears to have made this sighting just after Reno had halted his battalion on the open prairie just after clearing the fifty yards of timber west of Ford A and just before Reno reformed his command from column into line about a third of a mile west of Ford A. This would be at sometime between 1:15 and 1:20 p.m. as Custer's command was approaching Sharpshooter Ridge.

Gerard would also indicate that he saw Custer himself on the bluffs during the time that Reno deployed his troops on the valley skirmish line. This would be a time shortly after Lieutenant Varnum witnessed the Gray Horse Troop riding across the top of Sharpshooter Ridge as Reno halted and dismounted his battalion. My analysis is that Gerard very likely noticed a figure—or several figures—in buckskins standing on elevation 3411 while Reno deployed his skirmish line. Gerard, like De Rudio, assumed that this figure was Custer himself, but it is doubtful that anyone could recognize individual features at that distance, well over a mile, with the naked eye. I believe that it is more likely that Gerard actually saw Mitch Boyer and Curley standing on elevation 3411 at this time and simply mistook them for Custer and his officers. I do not believe the evidence supports the notion that Custer remained on the bluffs to witness Reno actually deploying his men to their skirmish line position at the Garryowen Bend of the river. Trumpeter Martin's testimony makes it clear that Custer and his command had left the bluffs prior to that event.

9b) <u>Fred Gerard</u> (*The Custer Myth*): not more than 4 rounds had been fired before they saw Custer's command dashing along the hills one mile to their rear.

<u>ANALYSIS</u>

Gerard in this account claims to see Custer's command riding north along the bluffs at the time Reno was deploying his battalion into his valley skirmish line. His reference to no more than four rounds having been fired at the time of this sighting references the opening shots of the valley fight being fired by either the hostiles or several of Reno's scouts—not from the troops of Reno's command, who had not yet formed into their skirmish lines at this time. This sighting by Gerard would be the same one made by Lieutenant Varnum as Reno halted and began to dismount his command. Gerard's distance of one mile to their rear—from the south side of the Garryowen Bend—references the general location of Sharpshooter Ridge.

10) Captain Edward McClernand (Chapter II (1866-91) *History of the 2nd Cavalry*) 2nd U.S. Cavalry: "Two horsemen were seen dashing down the valley. They were officers—Wallace and Hare, if the writer's memory is not at fault,—sent by Reno to tell of their desperate fight, and how the Indians seemed determined upon their extermination, until Gibbon's column appeared on the bluffs the day before. 'Where is Custer?' was then asked. They replied: 'The last we saw of him he was going down that high bluff towards the lower end of the village. We do not know where he is now.' They were told, 'We have found him.'

ANALYSIS

We now have five officers from Reno's battalion who reported spotting Custer's command on the bluffs moving northward. This would be Captain Moylan, Lieutenant De Rudio, Lieutenant Wallace, Lieutenant Varnum, and Lieutenant Hare.

OVERALL ANALYSIS

As we learn from Martin's account, Custer very likely ascended the bluffs well before Reno began his advance down the valley to charge the hostiles. It would appear from the participant accounts that Reno actually began his charge toward the hostiles about forty-five minutes after receiving his attack orders and separating from Custer near the Echleman Ford on Ash/Reno Creek, 1.2 miles east of Ford A. We learn from Martin's account that it took Custer about thirty minutes from when Reno left him—at about 12:45 p.m.—to reach the top of the bluffs at Sharpshooter Ridge—between 1:15 and 1:20 p.m. At this time, Reno had just finished his short pause to organize his command west of Ford A and was moving down the valley just before deploying his battalion from column to line formation about a third of a mile west of Ford A, as Captain Moylan tells us. This appears to be the location where most of Reno's men seem to have spotted Custer's command on the bluffs. By the time Reno had reached the vicinity of the Garryowen Bend of the river to dismount and deploy his battalion into a skirmish line, Custer's command would be seen by both Varnum and Gerard leaving Sharpshooter Ridge and riding down toward Medicine Tail Coulee. All other reported sightings of Custer after this were likely mistaken for the Crow or Ree scouts remaining behind on the bluffs.

CONCLUSION

Between 1:10 p.m. (H = 4 hours + 40 minutes) and 1:20 p.m. (4 hours + 50 minutes): Custer is spotted on the bluffs by members of Reno's command on the bluffs. At this time, Reno is approaching the point where he would reform his battalion from column into line formation about a third of a mile west of Ford A; Benteen's battalion was still watering at the eastern morass about 4.75 miles east of Ford A; the advance packs under Lieutenant Mathey were approaching the eastern morass about 5 or 5.5 miles east of Ford A; the main pack train and Rear Guard were now between 6 and 6.5 miles west of the divide or just under half the distance from the divide to the Little Big Horn River.

11. SERGEANT KANIPE'S RIDE TO THE PACK TRAIN

Most of our information about Sergeant Daniel Kanipe's ride to the pack train comes from Sergeant Kanipe himself in his many accounts given throughout the years. From his descriptions, and those of Packer John Frett and Lieutenant Edgerly, we can put together the essential pieces of his ride:

1) Packer John Frett (RCOI): Sometime between 11 a.m. and 1 p.m., the packs "were at the watering place near the tepee, the last tepee that was there before we got to the battlefield, the one with the dead Indian in it, when a sergeant came… and said we should hurry up, that General Custer was attacking the Indians."

ANALYSIS

John Frett's time of between 11 a.m. and 1 p.m. is certainly based on local sun time. This time range would convert into Headquarters watch time of between noon and 2 p.m. His reference to the packs being at the watering place near the tepee is a reference to the first or eastern morass, located about four and three-quarters miles east of Ford A, with the tepee referencing the first (eastern) lone tepee, located about 3.75 miles east of Ford A. On first reading, Frett's statement that this was the last tepee before they got to the battlefield might cause some to think he was referencing the second (western) lone tepee and the Hartung morass, but a further reading of Kanipe and Edgerly's account will dispel that notion. From Kanipe's statement, we can conclude that the head of the advance packs was at the first (eastern) lone tepee about 3.75 miles east of Ford A, while the rear elements of the pack train were still watering at the eastern morass about a mile farther east when Kanipe reached them there.

2) Lieutenant Winfield Edgerly (testifying at the RCOI, with my annotations in brackets): They went about one mile when they met up with Trumpeter Martin, but before they met up with Martin a sergeant from C Company [i.e., Kanipe] rode up and told Benteen Custer wanted him to bring up the packs. Edgerly thought this sergeant was either Kanipe or Hanley; Edgerly thought this was before they reached the [Hartung] morass, just a short time before Martin came up.

ANALYSIS

Lieutenant Edgerly tells us that Sergeant Kanipe met up with Benteen at some distance east of the Hartung morass, about a mile east of where they met up with Trumpeter Martin. As Benteen met up with Martin about 1.5 miles east of Ford A, Edgerly suggests that it was some distance over 2.5 miles east of Ford A when Kanipe met up with them, or just before Benteen's battalion reached the Hartung morass.

3a) Sergeant Daniel Kanipe (*I Fought With Custer, 81-82*): When he first saw Benteen, he was watering his horses.

ANALYSIS

Kanipe first saw Benteen's battalion several miles away as he approached the North Fork of Ash/Reno Creek. At this time, Benteen's battalion was seen watering their horses at the eastern morass, about four and three-quarters miles east of Ford A, and about three-plus miles east of Sergeant Kanipe's location by the North Fork. The time for this sighting would be about 1:30 p.m.

3b) Sergeant Kanipe (letter to Walter Camp): Kanipe met Benteen west of the watering place. The packs were two miles east of Benteen. He met Benteen a little west of the burning tepee.

ANALYSIS

Kanipe tells us that he met Benteen a distance west of the eastern morass and a short distance west of the first burning tepee. The eastern morass was 4.75 miles from Ford A, and the first (eastern) burning tepee was about 3.75 miles east of Ford A. Kanipe is suggesting that he met up with Benteen about two and three-quarters miles east of Ford A. He also indicates that the pack train was two miles east of Benteen, watering at the eastern morass about two miles east of where Kanipe met Benteen.

3c) Sergeant Daniel Kanipe (letter to Walter Camp, October 9, 1910): He met Benteen when the latter had just left the watering place, but when he first saw Benteen, his command was watering, and it was about a mile from where he met McDougall. He says he passed a small party of Rees—a half dozen or more—with ponies who passed ahead of him on his right about two hundred yards away before he reached Benteen. Camp notes that Kanipe met Benteen about one mile west of the lone tepee. Kanipe met the pack train at approximately where the lone tepee was. The packs were strung out about a mile, and Kanipe met McDougall about a half mile east of the lone tepee. Kanipe indicates that his route back took him across the North Fork of Ash/Reno Creek.

ANALYSIS

Kanipe says that he first saw Benteen watering his horses at the eastern morass, which would be about 1:30 p.m., just as Benteen was about to finish his lengthy water break about four and three-quarter miles east of Ford A. Kanipe would have been at or near the North Fork of Ash/Reno Creek when he made this first sighting of Benteen's battalion about three-plus miles east of his position. Camp notes that Kanipe met Benteen about a mile west of the first, or eastern lone tepee, which places their meeting point about 2.75 miles east of Ford A, or just as Benteen approached the Hartung morass. Kanipe then continues on to meet the head of the pack train at the first (eastern) lone tepee, about 3.75 miles east of Ford A. He continues riding east for another half mile until he meets up with Captain McDougall's Rear Guard about 4.25 miles east of Ford A, then continues on to the rear elements of the pack train still watering at the eastern morass about 4.75 miles east of Ford A. From Kanipe's time and distance estimates, it would appear to have taken him about twenty minutes from the time he saw Benteen's command watering at the eastern morass until he met up with him as he was approaching the Hartung morass about two miles farther west. It would take Kanipe another twenty minutes before he reached the rear elements of the pack train still watering at the eastern morass.

OVERALL ANALYSIS

Sergeant Kanipe leaves Custer's column as it reached the south end of Sharpshooter Ridge about 1:20 p.m., then rides back down to the North Fork of Ash/Reno Creek, arriving there about 1:30 p.m. where he spots Benteen's battalion about three-plus miles east of him watering his horses at the eastern morass. He crosses the North Fork and rides on eastward, first being passed by a group of Ree scouts with their captured Sioux ponies before reaching Benteen's battalion about 2.75 miles east of Ford A or just before reaching the Hartung morass at about 1:45 p.m. After briefly speaking to Benteen, he continues riding east until he meets the head of the advance packs near the first (eastern) lone tepee about a mile farther east, then continues riding until coming to Captain McDougall, about 4.25 miles east of Ford A. He does not stop there but continues riding back until he reaches the rearmost elements of the pack train, who were still watering their horses at the eastern morass about 4.75 miles east of Ford A, reaching them at about 2:05 p.m.

CONCLUSION

1:20 p.m. (H = 4 hours and 50 minutes): Sergeant Kanipe leaves Custer's column.

1:30 p.m. (H = 5 hours): Sergeant Kanipe first sees Benteen's battalion watering their horses at the eastern morass as he approaches the North Fork of Ash/Reno Creek.

1:45 p.m. (H = 5 hours and fifteen minutes): Sergeant Kanipe reaches Captain Benteen about 2.75 miles east of Ford A.

1:52 p.m. (H = 5 hours and 22 minutes): Sergeant Kanipe reaches the advance pack mules at the first (eastern) lone tepee 3.75 miles east of Ford A, where he is seen by packer John Frett.

2:02 p.m. (H = 5 hours and 32 minutes): Sergeant Kanipe reaches Captain McDougall about 4.25 miles east of Ford A.

2:05 p.m. (H = 5 hours and 35 minutes): Sergeant Kanipe reaches the Pack Train at the eastern morass.

12. RENO DISMOUNTS TO FORM A SKIRMISH LINE

How long did it take for Reno to advance down the valley and then dismount to form his skirmish line in the valley? This would effectively mark the start of Reno's valley fight. Here's what the participants have to say:

1a) Lieutenant Charles Varnum (RCOI): sees the Gray Horse Troop moving north on the bluffs as Reno halts his command to form his skirmish line.

ANALYSIS

Although Varnum does not give us a specific time for this event, he does indicate that Custer's five companies are moving over the highest point on the bluffs—likely Sharpshooter Ridge—just as Reno has halted his command and is dismounting to form up his valley skirmish line.

1b) Lieutenant Charles Varnum (RCOI): says it took Reno 15 to 20 minutes after crossing Ford A to reach his skirmish position at a fast trot—but also said it might have been longer.

ANALYSIS

Varnum's 15 to 20 minutes time for Reno to move down the valley until he reached his skirmish line position would include the time that Reno paused to dismount shortly after crossing the river at Ford A. Thus, Varnum is projecting a time beginning with Reno just crossing the river about 1:10 p.m. and arriving at his skirmish line position between 1:25 and 1:30 p.m., or perhaps a few minutes later.

2) Dr. Henry Porter (RCOI): it took Reno to advance down the valley to his skirmish line position between 15 to 20 minutes (p. 164, Graham).

ANALYSIS

Dr. Porter's 15 to 20 minutes time for Reno to move down the valley until he reached his skirmish line position would include the time that Reno paused to dismount shortly after crossing the river at Ford A. Thus, Porter is projecting a time beginning with Reno crossing the river at about 1:10 p.m., arriving at his skirmish line position between 1:25 and 1:30 p.m., and supporting Varnum's time estimate.

3) Fred Gerard (RCOI): it took Reno five to seven minutes to go down the valley to where he reached the skirmish line area…

ANALYSIS

Gerard indicates that it took Reno between five and seven minutes to charge down the valley to the skirmish line area, likely just after he deployed his battalion into line formation about a third of a mile west of Ford A (at 1:25 p.m.) thus reaching the skirmish line area about 1:30 or 1:32 p.m.

4) Lieutenant Luther Hare (RCOI): says it took Reno ten to fifteen minutes to go from Ford A to the skirmish line.

ANALYSIS

Lieutenant Hare's time estimates for Reno's advance down the valley likely begin just after his ten-minute pause to reorganize his command after crossing Ford A (at 1:20 p.m.), taking between ten and fifteen minutes to reach the skirmish line from that point, thus suggesting that Reno reached the general area of his skirmish line position between 1:30 and 1:35 p.m.

5) Captain Myles Moylan (RCOI) said that Reno advanced from Ford A one and a half miles to his skirmish line in either five, ten, or fifteen minutes, "I don't know the exact time." He claims that Reno stopped to form his skirmish line five to six hundred yards from "the point of timber".

ANALYSIS

Captain Moylan's time references likely refer to Reno taking about five minutes to reach the skirmish line after he begins his charge down the valley just after deploying his battalion into line formation a third of a mile west of Ford A or ten to fifteen minutes after Reno begins his advance following a pause to organize his command on the prairie just west of Ford A. This supports Gerard and Hare's time estimates depending on Reno's starting point, with Reno reaching the Garryowen Bend area about 1:30 to 1:35 p.m. Corporal Roy tells us that Reno's battalion would advance a short distance up to the "point of timber," which is another term for the Garryowen Bend of the Little Big Horn River. Moylan's distance of a mile and a half down the valley before Reno dismounted is contradicted by Varnum, De Rudio, and Reno himself, who all stated their distance estimates as just over two miles after crossing Ford A. Moylan is also contradicted by nearly all other participant accounts that Reno stopped his advance five hundred to six hundred yards from the point of timber. Thus Moylan's account in this regard is suspect and may have been influenced by his interpretation of the incorrect skirmish line position placed on the Maguire map at the RCOI.

6) Corporal Stanislas Roy - A Company (Camp Interview, *Custer in 76*, with my annotations in brackets): [Reno's battalion] dismounted about two hundred yards in advance [of the] timber [position] and [the] skirmish line [then] advanced to [the] point of timber [at the Garryowen Bend of the river].

ANALYSIS

Corporal Roy tells us that Reno dismounted his battalion about two hundred yards beyond the timber position and then advanced up to the point of timber at the Garryowen Bend of the river. Needless to say, there would be some delay in time from the moment Reno halted to dismount his command and the time it took to deploy his dismounted skirmish line several hundred yards in advance at the Garryowen Bend of the river with its right flank resting on what would later become the old Garryowen store location.

7a) Lieutenant George Wallace (RCOI): Question: What time do you fix for the fight in the timber? Wallace: I estimate it at about 2:30.

ANALYSIS

The recollections of other participants do not support Lieutenant Wallace's estimate of Reno's fight beginning at 2:30 p.m. As usual, his time estimate is off by almost one hour. According to the other participants, Reno's valley fight likely began about 1:30 p.m.

7b) Lieutenant George Wallace (RCOI, with my annotations in brackets): Question: How far was it from the point where Major Reno's command crossed the stream [i.e., at Ford A] to where he engaged the Indians? .... Wallace: It was a good mile and three-quarters before the first shot was fired, and two miles to the timber where he [i.e., Reno] dismounted.

ANALYSIS

Lieutenant Wallace indicates that Reno's advance down the valley from the crossing at Ford A to the point of dismount was about two miles. This would put Reno's dismount near the end of the Garryowen Bend of the river.

8) Private William Taylor - A Company (Camp Interview, *Custer in 76*): "on June 25 ... about 1 p.m. or a little later, we were nearing the Indian skirmishers on our ride toward the village ...." in about ten minutes we were heading down a long, but rather narrow valley .... we were then at a fast walk. Soon the command was given to "Trot." Then as little puffs of smoke were seen and the "ping" of bullets spoke out plainly, we were ordered to charge.

ANALYSIS

Private Taylor tells us that "about 1 p.m. or a little later," Reno's battalion neared the Indian skirmishers on their ride toward the village. Based on Headquarters watch time, his time estimate is remarkably accurate when one considers the actual time was closer to 1:30 p.m.

9) One Bull (Hunkpapa Sioux): thought it was about two p.m. white man's time (or Headquarters watch time) when Sitting Bull's camp was attacked.

ANALYSIS

One Bull's estimate of Reno's attack on the Hunkpapa camp being close to 2 p.m. was based on the time shown on a watch captured from one of General Crook's troopers at the Rosebud Fight six days earlier. As Reno would have dismounted to form his skirmish line at about 1:30 p.m., with the valley fight beginning a short time later, this watch time estimate that One Bull refers to is remarkably accurate.

10) Major Marcus Reno: (letter to General Terry, June 27, 1876, *The Little Big Horn, 1876*, by Overfield, p. 25): "I charged about 2 p.m., but meeting no support was forced back to the hills."

ANALYSIS

In this letter to Terry, Reno indicates a general time of 2 p.m. for his charge down the valley. Our other participants indicate that the charge actually occurred closer to 1:30 p.m. Thus, Reno's 2 p.m. time is a very rough estimate of the time instead of an exact recollection. The first shots from his valley skirmish line likely occurred about 1:40 p.m., or about twenty minutes shy of 2:00 p.m., so Reno may have been simply rounding off his estimate to the nearest hour.

11) Curley (General Scott interview 1919): Scott: How far down the valley did Reno get? Curley: The Sioux met Reno's command where Garryowen now is.

ANALYSIS

Curley supports the Vaughn location for the valley skirmish line instead of the alternative Pitsch location for the lines, which were well south of the Garryowen Bend.

12) Private Thomas O'Neill - G Company (Private O'Neill's letter to General Godfrey, Aug. 17, 1908, *Francis R. Hagner Collection*, Box 1, Manuscript Division, New York Public Library, with my annotations in brackets): We crossed the river at that place [i.e., Ford A]; there was some underbrush and fallen timber on the other side of the river, and as soon as we got clear of that timber we took up the trot in column of fours, and moved up the valley, what appears to me now to have been two miles.

ANALYSIS

Private O'Neill's recollection of Reno's battalion advancing about two miles down the valley from Ford A would bring Reno's advance to the vicinity of the end of the Garryowen Bend of the river.

OVERALL ANALYSIS

As the accounts of Reno's advance down the valley vary to a considerable degree, we must recognize that the likely cause of such is due to the differing starting points that each of our participants referenced. Varnum and Porter reference a time as long as twenty minutes after Reno crossed at Ford A, which indicates that they included the ten-minute halt just west of the crossing place to reorganize Reno's command. Their time estimates put Reno in the vicinity of where he halted his command to form his skirmish line at about 1:30 p.m. Hare and Moylan offer time estimates beginning just after Reno's first halt west of Ford A, and thus is about ten minutes shorter than Varnum's and Porter's. However, their times also put Reno in the vicinity of where he halted his command at about 1:30 p.m. Fred Gerard indicates a five- to seven-minute charge down the valley—also supported by Moylan—which suggests that his time estimate begins only after Reno had reformed his battalion into line formation—at 1:25 p.m.—about a third of a mile west of Ford A. This time estimate also places Reno's halt to form his dismounted skirmish line at about 1:30 or 1:32 p.m. Private Taylor's watch time estimate, along with One Bull's, both support a Reno advance toward the village at a time between 1 p.m. and 2 p.m., with Wallace's time, as usual, much later than all others.

At 1:30 p.m., Custer's command would depart from the bluffs near Sharpshooter Ridge, as seen by Lieutenant Varnum. Benteen's battalion would be just nearing the end of his lengthy water break at the eastern morass about 4.5 miles east of Ford A. The advance packs would just be arriving at the eastern morass, with the head of the main pack train about a quarter mile east of the junction of No Name Creek and Ash/Reno Creek.

CONCLUSION

1:30 p.m. (H = 5 hours): Reno halts his battalion about two hundred yards northwest of the timber position and dismounts his battalion.

1:40 p.m. (H = 5 hours and ten minutes): Reno would order the timber position cleared by First Sergeant Ryan's squad of ten men. At the same time, the rest of his battalion deployed into skirmish position. Then it advanced several hundred yards up to the point of timber at the Garryowen Bend where the old Garryowen store once stood at the narrowest point across the valley floor. The process would encompass approximately ten minutes of time, bringing Reno's valley fight to begin at 1:40 p.m.

Custer's command would be in Western Coulee at this time; Benteen's battalion would have left the eastern morass and was near the first (eastern) lone tepee location about 3.75 miles east of Ford A. The advance packs would be just finishing watering at the eastern morass, and the head of the main pack train and rear guard would have just passed the junction of No Name Creek along the main trail and approaching the eastern morass about 4.5 miles east of Ford A.

13. TRUMPETER MARTIN'S RIDE TO BENTEEN

The timeline for Trumpeter Martin's ride to Benteen with Custer's last message is discussed and analyzed from participant recollections and statements:

Colonel Benteen (RCOI):

Question: Then would you think it was about 3 o'clock when you met Major Reno on the hill?

Benteen: I think so. ...

Question: State at what time of day ... you received that order [i.e., from Trumpeter Martin] from General Custer.

Benteen: It was about 3 o'clock.

ANALYSIS

Colonel Benteen must realize he cannot simultaneously be in two locations—separated by over two miles. Benteen is found here making conflicting time estimates when he testified at the Reno Court of Inquiry that he united with Reno's battalion on the bluffs at 3 p.m. while at the same time also receiving Martin's message from Custer along Ash/Reno Creek over 2 miles away. Just as Benteen's claimed arrival on the bluffs at 3 p.m. is not supported by our primary sources, Benteen's alleged 3 p.m. meeting with Martin on Ash/Reno Creek is not supported by the evidence and should thus be rejected.

2) Private Golden (*Benteen-Goldin Letters*): Martin and myself went back to the point where he turned back with Custer's message to Benteen, which was not far from the point where the trail of the White Horse Troop left the main column and swung to the left and entered the valley by way of ... a dry creek bed. The main trail must have kept on northerly, back of the bluffs (p. 33).

ANALYSIS

Goldin appears to be describing to Benteen a point along Western Coulee (his "dry creek bed"), which would separate from the other four companies, who would continue moving northwards into North Medicine Tail Coulee just behind the Luce Ridge complex. At this junction in Western Coulee, the trail of E Company, serving as Custer's personal escort, moved from the command as Custer began his leader's recon shortly after Martin left on his ride to Benteen with Custer's last message. This is the point where Goldin is likely claiming as Martin's departure from Custer's command. It is interesting to note that Benteen never contradicted Goldin on this recollection, despite the twenty-plus years later allegation by Martin that Benteen supposedly identified his departure point as being at a point in a Coulee about six hundred yards (or a third of a mile) east of the river (which many assume to be at Ford B).

3) Lieutenant Godfrey (*Godfrey's Field Notes*): After watering, they continued their march very leisurely and not long after watering, a trumpeter brought a note from Colonel Cooke that the village was in front and to bring up the packs. We then increased our gait.

ANALYSIS

Godfrey tells us that not long after watering their horses—and this would be the second watering at the Hartung morass, although Godfrey is diplomatic enough not to point that fact out—Benteen meets up with Martin. This would be about a mile west of the Hartung morass and a quarter mile west of the second (western) lone tepee, about 1.5 miles east of Ford A.

4a) Lieutenant Edgerly (RCOI): They went about one mile when they met Trumpeter Martin. Benteen showed the message to Weir and Edgerly; before they met Martin a sergeant from C Company rode up and told Benteen Custer wanted him to bring up the packs. ... Edgerly thought this sergeant was either

Kanipe or Hanley; he thought the latter. Edgerly thought this was before they reached the morass, just a little time before Martini came up.

ANALYSIS

Edgerly tells us that Benteen met up with Martin about one mile west of the Hartung morass, about 1.5 miles east of Ford A. Edgerly also indicates that Benteen met Sergeant Kanipe just before they reached the morass. This would be the Hartung morass, about 2.5 miles east of Ford A, not the eastern morass, a further 2 miles east of the Hartung morass, or about 4.75 miles east of Ford A.

4b) Lieutenant Edgerly (RCOI): Edgerly heard Martin speaking to an orderly. "He was laughing and seemed very much elated, said it was the biggest village he ever saw, that they had found the Indians all asleep in their tepees, that Major Reno was charging it and killing everything, men, women and children." There was no halt or delay.

ANALYSIS

Edgerly indicates that Martin was heard to have told an orderly that they had struck the biggest village he had ever seen, that Custer had taken the village by surprise, and that Major Reno was charging the village with his battalion. This is a very important revelation, for it is the only indication we have that Martin mentioned that Reno's battalion had apparently separated from Custer's command to attack the village. Edgerly indicates that Benteen did not delay or halt his advance after receiving Martin's note.

5a) Trumpeter Martin (Camp Int., *Custer in 76*): says it took him 15-20 minutes to reach Benteen from [the] top of the ridge. [He] says he rode back 3/4 [of a] mile to reach the top of the ridge.

ANALYSIS

Martin gave various locations over the years as to the whereabouts of his departure point from Custer's command. These locations would include a) Medicine Tail Coulee - 600 yards east of Ford B; b) Medicine Tail Coulee - 3/4 mile east of Ford B; c) 3/4 of a mile north of Sharpshooter Ridge; & d) 600 yards north of Sharpshooter Ridge. So which of these 4 distinct locations is the correct departure point? We get a very strong clue from Camp's interview with Martin above. Basing calculations on Martin's starting point being his observation point on Sharpshooter Ridge, Martin tells Camp that he rode back 3/4 of a mile to reach the top of the ridge from which Custer first viewed the Indian village before Reno's attack. This distance estimate supports his earlier estimate made at the RCOI of a similar distance. At the RCOI, Martin also added 600 yards as his possible distance from this hill, but here in the Camp interview, he discards the conflicting 600 yards and sticks with his original 3/4 of a mile distance that he testified to at the RCOI. This is the distance we will use to calculate his time to return to the hill from whence Custer first viewed the village. Riding at a trot, it would take Martin about 7-8 minutes to reach the top of this hill—Sharpshooter Ridge).

5b) Trumpeter Martin (RCOI): says it took him about 45 minutes to reach Benteen from where he left Custer; he met Benteen 300 to 400 yards above the place where they watered, west of the lone tepee, which was not in sight; he claims he saw Reno's command fighting in the valley roughly between Sharpshooter Ridge and Reno Hill, then met Benteen about 15 or 20 minutes after seeing Reno fighting; he thought he traveled about 2 miles to reach Benteen; he says it took Benteen about 45 minutes after his meeting him to when they arrived on the bluffs; he says that McDougall was only 150 yards behind Benteen after delivering his note.

ANALYSIS

Martin contradicts his own testimony at the RCOI when he first states that it took him about 45 minutes to reach Benteen from where he left Custer, then later indicates he rode back 3/4 of a mile (at a trot) which would take him about 7-8 minutes to reach the hill where Custer had first viewed the village, then estimated it took him about 15-20 minutes more before he reached Benteen. This estimate puts Martin's total time for his ride between 22 to 28 minutes, which is a far more accurate time assessment than his previous 45 minutes. The latter time simply does not fit well with other participant recollections as to when Martin reached Benteen's command.

Now Martin gives us a very important clue that allows us to pinpoint with some degree of accuracy the location on Ash/Reno Creek where he met up with Benteen's command. He says he met Benteen 300 to 400 yards above—or east of—the place where they watered, which would likely be at the Echleman Ford on Ash/Reno Creek, located 1.2 miles east of Ford A, which was not in sight. This location allows us to determine that the distance from the top of the bluffs to the place where Martin meets Benteen a quarter mile east of the Echleman Ford to be between 2.5-3 miles, depending on where Martin rode down to Ash/Reno Creek. Martin claims it took him between 15 to 20 minutes to reach Benteen from this hill on the bluffs.

Martin tells us that he saw Reno's command fighting in the valley, which would likely be between 1:45 and 1:50 p.m. as he crossed over the top of the bluffs between Sharpshooter Ridge and Reno Hill, meeting up with Benteen 15 to 20 minutes later, or between 2:05 and 2:10 p.m. just after Benteen had finished his second watering at the Hartung morass. He would meet up with Benteen about a mile west of the Hartung morass and about 1/4 mile west of the second (western) lone tepee.

Martin then estimates his distance traveled to reach Benteen as about 2 miles, which is a fairly close distance if he were estimating his distance from Reno Hill to where he met Benteen east of the Echleman Ford, but well short of his actual distance from whence he left Custer near Western Coulee.

Martin then claims it took Benteen about 45 minutes after Martin delivered his message, to reach Reno on the bluffs. This time estimate is almost certainly false, as it would place Benteen's pace—which all other participants claim was increased—to a snail's pace of a 4 mph walk. Martin was likely attempting to say that it was about 45 minutes from whence he left Reno Hill on his ride back to Benteen until his return to the same point on the bluffs, but his language deficiencies may have contributed to his misunderstanding the question asked of him.

Martin claims that McDougall was only 150 yards behind Benteen after he delivered his note, but this is very likely a mistake on Martin's part. He mistook Captain McDougall and the Rear Guard with the advance pack mules under Lieutenant Mathey who had just about caught up to Benteen's battalion again after they had stopped for their second water break at the Hartung morass.

5c) Trumpeter Martin (*The Custer Myth*, Graham Int. 1922, with my annotation in brackets): When Martin reached the top of the hill [note: likely Sharpshooter Ridge] he looked down and saw Reno in action. He claimed it had been no more than 10 or 15 minutes since he was on the hill with Custer and now he saw lots of Indians and Reno on a skirmish line.

ANALYSIS

Martin indicates that upon his return to the top of the bluffs, it was no more than 10-15 minutes since he was on that spot with Custer. This indicates an advance and return of between 5 to 7.5 minutes each way at a trot, which is further proof that Martin did ride about 3/4 of a mile before being given his message for Benteen. When he returned to the top of Sharpshooter Ridge, he could see lots of Indians and Reno on his skirmish line. The time would be about 1:45 p.m. or 15 minutes after Martin left the bluffs with Custer at 1:30 p.m., just as Reno was halting to dismount his command by the Garryowen Bend of the river.

6) Private Windolph (*I Fought With Custer*): Trumpeter Martini figured it was maybe a mile on farther downstream, to the northward from our hill [i.e., Reno Hill], to the spot where he had been ordered to report to Benteen.

ANALYSIS

Private Windolph tells us that Trumpeter Martin indicated that he was given his message for Benteen about a mile north of Reno Hill. This distance would support Martin's RCOI testimony, indicating he had moved about three-quarters of a mile north of the hill before being sent back to Benteen. This observation point was roughly one-third mile north of Reno Hill, so Windolph's distance of about a mile north of Reno Hill would fit Martin's three-quarters of a mile estimate very nicely.

OVERALL ANALYSIS

If it took Martin seven to eight minutes to ride three-quarters of a mile back to the hill, then one must also factor in that it took Custer's command roughly seven to eight minutes to ride north of that hill to reach the position from which Martin was sent back. It took Martin about fifteen minutes from the time Custer left the hill from which he viewed the valley until Martin returned to it again on his way back to Benteen. In a previous section of the timeline, it was determined that Custer's command left Sharpshooter's Ridge at about 1:30 p.m., just as Reno was halting to dismount his command by the Garryowen Bend of the river in the valley. Thus, Martin returned to this hill at about 1:45 p.m., or roughly five minutes after the start of Reno's valley fight. Martin's ride across the top of the bluffs between

Sharpshooter Ridge and Reno Hill, where he observed Reno fighting in the valley, likely occurred between 1:45 and 1:50 p.m., during which time Reno's command was still on the skirmish line in the valley. Martin would see nothing of Reno's retreat into the timber at this time. Martin then reaches Benteen's command at 2:10 p.m. about a quarter mile west of the second (western) lone tepee, roughly 1.5 miles west of Ford A.

At the time Martin reaches Benteen at 2:10 p.m., Reno's battalion has fallen back to the timber and is just now preparing to retreat from the valley. According to Martin's recollection, the advance pack mules are only 150 yards behind Benteen's battalion at this point, with the head of the main pack train having just passed the first or eastern lone tepee about 3.75 miles east of Ford A.

Regarding Custer's command, we must now enter the realm of speculation, as our sources are hazy and contradictory. According to this model, Custer's command is divided in Medicine Tail Coulee, with four companies behind the Luce Ridge complex in North Medicine Tail Coulee about three quarters of a mile east of Ford B, while Custer himself is atop upper Boyer's Bluff with the Crow scouts, as the Gray Horse Troop waits at the foot of the bluff in Western Coulee below him. At this point, Custer begins to realize the imminent danger that presents itself to Reno's command.

<u>CONCLUSION</u>

1:38 p.m. (H = 5 hours and eight minutes): Trumpeter Martin leaves Custer's command about three-quarters of a mile north of Sharpshooter Ridge.

Between 1:45 and 1:50 p.m. (H = 5 hours and fifteen to twenty minutes): Martin reaches the top of the bluffs by 1:45 p.m. and views the Reno fight as he crosses the bluffs.

2:10 p.m. (H = 5 hours and forty minutes): Martin reaches Benteen's battalion 1.5 miles east of Ford A and delivers Custer's message to Benteen.

14. BOSTON CUSTER'S RIDE

A great deal of mystery surrounds where and when Boston "Boss" Custer left the main command and rode back to the rear, apparently to change to a fresh pony before entering combat with Custer's five companies. Where he did this and when he returned to the command has been debated for years. The participants of the battle give us several clues that clear up this mystery. Here's what they said:

1) <u>Trumpeter William Hardy</u>–A Company (Camp Interview, *Camp & Custer at the Little Big Horn*): Boss Custer was with the 5 companies when Reno separated, and Boss said he was going to where the fighting would be. Haddon had Boss Custer's extra pony. Boss had two Indian ponies ...

ANALYSIS

Trumpeter Hardy tells us that Boston Custer was with Custer's five companies when Reno's battalion separated from Custer's command. He does not specify where this separation point was. He could have been anywhere from the first lone tepee location 3.75 miles east of Ford A—where Reno's command first crossed over Ash Creek and then pulled ahead of Custer's five companies—all the way down to the Echleman Ford, 1.2 miles east of Ford A, when Reno left Custer's command to advance into the valley. Based on time distance factors, it would appear that Boston likely left Custer's command closer to the first lone tepee at some point between four and 3.5 miles east of Ford A, although it is possible that he left a short distance west of that point. There is too little information to speculate more about this score.

According to Hardy's account, Boss was heard to say that he was going to where the fighting would be, which would suggest that he may have been heard to utter this sentiment while A Company—in Reno's battalion—was passing ahead of Custer's command just a short distance west of the first lone tepee. Hardy's reference to a Private Haddon holding Boston Custer's spare Indian pony suggests that Boston's reason for leaving the command at this point was to ride back and secure a fresh mount before the command entered into combat.

It is important to note that there is no record of a Private Haddon in the 7th Cavalry. A number of battle students, uncomfortable with the account given by Private Hardy, attempt to discredit his entire recollection by focusing on this one faux pas. We call this a clear example of throwing the baby out with the bathwater—something which should never be done in historical research, but is a common tendency among amateur buffs in which a solitary mistake is seized upon to taint the entire credibility of the source. I would suggest that Private Hardy likely referred to Private William Hardden of D Company as the trooper holding Boston Custer's spare mount. A simple spelling error in the translation of his account would explain the discrepancy.

2) Saddler John Bailey – B Company (Camp Interview, *Custer & Company*): Bailey says that Boss Custer did not get back to the pack train. He said if he exchanged horses, he must have done it in advance of the pack train.

ANALYSIS

Saddler Bailey's statement again suggests that Boston Custer left Custer's five companies on the trail to change to a fresh mount. He indicates that Boston likely did this some distance ahead of the pack train, perhaps among the group of advance pack mules being led by Lieutenant Mathey.

3) Captain Thomas McDougall (Camp Interview, *Custer in 76*): says he saw and talked with him when Boston Custer came back to the pack train.

ANALYSIS

The above recollection by Captain McDougall, taken together with Saddler Bailey's statement, suggests that Captain McDougall was traveling some distance ahead of the main pack train, likely with the group of advance ammunition mules at this point on the trail.

4a) Private Fremont Kipp–D Company (*Walter Camp Papers,* Denver Public Library): said that the General had detached Boston Custer to act as a messenger between the other battalions and was with the main column from the start.

4b) Private Fremont Kipp–D Company (*Walter Camp Papers,* Denver Public Library): says Custer sent Boston Custer with a message to either Reno or Benteen, [or the pack train], and this [explains] how Boston was in the rear.

ANALYSIS

Private Kipp's recollection above supports Trumpeter Hardy's statement that Boston Custer was riding with the main column and was not traveling with the Pack Train, as some battle students believe.

5) Lieutenant Winfield Edgerly (The Gibson and Edgerly Narratives, *Vanishing Victory*, p. 39): Edgerly said that "He [Boston Custer] had stayed back with the pack train and was now hurrying up to join the General's immediate command. He gave me a cheery salutation as he passed and, with a smile on his face, rode to his death."

ANALYSIS

While Benteen's battalion is watering their horses at the eastern morass—4.75 miles east of Ford A—Lieutenant Edgerly sees Boston riding past their battalion there. He says that Boston was "now hurrying up," thus suggesting that Boston was riding past them at either a fast trot or a gallop. Edgerly assumes that Boston had been riding with the pack train since the divide, but this would be a false assumption, contradicted by the recollections of both Trumpeter Hardy and Private Kipp.

6a) Trumpeter John Martin (Camp Interview, *Custer in 76*): met Boston Custer on his ride back to the top of the bluffs shortly after receiving his message for Benteen. Boston pointed out to him that his horse was limping. This occurred halfway between "medium coulee and Weir Hill.".

6b) Trumpeter John Martin (Graham Interview, *The Custer Myth*): Martin says that just before he got to "the hill," he met Boston Custer riding at a run. Boston asked him where the General was, and Martin replied, "Just over that ridge."

ANALYSIS

Martin tells us that a few minutes after receiving his message for Benteen, or just before he got to the hill from where Custer first viewed the village fifteen minutes earlier, he met up with Boston Custer riding toward him at a run or gallop. I believe the hill that Martin refers to is at the southern end of Sharpshooter Ridge (in other accounts, Martin describes this hill as a ridge). Many battle students are confused by Martin's reference to "medium coulee" and "Weir Hill." I believe Medium Coulee is a reference to Cedar Coulee, as Martin has marked on Walter Camp's map. Weir Hill is not a reference to the Weir Peaks. I believe it refers to the south end of Sharpshooter Ridge, where Custer first viewed the Indian village and a location that both Captain Weir and D Company occupied shortly after reaching Reno's position on the bluffs. It was from this "hill" that D Company began their advance to the Weir Peaks.

Martin's reply to Boston's query (regarding the whereabouts of the General) as being over the next ridge was a reference to the eastern arm or loaf of the Weir Peaks, which blocked the view of Custer's command at that point. Martin marked the location of his meeting with Boston Custer on Walter Camp's map, indicating a location on the east slope of Cedar Coulee just shy of the top of Sharpshooter Ridge. From this position, neither Martin nor Boston could see into the area of Western or Medicine Tail Coulee hidden behind the eastern loaf of the Weir Peaks. Boston Custer would be about twelve to fifteen minutes from reaching Custer's command at that point.

7) Richard Roberts (from his monograph, p. 36): "He [i.e., Benteen] could have reached Custer, for Boston Custer came back and got a drink of water from one of Company K's men, who were with Benteen, under Captain Godfrey. I know this, for one of the wounded soldiers told me so, as he lay on his couch at the hospital in the upper post at Fort Lincoln. Boston got back and was found dead near his brother, the General."

ANALYSIS

Although Richard Roberts relates a secondary source account from one of the wounded troopers in the Fort Lincoln hospital, the account tends to confirm the previous primary source accounts in that Boston Custer did stop for a drink of water from a member of Benteen's battalion. Other sources place the location for this event at the morass.

OVERALL ANALYSIS

As the evidence remains sketchy, we can only speculate on the likely departure point of Boston Custer on his ride back to the pack train, either to deliver a message or, more likely, to change to a fresh mount that was being led by Private Hardden. Boston Custer likely left Custer's main command a short distance west of the first lone tepee shortly after Reno's battalion had crossed over Ash/Reno Creek and had taken the advance from Custer's five companies. This would be somewhere between 3.75 to 3.5 miles east of

Ford A at about 12:40 p.m. He would then ride back to meet up with the advance pack mules and meet with Private Hardden, who was leading his spare Indian pony along Ash/Reno Creek. He would meet up with the advance pack mules about six miles east of Ford A, or about one mile east of the junction of Ash/Reno Creek and No Name Creek at about 12:50 p.m. This would be between 2.25 and 2.5 miles at a lope of about ten to twelve miles per hour.

Boston would take about ten minutes to change his saddle and tack and converse briefly with Captain McDougall before riding out again at about 1:00 p.m., again, likely at a lope, witnessed by Lieutenant Edgerly passing Benteen's battalion—which has just reached the eastern morass—about ten minutes later at about 1:10 p.m. Boston, at this point, is 4.75 miles east of Ford A and about 4.25 miles east of Custer's command. It would take Boston, riding at a lope of about ten to twelve miles per hour, about fifteen minutes to reach the vicinity of the second (western) lone tepee, about 1.7 miles east of Ford A, at about 1:25 p.m. At that time, Custer's command would have reached the top of the bluffs by Sharpshooter Ridge and were paused there while Custer got his first view of the Indian village from its summit. Boston likely saw the command on these bluffs and swung to the north at this point to take a shortcut over the North Fork of Ash/Reno Creek and ride directly up the bluffs towards Sharpshooter Ridge.

It would take Boston about fifteen minutes to reach the top of these bluffs—a distance between 1.5 to two miles—reaching the top at about 1:40 p.m., just as the opening shots of Reno's valley fight commenced, and while Benteen's battalion had just passed the first (eastern) lone tepee 3.75 miles east of Ford A. Boston would then meet up with Martin on the northwest slope of Sharpshooter Ridge a short distance from the top, at a point in Cedar Coulee, at about 1:43 p.m. Boston would continue from that point reaching Custer's command about a mile and three quarters to two miles farther on, reaching them about ten to twelve minutes later at about 1:55 p.m.

At this point in time, General Custer was likely already engaged in his leader's recon (a personal reconnaissance of the terrain and tactical situation by the commander) at upper Boyer's Bluff, meaning that Boston Custer would have rejoined the other four companies in North Medicine Tail Coulee or perhaps met up with E Company in Western Coulee. It is clear from this timing and distance analysis that Boston Custer did not witness any portion of Reno's retreat. He would only have seen the very start of Reno's action in the valley as he crossed over the top of the bluffs across Sharpshooter Ridge. Still, he would have been able to relay the message to General Custer that Benteen was less than five miles from the Little Big Horn River and in position to support Reno's advance guard action in the valley.

CONCLUSION

12:40 p.m. (H = 4 hours and 5 minutes): Boston Custer leaves the main command about 3.75 to 3.5 miles east of Ford A or just west of the first (eastern) lone tepee.

12:50 p.m. (H = 4 hours and 20 minutes): Boston reaches the advance pack mules about 2.25 to 2.5 miles farther east—about six miles east of Ford A.

1:10 p.m. (H = 4 hours and 40 minutes): After a ten-minute delay in changing to a fresh mount, Boston rides back towards the main command, passing Benteen's battalion as they stopped for water at the eastern morass, 4.75 miles east of Ford A.

1:25 p.m. (H = 4 hours and 55 minutes): Boston reaches the vicinity of the second (western) lone tepee about 1.7 miles east of Ford A.

1:40 p.m. (H = 5 hours and 10 minutes): Boston reaches the top of the bluffs at Sharpshooter Ridge.

1:43 p.m. (H = 5 hours and 13 minutes): Boston meets up with Trumpeter Martin in Cedar Coulee on the west slope of Sharpshooter Ridge, a short distance northwest of the top of the ridge.

Between 1:50 and 1:55 p.m. (H = 5 hours and 20-25 minutes): Boston reaches Custer's command in either North Medicine Tail Coulee or Western Coulee.

15. THE DURATION OF THE VALLEY SKIRMISH LINE FIGHT

The length of time Reno's battalion stayed to fight on the valley skirmish line has been a subject of some debate among battle students. Reno had three companies, one of which was removed early on to redeploy in the timber position, leaving Reno with two companies to hold the valley skirmish line. The battle participants give us more details concerning these events below:

1) Major Marcus Reno (RCOI): says that G Company was on the skirmish line for fifteen to twenty minutes before they were brought into the timber. He also said (p. 502, Graham) that his remaining troops stayed on the skirmish line for twenty minutes before moving into the timber.

ANALYSIS

Reno implies that G Company was only brought into the timber position about five minutes before his remaining troop(s) fell back from the skirmish line. This time estimate does support the notion that A Company withdrew from the skirmish line about five minutes after G Troop was redeployed to the timber. According to Interpreter Fred Gerard's recollections, A Company likely fell back to a position not in, but close to, the timber, probably just behind a brow or small hillock on the northwest edge of the timber position. Reno's estimate of his valley skirmish line lasting for twenty minutes is supported by a number of other participants.

2) Lieutenant Charles Varnum (RCOI): G Company deploys in the timber after ten to fifteen minutes on the valley skirmish line. Varnum remained on the skirmish line for fifteen minutes before going to the timber. He said Reno's command did not remain in the timber position for more than ten to fifteen minutes. Reno stayed a total of thirty minutes from the formation of the valley skirmish line until they retreated from the timber.

ANALYSIS

Varnum tells us that G Company was sent to the timber position ten to fifteen minutes after the start of the valley fight, then remained in the timber for another ten to fifteen minutes, staying a total of thirty minutes before retreating from the timber, thus suggesting a valley skirmish line duration of between fifteen to twenty minutes for the remaining two companies.

3) Dr. Henry Porter (RCOI): judged the time from the formation of the skirmish line to the time the command left the woods at fifteen to twenty minutes, maybe a bit longer.

ANALYSIS

Dr. Porter's recollection of the entire valley fight lasting no longer than fifteen to twenty minutes is likely inaccurate. I would suggest that a more accurate interpretation of Dr. Porter's account would read thus: judges the time from the formation of the skirmish line to the time the command left for the woods at fifteen to twenty minutes, maybe a bit longer.

4a) George Herendeen (RCOI): estimated that the valley skirmish line lasted about fifteen minutes until it fell back to the timber. He also estimated that it was only twenty minutes from when the skirmish line was formed until it left the timber.

4b) George Herendeen (Camp Interview, *Custer in 76*): Reno's skirmish line lasted twenty minutes before they fell back.

ANALYSIS

Herendeen's time estimates suggest a fifteen-minute duration for the valley skirmish line fight and a five-minute stay in the timber position. Herendeen later modifies his estimate to twenty minutes for the skirmish line before it fell back to the timber position. This discrepancy of five minutes may indicate the difference in time from whence A Company left the skirmish line first to when M Troop fell back about five minutes afterwards.

5) Private Edward Pigford - M Company (Camp Interview, *Custer in 76*): says Reno's skirmish line was no sooner formed than it broke up.

ANALYSIS

Private Pigford's comment above suggests that the valley skirmish line did not last for any length of time.

6) Private William Taylor (*With Custer on the Little Big Horn*): The led horses, under the charge of Lieutenant Hare, were then taken into the woods for greater safety, "... Just how long we remained there I cannot say, but I shall never believe it was over fifteen or twenty minutes at the most."

ANALYSIS

Private Taylor's recollection of staying in the timber with the led horses for fifteen to twenty minutes suggests that this was the length of time he waited before seeing the rest of the command falling back to the timber position, suggesting a valley skirmish line fight of about fifteen to twenty minutes.

7) Private William Morris to M Company (letter to Robert Bruce, 5/23/1928, with my annotations in brackets): [the skirmish line action] it seemed to me at the time to be about fifteen minutes.

ANALYSIS

Private Morris' recollections reference fifteen minutes duration for the skirmish line before A Company likely fell back to the timber area.

8a) Fred Gerard (RCOI): claims the troopers fell back to the timber and remained on the brow fighting for about ten more minutes.

8b) Fred Gerard (Camp Interview, *Custer in 76*): Reno's skirmish line lasted about five minutes. Less than five minutes after entering the timber, the men began to bolt from it in disorder and spur their horses toward Ford A.

ANALYSIS

Gerard's confusing statements suggest that Reno's skirmish line lasted about five minutes before G Company was removed from it and taken into the timber position, likely when Gerard left the open plains and accompanied this troop back to the greater safety of the timber position. Gerard then tells us that troopers fell back to the brow (on the northwest edge of the timber position) and remained there fighting for about ten minutes longer. In a later interview with Walter Camp, he indicated that within five minutes of the last troopers entering the timber position, they ran for their horses and bolted from the timber. Gerard does not give us a time estimate for the length of the valley fight itself.

9) Lieutenant Carlo De Rudio (RCOI): the valley skirmish line lasted about ten to twelve minutes; he says the command left the timber after about another ten minutes; he estimates about twenty to thirty minutes from the beginning of the valley skirmish line until the time they left the timber.

ANALYSIS

De Rudio tells us that the valley skirmish line lasted about ten to twelve minutes before G Company was withdrawn from the line and taken into the timber position. This would also be when De Rudio left the skirmish line to retreat to the timber. He estimated they remained in the timber position for another ten minutes before the rest of the command fell back. They retreated from the valley shortly thereafter, spending twenty to thirty minutes in the valley fight.

10) Captain Myles Moylan (RCOI): claims that A Company remained on the skirmish line for forty minutes. He said G Company was sent into the timber after ten minutes to protect the horses. He then implied that A and M Troops remained on the valley skirmish line for another twenty-five to thirty minutes longer.

ANALYSIS

Captain Moylan's estimate of ten minutes before G Company was redeployed into the timber is supported by other accounts and is likely an accurate assessment. Moylan's forty minutes time estimate for the valley skirmish line very likely does not indicate the actual start of the valley fight but rather begins at the time Reno first dismounted his command, then moved forward to deploy at the Garryowen Bend of the river, as well as the time spent retreating to and fighting from the timber position.

11) Lieutenant Luther Hare (RCOI): It was thirty to forty minutes from the time Reno halted to when he left the timber (p. 239 and 262, Graham).

ANALYSIS

Lieutenant Hare, like Captain Moylan, above, indicates a total time for the Reno Valley fight as about 40 minutes, beginning with his halt to dismount his command at about 1:30 p.m.—referencing the actual start of the valley fight roughly ten minutes later after Reno's advance and deployment at the Garryowen Bend of the river. Hare's reference of thirty minutes likely refers to the actual time spent fighting in the valley before they retreated from the timber position.

12) Sergeant Ferdinand Culbertson – A Company (RCOI): the skirmish line lasted about thirty-five minutes; Reno spent five to ten minutes in the timber position before retreating from the valley.

ANALYSIS

Sergeant Culbertson's time estimates of five to ten minutes in the timber position are well supported by other participant accounts. His reference to a skirmish line lasting for thirty-five minutes is likely a reference to when Reno first halted to dismount his command and not at the actual start of the skirmish line action once they were deployed at the Garryowen Bend of the river. His reference to a total time of forty to forty-five minutes for Reno's valley fight—to the time Reno first halts to his retreat from the valley—supports the recollection of both Moylan and Hare.

13) Private Thomas O'Neill – G Company (Camp Interview, *Custer in 76*): Reno's valley skirmish line lasted no more than twenty minutes.

ANALYSIS

Private O'Neill, despite being a member of G Company, which left the skirmish line early, gives an accurate time estimate for the total length of the valley skirmish line from its opening shots until it retreats to the timber.

14) Stanislas Roy – A Company (Camp Interview, *Custer in 76*): says they were on the skirmish line for about twenty minutes.

ANALYSIS

Corporal Roy gives an accurate time estimate for the total length of the valley skirmish line from its opening shots until its retreat to the timber.

15) Little Voice – Hunkpapa (Hardorff, *Indian Views of the Custer Fight*): In less than forty minutes, the fight with Reno was over.

ANALYSIS

Little Voice estimates the Reno Valley Fight's duration to be less than forty minutes. The Participant Timeline suggests that this is an accurate estimate of the Valley Fight, beginning at roughly 1:30 p.m. and ending just before 2:10 p.m.

16) Captain Thomas French – M Company (from a private letter from General Charles Roe, 20 April 1906, to Dr. John M. Cooke, from the George Kush Collection) "Captain French told me himself that the fighting in the timber was inconclusive, with few casualties & only one man killed. He said that they (the troops) had not been engaged twenty minutes when Reno lost his head completely & gave the command to mount & then dismount, not once but three times in succession before screaming, 'It is all up with us boys. Follow me, it is every man for himself now.' "

ANALYSIS

According to the estimate given by Captain French to General Charles Roe, Reno's entire valley fight lasted about twenty minutes before he called for a retreat from the timber position. This time estimate supports those given by Lieutenant De Rudio, Lieutenant Varnum, Lieutenant Hare, Dr. Porter, Fred Gerard, George Herendeen, Corporal Stanislas Roy, Private O'Neil, Private Taylor, and Private Morris. Only Reno, Moylan, and Sergeant Culbertson gave a longer time estimate for the valley fight before the retreat began.

OVERALL ANALYSIS

Of the five participant estimates given for time elapsed before G Troop was taken off the valley skirmish line to redeploy into the timber, three of them fall within the range of ten to twelve minutes (Moylan, De Rudio, and Varnum), with the remaining two participants providing unsupported estimates of five minutes (Gerard) and twenty minutes (Reno). Focusing, therefore, on the three mutually supportive accounts above would suggest a time of about ten minutes after the start of the valley skirmish line deployment before G Troop was redeployed into the timber position, taking Lieutenant De Rudio along with them, and Lieutenant Varnum a few minutes later.

The participants tell us that the valley skirmish line, now composed of the remaining two companies of Reno's battalion—A and M Troops—along with a small detachment of G Co., lasted fifteen to twenty minutes before falling back to the timber position. We have eight participant time estimates that fall within this range, with two more suggesting it without explicitly saying so. As indicated before, Moylan, Hare, and Culbertson's extended time estimates for the valley fight likely include the additional ten minutes of time that included Reno's first halt and dismount of his command before they advanced several hundred yards dismounted to deploy their skirmish line at the Garryowen Bend of the river before the actual fighting began. The sources suggest that A Company fought on the skirmish line for about fifteen minutes before falling back to the timber. M Company fought for about twenty minutes before falling back to the timber last.

The participants all gave estimates of five to ten minutes for the duration of the fight at the timber once A and M Troops fell back from the valley skirmish line. It is likely that A Company, having fallen back first, may have fought for about ten minutes or so, partly behind the embankment edge of the timber position, before falling back to their horses. In contrast, M Troop may have fought for no more than five minutes upon reaching the northwest brow of the timber position.

We previously determined the start of Reno's valley fight at about 1:40 p.m., about ten minutes after Reno halted to dismount his battalion about a hundred yards southeast of the Garryowen Bend of the river at 1:30 p.m. The participants tell us that G Troop was taken off the valley skirmish line at about 1:50 p.m. with the remaining two troops (A and M Company) remaining on the skirmish line until 1:55 to 2:00 p.m. whence they fell back to the timber, first A Company followed shortly thereafter by M Company. The timber position was defended by these two troops for no longer than five minutes before they fell back to their horses and abandoned all efforts to defend their position at 2:05 p.m. Reno's battalion—more or less—leaves the timber position to retreat from the valley at 2:10 p.m., a duration of less than forty minutes for the Reno Valley Fight, as estimated by the Hunkpapa warrior Little Voice.

CONCLUSION

1:30 p.m. (H = 5 hours): Reno halts to dismount his command one hundred yards from the end of the Garryowen Bend of the river. Custer's command is just leaving the bluffs. Benteen is just leaving the eastern morass.

1:40 p.m. (H = 5 hours and 10 minutes): After Reno advances and deploys his battalion in a skirmish line across the valley floor at the Garryowen Bend of the river we come to the start of the valley fight. Custer has reached Western Coulee. Martin is riding back with his message for Benteen, about 3 minutes from passing Boston Custer. Benteen has just passed the first or eastern lone tepee location and is about 3.75 miles east of Ford A.

1:50 p.m. (H = 5 hours and 20 minutes): G Company is taken off the line and into the timber. Most of Custer's command is out of sight in North Medicine Tail Coulee behind Luce Ridge, while Custer has begun his leader's reconnaissance from Western Coulee with E Company as an escort and has now reached upper Boyer's Bluff. Benteen has just met with Sergeant Kanipe as he approached the Hartung morass three miles east of Ford A.

1:55 p.m. (H = 5 hours and 25 min.): A Company falls back to the timber embankment. Custer is soon spotted at a river crossing below by Private Peter Thompson. Boston Custer has reached Custer's four troops in North Medicine Tail Coulee. Benteen's battalion is watering again at the Hartung morass.

2:00 p.m. (H = 5 hours and 30 minutes): M Company falls back to the timber position. Custer has returned to the top of upper Boyer's Bluff and realizes Reno's command has fallen back to the timber position with no sign of Benteen's battalion having yet arrived in the valley to provide him with support. Benteen's battalion has just left the Hartung morass 2.5 miles from Ford A.

2:05 p.m. (H = 5 hours and 35 minutes): Reno's troops abandon their defense and race for their horses in the bottom. Custer has left upper Boyer's Bluff and rejoined E Company and are making their way down Western Coulee to the rest of his command in North Medicine Tail Coulee. Mitch Boyer arrives at upper Boyer's Bluff. Benteen's battalion has reached the second (western) lone tepee location 1.7 miles from Ford A.

2:10 p.m. (H = 5 hours and 40 minutes): Reno's battalion abandons the timber position and retreats from the valley. Custer has reached the junction of Western Coulee, Medicine Tail Coulee, and North Medicine Tail Coulee. Boyer witnesses the beginning of Reno's retreat from the timber. Benteen has just met Trumpeter Martin with Custer's last message about 1.5 miles from Ford A. The advance pack mules, under Lieutenant Mathey, carrying most of the ammunition, are fifteen yards behind Benteen's battalion at this point.

Appendix A: Participant Timeline of the Battle of the Little Big Horn

15a. INDIAN TIME ESTIMATES OF RENO'S VALLEY FIGHT

The following is a list of Indian time estimates on when Reno first began his attack in the valley, keeping in mind that almost all of these times referenced local sun time, about an hour earlier than the 7th Cavalry Headquarters watch time, with my annotations in brackets:

1) Standing Bear (Minneconjou/Oglala): Reno fight "began about noon." [or about 1 p.m. Headquarters watch time, referencing the time when Reno's battalion crossed the river at Ford A.]

2) Antelope: 10 a.m. [i.e., 11 a.m. Headquarters watch time]

3) Dewey Beard – aka. Iron Hail (Minneconjou): noon to 1 p.m. [Note: 1 p.m. to 2 p.m. Headquarters watch time. Beard's time estimate of Reno's valley fight beginning at some time between 1 p.m. and 2 p.m. watch time is an accurate time estimate of when Reno's fight started. The Participant Timeline suggests that the start of Reno's fight took place between 1:30 and 1:40 p.m.]

4) Crazy Horse (Oglala): 11 a.m. [i.e., noon Headquarters watch time].

5) Fears (Respects) Nothing (Oglala): 1 p.m. [or 2 p.m. Headquarters watch time; a fairly accurate time assessment based on the Participant Timeline.]

6) Flying Hawk (Oglala): 8 or 9 a.m. [9 or 10 a.m. Headquarters watch time]

7) Gall (Hunkpapa): before 2 p.m. [Gall's account is one of the few Indian accounts which likely reference a time based on Headquarters watch time, as opposed to local sun time. If so, it indicates a very close agreement with the Participant Timeline estimates.]

8) Hump (Minneconjou): noon [or 1 p.m. Headquarters watch time; suggesting that Hump's time estimate was based on Reno's crossing the river at Ford A.]

9) Kill Eagle (Blackfeet): noon [or 1 p.m. Headquarters watch time; suggesting that Kill Eagle's time estimate was based on Reno's crossing the river at Ford A.]

10) Lights (Minneconjou): 9:30 a.m. [or 10:30 a.m. Headquarters watch time]

11) Low Dog (Oglala): noon [or 1 p.m. Headquarters watch time; suggesting that Low Dog's time estimate was based on Reno's crossing the river at Ford A.]

12) One Bull (Hunkpapa): 2 p.m. [Note: I would speculate that One Bull's 2 p.m. time estimate is, like Gall's, one of the few Indian time estimates based on Headquarters watch time and not local sun time. If so, it is a close approximation to the time estimates suggested by the Participant Timeline as to the start of Reno's valley fight at 1:40 p.m.]

13) Red Feather (Oglala): before 10 a.m. [or before 11 a.m. Headquarters watch time]

14) Red Horse (Minneconjou): noon [or 1 p.m. Headquarters watch time; suggesting that Red Horse's time estimate was based on Reno's crossing the river at Ford A.]

15) Runs the Enemy (Two Kettles): 10 a.m. [or 11 a.m. Headquarters watch time]

16) Sitting Bull (Hunkpapa): 2 p.m. [Note: As with Gall and One Bull's accounts, these three Hunkpapa warriors all seem to reference a Headquarters watch time rather than a local sun time for their time estimates. If so, it is a close approximation to the time estimates suggested by the Participant Timeline as to the start of Reno's valley fight at 1:40 p.m.]

17) Waterman (Arapaho): 9 a.m. [or 10 a.m. Headquarters watch time]

18) White Cow Bull (Oglala): after midday [after noon or after 1 p.m. Headquarters watch time; this is an accurate time estimate referencing the time just after Reno's battalion had crossed the river at Ford A.]

19) Yellow Nose (Cheyenne/Ute): noon [or 1 p.m. Headquarters watch time; suggesting that Yellow Nose's time estimate was based on Reno's crossing the river at Ford A.]

20) Wooden Leg (Cheyenne): noon to 1 p.m. [Note: 1 p.m. to 2 p.m. Headquarters watch time. Wooden Leg's time estimate of Reno's valley fight, like Dewey Beard's, beginning at some time between 1 p.m. and 2 p.m. Headquarters watch time is an accurate time estimate of when Reno's fight started. The Participant Timeline suggests that Reno's fight began between 1:30 and 1:40 p.m.]

21) Red Hawk (Nicholas Ruleau Interview) (Oglala): 9 a.m. [or 10 a.m. Headquarters watch time]

22) Iron Hawk (Hunkpapa): 8 or 9 a.m. [9 or 10 a.m. Headquarters watch time]

23) Eagle Elk (Oglala): 8 a.m. [or 9 a.m. Headquarters watch time]

24) Two Eagles (Brulé): about 1 p.m. [or about 2 p.m. Headquarters watch time; a fairly accurate time assessment based on the Participant Timeline.]

25) Left Hand (Arapaho): 9 a.m. [or 10 a.m. Headquarters watch time]

26) Ice – aka. White Bull (Cheyenne): about 11 o'clock [or about noon Headquarters watch time]

27) White Bull (Minneconjou): 9 or 10 or 10:30 a.m. [or between 10 a.m. and 11:30 a.m. Headquarters watch time]

28) Soldier Wolf (Cheyenne): about noon [or about 1 p.m. Headquarters watch time, referencing the time when Reno's battalion crossed the river at Ford A.]

29) Four Woman (Hunkpapa): about noon [or about 1 p.m. Headquarters watch time, referencing the time when Reno's battalion crossed the river at Ford A.]

30) Lone Bear (Oglala): 9 a.m. [or 10 a.m. Headquarters watch time]

31) Goes Ahead (Crow): between 8 and 9 o'clock [or between 9 and 10 a.m. Headquarters watch time]

32) White Man Runs Him (Crow): gave a multiple of time estimates: about 9 or 10 o'clock [i.e., 10 or 11 a.m. Headquarters watch time]; about 9:15 o'clock [or 10:15 a.m. Headquarters watch time]; about 1 p.m. [or about 2 p.m. Headquarters watch time; only this last time estimate of 2 p.m. would support the time recollections of the Participant Timeline as it relates to the start of Reno's valley fight, which indicates it began between 1:30 and 1:40 p.m.]

33) Billy Cross (Two Kettles): 1 p.m. [Note: Billy Cross, a half-breed scout, was likely referencing a Headquarters watch time of 1 p.m. to mark the time that Reno's battalion crossed the river at Ford A, his likely reference point for the start of the valley fight, as this was actually the time when the Ree scouts were attacking the Sioux pony herds.]

34) Billy Jackson (Piegan): It was then about noon day. [Note: Jackson, a half-breed scout, is likely referencing a local sun time of noon as the start of the battle. Headquarters watch time would indicate a time of about 1 p.m.]

OVERALL ANALYSIS

Of the thirty-four Indian accounts listed above for the start of the Reno fight (based on Regimental Headquarters watch time), sixteen accounts refer to the start of the battle at noon or earlier in the morning. Nineteen Indian accounts—with White Man Runs Him falling into both categories; hence a total of thirty-five accounts instead of thirty-four—refer to the start of the battle at some time between 1 p.m. and 2 p.m., coinciding with the recollections of the majority of the 7th Cavalry battle participants, which suggest a starting time between 1:30 and 1:40 p.m. for the Reno valley fight. Of these more accurate time estimates, ten of the nineteen accounts suggest that the battle began about 1 p.m., which likely indicates the time of Reno's battalion crossing the river at Ford A. For these Indians, this crossing would indicate the actual start of the battle as this would be the time for the attack on the Sioux horse herds shortly afterward by Custer's Indian scout contingent. The remaining nine accounts indicate the start of the fight between 1 p.m. and 2 p.m., with six of those nine accounts specifying a time closer to 2 p.m. These time estimates support the valley fight starting between 1:30 to 1:40 p.m., resulting from a close examination of the 7th Cavalry participant recollections. Interestingly, no Indian accounts support Colonel William Graham's time estimation (*The Story of the Custer Fight* and *The Custer Myth*), based on Lieutenant George Wallace's time estimate, for the battle starting at about 3 p.m. in the afternoon.

CONCLUSION

Nineteen out of thirty-five Indian accounts support the evidence that the Reno valley fight began at some time between 1 and 2 p.m. Regimental Headquarters watch time. Ten of these Indian witnesses suggest the battle started with the Ree scouts' attack on the pony herd at about 1 p.m. In contrast, the remaining nine Indian accounts refer to the start of the Reno battle a short time later, with six accounts placing it closer to 2 p.m. These accounts support the start of Reno's attack on the village at 1:40 p.m.

16. RENO'S RETREAT FROM THE VALLEY

The following is an analysis of the participant accounts as to when Reno ended the valley fight and retreated across the Little Big Horn and onto the bluffs:

1a) Fred Gerard (RCOI): claimed it would take Reno's men about four minutes to retreat from the timber before they reached the top of the bluffs.... It was 2:10 p.m. when the command left the timber. Gerard knew this was the time because he looked at his watch; he said again that Reno left the timber at 2 p.m.

1b) Fred Gerard (*The Custer Myth*): Reno then gave the order: *"The Indians are taking us in the rear; mount and charge."* This was then about 1:30 p.m. ...

ANALYSIS

Gerard gives us the most accurate time assessment yet on the time of Reno's retreat from the timber, as he indicates that it is based on an actual watch time of 2:10 p.m. He later indicates a more general time of 2 p.m., but 2:10 p.m. actually fits the evidence provided by a number of independent sources quite nicely for the duration of the Reno valley fight. As far as Gerard's time estimates regarding Reno's retreat from the valley, we must factor in the realization that Gerard, like De Rudio, did not actually participate in this retreat, so their times are only a general estimate from what they could observe from a distance. Gerard suggests that it took the head of Reno's command only four minutes to reach the top of the bluffs after they left the timber position at 2:10 p.m., thus placing the head of Reno's command on the bluffs as early as 2:14 p.m. I seriously doubt this is likely. It is far more likely that the head of Reno's command reached the retreat crossing of the river by 2:14 p.m.

Gerard's second account, given in *The Custer Myth*, occurred years later and is less accurate than his original RCOI testimony. Gerard now uses a local sun time of *"about 1:30 p.m."* to mark Reno's retreat from the timber, which translates into about 2:30 p.m. Headquarters watch time for the retreat, which is a rather large generalization that better refers to the time that most of Reno's command actually reached the bluffs. Gerard's RCOI testimony, based on an actual Headquarters watch time reading of 2:10 p.m., is the better bet here, and is supported by a number of independent sources and fits into the overall evidence at hand, whereas 2:30 p.m. does not.

2) Lieutenant Charles Varnum (RCOI): Reno took six to eight minutes to retreat from the timber to the river crossing.... said it took twenty to twenty-five minutes from the time Reno retreated from the timber till he met Benteen on the bluffs.... He estimated the total length of Reno's valley fight as about thirty minutes.

ANALYSIS

If Reno left the timber at 2:10 p.m., as Gerard's reference to his watch time suggests, Varnum tells us that he reached the retreat crossing of the river at about 2:16 to 2:18 p.m. Varnum further tells us that it was a full twenty to twenty-five minutes from the start of Reno's retreat from the timber until he met up with Benteen on the bluffs; if Reno left the timber at 2:10 p.m., then he met up with Benteen on the bluffs between 2:30 and 2:35 p.m. Varnum also estimates the total length of Reno's valley fight as about thirty minutes. The Participant Timeline indicates that the valley fight began at close to 1:40 p.m. and ended at 2:10 p.m.—or about 30 minutes total.

3) Dr. Henry Porter (RCOI): estimated it took four to five minutes to go from the timber to the retreat crossing between a half to a mile away (p. 167, *The Custer Myth*). It was about one hour from when Reno got his attack orders until he reunited with Benteen on the bluffs. (p. 168, *The Custer Myth*).

ANALYSIS

If Reno left the timber at 2:10 p.m., as Gerard's reference to his Headquarters watch time suggests, then Porter suggests that Reno reached the river crossing at 2:14 to 2:15 p.m. He also claims that only one hour elapsed from the time that Reno got his attack orders (about 12:45 p.m. in the timeline) until he met Benteen on the bluffs, which would put the time of that meeting at 1:45 p.m., which clearly does not fit with other participant accounts. Thus, Dr. Porter is likely mistaken about the duration of the Reno valley fight and retreat to the bluffs as encompassing only one hour of time.

4) Lieutenant Carlo De Rudio (RCOI): claims to have seen three figures he thinks might be Custer, Cooke, and another individual—although they were more likely the three Crow or Ree scouts—about five minutes before Reno left the timber.... He says it took Reno's command about five to six minutes to get across the Little Big Horn in his retreat.... he estimates about twenty to thirty minutes from the beginning of the valley skirmish line (at 1:40 p.m. in the timeline) until the time they left the timber (at 2:10 p.m. according to Gerard's Headquarters watch time recollection)—or thirty minutes duration.... he says Reno left the timber somewhere between 2 and 3 p.m., he could not remember the correct minute, but said it was after 2 p.m.

ANALYSIS

According to Gerard's Headquarters watch time, De Rudio's sighting of the three figures on the bluffs would be about 2:05 p.m. He also indicates that it took Reno about five to six minutes to get his command across the river in his retreat out of the valley, suggesting that the head of Reno's command reached the east bank of the river at about 2:15 to 2:16 p.m. De Rudio's estimate of Reno's retreat from the valley as occurring sometime between 2 and 3 p.m. does support Fred Gerard's more accurate watch based time recollection of 2:10 p.m. marking the start of Reno's retreat from the timber.

5) Sergeant Ferdinand Culbertson (RCOI): it took Reno five to eight minutes to cross the Little Big Horn River.... From the time Reno left Custer at the tepee till he met Benteen on the bluffs was about one-and-a-half hours.... said the skirmish line lasted about thirty-five minutes; later saying that Reno spent five to ten minutes in the timber position before retreating from the valley, for a total time of forty to forty-five minutes in the valley before retreating.

ANALYSIS

Based on Gerard's 2:10 p.m. Headquarters watch time for Reno's departure from the timber, Culbertson suggests that it was 2:15 to 2:18 p.m. when Reno's battalion crossed the Little Big Horn river on its way up to the bluffs. He also claimed that it took one-and-a-half hours from the time Reno left Custer at the second (western) lone tepee (about 12:45 p.m.) until he met Benteen on the bluffs. The time elapsed would be closer to an hour and forty-five minutes according to the Participant Timeline. Culbertson does suggest a total time of forty to forty-five minutes for the valley fight, which fits well if his start of the valley fight was referenced at Reno's dismount and deployment of his battalion at 1:35 p.m. until its retreat from the timber at 2:10 p.m.—a thirty-five-minute duration. Of course, Culbertson may have included the time it took Reno to retreat from the valley in his time estimate for the valley fight.

6) Lieutenant Hare (RCOI): he estimated about one-and-a-half hours from when they left Custer at the tepee (about 12:45 p.m. in the timeline) to when Reno met Benteen on the bluffs (about 2:30 p.m. in the timeline).... he said it was thirty to forty minutes from the time Reno halted (at 1:30 p.m. in the timeline) to when he left the timber (at 2:10 p.m. according to Gerard's Headquarters watch time)—an actual duration of forty minutes. (p. 239 and 262, *The Custer Myth*).

ANALYSIS

The actual time between Reno leaving the second (western) lone tepee and then meeting Benteen on the bluffs—according to our Participant Timeline—would be slightly longer than an hour and a half, putting it closer to an hour and forty-five minutes. Yet, Hare's estimate of forty minutes from the time Reno halted his command to the time he left the timber appears to be very close to the forty minutes arrived at in our timeline.

7) Lieutenant Winfield Edgerly (*Edgerly Papers*): "When Reno left the bottom, Benteen's squadron was in sight, moving to his support."

ANALYSIS

Lieutenant Edgerly tells us that Benteen's battalion was in sight of the river by Ford A and moving to Reno's support in the valley when Reno's battalion left the bottom and began their retreat from the valley.

8) Private August Siefert (Camp Interview, *Custer in 76*): says that three Crow scouts met Benteen on the bluffs.

ANALYSIS

Private Siefert's recollection supports that of the Crow scouts themselves, who met Benteen's battalion some distance south of Reno Hill to direct them up the bluffs. It was likely these same three Crow scouts seen by De Rudio and mistaken for Custer and Cooke near elevation 3411 about five minutes before Reno's retreat from the valley.

9) Private Edward Davern (RCOI): says that Reno spent thirty to forty-five minutes from the time he deployed in the valley to the time he left the timber.... Reno reached the bluffs on his retreat between 2 and 3 p.m.

ANALYSIS

Private Davern tells us that thirty to forty-five minutes elapsed from the time he deployed his command in the valley (about 1:35 p.m. in the timeline) until Reno left the timber (at 2:10 p.m. in the timeline). This duration of time comes to thirty-five minutes, which fits well with Private Davern's recollection. Davern claims Reno's battalion reached the bluffs between 2 and 3 p.m., which supports the timeline estimate of a time between 2:20 and 2:45 p.m.

10) Major Marcus Reno (*Reno's Memoir*): says he fought in the valley for one and a half hours.

ANALYSIS

Reno's time estimate of an hour and a half for his valley fight is not supported by other participant accounts. It would be more accurate to say that it was an hour and a half from the time Reno received his orders to attack (about 12:45 p.m.) until his retreat from the valley (about 2:10 p.m.).

11) George Herendeen (RCOI): estimated it was only twenty minutes from when the skirmish line was formed until it left the timber.

ANALYSIS

Herendeen's recollection of a twenty-minute duration from when Reno first formed his skirmish line (at 1:40 p.m. in the Participant Timeline) until Reno left the timber is short by about ten minutes, as we base Reno's retreat from the timber at 2:10 p.m. according to Gerard's Headquarters watch recollection.

12) Captain Myles Moylan (RCOI, *Army and Navy Journal*): claims that the valley fight lasted forty minutes.

ANALYSIS

Moylan's time estimate of forty minutes for the valley fight would be very close to the actual time if he referred to the time Reno halted to dismount his command (at 1:30 p.m. in our timeline) and ended with Reno's retreat from the timber at 2:10 p.m.—a total of forty minutes.

13) Lieutenant George Wallace (RCOI, with my annotations in brackets): [The Reno valley fight began] sometime after half-past 2 o'clock.... I would say it was half past two when the fight commenced.... The valley fight lasted something like 3/4 of an hour .... it took Reno 3-5 minutes to cross the river in retreat.... Question: How much time elapsed from the time Lieutenant Cook delivered that order till Major Reno had been joined by Captain Benteen's command on the hill? Wallace: It must have been an hour and a half approximately.... somewhere about 4 o'clock or after [that Benteen joined Reno on the bluffs].

ANALYSIS

As usual, Lieutenant Wallace's time recollections are off by either an hour or an hour and a half and are contradicted by a number of other independent sources, including the Indians themselves. Wallace's recollection of the Reno valley fight starting at 2:30 p.m. is off by one hour, with the corrected time being closer to 1:30 p.m. if the start of his valley fight is marked at the time he halts to dismount his battalion. Wallace then gives us an accurate assessment for the duration of the valley fight, being about forty-five minutes, which is fairly close to correct if we base this on a 2:15 p.m. time for the head of Reno's battalion to reach the river on its retreat to mark the end of the fight—a time of forty-five minutes. He claims it took Reno three to five minutes to cross the river in retreat, placing the head of Reno's command on the east side of the river at 2:18 to 2:20 p.m., which compares favorably to other participant accounts. Wallace's estimation of an hour and a half from the time Reno received his attack orders (at 12:45 p.m. in the timeline) until he met Benteen's battalion on the bluffs (at 2:30 p.m. in the timeline) would be closer to an hour and forty-five minutes all told. Wallace's time estimate of a 4 p.m. junction of Reno and Benteen's battalions on the bluffs is contradicted by nearly every other participant account, including Benteen's.

14) Colonel John Gibbon (post battle report: 10/17/1876), with my annotations in brackets: Communication was now soon opened with Major Reno. His command was found intrenched [*sic*] upon the tops of several small ridges, their dead and living horses lying about them, ... We were informed that in this spot they had been surrounded by overwhelming numbers of Indians from the close of Major Reno's charge [to the rear] on the 25th (about 2 1/2 p.m.)....

Appendix A: Participant Timeline of the Battle of the Little Big Horn

ANALYSIS

According to Colonel Gibbon's post-battle report to General Terry, Major Reno ended his charge to the rear, or retreat out of the Little Big Horn valley at about 2:30 p.m., which appears to be about the same time Benteen's battalion arrived on the bluffs to join them.

OVERALL ANALYSIS

After fighting in the valley for thirty minutes, Reno's battalion leaves the timber position and retreats out of the valley at 2:10 p.m., based on Fred Gerard's Headquarters watch time recollection. It would take the head of Reno's battalion about four or five minutes to reach the river crossing at 2:14 to 2:15 p.m. and then ascend the bluffs by about 2:18 to 2:20 p.m. with the rest of his command straggling on afterward for the next thirty minutes or so, until about 2:45 p.m.

CONCLUSION

2:10 p.m.: Reno begins his retreat from the timber position. Custer is now nearing the junction of Medicine Tail Coulee, Western Coulee, and North Medicine Tail Coulee with his escort of E Company. Benteen has just met up with Trumpeter Martin with Custer's message to "Come on" and "Be quick" about 1.5 miles east of Ford A. The advance pack mules are only a hundred and fifty yards behind Benteen's battalion at this point. The three Crow scouts have left elevation 3411 and are now riding south of Reno Hill toward Ford A.

2:15 p.m.: The head of Reno's battalion has now reached the retreat crossing of the Little Big Horn River and is in the process of crossing to the east side. Mitch Boyer has left upper Boyer's Bluff and is riding down Western Coulee to alert Custer regarding Reno's retreat from the valley. Curley leaves Boyer and rides back up the bluffs away from the river.

2:20 p.m.: The head of Reno's command, including Reno himself and Captain Moylan a short distance behind, reach the top of the bluffs. Custer has now been alerted by Boyer as to Reno's plight and sends his escort E Company down to Ford B to demonstrate against the village. At the same time, Custer himself rides into North Medicine Tail Coulee to confer with his squadron commanders Keogh and Yates on this new development, leading to an immediate change of plans. Benteen's battalion has now reached the vicinity of Ford A and spots a portion of Reno's command retreating from the valley and up onto the bluffs. He meets the three Crow scouts who direct him north to junction with Reno's battalion on the bluffs.

17. BENTEEN'S BATTALION ARRIVES ON THE BLUFFS

Based on the participant accounts, it can be determined that Reno's retreat from the valley occurred at about 2:10 p.m., with the head of his battalion reaching the bluffs about ten minutes later or at 2:20 p.m. To support these time estimates, the participants should give us enough evidence to indicate the arrival of Benteen's battalion shortly thereafter. Here's what they had to say about this important event, along with an analysis of each of their accounts:

1) Major Marcus Reno (RCOI - *Army & Navy Journal* and Reno's After Battle Report): Benteen joined him on the bluffs at 2:30 [p.m.]

ANALYSIS

Major Reno gives a specific time of 2:30 p.m. for Benteen's arrival on the bluffs. This strongly supports a timeline estimate of Reno leaving the timber position and retreating from the valley at 2:10 p.m., as Fred Gerard so indicated, based on his Regimental Headquarters watch time.

2a) Lieutenant Charles Varnum (RCOI): Benteen arrives on the bluffs ten to fifteen minutes after Reno... Reno meets Benteen on the bluffs about an hour after he saw the Gray Horse troop (E Company) on the bluffs.... said it took twenty to twenty-five minutes from the time Reno retreated from the timber till he met Benteen on the bluffs.

ANALYSIS

The Participant Timeline indicates that Varnum's sighting of the Gray Horse troop (E Company) took place at about 1:30 p.m. Varnum now tells us that one hour elapsed from that time until Benteen joined Reno on the bluffs, putting the time for that event at 2:30 p.m., which supports Reno's recollection of the same in his own testimony at the RCOI and in his Post Battle Report. He also indicates that Benteen arrived on the bluffs about ten to fifteen minutes after Reno did. If 2:30 or 2:35 was the time Benteen arrived on the bluffs, then Varnum indicates that the head of Reno's command would have arrived there about 2:20 p.m. Varnum also tells us that it took twenty to twenty-five minutes from the time Reno retreated from the timber—at 2:10 p.m. according to the timeline—until he met Benteen on the bluffs. This would put Benteen's arrival on the bluffs at 2:30 or 2:35 p.m.

2b) Lieutenant Charles Varnum: (Camp Interview, *Custer in 76*): Benteen arrived on Reno Hill sometime between ten and twenty minutes after they [Reno's battalion] did.

Appendix A: Participant Timeline of the Battle of the Little Big Horn

ANALYSIS

Most of our independent participant accounts refer to a ten-minute span of time between Reno and Benteen's arrival on the bluffs. Thus, we suggest that Varnum's ten- or fifteen-minute estimate is closer to the truth than his twenty minutes.

3) General Terry (First *Post Battle Report* on June 27, 1876): refers to Benteen joining Reno on the bluffs "from about half-past two o'clock on the twenty fifth".

ANALYSIS

Although General Terry was not a witness to this event, he based his After Battle Report on interviews he held with surviving officers who were there. He concluded that Benteen joined with Reno on the bluffs about 2:30 p.m.

4a) Fred Gerard (RCOI–*Army and Navy Journal*): It was 2:10 p.m. when the command left the timber. Girard knew this was the time because he looked at his watch. He said again that Reno left the timber at 2 p.m.

4b) Fred Gerard (*The Custer Myth*): Reno then gave the order: "The Indians are taking us in the rear; mount and charge." This was then about 1:30 p.m. ... As soon as the hill was gained, Benteen and his command came up, and the demoralization of Reno's men affected his own men, and no attempt was made to go to Custer's aid. They remained where they were, though it was about 2 p.m. ...

ANALYSIS

Fred Gerard's two accounts conflict with each other in terms of time. In his first account, given only three years after the battle at the RCOI, Gerard tells us that Reno left the timber at 2:10 p.m.—a time he based on looking at his watch. Many years later, as an older man with, perhaps a fading memory, Gerard changes this time of Reno's retreat to about 1:30 p.m., with his joining with Benteen on the bluffs at about 2 p.m. This time sequencing does not fit in at all with our independent sources and must therefore be rejected. Gerard's earlier time estimates, given at the RCOI, support most independent participant accounts and are likely quite accurate as he bases them on a watch time.

5) Lieutenant Luther Hare (RCOI): Benteen arrived on the bluffs to join them in about fifteen minutes.

ANALYSIS

If Reno arrived on the bluffs at 2:20 p.m., Lieutenant Hare suggests that Benteen's battalion arrived to join them there about 2:35 p.m.

6) Sergeant Ferdinand Culbertson - A Company (RCOI): Benteen joined Reno on the bluffs five to ten minutes after Reno got there.

ANALYSIS

If Reno got to the bluffs about 2:20 p.m., then Sergeant Culbertson indicates that Benteen joined them there between 2:25 and 2:30 p.m.

7) Corporal Stanislas Roy - A Company (Camp Interview, *Custer in 76*): Benteen joined them on the bluffs in about ten minutes.

ANALYSIS

If Reno reached the bluffs at 2:20 p.m., Corporal Roy indicates that Benteen reached them there by 2:30 p.m.

8) Captain Frederick Benteen (RCOI): estimates his arrival on the bluffs at 3 p.m. but based that estimate on Lieutenant Wallace's 12:10 p.m. departure time for his flank march to the left shortly after crossing the divide.

ANALYSIS

Benteen's time estimate of 3 p.m. is undermined by his attaching its relevance to Lieutenant Wallace's flawed time estimate of 12:10 p.m. for the start of Benteen's flank march shortly after crossing the divide. Benteen is suggesting here that it took his battalion about two hours and fifty minutes to reach Reno Hill when he first left Custer near the divide. If we change Benteen's time of departure to a more logical 11:10 p.m., this would put his arrival on Reno Hill at about 2:00 p.m., which still does not fit with the testimony or recollections of other independent witnesses, as Reno would have still been fighting in the timber position at that time. That said, we must also factor in that elsewhere in Benteen's RCOI testimony, he claimed to have taken a total of three hours from whence he left Custer near the divide until he reached Reno on the bluffs. This would place his arrival time at closer to 2:10 or 2:15 p.m. which is closer to what other participants claimed.

9) Lieutenant George Wallace (RCOI): says that Benteen joined Reno on the bluffs at 3:45 p.m. He also indicated in his testimony that Benteen arrived on the bluffs at 4:00 p.m. Wallace said, "I don't think it was more than 10 minutes till Captain Benteen came up" to join them on the bluffs.

ANALYSIS

Wallace gives us two time estimates for Benteen's arrival on the bluffs: 3:45 p.m. and 4:00 p.m. Both of these time estimates conflict with the time recollections of the other participants regarding Reno's retreat from the timber, as well as all Indian time recollections. Wallace's times are usually off by at least an hour later than all others, but in this case, his time estimates are even more tardy—between an hour and fifteen minutes to an hour and a half. Pushing the time of Benteen's arrival on the bluffs back this amount of time would give the surviving officers of the 7th a handy excuse as to why no effort was made to support Custer's troops engaging the enemy in the north for as long as they did—a span of time that embraced nearly two hours before any serious effort was made to march north.

10) Private William Taylor (*With Custer on the Little Big Horn*, with our annotations in brackets): We had been there [on the bluffs] a very short time when we were joined by Captain Benteen.... This junction with all of Reno's command occurred at "2:30 p.m.", so Reno states.

ANALYSIS

It would appear that Private William Taylor was quoting his 2:30 p.m. for the arrival of Benteen's battalion on the bluffs from either Reno's or Terry's Post Battle Reports.

11) Private Patrick Corcoran – K Company (Camp Interview, *Custer in 76*): Benteen got to the bluffs before all of Reno's men did.

ANALYSIS

Private Corcoran tells us that Benteen's battalion reached the bluffs to join Reno there before all of Reno's men had escaped from the valley and rejoined their battalion. If Benteen's command reached the bluffs by 2:30 p.m., stragglers from Reno's battalion were still ascending the bluffs for a time afterward, likely until 2:45 or 2:50 p.m. Corcoran might also be referring to the later arrival on the bluffs of the group of troopers led by scout Herendeen.

12) Lieutenant Winfield Edgerly (1881 account in *The Custer Myth*): Benteen's battalion joined Reno's force fifteen to twenty minutes after Reno's retreat.

ANALYSIS

Based on other participant accounts, Edgerly's estimate would favor his fifteen-minute span of time over the longer twenty minutes. If Reno got to the bluffs at 2:20 p.m., Edgerly suggests that Benteen got there about 2:35 p.m.

13) Corporal John Hammon - G Company (Camp Interview, *Camp, Custer and the Little Bighorn*, with my annotations in brackets): Hammon says that Benteen came in and met Reno [on the bluffs] and [then] unsaddled.

ANALYSIS

The significance of Corporal Hammon's statement above lies with Benteen's decision to unsaddle upon reaching Reno's command. If Benteen ordered his battalion to unsaddle, it would indicate an intent to halt his progress and remain in position on the bluffs for an extended period of time—at least an hour or two—before moving out again, so as to prevent "heat bumps" or galling. This may be an indication of Benteen's frame of mind upon reaching the bluffs and may explain his inaction and decision to proceed no further until about two hours had passed after his arrival on the bluffs.

14a) Lieutenant Edward Godfrey (RCOI): guesses they (Benteen) reached Reno's bluff between 3 and 4 p.m.

14b) Lieutenant Edward Godfrey (Godfrey's narrative, *Custer's Last Battle*): my recollection is that it was half-past two when we joined Reno.

ANALYSIS

Lieutenant Godfrey gives two different times for Benteen's arrival on the bluffs. At the RCOI, he "guessed" that it was sometime between 3 and 4 p.m., which is one of the only participant accounts (aside from the questionable Windolph account) that supports Lieutenant Wallace's later time estimate and contradicts the recollections of nearly all other participants. However, in Godfrey's more carefully researched and published narrative of the Battle (i.e., his *Century Magazine* article, "Custer's Last Battle"), Godfrey relies on his field notes—which he apparently did not use during his RCOI testimony, as he makes clear he "guessed" at the time—to give us a more reliable and accurate time estimate of Benteen's arrival on the bluffs at 2:30 p.m., which supports the time estimates given by the majority of the participants who were there. Thus, Godfrey's 2:30 p.m. time is the more accurate of the two he gave.

15) Lieutenant Carlo De Rudio (RCOI): says Reno left the timber somewhere between 2 and 3 p.m. He could not remember the correct minute but said it was after 2 p.m.

ANALYSIS

Lieutenant De Rudio tells us that Reno's retreat from the valley occurred at some time after 2 p.m. and before 3 p.m. which supports our participant-based time of a 2:10 p.m. Reno departure from the timber position, with the arrival on the bluffs of his command between 2:20 and 2:40 p.m.

16) Private Edward Davern (RCOI): Reno reached the bluffs on his retreat between 2 and 3 p.m.

ANALYSIS

Private Davern tells us that Reno's battalion reached the bluffs sometime after 2 p.m. and before 3 p.m., which supports the participant-based time of a 2:10 p.m. Reno departure from the timber position and his command's arrival on the bluffs between 2:20 and 2:50 p.m.

17) Private Charles Windolph (*I Fought With Custer*): It was now somewhere around 3:30 in the afternoon [that Reno reached the bluffs].

ANALYSIS

Private Windolph's memoirs were written by Frazier and Robert Hunt and published in 1947. I strongly suspect that the times he gave for the events were heavily influenced by Col. Graham's timelines that were previously published in his book *The Story of the Little Big Horn* in 1926, as Windolph's times follow Graham's unique and popular times—based almost exclusively on those given by Lieutenant Wallace at the RCOI and thus contradicting the evidence provided by the great majority of the battle participants, both red and white. The Hunts reproduce and insert sections from Graham's book to supplement Windolph's own memoirs. Thus, our conclusion is that Windolph's times are not to be relied on.

18) White Man Runs Him (1913 interview with Frank Shively): regarding the meeting of the three Crows with Benteen's battalion previous to joining Reno on the bluffs, he said this: "His men were with him and the mules with the powder and bullets.... I do not know what time this was, but it was some time between 1 and 2 in the afternoon".

ANALYSIS

White Man Runs Him gives a local or sun time of "between 1 and 2 in the afternoon" when the three Crow scouts met Benteen's battalion just north of Ford A. This time would convert to between 2 and 3 pm regimental Headquarters watch time, supporting a Benteen arrival on Reno Hill at about 2:30 pm.

19) Lieutenant Edward McClernand's "Narrative of the Little Big Horn campaign," published in Gordon Harper's *The Fights on the Little Horn Companion*: "It is Godfrey's recollection that Benteen joined Reno about half past two, and that it was about 5 p.m., when the latter finally decided to advance to the support of his chief."

ANALYSIS

Lieutenant McClernand's times quoted above—given to him by Lieutenant Godfrey shortly after the battle—support the conclusion that Benteen's battalion arrived on Reno Hill at about 2:30 p.m.

OVERALL ANALYSIS

Of the participants listed above, only Lieutenant Wallace and Private Windolph (whose time estimates were likely derived from Wallace) offered us times for Benteen's arrival on the bluffs significantly longer than all others, placing the arrival at some time between 3:40 and 4 p.m. Godfrey at first supported Wallace in this estimate before revising his time estimate down to 2:30 p.m. in line with other participants. It was the time of 2:30 p.m. that he gave to Lieutenant McClernand shortly after the battle. Wallace and Windolph's time estimates are in a significant minority and can be safely rejected. By the same token, Benteen's time estimate, along with Gerard's later time estimates, suggesting an arrival time of 2 p.m., must also be rejected, as this would suggest too short a time for Reno's valley fight, said to have lasted at least thirty minutes before his retreat. The only time that fits for Benteen's arrival on the bluffs—the only time that does not conflict with the recollections of the remaining participants as to the time of Reno's valley fight and retreat—is that of 2:30 to 2:35 p.m., a time specifically mentioned by a number of the battle participants and a time supported by nine of the participant sources.

CONCLUSION

2:30 p.m.: Benteen's battalion arrived on the bluffs at this time. The head of Reno's command reached the bluffs at 2:20 p.m., with the rear elements not arriving until about 2:50 p.m. The advance pack mules under Lieutenant Mathey were just arriving on a hillock about a mile south of Reno Hill. The main pack train and Rear Guard have just passed the first (eastern) lone tepee and are now about 3.5 miles east of Ford A.

* On Custer's field: E Company is rapidly approaching the vicinity of Ford B at the mouth of Medicine Tail Coulee while Custer has rejoined his command in North Medicine Tail Coulee and is bringing both Keogh and Yates's squadrons up onto Luce Ridge into full view of the Indian camps below. The trumpeters of all four troops and Headquarters staff were assembled here at this time to prepare for the Keogh's squadron's advance down to the river with as much noise and fanfare as possible. (*Speculative)

18. LIEUTENANT HARE RIDES FOR THE AMMUNITION MULES

Shortly after Benteen arrived on the bluffs to join with Reno's battalion, Lieutenant Luther Hare was sent by Major Reno to hurry on the mules carrying the ammunition packs in the advance group of pack mules led by Lieutenant Edward Mathey. We will explore in this segment of the Participant Timeline when this event occurred, how long it took, and the distance involved:

1) Lieutenant Charles Varnum (Brininstool Interview): Lieutenant Hare was sent off for the ammunition packs soon after Benteen arrived.

ANALYSIS

The evidence will show that Hare was sent back for the ammunition packs within minutes of Benteen's arrival on the bluffs. Varnum's statement above supports this conclusion.

2) Major Marcus Reno (RCOI): claims he sent Lieutenant Hare for the ammunition mules after he returned from his foray down to see Lieutenant Hodgson's body (p. 504, Graham) .... said he was gone for thirty minutes to check on Hodgson's body.

ANALYSIS

Major Reno's foray down to see Lieutenant Hodgson took, according to his own testimony, about thirty minutes of time, thus Reno is saying that the earliest he could have sent Lieutenant Hare for the ammunition mules was about thirty minutes after Benteen reached the bluffs or even later than that. As we shall see from examining the other participant accounts, including Lieutenant Hare himself, this is not true. Reno's timing sequence above is entirely unsupported by any other participant who described this event and must be rejected as inaccurate. The evidence will show clearly that Lieutenant Hare was sent on his mission for the ammunition mules before—not after—Reno went down to the river to check on the body of Lieutenant Hodgson (p. 504, Graham).

3a) Lieutenant Luther Hare (RCOI, with my annotations in brackets): says that Reno sent him back to get the ammunition mules as soon as Benteen came up (p. 247, Graham). He was told to cut out the ammunition packs and send them on ahead. Also, to "find out where [the rest of] the pack train was and to get it up as soon as possible." He went back about one and a half miles to the pack train. He was gone and back in about twenty minutes. He did not wait for the ammunition packs. He reached the pack train and came back as fast as he could (p. 247-48).

ANALYSIS

Hare says his entire trip back and forth, to and from the pack train to Reno Hill, encompassed about twenty minutes. His distance estimate of reaching the packs about one and a half miles from Reno Hill is questionable, as that distance would likely put him down on the flats by Ash/Reno Creek and does not fit with other independent sources, although this distance does synchronize better with Packer Churchill's estimated distance of about two miles from Reno Hill, which would put the advance pack train down in the valley of Ash/Reno Creek rather than on the bluffs as described by Lieutenant Mathey.

3b) Lieutenant Luther Hare (Camp Interview, *Custer in 76*, with my annotations in brackets): Benteen came up [to join them] about ten minutes after Reno got to [the] top of [the] bluffs, and about ten minutes after this, Reno sent Hare to [the] packs. Hare's horse had been shot ... and he traded horses with Godfrey and started for [the] packs. [He] heard no firing in [the] direction of Custer before he started.

He met [the] packs north of the north fork of Sundance [i.e., Ash/Reno] Creek. A mile or not much more from Reno....—[He] was gone from Reno about twenty minutes. Just as he got back he heard firing in [the] direction of Custer and was told that previous firing down there was heard while he had been gone. Just as he got back he looked and saw Company D advancing toward Custer. They were some distance out but still in sight. [He] thinks the other six companies did not advance [towards the Peaks] until at least a full hour after Reno retreated up [the bluffs]. Reno sent him to go to Weir and tell Weir to connect with Custer. He found Edgerly out ahead but says [that] Edgerly did not hold his advanced position more than ten minutes.... [He] says [that] he (Hare) did not order any company back on his own authority.

<u>ANALYSIS</u>

In Hare's in-depth interview with Walter Camp, he reaffirms his twenty-minute trip to the advance pack mules and back. He also clarifies the distance to a more accurate distance of about a mile from Reno Hill. He mentions that he did not hear any firing from Custer's field before leaving, which places his departure time to just before the start of Custer's engagement down at Ford B, which was heard by so many participants on Reno Hill. Hare also tells us that when he returned to Reno Hill twenty minutes later, he noticed D Company moving ahead on their advance to the Peaks. They were some distance out but still in sight.

4) <u>Lieutenant Edward Godfrey</u> (Godfrey's narrative from his *Century Magazine* article, "Custer's Last Battle," with my annotations in brackets): We soon joined Reno's battalion [on the bluffs].... Very soon after this, the Indians withdrew from the attack .... Lieutenant Hare was ordered to go back and bring up the ammunition pack mules .... While waiting for the ammunition pack mules, Major Reno concluded to make an effort to recover and bury the body of Lieutenant Hodgson.

<u>ANALYSIS</u>

Lieutenant Godfrey supports Lieutenant Hare's recollections above and contradicts Reno's statement by claiming that Reno went down to check on Lieutenant Hodgson's body after he sent Hare off for the ammunition mules, not before.

5) <u>Sergeant Ferdinand Culbertson</u> (RCOI): Reno goes down to look for Lieutenant Hodgson shortly after sending Lieutenant Hare off for the ammunition mules; when they [note: Culbertson and several other troopers accompanied Reno on this foray to seek out Hodgson's body] returned to the bluffs, the ammunition packs were just arriving.

<u>ANALYSIS</u>

Sergeant Culbertson confirms that Reno went down to the river to check on Lieutenant Hodgson's body after sending Lieutenant Hare off for the ammunition mules. As Reno indicated that this foray took about thirty minutes, Culbertson, who accompanied Reno down to the river, indicates that the ammunition mules were just arriving on Reno Hill, about ten minutes after Lieutenant Hare got back to the hill.

6) Packer Benjamin Churchill (RCOI, with my annotations in brackets):

Question: Tell where it was the order [from Lieutenant Hare] came about sending the ammunition packs ahead? How far from that [lone] tepee?

Churchill: I think about half a mile after we passed it. [Note: This would be along the main Indian trail in Ash/Reno Creek about 1.2 miles east of the Little Big Horn river.]

Question: Were you on the trail or off it?

Churchill: On the trail.

Question: Tell what was done with the ammunition packs? Who went with them?

Churchill: Myself and a man named Mann took one—who took the other I don't know. He led the mule and I licked it up as fast as I could. We kept it going as fast as it could with two boxes of ammunition on [its back]. [Note: Churchill is referring here only to the two ammunition mules brought immediately to Reno Hill with Lieutenant Hare, not to the remaining ammunition mules in the advance group with Lieutenant Mathey.]

Question: How far did you go from that place till you reached the command [on Reno Hill]?

Churchill: About two miles. [Note: Lieutenant Hare put this distance at just over a mile to a mile and a half from Reno Hill.]

Question: How long did it take you to go over that distance? [Note: This distance would put the starting point down in Ash/Reno Creek, not on the bluffs as Mathey claimed.]

Churchill: I don't think it took us over eight or ten minutes. [Note: This would be a speed of about twelve miles per hour for these two mules. Mathey would claim to have soon followed Churchill to Reno Hill, arriving there in just over twenty minutes with his mules moving at a fast walk, or about four miles per hour, thus one of these two accounts must be mistaken as to distance and/or speed.]

ANALYSIS

Packer Churchill's account conflicts with the accounts given by Lieutenant Hare and Lieutenant Mathey as to the distance of the advance pack mules from Reno Hill when they received orders to send the ammunition mules on to Reno Hill. Churchill places this point about two miles south of Reno Hill down in the valley of Ash/Reno Creek about 1.2 miles east of the river (or a half mile west of the second, or western, lone tepee). He estimates that the two ammunition mules sent with Hare took eight to ten minutes to reach Reno Hill, moving at about twelve or thirteen miles per hour. On the other hand, Lieutenant Hare puts the distance at just over a mile—or a mile and a half in a different account—south of Reno Hill, or just on the southernmost bluffs overlooking Ash/Reno Creek and the Little Big Horn valley. Hare would support Churchill's estimated time of about eight to ten minutes for the mules to reach Reno Hill, but at that shorter distance, the mules would be moving about six or seven miles per hour if the distance was just over a mile, or about nine miles per hour if the distance were a mile and a half. These last two are much more reasonable speeds for the heavily laden ammunition mules than Churchill's twelve-plus miles per hour. Mathey would claim that it took him about twenty-plus minutes to reach Reno Hill with the rest of his advance mules moving at a fast walk (about five miles per hour), which would suggest that the distance from Reno Hill was closer to just over a mile rather than Churchill's two miles.

7) Lieutenant Edward Mathey (RCOI, with my annotations in brackets):

Question: From that point where you made the halt, how far were you from Major Reno's position on the hill, if you know?

Mathey: Where I made the halt, it was probably two or three miles. ...

Question: Did you meet any officers there with orders?

Mathey: After we started [moving again], I met Lieut. Hare, who said he wanted the ammunition; and I detached two mules from the train and ordered them to go with Lieut. Hare. ...

Question: State what was done about the pack train.

Mathey: We moved on till we got to Major Reno. (It) was at a pretty fast walk.

Question: After starting the ammunition pack mules with Lieut. Hare, how long was it till you arrived with the pack train? What time of day was it?

Mathey: Something less than half an hour; probably 20 minutes. I think it was about 3 o'clock when we reached Major Reno's command.

Question: How do you fix that as the time?

Mathey: I judge we started about 11 or 12 from where the long halt was made; [note: this would likely be a reference to Halt 2. The timeline would have the pack train crossing the divide shortly after 11 a.m.] We then marched six or seven miles to the morass [note: the morass would be closer to seven miles from the divide crossing], and about 3 miles from there to the [western lone] tepee; and from the tepee to where Major Reno was, was about 4 or 5 miles [note: the actual distance would be closer to three and a half miles]; and I think it would be about that time that we got there.

ANALYSIS

Lieutenant Mathey indicates that he first halted the advance pack mules about two or three miles from Reno Hill, which would be down on Ash/Reno Creek, a distance west of the second (western) lone tepee, which agrees with packer Churchill's recollection. He then indicates that he started the packs moving again, but does not state any distance covered. We can assume he moved about a mile and a half closer to Reno Hill up onto the bluffs to within about a mile of Reno Hill when he met up with Lieutenant Hare, when he halted the advance mules again, in order to send two of his ammunition mules ahead with Lieutenant Hare, while also sending a message back to Captain McDougall and the main pack train, several miles to his rear and still moving down Ash/Reno Creek. Mathey would then continue his advance with his packs until he arrived on Reno Hill about twenty minutes later, or between 3:10 and 3:15 p.m. The two mules he sent on in advance with packers Churchill and Mann would arrive at about 3:05 p.m.

OVERALL ANALYSIS

Lieutenant Hare tells us that he was sent back to hurry on the ammunition mules about ten minutes after Benteen arrived on the bluffs. We have previously established in the timeline a Benteen arrival on the bluffs at about 2:30 p.m. This would place Lieutenant Hare's departure from Reno Hill just prior to 2:40 p.m. He indicates that he returned back to Reno Hill—without waiting for the pack mules to accompany him—in about twenty minutes, placing his return to Reno Hill at about 3:00 p.m. Hare's estimated eight to ten minutes travel time each way is supported by Packer Churchill's similar time estimate of eight to ten minutes for the two advance ammunition mules to reach Reno Hill.

CONCLUSION

2:38 p.m.: Lieutenant Hare leaves Reno Hill and rides back to the advance pack mules under Lieutenant Mathey, now corralled on a small hill overlooking Ash/Reno Creek valley about a mile southeast of Reno Hill. He delivers his message to Lieutenant Mathey, who in turn sends a messenger back to the main pack train several miles to his rear and down on the flats between Ash/Reno Creek and the North Fork with instructions to hurry on to the bluffs. At this time, nearly all the warriors had withdrawn from Reno's position on the bluffs and are now concentrating on Custer's five companies in the north. E Company is now at the mouth of Medicine Tail Ford by Ford B-1 demonstrating there, while Keogh's squadron of I and C troops are rapidly advancing from Luce Ridge down to Butler Ridge with all their trumpeters making as much noise as possible on this advance, the likely purpose being to attract as much attention to his force converging in the area of Ford B to draw as many warriors off of the beleaguered Reno battalion and allow them to rejoin with either Benteen's battalion or the pack train rapidly approaching his position.

3:00 p.m.: Lieutenant Hare returns to Reno Hill after delivering his message for the ammunition mules under Lieutenant Mathey. He sees D Company advancing toward the peaks some distance out. Reno is still down at the river, checking on Lieutenant Hodgson's body. Lieutenant Mathey's advance pack mules are now moving towards Reno Hill and will reach that position within the next ten or fifteen minutes. However, the lead two ammunition mules would arrive within five minutes of Hare's arrival on the Hill.

Lieutenant Mathey's messenger has now reached Captain McDougall, who is now organizing the main pack train in a compact group with his rear guard protecting it from the front and rear with his two platoons of Company B. By this time, the Ford B demonstration is over, and Keogh's squadron is falling back up Deep Coulee towards Finley Ridge, with a squad of flankers moving up parallel with them on the high ground from Greasy Grass Hill to join them there. E Company is falling back from Butler Ridge to Luce Ridge to join the Yates squadron.

19. THE CUSTER FIGHT BEGINS

There has been much debate among battle students as to when the fight with Custer's five companies actually began. Here's what the participants tell us about this event:

1a) George Herendeen (RCOI): says he first heard firing from Custer (a great many volleys) less than thirty minutes after Reno left the timber. They were in the woods for twenty minutes when they heard firing downstream.

1b) George Herendeen (Camp Interview, *Custer in 76*): says he heard firing downriver about half an hour after Reno left the timber.

ANALYSIS

The Participant Timeline has established that Reno left the timber at about 2:10 p.m. according to Fred Gerard's Headquarters watch time recollection. Based on this foundation, George Herendeen tells us that Custer's Fight began about thirty minutes later, or about 2:40 p.m., or twenty minutes after Herendeen went back to hide himself in the timber at about 2:20 p.m.

2a) Fred Gerard (RCOI): Ten to fifteen minutes after Reno leaves the valley, Gerard hears more firing downstream. He heard three or four volleys of fifty to a hundred guns.

2b) Fred Gerard (Camp Interview, *Custer in 76*): about a half hour after Reno's troops left the timber, firing began downriver.

ANALYSIS

Fred Gerard tells us that Custer's fight begins about thirty minutes after Reno's troops left the timber. He previously testified that Reno left the timber at 2:10 p.m. Thus, he told Walter Camp that Custer's fight began at about 2:40 p.m. At the RCOI, Gerard says that the firing downstream (of three or four volleys) was heard ten to fifteen minutes after Reno left the valley (not the woods), indicating that most of Reno's command left the valley and regained the bluffs by 2:25 or 2:30 p.m.

3) Good Voiced Elk (Camp Interview, Camp, *Custer and the Little Bighorn*): says the Custer fight started half an hour after Reno retreats from the valley.

ANALYSIS

Good Voiced Elk agrees with both Herendeen and Gerard that the Custer fight began about thirty minutes after Reno's retreat from the valley. The timeline establishes Reno's retreat beginning at 2:10 p.m. Thus, the Custer fight begins at about 2:40 p.m.

4) Lieutenant Edward Godfrey (Godfrey's *Century Magazine* article, "Custer's Last Battle"): They heard Custer's firing downstream after Hare left for the ammunition packs and after Reno went down the bluffs to recover Hodgson's body.

ANALYSIS

Lieutenant Godfrey tells us that Custer's fight begins at some time after Lieutenant Hare left Reno Hill for the ammunition packs. Godfrey is likely mistaken about his recollection that the firing from Custer's field began after Reno went down the bluffs to check on Hodgson's body. Evidence is strong that Reno first had his altercation with Captain Weir when Weir vainly requested permission to advance shortly after hearing the firing from Custer's field. At that time, Reno had not yet left the bluffs to look for Hodgson's body. He would be gone for thirty minutes, and by the time of his return, Weir would be long gone on his unauthorized trek out to the Peaks with D Company.

5) Curley (*Helena Herald* Interview, July 15, 1876): the fight appears to have begun, from Curley's description of the location of the sun, about 2:30 p.m. or 3 o'clock p.m. and continued without intermission until nearly sunset.

ANALYSIS

Curley's time range of the Custer Fight beginning at some time between 2:30 and 3:00 p.m. fits well within the recollection of our participants and supports the notion that the Custer fight began about thirty minutes after Reno began his retreat from the valley, or about 2:40 p.m.

6) Private Davern (RCOI): Private Davern says he heard several volleys fired by Custer's command downriver about thirty minutes after he reached the bluffs and discussed it with Captain Weir.

ANALYSIS

Assuming Davern reached the bluffs about 2:15 p.m., his testimony puts the start of Custer's fight at about 2:45 p.m.

OVERALL ANALYSIS

All six of the participant accounts above support each other and indicate that Custer's fight began about thirty minutes after Reno began his retreat from the valley. This establishes the start of the Custer Fight at about 2:40 p.m.

CONCLUSION

2:40 p.m.: Custer's fight in the vicinity of Ford B begins several minutes after Lieutenant Hare left Reno Hill on his ride back for the ammunition mules. Lieutenant Mathey and the advance ammunition mules were corralled at this time on a small hillock about a mile south of Reno Hill. The main pack train and Rear Guard would be just west of the first (eastern) lone tepee and east of the Hartung morass at this time, about 3.5 to three miles east of Ford A.

* In Custer's wing, E Company has begun firing across the river by Ford B, while the Keogh squadron (I and C Company) is rapidly approaching the Ford area to support them there. Yates's squadron (F and L Company) is now deploying on Luce Ridge.

20. CAPTAIN WEIR AND D COMPANY LEAVE RENO HILL

When Captain Thomas Weir took D Company out on their advance toward Custer is a source of confusion for many battle students. To determine the truth, we must examine in detail what the participants had to say about this event:

1a) Captain Frederick Benteen (RCOI): estimated that Weir "sallied out" without orders about thirty minutes after Benteen's own arrival on the bluffs.

ANALYSIS

As Benteen's battalion arrived on Reno Hill about 2:30 p.m., Benteen estimates that Weir sallied out with D Company toward the Weir Peaks thirty minutes later at about 3:00 p.m.

1b) Captain Frederick Benteen (*Benteen's Narrative*): In an hour or less, the train came up (p. 186). About this time, I saw one of the troops of my battalion proceeding to the front without orders from me.

ANALYSIS

Years after the battle, Benteen wrote up his own narrative of the battle. In this narrative, Benteen now implies a change in the time of departure of D Company from Reno Hill. In his previous testimony at the RCOI in 1879 Benteen has Weir leaving thirty minutes after his own junction with Reno on the bluffs. Now, Benteen indicates that he only saw D Company moving to the front within an hour or less, or about the time of the pack train's arrival on the bluffs. As the advance packs did arrive on the bluffs between 3:05 and 3:15 p.m., Benteen is suggesting that Weir's D Company may still have been in sight on the Weir Peaks at 3:15 p.m.

2a) Lieutenant Edward Godfrey (RCOI): Weir left on his advance shortly before the arrival of the ammunition packs on the bluffs.

2b) Lieutenant Edward Godfrey ("Custer's Last Battle" article in *Century Magazine*, with my annotations in brackets): Captain Weir and Lieutenant Edgerly ... heard the firing [i.e., the volleys from Custer's field], became impatient at the delay and thought they should move down that way if they should be permitted.... Edgerly, seeing Weir going in the direction of the firing, ... started down the [Cedar Coulee] ravine with the troop. Weir, from the high point [later named in his honor] ... signaled for him to change his direction, and Edgerly went over to the high point [i.e., the Weir Peaks] where they remained, not seriously molested until the remainder of the troops marched down there....

ANALYSIS

The timeline previously established the arrival on Reno Hill of the advance ammunition pack mules under Lieutenant Mathey at about 3:05 p.m., thus Godfrey indicates that Weir left Reno Hill at "shortly before" 3:05 p.m. Godfrey also indicates that D Company remained out on the Peaks until "the remainder of the troops marched down there...."

3) Dr. Henry Porter (RCOI): is "pretty positive" that Weir left downriver before the pack train arrived.

ANALYSIS

Dr. Porter supports the recollection of Lieutenant Godfrey that Weir left on his advance before the arrival of the advance pack mules on Reno Hill. The timeline indicates the arrival of the advance pack mules at about 3:05 p.m.

4) Private George Wylie – D Company (Camp Interview, *Custer in 76*): soon after Benteen joined Reno on the hill, Captain Weir started out to reach Custer.

ANALYSIS

Private Wyle, a member of D Company himself, indicates that his troop left Reno Hill shortly after Benteen's battalion arrived there, thus favoring an earlier, rather than later departure of this company to the Weir Peaks.

5) Corporal Stanislas Roy – A Company (Camp Interview, *Custer in 76*): soon after Benteen came up, Weir went out.

ANALYSIS

Corporal Roy supports the recollection of Private Wylie that D Company moved out on their advance soon after Benteen's arrival on the bluffs.

6) Captain Myles Moylan (RCOI): claims he saw Weir's company head downstream about the time the packs arrived on the bluffs one-half to three-quarters of an hour after they got onto the bluffs.

ANALYSIS

Captain Moylan indicates that he saw Weir's company conducting their advance downstream about the time that the advance ammunition packs arrived on the bluffs. The timeline has established that time at about 3:05 p.m. Moylan further indicates that this time would be between thirty to forty-five minutes after Reno's battalion got onto the bluffs. As we have established the head of Reno's troops arriving on the bluffs at about 2:20 p.m., Moylan is confirming that the advance pack mules under Lieutenant Mathey arrived on the bluffs about forty-five minutes later at 3:05 p.m. He indicates that Weir would have left Reno Hill at some time between 2:50 and 3:05 p.m., or thirty to forty-five minutes after Reno's command first reached the bluffs.

7) Sergeant Ferdinand Culbertson – A Company (RCOI): claims that Weir left twenty minutes after the arrival of the first ammunition mules at Reno Hill.

ANALYSIS

Sergeant Culbertson's time estimate above conflicts with the majority of our participant accounts regarding the time of Weir's departure from Reno Hill. Most other accounts relate Weir leaving with D Company prior to the arrival of the advance pack mules under Lieutenant Mathey, who likely did not arrive at Reno Hill before 3:05 p.m. However, Culbertson claims that he saw what he believed to be Weir or his troop leaving about twenty minutes later, placing the departure time at about 3:25 p.m. To resolve this apparent conflict, we suggest that Culbertson actually witnessed the departure of Captain French's M Company on its way out to join Weir's D Company out on the Peaks, and mistook the former for the latter.

8) Private Edward Davern (RCOI): claims that Weir left on his advance over an hour (actually at some time between one and two hours) after the arrival of the advance packs.

ANALYSIS

Private Davern claims that Weir began his advance to the Peaks between an hour and two hours after the arrival of the advance packs on Reno Hill. As the advance packs did not arrive on Reno Hill until 3:05 p.m., Davern suggests that Weir did not begin his advance north until between 4:00 and 5:00 p.m. This time estimate conflicts with all other participants, yet it matches roughly the time that the rest of Reno's battalion began their advance toward the Weir Peaks to join D Troop there. Davern was likely confusing Weir's advance to the Peaks with the later general advance of Reno's remaining troops to join Weir there. Captain Weir himself may have left the Weir Peaks and ridden back to Reno Hill to lead the rest of Reno's command out to his advanced position being held by Edgerly and D Company out on the Peaks. Thus, Davern may well have seen Weir at this time, rather than D Company, moving out to lead the advance to the Weir Peaks.

9a) Lieutenant Luther Hare (RCOI): said that Weir had left on his advance while Hare had gone to the pack train. He also said it was one-and-a-half hours from the time Weir advanced until they began to seriously engage the Indians.

ANALYSIS

Lieutenant Hare supports the recollections of Godfrey and Moylan that Weir left on his advance during the time that Hare had ridden back to the advance pack mules under Lieutenant Mathey. This would place Weir's departure from Reno Hill at some time between 2:40 and 3:00 p.m. One can probably narrow that timeframe down to between 2:45 and 2:55 p.m. Lieutenant Hare also mentions an important time reference above when he claimed that D Company was not seriously engaged out on the Peaks for the next one and a half hours, a time in which the Indians were apparently fully engaged in fighting Custer's five companies. As D Company would not have arrived and deployed on the Weir Peaks until roughly forty-five minutes after Benteen arrived on the bluffs, it suggests that the Custer fight would have lasted for roughly two hours and fifteen or twenty minutes.

9b) Lieutenant Luther Hare (Camp Interview, *Custer in 76*, with my annotations in brackets): says he was gone [from Reno Hill] about twenty minutes [on his ride back to the advance ammunition mules]. Just as he got back [to Reno Hill at about 3:00 p.m. according to the Participant Timeline] ... he saw D Company advancing toward Custer. They were some distance out but still in sight. Reno [after he returned to the bluffs—about 3:15 p.m. according to the timeline—from his trip to the river to check on Lieutenant Hodgson's body] sent him out to Weir to tell him to connect with Custer. He found Edgerly out ahead [with D Company] but Edgerly did not hold his advance position more than ten minutes. [Note: this would likely have been between 3:30 and 3:40 p.m.] He said M, K, and H companies were strung out along the bluffs behind D Company parallel with the river, but no other company was quite up to D Company's [advanced] position.

ANALYSIS

Here Lieutenant Hare gives us more specific information on Weir's likely time of departure from Reno Hill. He confirms that D Company left Reno Hill during his ride back to the advance ammunition mules, and were still within sight upon Hare's arrival back on Reno Hill, which we have at 3:00 p.m. This would suggest that D Company actually left Reno Hill no less than five or ten minutes earlier—this would suggest a time of 2:55 p.m. at the latest.

10a) Lieutenant Winfield Edgerly (RCOI, with my annotations in brackets): says that Weir and D Company moved downstream about thirty to thirty-five minutes after reaching Reno on the bluffs (p. 393, RCOI - Graham). Said it was about forty-five minutes for Benteen and his command [to advance—probably only a reference to the arrival on the Weir Peaks of Captain French's M Troop at that time]. As they joined Reno on the bluffs, they heard heavy firing downstream, by volleys, almost immediately (within five minutes of their arrival on the bluffs). He discussed this with Weir and their First Sergeant for several minutes. Then Weir went off *"walking up and down rather anxiously"* (p. 393, Graham RCOI). Weir returned after ten minutes and asked Edgerly if he was willing to bring D Company forward if the rest of the command would not move. He said he would do so. Weir said he would ask permission from Reno or Benteen. Soon after, he came back, called for his orderly, mounted and went off; then Edgerly mounted up D Company and followed Weir, who was keeping up on a ridge while D Company went in a sort of valley [note: likely Cedar Coulee]. He says that D Company was deployed on a spur on the right loaf of the Weir Peaks about three-quarters to one mile from [southeast of] Ford B.

10b) Lieutenant Winfield Edgerly (Edgerly's Narrative): claims that only *"M and K Troops were sent to our support"* on the Weir Peaks.

10c) Lieutenant Winfield Edgerly (Camp Interview, *Custer in 76*): after D Company advanced to the peaks, Lieutenant Hare arrived with orders to try to open communications with Custer; then Hare returned to Reno. He then claims that Troops M, K, and H arrived on the Peaks.

ANALYSIS

Lieutenant Edgerly confirms Benteen's original estimate of D Company leaving for the Peaks about thirty to thirty-five minutes after their arrival on Reno Hill. As the Participant Timeline established that Benteen's battalion arrived on Reno Hill about 2:30 p.m., Edgerly indicates that D Company began their advance about 3:00 or 3:05 p.m. You will note also how Edgerly mentions only three additional troops ever joining them on the Weir Peaks: M, K and H Troops—and always in that particular order. Lieutenant Hare also mentioned these three troops deployed on the Weir Peaks, and in that very same order. This is very likely no coincidence, but a direct reference to the order of arrival of these tree troops on the Weir Peaks. M Company arrived there first, followed later by K Company and finally joined by H Company.

11) Private Charles Windolph (*I Fought With Custer*): It was now maybe 4:30 p.m. and the sun was fairly high in the sky.... I remember that Captain Weir suddenly rode off to the north alone, and a minute later, Lieutenant Edgerly, second in command of "D," followed with the whole troop. The pack mules were coming up about this time .... As I recall, Reno had seven wounded men, some of them in pretty bad shape.

ANALYSIS

We strongly suspect that Windolph's 4:30 p.m. time comes more from the authors Robert and Frazier Hunt than from Windolph himself. Aside from the conflict with so many other participant accounts, we have a few additional clues. First, Windolph's comment that when Weir left Reno Hill "the sun was fairly high in the sky". This was likely the best Windolph could give us timewise, and at 4:30 p.m. the sun is certainly not fairly high in the sky. The second clue given us by Windolph is that when Weir left Reno Hill, "the pack mules were coming up about this time." This description would coincide with the arrival on Reno Hill of Lieutenant Mathey and his advance ammunition mules, which Lieutenant Hare and Mathey indicated occurred about 3:00 p.m. or a short time afterwards. Private Windolph's memoirs were written by Robert and Frazier Hunt and published in 1947. I believe that the times he gave for the events were heavily influenced by Colonel William Graham's timelines that were previously published in his book *The Story of the Little Big Horn* in 1926, as Windolph's times follow Graham's unique and popular times estimates—based almost exclusively on those given by Lieutenant George Wallace at the Reno Court of Inquiry, and thus contradicting the evidence provided by the great majority of the battle participants, both red and white. The Hunts reproduce and insert sections from Graham's book to supplement Windolph's own memoirs, thus my conclusion is that Windolph's times are not to be relied on.

12) Private Jacob Hetler, Company D, from his Memories of the Little Big Horn, *"Indian War Veterans"* Jerome A. Greene, 2007:

(after reaching Major Reno)

"For an hour and a half after that, Captain Thomas B. Weir of my company asked Major Reno for permission to go to after Custer, and this was granted. We had heard firing and knew there was trouble and we hoped to arrive in time to aid Custer, but as the world knows now, we were too late. We were badly outnumbered when we finally came in sight of the scene of the massacre, and were fortunate to find a way to retreat back to the main company."

ANALYSIS

Private Hetler's comments above might initially be interpreted as indicating that Company D did not begin their advance to the Weir Peaks until an hour and half had elapsed from the time Captain Weir first asked Reno to advance, but this interpretation would conflict with the preponderance of primary source accounts which indicate otherwise. An alternative interpretation, and one that does not conflict with the other accounts, indicates that Private Hetler was not referring at all to Company D's initial advance out to the Weir Peaks, but rather, its later advance from the Weir Peaks down to the ridge overlooking Medicine Tail Coulee, which would have occurred about an hour and a half after Weir first asked—and was refused—permission from Reno to advance to Custer. As Company D began its advance to the Peaks a few minutes before 3 p.m., Private Hetler's account would indicate that his Company did not begin its move beyond the Weir Peaks to its advanced position to a ridge above Medicine Tail Coulee until roughly 4:20 p.m.

OVERALL ANALYSIS

From the twelve participants listed above, all but four of them indicate that D Company began their advance prior to the arrival of the advance ammunition mules on Reno Hill. As the latter occurred at about 3:05 p.m., one can safely conclude that D Company began their advance prior to 3:05 p.m. We can move this time ahead further with Lieutenant Hare's statement that he saw D Company out on their advance as he returned to Reno Hill, a time which I have placed at about 3:00 p.m. Thus it is very likely that D Company actually left the vicinity of Reno Hill—they were likely deployed on the south end of Sharpshooter Ridge and just west of it at this time, about a quarter mile north of Reno Hill—at about 2:55 p.m., or roughly twenty-five minutes after the arrival of Benteen's battalion at Reno Hill. Sergeant Culbertson's recollection of seeing Weir leave twenty minutes after the arrival of the advance pack mules would suggest that he mistook the likely departure of M Troop at about that time (3:25 to 3:35 p.m.) for Captain Weir's troop. Private Davern, likewise, mistook the later advance of K and H troops, possibly accompanied by Weir himself, who would have begun their advance at the time he suggested. Private Hetler's apparently conflicting comments will confuse many students unless understood in its proper context, referring to D Company's second advance beyond the Weir Peaks, and not its initial advance from Reno Hill.

CONCLUSION

2:55 p.m.: Captain Weir and D Troop begin their advance to the Weir Peaks from near the south end of Sharpshooter Ridge.

3:10 p.m.: Weir and Company D arrive on the Weir Peaks just as the advance ammunition mules under Lieutenant Mathey were arriving on Reno Hill. The main pack train and Rear Guard under Captain McDougall would be about 1.5 miles east of the Little Big Horn River just west of the second (western) Lone Tepee. Custer's second volleys were reportedly heard by both Godfrey and Davern at this time and we would speculate that this was because fighting was taking place on Luce and Blummer Ridges at this time, with the squadron previously sent to the river now falling back up Deep Coulee.

3:15 p.m.: Major Reno has returned to the bluffs from his trip down to the river and has issued orders for Lieutenant Hare to ride to the Peaks and order Captain Weir to make contact with Custer if possible.

3:25 p.m.: Lieutenant Hare arrives at the Weir Peaks and delivers his orders to Captain Weir.

4:20 p.m.: D Company begins its second advance north of the Weir Peaks to a ridge overlooking Medicine Tail Coulee.

21. LIEUTENANT MATHEY AND THE AMMUNITION PACKS ARRIVE ON THE BLUFFS

Lieutenant Mathey and the advance group of mules carrying ammunition arrive on the bluffs to join Benteen and Reno about forty-five minutes to an hour before the main pack train did. Here's what the participants had to say about their time of arrival:

1) Major Marcus Reno (RCOI, *Army and Navy Journal*): claimed the pack train arrived in about an hour.

ANALYSIS

Reno suggests that the pack train arrived on the bluffs in about an hour, presumably a reference to his own arrival on the bluffs at 2:15 or 2:20 p.m. In this case, Reno would be referring to the arrival of the advance packs at about 3:15 p.m.

2a) Captain Frederick Benteen (RCOI): claims that "the advance packs, containing the ammunition" arrived on Reno Hill over an hour after he got to the bluffs. (p. 383, Graham)

2b) Captain Frederick Benteen (Benteen's Narrative): "Reno dispatched Lieutenant Hare ... to go to the train and hasten it along. In an hour or less the train came up."

ANALYSIS

Benteen tells us that the advance packs arrived on the bluffs over an hour after his arrival there at 2:30 p.m., thus suggesting an indefinite arrival time between 3:30 to perhaps 3:45 p.m. As we shall see later, 3:45 p.m. would be the arrival time of the advance elements of the Rear Guard escort for the main pack train traveling under Captain McDougall. Benteen, either intentionally or unintentionally, merged the arrival times of the advance packs together with the main pack train.

In Benteen's narrative, he gives a slightly different assessment for the arrival of the advance pack mules, now claiming they arrived on the bluffs in an hour *or less* after Lieutenant Hare was sent off to hasten them along. The departure time for Hare's ride is at 2:38 p.m., thus suggesting an arrival time for the advance packs, according to Benteen, at 3:40 p.m. or some time less than that. If we compare it to his earlier statement made at the RCOI, Benteen is suggesting that the advance pack mules arrived on Reno Hill between 3:30 and 3:40 p.m.

3) Lieutenant George Wallace (RCOI): Question: How long after Benteen came up did the pack train come up? Wallace: It must have been an hour.... Question: How long after Captain Weir moved out did the pack train come up? Wallace: I think the train came up soon after he started.

ANALYSIS

Wallace first estimates the arrival of the pack train on the bluffs about an hour after Benteen arrived on the bluffs, placing the time for this arrival at about 3:30 p.m., and thus supporting Benteen's recollection of its arrival time. Then Wallace somewhat contradicts this statement by claiming that the packs arrived soon after Weir started out with D Company on their trek to the north. However, evidence strongly suggests that Weir left Reno Hill about thirty minutes after Benteen's arrival there, not an hour later. Of the two answers Wallace gave above, we believe his second response was the more accurate assessment of the arrival of the advance packs: soon after, Weir left Reno Hill on his ride out to the peaks.

4a) Trumpeter John Martin (RCOI): says it took the pack mules about ten to fifteen minutes to come up.

4b) Trumpeter John Martin (Camp Interview, *Custer in 76*): says the packs got up to Reno Hill ten or fifteen minutes behind Benteen.

ANALYSIS

Martin's recollection of the advance packs reaching Reno Hill ten to fifteen minutes after Benteen's arrival—placing its arrival time there at between 2:40 and 2:45 p.m.—is not supported by the recollections of any other participant that day and conflicts with other evidence indicating Lieutenant Hare's departure from Reno Hill only a few minutes before that time on his ride back to the advance packs located about a mile south of Reno Hill. Martin's recollection here must be seriously questioned—unless Martin was referring to Lieutenant Mathey's advance pack mules leaving Ash/Reno Creek and moving up to the bluffs about ten or fifteen minutes after Benteen left the creek and moved up to the bluffs. Mathey would halt and remain on a hillock over a mile away for another fifteen or twenty minutes before moving on to Reno Hill.

5) Private William Taylor - A Company (*With Custer on the Little Big Horn*, with my annotations in brackets): "A very short time afterwards [i.e., after Benteen's junction with Reno on the bluffs about 2:30 p.m.] Captain McDougall with B Troop, escorting the pack train, came along and joined us...."

ANALYSIS

Private Taylor's recollection of an early McDougall arrival on the bluffs "a very short time" after Benteen's arrival there is not supported by other participant accounts, including those of McDougall himself. Taylor likely confused the arrival of the advance pack mules under Lieutenant Mathey with those of the main pack train under Captain McDougall. The latter group would not arrive on the bluffs for another three-quarters of an hour. The possible source of Taylor's confusion may lie in the strong likelihood that B Company's pack mules were actually traveling in the advance group under Lieutenant Mathey. This comes from the recollection of Private Augustine DeVoto. It is possible that Taylor recognized several members of B Troop at this time and mistook them for the arrival of their company at that time.

6) Dr. Henry Porter (RCOI): says the pack train arrived between thirty and sixty minutes after Benteen arrived on the bluffs (p. 168, Graham), but isn't sure of this.

ANALYSIS

Dr. Porter gives us a rather indistinct estimate for the arrival of the pack train at some time between 3:00 p.m. and 3:30 p.m. after Benteen's 2:30 p.m. arrival on the bluffs. This thirty-minute range of time is, I believe, reliable, based on other participant recollections. Lieutenant Mathey's advance pack mules arrived on Reno Hill at some time after 3:00 p.m. and sometime before 3:30 p.m. This time frame supports Lieutenant Hare's recollection of returning to Reno Hill twenty minutes after leaving it, placing his return time there at about 3:00 p.m. with the advance pack mules not far behind.

7a) Private Edward Davern (RCOI): says the pack train arrived on the bluffs shortly after he arrived on the bluffs.

7b) Private Edward Davern (RCOI): Question: You say something about the advance of the pack train. What did it have? Davern: Ammunition.

ANALYSIS

Private Davern's recollection of the advance pack mules arriving shortly after his (i.e., Reno's) arrival on the bluffs is supported by other participant accounts of the advance pack mules arrival just after the return of Lieutenant Hare on Reno Hill about 3 p.m. Davern also tells us that the advance pack mules were carrying ammunition.

8) Captain Myles Moylan (RCOI, with my annotations in brackets) claims the packs arrived on the bluffs one-half to three-quarters of an hour after they [note: either Reno's or Benteen's battalion] got to the bluffs.

ANALYSIS

Given that the head of Reno's battalion—including Moylan himself—reached the bluffs at about 2:20 p.m., Moylan indicates that the packs then arrived from half to three-quarters of an hour later, or between 2:50 to 3:05 p.m. If Moylan was referencing Benteen's arrival on the bluffs, then we can shift his time range back about ten minutes to between 3:00 and 3:15 p.m., which coincides more closely with the recollections of Varnum, Godrey, and Mathey. His thirty-minute estimate likely referred to the two advance ammo mules arriving on Reno Hill following Lieutenant Hare. His forty-five-minute estimate likely refers to the subsequent arrival of Lieutenant Mathey's remaining ammunition mules.

9) <u>Lieutenant Charles Varnum</u> (RCOI): the pack train came up to the bluffs about 3/4 of an hour, more or less, after Benteen did and came up from a different direction than Benteen.

<u>ANALYSIS</u>

Lieutenant Varnum's recollections, as usual, are remarkably accurate and reliable. He estimates three-quarters of an hour span of time, more or less, for the arrival of the pack mules after Benteen's arrival on the bluffs. As we have established that Benteen arrived on the bluffs at 2:30 p.m., Varnum indicates that Mathey's advance pack mules have arrived there by 3:15 p.m.

10) <u>Lieutenant Edward Godfrey</u> (RCOI): claims the ammunition packs arrived thirty to forty-five minutes after Benteen arrived on the bluffs.... the ammunition packs arrived soon after Weir left on his advance.

<u>ANALYSIS</u>

Lieutenant Godfrey's time estimate closely reflects that of Moylan and Varnum. If Benteen arrived on the bluffs at 2:30 p.m., as our participant accounts tell us, then the first of the ammo mules arrive on Reno Hill about 3:00 p.m. with the remaining packs under Lieutenant Mathey arriving by about 3:15 p.m. Godfrey's comments above also tell us that Captain Weir moved out on his advance a short time before the arrival of these packs, or just under thirty minutes after Benteen's arrival on the bluffs. Godfrey also indicates that Lieutenant Mathey's advance pack mules carried the ammunition packs.

11) <u>Lieutenant Luther Hare</u> (Camp Interview, *Custer in 76*, with my annotations in brackets): Benteen came up [to join them] about 10 minutes after Reno got to [the] top of [the] bluffs, and about 10 minutes after this, Reno sent Hare to [the] packs. Hare's horse had been shot ... and he traded horses with Godfrey and started for [the] packs. [He] heard no firing in [the] direction of Custer before he started. He met [the] packs north of the north fork of Sundance [i.e., Ash/Reno] Creek. A mile or not much more from Reno.... [He] was gone from Reno about 20 minutes. Just as he got back he heard firing in [the] direction of Custer and was told that previous firing down there was heard while he had been gone. Just as he got back he looked and saw Company D advancing toward Custer. They were some distance out but still in sight.

<u>ANALYSIS</u>

In Hare's in-depth interview with Walter Camp, he reaffirms his twenty-minute trip to the advance pack mules and back. He also clarifies the distance to a more accurate distance of about a mile from Reno Hill. He mentions that he did not hear any firing from Custer's field before leaving, which places his departure time to just before the start of Custer's engagement down at Ford B, heard by so many participants on Reno Hill. Hare also tells us that when he returned to Reno Hill twenty minutes later, he noticed D Company moving ahead on their advance to the Peaks. They were some distance out but still in sight.

12) Lieutenant Edward Mathey (RCOI): after sending ahead the two ammunition mules, he claimed he reached Reno Hill with the advance mules in less than thirty minutes—probably twenty minutes—at a fast walk. He thought it was about 3:00 p.m. when he reached Reno Hill. The two ammunition mules arrived earlier.

ANALYSIS

Lieutenant Mathey gives us more detailed information on which to base our timeline for the arrival of his advance ammunition mules. He tells us that he first sent ahead the two ammunition mules following Lieutenant Hare's directive, then followed shortly thereafter with his own group of pack mules, arriving on Reno Hill about twenty minutes after moving out at a fast walk (about four to five miles per hour). As his distance from Reno Hill was only a mile or so to the south, it would take him about twenty minutes to arrive. If Mathey moved out at 2:45 p.m., the head of his pack train would arrive on Reno Hill about 3:05 p.m.—or about 3:00 p.m. as Mathey himself recollected.

13) Private Charles Windolph – H Company (*I Fought With Custer*, with my annotations in brackets): Reno and Benteen and two or three of the officers held a little conference, and we saw Lieutenant Hare, who had had charge of Reno's Indian scouts, suddenly mount Godfrey's horse and head back down our trail to the southward. Maybe it was fifteen or twenty minutes, or possibly a half hour, before he came up at a trot with several pack mules, loaded with ammunition boxes. The rest of the dozen ammunition mules slowly dribbled in, and before long the pack train itself came up. We now had 24,000 rounds of rifle ammunition, food and a new force of not less than 130 armed men. Except for an odd hostile shot, fired from a distance, there was no firing near us now. The Indians apparently had left for the north. We could hear the sound of distant firing echoing down through the hills and valleys from that direction. Custer must be down there.... I remember that Captain Weir suddenly rode off to the north alone, and a minute later, Lieutenant Edgerly, second in command of "D," followed with the whole troop. The pack mules were coming up about this time and there was a lot of speculating going on. As I recall, Reno had seven wounded men, some of them in pretty bad shape....

ANALYSIS

Private Windolph supports other accounts, which place Lieutenant Hare's arrival back on Reno Hill with the two advance pack mules in about twenty minutes—or close to 3:00 pm. It is important to note that Windolph then claims that "the rest of the dozen ammunition mules slowly dribbled in." According to other participants, this would be between 3:05 and 3:15 p.m. Windoph indicates that these dozen or so ammunition mules arrived on Reno Hill before the rest of the main pack train arrived ("before long the pack train itself came up). Windolph also indicates that Captain Weir and D Company rode off to the Weir Peaks about the time that the [ammunition] "pack mules were coming up."

OVERALL ANALYSIS

There is a general split in the time estimates for the arrival on Reno Hill for the advance packs traveling under Lieutenant Mathey. Captain Benteen and Lieutenant Wallace both suggest an arrival time of about an hour after Benteen arrives on the bluffs (at 2:30 p.m.). These two participants indicate an arrival time for these pack mules at about 3:30 p.m. at the earliest. Eight other accounts contradict these two estimates. Trumpeter Martin gives us a time estimate that is simply too short—ten to fifteen minutes after Benteen's arrival—and is not supported by any other participants and can thus be safely disregarded as unsupportable. Our remaining participants—Moylan, Godfrey, Varnum, Mathey, and Windolph—all indicate an arrival time of thirty to forty-five minutes after the arrival of Benteen. Dr. Porter's broader time range would support these estimates, along with Major Reno and Private Davern, and with Hare indicating an even shorter time for the arrival of the first two ammunition mules following behind him. I would suggest that these five accounts, supported by four others, are the more accurate assessment of when the advance packs under Lieutenant Mathey arrived on the bluffs. The two advance pack mules following Lieutenant Hare would arrive on Reno Hill by about 3:00 p.m., shortly after Weir's departure on his advance to the Peaks, with Lieutenant Mathey's remaining pack mules likely arriving between 3:05 and 3:15 p.m.

CONCLUSION

2:45 p.m.: Mathey sends ahead his two strongest and fastest mules after receiving Lieutenant Hare's message to hurry up the packs. Mathey starts out with the remainder of his pack mules shortly thereafter.

3:00 p.m.: The two lead ammunition mules arrive on Reno Hill, shortly after Captain Weir and D Company have left on their advance to the Peaks.

3:05 to 3:15 p.m.: Lieutenant Mathey and his remaining pack mules from the advance group of packs arrive on Reno Hill with the rest of the ammunition mules, about the same time as Major Reno's return to the bluffs—at 3:15 p.m.—after checking on Hodgson's body.

22. CUSTER'S INITIAL VOLLEY FIRE

Here are the participant recollections of hearing the first volleys fired by Custer's wing:

Appendix A: Participant Timeline of the Battle of the Little Big Horn

1a) George Herendeen (RCOI): heard volley fire from Custer less than half an hour after Reno left the timber (he heard a great many volleys). They were in the woods for twenty minutes when they heard firing downstream. The heavy firing lasted for about an hour.

1b) George Herendeen (Camp Interview, *Custer in 76*): heard firing downriver about half an hour after Reno's troops left the timber. He heard volleys fired in the early part of the fight. There were a great many volleys, and there were scattering shots between them. He thinks Custer's battle did not last over an hour.

ANALYSIS

Herendeen claims to have heard volley fire from Custer less than a half-hour after Reno left the timber. As the time for Reno's retreat from the timber was established at 2:10 p.m. Department Headquarters watch time, Herendeen would indicate hearing volley fire from Custer's command about 2:40 p.m. or less. Herendeen had returned to the woods by this time—about 2:20 p.m.—after his aborted attempts to escape and had been there for about twenty minutes when he heard a great many volleys fired by Custer's troops. Herendeen can hear the heavy firing lasting for one hour, or until about 3:40 p.m.

2) Lieutenant Charles Varnum (RCOI): hears firing from Custer's command shortly after Benteen joins them on the bluffs. He implies hearing two volleys.

ANALYSIS

Varnum's recollection fits in well with Herendeen's recollections, with two of Custer's volleys being heard "shortly after Benteen joins them on the bluffs." As Benteen joined Reno on the bluffs at about 2:30 p.m., Custer's volleys would begin about ten minutes later—about 2:40 p.m.

3a) Lieutenant Luther Hare (RCOI, *Army and Navy Journal*): heard two distinct volleys before he left on his ride back to the pack train.

3b) Lieutenant Luther Hare (RCOI): Just after Benteen came up and before he left for the packs, Godfrey called his attention to firing downstream. He heard two distinct volleys. (p. 248, Graham RCOI)

3c) Lieutenant Luther Hare (Camp interview, *Custer in 76*): claims he heard no firing from Custer's field before he started back for the ammunition packs.

ANALYSIS

Although this account supports those of both Varnum and Herendeen, Hare would contradict himself many years later in an interview with Walter Camp and deny that he heard any volleys or firing from Custer's field before riding back to the advance pack mules—a time we have placed at 2:38 p.m.

4) Lieutenant Winfield Edgerly (RCOI): as they joined Reno on the bluffs they heard heavy firing downstream, by volleys, almost immediately. He discussed it with Weir and their first sergeant. He says the firing they first heard on arriving on the bluffs lasted for about forty-five minutes. (p. 393, (Graham RCOI) He claims they heard firing from downstream within five minutes (or less) of arriving on the bluffs.

ANALYSIS

Upon reaching the bluffs, D Company was deployed the farthest north of Reno Hill, on the southern end of Sharpshooter Ridge, and just west of the bluffs. Within five minutes of deploying on this ridge, Edgerly and Weir hear volleys fired from Custer's field. If D Company deployed on Sharpshooter Ridge at 2:35 p.m., this volley fire would have been heard about 2:40 p.m., supporting the recollections of Varnum, Herendeen, and Gerard. Edgerly claims that this heavy firing lasted about 45 minutes, or until 3:25 p.m. It is interesting to note here that D Company would be on the Weir Peaks by that time, yet Edgerly makes no mention of hearing any heavy firing from Custer's command while out on the Peaks.

5) Private Thomas O'Neill - G Company (Camp Interview, *Custer in 76*): after Reno retreated from the valley, he heard firing from Custer's field—at one time he heard three volleys—and thought the firing ended from that direction about 5 p.m.

ANALYSIS

Private O'Neill does not give a specific time for the start of Custer's volley firing, but mentions hearing 3 volleys, with the firing on Custer's field lasting till about 5 p.m. This supports Gerard's recollections for the length of the fighting on Custer's field.

6a) Fred Gerard (RCOI): 10-15 minutes after Reno leaves the valley, he hears 3 or 4 volleys of 50 to 100 guns. He heard continuous fire moving downstream. After the first few volleys, it was 20-25 minutes before the firing became general. The heavy, continuous gunfire further downstream lasted about 2 hours. He estimated that he heard the heavy firing between 3 and 4 p.m.

ANALYSIS

Gerard tells us that ten to fifteen minutes after Reno left the valley (or ascended the bluffs) he heard three or four volleys fired by one or two companies. The head of Reno's command would ascend the bluffs between 2:15 and 2:20 p.m.; if we use this as a base, it would place Gerard's timing sequence in general support of Varnum, Herendeen and Edgerly's recollections, but with a slightly earlier start at about 2:35 p.m. Gerard also tells us that the firing from Custer's field lasted for about two hours and twenty to twenty-five minutes, which supports the recollection of Private O'Neill.

6b) Fred Gerard (Camp Interview, *Custer in 76*): about half an hour after Reno's troops retreated from the timber, firing began downriver. Before this, there had been some general firing; then, he heard volleys and thought Custer was coming nearer. He heard two volleys fired, which he believed were fired down at the river, about thirty to forty carbines in each volley.

ANALYSIS

Gerard now clarifies that he heard Custer's volleys fired about thirty minutes after Reno left the timber, which he had noted occurred at 2:10 p.m., thus placing the time for these volleys at about 2:40 p.m. Gerard believes the volleys were fired by one company down at the river—which would likely be E Company at Ford B. The general firing he heard before these volleys was likely sporadic fire directed at the troopers from various warriors after spotting Custer's command approaching the river.

7) Lieutenant De Rudio (RCOI, with my annotations in brackets): says he heard Custer's fighting downriver about twenty minutes after Reno left the timber. Reno was not on the hill for ten minutes when the firing started downstream. He heard "immense volleys" of firing, and most of the Indians left Reno; some went down the river while others went up to the highest bluffs. When the heavy firing was heard from downstream, the Indians left Reno. The firing from downriver lasted about one and a half hours. The Indians watched that command [i.e., Benteen's battalion on the bluffs] for about ten minutes, then the heavy firing began at the lower end of the village and the Indians began leaving Reno.

ANALYSIS

De Rudio says it was about twenty minutes after Reno left the timber—placing the time at about 2:30—when Custer's firing was first heard downstream; he says this was about ten minutes after Reno had reached the top of the bluffs or about 2:20 p.m. He would later revise this time estimate by claiming that the Indians watched Benteen's battalion on the bluffs for about ten minutes before leaving to attack Custer after hearing firing from the north, thus placing the start of Custer's volley fire at about 2:40 p.m., supporting the recollections of Varnum, Herendeen, Gerard, and Edgerly. De Rudio claims this firing lasted for about an hour and a half, ending at about 4:10 p.m.

8a) Lieutenant Edward Godfrey (*Godfrey's Field Notes*): Upon meeting Reno's command on the bluffs, they heard volley fire and the rattle of guns.

8b) Lieutenant Edward Godfrey ("Custer's Last Battle," *Century Magazine*): after Hare left for the ammunition mules, they heard firing downriver in the direction of Custer's command .... We heard 2 distinct volleys which excited some surprise .... I have little doubt now that these volleys were fired by Custer's orders as signals of distress and to indicate where he was.

ANALYSIS

Godfrey's account supports the recollections of Varnum, Edgerly, Herendeen, and Gerard. Godfrey indicates that this volley firing occurred shortly after Lieutenant Hare departed on his ride back to the advance pack mules at about 2:38 p.m. Godfrey believes that the volleys were fired to signal to the rest of the regiment the location of Custer's wing, now engaging the enemy.

9) Private Edward Davern (RCOI): heard volleys about thirty minutes after he reached the bluffs.

ANALYSIS

As we cannot determine the time that Davern actually reached the bluffs, we cannot determine the time he heard volleys fired from Custer's field "30 minutes after."

OVERALL ANALYSIS

The recollections of Herendeen, Varnum, Hare, Edgerly, Gerard, De Rudio and Godfrey all support the premise that the initial reports of firing of volleys from the Custer field occurred at about 2:40 p.m. Headquarters watch time. This should mark the start of the Custer fight at or near Ford B.

CONCLUSION

2:40 p.m.: The initial volleys from Custer's command are heard by several officers, troopers, and scouts on the bluffs and those left behind in the timber position in the valley. Gerard indicates that these volleys were fired by one company—this would likely be E Company—near the river at Ford B. The result of this volley fire was the immediate departure of nearly all the warriors from Reno and Benteen's command within about ten minutes, who were now riding hard to confront Custer's command demonstrating on the east bank of Ford B.

23. CUSTER'S SECOND VOLLEY FIRE

Here are the participant's descriptions of hearing the second set of volleys fired by Custer's wing:

1) Lieutenant Edward Godfrey (RCOI): Heard two distinct volleys fired shortly after Hare returned from the pack train.

Appendix A: Participant Timeline of the Battle of the Little Big Horn

ANALYSIS

According to the Participant Timeline, Lieutenant Hare took about twenty minutes for his ride to the pack train and return to Reno Hill, placing his return there at about 3:00 p.m. Thus, Lieutenant Godfrey tells us that he heard two distinct volleys fired by Custer's troops shortly after 3:00 p.m., or about twenty to twenty-five minutes after the first set of volleys were reportedly heard.

2) Fred Gerard (RCOI): After he heard the first set of volleys fired by Custer's command, it was twenty to twenty-five minutes before the firing became general. The heavy continuous gunfire farther downstream lasted about two hours. He estimated the heavy firing he heard occurred between 3 and 4 p.m.

ANALYSIS

Gerard tells us that he heard a gap of about twenty to twenty-five minutes between the initial volleys he heard fired until the firing became heavy, continuous gunfire. Although Godfrey does not mention hearing any additional volleys, he does notice the start of heavy firing at this time. The participants tell us that the initial volleys of Custer's command started about 2:40 p.m., and Gerard's recollections support this time estimate. Thus, the start of Gerard's heavy firing would have started twenty or twenty-five minutes later, or at 3:00 or 3:05 p.m., thus supporting Godfrey's recollections.

3) Lieutenant Winfield Edgerly (RCOI): Says the firing they first heard when arriving on the bluffs lasted about forty-five minutes.

ANALYSIS

Although Edgerly, like Gerard, does not mention hearing any volleys at this time, he does reference hearing firing that lasted for about forty-five minutes, covering the same period of time when a second set of volleys were reportedly heard by Godfrey, or lasting until about 3:30 p.m.

4) George Herendeen (Camp Interview, *Custer in 76*): He heard volleys fired in the early part of the fight. It was a great many volleys with scattering shots between them. He thinks Custer's battle did not last over an hour.

ANALYSIS

Herendeen hears a great many volleys fired in the early part of the fight and hears the firing continue for just under an hour, supporting the recollection of Edgerly's timeframe and including the period when Godfrey heard volleys fired by Custer's troops, hearing firing that lasted until about 3:30 p.m.

5) Lieutenant Carlo De Rudio (RCOI): The firing downriver lasted about an hour and a half.

ANALYSIS

Lieutenant De Rudio hears firing from Custer's position that lasts for an hour and a half, thus encompassing and supporting the time that other participants, closer to the firing, describe hearing volleys fired from Custer's command.

6) Lieutenant Luther Hare (Camp Interview, *Custer in 76*): Says that just as he got back to Reno Hill (twenty minutes after leaving for the pack train) he heard firing in the direction of Custer and was told that previous firing from down there was heard while he had been gone.

ANALYSIS

According to the Participant Timeline, Lieutenant Hare tells Walter Camp that just as he returned to Reno Hill from his ride to the pack train, at about 3:00 p.m., he heard firing from Custer's direction. This supports the recollection of Godfrey's time estimate of hearing volleys fired from Custer's troops during this period. Hare also mentions that others told him that firing—he does not mention volleys, but other participants do—was heard previously while he had been gone from the hill on his ride back to the packs. This would be a reference to the initial volleys heard fired from Custer's command about 2:40 p.m., just a few minutes after Hare departs from Reno's Hill.

7) Packer Benjamin Churchill (RCOI - *Army and Navy Journal*): [for] about 1 1/2 hours after he reached the bluffs he heard firing from Custer's field. He heard 4 or 5 volleys at that time—likely about 3 pm—about 2 1/2 to 3 miles downstream. He thought the firing lasted about 1 1/2 to 2 hours before the Indians came back in force to Reno's position on the bluffs. [Note: Churchill would later amend his RCOI testimony to read: "I heard the volley firing down the river soon after we arrived on the hill with the packs."]

ANALYSIS

Packer Churchill first arrived on the bluffs at about 3:00-3:05 p.m. with the two lead mules that followed in the wake of Lieutenant Hare. He tells us that he then heard four or five volleys fired from Custer's field two and a half to three miles to the north, for between one and a half to two hours (or until 4:30-5:00 p.m.) until the Indians came back in force to Reno's position on the bluffs [i.e., to the Weir Peaks]. Churchill's amended testimony indicates that he heard the volley firing from Custer's wing at sometime between 3:05 and 3:15 p.m. or a short time later.

OVERALL ANALYSIS

One participant, Lieutenant Godfrey, heard a second set of volley fire from Custer's command about twenty-five minutes after the first set of volleys were heard fired. This recollection is supported by six other participants who indicated hearing heavy firing from Custer's field during this time. Gerard and Hare's accounts indicate this occurred about twenty or twenty-five minutes after the initial volleys were heard. Other participants describe hearing this firing lasting at least forty-five minutes or longer—depending on our source—after the first set of volleys were fired. As none of the participants' accounts from the timber position describe hearing volleys fired during this second phase of firing from Custer's field, the volleys that were reportedly heard from Godfrey instead were likely fired by a portion of Custer's command that were deployed farther away from the river and Ford B—and closer to the bluffs—and thus beyond the hearing range of those left in the timber position, who had the disadvantage of intervening bluffs to mask the sound of this volley firing. None of the other officers on Reno Hill, including Edgerly, specifically mentioned hearing volley firing. It is possible that their motives for not doing so may be entirely different from those of the participants in the timber, or possibly Godfrey was simply mistaken about hearing volleys at this time.

CONCLUSION

3:05 p.m.: The second set of volleys are heard fired from Custer's troops, likely from the Luce/Blummer Ridge complex further back from the river, and likely fired at warriors entering Medicine Tail Coulee or North Medicine Tail Coulee from the bluffs on the north side of the Weir Peaks, or possibly from Wolf Tooth's small band of warriors east of Blummer and Luce Ridge.

3:05-3:30 p.m.: The firing during this phase is described as heavy and likely takes place on Luce and Blummer (i.e., Nye-Cartright) Ridge. It ends about 3:30 p.m., which likely indicates the withdrawal of this squadron of Custer's command further north across Deep Coulee on their way towards Calhoun Hill.

24. CUSTER'S LAST VOLLEYS

According to the battle participants, additional volleys fired at a later time during the battle were reportedly heard from Custer's field. Here are their accounts of when these shots were heard along with other supporting accounts of hearing sounds of his battle:

1a) Captain Myles Moylan (RCOI – *Army and Navy Journal*): heard volley firing from Custer's field about an hour after reaching Reno Hill and commented on it to Captain Thomas McDougall.

1b) Captain Myles Moylan (RCOI): claims to have heard firing downstream—faint volley firing—after the arrival of the pack train and Rear Guard. He discussed this with Captain Thomas McDougall.

ANALYSIS

In the first account, reported by the *Army and Navy Journal*, Captain Moylan describes hearing what amounts to a third set of volleys fired by Custer's command about an hour after he arrived on the bluffs. As Moylan was one of the earliest officers to arrive on the bluffs at about 2:20 p.m., these volleys he heard fired would have been about fifteen minutes or so after the volleys described by Lieutenant Godfrey and Private Davern and would have been fired at about 3:20 p.m. However, other accounts from the RCOI indicate that Moylan only heard these volleys fired after the arrival of the main pack train and Rear Guard, and he discussed hearing these volleys with Captain McDougall, which places the time of these volleys a good deal later than an hour after Moylan reached the bluffs. As McDougall indicated, he did not arrive on Reno Hill before 3:45 p.m.; this would be the earliest time these volleys could have been fired. However, we can place the time of these volleys back even further—likely at or just after 4:00 p.m., as Moylan indicated that it occurred after the arrival of the pack train on Reno Hill.

2a) Captain Thomas McDougall (RCOI): heard firing off to his right (as he was coming up the bluffs just after crossing the North Fork). He says he heard two volleys fired about fifteen minutes before reaching Reno Hill.

ANALYSIS

Captain McDougall tells us that he heard two volleys fired by Custer's command as he was ascending the bluffs to Reno Hill. The time for this event would be somewhere between 3:30 and 3:45 p.m. The volleys heard fired by McDougall were apparently not the same faint volleys heard and reported by Captain Moylan, who indicated that the pack train had already joined them atop Reno Hill at the time they were heard. McDougall's report of hearing volley fire occurred a good fifteen minutes before the pack train arrived on the bluffs.

2b) Captain Thomas McDougall (Camp Interview, *Custer in 76*): On arriving on Reno Hill, he heard firing from Custer's field and asked Godfrey if he heard it. Godfrey, who was partially deaf, said he believed he did.

ANALYSIS

Captain McDougall now reports hearing additional fire—although not necessarily volleys—from Custer's field upon his arrival on Reno Hill about fifteen minutes later. The timeframe for this would be between 3:45 and 4 p.m. and synchronizes very well with the faint volleys heard by Captain Moylan shortly after McDougall arrived on the bluffs. Moylan discussed this with McDougall, who, in turn, discussed hearing this with Lieutenant Godfrey, who, despite being partially deaf, also claimed to have heard them.

3) Fred Gerard (RCOI): estimated that the heaviest firing he heard occurred between 3 and 4 p.m.

ANALYSIS

Gerard's recollections of hearing heavy firing between 3 and 4 p.m. supports the accounts of both McDougall and Moylan who reported hearing volleys fired during this timeframe (between 3:30 and 4:00 p.m.)

4) Lieutenant Carlo De Rudio (RCOI): The firing from downriver lasted about one and a half hours.

ANALYSIS

De Rudio's account of hearing firing from Custer's field for 1.5 hours supports the accounts reported above from Moylan and McDougall, whose reports of volley firing would fall within De Rudio's timeframe.

5) Lieutenant Edward Godfrey (*Custer's Last Battle, Century Magazine*): McDougall came up with the pack train and reported the firing he heard when he reported to Reno.

ANALYSIS

Lieutenant Godfrey validates McDougall's recollections of hearing firing from Custer's field, which he reported to Major Reno upon his arrival on the bluffs.

6) Packer Benjamin Churchill (RCOI – *Army and Navy Journal*), with my annotations in brackets: [for] about an hour and a half after he reached the bluffs, he heard firing from Custer's field. He heard four or five volleys at that time [note: likely about 3 pm] about two and a half to three miles downstream. He thought the firing lasted about an hour and a half to two hours before the Indians came back in force to Reno's position on the bluffs.

ANALYSIS

Packer Churchill first arrived on the bluffs at about 3:00 p.m. with the two lead mules that followed in the wake of Lieutenant Hare. He tells us that he then heard four or five volleys fired from Custer's field 2.5 to three miles to the north with the firing lasting between 1.5 to two hours (until 4:30 - 5:00 p.m) until the Indians came back in force to Reno's position on the bluffs [i.e., at the Weir Peaks].

7) Private Edward Davern (RCOI): says that after reaching the top of the bluffs, he heard Custer's firing downstream "occasionally for about an hour and a half."

ANALYSIS

Private Davern hears the firing from Custer's field from roughly 2:40 to 4:10 p.m., suggesting that he could hear the action taking place in the Keogh sector before the fighting moved farther downriver and out of his hearing range.

8) Private Thomas O'Neill – G Company

(Walter Camp's interview with Private Tom O'Neill on Oct. 12, likely in 1919, published in Kenneth Hammer's *Custer in 76*, with my annotations in brackets):

During the afternoon, they heard firing in [the] direction of Custer, at one time 3 volleys, and O'Neill thinks [that] all [the] firing in [the] direction [of Custer was] over [by] about 5:00 pm [note: either sun time or HQ watch time. The Participant Timeline indicates that the Custer Fight ended about 5:20 p.m. HQ watch time] as nearly as he can estimate.

OVERALL ANALYSIS

The participants described two additional periods of volley fire heard from Custer's command, each separated from the other by about fifteen or twenty minutes. The first set of volleys, reported by Captain McDougall as he was ascending Reno Hill with the Rear Guard, occurred about fifteen minutes before he arrived on the bluffs. The earliest time we can place for this event would be 3:30 p.m., while the latest would be about 3:45 p.m. The second set of volleys reported by Captain Moylan occurred just after the arrival of McDougall and the pack train itself, which would be at or shortly after 4:00 p.m. Godfrey also heard about this firing, and it was reported to Major Reno. Packer Churchill's recollections of firing from Custer's field, lasting until 4:30 or 5:00 p.m., support that of Private Thomas O'Neill. Although it is difficult to distinguish precise times here, it is clear that Custer's command was still alive and defending itself with volley fire at the time of the arrival of the pack train and Rear Guard on Reno Hill.

As these volleys are reported to be faint, it is very likely that they were fired from either Finley Ridge or Calhoun Hill by elements of Keogh's battalion deployed on these positions.

CONCLUSION

3:30 to 3:45 p.m.: The third set of volleys fired from Custer's command are reportedly heard by Captain McDougall as he was ascending the bluffs to Reno Hill with the pack train and Rear Guard.

4:00 to 4:10 p.m.: The fourth set of volleys fired by Custer's command is reportedly heard by Captain Moylan, McDougall, and Godfrey and reported to Major Reno.

25. M TROOP LEAVES FOR THE WEIR PEAKS

Determining when the rest of Reno's troops advanced to the Weir Peaks has been an ongoing mystery that has never been fully solved since the time of the battle. This is due primarily to the efforts of the officers to obfuscate when and how this advance took place. Most of the officers at the Reno Court of Inquiry gave the impression that this advance took place all at one time and involved all of the remaining six companies and the pack train. Years later, in private correspondence, narratives, and interviews with Walter Camp, an entirely different story emerged that indicated the advance to the Weir Peaks was made, in fact, in a piecemeal fashion, with the likelihood that no more than three of the six remaining troops ever made it out to the Peaks to join D Company there. Research into the primary sources indicates that M Company was the first, after D Company, to leave Reno Hill and advance to the Peaks. Only Private Morris' account states this openly, but we have enough circumstantial evidence to support such a conclusion. Here is what the participants had to say about the advance of the first portion of Reno's command to the Weir Peaks:

1) Captain Frederick Benteen (*Benteen's Narrative*, with my annotations in brackets): I followed [Weir's D Company] with the remaining two companies of my battalion [i.e., K and H companies on] the trail that Weir had "sallied on" ... some ten minutes later perhaps.

ANALYSIS

Captain Benteen tells us that K and H companies followed Captain Weir out to the Weir Peaks about ten minutes after D Company began their advance, placing Benteen's departure from Reno Hill at about 3:05 p.m. This time estimate is patently absurd and is a glaring example of Benteen's willingness to engage in blatant falsehoods to cover up the truth of this episode. No other independent accounts support Benteen's claim that he brought H and K companies out to the Peaks ten minutes after Weir left Reno Hill. Captain McDougall indicated that both Benteen and Godfrey were still on Reno Hill at the time the Rear Guard arrived there at 4 p.m., and Lieutenant Godfrey supported this recollection in his own published article on the battle, indicating that both he and Benteen were still on Reno Hill at 4:20 p.m., or roughly two hours after Benteen first arrived on the bluffs.

2) Private Jean Gallenne – M Company (letter to Father Genin, 7/5/76, with my annotations in brackets): We have about forty-five minutes respite [after M Company reaches the bluffs before moving out again to the Weir Peaks].

ANALYSIS

Private Gallenne indicates that M Company had a respite of about forty-five minutes before moving out again from Reno Hill. If Gallenne is basing this on the 2:30 p.m. junction with Benteen's battalion on the bluffs, he is suggesting that M Troop then departed Reno Hill for the Peaks about forty-five minutes later, or about 3:15 p.m. If, however, he is basing the start when the final members of M Company ascended the bluffs—about 2:45 p.m.—then he is suggesting that M Troop departed from Reno Hill about 3:30 p.m.

3a) Lieutenant Winfield Edgerly (RCOI, with my annotations in brackets): said it was about forty-five minutes [after D Company left on its advance] for Benteen and his command [to come up].

ANALYSIS

Lieutenant Edgerly gives us the impression that Benteen and his battalion came up to the Weir Peaks about forty-five minutes after D Company left Reno Hill on its advance. As D Company likely left Reno Hill at about 2:55 p.m., Edgerly is suggesting that H and K companies reached the Weir Peaks about 3:40 p.m. The problem is that we know this simply cannot be true, as both Godfrey and McDougall tell us that both Benteen and Godfrey were still on Reno Hill until after the arrival of the pack train and Rear Guard after 4 p.m. So how do we explain this apparently conflicting evidence? I believe Edgerly was being coy here. He gives us a clear impression that both Benteen and Godfrey arrived on the Peaks forty-five minutes after his own company did, but this is simply not correct. I believe Edgerly was actually indicating the arrival of the second troop on the Peaks about forty-five minutes after D Company got there. This second troop was M Company. Edgerly wishes to give the impression that the arrival of this second troop actually was part and parcel of Benteen's later advance to the Weir Peaks. In point of fact, it was not. M Troop was never a part of Benteen's battalion, and Captain French apparently moved down to the Weir Peaks entirely of his own volition. This is a fact the surviving officers of the 7th Cavalry went out of their way to obfuscate at the Reno Court of Inquiry.

3b) Lieutenant Winfield Edgerly (Camp Interview, *Custer in 76*): says that Troops M, K and H arrived on the Weir Peaks. He says they must have followed within thirty minutes of the time D Company left Reno Hill. He says that he was on the second ... peak (farthest north). French's Troop came up next to them [on the western ridge]; Godfrey came next [behind them on the Peaks], then Benteen [on the west side]. [After all three companies—M, K, and H—arrived on the Peaks] Edgerly advances D Company farther down along the ridge (loaf), dismounted, and returned fire at Indians on a peak in the direction of Ford B. He said the firing lasted about half an hour before they were ordered back to Reno's position.

ANALYSIS

Lieutenant Edgerly informs us that only three more troops ever reached and deployed on the Weir Peaks—M, K, and H. This contradicts the accounts of other officers who suggested that they or their companies had reached the Weir Peaks. It is very likely that only four troops ever reached and deployed on the Weir Peaks. This suggests another part of the coverup engaged in by the 7th Cavalry officers at the Reno Court of Inquiry. In his account to Walter Camp above, Edgerly claims that these troops must have left Reno Hill about thirty minutes after D Company left, placing their departure at about 3:25 p.m.

The problem here is that we know from Godfrey and McDougall's account that K and H companies did not leave Reno Hill until after the arrival of the pack train and Rear Guard on the bluffs and that did not occur anywhere close to 3:25 p.m. That leaves only M Troop, who could have left that early. Edgerly continues to be coy about this episode to Walter Camp, still promoting the fiction that M, K, and H companies all moved out to the Weir Peaks at the same time. Edgerly does give us a clue to the actual order of arrival for these three companies, when he chooses to say M, K, and then H Company last. That was very likely the exact order of their arrival on the Weir Peaks.

4a) Lieutenant Luther Hare (RCOI, with my annotations in brackets): After returning from his ride to the [advance] pack train, Reno sent him to Weir to have him try to open communications with Custer. He gave the message to Weir and then headed back. He met the command coming downstream (Graham, p. 249). He estimated it was about forty-five minutes from the time Benteen arrived [on the bluffs] to the time the command headed toward Weir.

ANALYSIS

Lieutenant Hare tells us that after his return to Reno Hill from delivering his message to the advance pack mules—this would be about 3 p.m.—he was sent by Major Reno with a message for Captain Weir out on the Peaks. Hare would not have been given this message until at least 3:15 p.m. when Reno returned to the bluffs after his trip down to look for Hodgson's body. After delivering his message to Weir—at about 3:30 p.m.—he then returned back to Reno Hill, passing a portion of the command heading out to the Weir Peaks. Hare estimated that this command left Reno Hill about forty-five minutes after the arrival of Captain Benteen on the bluffs. As Benteen's arrival on the bluffs was at about 2:30 p.m., Hare suggests that this part of Reno's command advancing to the Weir Peaks to support Weir's advance left Reno Hill about 3:15 p.m. This, however, despite supporting Private Gallenne's time estimate of a forty-five minute respite before M Troop moved out again, appears to be too short a time span for M Troop to begin their advance, as Hare himself could not have left Reno Hill much before 3:15 p.m. It is more likely that Edgerly's time estimate of thirty minutes after D Company left—or about 3:25 p.m.—is a more accurate indicator for M Troop's departure from Reno Hill. Like the other surviving officers of the 7th, Hare chooses to obfuscate the truth by substituting the word "command" in place of "Troop" in describing this advance to the Weir Peaks at this time.

4b) Lieutenant Luther Hare (Camp interview, *Custer in 76*): Thinks the other six companies did not advance until "at least" a full hour after Reno retreated up the bluffs.

ANALYSIS

In Hare's latter interview with Walter Camp years later, he continues the fiction that all six of Reno's remaining companies—minus D Company—advanced together to the Weir Peaks. This clearly did not happen, yet the surviving officers continued to promote this false notion. Hare does give us a solid clue, though it was at least a full hour—and possibly a bit longer—after Reno retreated up the bluffs before the first company advanced to support D Company out on the Weir Peaks. The Participant Timeline established the head of Reno's battalion reaching the bluffs at about 2:15 to 2:20 p.m., with the remaining troopers arriving within the next twenty-five or thirty minutes. Thus Hare's time estimate for the departure of the first troops to the Weir Peaks would likely occur at some time between 3:15 and 3:45 p.m.

5) Sergeant Ferdinand Culbertson – A Company (RCOI): claims Weir left for the Peaks twenty minutes after the arrival of the first ammunition mules. He claimed the command moved out toward the Weir Peaks just under an hour after hearing Custer's first volley just after Benteen arrived on the bluffs.

ANALYSIS

Sergeant Culbertson claims that Captain Weir left for the Peaks twenty minutes after the arrival of the first ammunition mules. As the first ammunition mules arrived at about 3:05 p.m., this would place Weir's departure from Reno Hill at about 3:25 p.m. The problem with this is that other participants clearly stated that Weir and D Company left Reno Hill before the arrival of the ammunition mules. So, how do we reconcile Culbertson's contradictory account? I believe the answer lies in the fact that at the time of D Company's advance to the Weir Peaks, Culbertson had left with Reno to go down to the river to find Lieutenant Hodsgon's body. He did not witness Weir and D Company leave for the Peaks. Upon his return to the bluffs about twenty-five minutes later, he witnessed the advance of a single company to the Weir Peaks. This company was not—could not have been—D Company, as D Company was already on the Peaks, as other witnesses, including Lieutenant Edgerly himself, so stated. Culbertson very likely witnessed the second troop move out to the Peaks, a company he thought was Weir's D Troop, but was, in fact, Captain French's M Troop. If this interpretation is correct, then Culbertson is telling us that the second company to move out to the Weir Peaks left twenty minutes after the arrival of the first ammunition mules on Reno Hill. This would place the departure of this company at about 3:25 p.m. This second company to leave for the Weir Peaks was very likely M Troop.

6) Private William Morris – M Company (letter to Robert Bruce, May 23, 1928, with my annotations in brackets): ... we [i.e., Company M] moved up [to the Weir Peaks], Captain French leading to the top of the highest bluff where Captain Weir [was] and Lieutenant Godfrey went [to] later .... Benteen, with his battalion, reached us about an hour and a half after we were on the hill [i.e., Weir Point], and Captain McDougall, with the rear guard, arrived about the same time.

ANALYSIS

Private Morris, in this private letter above, confirms the thesis that Captain French's M Troop arrived on the Weir Peaks well before the arrival there of Captain Benteen and his battalion (H and K Troops). Morris indicates that Benteen did not arrive on the Peaks for about an hour and a half after M Troop got there, thus giving the lie to the surviving officers of the 7th who went out of their way to create the false impression that M, K, and H Troops all moved out to the Weir Peaks together. Private Morris confirms that M Troop was led out to the Peaks by Captain French on his own volition and was never a part of Benteen's general advance, which occurred at a much later time after the arrival of the pack train and Rear Guard on Reno Hill. Morris also indicates that Godfrey's K Company arrived at the Peaks before H Company, again confirming the premise that both Edgerly and Hare's reference to M, K, and H Company's arrival on the Peaks actually referred to their specific order of arrival there.

7) Private Edward Pigford—M Company (Camp Interview, *Custer in 76*, with my annotations in brackets): after Company M got on the Weir Peaks, they looked in the direction of Custer [or Battle] Ridge and saw Indians firing from a big circle; but it gradually closed until they seemed to converge into a large black mass on the side hill toward the river [note: likely a reference to Finley Ridge] and all along the [Battle] ridge. He thinks that what they saw were the last stages of the fight.

ANALYSIS

Private Pigford appears to be describing the action leading to the destruction of the Keogh battalion on Finley Ridge, Calhoun Hill, and Battle Ridge. As Pigford was a member of M Troop, this action could only have taken place at some time after the arrival and deployment of M Troop on the Weir Peaks. If the participant accounts are correct that M Troop was the second company to leave Reno Hill about thirty minutes after D Company, the time of its departure from Reno Hill would be about 3:25 p.m., and their arrival on the Weir Peaks would be about 3:40 p.m. The action that Pigford describes above would likely have occurred at some time after 3:40 p.m.

OVERALL ANALYSIS

Based on the participant accounts listed above, it is clear that at least some portion of Reno's command moved out to the Weir Peaks within thirty minutes of D Company's departure. Through a process of elimination, we can surmise that M Troop was this second troop to move out—independently—to the Weir Peaks. They likely left Reno Hill about thirty minutes after D Troop—or twenty minutes after the arrival of the first ammunition mules on Reno Hill—at about 3:25 p.m., as our participants describe, and arrived on the Peaks about 3:40 p.m., passing Lieutenant Hare on his way back to Reno Hill. The destruction of the Keogh battalion, described so well by Private Pigford of M Company, likely did not occur at 3:40 or 3:45 p.m. as this was the time referenced by McDougall as hearing volley fire from Custer's command. Thus, Keogh's command would have been destroyed at some time after 3:45 p.m. From the evidence given to us so far, a more accurate time for this event to have taken place cannot be determined, and we must leave this question for a future discussion. All references by the surviving officers of the 7th to multiple units moving out to the Weir Peaks at this time are very likely an intentional diversion from the truth.

CONCLUSION

3:25 p.m.: Captain French leads M Troop on its advance to the Weir Peaks.

3:40 p.m.: M Troop arrives on the Weir Peaks and deploys on the Peaks just west of D Company deployed on a small spur just east of the eastern loaf of the Peaks. At this time, Captain McDougall's first platoon of the Rear Guard has nearly reached the summit of Reno Hill and can clearly hear volley fire from the direction of Custer's command about four to 4.5 miles to the north.

26. CAPTAIN MCDOUGALL AND THE PACK TRAIN ARRIVE ON RENO HILL

There is much confusion as to when the main pack train and Rear Guard under Captain McDougall arrived on Reno Hill. This confusion is due primarily to the efforts, I believe, of the 7th Cavalry officers at the Reno Court of Inquiry and later to confuse the fact that the pack train was actually divided into two groups: most of the ammunition mules were in the advance group led by Lieutenant Mathey with the remaining pack mules arriving 45 minutes to an hour later. The following are the participant accounts that describe the later arrival of the main pack train and Rear Guard under the command of Captain McDougall:

1) Private Patrick Corcoran – K Company (Camp Interview, *Custer in 76*): says that McDougall got up to the bluffs with the packs about an hour after Benteen did.

ANALYSIS

As Benteen's battalion arrived on the bluffs at about 2:30 p.m., Private Corcoran indicates that McDougall and the packs arrived about an hour later at about 3:30 p.m.

2) Private William Taylor (*With Custer on the Little Big Horn*, with my annotations in brackets): "A very short time afterwards [i.e., after Benteen's arrival on the bluffs], Captain McDougall with B Troop, escorting the pack train, came along and joined us ...."

ANALYSIS

Private Taylor very likely mistook the earlier arrival of Lieutenant Mathey and the advance pack mules—who arrived on the bluffs about thirty-five minutes after Benteen did—with that of Captain McDougall's arrival with the Rear Guard and packs, which occurred much later—certainly not *"a very short time"* after Benteen got to the bluffs.

3) Lieutenant George Wallace (RCOI): It was well on to an hour before the pack train came in.

ANALYSIS

Although Lieutenant Wallace does not reference what event he anchors his time to, presumably it was the arrival of Benteen's battalion on the bluffs at 2:30 p.m., thus suggesting a 3:30 p.m. arrival of the pack train on the bluffs, similar to the estimate given by Private Corcoran above.

4a) Captain Thomas McDougall (RCOI): said he reached Reno Hill at 3:45 or 4:00 p.m. and heard two volleys fired, which he reported to Reno. He thought the volleys were four to four and a half miles away. He said he spoke to Benteen on the bluffs after he arrived there, and then he threw out a skirmish line.

4b) Captain Thomas McDougall (Camp Interview, *Custer in 76*, with my annotations in brackets): says he got to Reno Hill about 4:00 p.m. On arriving at Reno Hill, he heard firing and asked Godfrey if he heard it ... McDougall said, "I think we ought to be down there with him [Custer]" and went to where Reno and Benteen were talking and expressed the same opinion.

ANALYSIS

Captain McDougall, in the two accounts above, gives us a fifteen-minute time range for his arrival on the bluffs between 3:45 and 4:00 p.m. This was likely the time of arrival for his advance platoon of the Rear Guard, including himself personally. The head of the pack train itself likely did not reach the top of the bluffs until about 4:00 p.m. McDougall had reported hearing volleys as he ascended the bluffs fifteen minutes earlier, as well as additional firing on reaching the bluffs—estimated at four to 4.5 miles north of their position on the bluffs which he reported to Major Reno. He also mentioned conversing with both Godfrey and Benteen at this time, indicating that both of those officers had not yet left the bluffs to advance to the Weir Peaks. It is well to keep this in mind when reading other accounts that infer that Benteen and Godfrey began their trek to the Weir Peaks before McDougall got to the bluffs.

5) Lieutenant Charles Varnum (Camp Interview, *Custer in 76*): says the pack train arrived at 4 p.m. or later.

ANALYSIS

Lieutenant Varnum supports the time recollection of Captain McDougall by telling us that the pack train arrived on the bluffs at 4:00 p.m. or a bit later. It is likely that the head of the pack train arrived at 4:00 p.m. with the last mules arriving there about twenty minutes later.

6) Goes Ahead (*The Vanishing Race*): about 4 or 5 o'clock the pack train came up ....

ANALYSIS

The Crow scout Goes Ahead supports Varnum and McDougall's recollection that the pack train arrived on the bluffs about 4:00 p.m. or a bit later.

7) Billy Cross (*NY Times* Interview, July 4, 1876): All the companies with the pack train were on the hills when I left about 4 o'clock Sunday afternoon.

ANALYSIS

Scout Billy Cross tells us that he left Reno Hill at about 4 p.m. and that the pack train was apparently on the bluffs—or nearly so—at that time, thus supporting the recollection of Goes Ahead, Varnum and McDougall.

8) Lieutenant Edward Godfrey ("Custer's Last Fight," *Century Magazine*): McDougall came up with the pack train and reported the firing when he reported his arrival to Reno. I remember distinctly looking at my watch at 20 past 4 and made a note of it in my memorandum book, and although I have never ... been able to recall what ... happened at that time, it was some important event before we started down the river. It was my impression, however, that it was the arrival of the pack train.

ANALYSIS

Lieutenant Godfrey's recollections support those of McDougall, Varnum, Goes Ahead, and Billy Cross. He apparently looked at his watch and noted the time of 4:20 p.m., an impression he believes referred to the arrival of the pack train—likely the last of the pack mules and second platoon of the Rear Guard—reaching the bluffs at that time. It is important to note here that Godfrey confirms McDougall's further recollection that both Benteen and Godfrey were still on Reno Hill at 4:20 p.m. and had not yet started downriver for the Weir Peaks. Godfrey, in referencing this time (4:20 p.m.), specifically states that it was *"before we started down the river"*. It is well to keep this in mind when reading other accounts that infer an earlier departure to the Weir Peaks by Benteen's battalion. McDougall and Godfrey tell us it just was not so.

9) Lieutenant Edward McClernand ("Narrative of the Little Big Horn Campaign," published in Gordon Harper's *The Fights on the Little Horn Companion*): McDougall now came up with his troop and the pack train and reported the firing he had heard in the direction Custer had taken. There is fair reason to believe this was near 4:20 p.m.

ANALYSIS

Lieutenant McClernand's time estimate of McDougall and the pack train's arrival on Reno Hill at 4:20 pm likely derives from Godfrey's notation of 4:20 pm in his field notes. This time estimate likely refers to the last of the pack train and rear guard arriving on Reno Hill shortly before they departed towards the Weir Peaks.

OVERALL ANALYSIS

Of the nine participant accounts listed above, two of them suggest that the pack train arrived about an hour after Benteen reached the bluffs (at 2:30 p.m.), thus placing the packs there at about 3:30 p.m. Private Taylor's account appears to refer to the arrival of the advance packs, despite his mentioning McDougall and the Rear Guard, which makes sense as no other account supports the notion that McDougall arrived on the bluffs "very shortly after" the arrival of Benteen there. The remaining six accounts, including those of McDougall himself, indicate that the advance elements of the Rear Guard may have reached Reno Hill as early as 3:45 p.m. with the head of the pack train arriving about 4:00 p.m. and the rear elements and second platoon of the Rear Guard not arriving there until 4:20 p.m., likely due to difficulties in crossing over the North Fork of Ash/Reno Creek just before ascending the bluffs to Reno Hill.

CONCLUSION

3:45 p.m.: Captain McDougall and the first platoon of the Rear Guard reach Reno Hill in advance of the main pack train ascending the bluffs. McDougall hears volley fire from Custer's field about four-plus miles to the north between 3:30 and 3:45 p.m. At this time, Custer's command has left Luce and Blummer Ridges, and all five companies are now on the high ground north of Deep Coulee.

4:00 p.m.: The head of the main pack train arrives on Reno Hill. The rest of the pack train has just finished crossing the North Fork of Ash/Reno Creek just southeast of Reno Hill.

4:20 p.m.: The rearmost elements of the pack mules and the second platoon of the Rear Guard arrive on Reno Hill. According to Lieutenant Godfrey, K and H companies are still on Reno Hill at this time and have not yet begun their advance to the Weir Peaks.

27. THE REST OF THE REGIMENT LEAVES FOR THE WEIR PEAKS

So when did the remaining five companies of Reno's command leave for the Weir Peaks to join D and M Troops there? Here's what the participants tell us:

1) Private Edward Davern (RCOI, with my annotations in brackets): says Reno's command began its advance to the Weir Peaks "about 2 hours after the advance of the pack train came up." He claims that Weir left [for the Peaks] an hour or two after the arrival of the first packs.

ANALYSIS

Private Davern tells us that Reno did not begin his advance towards the Weir Peaks until about two hours after the arrival of the advance pack mules. Davern here would be referencing the time of near 3 p.m. HQ watch time as his starting time (the arrival of the advance packs at Reno Hill) and is thus saying that Reno's remaining five companies began their advance toward the Weir Peaks at about 5:00 p.m. He also mentions that Captain Weir was with them just a short period before this time. As we know from other participants (including Lieutenant Edgerly himself) D Company had already advanced out to the Weir Peaks about thirty minutes after Benteen's arrival on the bluffs—at about 3:00 p.m. on the timeline. We are then left with several possibilities to explain this conundrum. a) Davern was mistaken about seeing Weir at this time; b) Company D returned from the Weir Peaks earlier and then rode out again a second time; or c) Captain Weir rode back alone to Reno Hill from the Weir Peaks in an attempt to spur on the remaining companies to advance to the Weir Peaks to support D and M Troops already out there. I favor c) as the most likely explanation of these three choices. Davern refers to the start of this advance as sometime between one to two hours after the arrival of the first pack mules to Reno Hill. As the first mules arrived just after 3:00 p.m., this would support the advance of Reno's command at about 4:30 p.m., or one and a half hours after the arrival of the first pack mules. As Reno then halted the command for a period of time to await the arrival of Herendeen and eleven soldiers from the valley, it was closer to 5:00 p.m. before he once again began his advance towards the Peaks, or roughly two hours after the arrival of the advance pack mules at 3 p.m.

2a) Lieutenant Charles Varnum (RCOI): Question: "How long did Reno remain on the bluffs before the column moved downriver?" Varnum: "About 1 1/2 hours before it started.

ANALYSIS

Lieutenant Varnum is likely referencing his time from the arrival of the first ammunition packs on the hill, a time we have marked at just after 3:00 p.m. An hour and a half from that time would mark Reno's advance to the Peaks at about 4:30 p.m. Captain McDougall gave the same time reference for how long Reno remained on the bluffs before moving downriver.

2b) Lieutenant Charles Varnum (Camp Interview, *Custer in 76*, with my annotations in brackets): says it was 5:30 p.m. or later when the three companies [note: a likely reference here to B, G, and A Troops] moved down to support Weir. Varnum was out ahead with these advance companies [note: a reference here to Troops D, M, K, and H] and said they were not hard pressed when they fell back [to Reno Hill].

ANALYSIS

Lieutenant Varnum is likely incorrect about his time of 5:30 p.m. for the advance of the final three companies of Reno's command (B, G, and A) to the Weir Peaks, as other accounts from a number of participants indicate that 5:30 p.m. was the approximate time these same three companies returned to Reno Hill. Due to this conflict, it is more likely that Varnum was simply mistaken about his estimate of 5:30 p.m. as the start of the advance to the Weir Peaks. Varnum also tells us here that he was with the advance companies (D, M, K, and H) out on the Weir Peaks. Varnum very likely arrived there about the same time as Benteen and Godfrey at about 5:00 p.m. Of course, it is also possible that Varnum was simply adding an hour to his clock time estimates based on his acceptance of the problematic Wallace's noon crossing of the divide. If so, then Varnum's actual departure time for the Peaks would be closer to 4:30 p.m. HQ watch time.

3) Lieutenant Edward Mathey (RCOI): they began to move downstream about half an hour after the packs reached the bluffs.

ANALYSIS

Lieutenant Mathey's comments above could be read two ways, depending on whether he was referencing the arrival of his own advance ammunition pack mules to Reno Hill at about 3:00 p.m. or the arrival of the main pack train under Captain McDougall at about 4:00 p.m. If Mathey was referring to his own earlier arrival on the bluffs, then his reference to an advance thirty minutes later would likely refer to M Troop's departure from Reno Hill to the Weir Peaks, which would have occurred about 3:30 p.m. If, however, Mathey's comments referred to the arrival of the main pack train at about 4:00 p.m., then he is telling us that Reno began his advance to the Weir Peaks about thirty minutes later, or at about 4:30 p.m.

4a) Captain Thomas McDougall (RCOI): said he spoke to Benteen on the bluffs after he arrived there; he says the command moved downriver about thirty minutes after his arrival.

ANALYSIS

Captain McDougall gives us two important pieces of information here. First, he indicates that Benteen (and also Godfrey—although not mentioned in this account) were still on Reno Hill when his pack train arrived there at 4:00 p.m. They had not yet departed for the Weir Peaks. Then he supports Mathey's recollection that Reno's command began to advance towards the Weir Peaks about thirty minutes after his arrival there, placing this time at about 4:30 p.m., which supports the recollections of Mathey, Varnum and Davern.

4b) Captain Thomas McDougall (Camp Interview, *Custer in 76*, with my annotations in brackets): They remained on Reno Hill one and a half hours before the whole command started to find Custer. He still could see D Troop out on the bluffs ahead. A Company was the last to move out with the wounded. Benteen's men [i.e., H and K Troops] moved out by file on the ridge parallel to the river. The pack train merely got started before it was ordered to fall back [to Reno Hill] .... He let Moylan use a platoon of B Company to help move the wounded just as they were ordered to fall back.

ANALYSIS

Captain McDougall indicated that Reno's command remained inactive on Reno Hill for about one and a half hours before they began their advance to the Weir Peaks at about 4:30 p.m.—or thirty minutes after the arrival of the pack train on the bluffs. It would appear that McDougall is referencing the start of his time sequence above to the arrival of the advance ammunition mules under Lieutenant Mathey at about 3:00 p.m. He also claims to have been able to see a troop which he supposed was D Company silhouetted on the Weir Peaks after his own arrival on the bluffs, but as D Company was likely deployed farther down the northern slopes of the Weir Peaks, it is more likely that McDougall spotted troopers from M Company at the top of the Weir Peaks at this time.

5) Captain Myles Moylan (RCOI): soon after the arrival of the packs, the order was given to prepare to move downstream.

ANALYSIS

Although Captain Moylan does not give us any specific time, he does affirm that the move down to the Weir Peaks by Reno's command occurred soon after the arrival of the pack train on the bluffs, which McDougall tells us occurred about 4:00 p.m. Thus, Moylan's comments would support those of Varnum, Mathey, McDougall and Davern who all indicate a 4:30 p.m. advance to the Weir Peaks.

6a) Trumpeter John Martin (RCOI): Company H did not move down towards Custer until after some men from the bottom arrived (Herendeen's group) and the rest of the pack train arrived.

Question: Can you tell how long it was after the packs moved up, till the command moved down the river?

Martin: I think about an hour and a half. We waited for some men from the bottom and then moved out together.

Question: You came up with your company to where Major Reno was on the hill?

Martin: Yes sir.

Question: When you got there what did you do?

Martin: I did nothing. I was on my horse behind Captain Benteen. I was his orderly.

Question: How long were you there before the command started to move out all together?

Martin: It was about an hour and a half.

ANALYSIS

Trumpeter Martin tells us two things here. Captain Benteen and Company H did not leave for the Weir Peaks until after the arrival of Herendeen's group from the valley and only after the arrival of the "rest of the pack train" onto Reno Hill. Martin then goes on to explain that it was an hour and a half after the packs arrived before they moved out to the Weir Peaks. We believe Martin is referring to the arrival of the advance pack mules carrying the ammunition which arrived shortly after 3:00 p.m. Thus, Martin's hour and a half delay would start at about 3:00 p.m. and take us to a 4:30 p.m. departure time from Reno Hill, with Benteen pausing again for another fifteen or twenty minutes until the Herendeen group arrived to join them on the bluffs.

6c) Trumpeter John Martin (Camp Interview, *Custer in 76*): says it was one and a half hours after McDougall reached Reno Hill before the whole command advanced north.

ANALYSIS

Trumpeter Martin agrees with both Varnum and McDougall that Reno's command remained inactive on the bluffs for an hour and a half before moving north to join with Custer. Although Martin claims the advance occurred ninety minutes after the arrival of McDougall, he has likely mistaken McDougall here for Lieutenant Mathey and the arrival of the advance ammunition mules, which arrived on the bluffs shortly after 3:00 p.m. An hour and a half later, at about 4:30 p.m., Reno began his general advance towards the Weir Peaks, as Martin and other participants indicate.

7) Private August Seifert—K Company (Camp Interview, *Custer in 76*): says Reno began his move to Custer at 5 p.m.

ANALYSIS

Private Seifert's comments to Walter Camp tell us that Reno began his advance toward the Weir Peaks at about 5 p.m. despite most of our other participant accounts indicating an advance about thirty minutes earlier. I believe Seifert's comments above can be explained more easily once it is understood that Reno halted his advance north upon seeing Herendeen's group moving out of the valley to join them on the bluffs. There was a subsequent delay of about thirty minutes before they reached Reno's command on the bluffs, made their report to Reno, resupplied with ammunition, and had their wounded sergeant properly taken care of before Reno began his advance to the Peaks once again. As Private Seifert tells us, it was now about 5:00 p.m.

8a) Lieutenant Edward Godfrey ("Custer's Last Battle," *Century Magazine*): I remember distinctly looking at my watch at 20 past 4 and made a note of it in my memorandum book, and although I have never ... been able to recall what ... happened at that time, it was some important event before we started down the river. It was my impression, however, that it was the arrival of the pack train. It was about this time that 13 men and a scout named Herendeen rejoined the command ....

ANALYSIS

Lieutenant Godfrey gives us some valuable information above. For one thing, he confirms that Benteen's remaining battalion (H and K companies) had not yet left Reno Hill for the Weir Peaks as of 4:20 p.m. He also tells us that his impression was that the pack train—very likely the tail end of it, as McDougall and Hare referenced the head of it arriving twenty minutes earlier—had arrived on Reno Hill by 4:20 p.m. Godfrey also indicates that Herendeen's group of thirteen survivors from the valley fight were seen by the command leaving the valley and heading for the bluffs to rejoin the command about this time as well. These comments by Godfrey would support the recollections of McDougall, Martin, Varnum, Mathey, and Davern that Reno's command began their advance to the Weir Peaks at about 4:30 p.m.

8b) Lieutenant Edward Godfrey ("Custer's Last Battle," *Century Magazine*): about 5 o'clock the command moved down towards Custer's supposed whereabouts, intending to join him.

ANALYSIS

Lieutenant Godfrey supports the recollection of Private Seifert that it was about 5:00 p.m. when Reno's remaining three companies and pack train resumed their advance towards the Weir Peaks, shortly after the arrival on the bluffs of Herendeen and his group of survivors from the valley fight. At 5:00 p.m., Benteen and Godfrey had likely just arrived at the Peaks, while the rest of Reno's command followed a distance behind.

8c) Lieutenant Edward Godfrey ("Custer's Last Battle," *Century Magazine*, *The Custer Myth*, p. 149, with my annotations in brackets): We thought he [i.e., Reno] ought to go [to Custer's support], his attention was called to the firing on Custer's field; it was suggested that he go; he was [first] waiting for the ammunition packs to replenish the ammunition; then he waited for the delayed pack train. We started and marched as far as the first ridge [note: about a quarter of a mile north of Reno Hill] when we beheld the dismounted men [of Herendeen's party] crossing the valley, men that had been left [behind] when he [i.e., Reno] made his retreat. We halted and waited for them to join, to be mounted and have their ammunition replenished; then we advanced to the high point [note: called the Weir Peaks] where we could overlook the Custer battlefield. We saw the commotion on that field when the Hostiles started in our direction. Three troops, D, K and M were then ordered to take positions and hold their ground [on the Peaks]; with the other four [troops] and the pack train Reno returned to our first position on the bluffs, nearly two miles away [note: from D Company's deployment on the forward slopes of the Weir Peaks]. Then he ordered the troops in position [on the Weir Peaks] to fall back at once.

ANALYSIS

Godfrey tells us several things here. First, he tells us that Reno was informed of the firing heard from Custer's field, something Reno denied in his testimony at the Reno Court of Inquiry. He also tells us that the pack train was, indeed, divided into two groups; an advance group of packs carrying the ammunition mules and the main or "delayed" pack train which arrived later. He also tells us that both his own company as well as Benteen's began their advance towards the Weir Peaks while Reno ordered the general advance that involved troops G, B, A. and the pack train. It is important to note here that neither Benteen nor Godfrey left Reno Hill before the other companies. Then Godfrey indicates that Reno's entire advance halted about a quarter of a mile north of Reno Hill—at the first ridge—to await the arrival of Herendeen's group, now seen "crossing the valley"—not simply climbing up the bluffs. He then indicates another delay in getting Herendeen's group of survivors mounted and resupplied with ammunition—and the wounded likely cared for as well—before the general advance was once again made. It is my contention that Benteen and Godfrey both took their companies out to the Weir Peaks during this delay in getting Herendeen's group re-equipped.

9) Private William Morris—M Company (letter to Robert Bruce, May 23, 1928, with our annotations in brackets): Benteen, with his battalion, reached us about an hour and a half after we were on the hill [i.e., Weir Point], and Captain McDougall with the rear guard arrived about the same time.

ANALYSIS

Private Morris was in M Company, and, as indicated in that section of the timeline, had arrived on the Weir Peaks at about 3:40 p.m. Morris now tells us above that Benteen's battalion—referring here to Companies K and H—arrived on the Weir Peaks about an hour and a half later, or at about 5 p.m. or a few minutes later, accounting for the short delay of these two troops as they waited for the arrival of Herendeen's group from the valley floor. Morris also indicates the arrival of Captain McDougall and the rear guard on the Weir Peaks at about the same time. Although the pack train itself, and the second platoon of B Company would never arrive on the Weir Peaks, it is possible that Captain McDougall, moving out ahead of Reno's column with his first platoon of B Company, reached the Peaks just after Benteen and Godfrey did.

10a) Captain Frederick Benteen ("Benteen's Narrative," with my annotations in brackets): About this time, I saw one of the troops of my battalion [i.e., D Troop] proceeding to the front without orders from me. Upon this, I followed with the other 2 troops [i.e., H & K Troops—you will note that M Troop is not included here by Benteen.] Major Reno, having his trumpeter sound the "Halt" continuously and assiduously, but I had to get in sight then of what I had left my valley-hunting mission for. On reaching the highest bluff in the vicinity [i.e., Weir Point], I saw what I estimated to be about 900 Indians in the valley from which Reno had just been driven; the officer [Captain Weir] who preceded me on his "own hook" with his [D] troop had gone down a gorge with it; Indians were riding around the bluff on either side of this troop signaling; I then formed a troop [i.e., M Troop] dismounted at right angles with the river, and one [i.e., K Troop] on the bluff parallel with the river, so that if Custer's [or Keogh's] forces were near ... our position would be defined. Just then, up the gorge, pell-mell, came the troops which the commander [Captain Weir] had so fearlessly sallied forth with, and the preparation I had made allowed he and his troops to come down to a less frantic style of going. ... slowly, we fell back to about just the very point where Reno reached the bluffs ... the fight lasted ... till dark on the 25th. ...

10b) <u>Captain Frederick Benteen</u> ("The Benteen-*Goldin Letters*," with my annotations in brackets): I followed [Weir's D Company] with the remaining two companies of my battalion [i.e., H and K—you will note there that Benteen does not include M Troop, which was already out on the Peaks by this time] the trail that Weir had "sallied on," he having no orders to proceed—some 10 minutes later perhaps; anyway, just as soon as I learned of his insubordination and unauthorized movements. While enroute to the highest point [i.e., Weir Point] ... Major Reno kept his trumpeter pretty busily engaged in sounding the "Halt" for the purpose of bringing my command to a stand. [Note that Benteen still regards himself in command of his own battalion, and not under Major Reno's authority at this time.] However, I paid no heed whatever to the signal, but went to the highest point of bluffs [i.e., the Weir Peaks], the battalion [K and H Troops] being in column of fours. On arriving at [this] elevation, I then had my first glimpse of the Indian village from the height. ... I saw enough to think that ... this time we had bitten off quite as much as we would be able to well chew. ... Reno had then gotten up to the point where I was. [Note: It is doubtful that Reno ever reached the Weir Peaks.] However, I ordered French to put his [M] troop on a line on the [western] bluff, which was at right angles with ... the river, and then for the purpose of showing [Custer or Keogh] where our command was, if there were any other bodies of our troops in sight.

Weir's [D] troop then "showed up" midway in [the] canyon coming in hot haste toward us, and, as it seemed, myriads of howling red devils behind him. [Note: this account of the retreat from the Weir Peaks differs significantly from those of other participants including those of Lieutenant Edgerly and 1st Sergeant Harrison, both of D Company, who give a contradictory account of this withdrawal, which did not occur until Benteen himself was long gone from the Peaks.]

Of course, I knew that Reno's retreat from the bottom had been a rout—a panic—and I made up my mind then and there that there was no necessity of having a repetition of [the] same. So I ordered French to dismount his troop, keep his dismounted men on the bluff, send his led horses to [the] rear, let Weir's troop through, then slowly fall back with his dismounted men. ... [Note: Benteen here gives the proper formula for withdrawing in the face of large numbers of mounted hostiles. Contrast this method of retreat with the one engaged in by Reno in his retreat from the valley.] Well, French "weakened" (no doubt of it!) Twas then I ordered Godfrey to take position on the point he writes of. [Note: Godfrey denied that Benteen had ordered him to do this, making it clear that his rear guard action was done on his own volition and without orders.]

ANALYSIS

In the above narrative, we can clearly see examples of Benteen's willingness to engage in outright falsehoods (his claim to have followed Weir out to the Peaks ten minutes after D Company left Reno Hill; his fanciful account of the withdrawal of D Company from the Weir Peaks; his phantom orders to Godfrey to form a dismounted rear guard to cover this withdrawal), yet despite these blatant attempts at securing credit where none is due and denigrating the actions of others, Benteen does give us some valuable information regarding the advance of his battalion to the Weir Peaks. First we learn that Benteen does not include M Troop in his advance, but clearly states that he advanced with only the remaining two troops of his battalion. He thus eliminates any of the remaining companies left under Reno's command in this advance. Benteen also states that during his advance to the Peaks, Reno attempted to order him to halt by having "his trumpeter sound the 'Halt' continuously and assiduously". This would suggest that Reno, after beginning his belated advance towards the Weir Peaks, suddenly found a need to halt his command before getting very far—no doubt due to the observation of Herendeen's group of survivors spotted making their way out of the valley and up the heights to join Reno on the bluffs. It would appear that Benteen, after a short halt to await the arrival of this group, began his advance along with Godfrey once again to reach the Weir Peaks when they lost patience with Reno's unwillingness to advance at a faster pace, likely due now to refitting and resupplying these stragglers from the valley and perhaps arranging for medical treatment or examination of Sergeant White's wounds. Reno attempted to stop Benteen and Godfrey from continuing their subsequent advance out to the Peaks, which the latter accomplished at about 5 p.m. or very shortly thereafter. It is likely that Reno himself began his own advance with the remaining three companies and pack train at about the same time that Benteen and Godfrey reached the Weir Peaks.

11) Lieutenant Winfield Edgerly (RCOI, with my annotations in brackets): Said it was about forty-five minutes for Benteen and his command [to advance—probably a reference to the arrival on the Weir Peaks of Captain French's M Troop at that time].

ANALYSIS

Since Edgerly estimated D Company's advance to the Peaks as occurring about thirty minutes after Benteen arrived on the bluffs, his further comment above that it was about forty-five minutes for Benteen and his command to advance to the Peaks suggests that Edgerly wishes us to believe that Benteen started his advance at about 3:15 p.m. Unfortunately for Edgerly's memory, this suggested timeframe is simply not supported by any of our other witnesses other than Benteen himself. It is directly contradicted by at least eight other participant recollections, which strongly suggests that this estimated time is invalid, unless he is actually referring to his own D Company (as part of Benteen's battalion) arriving on the Weir Peaks about 3:10-15 p.m.

12) Lieutenant George Wallace (RCOI, with my annotations in brackets): Question: How long was the particular column [moving down to the Weir Peaks] you speak of? Wallace: There were 3 companies [B, G, and A] marching in column of two's .... Question: Give the approximate length of the column. Wallace: 150 to 200 yards. Question: Were you moving with the column? Wallace: I was in the column towards the rear. Question: How far did you move? Wallace: Till the head of the column halted, which was probably three-quarters of a mile or a mile.

ANALYSIS

Lieutenant Wallace indicates in his account above that he was traveling out towards the Weir Peaks with the last three companies of Reno's command—B, G, and A Troops—and that Wallace was towards the rear of the column. He then goes on to state that these last three troops moved no farther than three-quarters of a mile to a mile towards the Weir Peaks, or between a half to three-quarters mile short of the Weir Peaks before halting their advance and then returning back toward Reno's corral. Thus, it is very likely that Lieutenant Wallace, contrary to his own testimony at the RCOI, never reached the Weir Peaks before the column was turned back to retreat to Reno Hill.

13) Sergeant Daniel Kanipe (Camp Interview, *Custer in 76*): was with B Company on its advance to the Weir Peaks. They only got about 1/2 way (says three-quarters of a mile) before returning to Reno Hill. They stayed out, all told, about thirty minutes.

ANALYSIS

Sergeant Kanipe indicates that he moved out towards the Weir Peaks with B Company—likely its second platoon—detailed to help A Company with the wounded. He further indicates that they got only about halfway down to the Weir Peaks, or about three-quarters of a mile north of Reno Hill before they turned around and returned to Reno Hill. It is likely that Kanipe, like Lieutenant Wallace, never reached the Weir Peaks. He claims they stayed out on this advance, all told, about thirty minutes, suggesting that Kanipe begins this timeframe with Reno's second advance north at about 5:00 p.m. shortly after the arrival of Herendeen's group of survivors onto the bluffs. Thirty minutes later, at 5:30 p.m., the tail end of Reno's command (B, G, A Troops, and the pack train) arrive back on Reno Hill.

14) Private William Taylor – A Troop (*With Custer on the Little Big Horn*, with my annotations in brackets): For over 2 hours after the arrival of Captain Benteen's command, we remained there on the bluffs, unmolested in any manner .... we had heard firing off in the general direction Custer was supposed to have gone. *"Why don't we move?"* was a question asked by more than one. The 3 troops [M, A and G] that had been engaged in the valley were, it is true, somewhat demoralized, but that was no excuse for the whole command to remain inactive [for two hours]. A few of our men [about seven] had been wounded, but none so seriously that they could not ride with the pack train. ... soon after the party of 12 [i.e., Herendeen and the survivors from the valley] rejoined us, and somewhere near 5 o'clock, we

had orders to "fall in", and mounting our horses, we started in the direction that Custer was believed to have gone. ... Soon they [i.e., the Indians from Custer's field] began to come in our direction, and our advance troop [D] under Weir was attacked. The whole command was then ordered to retire and we returned to the place we had left a short time before .... This was, according to my belief, about 6 o'clock.

ANALYSIS

Private Taylor gives us some very important information surrounding Reno's advance towards the Peaks in his account above. First he tells us that Reno's command remained for two hours on the bluffs before moving north to attempt to reach Custer. Thus, Taylor confirms other accounts in suggesting that Reno remained inactive on the bluffs from the time Benteen joined them at 2:30 p.m. until they began a general advance north two hours later at about 4:30 p.m. Reno would halt his advance upon spotting Herendeen's group of survivors leaving the valley, after which Taylor informs us that soon after Herendeen's group from the valley rejoined them on the bluffs—somewhere near 5 p.m.—they had orders to "fall in," and mounting their horses, they started out again in the direction that Custer was believed to have gone. As Taylor was traveling with A Company, his comments refer to the last three troops—G, B, and A—moving out towards the Weir Peaks at 5 p.m. about the same time that Benteen and Godfrey arrived at the Weir Peaks. Taylor then tells us that soon after 5:00 p.m., the Indians, having finished fighting Custer's command in the north, began to come in their direction, and D Company was attacked on the northern slopes of the Weir Peaks. He then says that the whole command—actually only companies A, B, G, and H—were ordered to retire back to Reno Hill, with the final companies arriving there a bit later, according to Taylor, about 6 o'clock.

15) Private Charles Windolph (*I Fought With Custer*): I believe it must have been around 5 o'clock when Reno and Benteen ordered the whole outfit to move northward, in the general direction Captain Weir and his troop "D" had taken a good half hour before.

ANALYSIS

Private Windolph's memoirs were written by Frazier & Robert Hunt and published in 1947. We strongly suspect that the times he gave for the events were heavily influenced by Colonel William Graham's timelines that were previously published in his book *The Story of the Little Big Horn* in 1926, as Windolph's times follow Graham's unique and popular times—based almost exclusively on those given by Lieutenant Wallace at the Reno Court of Inquiry—contradicting the evidence provided by the great majority of the battle participants, both red and white. The Hunt's reproduce and insert sections from Graham's book to supplement Windolph's own memoirs. Thus, my conclusion is that Windolph's times are not to be relied on. It is likely that Windolph was basing his 5:00 p.m. advance of Reno's troops on his mistaken notion that Weir left Reno Hill at 4:30 p.m. This time is not supported by other participant accounts.

OVERALL ANALYSIS

Of the 10 time estimates given by the participants above, those of Davern, Martin, Mathey, McDougall, Godfrey and Taylor support a departure time near 4:30 p.m. for Reno's general advance towards the Weir Peaks. After halting to wait for the arrival of the Herendeen group of survivors from the valley, it was not until near or after 5 p.m. that several of Reno's companies in this advance actually reached the Peaks. Companies G, A, and half of B, along with the pack train, never reached the Peaks before returning to Reno Hill. The Edgerly and Benteen accounts are the only ones that support an earlier departure from Reno Hill for H and K Companies, but as they are contradicted by the remaining accounts, particularly those of Godfrey and McDougall, they are not accepted as valid.

CONCLUSION

4:25 p.m.: Reno's remaining five companies on Reno Hill begin their advance towards the Weir Peaks with the intent to join with Custer's command in the north.

4:35 p.m.: Reno's command, having moved about a half to a quarter mile downriver along the bluffs, halts their advance upon seeing Herendeen's group of survivors making their way out of the valley towards their position on the bluffs.

4:45–4:50 p.m.: Herendeen's group ascends the bluffs and joins with Reno's command. Benteen and Godfrey move out with H and K Troops to the Weir Peaks while Reno orders his trumpeter to sound the Halt to prevent their departure from the rest of his command.

5:00 p.m.: Godfrey and Benteen arrive on the Weir Peaks with K and H Companies—in that order—to join with D and M Troops who were already deployed on the Peaks. After having resupplied the survivors from the valley and inspected and tended to Sergeant White's wounds, Major Reno has now begun his advance to the Peaks with the remaining Companies under his command (G, A, the second platoon of B, and the pack train). Captain McDougall and the first platoon of B would likely have moved out to the Weir Peaks at a trot at this time, leaving Reno's slow-moving command behind, and arriving at the Peaks a few minutes after Benteen.

5:15p.m.: Reno begins his retreat back to his corral on the bluffs. This would involve companies A, G, B, and H along with the pack train while Troops D, M, and K remained out on the Weir Peaks a while longer to engage the hostiles and give time for Reno and Benteen to locate a defensive position on the bluffs for the command.

28. HERENDEEN'S GROUP ARRIVES ON THE BLUFFS

Scout George Herendeen led a small group of a dozen or less troopers abandoned by Major Reno in his retreat from the valley up onto the bluffs to rejoin their command there. Here we examine what the participants themselves tell us about the timing surrounding this event:

Appendix A: Participant Timeline of the Battle of the Little Big Horn

1) George Herendeen (Camp Interview, *Custer in 76*, with my annotations in brackets): says he stayed in the timber with the troopers about two hours before leaving; They left the timber after seeing only a few Indians in the valley. They walked—they did not run—in skirmish order [out of the valley] fording [the river] some distance below Reno's retreat ford and went up the bluffs farther north from where Reno retreated and farther north from where [Dr.] DeWolf was killed. As they got up the bluffs, the Indians now came up the valley to attack Reno. As they neared the top of the bluffs, they met Billy Cross with the Ree scouts mounted and coming down [the bluffs] for water. They met Reno's advance at the top of the bluffs just as they stopped and [shortly afterward] fell back [to Reno Hill]. Twenty minutes later would have been too late for them to rejoin with Reno [note: due to the Indians arriving on the bluffs and the flats below to cut them off].

ANALYSIS

Herendeen tells us that he remained hidden in the timber with the eleven or so troopers for about two hours before they left the timber and made their way out of the valley and up the bluffs to rejoin with Reno. Assuming Herendeen references his time shortly after Reno's retreat from the valley and his own return to the timber, we can likely begin this two-hour stretch of time at about 2:15 p.m. Thus, at about 4:15 p.m., Herendeen's group walks on foot out of the valley about three-quarters of a mile from where they would cross the river and ascend the bluffs. At the top of the bluffs, somewhere north of the ridge where Dr. DeWolfe was killed north of Reno Hill, his group met Reno's command, which had paused during their advance to the Weir Peaks to await the arrival of his group on the bluffs, which would have likely occurred at some time around 4:45 p.m.

2) Lieutenant Edward Godfrey (*Custer's Last Battle, Century Magazine*, with my annotations in brackets): It was about this time [i.e., 4:20 p.m. according to Godfrey's watch] that thirteen men and a scout named Herendeen rejoined the command.

ANALYSIS

Our interpretation of Lieutenant Godfrey's above passage is that Herendeen's group was seen in the valley making their way out to rejoin their command at about 4:20 p.m.

3) Private William Taylor—A Company (*With Custer on the Little Big Horn*, with my annotations in brackets): In the latter part of the afternoon, not far from half-past four, a squad of 12 of our comrades were seen coming up the bluffs.... The party that escaped consisted of George Herendeen, a scout, 6 soldiers of G Troop, 3 of M, and 2 of A. ... Soon after the party of 12 rejoined us, and somewhere near 5 o'clock we had orders to "fall in", and mounting our horses, we started in the direction Custer was believed to have gone.... Soon they [the Indians] began to come in our direction and our advance troop

under Captain Weir was attacked. The whole command was then ordered to retire and we returned to the place we had left a short time before.... This was, according to my belief, about 6 o'clock.

ANALYSIS

Private Taylor tells us that at about 4:30 p.m., Herendeen's group was seen coming up to join them on the bluffs. He then indicates that at about 5:00 p.m., they had orders to "fall in" and move out again towards the Weir Peaks. Then Indians were seen coming in their direction, and soon afterward, Captain Weir and D Troop were attacked out on the advance ridges of the Weir Peaks. At this time, the command then retired from the advance and returned to Reno Hill, with most of the troops—very likely all except for K Troop acting as a Rear Guard—having returned to the Reno Hill position by about 6:00 p.m. Regiment Headquarters watch time.

4) Billy Cross (*NY Times* Interview, July 1876): left Reno's bluff about 4 p.m. (Herendeen says he met him going down the bluffs as he was ascending.)

ANALYSIS

Billy Cross's recollection of leaving Reno's command on the bluffs at about 4 p.m. coincides closely with Herendeen recollection of meeting him on the way up the bluffs. Both Godfrey and Taylor would put this time closer to 4:30 p.m., or a bit later.

5) Private Henry Petring—G Company (Camp Interview, *Custer in 76*): says he left the timber with Herendeen's group. He says they left the timber near sundown to reach Reno's command on the bluffs.

ANALYSIS

Private Petring's assertion that Herendeen's group—of which he was a part—did not leave the timber until sundown (or about 7:00 p.m.) is simply not accurate. We must dismiss this recollection as unsupportable from any other account.

6) Captain Myles Moylan (RCOI): after about an hour the men left behind in the timber rejoined them.

ANALYSIS

Assuming Moylan is referring above to the Herendeen group, we cannot use his time reference as he does not reference the starting time of his reference. He simply says "after an hour ..." One might ask, an hour after what? Reno's arrival on the bluffs? Benteen's arrival? The advance packs? We really have no idea, thus Moylan gives us very little information on when this event took place.

7) Private Hugh McGonigle – G Company (Camp Interview *Custer in 76*, with my annotations in brackets): says he was with Herendeen's group. Says he remained hiding in the brush about an hour [before heading out of the valley to rejoin Reno's command on the bluffs].

ANALYSIS

Private McGonigle contradicts Herendeen in estimating their time of hiding in the timber at about an hour before leaving to rejoin Reno on the bluffs. Herendeen claimed it was about two hours. Of the two, Herendeen's time is more likely to be correct, as it correlates more closely with the time estimates given for Reno's advance towards the Weir Peaks, which, according to Lieutenant Godfrey, did not begin until some time after 4:20 p.m.

8) Trumpeter John Martin (RCOI): claims that H Company did not move down to the Weir Peaks until after the arrival of some men from the bottom....

ANALYSIS

Trumpeter Martin tells us that Benteen's two companies (H and K) did not move down to the Weir Peaks until after the arrival on the bluffs of Herendeen's group.

9) Private John Sivertson—M Company (Camp Interview, *Custer in 76*): says he was the first man to wade through the Little Big Horn river going back with Herendeen's group about 4:30 p.m.

ANALYSIS

Private Sivertson, traveling with Herendeen's group, indicates that he crossed the Little Big Horn river about 4:30 p.m., which suggests that Herendeen's group actually reached Reno on the bluffs at some time between 4:40–4:45 p.m. His account supports that of Lieutenant Godfrey and Private Taylor and better clarifies Billy Cross's account.

10) Private Charles Windolph (*I Fought With Custer*): Of nineteen men left behind, seventeen crossed the river and reached Reno Hill on foot within two hours.

ANALYSIS

Private Windolph recalls that Herendeen's group reached the bluffs to rejoin with Reno's command within two hours of Reno's retreat out of the valley. If most of the tail end of Reno's command had attained the bluffs between 2:30 and 2:40 p.m., then Windolph's two-hour estimate would put Herendeen's survivors from the valley fight—about four or five less than the figure Windolph gives us—on the bluffs between 4:30 and 4:40 p.m., which fits nicely with the time estimates of five other accounts indicating the Herendeen group's arrival at some time after 4 p.m.

OVERALL ANALYSIS

From the accounts listed above, we can dismiss Private Petring, McGonigle and Moylan's time references as being either unsupported or not clear enough to determine a time. The same might be said for Martin's comments, whose principal value is in telling us that Benteen delayed his advance to the Weir Peaks until after Herendeen's group arrived on the bluffs. The remaining six accounts indicate that Herendeen's group arrived on the bluffs sometime after 4:00 p.m. Herendeen himself says he remained in the valley for about two hours, placing his departure time at about 4:15 p.m. according to the timeline. Godfrey, Taylor, and Sivertson confirm that Herendeen was seen at about 4:30 p.m. Sivertson clarifies this by saying he crossed the Little Big Horn river at this time and that he was the first man in Herendeen's group to do so. Herendeen's group, helping along a wounded man, would likely have surmounted the bluffs about fifteen minutes later.

CONCLUSION

4:15 p.m.:: Herendeen's group was first seen in the valley by elements of Reno's command, shortly after they began their advance along the bluffs towards the Weir Peaks.

4:25 p.m.: Herendeen's group crosses the river just downriver from Reno's retreat crossing.

4:35 p.m.: Herendeen's group, ascending the bluffs, passes Billy Cross and his party of Rees as the latter group was descending to the river to water their horses.

4:45 p.m.: Herendeen's group arrives on the bluffs north of the ridge where Dr. DeWolfe was killed. They are then outfitted with more cartridges for their weapons, and medical attention is given to Sergeant White while they quickly debriefed and reported the situation in the timber to Major Reno.

4:50 p.m.: Benteen and Godfrey move out with H and K Troops to the Weir Peaks, defying Reno's orders to halt and await the advance of the rest of the command.

5:00 p.m.: Major Reno resumes his advance towards the Weir Peaks.

29. HOW LONG D COMPANY REMAINED OUT ON THE PEAKS

Perhaps one of the least understood aspects of the battle was the length of time that D Company spent out on their advance to the Weir Peaks. There is no real mystery over this length of time. From earlier investigations and analyses of the battle participants, we learned that D Company moved out on their advance to the Peaks at about 2:55 p.m., or just a few minutes before Lieutenant Hare returned to Reno Hill from his two-mile round trip ride to the advance pack mules carrying the ammunition. D Company would likely have arrived on the Weir Peaks at about 3:10 p.m. based on time/distance factors. In my analysis of that event, I have also determined the likely time of D Company's withdrawal to Reno Hill, placing the watch time for this at about 5:45

p.m. Thus, simple math tells us that D Company remained out on the Peaks for roughly two and a half hours before being recalled back to Reno Hill. Can this timeframe be supported from the participant accounts? Despite the apparent efforts made by several officers to cover up this part of the battle, we are left with a number of significant statements that allow us to shed some light on the truth of this matter:

1) Lieutenant Winfield Edgerly (Carroll, *The Gibson and Edgerly Narratives*, p. 11, with my annotations in brackets): We stayed out there [on the Weir Peaks] about 2 hours, according to my recollection ....

ANALYSIS

Lieutenant Edgerly gives us a very modest estimate of two hours duration, which is actually about thirty minutes less that what I calculate to be the actual time his company spent out on the Peaks.

2a) Lieutenant Charles Varnum (Carroll; *Custer's Chief of Scouts*, p. 144): Varnum (one of the last officers to reach the Weir Peaks) would testify that about one and one-half hours were consumed from the time he reached Weir's company on the point until they returned to the defensive position on Reno Hill.

ANALYSIS

Lieutenant Varnum would have arrived on the Weir Peaks sometime between 5:00 and 5:15 p.m. He then claims that one and one-half hours would go by from the time he arrived on the Peaks (5:00 to 5:15 p.m.) until all the companies had returned to their defensive position on Reno Hill. Varnum tells us that K Company—the last company to return to Reno Hill from the Weir Peaks—got into the defensive corral there between 6:30 and 6:45 p.m., which supports the time given us by Lieutenant Godfrey for the arrival of K Troop on Reno Hill.

2b) Lieutenant Charles Varnum (Camp Interview, *Custer in 76*): He thinks [it] must have been 5:30 p.m. or later when the three companies moved down to support Weir.

ANALYSIS

Lieutenant Varnum estimates that it was close to 5:30 p.m. or later when the three companies (likely G, B, and A) moved down to support Weir. I believe Varnum was off by fifteen to thirty minutes in his time estimate above. Reno's last three companies likely had halted their advance to the Peaks by about 5:15 p.m.. They began to fall back shortly afterward to Reno Hill reaching the latter at about 5:30 p.m., but Varnum would not know this as he was, at this time, out on the northern slopes of the Weir Peaks himself and well out of sight of Reno and these last three troops moving along the bluffs.

3) Lieutenant Luther Hare (RCOI): said it was one and a half hours from the time Weir advanced until they began to seriously engage the Indians.

ANALYSIS

Lieutenant Hare also mentions an important time reference above when he claimed that D Company was not seriously engaged out on the Peaks for the next one and a half hours, a time in which the Indians were apparently fully engaged in fighting Custer's five companies. This supports Lieutenant Edgerly's estimate of being on the Peaks for about two hours with only the last thirty minutes engaged in skirmishing with the Indians. In other words, D Company remained on the Peaks for at least one and a half hours before engaging with the enemy.

4) Captain Thomas Weir (From an unpublished letter written by Captain Weir on 22 Oct. 1876 (from the George Kush Collection): "With our cherished ones deliverance within our grasp we waited breathless two hours, for the order that never came."

ANALYSIS

Captain Weir supports Lieutenant Edgerly's two-hour duration of D Company's stay out on the Weir Peaks waiting for orders for a general advance to support Custer's wing that never came.

OVERALL ANALYSIS

The four officer accounts above support this timeline's conclusion that D Company remained on the Peaks from roughly 3:15 p.m. to 5:45 p.m., a span of time amounting to two hours and thirty minutes. The warriors engaged D Company during the final thirty minutes of their stay out on the Peaks, beginning at about 5:15 p.m.

There is much confusion as to the sequence of events during D Company's stay out on the Peaks as several of the officers went out of their way to shorten and condense the actual time spent out on the Peaks by both D and M Troops.

CONCLUSION

3:10-15 p.m.: Lieutenant Edgerly and D Company arrive on the Weir Peaks.

3:40 p.m.: M Troop joins D Troop on the Weir Peaks.

5:00 p.m.: K Troop followed by H Troop join D and M Troops on the Weir Peaks.

5:10 p.m.: H Company leaves the Weir Peaks; the warriors begin to move towards the troops seen on the Weir Peaks.

5:15 p.m.: D Company engages in long range fire with a group of Indians on the lower ridges of the northern Peaks. Reno halts the advance of Cos. B, G, A, and the pack train.

5:20 p.m.: Reno begins a retreat back to Reno Hill.

5:35 p.m.: H, A, G. and B Troops as well as the Pack Train return to Reno Hill and begin to deploy there.

5:35 p.m.: K Company leaves the Weir Peaks after being recalled by Lieutenant Hare acting on orders from Major Reno, with M and D Troops acting as a Rear Guard.

5:40 p.m.: M Company leaves the Weir Peaks. Captain French advises Lieutenant Edgerly to do likewise.

5:45 p.m.: D Company leaves the Weir Peaks.

5:50 p.m.: Trooper Vincent Charley is shot down about 650 feet (or just over 200 yards) south of the Weir Peaks. Lieutenant Godfrey dismounts K Troop about 500 yards north of Reno Hill to provide cover fire for M and D companies withdrawal from the Peaks.

5:55 p.m.: M Troop arrives at Reno Hill. D Troop passes K Troop on their retreat from the Peaks.

6:00 p.m.: D Troop arrives at Reno Hill.

6:00 - 6:40 p.m.: K Troop engages in their dismounted Rear Guard skirmish action until repeatedly recalled to fall back to Reno Hill.

6:45 p.m.: First Sergeant Winney of K Troop is killed at Reno Hill, marking the general start of heavy skirmish firing directed at Reno Hill, which continues till about 9 p.m.

30. RENO'S COMMAND RETURNS TO RENO HILL

There are varying accounts of when the surviving troops of the 7th Cavalry returned back to Reno Hill later that afternoon. Lets see if we can make some sense of these differing accounts:

1a) Captain Frederick Benteen (RCOI): claims they set up their defense and were attacked at 5:30 p.m. or a bit later, and were under attack till it was too dark to see.

ANALYSIS

Benteen tells us that Reno began to set up his defense on his hill at 5:30 p.m. with the lead elements of his command. Benteen also claims they were attacked at the same time or a bit later. Most of the other participants put the initial attack at Reno Hill to be nearer to 6 p.m. or later. We believe Benteen was correct in estimating the initial attack there to be "a bit later" than the 5:30 p.m. time he first suggested.

1b) Captain Frederick Benteen (RCOI – *Army and Navy Journal*): says they fought the Indians for three hours.

ANALYSIS

As stated by nearly all accounts, the fight on Reno Hill ended about 9 p.m. on June 25th. Benteen suggests that the fight actually began about 6 p.m., which may or may not include the fighting during the Rear Guard action of K Company from the Weir Peaks.

2a) Major Marcus Reno (RCOI): implies that they fought at his hilltop defense from 6 to 9 p.m. or until dark.

2b) Major Marcus Reno (RCOI - *Army and Navy Journal*): suggests that his hilltop defense lasted from 6 to 9 p.m.

2c) Major Marcus Reno (*Reno's Memoirs*): The fight on Reno Hill ended at 9 p.m.

ANALYSIS

Major Reno would agree with Benteen's time estimates that the fight on Reno Hill took place between 6 and 9 p.m. Again, Reno may or may not have included the Rear Guard action of K Company during the retreat from the Weir Peaks.

3) Capt. Thomas McDougall (RCOI): The fight ended at 9 p.m.

ANALYSIS

McDougall agrees with Benteen and Reno that the fight at Reno Hill ended about 9 p.m.

4) Lieutenant George Wallace (RCOI, with my annotations in brackets): It was somewhere between 5 and 6 o'clock [when the command returned to Reno Hill] .... they defended Reno Hill from 6 p.m. till 9 p.m.

ANALYSIS

Wallace tells us that it was between 5 and 6 p.m. when Reno's command returned to the corral on Reno Hill. His statement above suggests that elements of Reno's command reached Reno Hill between 5:30 p.m. and 6:00 p.m., about thirty minutes before the fight actually began against Reno's troops. Wallace agrees with Benteen and Reno that the fight lasted from 6 to 9 p.m., but again, it is not clear if he was including Godfrey's Rear Guard action in falling back from the Weir Peaks.

5) Packer Ben Churchill (RCOI): he got back to Reno's corral about thirty minutes before the Indians attacked them there.

ANALYSIS:

Packer Churchill, traveling with the pack train, indicates that he arrived at Reno Hill about thirty minutes before the Indians attacked them there. This would suggest a 5:30 p.m. arrival on Reno Hill if the Indians began their attack at 6 p.m., as Benteen, Reno, and Wallace so indicated. What is not clear is if Churchill was referring to the Indians' attack at Reno Hill or if he was referring to the fighting engaged in by Godfrey's K Troop as it fell back from the Weir Peaks. I believe he was most likely referring to the latter.

6) Captain Myles Moylan (RCOI): after returning from their aborted advance, the action continued at Reno Hill until after dark.

ANALYSIS

Moylan suggests that the fight at Reno Hill ended about 9 p.m.

7) Lieutenant Charles Varnum (RCOI): believes they reached Reno Hill about 5:30 p.m.

ANALYSIS:

Lieutenant Varnum agrees with Benteen that the lead elements of Reno's command reached Reno Hill by about 5:30 p.m.

8) George Herendeen (RCOI): Reno retreats to his position on the bluffs at 5 p.m. and fights till it is too dark to see.

ANALYSIS:

Herendeen tells us that Reno begins his retreat to his corral at about 5 p.m., likely a local sun time estimate, which would convert to about 6:00 p.m. Regimental Headquarters watch time, and then fought the Indians until it was too dark to see—which would be about 9 p.m.

9) Sergeant Ferdinand Culberston - A Company (RCOI): the Indians attacked them on the bluffs around 5 or 5:30 p.m.

ANALYSIS

Sergeant Culbertson is likely referring to the time when the Indians began their attack on the three troops still out on the Weir Peaks at this time, and not the start of the fight at Reno Hill.

10) Private Edward Davern (RCOI): the fight that evening at Reno Hill lasted about two hours and ended a little before sundown.... Reno's command got back an hour before sundown.

ANALYSIS:

There are a number of problems with Davern's recollections above. The first of which is that he is the only participant to claim that Reno's fight ended a little before sundown, or that Reno's command returned to Reno Hill an hour before sundown. In the latter case, he very likely meant to say that Reno's command returned to Reno Hill at about an hour before dusk—(as dusk is at 7 p.m. Headquarters watch time, Davern would suggest that Reno's command returned to the Hill about 6 p.m.), but if so, then the fight would not have ended as he described unless he meant to say too dark to see. Most of our participants indicate the fight lasted about three hours, yet Davern shortens it to about two hours, indicating that he was likely only referring to the action on Reno Hill, which began about 7 p.m. and did not include Godfrey's Rear Guard action falling back from the Weir Peaks.

11a) Lieutenant Winfield Edgerly (Camp Interview, *Custer in 76*): says that the fight on Reno Hill that evening began about dusk when First Sergeant Winney of K Troop was killed on Reno Hill.

ANALYSIS:

Edgerly tells us that the fight at Reno's Hill began about the time that First Sergeant Winney was killed. As First Sergeant Winney was a member of Godfrey's K Company, Edgerly is telling us that the fight began shortly after the arrival of K Company (the Rear Guard on the retreat from the Weir Peaks) on Reno Hill. We believe K Troop's arrival at Reno Hill to be about 6:40 p.m. or a bit later, indicating that First Sergeant Winney was killed at some time between 6:40 and 7:00 p.m.—or as Edgerly says, "about dusk."

11b) Lieutenant Winfield Edgerly (Edgerly's Narrative): the fight on the bluffs lasted from 5 p.m. to 9 p.m.

ANALYSIS:

Edgerly is likely including D Troop's fighting at the Weir Peaks and the Rear Guard action of K Company in his 5 p.m. estimate, just as Sergeant Culbertson did. His comment above would indicate that the fighting at Reno Hill did not begin until First Sergeant Winney's death after K Troop had fallen back to that position at some time between 6:40 and 7 p.m.

11c) Lieutenant Winfield Edgerly (RCOI): claims the Indians began firing at them within a minute of the troops getting in position on Reno Hill; he said it was quite dark at 9 p.m.; he claims that the Indians fired one and a half hours more after dark.

ANALYSIS:

Edgerly tells us that the Indians began firing at them within a minute of K Company returning to Reno Hill. He confirms that it was quite dark by 9 p.m., which is powerful evidence that the command was using Headquarters watch time based on the time at their Department Headquarters in St. Paul, Minnesota. He then claims that the Indians fired for "1 1/2 hours more after dark"; he very likely means that the Indians fired for one and a half hours more after *dusk*, indicating that the heaviest firing likely occurred either between 7:00 and 8:30 p.m. or between 7:30 p.m. and 9:00 p.m. when it became too dark to see.

12a) Lieutenant Edward Godfrey (RCOI): claims that his company got back to Reno Hill after 6 p.m. and that the Indians attacked right away, and lasted till dusk.... He says that K Company got back to Reno Hill about fifteen to twenty minutes before dark.

ANALYSIS

Godfrey tells us that K Troop got back to Reno Hill at some time after 6 p.m. with the Indians beginning their attack at Reno Hill right afterward, with their attack lasting until dusk. The latter is an obvious error. Godfrey likely meant that the attack began at dusk and ended when too dark to see. Godfrey clarifies this when he says that K Company got into Reno's corral about fifteen to twenty minutes before dark. He likely means 'dusk' here, not dark, meaning that K Company got back to Reno Hill after its Rear Guard action covering the retreat of the command from the Weir Peaks at about 6:40 to 6:45 p.m., or fifteen to twenty minutes before dusk at 7 p.m.

12b) Lieutenant Edward Godfrey ("Custer's Last Battle," *Century Magazine*): K Company fell back to Reno Hill at about 7 o'clock. The firing continued till nearly dark (between 9 and 10 o'clock), although after dusk, but little attention was paid to the firing, as everybody moved about freely.

ANALYSIS

Lieutenant Godfrey clarifies his comments made at the RCOI here. He indicates that K Company returned to Reno's corral at about 7 p.m. At the Reno Court of Inquiry, he indicated that K Company arrived fifteen to twenty minutes before dark—which I take to mean 'dusk'—placing his arrival on Reno Hill at about 6:40 or 6:45 p.m. Lieutenant Edgerly supports this time when he indicates that First Sergeant Winney was killed about dusk—or about 7 p.m.—and that coincided with the commencement of the Indian attack at Reno Hill. Sergeant Davern also supports this recollection by claiming that the fighting at Reno Hill lasted about two hours—or roughly from 7 p.m. to 9 p.m.

13) Private William Taylor – A Company (*With Custer on the Little Big Horn*): The whole command was then ordered to retire and we returned to the place we had left a short time before.... This was, according

to my belief, about 6 o'clock, and that is the time set by Major Reno in his report. Capt. Godfrey seems to think it was a little later, but he must be mistaken, for it certainly never took us 2 hours to travel a mile and return to our starting place.... The firing ... was continuous on the part of the Indians until dusk, and after that, a few scattering shots, ceasing altogether about 9 o'clock.

ANALYSIS

Private Taylor tells us that most of Reno's command had fallen back to Reno Hill at about 6 p.m. and bases that time on Reno's after-battle report. He notes that Godfrey indicated a later arrival time of 7 p.m. and disputed that by saying it did not take them two hours to advance and retreat about a mile all told (note: Taylor's A Company was the last company to move out to the Weir Peaks and did not get more than halfway down before Reno turned back to Reno Hill). By this statement, Taylor is likely telling us that Reno began his retreat back to the corral a short time after 5 p.m.—or just shy of two hours before Godfrey's estimated time of a 7 p.m. arrival at the Hill. It is very likely that Taylor's A Company was the first to arrive back at Reno Hill at about 5:35 p.m. with the rest of the command falling back at some time between 5:35 and 6 p.m. Only K Company—engaged in their dismounted Rear Guard action—arrived significantly after the others, getting back to Reno Hill about fifteen or twenty minutes before dusk—or about 6:40 p.m.

14) Private Charles Windolph – H Company (*I Fought With Custer*), with my annotations in brackets: We'd gone less than a mile when we [i.e., H Company] got in sight of Weir's troop. Way off to the north you could see what looked to be groups of mounted Indians. There was plenty of firing going on. Pretty soon it looked as if the Indian masses were coming towards us.... Here we were stretched out all over the hell's half acre, a troop [likely D Company] on this hill knob, another [likely M Company] in this little valley and over there a third troop [likely K Company]. Behind [the Weir Peaks, at least 1/2 mile or more], at a slow walk, came the pack trains, the [10] wounded men and the rear guard. Reno and Benteen both sensed the danger and ordered a withdrawal. The advance troops [i.e., D, M & K Co's.] were dismounted and fought as skirmishers. Soon the Indians were pressing hard, and it was only good luck and the hard courage of Lieutenant Godfrey's [K] troop, fighting stubbornly on foot, that kept disaster from overtaking us. We were able to regain Reno Hill while Troop "K" kept back the Indians until the men, retreating slowly, got close enough to the hilltop to make a dash for it. It was now possibly 6:30 in the afternoon, with 3 hours of daylight still to go.

ANALYSIS

Private Windolph in his account only mentions seeing four troops deployed on the Weir Peaks before Reno and Benteen called for a retreat back to Reno Hill. The remaining three troops were apparently still advancing at a slow walk when Reno and Benteen decided to return to Reno Hill, leaving three troops as a rear guard to delay the Indian attack on the command until that position was reached. He indicates that all the troops had returned to Reno Hill by about 6:30 p.m., whereupon the fight at Reno Hill would commence and carry on for the remaining three hours of daylight, or until about 9:30 p.m.

CONCLUSION

Reno begins his retreat, as indicated by scout Herendeen and implied by Private Taylor, at about 5 p.m. or a short time after, likely about 5:20 p.m. It is important to note that Reno began this retreat with his rearmost troops (A Company, B Company, the pack train, H Company, and G Company) without bothering to inform his three remaining companies out on the Weir Peaks until the four troops above had reached Reno Hill and began to deploy themselves there. Reno likely wished to keep D, M, and K Troops out on the Weir Peaks as a Rear Guard to occupy the rapidly approaching warriors until he could secure a place to defend at Reno Hill. Once this was accomplished—likely at some time between 5:30 and 5:45 p.m.—Lieutenant Hare was sent back to the Peaks to order the remaining troops there to fall back to their new position a mile and a quarter to the south. K Troop left the Peaks first, followed by M Troop and then D Troop. The latter two troops rode rapidly to the rear, reaching Reno Hill at about 6 p.m., as Private Taylor and most other accounts indicated. K Company, however, dismounted and assumed the role of a Rear Guard and continued to engage the warriors while slowly falling back to Reno Hill. Godfrey was in no hurry to fall back and began to deploy part of his company across the mouth of Cedar Coulee when Reno, running out of patience, ordered him once again to fall back immediately to their position. Godfrey now falls back to Reno Hill, reaching that position at about 6:40 p.m., fifteen or twenty minutes before dusk. Lieutenant Edgerly supports this time when he indicates that First Sergeant Winney was killed about dusk—or about 7 p.m.—and that coincided with the commencement of the Indian attack on Reno Hill. Sergeant Davern also supports this recollection by claiming that the fighting at Reno Hill lasted about two hours—or roughly from 7 p.m. to 9 p.m.

5:20 p.m. (H + 8 hours and 50 minutes): Reno begins his retreat from his advance towards the Weir Peaks.

5:35 p.m. (H + 9 hours and 5 minutes): A, B, G, H companies and the pack train arrive at Reno Hill.

5:45 p.m. (H + 9 hours & 15 min.): The remaining troops on the Weir Peaks are recalled.

6 p.m. (H + 9 hours and 30 minutes): Troops M and D arrive at Reno Hill.

6:40 p.m. (H + 10 hours and 10 minutes): K Troop arrives at Reno Hill.

6:45 p.m. (H + 10 hours and 15 minutes): The fight at Reno Hill begins.

9:00 p.m., with sporadic firing continuing until 9:30 p.m. (H + 10 hours and 30+ minutes) : The fight at Reno Hill ends.

31. THE DURATION OF THE CUSTER FIGHT

There has been much debate among Little Big Horn battle students concerning the actual length of the fighting engaged in by Custer's five companies before their ultimate destruction. Here is what the participants who were there had to say about its length and the time the fighting ended, along with my analysis:

1) Major Marcus Reno ("Reno's Narrative"): Custer's command was destroyed before Benteen arrived on the bluffs.

ANALYSIS

Major Reno's assertion above is false. The primary source evidence we have from multiple participant accounts is that the Custer fight did not begin until after Benteen arrived on the bluffs to join with Reno's battalion. Benteen's battalion arrived on the bluffs at about 2:30 p.m. Thus, Reno is telling us here that Custer's entire command of 210 men had already been destroyed by that time. This, of course, contradicts the numerous reports of volley fire heard by elements of Reno's command over the next hour or longer, as well as the two-hour or more delay before the hostiles returned to attack his own position on the bluffs. In light of the weight of contradictory evidence, we can only assume that Reno must have been aware of the absurdity of his estimate above. That he would seriously make such an estimate publicly would suggest that Reno was either delusional or could not bring himself to face the truth.

2) George Herendeen (RCOI): heard firing from Custer's position, which lasted about one hour. He also added that he remained in the brush for about two hours, not leaving the valley until the firing from Custer's field died down.

ANALYSIS

Herendeen tells us that he could hear firing from Custer's field for only one hour, yet also says that he remained in the brush about two hours before leaving the valley when the firing from Custer's field died down, thus presenting us with a contradiction in his account as far as the duration he attributes to the Custer fight. Was it one hour or two? One explanation would be that Herendeen's hearing was, like Lieutenant Godfrey's, limited in scope. He likely could only hear the initial action at Ford B, Luce and Blummer Ridges, and possibly some of the action around Finley Ridge or Calhoun Hill as well, but little or nothing beyond that point. The rest of the troopers with him likely had significantly better hearing, which kept them hiding in the brush for at least another hour before making an attempt to venture out of the valley to rejoin with Reno's command on the bluffs at a time when the hostiles were still fighting Custer's command in the north.

3a) Fred Gerard (RCOI – *Army and Navy Journal*): The firing on Custer's field lasted about 2 hours.

3b) Fred Gerard (RCOI): After the first few volleys, it was twenty to twenty-five minutes before the fighting became general. The heavy, continuous gunfire lasted about two hours.

ANALYSIS

Gerard tells us that the heavy, continuous gunfire from Custer's field lasted about two hours. He also tells us that this heavy gunfire did not begin until about twenty to twenty-five minutes after he heard the first few volleys fired, which I would interpret as being fired in the vicinity of Ford B. Thus, according to what he could hear, Gerard is telling us that Custer's Fight lasted at least two hours and twenty or twenty-five minutes.

4a) Private Thomas O'Neill – G Company (Camp Interview, *Custer in 76*): he heard Custer's firing (including three volleys) after Reno left the valley. The firing continued until 5 p.m.

4b) Private Thomas O'Neill – G Company (*Troopers With Custer*, p. 136): We had listened to terrific firing both up and down the river [i.e., from the valley timber position], and realized that the loss must have been heavy. For myself, I thought that when the firing had ceased in the direction Custer had gone, it meant that he had retreated to Major Reno's position. I think it was about 5 o'clock when we heard the last shots from Custer's direction; but Major Reno's command kept up their firing at intervals.... But along about dark, all firing from that direction also ceased.

ANALYSIS

As established from the Participant Timeline, Reno's retreat from the valley occurred shortly before 2:10 p.m., based on Gerard's watch time. Private O'Neill tells us that sometime after Reno left the valley, the Custer fight began. In a previous section, the start of the Custer fight was determined to occur at about 2:40 p.m. Private O'Neill tells us that the firing from Custer's field lasted until 5 p.m., a duration of two hours and twenty minutes, which supports Gerard's time estimate perfectly.

5) Lieutenant Charles Varnum (RCOI – *Army and Navy Journal*): from the time Reno reached the bluffs about 2 hours passed before he began defending against the whole force of Indians.

ANALYSIS

I believe Lieutenant Varnum is telling us that by the time most of Reno's battalion had ascended the bluffs—likely about 2:30 p.m. when Benteen's battalion had joined them—about two hours would pass before the whole force of Indians began to return to confront Reno's advance troops deployed out on the Weir Peaks. According to Varnum, the beginning of the action on the Weir Peaks—primarily against D Company's forward-deployed troopers— would begin shortly after 4:30 p.m., or roughly one and a half hours after D Troop first arrived out on the Peaks.

6) Lieutenant Carlo De Rudio (RCOI): The Indians returned to attack Reno about twenty minutes after most of the firing died out downstream. De Rudio also claimed to hear firing downstream for one and a half hours starting about ten minutes after Reno arrived on the bluffs.

ANALYSIS

Lieutenant De Rudio tells us that about ten minutes after most of Reno's battalion had reached the bluffs the start of the Custer Fight began. As most of Reno's men had reached the bluffs by 2:30 p.m., the same time that Benteen's battalion arrived there according to the Participant Timeline, ten minutes later would place the start of the Custer Fight at about 2:40 p.m. De Rudio could hear firing going on for another hour and a half, ending about thirty minutes earlier than the time reported by Gerard and O'Neill, meaning that De Rudio could hear no more firing from Custer's field after 4:10 p.m. He then reports a lull in the sounds of battle for about twenty minutes until he hears the fighting start up again on the bluffs against Reno's command. As this would bring us to a time of 4:30 p.m., De Rudio is likely describing the skirmishing now taking place between D Company and the hostiles on the Weir Peaks.

7) <u>Trooper John Henley</u> – B Company (Camp Interview, *Custer in 76*): says Sitting Bull told him at Standing Rock (he knew him well and could speak fluent Sioux) that the Custer fight lasted about two hours.

<u>ANALYSIS</u>

Sitting Bull concurs with Gerard, O'Neill, and Varnum on the length of the Custer Fight, which was about two hours in duration once Custer reached and deployed on Battle Ridge about 3 p.m., thus suggesting that the Custer Fight ended about 5 p.m.

8) <u>Left Hand</u> – Arapaho (Graham, *The Custer Myth*): claimed Custer's fight started at 12 noon and ended at 3 p.m.

<u>ANALYSIS</u>

Left Hand tells us that the Custer Fight started at about 12 noon, or 1 p.m. Headquarters watch time and ended about 3 p.m., (4 p.m. Headquarters watch time). Although Left Hand's clock times are early by about an hour and a half, he does estimate the Custer Fight at about three hours duration. If we transcribe his inaccurate clock times into the participant times for the start and end of the battle, we would end up with the Custer fight lasting from about 2:30 p.m. until about 5:30 p.m.

9a) <u>Wooden Leg</u> – Northern Cheyenne (Marquis, *A Warrior Who Fought Custer*): the long-range fighting took about an hour and a half.

9b) <u>Thomas Marquis's unnamed Cheyenne Accounts (Marquis, *Keep the Last Bullet for Yourself*)</u>: for about an hour and a half the fighting was slow, with little damage done to the Indians and none ... done to the soldiers.

<u>ANALYSIS</u>

Although Wooden Leg does not give us a time that encompasses the total duration of the fight, he does tell us that the initial long-range fighting took up about an hour and a half before the final melee fighting ended the battle. This would suggest a duration of fighting lasting about two hours in duration. If he is referencing the start of the fighting at 2:40 p.m., as per the Participant Timeline, then Wooden Leg is saying that the close-range melee fighting against Keogh's battalion began about 4:10 p.m.

10) Seven unidentified Sioux: (Seven Anonymous Warrior's Accounts of the Little Big Horn Battle," *Chicago Tribune*, 1876) After the Medicine Tail Ford episode, Custer's fight on the bluffs lasted about an hour.

ANALYSIS

These seven Sioux warriors estimated a duration of about an hour after the Ford B episode. I believe they were referencing the duration of the fighting at the Keogh sector, which would have ended about 4:00 p.m. or a bit later, as opposed to the entire Custer field.

11) American Horse – Northern Cheyenne (The Grinnell Collection, Item 497, *"Notes on the Custer Fight,"* Braun Research Library, Southwest Museum, Los Angeles, CA): Thinks the fight ended about 2 or 3 o'clock.

ANALYSIS

American Horse thought the battle ended about 2 or 3 p.m., which equates to 3 or 4 p.m. Headquarters watch time. We speculate that he was referring to the Keogh sector of the fighting here, which would have ended about 4:00 p.m. or a bit later.

12) He Dog – Oglala (Scott 1919 Interview, *Lakota Recollections*): It was about two hours from Reno's attack until all was over.

ANALYSIS

He Dog suggests that the battle lasted for two hours duration from the Reno's attack in the valley. We suspect that He Dog was referencing a starting time of about 2 p.m.—the height of the Reno valley fight—as opposed to its actual start at about 1:40 p.m., thus He Dog claims that the Custer Fight ended about 4:00 p.m. We would speculate that he was referring to the fighting around the Keogh sector only, not the entire Custer field.

13) Respects (or Fears) Nothing – Oglala (Eli Ricker 1906 Interview, *Lakota Recollections*): the battle began about 1 p.m. and lasted till 4 p.m. based on a watch the Indians recovered from one of Crook's soldiers at the Rosebud fight. (Respects Nothing then went outside and correctly pointed to the sun time that the battle began—which equated to approximately 1 p.m.).

ANALYSIS

Respects (or Fears) Nothing tells us that the battle (likely a reference to the Reno valley fight) began about 1:00 p.m. based on a watch time taken from a captured watch from one of Crook's soldiers at the Battle of the Rosebud six days earlier. This watch time would have been set to Montana sun time, or roughly one hour earlier than St. Paul watch time, which the 7th Cavalry watches would have been set to, thus Respects Nothing's 1 p.m. start of the Reno fight would actually translate into a 2 p.m. watch time based on regimental Headquarters time. The Participant Timeline indicates that the Reno valley fight actually started about 1:40 p.m., about twenty minutes earlier, but Respects Nothing is likely estimating his time to the nearest hour. He also indicates that the Custer Fight ended about 4:00 p.m., which translates into 5:00 p.m. Regimental Headquarters watch time, for a duration of three hours from the start of Reno's fight until the end of the Custer fight. Respects Nothing's estimate of a 5:00 p.m. HQ watch end time for the Custer Fight agrees with the time estimates made by Fred Gerard, Private O'Neill, and Sitting Bull.

14) Jack Red Cloud – Oglala (Dixon, *The Vanishing Race*): we fought this battle from daylight up until 3 o'clock in the afternoon, and all of the white men were killed. I think that Custer was a very brave man to fight all these Indians with his few men from daylight until the sun was almost going down.

ANALYSIS

Jack Red Cloud tells us that the battle began at daylight, which is simply not true, and ended at a sun time of 3 p.m., or a Regimental Headquarters watch time of 4:00 p.m., which I would interpret as the end of the fighting in the Keogh sector. His later reference to the fighting ending when the sun was almost going down would indicate the subsequent attack on the Reno Hill position afterward.

15) Flying Hawk – Oglala (Eli Ricker 1907 Interview, *Lakota Recollections*): The battle lasted about 4 or 5 hours, starting with Reno's attack in the valley at 8 or 9 a.m. and the Custer Fight ending at 1 or 2 p.m.

ANALYSIS

Flying Hawk's times are too early by about five hours. If we add five hours to his estimated clock times, we have Reno's attack in the valley starting between 1:00 and 2:00 p.m. (the Participant Timeline indicates a time of 1:40 p.m.) with the Custer Fight ending at 6 or 7 p.m., which is a time much later than any other Indian account. This time does indicate the approximate start of the Reno hilltop fight later that evening, so perhaps Flying Hawk is referencing the start of the Reno hilltop fight as the approximate end of the Custer fight.

16) Red Hawk – Oglala (Eli Ricker 1906 Interview, *Lakota Recollections*): has the Reno fight starting at about 9 a.m. with the Custer Fight ending about 12 noon.

ANALYSIS

Red Hawk tells us that the Reno valley fight began at 9 a.m. sun time or about 10 a.m. Headquarters watch time, with the Custer fight ending about three hours later at noon sun time or 1 p.m. Headquarters watch time. Red Hawk's clock times are early by about 4 hours according to the Participant Timeline, thus by adding 4 hours to the starting and ending times, we have a starting time of about 2 p.m. for the Reno fight (the Participants Timeline would say 1:40 p.m., or about 20 minutes earlier) and an ending time of about 5 p.m. Headquarters watch time, which supports the time estimates made by Gerard, O'Neill, Sitting Bull, and Respects Nothing.

17) Iron Hawk – Hunkpapa (Eli Ricker 1907 Interview, *Lakota Recollections*): said the Reno fight in the valley began about 8 or 9 a.m. with the entire battle (Reno and Custer's) lasting between three to four hours.

ANALYSIS

According to the Participant Timeline, Iron Hawk's start of the Reno Fight is too early by about five hours. If we adjust his times forward five hours from his estimated 8:30 a.m. start, we have the Reno valley fight beginning at about 1:30 p.m. and the Custer Fight ending about three and a half hours later at 5 p.m., which agrees with the time estimates of Gerard, O'Neill, Respects Nothing, Sitting Bull, and Red Hawk.

18) White Bull – Minneconjou (*Walter S. Campbell Collection*, University of Oklahoma): Estimates that the Reno valley fight started at either 9 a.m., 10 a.m., or 10:30 a.m. while the Custer Fight ended about noon. If we adjust his sun times to Regimental Headquarters watch times, we get the Reno valley fight starting at either 10, 11 or 11:30 a.m. and the Custer Fight ending at about 1:00 p.m. It appears that White Bull's clock estimates are early by about three hours based on the Participant Timeline. If we add three hours to the starting and ending times, we get the start of the Reno valley fight at some time between 1:00 and 2:00 p.m. and the Custer Fight ending about 4:00 p.m., which, like the accounts of Red Cloud, He Dog, and American Horse, appear to be describing the approximate time of the destruction of the Keogh battalion.

19) Lone Bear – Oglala (Walter Camp 1909 Interview, *Lakota Recollections*): The battle lasted from about noon until 4 o'clock.

ANALYSIS

Lone Bear tells us that the Custer fight started about noon, or 1:00 p.m. Regimental Headquarters watch time (he has the Reno valley fight starting at some time between 9 and 11 a.m. or between 10 and 12 noon Headquarters watch time), which is about three hours too early based on the Participant Timeline. If we add three hours to his starting times, he would have the Reno valley fight starting between 1 and 2 p.m. (the Participant Timeline indicates it was about 1:40 p.m.) and the Custer Fight starting about 3:00 p.m. Headquarters watch time, which is about the time that Custer's command reached and deployed on Battle Ridge. Lone Bear then says the Custer Fight lasted until 4 p.m., which converts to about 7:00 p.m. Headquarters watch time once we factor in the additional three hours, which suggests that, like Flying Hawk, Lone Bear was referencing the start of the Reno Hilltop fight as the end of the Custer fight.

20) Two Eagles – Brulé (Sewell B. Weston 1908 Interview, *Lakota Recollections*): placed the start of the battle—i.e., Reno's valley fight—at about 1 p.m. with the Custer Fight ending about 5 p.m.

ANALYSIS

Most Indian accounts refer to sun time estimates, however, it is possible that Two Eagles is actually referring to regimental Headquarters watch time in his estimates. The Participant Timeline tells us that the Reno valley fight began about 1:40 p.m., and a good number of participants (Gerard, O'Neill, Respects Nothing, Red Hawk, Sitting Bull, and Iron Hawk) tell us that the Custer fight ended at about 5:00 p.m., thus Two Eagles' time estimates of a fight lasting from 1 p.m. to 5 p.m.—a total of four hours duration for both the Reno valley fight and the Custer Fight—may well be the most accurate times given by the participants. If, however, Two Eagles was referring to sun time, as most Indian accounts do, then we would add one hour to his time estimates, giving us a time frame of 2 p.m. for the start of the Reno valley fight and a 6 p.m. end of the Custer Fight.

21) Private Peter Thompson – C Company (*Peter Thompson Narrative*): to my certain knowledge the battle lasted three hours.

ANALYSIS

Private Peter Thompson states emphatically that he is quite certain that Custer's fight lasted for three hours. This would encompass a time of from 2:30 p.m. until about 5:30 p.m., just after the first three companies of Reno's command returned to Reno Hill from their foray out toward the Weir Peaks. Thompson's time estimate of a three-hour duration for the Custer Fight agrees very closely with that of the Arapaho Left Hand.

22) Packer Benjamin Churchill (RCOI)—the only thing I could go by was the time the Indians came back on us on the hills. I think probably it [i.e., the firing from Custer's field] lasted an hour and a half or two hours.

ANALYSIS

The Timeline previously established that the advance packs group (where Churchill was traveling) arrived on Reno Hill about 3:10 to 3:15 p.m. Churchill thus claims that the firing on Custer's field lasted between 1.5 to two hours before the Indians came back to attack the troops out on the hills, likely the Weir Peaks. Thus, Churchill's testimony suggests that the attack on the Weir Peaks by the mass of warriors began sometime between 4:40 and 5:00 p.m., keeping in mind that Churchill himself likely never reached the Weir Peaks to make this observation, as he would have been traveling in the rear with the packs. His account does support that of Private O'Neill who likewise suggested an end to the Custer fight at about 5 p.m. We believe Churchill's two-hour estimate is closer to the truth than his hour-and-a-half estimate. This two-hour timeframe for the Custer fight does not include the first twenty or thirty minutes of Custer's fight in the vicinity of Ford B and Luce Ridge, which Churchill apparently could not discern, being about a mile or so south of Reno Hill at the time.

23) Private Edward Davern (RCOI): Q. How long did the firing last that you heard down the stream? Davern: I heard it occasionally for about an hour and a half.

ANALYSIS

Private Davern hears the firing from Custer's field lasting about an hour and a half—or from roughly 2:40 p.m. until 4:10 p.m., suggesting the duration of the Keogh fight on Battle Ridge. Since other participants estimated a longer period of fighting before the end of the battle, it is likely that the remaining fighting on Battle Ridge occurred farther to the north and out of range of Davern's hearing.

24) Dry Hide, Red Cloud, and Howling Eagle – Cheyennes: said the fight lasted less than an hour.

ANALYSIS

These Cheyenne estimates were second-hand reports as they were not participants at this battle. As their estimate conflicts with the greater majority of the actual participants of the battle, they can be safely dismissed.

25) Charles Corn – Oglala (letter to Pvt. William Taylor, 7th Cavalry, *Indian Views of the Custer Fight*): said the battle lasted about two hours.

ANALYSIS

Charles Corn's two hour estimate for the length of the battle would likely refer only to the Custer fight. If so, then this estimate agrees with most other Indian estimates.

26) Eagle Bear – Oglala (Hunt, *I Fought With Custer*): I do not know how long it lasted. I think maybe it was one hour and a half....

ANALYSIS

Eagle Bear's analysis of the length of the fight corresponds well with those describing the likely duration of the Keogh fight.

27) Lieutenant Luther Hare, (Walter Camp 1910 interview, published in Kenneth Hammer's *Custer in 76*): While out in the advance with Company D, the Indians were thick over on Custer ridge and were firing and at that moment, Hare thought Custer was fighting them.

ANALYSIS

Lieutenant Hare informs us that he believed Custer's wing—or likely Keogh's battalion—was still actively fighting the hostiles from what he could observe from the heights of the Weir Peaks. This statement contradicts the notion held by some students that the fighting on Battle Ridge ended shortly after D Company arrived on the Weir Peaks.

28) Billy Jackson – (Interview with the *Anaconda Standard*, June 19, 1898): He says of the firing when Custer's battle began: "... the fighting in the lower end of the camp was hot and heavy. The sound of the firing increased steadily until it became a roar, and then it died gradually away until there was only the scattered reports of single shots. All this took place in the space of two hours."

ANALYSIS

Billy Jackson estimates the length of the Custer Fight at about two hours duration. His estimate concurs with Gerard, O'Neill, Varnum, and Sitting Bull on the length of the Custer fight, which was about two hours in duration once Custer reached and deployed on Battle Ridge about 3 p.m., thus suggesting that the Custer fight ended about 5 p.m.

29) Private William Morris – M Company, (published by the *New York Herald* on August 4th, 1895, with our annotations in brackets): "When we had mounted the highest hill [i.e., Reno Hill] we were joined by Benteen, with the pack train, and about this time the Indians left us and went to the other side of the village, which was plainly discernible from our position. They were absent for about three hours,

during which [time] Custer fell, but we were in ignorance of it. They returned at half past five and attacked us in full force."

ANALYSIS

Private Morris claims in his statement above that the Indians left Reno's position around 2:30 pm and returned to engage them on the bluffs at about 5:30 pm, giving a rough time estimate of three hours, or just less considering the distance traveled, for the Custer fight.

OVERALL ANALYSIS

Of the twenty-nine accounts listed above, twelve participants estimated or strongly suggested that the Custer Fight ended about 5:00 p.m. or later, and fourteen participants indicated the Custer fight lasted between two to three hours in duration. We can compare this to three participants who estimated the Custer fight only lasted about an hour, and four who claimed it lasted an hour and a half. It is very possible that these seven participants were referring specifically to the Keogh sector of the field, which ended earlier than the action on Last Stand Hill and beyond, out of hearing range of those participants. The same thing can be said for the handful of participants who estimated the battle ended earlier than 5 p.m. These Indian participants likely left the battlefield during or immediately after the final destruction of the Keogh battalion, and for all intents and purposes, the battle was over for them at that time. It is apparent that the majority of the participant accounts tell us that the Custer fight lasted somewhere between two to three hours in duration, likely two hours and thirty to forty minutes, with the fighting on Custer field ending about 5:10 p.m. or a short time afterward—perhaps 5:20 p.m.

CONCLUSION

The Custer fight lasted about two hours and thirty or forty minutes. The fighting on Custer field likely ended at some time between 5:10 and 5:20 p.m.

Between 5:10 and 5:20 p.m. (between H + 8 hours and 40 to 50 minutes): The fighting on Custer's field ends.

Appendix B: United States Seventh Cavalry Casualties at Little Big Horn

Ever since the battle, there have been low-voltage debates on the casualties suffered at Little Big Horn. Lieutenant James Bradley said 214 died on the Custer Battlefield; Captain Benteen said 212; and still others, 211. Nathanial Philbrick, author of *The Last Stand,* bandied a figure of approximately 215. General Terry had been making his way to the mouth of the Little Big Horn having told Custer that he would have Gibbon's column at that destination by the morning of June 26. But that morning he was still forty miles away. The country they were traveling was rough territory populated by numerous hills, deep ravines, and, according to Lieutenant Bradley, destitute of water. Bradley and his Crow scouts were riding several miles ahead of Terry's main force.

Returning to the command, they reoriented Terry's march. Bradley set off again with his scouts, this time accompanied by the Crow scout Little Face, who was said to know every inch of ground in that region. Soon, Gibbon's Crow scouts encountered three of Custer's Crow scouts, White Man Runs Him, Goes Ahead, and Hairy Moccasin with dreadful news. They were told the Sioux had defeated the 7th, and all but a small portion had perished. Those few who survived were now besieged on a hill. Bradley's Crows believed it and began howling in the most aggrieved manner. Bradley repeated all this to Terry, who still viewed a defeat of so fine a regiment at the 7th almost incomprehensible. With stops and starts, they trudged on, the sights deeply discomforting, with heavy smoke seen on the horizon to the south. Large pockets of Indian warriors were first mistaken for the U.S. cavalry because they were wearing soldier's uniforms and carried army guidons. Bradley and eleven of his men scouted on the east side of the Little Big Horn amid its sloping hills. Gibbon's advance had run into signs of a large Indian village along the west bank. The grass had been burned black and still smoldered—a sign of an Indian retreat. A ring of horses surrounded two tepees that housed dead warriors. A battle had clearly been fought; trepidations grew for the fate of the regiment. The next sights did nothing to calm those fears. A village was strewn with Indian accouterments for two or three miles. Much more ominous was the gruesome discovery in the village of army equipment, a soldier's underwear, a buckskin jacket, a scattering of clothes, and most ominous—the decapitated heads of three white men.

Bradley reported the grim news to the main command. His voice, fraught with emotion and pain, said, "I have a sad report to make. I have counted 197 bodies lying in the hills." Someone dared to ask what everyone was thinking, "White men?" "Yes," said Bradley. "White men." Indeed, white men were stripped, scalped, with most naked and horribly mutilated lying all along the crest and either side of a hill for more than a mile.

Custer's Last Stand Demystified

The burial details would bury 204 men. Each officer's grave was marked by his name on a piece of paper in an empty cartridge shell, which was then hammered into the top of a stake. The enlisted men were simply buried wherever enough earth was found to partially cover them. Some remains of their dead comrades were not discovered until years and even decades later. There was, of course, the controversy of how many bodies were interred in Deep Ravine. Three of Custer's officers were never identified, Harrington, Sturgis, and Porter—though the head of young Sturgis was hideously discovered in the village, and Porter's buckskin jacket containing a bullet hole was also there. Not a trace of Lieutenant Harrington was ever discovered. His name is known but he remains the Little Big Horn's Unknown Soldier.

Of course, others of the 7th died, but they did not fight alongside Custer that day. They also must be added to the total. The first reports gave a total of 262 fatalities. Others would later report one additional fatality, making it 263 who died in the two-day battle. The 263 number is likely correct. Five others would succumb to their wounds, making the total dead in the 7th for this engagement 268. There were 52 wounded who were transported by the *Far West*, a 190-foot steamer docked at the mouth of the Little Big Horn, and then transported to Fort Abraham Lincoln. Another twelve were only slightly wounded and remained with the command. Twenty-one of the men were so severely wounded they needed to be delicately carried, and it is a wonder that more did not die. The total casualties for the 7th at the Little Big Horn were 334 when counting all the wounded. The total outfit numbered 655, translating into a staggering 50 percent-plus casualty rate. More wounded would have given up the ghost had it not been for the skill of Captain Grant Marsh, who made extensive preparations to receive the wounded. He piloted the *Far West* along the shallow Little Big Horn River and around its sandbars and small islands. Then he navigated the steamer into the Yellowstone River, reaching Fort Lincoln in record time. The number of Indian casualties is unknown, and getting an accurate fix on that number is almost impossible. It is safe to say, however, that the number was significantly less than the 7th suffered.

Most of the remains of the soldiers who died with Custer have been accounted for and identified. Several officers from Custer's respective commands who fought on the east side of the river, such as Lieutenant Henry Harrington, Lieutenant James Porter, and Lieutenant James Sturgis, were the only officers not accounted for. Blood-soaked clothing later found in the Indian village strongly points to the possibility that Porter and Sturgis were wounded during the battle but killed just afterward (the latter's three-letter initialed jacket and shirt, as well as his one-piece blood-stained underwear buttoned to the top, indicated decapitation). At the same time, the circumstances surrounding the fate of Harrington remain unknown and more speculative, other than the fact he did not survive. Sergeant John Ryan remembers seeing Sergeant O'Hara's detached head in the village with the heads of two unidentified soldiers (with one of them possibly being Sturgis) tied together with wires and suspended on a lodge pole in one of the Indian lodges with the hair all singed off.

So, how many died with Custer? The best evidence shows that the number of soldiers under his direct command who died was 210. In the following list are the 7th's dead and wounded at the Little Big Horn.

Appendix B: United States Seventh Cavalry Casualties at Little Big Horn

They are listed by companies and, if the information was available, where on the battlefield they were killed or wounded:

Casualties of the Battle of Little Big Horn—From the research of noted Little Big Horn scholar Frederick C. Wagner.

Killed In Action And Died of Wounds—Total killed: 258 plus 10 civilians. Those who died of wounds are listed in italics.

Headquarters—(6)

1. LTC George Armstrong Custer: Body identified on Last Stand Hill. Marker 109.
2. 1LT William W. Cooke: Body identified on Last Stand Hill. Marker 102.
3. 1LT Dr. Geroge Lord: Body identified on Last Stand Hill. Marker 98.
4. SGM William Sharrow: Body identified over a mile north of Last Stand Hill.
5. CTMP Henry Voss: Body identified on Last Stand Hill. Marker 106.
6. Dr. James DeWolf: Killed during Reno's retreat.

Company A—(9)

7. CPL James Dalious: Killed during Reno's retreat.
8. CPL George H King: *Died of wounds on July 1, 1876.*
9. PVT John E. Armstrong: Believed to have been killed approaching the village during Reno's advance in the valley.
10. PVT James Drinan (also spelled Drinnan): Killed during Reno's retreat.
11. PVT James McDonald: Killed during Reno's retreat.
12. PVT William Moodie: Killed during Reno's retreat.
13. PVT Richard Rollins (aka Rawlins): Killed during Reno's retreat.
14. PVT John Sullivan: Killed during Reno's retreat.
15. PVT Thomas P. Sweetser (aka Switzer): Killed during Reno's retreat.

Company B—(3)

16. 2LT Benjamin "Benny" Hubert Hodgson: Killed during Reno's retreat.
17. PVT Richard B. Dorn (aka Doran): Killed on Reno Hill.
18. PVT George Brown Mask (or Mack) [Hodgson's orderly: Killed during Reno's retreat.

Company C—(39)

19. CPT (Bvt LTC) Thomas Ward Custer: Body identified on Last Stand Hill. Marker 70.
20. 2LT Henry Moore Harrington: Body never found.

21. 1SGT L. Edwin Bobo: Body identified in Keogh's Swale. Marker 190.
22. SGT George August Finckle (also spelled Finkle or Finkel): Body identified on Finley Ridge. Marker 131.
23. SGT Jeremiah Finley: Body identified on Finley Ridge. Marker 139.
24. CPL John Foley: Body identified off reservation on Butler Ridge extension. Marker 260.
25. CPL Henry Eldon French
26. CPL Daniel Ryan
27. TMP Thomas J. Bucknell
28. TMP William Kramer
29. BSM John King
30. SAD George Howell (aka Hamel or Hammel)
31. PVT Alfred (Fred) Ernest Allan (or Allen)
32. PVT John Brightfield
33. PVT Christopher Criddle
34. PVT George Eiseman
35. PVT Gustave Engle
36. PVT James Farrand
37. PVT Patrick Griffin
38. PVT James Hathersall
39. PVT John Lewis
40. PVT August Meyer
41. PVT Frederick Meyer (also Mayer, Meier)
42. PVT Edgar Phillips
43. PVT John Rauter
44. PVT Edward Rix
45. PVT James Henry Russell
46. PVT St. John Ludwick
47. PVT Samuel S. Shade
48. PVT Jeremiah Shea
49. PVT Nathan Short: Probably killed in Calhoun Coulee at Marker 128.
50. PVT Alpheus Stuart
51. PVT Ygnatz Stungewitz: Body identified on Last Stand Hill. Marker 84.
52. PVT John Thadus
53. PVT Garrett H. Van Allen (aka Garrett H. Niver)
54. PVT Oscar T. Warner (also listed as Warren and Oscar L. Warner)
55. PVT Willis B. Wright: Body identified in Keogh's Swale.
56. PVT Henry Wyman.
57. PVT James C. Bennett: *Fatally wounded on Reno Hill, died on July 5, 1876.*

Appendix B: United States Seventh Cavalry Casualties at Little Big Horn

Company D—(3)

58. FAR PVT Vincent Charley: Killed on retreat from Weir Point.
59. PVT Patrick M. Golden: Killed on Reno Hill.
60. PVT Edward Housen (aka Edward Dellienhousen): Possibly killed while carrying water.

Company E—(39)

61. 1LT (Bvt MAJ) Algernon Emory "Fresh" Smith: Body identified on Last Stand Hill. Marker 74.
62. 2LT James (Jack) Garland Sturgis
63. 1SGT Frederick Hohmeyer: Body identified in Deep Ravine. Marker 25.
64. SGT William Batine James
65. SGT John S. Ogden: Body identified in Deep Ravine. Marker 1.
66. CPL George C. Brown: Body identified in Deep Ravine. Marker 20.
67. CPL Thomas P. Hagan (aka Eagan)
68. CPL Henry S. Mason
69. CPL Albert H. Meyer: Body identified in Deep Ravine. Marker 21.
70. TMP Thomas McElroy (listed elsewhere as McElvay)
71. TMP George A. Moonie
72. PVT William H. Baker
73. PVT Robert Barth
74. PVT Owen Boyle
75. PVT James Brogan
76. PVT Edward Connor
77. PVT John Darris
78. PVT William Davis (aka Dawes)
79. PVT Richard Farrell: Body identified in Deep Ravine. Marker 5.
80. PVT John Heim
81. PVT John Henderson
82. PVT Sykes Henderson
83. PVT John S. Hiley (aka Forbes, John Stuart Stuart (Double Stuart is correct)
84. PVT William Huber (also Heiber). Body identified in Deep Ravine. Marker 22.
85. PVT Andrew Knecht: Body identified in Deep Ravine. Marker 23.
86. PVT Patrick Edward O'Conner
87. PVT William Henry Rees: Body identified in Deep Ravine. Marker 19.
88. PVT Edward Rood
89. PVT Henry Schele (aka Schoole)
90. PVT William Smallwood
91. PVT Albert A. Smith

92. PVT James Smith (1st named)
93. PVT James Smith (2nd named)
94. PVT Benjamin F. Stafford
95. PVT Alexander Stella (aka Stern)
96. PVT William A. Torrey (aka Tarr)
97. PVT Cornelius Van Sant (aka Vaugant)
98. PVT George P. Walker (aka George P. Weldon)
99. PVT Herod T. Liddiard *Fatally wounded on Reno Hill, died on June 27, 1876.*

Company F—(38)

100. CPT (BVT LTC) George Wilhelmus Mancius Yates: Body identified on Last Stand Hill. Marker 108.
101. 2LT William Van Wyck Reily (aka Reilly and Reilley): Body identified on Last Stand Hill. Marker 110.
102. 1SGT Michael Kenney: Body identified in The Basin area. Marker 45.
103. SGT Frederick Nursey.
104. SGT John H. Vickory (also Vickery; aka Groesbeck): Body identified on Last Stand Hill. Marker 104.
105. SGT John R. Wilkinson.
106. CPL John Briody: Body identified in The Basin area, Marker 257. Also identified over a mile north of Last Stand Hill near Sharrow. The latter likely correct.
107. CPL Charles Coleman
108. CPL William Teeman (Tieman; aka William A. Adams): Body identified on Last Stand Hill. Marker 56.
109. TMP Thomas N. Way.
110. BSM James R. Manning
111. FAR Benjamin Brandon (aka Robert Nelson)
112. PVT Thomas Atcheson
113. PVT William Brady
114. PVT Benjamin Franklin Brown
115. PVT William A. Brown: Body identified west of the river, across Ford C, opposite the mouth of Deep Ravine, in the village. No marker.
116. PVT Patrick Bruce
117. PVT Lucien Burnham
118. PVT James Carney
119. PVT Armantheus D. Cather
120. PVT Anton Dohman (also Dorman or Dohlman): Body identified on Last Stand Hill. Marker 72.
121. PVT Timothy Donnelly: Body identified in The Basin area. Marker 2.

122. PVT John Gardiner (also William Gardner), body identified over a mile north of Last Stand Hill near Sharrow and Briody.
123. PVT George W. Hammon
124. PVT John P. Kelly (also listed as Kidey), body identified on Last Stand Hill.
125. PVT Gustav Klein (aka Heinrich Klein and Henry Klein): Body identified on Last Stand Hill. Marker 55.
126. PVT Herman Knauth (also Krianth)
127. PVT William H. Lerock (also LeRoche): Body identified on Last Stand Hill. Marker 94.
128. PVT Werner L. Liemann (also Luman, Lunden, or Lieman): Body identified on Last Stand Hill. Marker 105.
129. PVT William Lossee (also Loose or Losse)
130. PVT Christian Madsen (also Madison)
131. PVT Francis E. Milton
132. PVT Joseph Monroe
133. PVT Sebastian Omling (aka Smeling)
134. PVT Patrick Rudden
135. PVT Richard D. Saunders
136. PVT Francis W. Sicfous
137. PVT George A. Warren (also Wanew)

Company G—(14)

138. 1LT Donald McIntosh: Killed during Reno's retreat near the retreat crossing.
139. 1SGT (acting) Edward Botzer: Killed during Reno's retreat.
140. SGT Martin Considine: Killed during Reno's retreat.
141. CPL Otto Hagemann: Killed during Reno's retreat.
142. CPL James Martin: Killed during Reno's retreat.
143. TMP Henry C. Dose: Body identified off reservation near or south of Deep Coulee. No marker number.
144. FAR Benjamin J. Wells: Killed during Reno's retreat
145. SAD Crawford Selby: Killed during Reno's retreat.
146. PVT John J. McGinniss: Killed during Reno's retreat.
147. PVT Andrew J. Moore: Killed on Reno Hill
148. PVT John Rapp: Killed during Reno's retreat.
149. PVT Benjamin F. Rogers: Killed during Reno's retreat.
150. PVT Henry Seafferman: Killed during Reno's retreat.
151. PVT Edward Stanley: Killed during Reno's retreat.

Company H—(4)

152. CPL. George Lell: Killed on Reno Hill.
153. PVT William Montell George: *Died of wounds at the Powder River Depot on July 3, 1876.*
154. PVT Julien D. Jones: Killed on Reno Hill.
155. PVT Thomas Meader: Killed on Reno Hill.

Company I—(39)

156. CPT. (BVT LTC) Myles Walter Keogh: Body identified in Keogh's Swale. Marker 199..
157. 1LT James Ezekiel Porter: Body never identified.
158. 1SGT Frank E. Vardon: Body identified in Keogh's Swale. Marker 185.
159. SGT James Bustard: Body identified in Keogh's Swale. Marker 179.
160. CPL George C. Morris
161. CPL Samuel Frederick Staples
162. CPL John Wild: Body identified in Keogh's Swale. Marker 178.
163. TMP John McGucker
164. John W. Patton: Body identified in Keogh's Swale. Marker 180.
165. BSM PVT Henry Allen Bailey: Body identified in Keogh's Swale. Marker 174.
166. PVT John D. Barry
167. PVT Joseph F. Broadhurst
168. PVT Thomas Connors
169. PVT Thomas Patrick Downing
170. PVT Edward C. Driscoll: Body identified on Last Stand Hill. Marker 64.
171. PVT David C. Gillette
172. PVT George H. Gross
173. PVT Adam Hetesimer
174. PVT Edward P. Holcomb
175. PVT Marion E. Horn
176. PVT Patrick Kelly (or Kelley; aka Edward H. Kelly)
177. PVT Frederick Lehman
178. PVT Henry Lehmann
179. PVT Edward W. Lloyd: Body identified in Keogh's Swale. Marker 181.
180. PVT Archibald McIlhargey (also McEllargey): Body identified on Last Stand Hill. Marker 103.
181. PVT John Edward Mitchell: Body identified on Last Stand Hill. Marker 66.
182. PVT Jacob Noshang:
183. PVT John O'Bryan
184. PVT John Parker: Body identified on Last Stand Hill. Marker 59.
185. PVT James Felix Pitter

186. PVT George Post
187. PVT James Quinn
188. PVT William Reed
189. PVT John W. Rossbury
190. PVT Darwin L. Symms
191. PVT James E. Troy
192. PVT Charles Van Bramer
193. PVT William B. Whaley
194. PVT David Cooney (aka Corey): *Fatally wounded on Reno Hill, died of wounds on July 20, 1876.*

Company K—(5)

195. 1SGT DeWitt Winney: Killed on Reno Hill.
196. SGT Robert H. Hughes: Body identified in The Basin area. Marker 7. Also identified at the top of Last Stand Hill, near Custer.
197. CPL John J. Callahan: Body identified on Last Stand Hill. Marker 73.
198. TMP Julius Helmer: Killed on Reno Hill.
199. PVT Elihu F. Clear (also seen listed as Clare): Killed during Reno's retreat on the east side of the river.

Company L—(46)

200. 1LT James Calhoun: Body identified on Calhoun Hill. Marker 156.
201. 2LT John Jordan Crittenden: Body identified on Calhoun Hill. Marker 157.
202. 1SGT James Butler: Body identified in or near Deep Coulee, off reservation along the Butler Ridge-Nye-Cartwright extension. Marker B33.
203. SGT William Cashan (or Cashen)
204. SGT Amos B. Warren
205. CPL William H. Gilbert
206. CPL William H. Harrison
207. CPL John Seiler
208. TMP Frederick Walsh
209. BSM Charles Siemon
210. FAR William H. Heath
211. SAD Charles Perkins
212. PVT George E. Adams
213. PVT William Andrews
214. PVT Anthony Assadaly

215. PVT Elmer Babock
216. PVT John Burke (aka, Oscar F. Pardee): Body identified on Last Stand Hill. Marker 76.
217. PVT Ami Cheever (aka Ami Chreer, also Andrew Hester)
218. PVT William B. Crisfield
219. PVT John F. Duggan (aka John L Crowley)
220. PVT William Dye
221. PVT James J. Galvan (aka Michael J. Miller; also Calran)
222. PVT Charles Graham: Body identified in Keogh's Swale. Marker 160.
223. PVT Henry Hamilton
224. PVT Weston Harrington: Body identified in The Basin area. Marker 54.
225. PVT Louis Hauggi (also Hange, Hanggie, and Haugge)
226. PVT Francis Thomas Hughes: Body identified on Last Stand Hill. Marker 76.
227. PVT Thomas G. Kavanaugh
228. PVT Louis Lobering
229. PVT Bartholomew Mahoney
230. PVT Thomas E. Maxwell
231. PVT Charles McCarthy: Body identified on Last Stand Hill. Marker 79.
232. PVT Peter McGue
233. PVT John Miller
234. PVT David J. O'Connell
235. PVT Christian Reibold
236. PVT Henry Roberts
237. PVT Walter B. Rogers
238. PVT Charles Schmidt
239. PVT Charles Scott
240. PVT Bent Siemonson
241. PVT Andrew Snow
242. PVT Byron L. Tarbox
243. PVT Edmond D. Tessier
244. PVT Thomas S. Tweed: Body identified on Last Stand Hill. Marker 80.
245. PVT Johann Michael Vetter

Company M—(13)

246. SGT Miles F. O'Harra (also spelled O'Hara): *Mortally wounded in the valley fight.*
247. CPL Henry M. Scollin (or Scullin; real name was Cody): Killed during Reno's retreat.
248. CPL Frederick Stressinger (also Streing and Stringer): Killed during Reno's retreat.
249. PVT Frank Braun (also Baum and Brunn): *Died of wounds at Fort Abraham Lincoln on October 4, 1876, the final fatality of the Little Big Horn battle.*
250. PVT Henry James Turley (aka Tenley and Farley): Killed during Reno's retreat.

251. PVT Henry Gordon: Killed during Reno's retreat.
252. PVT Henry Klotzbucher: Killed in the timber.
253. PVT George Lorentz: Killed in the timber.
254. PVT William D. Meyer (also seen spelled Myer and Myers): Killed during Reno's retreat.
255. PVT George E. Smith: Killed during Reno's advance in the valley.
256. PVT David Summers: Killed during Reno's retreat.
257. PVT James J. Tanner (aka Jacob Henry, Gebhart): Killed during Reno's retreat.
258. PVT Henry Carl Voigt (also spelled as Voyte, Voight, Voyt)

Auxiliaries—(10)

259. Harry Armstrong "Autie" Reed: Body identified below Last Stand Hill. Marker 82.
260. Marcus Henry "Mark" Kellogg: Body identified off reservation northwest of Last Stand Hill.
261. Isaiah Dorman (aka "Teat;" also Darman and Darwin): Killed during Reno's retreat.
262. Bloody Knife: Killed in the timber.
263. Boston Custer: Body identified below Last Stand Hill in the Basin. Marker 83.
264. Charles Alexander Reynolds: Killed during Reno's retreat.
265. Michel "Mitch" Boyer (sometimes spelled Bouyer): Body identified in The Basin area. Marker 33.
266. Frank C. Mann: Killed on Reno's Hill.
267. Bob-tail Bull (Hocus -ta-ris; Hucus ta-nix) (SGT): Killed during Reno's retreat.
268. Little Brave (aka Little Soldier and Stub, also Bear's Trail): Killed during Reno's retreat.

Wounded At the Little Big Horn—Total wounded: 61 plus 3 civilians

Company A (9)

1. LT Charles Varnum: Slightly wounded on Reno's Hill.
2. ISG William Heyn: Wounded in the valley fight.
3. PVT Thomas Blake
4. PVT Jacob Deihle: Wounded on Reno Hill.
5. PVT Samuel James Foster: Carried as sick on the June returns.
6. PVT Frederick Holmstead: Wounded in the valley fight. Carried as sick on the June returns.
7. PVT William David Nugent: Wounded while carrying water back to the hilltop.
8. PVT Francis M. Reeves: Wounded leaving the timber. Carried as sick on the June returns.
9. PVT Elijah T. Strode: Wounded on the way up to Reno Hill. Carried as sick on the June returns.

Company B—(5)

10. SGT Benjamin C. Criswell
11. SGT Thomas Murray
12. CPL Charles Cunningham.
13. CPL William M. Smith
14. PVT James Pym

Company C—(3)

15. PVT Peter Thompson
16. PVT John B. McGuire, Jr.
17. PVT Alfred Whitaker

Company D—(3)

18. PVT Jacob Helter
19. PVT Joseph Kretchmer
20. PVT Patrick McDonnell

Company E—(2)

21. SAD William M. Shields
22. SGT James T. Riley

Company F—None

Company G—(6)

23. PVT James P. Boyle: Wounded on Reno Hill.
24. PVT John Hackett: Wounded on Reno Hill.
25. PVT John McVay: Wounded on Reno Hill.
26. PVT John Morrison
27. PVT Henry Petring: Wounded on Reno Hill.
28. PVT Charles W. Campbell: Wounded while attempting to get water.

Appendix B: United States Seventh Cavalry Casualties at Little Big Horn

Company H—(19)

29. CPT F.W. Benteen: Slightly wounded on Reno Hill.
30. 1SGT Joseph McCurry
31. SGT Patrick Connelly
32. SGT Thomas F. McLaughlin
33. SGT John Pahl
34. CPL Alexander B. Bishop
35. TMP William Ramell
36. SAD Otto Voit
37. PVT P. Henry Bishley
38. PVT Charles H. Bishop
39. PVT Henry Black
40. PVT John H. Cooper
41. PVT William Farley
42. PVT Thomas Hughes
43. PVT Jan Moller
44. PVT John J. Phillips
45. PVT Samuel Severs
46. PVT William C. Williams
47. PVT Charles A. Windolph

Company I—None

Company K—(2)

48. SAD Michael P. Madden
49. PVT Patrick Corcoran
50. PVT Max Mielke: Carried as sick on the June returns.

Company L—(1)

51. PVT Jasper Marshall

Company M—(10)

52. SGT Patrick Carey: Wounded in the timber.
53. SGT Henry White: Wounded in the timber.
54. PVT James Wilber Darcy: Wounded either on the hilltop or while trying to get water.

55. PVT John H. Meier
56. PVT William Ephraim Morris: Wounded while climbing the bluffs.
57. PVT Daniel J. Newell: Wounded during the retreat.
58. PVT Edward D. Pigford: Wounded during the retreat.
59. PVT Roman Rutten: Probably wounded on Reno Hill.
60. PVT Thomas B. Varner: Wounded on Reno Hill.
61. PVT Charles Theodore Wiedman: Wounded on Reno Hill.

Civilians and Scouts—(3)

62. John C. Wagoner: Packer, wounded on Reno Hill.
63. White Swan: Wounded during Reno's retreat.
64. Goose: Wounded during Reno's retreat.

Appendix C: Glossary

Many different geographical landmarks on the battlefield have been given adopted names. To help readers establish their bearings, this glossary will help in identifying these sites, their location relative to the battlefield, and, in some cases, describing the events that invested it with significance:

1. Ash/Reno Creek: A western flowing creek between the divide and the Little Big Horn. Custer and Reno rode down this creek to the Little Big Horn Valley.
2. Basin Area: A low-lying area south of Cemetery Ridge and west of Last Stand Hill occupied by a portion of Custer's wing during the battle.
3. Basin Ridge: A low ridge within The Basin area that may have been occupied by a detachment of troopers, which may have formed what later became known as the South Skirmish Line—a position identified by noted Little Big Horn author Gregory Michno.
4. Battle Ridge: The long ridge between Calhoun Hill and Last Stand Hill. The Keogh battalion occupied this position during the battle.
5. Battle Ridge Extension: The northern extension of Battle Ridge north of Last Stand Hill.
6. Blummer Ridge: See Nye-Cartwright Ridge.
7. Boyer's Bluff: A long ridge just east of the Little Big Horn River and just south of Ford B-1. Named after the scout Mitch Boyer, who viewed the Reno valley fight from that location.
8. Busby Camp: About 12 miles from the divide. This was the last camp of the 7th Cavalry on Rosebud Creek on the evening of June 24th, near the present town of Busby, Montana.
9. Calhoun Coulee: A large ravine just west of Calhoun Hill and Battle Ridge and east of Finley Ridge that runs in a general west then north to its junction with Deep Ravine.
10. Calhoun Hill: A large hill at the southern end of Battle Ridge occupied by L Company during the battle. It is named after Lieutenant James Calhoun, commander of L Company and brother-in-law of George Armstrong Custer, who died there.
11. Cemetery Flats: A relatively flat area on the western end of Cemetery Ridge just east of the Little Big Horn River.

12. Cemetery Ridge: A long ridge just northwest of Last Stand Hill, today it is occupied by a National Military Cemetery.
13. Chalk (or White) Bluffs: A series of elevations lying between Ash/Reno Creek and the North Fork four miles east of the Little Big Horn River.
14. Chetish Mountains: See Wolf Mountains.
15. Crazy Horse Ridge: A long ridge just east of Battle Ridge and Keogh's Swale occupied by Indian warriors led by the Lakota warrior Crazy Horse.
16. Crow's Nest: A high hill near the divide of the Wolf Mountains where Custer's Crow scouts first saw smoke from early morning campfires. The precise location ofis still debated today.
17. Deep Coulee: A large ravine running in a general southwest direction to its mouth near Ford B-2.
18. Deep Ravine: A large ravine running in a general westerly direction culminating at Ford C. It is believed that as many as twenty-eight troopers were killed here at the end of the battle and remain entombed under tons of dirt.
19. The Divide: A location at the top of the Wolf or Chetish Mountains that marks the line between the Rosebud Valley and the Little Big Horn Valley.
20. Eastern Morass: A watering slough used by Benteen's battalion, located about 4.75 miles east of the Little Big Horn River.
21. Elevation 3411: A point along the bluffs about a quarter mile north of Reno Hill, likely used as an observation point into the Little Big Horn Valley by Custer's Crow scouts during Custer's advance along the bluffs. A position identified by Little Big Horn author by Frederick C. Wagner III.
22. Echleman Ford: Located along Ash/Reno Creek about 1.25 miles east of Little Big Horn River. It is believed that this is where Custer watered his horses before he climbed the bluffs with his five companies. It is also where Reno's battalion separated from Custer's wing on their advance to contact with the hostiles.
23. Finley Ridge: An extension of Calhoun Hill running in a general east–west direction between Calhoun Hill and the Little Big Horn River. A platoon of C Company troopers made another last stand here.
24. Ford A: Near the mouth of Ash/Reno Creek and where Major Reno crossed to first enter the valley immediately preceding his charge toward the Indian village.
25. Ford B-1: At the mouth of Medicine Tail Coulee.
26. Ford B-2: At the mouth of Deep Coulee—several hundred yards northwest of Ford B-1.
27. Ford C: Also known as Crazy Horse Ford, which is at the mouth of Deep Ravine.

28. Ford D-1: Near the northwest corner of Cemetery Ridge.
29. Ford D-2: A point farthest north and farthest west near the old Willy Bends place along the Little Big Horn River.
30. Garryowen Bend: The location of Major Reno's valley skirmish line at the westernmost loop of the Little Big Horn River along Major Reno's advance down the valley.
31. Gerard's Knoll: A small hillock just north of Ash/Reno Creek about 1.7 miles from the Little Big Horn River, where interpreter Fred Gerard spotted a band of hostile warriors before the battle. It was a short distance from the western lone tepee.
32. Greasy Grass Hill: A prominent hill along the Little Big Horn River just north of Ford B-2.
33. Greasy Grass Ridge: The first line of bluffs east of the Little Big Horn River, west of Battle Ridge, north of Deep Coulee, and south of Deep Ravine. This was an Indian firing position during the battle.
34. Halt One: About 4.5 miles east of the divide along Davis Creek between the Rosebud and the Little Big Horn valleys. Custer rested the regiment in this area.
35. Halt Two: About a quarter mile east of the divide between the Rosebud and the Little Big Horn valleys where Custer again halted his advance.
36. Harrington Ridge: A shallow ridge just north of Finley Ridge and upper Calhoun Coulee likely occupied by a detachment of C Company.
37. Hartung Morass: A watering slough used by part of Benteen's battalion, located about 2.5 miles east of the Little Big Horn River.
38. Henryville: An area where a large concentration of expended Indian cartridge shells from Henry rifles were found a short distance southeast of the summit of Calhoun Hill.
39. Horse Holder Ravine: A ravine at the southernmost end of Keogh's Swale where the horses of L Company were being held.
40. Keogh's Swale: A low-lying ravine or valley east of Battle Ridge and west of Crazy Horse Ridge, occupied by elements of Keogh's Battalion during the battle.
41. Last Stand Hill: A hill at the northern end of Battle Ridge where forty-two men, including Custer, perished.
42. Luce/Blummer Ridge Complex: A series of high ridges about two-thirds of a mile east of the Little Big Horn River between Medicine Tail Coulee and Deep Coulee. Named after Edward Luce, a former superintendent of the battlefield, and Joseph Blummer, a local resident.
43. Mathey's Hillock: A small hill temporarily occupied by Lieutenant. Mathey and the advance ammunition packs just over a mile south of Reno Hill and north of Ash/Reno Creek.

44. Medicine Tail Coulee: A large ravine running on a general northwest direction north of the Weir Peaks and culminating at Ford B-1.

45. No Name Creek: A tributary of Ash/Reno Creek running in a general northwest direction to its junction with Ash/Reno Creek about five miles from the Little Big Horn River. About a quarter mile east of the Eastern Morass. This tributary was used by Benteen's battalion on its return to the main Indian trail from its flank march.

46. Northern Flats: An area of flat ground between Ford D-1 and D-2 where Custer may have been planning a defense in the event that Benteen or Reno failed to arrive on the battlefield.

47. North Fork of Ash/Reno Creek: A tributary of Ash/Reno Creek that runs in a general southwest direction, with its junction at Ash/Reno Creek about three-quarters of a mile east of the Little Big Horn River.

48. Nye-Cartwright Ridge: Also known as Blummer Ridge, a ridge about two-thirds of a mile east of the Little Big Horn River between Medicine Tail Coulee and Deep Coulee. The ridge was named after Lt. Colonel Elwood Nye, a veterinarian who helped stimulate work on a horse bone grave found on the battlefield, and Cartwright, who owned property in the area. Numerous shell casings were also found in the vicinity, likely from a firefight between the soldiers and Wolf Tooth's band who arrived from the east.

49. Reno Hill: A defensive position named after Major Marcus Reno along the bluffs used by the battalions of Major Reno and Captain Benteen, as well as the pack train and rear guard on June 25 and 26.

50. Sharpshooter Ridge: A high ridge where Custer first saw a portion of the Little Big Horn Valley and Indian village therein. Later used as a firing position by warriors firing at Reno Hill. It is located about one-third mile north of Reno Hill.

51. Shoulder Blade Creek or Box Elder Creek: A large ravine running from the western foothills to the valley floor between the Hunkpapa camp and the Garryowen Bend of the river.

52. South Fork of Ash/Reno Creek: This tributary runs in a general northwest direction, with its junction at Ash/Reno Creek about four miles east of the Little Big Horn River.

53. South Skirmish Line: Identified by historian Charles Kuhlman as located in The Basin area west of Last Stand Hill, south of Cemetery Ridge, and north of Deep Ravine. Might have been occupied by a detachment of F Company during the battle.

54. Squaw Creek: Believed to be the place where many of the Indian families gathered as they left the village after Major Reno's charge down the valley. It was located on the west side of the Little Big Horn between Ford B and Ford D-1.

55. Thompson's Ford: Private Peter Thompson mentioned this ford in his narrative of the battle. It crosses the Little Big Horn River over a half mile south or upriver from Ford B-1.

56. Upper Boyer's Bluff: An area along the westernmost ridge of the bluffs between the Weir Peaks and Medicine Tail Coulee well south of Ford B-1 and opposite Thompson's Ford.

57. Upper Calhoun Coulee: The easternmost portion of Calhoun Coulee lying between Finley Ridge and Harrington Ridge just west of Battle Ridge.

58. Varnum's Lookout: An observation point on a high hill south of Ash/Reno Creek used by Lieutenant Varnum to locate the Indian village just prior to the battle. This location was identified by noted Little Big Horn author Bruce Liddic.

59. Western Coulee: A large ravine running on a generally north-south axis just east of Boyer's Bluff. This coulee was likely used by Custer's five troops as they advanced north behind the bluffs.

60. White Bluffs: See Chalk Bluffs.

61. Willy Bends Place: An area along the Little Big Horn River near Ford D-2 at the farthest northern and western extremity of the battlefield, named for a local Crow Indian who once resided there.

62. Wolf Mountains: Also known as Chetish Mountains, they separate the Rosebud Valley from the Little Big Horn Valley.

Endnotes

Chapter 1
(Pages 11–15)

1. For information regarding Custer's departure from the Yellowstone, we relied on James Willert: *Little Big Horn Diary*; (El Secundo, California: Upton, 1997), 211. Willert's book is comprehensive and useful, although he has a definite bias against Custer. His description of the 7th's departure and the observations of Terry and Gibbon are interesting considering future events. Supplementing this account, we relied on Edward Godfrey's "Custer's Last Battle," published in *Century Magazine,* Vol. XLIII, no. 3, 1892), and finally, Godfrey's letter to Edgar S. Paxon dated January 16, 1896, which is found in W.A. Graham's *The Custer Myth:* (Mechanicsburg, PA: Stackpole Books, 2000), 345. Godfrey wrote that for the 1876 Campaign, Custer carried a Remington sporting rifle with an octagonal barrel, two self-cocking English, white-handled pistols with a ring in the butt for the lanyard, a hunting knife in a beaded sheath, and a canvas cartridge belt.

2. See Hunt, Fraser, and Robert Hunt, *I Fought With Custer*, (New York: Charles Scribner, 1947), 70, about his two sergeant color guards on the parade ground at the Yellowstone.

3. Numbers regarding the exact strength of the regiment are all over the map. Lost muster rolls and how many soldiers were left behind on Grant Marsh's *Far West* steamer have muddied the waters in determining the strength of the 7th Cavalry when it left the Yellowstone River. While conceding that there can be no definitive number, we have relied on Frederic C. Wagner III's estimates. Wagner, author of *Strategy of Defeat,* was conspicuously attuned to numbers and exhaustively researched numbers on both sides of the battle. Following numerous conversations with Wagner in 2018, the authors have determined that his count of a total of 655 in the command, 607 uniformed, is as good an educated guess as there is extant.

4. W.A. Graham: *Reno Court of Inquiry*: Abstract 23: (Harrisburg PA.: Stackpole, 1953).

5. Quote about whipping all the Indians on the Plains, see J.R. Perkins: *Trails, Rails and War: The Life of General G.M. Dodge*, (Indianapolis: Bobbs-Merrill, 1929).

6. That Custer was true as steel and was depended on for great things, see quote from Jacob Greene, see Brady, *Indian Fights and Fighters*, 392.

7. Custer singlehandedly capturing six rebel flags and his zest for war near Williamsburg and Chickahominy, see Andrew Paul Hutton, *The Custer Reader*, 10–11, also O. S. Barrett, *Reminiscences, Incidents, Marches, and Camp Life of the Old Fourth Michigan Infantry in War of Rebellion, 1861 to1864* (Detroit: W. S. Ostler, 1888).

8. Astonishment about Terry sending this wild man and insufferable ass down the Rosebud, see James Brisbin: Letter to *Los Angeles Times* November 8, 1884.

9. Colonel John Gibbon: "Last Summer's Expedition Against the Sioux," (*American Catholic Quarterly*, April 1877), 293. This rather jocular admonition telling Custer not to be greedy on Gibbon's part was rich. Gibbon had been doing all he could on this campaign not to attack the Indians. We reference when his scouts first located the Indian village on May 17, at the Tongue River, then again on the Rosebud, May 27. Gibbon remained immobile on the Yellowstone's north bank for so long that a disgusted Dr. Paulding (see "A Surgeon at the Little Big Horn: The Letters of Dr. Holmes Paulding," edited by Thomas R. Buecker, 139) wrote his mother: "After laying there for 10 days our genial Commanding Officer did nothing, despite the Indians showing themselves every day as if they knew what a harmless command they were dealing with."

10. On William Ludlow's observations regarding Custer, see Fred Dustin: *The Custer Tragedy*, (Reprint, El Secundo CA, Upton and Sons, 2011), 202–203.

11. Custer's abilities as a soldier, see General A.B. Nettleton: "Custer's Characteristics," *Army and Navy Journal*, (New York: Publication Office, July 29, 1876) 818.

12. Jacob L. Greene's assessment of Custer, see Cyrus Townsend Brady: *Indian Fights and Fighters*. (New York: McClure, Philips, 1904) 391–392.

13. Reasons for Terry's beliefs that the Indians were in the vicinity of the Little Big Horn was because of the sightings of "many smokes," see Robert Hughes, *Campaign Against the Sioux in 1876*, 36.

14. Description of the scene on the *Far West* of Custer's projected route see Letter by Brisbin to Godfrey in his January 1, 1892, letter, Carroll, *The Two Battles of the Little Big Horn*, 141–142, see also Brisbin's letter to *Los Angeles Times*, November 8, 1884.

15. That Terry was supporting Custer with his 400-man infantry, see Lee Noyes, "A Dispatch from the Battlefield, *Research Review* 18, no.2, Summer 2004.

16. For the plan inferring that Custer would allegedly meet Terry on June 26 near the mouth of the Little Big Horn, see Graham, *The Custer Myth*, 261. The verb inference is used here because this rendezvous mentions this only after the fact, which by no means makes the story of a potential conjunction illegitimate or apocryphal. Traditionally, it has been extrapolated from Terry's orders that Custer was to make a junction with his Commanding Officer at the mouth of the Little Big Horn on June 26. Still, the orders do not explicitly say this, nor does it even implicitly suggest it. Custer could have, based on the written communication, met up with Terry after his fifteen-day itinerary on July 6 or 7. Interestingly, not even Varnum, Custer's Chief of Scouts, who might be expected to know such intelligence, gives a single intimation of any knowledge for such an arrangement.

17. Terry gave a specific date to let Custer know where he might be found, see Edgar Stewart, *Custer's Luck*, 250.

18. Bradley's quote about Custer seizing all the laurels can be found in his Journal entry dated June 21, 1876.

19. For John Ryan's comments about the regiment never looking better, see Barnard, *Ten Years With Custer*, 285, 287.

20. For Lt. Gibson's comment about horses being jaded, see Graham, *Custer Myth*, 138.

21. Slaper saying night march and lack of grain fatigued horses, see Brininstool, *Troopers with Custer*, 46.

22. Varnum stating horses played out, see Carroll, *I Varnum: Custer's Chief of Scouts*, 115.

23. Godfrey on the condition of the horses, see Nichols, *Reno Court of Inquiry*, 495.

24. Wallace on tired horses, ibid., 332.

(Pages 16–21)

25. Benteen first mentions fatigue of horses when climbing bluffs to Reno Hill, see ibid., 406.

26. See Lieutenant Edward McGuire: 1876 Report in Fay, *Military Engagements*, 30.

27. For Dr. Porter's horses in the regiment being in good condition, see Nichols, *Reno Court.* 189.

28. Mounts doing well, see Frazer and Hunt, *I Fought With Custer: The Story of Sergeant Charles Windolph: Last Survivor of the Little Big Horn Battle*, 73.

29. Herendeen, horses galloped in good style, see Nichols, *Reno Court*, 252.

30. Slaper stating that horses were made to ride around hills, See Brininstool, *Troopers with Custer*, 43.

31. Report of Custer's brutal disregard of animals, see Lieutenant Colonel Elwood Nye: *Marching with Custer*, 13.

32. Davern: Horses in tolerable condition, see Nichols, *Reno Court*, 232.

33. Oscar Long, 5th U.S. Infantry: Military Report to Frank Baldwin, June 27, 1878, Bancroft Library, CB 939, Box 1, University of Berkley.

34. That defeat is rarely the result of a wearied offense, see Comments by General J.B. Fry on Godfrey's Narrative of the Custer Fight, *Centennial Magazine*, January 1892, published in Graham's *Custer Myth*, 154.

35. Example of Colonel Henry of 9th Cavalry, see ibid., 154.

36. Regarding the possible defectiveness of the 7th Cavalry firearms, see Douglas D. Scott and Richard Allan Fox, *Archeological Insights into the Custer Battle*,1987, 113; see also, Paul L. Hedren, "Carbine Extraction Failure at the Little Big Horn: A New Examination," *Military Collector and Historian* Volume 2, 1973, 66–68.

37. The 7th Cavalry firearms not being commensurate with Indian weapons as being false, see *Ibid.*

38. Benteen the army pays you to get shot at, see Frazer and Hunt, *I Fought With Custer*, 77.

39. For Moylan's comments on the battle readiness of the 7th, see Nichols, *Reno Court*, 43.

40. For John Ryan's view of the 7th's combat readiness, see Barnard, *Ten Years With Custer*, 273. Ryan's criticism of the regiment's preparedness is questionable. The 7th was probably at its summit during the Washita Campaign of 1868. However, confidence and military capability were still high, and even Ryan, in his later correspondence with Elizabeth Custer and Walter Camp, said that he had never been disposed to give such criticisms.

41. For Reno as drill master, see Fort Lincoln, Circular 1 and Circular 24.

42. That soldiers were malingerers and bounty jumpers, see Beadle, *Western Wilds and the Men Who Redeem Them*, Cincinnati, Jones Brothers. (1881), page 550.

43. For a brief but solid discussion on soldier readiness in the 7th, see Frederic Wagner, *Strategy of Defeat*, 4. See also Robert Utley, *Cavalier in Buckskin*, 113.

44. That recent recruits numbered only 10 percent, see Robert Utley, *Cavalier in Buckskin*, 113.

45. Nothing could happen to the 7th, see Frazer and Hunt, *I Fought With Custer*, 17.

46. Windolph's jaunty description of serving with the 7th, ibid., 53.

47. For Boyer, next to Jim Bridger, being the best scout on the frontier, see Edgar Stewart, *Custer's Luck*, 105.

48. For Herendeen's background as a so-called "Jack of All Trades" scout, see Donovan, *Terrible Glory*, 182–183.

49. For Dorman's history and plans, see Stanley Vestal, *New Sources of Indian History*, 339; also see Camp, 712.

50. On Lonesome Charley Reynolds, see Libby, *The Arikara Narrative of Custer's Campaign*, 168; also see Frazer and Scott, *I Fought With Custer*, 68.

(Pages 22–27)

51. The Arikara leaving on the campaign with Custer, see Libby, *Arikara Narrative*, 59, 139; also see Custer, *Boots and Saddles*, 218.

52. Crows as enemies to the Sioux, their history and numbers on the Plains, see Peter Nabokov, *Two Leggings*, 155. See also Mark H. Brown, *The Plainsman of the Yellowstone*. Custer wrote how the Indian scouts were a powerful ally and how ancient hatred between the tribes could be exploited, see

Custer, *My Life on the Plains*, 271. The Crows were also happy to ride with Custer, stating that he never, no matter how onerous, ever abandoned a trail, see Custer, *Boots and Saddles*, 275.

53. Reno was cited in the Civil War for gallant and meritorious service in action at Kelly's Ford in Virginia on May 1863, at the battle of Trevilian Station on June 1, 1864, and at Darbytown on July 28, 1864.

54. Tom Custer's heroics in the Civil War earned him two Medals of Honor even though his older brother still overshadowed him. See Jeffery M. Wert, *The Controversial Life of George Armstrong Custer,* 219–220.

55. There is a plethora of material on George Armstrong Custer's life and military career. A sample of the works the authors consulted include but are not limited to George A. Custer: *The Civil War Memories of George Armstrong Custer*; Custer: *My Life on the Plains*; Elizabeth B. Custer, *Boots and Saddles*; Robert Utley, *Cavalier In Buckskin*; Frederick Whittaker: *George Armstrong Custer and the American Military Frontier: A Complete Life of George A. Custer*; Jeffery D. Wert: *The Controversial Life of George Armstrong Custer*; Louise Barnett: *Touched by Fire: The Life and Death and Mythic Afterlife of George Armstrong Custer*; T.J. Stiles: *Custer Trials: A Life on the Frontier of a New America* (probably the best and most definitive of the Custer biographies though it is mildly marred by a subpar treatment of the Battle of Little Big Horn); Marguerite Merington, *The Custer Story*.

56. For an early intelligence report regarding the numerical strength of hostile tribes Custer could face, see Mark Kellogg's dispatch dated July 11, 1876, found in Graham, *The Custer Myth*, 237.

57. In an Officer's Call, Custer estimated an enemy fighting force of 1,500, see E.S. Godfrey *Buffalo Days in Carroll, A Custer Chrestomathy*, 71.

58. For Gerard's higher estimate of hostiles, see Nichols, *Reno Court*, 85.

59. Custer's march up the Rosebud Valley is taken from *The Field Diary of Lieutenant Edward Settle Godfrey*, Edited by Edgar I. Stewart and Jane R. Stewart, including the regiment's activities at the site of Sitting Bull's Sun Dance, 9–10; also Graham, *The Custer Myth*, recounts how the regiment passed a number of abandoned camps making halts at each one studying pony droppings, to determine the age of the droppings, the tracks and lodge trails as a means of estimating the size of the Indian population, 135–136.

60. Varnum's comment on one continuous village can be found in a letter from Varnum to Walter Camp and found in Richard Hardorff's *On the Little Big Horn, Walter Camp*, 71.

61. Indian campfires still smoldering, see Stewart, *Custer's Luck*, 265.

62. Burkman's exchange with Custer concluding "not too many to lick," see Glendolin Damon Wagner, *Old Neutriment*, 144–145.

63. Not sounding the trumpets can be found in Graham, *The Custer Myth*, 135.

64. Managing the dust on the trail and the decision to carry no sabers, see Barnard *Ten Years with Custer*, 288–289.

65. Herendeen telling Custer how he and a number of civilians held off hundreds of Indians, see Hammer, *Custer in '76*, 220.

66. Bloody Knife's words to Custer vary from account to account and are not always attributed to him, probably because they were hand-signaled. Whether it was he or others who commented on the size of the village, they were all pessimistic about Custer's chances. According to David Evans in *Custer's Last Fight*, 193, Bloody Knife, speaking to the sun, said, "I shall not see you go down behind the mountains tonight."

67. Custer's exchange with Red Star about attacking the village, see Graham, *The Custer Myth*, 15, 33.

68. A note here on premonitions. The Little Big Horn saga abounds with them from the time the 7th marched out of Fort Lincoln. Some of them even had a quasi-religious imagery, as when witnesses attested that they saw a vision of soldiers marching into the sky. A harbinger of doom, see Elizabeth B. Custer, *Boots and Saddles*, 218. A possible explanation for this phenomenon is that when the sun burned through the morning mist, it created the illusion (visions of this sort were not terribly uncommon on the Western plains) of giving the impression that half the cavalry line was marching into the sky. There were other omens, and in General Edward Godfrey's narrative, he states that on the night before the battle, General Custer's Headquarters flag is blown down by a southerly breeze. Being near the flag, Godfrey relates how he picked it up and planted it again, only to have the wind blow it down a second time before he secured it by boring the staff into the ground supported by sagebrush. Godfrey forgot the incident until Lieutenant George Wallace, who witnessed it, brought it up, seeing it in retrospect as a premonition of disaster. Godfrey agreed, see Graham, *The Custer Myth*, 135–136.

69. Captain Charles King's remarks on the unique fitness and battle readiness of the 7th Cavalry, see King, *Campaigning with Crook: The Meeting of Terry and Crook,* 72.

70. Regarding the jamming of carbines, see Paul Hedren: "Carbine Extractor Failure at the Little Big Horn: A New Examination," *Military Collector and Historian*, Volume 25, 1973, 32; also, Scott, Fox, Connor, Harmon, *Archeological Perspectives on the Battle of the Little Big Horn*, 114.

71. For Edgerly's testimony on the character and fighting fitness of the officers and men of the regiment, see Nichols, *Reno Court*, 454–455.

Chapter 2

(Pages 29–32)

1. For Custer interpreting his orders to the point of disobedience, see *New York Herald*, July 11, 1876, as reported in Graham, *The Custer Myth*, 237.

2. Living on horsemeat, see Godfrey, *Custer's Last Battle*, 14.

3. General Terry's well-written and much-interpreted set of orders to Custer can be found in any number of books. We used David Evans, *Custer's Last Fight*, Appendix 4, 455.

4. Stewart, *Custer's Luck*, 249–250. The reputed affidavit of Mary Adams, Custer's cook, in which she stated that she overheard Terry's last words to Custer, stating that he should use his own judgment and do what he thinks best if he strikes the trail. General Miles first mentioned it in his 1896 autobiography, *The Personal Observations of General Nelson A. Miles,* in which he claims to have an affidavit from an unidentified individual (believed to be Mary Adams) who overheard Terry's final conversation with Custer. Verification of this would absolutely exonerate Custer from disobeying orders. Terry, an honorable man, never admitted or even hinted that such a conversation took place. It is the author's contention that the story has enough supporting evidence and somewhat echoes Terry's orders/instructions to Custer and is, therefore, worthy of consideration. Those who wish to look deeper into the matter should consult John S. Manion's *Terry's Last Statement to Custer,* which definitively verifies Mary Adams's affidavit (El Secundo, CA: Upton and Sons, 2000), 62; and also see Francis Taunton's *The Mystery of Miss Adams*, (Brand Book No. 2, Summer 1991). Finally, see Barnard, *Custer's First Sergeant John Ryan*, 174.

5. General James B. Fry's commentary on the Little Big Horn fight, first published in the *Army and Navy Journal*, July 22, 1876, and in his book *Military Miscellanies*, (1889), 506–511, is excellent in parsing the language of Terry's order in context with the vernacular of the Army in 1876.

6. Terry used Reno for the June 11 scout because he feared Custer would ride straight to the Indian encampment, see Graham, *The Custer Myth*, 230.

7. For Kellogg writing that Custer declined the scouting order that Reno readily accepted, see *New York Herald*, June 21, 1876.

8. For Reno's special orders dated June 10, 1876, and signed by General Terry, see Willert, *Little Big Horn Diary*, 118.

(Pages 33–36)

9. Terry underscored his displeasure with Reno when a reporter, Charles S. Diehl, inquired about Reno's disobedience on his scouting foray, Terry curtly shot back, "That is a matter of record." See Oliver Knight, *Following the Indian Wars*, (Norman, University of Oklahoma Press, 1963), 288.

10. That Terry gave Custer carte blanche, see *Varnum, Custer's Chief of Scouts*, 24.

11. Gatling guns did not offer the advantage they seemed to promise. They gave Reno much trouble during his scouting, forcing him to abandon them, see Hammer, *Custer in '76*, 53. General Miles also believed that Gatling guns were useless when fighting Indians.

12. That Custer was at liberty to attack first, see "James H. Bradley: Journal of the Sioux Campaign Under the Command of General John Gibbon," recorded June 21, 1876, (Contributions to the Historical Society of Montana (1903), 218.

13. The 7th Cavalry's overwhelming concern to prevent the Indians from escaping, see Gibbon, *Last Summer's Expedition Against the Sioux*, 22.

14. Custer wanting Varnum to investigate a trail Godfrey putatively discovered, Carroll, *I Varnum*, 59–60.

15. Regarding any rendezvous with Gibbon's forces can be readily dismissed as fantasy. Never was there any mention of a junction with Gibbon on June 26 or at any other time. Custer was ordered to report not later than the expiration of the time for which the troops were rationed. This would give Custer fifteen days from the date of his departure or on July 7. See Graham, *Custer Myth*, 130.

16. Stewart, *Custer's Luck*, 244. The Crows were driven from that very land by the Sioux and hence knew the ins and outs of the territory.

17. General Miles arguing that Custer did not disobey orders and that once the respective commands were seventy-five to one hundred miles apart, logically the responsibility of acting on the offensive or defensive rested upon Custer alone, see Brady, *Indian Fights and Fighters*, 364.

18. For the likelihood of Indians permitting two different columns of soldiers to move around the country, see *Ibid.*, 364.

19. The fortuitous circumstances of finding large herds of antelope on the foothills west of the Big Horn and in the neighborhood of the mouth of the Little Big Horn caused the Indians to go downstream instead of upstream.See Charles Kuhlman: *Legend into History*, 38.

20. Pretty White Buffalo said that the Sioux did not expect a battle. See James McLaughlin, *My Friend the Indian* (Seattle, Superior Publishing Company, 1970), 43–44.

21. For Gall stating that the attack caught him and all the Indians unprepared, see Graham, *The Custer Myth*, 92. See also Marquis, *Custer on the Little Big Horn*, 110.

(Pages 37–42)

22. For speculation if the Indians were not found at the Little Big Horn but rather at the Rosebud or Powder River was a quote from an unidentified officer of the regiment present at the battle (most likely Captain Edward Godfrey) and recounted at a reunion of the survivors ten years later, see *The Winona Times*.

23. Chief Sitting Bull was caught so off guard by the attack that he was ready to negotiate, see Hardorff: *Indian Views of the Custer Fight* (Spokane, A Source Book: Norman University of Oklahoma Press, 2004), 182.

24. That it was best to forego an Indian attack and attack later, see Brady, *Indian Fights and Fighters*.

25. It should be noted that while Brisbin was correct that the 7th would take the lead and be the striking force, he was, nonetheless, Terry's brother-in-law and did all he could to shield Terry from any blame for the disaster. Custer and Brisbin did not get along, which is one of the reasons why he refused

the four companies of Brisbin's 2nd Cavalry, and why Brisbin refused to serve under Custer's direct command, see Stewart, *Custer's Luck*, 242.

26. That Terry would have harnessed the 7th's reputed zeal had he led the expedition into Indian territory, see John Bailey, Pacifying the Plains: General Alfred Terry and the Decline of the Sioux (Westport, Connecticut, Greenwood Press, 1979), 156.

27. For comments that, had Custer just obeyed Terry's orders, all would have been well, see Colonel John Gibbon's "Last Summer's Expedition against the Sioux," *American Catholic Quarterly Review*, April and October 1877. Also, Captain Benteen's July 4th letter to his wife, Graham, *The Custer Myth*, 300.

28. E.A. Brininstool: *Troopers with Custer* (Mechanicsburg PA. Stackpole, 1952), 278–279. Conversations revolving around Custer's impulsiveness and overconfidence waxed heavily among some of the officers, most conspicuously Gibbon and Brisbin. Brisbin pointed out, or so he maintains, that based on Lieutenant James Bradley's intelligence, the Indians had three thousand fighting men and did not seem to have many squaws or children with them. All this, Bradley asserted, Custer knew, "for I told him about it and cautioned him to be careful." He further states that he advised General Terry to put all the cavalry together and to command the exhibition himself. In failing to take the initiative, Terry underrated himself and overrated Custer. Terry demurred, stating he had little experience in Indian fighting. Terry finally settled on having Brisbin inquire of Custer about Brisbin going with him and that if Custer thought well of uniting the two columns, then he (Terry) would go in command. "This," said Brisbin, "is how the 2nd Cavalry came to be offered to Custer." Custer saw all of this as a ploy by Brisbin to wrest command from himself and later that evening said to John Carland, 6th Infantry, that "The 7th can handle anything it meets …" That "this is to be a 7th Cavalry battle, and I want all the glory for the 7th Cavalry there is in it," (See Brininstool, *Troopers with Custer*, 278–279.

29. For Terry's favorite motto of "too much zeal" see Hughes, *Campaign Against the Sioux*, 43.

30. Brisbin's scathing remarks years later about Custer violating both the spirit and letter of Terry's orders, see Brady, *Indian Fights and Fighters*, 377.

31. For Godfrey's discussion on whether "desires" and wishes" has the full force of a direct order, see Brady, *Indian Fighters and Fights*, 389.

32. For a brief discussion on Custer's orders and Tullock Creek, see ibid., 379.

33. Varnum was emphatic on this point. Two decades later, he insisted to Robert Hughes that in all his instructions from Custer regarding scouting, he never heard of Tullock's Creek and any examination of it was made without his knowledge and by someone not under his control. See Graham: *Story of the Little Big Horn* (Harrisburg PA. 1959), 31.

34. For further discussion on Cullock Creek, see Brady, *Indian Fights and Fighters*, 379.

35. Herendeen's exchange with Custer about scouting Tullock's Creek at this juncture, telling him it was an inopportune time to do so and Custer agreeing by saying, "All right. I could wait." See Gray, *Centennial Campaign*, 156.

36. Custer telling Herendeen it was no longer necessary to scout Tullock Fork on June 25th, and the scout believing that Custer wished to fight the Indians alone without any help or interference from anyone see Hammer, *Custer in '76*, 221.

37. For information attesting that Custer initially planned to attack on the morning of June 26, see Brady: *Indian Fights and Fighters*, 379.

38. Gerard saying that from the Crow's Nest they knew there were no Indians in that part of the country, see Hammer, *Custer in '76*, 231.

39. Edgerly saying it would have been absurd to scout Tullock's Creek, see *The Custer Myth*, 336.

40. *George Armstrong Custer, My Life on the Plains* (Norman: University of Oklahoma Press, 1962). "Faint heart never won fair lady." This epitaph, believed to have been written by Custer on June 21, 1876, appeared in the *New York Herald*, July 11, 1876, quoted in Graham's *The Custer Myth*, 234–235.

41. Custer's quote that he was fortunate from the time of his birth, see Louise Barnett: *Touched by Fire: The Life, Death, and Mythic Afterlife of George Armstrong Custer* (New York: Holt, 1996), 59–60.

42. This dramatic line regarding "audacity" concludes a speech by the French Revolutionary Georges Danton (1759 –National Assembly on September 2, 1792, and reported by the *National Gazette* or Le Moniteur Universel on September 4, 1792.

Chapter 3

(Pages 43–46)

1. Literally thousands of lodgepoles dragged along by the Indians scarred the earth. Numerous campfires still smoldered from abandoned sites. See Stewart's *Custer's Luck*, 265.

2. Regarding Jackson's observations, see James William Schultz, *William Jackson, Indian Scout* (Boston: Houghton Mifflin, 1926).

3. Regarding Custer's awe upon sighting the mountains, see *Custer, My Life on the Plains*, 26.

4. For the sights and conditions of Custer's trail to the divide, see Donavan, *Terrible Glory*, 192.

5. About Custer writing that he was hopeful of achieving great things, see Elizabeth B. Custer, *Boots and Saddles*, 275.

6. That the scouts were directed to go without sleep and travel all night to find the Sioux camp, see Graham, *The Custer Myth*, 32.

7. Regarding riding to the Crow's Nest, see Carroll, *I Varnum*, 87; Libby G. Orin, *The Arikara Narrative of Custer's Campaign and the Battle of the Little Big Horn*. (Norman: University of Oklahoma Press, 1998). Lieutenant Varnum related that his journey to the Wolf Mountains commenced when General Custer told me that the Crow scouts reported that between us and the Little Big Horn was a high hill where the Crows watched the Sioux. Custer said that the Crows were going at once and that he wanted an intelligent white man to go with them. Custer agreed to let Varnum take Charley Reynolds, Mitch Boyer, as well as five Crows and eight to ten Rees. See the letter Varnum wrote to Walter Camp on April 14, 1909, from Boise, Iowa, in Hammer, *Custer in '76*. The nomenclature of Crow's Nest came about because the spot reminded Varnum of a similar location at West Point Military Academy also dubbed Crow's Nest.

8. The eyesight of Indians was almost preternatural. Charles Windolph, with just a bit of exaggeration to make his point, stated that Indians could see better with their naked eye at that time in the morning than the officers could in broad daylight with high-powered field glasses. See *I Fought with Custer*, 73.

9. Red Star, seeing the light smoke indicating an Indian village and Charley Reynolds acknowledging the same thing, see *The Custer Myth*, 32.

10. Varnum sending scouts to fetch Custer, see Carroll, "General Custer and the Battle of Little Big Horn," *The Federal View* (Bryan Texas: Guidon Press, 1978), 65.

11. Interpreter Dorman telling Burkman that important news has been brought in by the scouts for Custer, see Glendolin Damon Wagner *Old Neutriment*,(New York: Sol Lewis, 1973), 149.

12. Custer riding to the Crow's Nest, see Graham, *The Custer Myth*, 32.

13. Varnum meeting and apprising Custer at the bottom of the hill on what the scouts had seen from the Crow's Nest, see Letter from Varnum to Albert Johnson, November 13, 1929, (Dustin Collection: Custer National Battlefield).

14. Worms crawling in the grass is one of the more memorable metaphors underscoring the vast number of ponies. Also, Carroll, *I Varnum*, 14. One can only make an educated guess regarding the size of the pony herd. Factoring in what we know about the size of the village and the importance of horses to the Indian way of life, a range of 10,000 for the low and 20,000 for the high seems about right. Gerard was an experienced Indian scout with an educated eye for the frontier. He noted that he could make out "a pony herd on the hills or tableland beyond the stream, as he could distinctly see a large dark spot or mass and even see the rising dust "from which I concluded that I was looking at a herd of ponies being driven," see Hammer, *Custer in '76*.

15. Custer, partially because of the haze, could not make out what the Indian scouts were so certain they saw, see Carroll, *I Varnum*, 63–64.

16. The Crows pointing out a large pony herd to Varnum and told him, "Look for worms on the grass." Ibid., 87.

17. That Custer, after his first sighting of the valley, thought that there was no Indian camp, see Liddic, *Vanishing Victory*, 26.

18. Camp Interview with Varnum, (Undated, Camp Collection, BYU). There has been an interesting debate about whether Custer returned to the Crow's Nest for a second look. Varnum rejected the possibility, see *I Varnum*, 102. De Rudio disagreed, stating Custer did go up a second time and with his glasses, which apparently went with Custer to his death, never to be recovered (See Hardorff *On the Little Big Horn with Walter Camp* (El Secundo, CA: Upton and Sons, 2002), 100–101. Lieutenant Hare agreed with De Rudio that he did go up a second time, Hammer, *Custer in '76*, 64. James Willert quoting the controversial Curley, who he had more faith in than most, stated that Custer's "second look" occurred from the top of the slope north of Davis Creek. See Willert *Little Big Horn Diary*.

19. That Custer planned to reconnoiter the enemy's position during daylight, see Graham, *The Custer Myth*, 33.

20. See *New York Times* on the William Belknap scandal dated March 30, 1876 and *Chicago Daily Tribune*, dated March 29, 1876; Robert C. Prickett, "The Malfeasance of William Worth Belknap," *North Dakota History* 17, no. 1, 1950; See also Willard, *The Black Hills Trails* (Rapid City, S.D.: Rapid City Journal Company, 1924), 133; Robert M. Utley: *Cavalier In Buckskin: George Armstrong and the Western Military Frontier* (Norman: University of Oklahoma Press, 1988), 152–154, provides a good, albeit brief, account of Custer's testimony before Congress in the Spring of 1876. These scandals would implicate President Grant's brother, his brother-in-law, as well as other associates in his administration. Custer's testimony before the House Committee of Military Expenditures implicated Secretary of War Belknap, Orvil Grant, the President's brother, and the corruption among Indian and Army traders. Grant was incensed by the bad publicity and how it tainted his Administration and his legacy. He was also outraged by Custer's very publicly anti-Administration bent. His relationship with the president had already been suffering when Custer had Captain Fred Grant, the president's son, arrested for public drunkenness.

21. Camp, Brigham Young University, roll 5, box 6. See also Richard Hardorff's *Hokahey! A Good Day to Die: The Indian Casualties of the Custer Fight*, which furnishes a useful account of Sergeant Curtiss detail as it came upon the Indians. Up to that point, there was disagreement between Custer and the Crow scout Half Yellow Face about whether they had been seen. Gerard said that Custer was sure that they had not been seen by the Sioux. See Graham, *Custer Myth*, 33. At the last officer's call, Custer proposed crossing the divide before daylight, concealing the regiment, and launching a surprise attack on Monday, June 26, at dawn. See Carroll, *The Federal View*, 65.

22. Custer telling Herendeen there were no Indians in the upper reaches of Tullock Creek, see Cyrus T. Brady: *Indian Fights and Fighters* (Lincoln and London: University of Nebraska Press, 1971), 372.

23. Custer seeing the futility of sending a scout through the valley to inform Terry of the situation after he made his attack, see Camp, Brigham Young University, roll 5, box 6, no. 511.

(Pages 47–51)

24. Herendeen made this comment about Tullock Creek early on June 25, in the hollow east of the divide. Herendeen impressed Custer from when they first met on the *Far West* steamer at the mouth of the Rosebud. He was summoned by General Terry to a conference where he also saw Custer, Gibbon and Brisbin. Gibbon had been impressed by the frontiersman who had been attached to the Montana column. Looking at a map spread out on the table, Custer pointed out the Tullock Creek area, west of the Rosebud, and asked if he knew that area. When the scout assented, Custer unhesitatingly replied, "You are the man I want." See Hammer, *Custer in '76*, 221. So Custer did not forget or ignore Terry about Tullock Creek. The situation had changed making Terry's order to scout that area for Indians no longer necessary and, indeed, a waste of time since they could see there were no Indians there. While it can be argued that Custer technically disobeyed Terry's order by not sending Herendeen it would be all but impossible, given the larger and changing picture, to make Custer culpable.

25. Custer, believing he was discovered and would attack without delay, see Edward Godfrey, *Custer's Last Battle, 1892*, (Reprint, Olympic Valley, CA.: Outbooks, 1976), 21.

26. General Carl Von Clausewitz: A Prussian military commander and theorist, 1780–1831. See his *On War*, edited and translated by Michael Howard and Peter Paret (Princeton University Press, 1976). Valuable and indispensable regarding the dynamics of war, no matter the period under study.

27. Custer ordering Keogh to kill all the Indians regardless of sex and age, see John B. Kurt, Hamilton Cox, and Brian Pohanka: *The Life and Legend of an Irish Dragoon in the 7th Cavalry* (El Secundo CA., Upton and Sons, 1999), 108.

28. Regarding how the 7th looked that day, we relied on Graham, *Custer Myth*, 345–346; Charles Windolph's description in *I Fought with Custer*, 5; Colonel T.M. Coughlin *The Last of Custer's Lieutenants* (Bryan, Texas), in which Varnum describes the soldiers' hats, their shapes being mostly irregular and rolled up on both sides, others just flat and still others turned with the brim up in a Napoleonic appearance.

29. This was the distance between columns of four that Charles Windolph observed. See *I Fought with Custer*, 78.

30. Regarding Custer's apparel, Godfrey in a letter to Edgar S. Paxson described Custer wearing a broad-brimmed, whitish-gray hat with a low crown and a fringed buckskin suit, the blouse of which was double-breasted with military buttons "… he carried as personal arms a Remington Sporting rifle with octagonal barrel, two self-cocking English, white-handled pistols with a ring in the butt for the lanyard, a hunting knife in a beaded sheath, and a canvas cartridge belt." Again, see Graham, *Custer Myth*, 345–346.

31. Custer's impulsiveness is overplayed by too many who have studied and written about this battle. Nathaniel Philbrick states that at Washita, Custer probably intended to also attack the huge village to the east if the Indian scout Ben Clark had not dissuaded him. See Philbrick, *The Last Stand*, 260. The

authors, however, have combed Clark's reports as well as every reference we could find on Washita and have found not a single reference supporting this claim.

32. Burkman's comment that the very idea of the 7th being whipped was impossible, see Wagner, *Old Neutriment*.

33. Custer cautioning Benteen, ironically enough, about moving too fast, see Carroll, *Benteen–Goldin Letters*, 268–269.

34. Edgerly speaking of the assigned divisions of the troops as prescribed by Custer and Cooke, see Nichols, *Reno Court*, 439. For a definitive breakdown of this division, see Larry Sklenar, *To Hell With Honor*, 114–119.

35. Reno surprise at taking command of three companies at the divide, see Lloyd J. Overfield II, *The Little Big Horn 1876: The Official Communications, Documents and Reports*, 43.

36. Carroll, *Benteen–Goldin Letters*, 168–169. Benteen discusses how he was ordered with three troops to move to the left to a line of bluffs about two miles away.

37. Nichols, *Reno Court*, Edgerly testimony, 43, mentions that Benteen would go off to the left away from the regiment, but he never heard of any plan to reunite the regiment. See also Graham, *Reno Court*, 388.

38. For Custer's attack on Black Kettle's village and how its success influenced the thinking at Little Big Horn, see John Carroll, "General Custer and the Battle of Washita," *The Federal View* (Bryan, Texas: Guidon Press, 1978).

39. For a discussion regarding criticism that in dividing the regiment, Custer was tying himself down with only part of it, when he should have been overseeing the entire regiment to coordinate one part with another, see Colonel T.M. Coughlan, "The Battle of Little Big Horn, *Cavalry Journal* (Vol. 43, No. 181, January–February, 1934), 17.

40. Graham gives a good explanation of what might otherwise seem to be a cavalier dismissal on Custer's part to accept the additional force of Gibbon's 2nd Cavalry. It was not only because, as Custer believed, that the 7th could defeat any force the Indians could put against it, but with the regiment acting alone, there would be harmony, whereas another cavalry unit would be sure to cause friction and jealousy. This is also discussed in an earlier note. See Graham, *The Custer Myth*, 134.

41. Benteen saying that he believed that Brisbin's four companies offered by Terry but rejected by Custer would be missed, see Evan Connell, *Son of the Morning Star*, 134. Of course, had Custer accepted those four companies, he would have lost independent command of the strike force.

42. That the 7th, according to Charles Windolph, was the greatest cavalry unit in the U.S. Army, see Hunt, *I Fought with Custer*, 56.

43. Boyer telling Custer, "if you don't find more Indians in that valley than you ever saw, you can hang me," see Carroll, *I Varnum*, 63–64.

44. Going home by a road we don't know, see James Willert, *William Jackson, Indian Scout* (Boston, MA: Houghton and Mifflin & Co., 1926).

45. Charles Windolph, quoting Benteen cautioning Custer about keeping the regiment together, can be found in Charles K. Mills, *Harvest of Barren Regrets* (Glendale, CA: Arthur H. Clark Co., 1985), 52-53.

46. Custer was so concerned about the village breaking up that he sent flankers to the right and the left, for nothing must leave the trail without him knowing it. Gray, *Centennial Campaign*, 161.

(Pages 52-55)

47. Carroll, *Benteen–Goldin Letters*, 181. "Benteen went to his grave believing that Custer neither expected nor desired that I should have the advance of the regiment."

48. Benteen saying that the ground was too rugged and uneven for the Indians to go in that direction, see Nichols, *Reno Court of Inquiry*, 404.

49. Charles Roe's belief that Custer purposely kept Benteen out of the upcoming fight, see Hammer, *Custer in '76*, 249. Interview with Charles F. Roe, December 8, 1910.

50. Walter Camp believed that Terry's order to keep feeling to the south caused Custer's fatal mistake of dividing the regiment; see ibid., 261.

51. Edgerly speculating on what Custer had envisioned for what would be Benteen's line of attack, see Nichols, *Reno Court*, 457.

52. Daniel Newell overhearing Custer and French speaking about the number of tepees, see Tom O'Neil, *Sagas of the Greasy Grass*. (Brooklyn, New York: Arrow and Trooper, 1992), 26–27.

53. Kuhlman's belief that Benteen's battalion was a "Reconnaissance in force," see Kuhlman, *Legend into History*, 81.

54. Low Dog stating the Indians believed the village too strong to be attacked, see Graham, *Custer Myth*, Low Dog interview, 75.

55. That many warriors were asleep or away hunting, see Thomas Marquis, *Custer on the Little Big Horn*. Edited by Paul Hutton, (Lodi, CA: Kain Publishing, 1969), 82-84.

Chapter 4

(Pages 57–59)

1. That Custer was flying by the seat of his pants, see Michael N. Donahue, *Where the Rivers Ran Red: The Indian Fights of George Armstrong Custer*, (San Jose Publishing Group, 2018), 56.

2. Regarding the immutable lore of war, see Lord Wolseley, see Garnet Wolseley: *The Story of a Soldier's Life*, Volumes 1 and 2, published in 1903 and 1904. (Westminster: A. Constable & Co., Ltd).

3. Some authorities have questioned if Custer had brought his hounds along on this campaign. It would not be unprecedented, since Custer was known to bring hounds along on campaigns, which he used for hunting game. There is even documentation of Custer bringing forty hunting dogs on one of his campaigns. There is no reason for John Burkman, an eyewitness and an intimate of the General, to lie about this, and being a primary source, the authors find it easy to accept his testimony that the hounds Blucher and Tuck came along with him on his last campaign, see Wagner, *Old Neutriment*, 152.

4. The Canadian William Cooke may well have been the most popular man in the regiment. Affection for him was not just reserved for his fellow soldiers. Years later, when Walter Camp was interviewing Soldier, an Arikara Indian, who was in Reno's column, he showed him a picture of Lieutenant Cooke. Moved, the scout kissed the picture, saying he was a lovable man, his very breath being nothing but kindness. See Camp, Folder 54, Brigham Young University.

5. Regarding the trumpeter John Martin being called "Dry Martini", see Theodore Goldin to E.A. Brininstool, December 3, 1993, Brininstool Collection. Giovanni Martini, who famously carried Custer's last message, is often referred to by his Italian birth name. But once he had emigrated to America in 1873, he anglicized it to John Martin. It is how he always referred to himself. His tombstone is also inscribed John Martin.

6. Burkman telling Autie Reed not to go with the lead columns, see Wagner, *Old Neutriment*, 152.

7. On a Northern Pacific snowbound train, Kellogg was immediately taken by Custer's dash and charisma from the time he first met him. See Barnard, "I Go With Custer: The Life and Death of Mark Kellogg," (Bismark ND: *Bismark Tribune*, 1996), 207.

8. Cooke rode to Mathey, who was leading the pack train, once and then a second time to keep the mules off the trail in an attempt to contain clouds of dust. (See Hammer, *Custer in '76*), 78. Some mules proved troublesome as each was carrying some three hundred pounds of equipment. Moreover, most of the mules were not regular pack mules, but animals that had been taken out of the wagon teams and were not accustomed to this kind of drudgery. Many were poorly packed by inexperienced packers, the packs slipping off their sore backs, and they suffered long stretches, as much as twenty-four hours, without water. (See Hunt, *I Fought with Custer*, 81.) But the pack mules were an indispensable and necessary part of the campaign. Custer paid careful attention to them, making a list and putting them in order (each troop had to pack its own mules) of those companies that did the best job in packing and assigned marching orders in accordance with that list. It was all incentive-oriented, for the troops who did the worst job were assigned to the rear and had to eat the alkaline dust of the column. (See ibid., 73)

9. That Reno rode on the left side of Ash Creek and Custer hugged the right, see Camp Collection, BYU, undated. This came from Camp's interview with Daniel Kanipe in which Custer's path toward the village can be reasonably intuited from the description he offers.

10. Scouts being a beehive of energy riding back and forth, see Donovan, *A Terrible Glory*, 213.

11. Hairy Moccasin confirmed what Varnum had later told Custer that he saw the village and it was still standing, see Coughlin, *Battle of the Little Big Horn*, 48. For Varnum, see Graham *Reno Court*, 121-122. Although Varnum now voiced the much repeated "more Indians than I ever saw," he also stated that he did not get as good a look as he would have liked because of the bends in the river and the timber on the left bank. There are a few others, however, who have questioned whether Varnum really saw the main village in the same way they questioned whether Gibson really saw the Valley of the Little Big Horn during the Benteen scout. The authors see no good reason to doubt Varnum, one of Custer's most capable and reliable scouts. See Carroll, *I Varnum*, 64.

12. An anxious Custer riding out in front with his scouts to assess matters see Hunt, *I Fought With Custer*, 85.

13. Benteen saying that having a plan at this point was premature, see Graham, *Reno Court*, Abstract 149.

14. Custer pointedly telling the Indian scout, Half Yellow Face, that the regiment had not been spotted, see Graham, *The Custer Myth*, 33.

(Pages 60–62)

15. A few battle historians have expressed skepticism about a hastily abandoned village. The authors do not count themselves among them. Custer was heard to tell Captain French that he saw a dozen or so tepees still standing on Ash Creek after he returned from the Crow's Nest on the divide. Additionally, there are multiple attestations from reliable eyewitnesses stating this hurriedly abandoned village existed without any primary source contradicting their testimony. These include 1) Sergeant Daniel Kanipe, 2) Pvt Thomas O'Neill, 3) Pvt. Peter Thompson, 4) Pvt. Daniel Newell, 5) Wooden Leg, 6) Scout George Herendeen, 7) Lt. Charles De Rudio, 8) First Sergeant John Ryan. See citation link: https://thelbha.proboards.com/thread/1315/evidence- satellite-village-Ash-Creek, 16. Also see Stewart, *Custer's Luck,* 323–324.

16. Custer with five companies, with Reno on the left moving at a trot and gallop in columns of twos toward the small Indian encampment, see Hammer, *Custer in '76.*

17. For a discussion of the possible identity of the Sans Arc warrior found in the tepee, see Richard Hardorff: *Cheyenne Memories of the Custer Fight*,153. Also, Hardorff was of the opinion, as were most Little Big Horn historians at that time, that there was only one lone tepee. The evidence that there were, however, two lone tepees is substantial.

18. For more information on the lone teepee, see also "The Lone Tepee along Reno Creek" (The 12th Annual Symposium Custer Battlefield Historical & Museum Association, 1998), 1–11.

19. That Young Hawk was the Crow scout who slashed the lone tepee open, see Libby *The Arikara Narrative*, 92.

20. For Indian rituals regarding dead warriors, see Joseph Dixon, who noted that, to the Indian mind, all signs are symbolic. Every single act of his life carries some ceremonial function, a vast, glittering mosaic of rite and ritual. This will, of course, include death. The Indian, for example, buries

their dead on a higher elevation, nearer to the spirit world, and so was the case with the Sans Arc warrior. See Joseph Dixon, *The Vanishing Race*, 27, 29.

21. Custer's order to set the tepee on fire can be found in Hardorff, *On the Little Big Horn with Walter Camp*, 6.

22. Custer setting fire to the tepee to alert the hostiles to the regiment's presence is, candidly, ludicrous, but the authors cite it for informational purposes. See Daniel O. Magnussen: *Peter Thompson's Narrative of the Little Big Horn Campaign, 1876* (Glendale CA: Arthur H. Clark, 1974), 78.

23. Custer ordering Reno to cross over to the north side of the creek and fall behind him, see Carroll, *Federal View*, 66; Charles Kuhlman, *Legend into History*, 51.

24. For Reno's actions during the battle of Gaines Mills, see John Upton Terrell and George Wilton, *Faint the Trumpet Sounds*, 36-37.

25. Custer speaking of his early friendship with Reno, see his letter to Libbie Custer January 14, 1869, List 152, American Historical Autographs.

26. Reno discharging the Arikara interpreter, Gerard, because he thought he was stealing from the government, see Graham, *Reno Court*, 223.

27. Reno's run-in with Weir, see Nichols, *In Custer's Shadow*, 116 -120.

28. Gerard's sighting from the knoll of running Indians occurred only moments after Custer summoned Reno and his column over with a wave of his hat. See Nichols, *Reno Court*, 43.

29. That the small band of fleeing Indians also saw Custer's command, see Graham *Reno Court*, 236.

30. Both Herendeen and Gerard were in earshot of this order and repeated it as such. See Nichols, *Reno Court of Inquiry*, see the testimony of Gerard and Herendeen.

31. Reno stated that before he was directed to take command of Companies M, A, and G, he had never received any orders and that all he had heard at the last officers' meeting was that Custer said that the Indian scouts had spotted a large Indian village from the top of the mountain. See ibid., 560.

32. For Reno's almost undisguised uneasiness of Benteen's departure from his wing, see Nichols, *Reno Court*, 423.

33. That the wings at the divide were reconfigured with officers, regardless of their companies, to support the leadership of those columns, see Sklenar, *To Hell with Honor*, 117.

(Pages 63–65)

34. Theodore A. Ediger and Vinnie Hoffmann: "Some Reminiscences of the Battle of Washita (*Chronicles of Oklahoma* quotes, also *Journal of Southern History*, Vol. 21, No. 4, November 1955), 571–581, quotes a Cheyenne girl mournfully remembering how the dying ponies would moan loudly like human beings.

35. That most of the Indians were either sleeping or cooling themselves in the river, see Livey *Irikira Narratives*, 121-122.

36. It should be noted that the fleeing Indians were not, as was then believed, from the large encampment, which would soon be encountered across the river. See Robert Utley, *The Reno Court of Inquiry: The Chicago Times Account*. (Fort Collins, CO: Old Army Press, 1972), 276.

37. Reno testified that these were Cooke's exact words. See Carroll, *The Federal View*, 103. We also believe that Custer, along with Cooke, as Gerard and Herendeen later maintained, also ordered Reno to the valley. Hare also thought Custer repeated Cooke's orders to Reno see Hammer, *Custer in '76*, 9. That both would say essentially the same thing to Reno, albeit at different times before he moved ahead, is convincing. See Graham, *Reno Court*, 500. But this was contradicted by Dr. Henry Porter, who said that Reno asked if Custer would support him. Cooke said yes, but Reno again asked about being supported, and again Cooke replied in the affirmative, see Graham, *Reno Court*, 76. Reno would later mount a vigorous defense against criticism that he was reluctant, as a calumny against him and the gallant soldiers of his command. See "Reno's Reply to General Rosser," *New York Herald*, August 8, 1876, see Graham, *The Custer Myth,* 226-227. The two would exchange several critically worded letters about Reno's conduct during the battle.

38. Davern of F Company was Reno's orderly and, therefore, right next to him when the order was given from Cooke, who ordered him to take three companies and drive everything before him, adding only that Cooke also said that Colonel Benteen will be on your left with the same instructions. See Graham, *Reno Court*, 286, though Reno, in a letter published in the *New York Herald* nearly six weeks later, says nothing about Benteen. See Graham, *The Custer Myth*, 226.

39. Here Martin agrees with Davern that Cooke mentions Benteen in his order to Reno to advance into the valley, see Colonel William A. Graham, "Come On—Be Quick! Bring Packs!" (*The Cavalry Journal*, Vol. 32, no. 132, July 1932), 306.

40. Reno's remark, "Let us keep together," to which Custer responded by his jaunty way of lifting his broad-brimmed hat, see Hunt, *I Fought With Custer*, 165.

41. We believe that Custer did lift his hat in this manner as it seems consistent with Custer's personality and his innate assuredness at the precipice of battle, not to mention primary source support for it, see *New York Herald* (New York, August 8, 1876); Graham, *Custer Myth*, 228.

42. There is no reason not to believe Kanipe, though he is the only source who says that Custer and Reno rode a stretch together talking to one another, see Hammer, *Custer in '76*, 92.

43. : If Custer had any plan, it was to strike the village whether it was standing or fleeing. Charles Kuhlman's incisive analysis of why this is the case has never been surpassed, see Graham, *Reno Court*, 213; Kuhlman, *Legend into History*, 53–54. As long as the Indians were free to hang on, Kuhlman wrote, to the rear and flanks of the troops, dodging from cover to cover, the superior organization, tactics, and long-range weaponry the army had availed them nothing. The only way to deprive the Indians of this advantage was to move against their camps, compelling them to concentrate on the defense of their

non-combatants and property, leaving the initiative and the freedom to maneuver to the troops. This was one of the largely unexamined reasons why Crook, once driven from the Rosebud, did not regroup and pursue the Indians as he simply would have been driving the Indians from one position to another in a fruitless pursuit, taking him farther away from his base, which was but a day's march from the Rosebud.

44. For Varnum testifying having seen Custer for the last time and that he was moving at walking speed, see Graham, *Reno Court*, 237.

45. For Cooke conversing with Reno on the way to Ford A, encouraging Reno and also allowing Cooke and Keogh to make a personal reconnaissance of the river crossing, see *New York Herald*, August 8, 1876, and Graham, *The Custer Myth*, 228.

(Pages 66–70)

46. Reno ignored Gerard's alarms, at least as they concerned him (he did not ignore it in a general sense as he did send a messenger to Custer about Gerard informing him of this development) and continued forming the troops in 3 columns as they straggled out of the river and onto dry, open ground. Reno's Indian scouts had ridden out ahead down the valley, shouting about enemy warriors on the move. At the same time, some warriors were heading for the Indian scouts who were attempting to run off and capture the pony herd.

47. For McIlhargey being sent as a messenger, see Hammer, *Custer in '76*, 232. For Reno dispatching the cook John Mitchell as a messenger, see Graham, *Reno Court*, 500. Both messengers died with Custer.

48. For Goldin writing about his account of being a messenger sent by Custer to Reno, see *Janesville Gazette* (Janesville, WI, July 8, 1886). This was a topic that Custer researcher Fred Dustin first accepted and then went back and forth on whether Goldin was telling the truth, see Dustin, *The Custer Tragedy*, 149–150. Of course, Custer may have sent Goldin as a messenger, but it is questionable considering his tender years (age 17) and that he was a tyro. Whatever the case, that note never reached Reno. Goldin was a notorious and well-known fabricator. This pattern of lying continued, and Goldin even stated that he was at the Reno Court of Inquiry in Chicago and spoke to Captain Weir and French, even though Weir had, by that time, been dead for two years and French was confined to quarters undergoing his own Court-martial.

49. This was the account that Kanipe gave to Camp as the reason Custer changed his plans and moved up the bluffs instead of into the valley, see Camp, Indiana University Library, Box 2, Folder 7. We should note again Camp's extensive research on the battle and his determination to seek out and interview every eyewitness he could find. Camp interviewed more than 60 survivors of Reno's command and more than 150 Sioux and Cheyenne.

50. For Custer saying that you needed a racehorse to catch Indian horsemen on the plains, see *Custer, My Life on the Plains*.

51. Weibert states that Custer watered his horses at the North Fork, see Weibert, *Sixty-six Years In Custer's Shadow*, (Billings MT: privately printed, 1985), 1–19. However, the North Fork is practically

bone dry in the summer months. In all likelihood, the battalion did not water some 220 mounts at that location. The authors have concluded that Custer's watering of the horses was at the Echleman Ford of Ash Creek, where there was water at that time of year.

52. For Custer being sighted waving his hat, see Donahue, *Where the Rivers Ran Red*, Appendix 5, 25. Little Big Horn Park Ranger Michael Donahue conducted two site tests, one in 2006 and another in 2011, with Park Rangers Steve Littleton and Tom Smith, and the latter test with participant Carolyn Bennett. Donohue positioned one ranger on Weir Point and another on Sharpshooters Ridge. Donahue stood in the timber area where De Rudio testified that he saw Custer waving his hat and recognized them by their dress. Equipped with radios, he asked the rangers to wave their hats. Donahue, who has excellent long-sighted vision, could barely make out the rangers, and it was impossible to see anyone waving a hat. It appears that while De Rudio may have well seen the figures on those hills, he took the liberty of dramatically stating it was Custer.

53. Regarding Custer's secrecy on his military plans in a campaign, see Elizabeth B. Custer, *Boots and Saddles*. (New York: Harper and Brothers, 1885), 116.

Chapter 5

(Pages 75–78)

1. Whatever distance Benteen was when he last saw Custer's column going at a dead gallop (that Custer was riding the horses that hard so early and on such a hot day is utterly implausible) is really beside the point. What really mattered was that the order given to Benteen showed that Custer was skeptical that the Indians were all gathered in one village, see Kuhlman, *Legend into History*, 92; Nichols, *Reno Court*, 404.

2. Gibson and six troopers to ride three hundred yards in front, see Hammer, *Custer in '76,* 80. The formation and distances of Gibson's detachment and the rest of the wing can be found in Hunt, *I Fought With Custer*, 80.

3. About Gibson going to the top of The Bluffs four separate times, see Graham, *Reno Court Abstract*, 157.

4. Custer sent Trumpeter Henry Voss and another messenger, William Sharrow, to Benteen as he trekked southward, illustrating the urgency he felt about gathering intelligence about what lay to the south see Nichols, *Reno Court*, 403.

5. Here Benteen describes for the court his encounter with Sergeant Major Sharrow and sending a detachment in advance of his wing, see Nichols, *Reno Court*, 403.

6. Gibson tells Benteen, after looking down the valley and seeing no Indians, that it would be useless to go any farther with even the troopers thinking it was a "wild goose chase." See Windolph, *I Fought With Custer*, 80.

7. Differing opinions regarding Benteen's scout, see Kuhlman, *Legend into History*, 81. Both its subject and extent, says Kuhlman, are in dispute.

8. Edgerly's view on Benteen's scout, see ibid., 81.

9. For Godfrey's view, see ibid., 81.

10. For Benteen's dissatisfaction about his scouting mission and its senselessness, and that, by turning back, his action amounted to a violation of his orders, see Nichols, *Reno Court* 431.

11. Benteen's letter to his wife dated July 2, 1876, see Graham, *The Custer Myth*, 187.

12. For Lieutenant Gibson's letter to his wife dated July 4, see Gray, *Custer's Last Campaign: Mitch Bouyer and the Little Big Horn Reconstructed*, 259. Supporting this is an account given by Gibson to the *New York Evening Post* on February 20, 1897, wherein he states that Benteen was to search for the Valley of the Little Big Horn as soon as possible.

13. See for another indication that Benteen knew that Custer wanted him to rejoin the command, see Graham, *Reno Court*, 380.

14. For Benteen saying he had no orders to reunite with anyone, see Graham, *Reno Court Abstract*, 147-149.

15. Benteen letting down his guard before the Reno Court, admitting that there was nothing in Custer's orders that would imply that he was not to rejoin the command. See Graham, *Reno Court*, 380.

16. Letter to Godfrey by Gibson telling him that on his second ascent, even with the aid of glasses, he saw not a living thing down the valley. This was an account from Colonel Robert P. Hughes, Aide de Camp to General Terry, in his article, "The Campaign Against the Sioux in 1876," published as Note 10, 130, in Graham, *The Story of the Little Big Horn*, wherein he quotes from a letter from Francis Gibson to General Godfrey dated August 8, 1908.

(Pages 79–81)

17. Godfrey says the pace of the scout was rapid and the distance twelve to fifteen miles, see Nichols, *Reno Court*, 479; Godfrey, *Custer's Last Battle*, 22.

18. Windolph's impressions of Benteen's scout, and that they rode over rough and hilly country, which was hard on the horses, see Hunt, *I Fought with Custer*, 80.

19. For Roe arguing that Benteen should have been with Headquarters Staff and not leading a scout, see Hammer, *Custer in '76*, 249.

20. Custer's quote about where water is found, so are mosquitoes and gadflies; see *Custer, My Life*, 18.

21. Chip Creighton said the alkaline water was enough "to burn the hide off your tongue," see Usher Burdick: *The Army Life of Charles Chip Creighton*, (Paris MD: National Reform Associates, 1937), 27.

22. The quote "What is the old man keeping us here for so long?" can be found in Hammer, *Custer in '76*, 75.

23. Here Benteen describes how he reached the "boggy place" and watered the animals for fifteen minutes—in reality more like twenty-five minutes, according to the Participant Timeline, see E.A. Brininstool, *Troopers with Custer*, (Harrisburg, PA: The Stackpole CO., 1952), 77.

24. Boston was employed in the Seventh as a guide and probably rode with the vanguard of the cavalry. Regarding the filial connection of the 7th in his book *Cavalier In Buckskin*, the historian Robert Utley called it "an army full of assiduous nepotism," see Ibid., 69.

25. Edgerly remembered that Boston, now returned from the pack train, was full of energy and anticipation to join his older brothers, see *The Gibson–Edgerly Narratives*, 5.

26. Impressions from the trail after Boston left on his own, including trouble with the mules, can be derived from several sources, including Stewart, *Custer's Luck, The Gibson-Edgerly Narratives*, Graham, *The Custer Myth*, Carroll, *Custer's Chief of Scouts*.

27. Private Coleman remembering the difficulty with packing and repacking the mules, see Liddic, *I Buried Custer*, 17.

28. This would not be Weir's last defiance of a battalion commander that day. He would become even more defiant with the other battalion commander on Reno Hill, Major Marcus Reno—both times because Weir was anxious to ride to where he heard firing, see Hammer, *Custer in '76*, 75.

29. For additional perspectives regarding the pack mules, see Edward S. Godfrey, *Custer's Last Battle*. In W.A. Graham's *Custer Myth: A Source Book of Custeriana*, (Mechanicsburg, Pa.: Stackpole, 2000), 125–149.

30. The death of the Indian boy Deeds and some of the women and children, see Richard Hardorff: *Hokahey! A Good Day to Die! The Indian Casualties of the Custer Fight*, (Spokane, WA: Arthur H. Clark Company, 1993), 17-30.

31. An early newspaper account stating that ten Indian women were killed, later amended to six women and four children. George Herendeen was emphatic that our men did not kill any squaws, but the Indian scouts did. See: Graham, *The Custer Myth*, 90; also see *The Bismark Tribune* July 7, 1876; Hardorff's *Hokahey! A Good Day to Die!*, 32–34; also Libby, *The Arikara Narrative*, 151.

32. Benteen's sarcastic remark about a "good Indian" see Brininstool, *Troopers with Custer*, 77.

33. Benteen believed the order to bring the pack trains did not pertain to him and sent Kanipe back to the rear, see Graham, The Reno Court, 357, 410.

34. For Kanipe's breezy remark "We got them boys"—"we're licking the stuffing out of them," whereupon everyone erroneously inferred, that the Indians were on the run, see Graham, *The Custer Myth*, 140.

(Pages 82–85)

35. Martin was directed to follow the command's back trail because both Custer and Cooke knew about how long it would take the Trumpeter based on the distance of the dust cloud, which they had observed some ten minutes earlier, see Graham, *Reno Court*, 347.

36. Martin describing at what location he left Custer's command, see Graham, *Reno Court*, 342, 347.

37. Martin seeing Custer's command galloping down into the ravine see Graham, *The Custer Myth*, 290.

38. Indians shooting at Martin, see Graham, *The Custer Myth*, 290. His horse being hit and the blood splattering on him, see Hammer, *Custer in '76*, 101.

39. Martin meeting up with Boston Custer, see Camp: IU, Box 3, Folder 3, 130–132.

40. Martin telling Boston the whereabouts of the Custer command. Martin also cautions about Indians in the vicinity, but the younger Custer gallantly replies he is going anyway. See Graham, *The Custer Myth*, 290; and Camp IU, box 3, Folder 3, 130–132.

41. Martin recounting locating Benteen and waving his hat, see Graham, *The Custer Myth*, 290.

42. Martin, who Benteen believed was just as much cut out for a cavalryman as a king, telling Benteen the Indians were skedaddling, see *Benteen–Goldin Letters* also Hammer, *Custer in '76*, 101. Others would come to think differently about Martin. After Martin's intrepid ride, Charles Windolph said that they never mocked him again. (See *I Fought with Custer*, 83).

43. Benteen noting that the Cooke message was not dated, see Graham, *Reno Court*, 357.

44. For Martin saying how the Indians were being killed left and right reinforced Kanipe's comment about the Indians being badly punished, although Martin, with his broken English, almost certainly never used a polysyllabic verb like skedaddle. Martin claimed he just told Benteen that Custer was three miles distant, and he attacked the village, see Hammer, *Custer in '76*, 101.

45. That Kanipe's note should be directed to McDougall for the order did not pertain to Benteen, see Graham, *Reno Court,* 357, 410.

46. Kudos to Frederic Wagner for sleuthing out the weaknesses of Kanipe's story, see Wagner, *Strategy of Defeat*, 107.

47. Wagner theorizes what Kanipe would have said if Custer survived, see ibid., 108.

48. Neither McDougall nor Mathey said they ever received any order from Custer or notification from anyone to hurry up the pack train. Wagner argues, persuasively, that if the Custers survived, it would have been easy enough for Kanipe to claim his horse broke down and that he was doing his duty as any good NCO might to help bring up the packs. Ibid., 108.

49. For DeVoto's statement on the matter, see Walter M. Camp Papers, Crow Agency, MT.: Custer Battlefield National Monument, no.1138c.

50. Benteen explains how a mile or so after he was at the tepee with a dead Indian, he encountered a sergeant telling him to bring the pack trains to the front with great rapidity, see Carroll, *The Federal View*, 106.

51. Benteen said the order did not pertain to him and pointed out where McDougall was to which the order pertained, see Graham, *Reno Court*, 357, 410; Camp, IU, box 2, folder 8.

52. McDougall was effusive in his praise, saying in his testimonial that he had known Kanipe since 1873. See Bill Boyes, "Tarheel Survivor of Custer's Last Stand," *Research Review*: The Journal of Little Big Horn Associates, Volume 7, no. 2, June 1993, 30.

53. Bruce Liddic's questionable conclusion in endorsing Kanipe's story that he carried an order from Tom Custer because the pack train had ceased to follow the main trail was taken from *Vanishing Victory*, based on Walter Camp Interview, BYU, roll 1, box 1, folder 12. See 71.

54. Benteen stating that if he is to be of service to Custer, he better not wait for the packs, see Hammer, *Custer in '76*, 54.

(Pages 86–90)

55. Although he never names him, it is obvious that the sergeant whom Benteen was directing to McDougall was Kanipe. The reason Benteen says the pack train was seven miles behind was to give the court the impression that he had to wait longer for the ammunition they were carrying than he actually did. See Nichols *Reno Court*, 404.

56. Cooke telling Martin to go Benteen, see Nichols, *Reno Court*, 390.

57. For the view that Benteen moved at a reasonable pace in returning to the main trail, see Gray, *Centennial Campaign*, 179.

58. On the proper gait of a cavalry battalion, see Clair Conzelman's analysis: https://thelbha.proboard.com/thread/933/cavalry-unit-march-rates?. Conzelman is a graduate of the U.S. Military Academy at West Point. Held two company commands in the U.S. Army; served with the 1st Cavalry Division at Fort Hood; Attended Cavalry School at Fort Knox. Member of a riding team; currently owns a horse ranch in Louisville, Kentucky. Widely considered an expert on horses and cavalry tactics.

59. Benteen's testimony regarding Custer moving at a gallop implying that this pace resulted in a greater distance between Custer's command and his own is misleading. What Benteen observed was an accordion effect—a kind of optical illusion accentuating the gap between companies.

60. Regarding the importance of determining the distance based on where Boston met the General (about forty minutes at a trot) from the point where Benteen encountered Martin is not, save for informational purposes, significant. The authors continue to emphasize that Benteen was not going over The Bluffs to reach Custer but was heading directly to the valley where the Indian village was. Benteen said he was poised to cross the river at Ford A, which is entirely believable since one of his companies, Company D, was already down at the crossing. In connection with this, some students of the battle still

find it hard to believe that Martin did not tell Benteen what side of the river Custer was on. It is not difficult to believe that at all. Martin was told that if he was not coming back to Custer's wing, he was to stay with Benteen, which indisputably shows that everyone knew, including Custer and Benteen, that Benteen was going to the valley. Colonel William Graham, who published the highly regarded "*The Custer Myth,"* erroneously contributed to the myth that Benteen was to support Custer via the bluffs.

61. Kuhlman stated that the significant thing to note here is that the Martin statement at the Court of Inquiry stated that the order specifically called for the ammunition packs, not the whole pack train to which an order to hurry forward had been sent just before the command reached Sharpshooter Ridge, see Kuhlman, *Legend into History*, 163.

62. General Hamilton Hawkins stating that "packs" always means the pack train—the entire pack train, see Letter to Charles Dubois, May 18, 1954, Moses Collection.

63. Benteen's testimony that by the time he received Martin, he thought Custer's command was dead can be found in Nichols, *Reno Court*, 430.

64. Benteen's discussion with Terry informing him that Custer's command was destroyed and Benteen's denial of it can be found in Hammer, *Custer in '76*, 249–250.

Chapter 6

(Pages 91–95)

1. Sergeant James Wilson recorded in his Diary the temperature on the steamer *Far West* on June 25, 1876, at 91 degrees Fahrenheit, see Michael J. Koury: *Diaries of the Little Big Horn*.

2. Almost every author and battle scholar tends to get the time wrong for the start of the battle. We think that Frederic Wagner did the hard chair time and the necessary legwork to get it right, see Wagner, *Strategy of Defeat*, 2–3.

3. William Taylor, seeing Reno imbibe, can be found in Camp, BYU, roll 1, box 1, folder 1. Reno was certainly imbibing the spirits even though he claimed that he did not take a drink until after midnight, June 25 (See Graham, *Reno Court*, 511). There are multiple attestations of those who saw him drinking from his whisky-filled flask throughout the day. While it is true that Private William Slaper of A Company asserted that he saw no sign of drunkenness in Reno (See Brininstool, *Troopers with Custer*, 59) does not mean Reno was not drinking. Others emphatically disagree that Reno was not imbibing. De Rudio observed Reno drinking from a whisky flask. (See Camp, Box 8, folder 4.) Private William Taylor made the same mental notation (See Camp, Brigham Young University, roll 1, box 1, folder 1.) We believe Reno had started drinking but was not yet drunk, or at least obviously drunk. Some individuals can hold their liquor better than others, and Reno, no stranger to alcohol, may well have been one of those.

4. Hare and other scouts going ahead down the valley ahead of Reno's command, see Hammer, *Custer in '76*, 65.

5. Cook and Keogh rode with Reno for some time, but at the river crossing, Reno said he lost track of them, see interview in *New York Herald*, August 8, 1876, and also Graham, *Custer Myth*, 139.

6. After his much-criticized scout, Reno was relegated to obscurity until Custer first divided his regiment, causing the major to complain. He was mostly kept in the dark until he was placed in command of three companies. See Graham, *Reno Court of Inquiry*, 211.

7. Companies M, A, and G crossing the stream, see Stewart, *Custer's Luck*, 331.

8. Trumpeter William Hardy saw Cooke on the eastern bank, watching the battalion's progress, see Hammer, *Custer in '76*, 148. See also Camp IY. Folder, Box 10.

9. Private James Wilbur of M Company remembered that some were galloping too fast as they approached the ford, and Cooke told them not to run the horses like that can be found in Hammer, *Custer in '76*, 148.

10. Reno delaying the charge, probably giving time for Custer's promised support to arrive, see Graham, *Reno Court Abstract*, 89.

11. Reno chastising De Rudio can be found in Hammer, *Custer in '76*, 84.

12. Interestingly, Keogh is never mentioned by anyone as leaving the river, although he clearly did. There is a good chance he may have even departed before Cooke.

13. Gerard's alarms about the Sioux coming can be found in Camp BYU Notes, Reel 5, 463, and Graham, *The Custer Myth*, 263. The departure of Private McIlhargey, see Nichols, *Reno Court*, 561.

14. Reno ignoring Gerard at Ford A, see Graham, *Reno Court Abstract*, 223.

15. Determining the numbers involved in the charge down the valley is not an enviable task. Reno, give or take, had about 175 men with him. Some 153 charged down the valley, all but fifteen uniformed. The others rode to the valley to either capture or stampede Indian horses. Reno falsely claimed he only had 112 men, an obvious attempt to make his task of charging such a big village with so few men even more daunting than it was.

16. For Reno slurring his words, see Taylor, *With Custer on the Little Big Horn*, 36. Again, the authors stress that we do not believe Reno was drunk during the charge, though he had clearly imbibed. Custer, who religiously abstained from alcohol, had recently been with Reno just prior to ordering him to charge the village. At one point, according to Sergeant Kanipe, they rode side by side, talking for a stretch. Custer would have certainly picked up any signs of palpable inebriation and, if he had, would not likely have sent Reno off with three companies chasing hostiles.

17. Reno shouting, "Stop that noise," see Camp, BYU, Box 6, Folder 2.

18. For Private Sivertsen stating Reno's bravado in urging on the charge, see Coffeen, *Tepee*, Book 2, 79.

19. Describing the terrain of cottonwood trees and bullberry bushes as they moved toward the village and the job of the flankers. See Barnard, *Ten Years With Custer*, 29.

(Pages 96–100)

20. Disposition of the scouts during the charge can be found in Carroll, *Chief of Scouts*, 89.

21. Account of Moving Robe registering shock that they were being attacked. See Hardorff, *Lakota Recollections*, 92.

22. Sitting Bull escorted the women and children to the western foothills, see Hardorff, *Indian Views*, 148–160.

23. The Lakota Eagle Ben stating none of the Indians expected an attack, see Vestal, *Warpath: The True Story of the Fighting Sioux Told in the Biography of Chief White Bull*. Lincoln NB: University of Nebraska Press, 1984.

24. For Crazy Horse racing back to the village, exhorting his warriors to kill all the soldiers, see Hardorff, *Lakota Recollections*, 40.

25. Two separate times, portions of Custer's five companies were spotted on the eastern heights above the river. The first sighting came a minute or two after Reno began his advance into the valley, as reported by some seven soldiers in Reno's battalion. Approximately ten minutes later, just before the battalion had dismounted to form a skirmish line, there was a second sighting by Lieutenant Charles Varnum, who first saw part of E Company "some little distance…from the point where Major Reno was when he struck the hill." (See Graham, *Reno Court*, 147.) Fred Gerard would claim to have seen Custer's men on the bluffs somewhat later. However, the small figures on the bluffs he had mistaken for Custer's battalion were likely the Ree scouts since, according to the participant timeline, Custer was well beyond that point downriver, and the scouts were upriver where Gerard likely spotted them. Lieutenant De Rudio saw Custer at the highest point on the right bank about a thousand yards from where he was, see *ibid.*, 291—De Rudio's claim that he saw Custer and Cooke on the bluffs will be discussed later. Private Daniel Newell saw Custer's column on the height as they charged down the valley (See, Carroll, *The Sunshine Magazine*, 10); Reno, though inconsistent, does allude to seeing Custer on the bluffs and knowing then that his support would come elsewhere rather than from the rear as he expected (Reno's thoughts on this is recorded in *The Federal View*, 103–104). See also Francis Gibson's *New York Evening Post*, February 20, 1897. In that article, Gibson quotes Reno as saying that he "last saw Custer on the crest of the hill we were then on (Reno Hill)—but that his troops must have been behind the slope, as he did not see them. Again, much closer to the event, Reno is asked in an interview by the St. Paul Pioneer Press, January 13, 1879, whether Custer was aware of his (Reno's) engagement with the Indians, wherein he replied that Custer could see from the eminence where he halted for a brief time. It seems that Reno had at least heard Custer was on the bluff, but the weight of evidence strongly indicates that he also caught a glimpse of them; Lieutenant Luther Hare saw Custer's troopers on the bluffs (see Graham, *Reno Court*, 137); Lieutenant George Wallace also saw soldiers on the eastern heights, when on June 27 he responded to an inquiry from Lieutenant Edward McClernand, an officer with General Gibbon's column about where was Custer; he replied, "the last we saw of him he was going along that high bluff" [upon which

he pointed in the general direction to a point on the bluffs downstream from the position where he had located Reno], "toward the lower end of the village." Wallace even added that he waved his hat at us (see McClernand, *With the Indian and the Buffalo in Montana*, 27). The reader is encouraged to consult Hammer, *Custer in '76*, 112, who also adds Brinkerhoff, O'Neill, Petring, and Rory.

26. Regarding messenger John Mitchell, see Graham, *Reno Court*, 500.

27. Evidence that Reno, by his own admission, understood that Custer was now intending to support him not from the rear but by moving farther down the stream and attacking the village in flank. (See Carroll, *Federal View*, 103–105.)

28. For Reno checking the formation, see Nichols, *Reno Court*, 105; bugle calls during the charge, see Nichols, *Reno Court*, 277, and Hodgson's bringing G Company into line, *ibid*., 561.

29. If the attacking troopers fired any shots, they must have been of the scattering sort at best. Varnum did not believe that Reno's command fired any shots until they dismounted and formed a skirmish line. See Carroll, *Chief of Scouts*, 121–122.

30. For reactions of the Indians in and around the village when they realized they were under attack, see Michno, *Lakota Noon*, 35-41. Therein are the accounts of Wooden Leg, Gall, White Cow Bull, Two Moons, Iron Hawk, and He Dog. See also James McLaughlin, *My Friend the Indian*, 43–44.

31. For Pretty White Buffalo remembering the warriors were not anticipating a battle and were attacked from a direction they did not expect, see Michno, *Lakota Noon*, 45.

32. That many in the village were still asleep, see Hardorff, *Lakota Recollections*, 99-100.

33. Lieutenant Winfield Scott Edgerly thought Reno made a mistake not to run through the village or at least the Hunkpapa circle at the most southern end of the camp (See Lt. O.L. Hein, *Memories of Long Ago By An Old Army Officer* (NY, NY, G.P. Putnam and Sons, 1925), 145. And this criticism is not limited to soldiers. Pretty White Buffalo was representative of many of the Indians that day. Mrs. Spotted Bull, who was in the Indian encampment, believed striking the village would have had a devastating effect. Not doing so robbed the soldiers of victory (See *My Friend the Indian*, (Boston, MA, Houghton Mifflin Co., 1926), 170.

34. With all the dust, Varnum certainly saw plenty of Indians, but not necessarily in front of him. See Carroll, *Chief of Scouts*, 65.

35. It is a revelation to many that Indian villages, even larger ones, could be rolled up by a smaller but effective contingent of cavalry making a charge. Examples can be cited: At the battle of Summit Springs in Colorado in July 1869, a 300-man contingent of the 5th Cavalry and a passel of Pawnees crashed into a sleeping camp of five hundred Cheyenne Dog soldiers and routed them with only one man slightly wounded, see Michno: *Encyclopedia of Indian Wars*, 258–259. Ranald S. Mackenzie's 4th Cavalry in September 1872 led five companies against a village of 262 lodges, successfully overcoming the village with a loss of four troopers. Of course, in the first instance, Reno did not have three hundred men and the village he was attacking was at the Little Big Horn was at least four times larger. In the second instance, unlike Mackenzie, Reno had only three companies, not five.

Additionally, at the Powder River Fight on March 17, 1876, one company of forty-seven men charged into and routed an Indian village of over one hundred lodges.

36. For Mrs. Spotted Horn Bull's reaction to the attack, see *My Friend the Indian*, 170.

37. This is a hunting expression wherein holding a leg would have been tantamount to holding a position in the valley. Godfrey's narrative was first published in *Century Magazine*, January 1892. The narrative was later published in Graham, *The Custer Myth*, 125–149, where he details Reno's failures in leadership. This would not be the only time Godfrey would point an accusing finger at Reno. On January 18, 1929, in a letter written by Godfrey to Major C.H. Asbury, he again notes Reno's faults and errors in the valley fight.

38. That Reno had no dash and the attack was spiritless, see Charles King, "Custer's Last Battle," *Harpers*, August 1890.

39. That Reno was thrown on the defensive after he dismounted; see Graham, *Reno Court*, 501.

40. De Rudio stating that if they advanced another five hundred yards, they would have been butchered, see *ibid*, 282.

41. Reno's explanation on why he called a halt, see Graham, *The Custer Myth*, 173.

42. Varnum testified that there was no ravine or ditch—it was all open prairie until way down. See Graham, *Reno Court*, 127. Battle scholar Bruce Liddick scoffed at Reno's explanation, stating he should have used good horsemanship to avoid the ravine, see Liddick, *Vanishing Victory*, 66.

43. For Reno claiming that he was fighting everyone and his uncle that day, see Carroll, *The Federal View*, 103–104.

44. Herendeen on the number of Indians, see Graham, *Custer Myth*, 264.

45. Regarding Indians in the large ravine, see Hammer *Custer in '76*, 66, Utley, *The Reno Court of Inquiry*, 402, and Bruce Liddic's excellent analysis in *Vanishing Victory*, 63.

46. Hare agreed with Herendeen that there were two hundred, maybe more, constantly in Reno's front. See Graham, *Reno Court*, 32, 238.

47. Stating that Indians only began pouring out of the ravines after Reno halted and not before. See *ibid.*, 237.

(Pages 101–104)

48. Rutten's refractory horse, see Hammer, *Men with Custer*, 352. Regarding Meier, Smith, and Turley, see Brady, *Indian Fights and Fighters*, 402; also see *Men with Custer*, 342.

49. For teepees catching fire, see Kill Eagle's (Blackfeet Sioux) account of the Little Big Horn battle (Condensed in an interview with Captain R.E. Johnson at Standing Rock Agency, given September 17, 1876, published in the *New York Herald*, October 6, 1876, and in Colonel William Graham's *The Custer Myth*, with our annotations in brackets: "At this time, Sitting Bull's men set fire to the prairie around my lodges and burnt some of my lodges up." A number of Indian accounts mistakenly believed

that Reno's soldiers set fire to some of the Indian lodges in the village, but Kill Eagle places the blame on the Hunkpapa warriors themselves.

50. For Hodgson giving the order to halt first to Company G, while Reno gave the orders to Companies A and M, see Graham, *Reno Court*, 501.

51. Wallace's estimate that they were a hundred yards from the closest tepees is wildly inaccurate. (See Graham, *Reno Court*, 23. Porter's estimate in Nichols, *Reno Court*, 196.

52. Reno's detractors have made much of what they see as Reno's timidity, pointing out this marked the only time of nineteen documented attacks on Indian villages between 1868 and 1876 where the troops failed to overcome a sitting encampment. See Jay Smith, *A Hundred Years Later*, 105. It is also true, notwithstanding that it was known one could often stampede a large village as easily as a small one, while Reno deserves the preponderance of blame which we cannot ignore that this was an especially large village. The authors believe that while Reno surely deserves the greater proportion of blame for the battle's catastrophic outcome, it is an overreach to unduly castigate him considering all the factors for arresting the charge.

53. Ryan clearing out any possible hostiles in the timber could be found in Dustin, *The Custer Tragedy*, 112.

54. The dismounted troops advancing on foot seventy-five to a hundred yards, see Nichols, *Reno Court*, 25, 91, 222, 314.

55. Mounts being moved into the glade of woods can be found in Dale T. Schoenberger, "Custer's Scouts," Helena, Montana, *Montana Magazine*, Spring, 1966. Also see Magnum, *Reno's Battalion*, 5.

56. These mounds were created by prairie dogs, herbivorous burrowing, ground squirrels native to the grasslands of North America, and the dirt they dug up provided minimal protection from gunfire.

57. For John Ryan believing that five hundred warriors were coming from the direction of the village, see Barnard, *Ten Years with Custer*, 293.

58. Regarding Varnum's observation on the visibility of the tepees at the point of dismount, see *On the Little Big Horn* with Walter Camp: A Collection of Walter Camp's Letters, Notes on Custer's Last Fight, 101.

59. For the early actions of Reno and French on the skirmish line, see Brady, *Indian Fights and Fighters*, 402.

60. Men being in good spirits on the skirmish line, see interview with Thomas O'Neill, Folder 83, Brigham Young University Library, and Hammer, *Custer in '76*, 106.

61. That every fourth trooper held the horses; see Nichol's *Reno Court*, 342 and Graham, *Custer Myth*, 384.

62. For the prescribed five-yard interval between troopers, see Custer, *My Life on the Plains*, 251.

63. About the soldiers bunching up, see Kennedy, W.J.D., "A Captain of Chivalric Courage," *English Westerners*, Vol. 25, no. 1 and 2, 1987/1988, 13.

64. For an excellent analysis of the skirmish line location, see Bill Rini, "The Reno Skirmish Line Locations," *Research Review: Journal of the Little Big Horn Associates*, Vol. 34, 2020, 31. See also Jesse Vaughn's *Indian Fights: New Facts on Seven Encounters* (Mormon: University of Oklahoma Press, 1966), 147–150.

65. For the unearthing of army cartridges near the skirmish line, see *ibid.*, Rini, 31 and Douglas Scott, *Uncovering History*, 57.

66. For Lieutenant Maguire drawing a dotted line representing Reno's skirmish line, see Nichols, *Reno Court of Inquiry*, 14.

67. For Lieutenant George Wallace's testimony on the Reno skirmish line, see Nichols, *Reno Court of Inquiry*, 23.

68. For Hodgson walking up and down the line encouraging the men to keep cool, and fire low, see Barnard, *Ten Years With Custer*, 293.

69. See the testimony of Charles Varnum. As far as the firing, in general, was concerned, it was more moderate, see Nichols, *Reno Court of Inquiry*, 174. Sergeant Ferdinand Culbertson said most of the men told him they had fired sixty rounds (each soldier had one hundred rounds—a few siloed them in their prairie ammunition belts, but some of the soldiers carried the ammunition in cartridge boxes when away from their horses), although he himself expended only twenty-one cartridges. For Culbertson's remarks, see Nichols, *Reno Court*, 376, 382, regarding cartridge boxes versus ammunition belts, see Brown and Willard. For A Company running out of ammunition, see Graham, *Reno Court*, 106; The Black Hills Trails, 159; Moylan thought each soldier on the skirmish line fired fifty rounds (see Camp, IU Notes, 112); Gerard and De Rudio testified somewhere between thirty and forty rounds of ammunition were fired per soldier (see Nichols, *Reno Court*, 104, 344).

(Pages 105—110)

70. For Indians riding around the left flank, see Nichols, *Reno Court*, 88.

71. The Indians' skill on horseback was unparalleled. "No race of men," Custer wrote, "not even the famous Cossacks, could display more wonderful skills of horsemanship than the Indian warriors in their native plains on his well-trained war pony." See Custer, *My Life on the Plains*, 251.

72. For an insightful analysis of the disintegration of the skirmish line, see Sklenar, *To Hell with Honor*, 177.

73. Newell's exclamation of the troopers behaving like sheep can be found in *Sunshine Magazine*, Bryan, Texas, N.P., 1979, 10.

74. Moylan testified that Reno had gotten information that the Indians were coming up the left bank of the river, threatening the horses, see Nichols, *Reno Court*, 216.

75. Horses being held in an open glade, see Brininstool, *Troopers with Custer*, 99.

76. For Ryan's assessment of French's quality as an officer, see Barnard, *Ten Years With Custer*, 257.

77. Regarding Sergeant O'Hara's death, see Liddic, *Camp on Custer*, (Spokane WA: Arthur H. Clark Co., 1995), 86; Barnard, *Ten Years With Custer*, 294.

78. Indian scouts driving the ponies over the ford and up Ash Creek, but were overwhelmed by Sioux moving to the left of Reno's line. The attempt to disperse the Indian pony herd was an abysmal failure. The Arikara would claim that they failed because Reno was unsuccessful in holding their line of defense, releasing swarms of Lakota horsemen that swept down upon them, see Libby, *The Arikara Narrative*, 12.

79. Crow King's quote can be found in Hardorff, *Lakota Recollections*, 94–95.

80. The scout Gerard stating that ten men placed in the timber could have stopped the Indians at that juncture; see Graham, *Reno Court*, 215.

81. Regarding Reno's lack of aggressive and controlling leadership, see McClernand, *With the Indians and the Buffalo in Montana: 1870–1878, (Glendale, CA*: Arthur H. Clark Co., 1969), 72.

82. Benteen explains why, with just 120 men, the defense of the Timber area could have lasted five to six hours, see Graham, *Reno Court*, 366.

83. For Moylan's reasons for leaving the timber, see Nichols, *Reno Court*, 231.

84. Red Feather remarks that the warriors could not get to the soldiers while in the woods, see Hardorff, *Lakota Recollections*, 83.

85. Windolph remarking about how to compensate for the lack of manpower, see Frazier and Hunt, *I Fought With Custer*, 101.

86. Reno stating that six hundred to seven hundred men to defend the timber is preposterous. See Nichols, *Reno Court*, 564.

87. Ryan said this to no one in particular and ten years earlier said that he said this only after he saw Indians penetrating the woods. See *Hardin Tribune*, Hardin, MT., June 22, 1923, and Barnard, *Ten Years With Custer*, 293. Ryan saw that the soldiers had allowed the Indians to gain the higher ground around the perimeter of the timber.

88. The passage about French's coolness under fire, see Brininstool, *Troopers with Custer*, 55.

89. Jackson, who discouraged a retreat to Ford A, (see Nichols, *Reno Court of Inquiry*, 53, 249), also apparently had a role in convincing Reno that sending another messenger to Custer was hopeless, see *ibid*, 211.

90. Regarding the death of Private George Lorenz, see *The Hardin Tribune*, Hardin, MT, June 22, 1923.

91. Retreating soldiers left the timber open for Indian infiltration. See Brininstool, *Troopers with Custer*, 131.

92. Bloody Knife being shot dead while mounted in the timber. See Hammer, *Custer in '76*, 223.

93. While in the wooded area, the authors contend that Reno's serious drinking had begun and was the start of his growing inebriation. (See Hammer, *Custer in '76*, 232).

94. For an account of Reno's contradictory commands regarding mounting and dismounting in the timber, see Nichols, *Reno Court*, 255.

95. Varnum relating that Reno's retreat was not disorganized from the head of the column but from the rear, see Graham, *Reno Court*, 156.

96. Giving up their ammo ungraciously, see Rickey, *Forty Miles a Day on Beans and Hay,* 291.

97. . G Company not hearing the order to mount and in a panic and a dead run, see Nichols, *Reno Court*, 336, Graham, *Reno Court*, 301. It should also be noted that General Nelson A. Miles remarked that the Indians would have had to abandon their village if the troops had not retreated, see Nichols, *Reno Court*, 223.

98. For how his comrades were left in confusion in the woods, see Taylor, *With Custer on the Little Big Horn*, 41.

99. For Charley Reynold's death, see Graham, *Reno Court*, 79, also Hammer, *Custer in '76*, 223.

100. For Dorman's death, see Hardorff, *Lakota Recollections of the Custer Fight*, 102; also Hammer, Custer in '76, 223–224; Dixon, The Vanishing Race, 173.

(Pages 111–115)

101. Gerard's comments on Reno losing his head can be found in Hammer, *Custer in '76*, 233.

102. For another account of Reno's retreat, see Thomas B. Marquis, *Wooden Leg: A Warrior who fought Custer*, 220–221.

103. Taylor remembering the warriors as a whooping, howling mass, see William O. Taylor: *With Custer on the Little Big Horn*, 41.

104. Indian descriptions, such as American Horse saying that it was like a Buffalo chase, can be found in Hardorff, *Cheyenne Memories*, 28.

105. For soldiers trying to surrender to the Indians, see Hardorff, *Lakota Recollections of the Custer Fight*, 53.

106. Colonel John Gibbon stating "once I dis-mount I never remount," see John Gibbon, *Arms to Fight Indians*, 15.

107. For the death of McIntosh, see Hardorff, *Battle Casualties*, 131, Stewart, *Custer's Luck*, 375. Hardorff, *Battle Casualties II*, Karen L. Daniels and Elden E. Davis, *That Fateful Day: Eight More With Custer* (Howell, MI: Power River Press, 1992).

108. Indians riding in all directions naked as the day they were born, *ibid.*, 48. See also Thomas B. Marquis, *Keep the Last Bullet for Yourself*, 54.

109. This dozen or so men attempted to give cover to the panicked retreat on the west bank, and though they didn't survive, their heroic actions undoubtedly saved others. See Nichols, *Reno Court*, 405.

110. Brave Bear saying that mounts hitting the water sounded like cannon fire can be found in Hardorff, *Indian Views*, 84.

111. Regarding the steepness of the eastern side of the bank, see Graham, *Reno Court*, 167.

112. Information regarding DeWolf's death, see Nichols, *Reno Court*, 351.

113. Description of Hodgson's death can be found in Daniels and Davis, *That Fatal Day*, 7–8.

114. Morris' ordeal at the crossing can be found in Barry C. Johnson's *A Captain of Chivalric Courage*, (Brand Books 25, nos. 1 and 2, 1987–1988), 14. For every man for himself quote, see Neil C. Mangum, "Reno's battalion at the Little Big Horn," *Volume 2 of Greasy Grass*, May 1986, 6.

115. For American Horse's observation that the soldiers escaped only because the Indians did not follow, see George Bird Grinnell, *The Fighting Cheyennes*, (New York: Charles Scribner, 1915).

116. Hare essentially agreeing with American Horse can be found in Graham, *Reno Court*, 91.

117. French's homicidal thoughts toward Reno because of his hasty retreat can be found in Graham, *Custer Myth*, 341–342 and W. J. Ghent: "Varnum, Reno and the Little Big Horn," *Winners of the West*, Volume 13, 35, April 30, 1936. French is interesting in what he told who about the battle. He sent two sets of letters about Little Big Horn, one to Adjutant Cooke's mother in Canada and another set to an uncle. The letters to Cooke's mother emphasized Reno's dereliction as he sought to commiserate with her, while the letters to his uncle defended the honor of the regiment, seeing that the stain of one soldier was a stain on all. Regarding the 7th's prospects in the battle, he said that if unsupported, the command had as much of a chance as a flea in a tar barrel. No doubt seeing a stain on one soldier as a stain on all.

118. Hare calling himself a fighting son of a bitch and telling the soldiers not to run like with curs, see Daniel and Davis, *That Fatal Day*, 27.

119. Godfrey explains why he believes that Custer's command would not have been rescued under Reno's leadership, since at no time during the battle did his conduct inspire confidence. Graham, *Custer Myth*, 149.

120. Marcus Reno and the judges being appareled in full dress uniform, see Robert Utley, *Reno Court of Inquiry*, 420–421; *Washington Post*: February 10, 1879.

121. Reno stating he couldn't have done more for Custer and his command if his own brothers had been in that column, see Utley, *Reno Court of Inquiry, 420–421*; *Washington Post*: February 10, 1879.

122. That he had no confidence in Custer's ability as a soldier, see *ibid.*

Chapter 7

(Pages 121–127)

1. That Custer would attack the enemy from multiple directions, see Arthur C. Unger, *Custer's First Messenger! Debunking the Story of Sergeant Daniel A. Kanipe?* (Washington, NC: Custer 1876 Publisher, 2011), 26.

2. American military attacking Indian villages, see Gregory Michno, *Encyclopedia of Indian Wars* (Missoula, Montana: Mountain Press, 2003), 356.

3. For trouble with the mule packs, see Bruce Liddic, *I Buried Custer*, (College Station TX, Creative Publisher Company, 1979), 17 and Graham, *Reno Court*, 471.

4. For not informing Reno of his change of plans, see Hammer, *Custer in '76*, 188. After all, Reno was second in command and had sent two messengers that the Indians were coming out in force. Sending Stabbed would satisfy this delinquency, but it is more anecdotal than factual that Custer sent Stabbed to Benteen. Reno had been promised by Custer that he would be supported by the entire outfit. We do not know what was said between them in the moments preceding Reno's departure; Kanipe had seen them riding together and conversing. While Custer had a noted failure to fully apprise his subordinates of his intentions, it is also possible that in this case, as stated by Bruce Liddic, Custer believed that he could execute an attack well before any messenger would be able to get through. (See Liddic, *Vanishing Victory*, 53.) The subject of Theodore Goldin as a possible messenger was discussed earlier, and his chronic untruthfulness seriously undermines his claims. We also have Sergeant Thomas Harrison of D Company, who wrote that Private Henry Jones of I Company was sent by Custer to Reno. This claim, however, was denied by Jones himself, pointing out that he was assigned to the pack train and nowhere near Custer. See Camp, BYU, roll 1, box 1, folder 20.

5. The Crows, happy to go with Custer, can be found in Custer's letter to Elizabeth, published in *Boots and Saddles*, 275.

6. "That man will stop at nothing" reference, see Graham, *Custer Myth*, 18.

7. Varnum's inquiry to Custer about where Reno was going, see Carroll, *I Varnum*, 64.

8. The men cheering Custer can be found in Hammer, *Custer in '76*, 100; Graham, *The Custer Myth*, 290.

9. Rancher Hank Weibert, author of *Sixty-Six Years in Custer's Shadow,* maintained that Custer watered his horses at the North Fork. But battle student Bill Rini has been out by the North Fork on June 25 numerous times and has also found the fork virtually bone dry. Rini has also spoken to other ranchers and denizens in the area, and everyone residing in the area agrees that there simply was not enough water in the North Fork for more than two hundred thirsty mounts.

10. For Martin riding right behind the General, see Graham, *Custer Myth*, 289.

11. Kanipe stating that Custer rode up the bluffs in that particular formation can be found in Hammer, *Custer in '76*, 92; Donovan, *Terrible Glory*, 220.

12. Martin remembers that Custer was in a big hurry; see Graham, *Custer Myth*, 249; Hammer, *Custer in '76*, 92, 156.

13. Custer studying the valley floor for ten minutes can be found in Martin's testimony; see Graham, *Reno Court*, 349.

14. Martin seeing nothing but dogs barking, see Hammer, *Custer in '76*, 100; Graham, *Custer Myth*, 290.

(Pages 128–131)

15. Low Dog's comments about warriors falling back after the women and children were safe can be found in Graham, *Custer Myth*, 179.

16. Regarding that the Indians were caught napping, William Graham argued that Martin had only been in the United States for three years and was not well versed in that American idiom and probably mistook the intent of what was being said. Others, however, have persuasively argued that Custer did not mean the Indians were sound asleep when he said, "We caught them napping," with napping signifying catching the Indians unaware as opposed to snoozing—although there is no doubt that some Indians were indeed sleeping in late after a long night of festivities. See Graham, *Custer Myth*, 290.

17. Little Soldier said that many of the warriors were out hunting (and this was also alluded to by Mitch Boyer) so early on in the fight that the early defense of the village depended on boys as young as 14, who were raising dust to slow down Reno's charge. See Hardorff, *Lakota Recollections*, 175; Graham, *Custer Myth*, 13, 39.

18. That Custer made sure he knew where Benteen's column was before he had gone downstream, see John Carroll, Editor: *The Benteen–Goldin Letters on Custer and His Last Battle*, (Lincoln and London: University of Nebraska Press, 1974), referencing letter of Goldin to Albert W. Johnson, 48–49.

19. Hank Weibert had spent six and a half decades traveling all over the backcountry on horseback in this area and, knowing the terrain features as well as he did the back of his hand, plausibly asked why Custer would take this route: "There is nothing but rough going down there." While it is true that Custer himself may not have known the terrain and proceeded all the way down Cedar Coulee by way of ignorance, in all probability, Boyer and the Crows probably did know the lay of the land in that vicinity and would have counseled Custer against taking such a route. See Weibert, *Sixty-Six Years in Custer's Shadow*, 46.

20. For a good analysis that Custer did not travel all the way down Cedar Coulee, see Donahue, *Where the Rivers Ran Red*, Appendix 4: Field Research on the Trail of George Custer and John Martin, 273–274.

21. That Custer cut over toward the west to Western Coulee leading to the river, see *ibid.*, 273.

22. Martin telling Benteen where he left with Cooke's note approximately six hundred yards from the river, see Hammer, *Custer in '76*, 105.

23. The Benteen sketch was probably the best of the early maps drawn on the battlefield detailing Custer's line of march. Although the map does not show the Weir Peaks, it can be reasonably assumed that Custer never went over the Peaks. Benteen's scale is also impressively accurate. As to Custer's line of march, Benteen wrote, "When Reno crossed the river to charge the Indians, Custer promised to support him,—instead, he went to his own death by the line marked, as per margin." See Donahue: *Drawing Battle Lines* (El Segundo, California: Upton and Sons, Publishers, 2009), 66.

24. Despite Topographic errors, less than optimum surveying conditions, and lacking the luxury of time to double check his work (never dreaming he would be called to testify on his map at a court of inquiry), Lieutenant Edward Maguire in this early map still manages to successfully depict Custer's advance as it passes Reno's defense area and angles down to the river in the area commonly known today as Medicine Tail Coulee. The Benteen and the Maguire maps, as Donahue acknowledges, were probably drawn only days apart, shortly after the battle. There would also be a third map drawn by Sergeant G. Becker under the supervision of Lieutenant Maguire, which, although undated, is generally believed to have been created some ten weeks after the battle. It was made to enhance and improve the notations of the earlier maps. Despite some slight differences, Maguire erroneously showed Custer going over the top of Weir Peaks (Maguire's earlier map does not even label the Weir Peaks, which cannot be discerned on the map—although a later map shows a partial rendering of them) the maps, when taken together, do provide a plausible narrative of the trajectory of Custer's march to Medicine Tail Coulee via Western Coulee as well as supporting Martin's contention that Custer never rode over the Weir Peaks. Regarding Edward Maguire's map, see Donahue, *Drawing Battle Lines*, 31–36.

25. Martin testifying that Custer climbed the high ridge (Sharpshooter Ridge), see Nichols, *Reno Court*, 394.

26. Colonel John Gibbon also believing that Custer rode to Sharpshooter Ridge, not the Weir Peaks, see "Last Summer's Expedition Against the Sioux," *American Catholic Quarterly*, April 1877.

27. Captain Thomas French's statement that Custer's men "rode over the saddle," see unpublished letter cited in Donahue's *Where the Rivers Ran Red*, Appendix 4, 273.

28. Custer supposedly wanting to be seen on the bluffs, see Sklenar, *To Hell with Honor*, 268.

29. Stating that there was no physical evidence that Custer had gone over the top of the Weir Peaks, see Winfield Edgerly's testimony in Nichols *Reno Court*, 453.

Chapter 8

(Pages 133-142)

1. Walter Camp's interview with Trumpeter John Martin on October 24, 1908, published in Hammer, *Custer in '76,* 100–101: It was probably approaching Sharpshooter Ridge that Martin heard heavy firing to his right or westward toward the valley. This was almost certainly firing from Reno's battalion and not Custer's command. Afterward, quite a few years later, he came to surmise that the firing he heard was at Ford B or the start of the Custer fight. But this was an erroneous conclusion. When Martin got to the top of Sharpshooter Ridge, he looked down and saw Indians charging like a swarm of bees toward the river, waving buffalo hides. This was likely a reference to seeing Indians driving their horses off the upper foothills down into the valley roughly opposite Ford B. Martin would mistakenly believe that the objective of these Indians would be Ford B, as opposed to the more likely objective to drive in the horse herds to the safety of the village. At the same time, he saw what appeared to be Custer retreating up the open country in the direction of the battlefield. In actuality, Custer's command was moving northeastwards into north Medicine Tail Coulee behind the Luce Ridge Complex, keeping well out of sight of the warriors. He said that he did not reveal this new stated version of events at the Reno Court of Inquiry because he was not asked the question.

2. Indians across the river waving blankets, see *ibid.* 101.

3. Bobtail Horse's account was given to George Bird Grinnell on September 27, 1908, and was published in Father Peter Powell's *People of the Sacred Mountain,* 1020.

4. This is from a letter by General Charles Woodruff, who was a lieutenant serving under Colonel Gibbon during the 1876 campaign against the Sioux, that was written to Walter Camp on November 17, 1910, regarding the time between Reno's charge in the valley and the Custer fight at Medicine Tail Coulee. See Walter Camp Collection, LBBNM, with our annotations for clarification and explication in brackets:

> Dear Mr. Camp:
>
> I was glad to answer your questionnaire, I have been intending to write you concerning [Theodore] Goldin—his statement, if true, would damn Reno, for if he did carry a message to Reno, it must, of necessity, have been in effect, "Push through and I'll strike them in the flank."—In regard to time, and I agree with you that the most important of the unknown facts is to account for what <u>Custer was doing during the period between leaving thick Ash or Sundance Creek (i.e., Ash/Reno Creek) and at appearing Ford B</u>—I think it is sure that he halted for over a half hour somewhere. [Note: The Participant Timeline would agree with General Woodruff's statement, however, placing Custer's halt at closer to an hour's delay before attacking at Ford B.]

> I think it was Hall that led General Godfrey into what was, I think, certainly an error about the route. [Note: A reference to General Godfrey's belief that Custer took all 5 troops directly across Luce Ridge to Calhoun Hill, effectively bypassing Medicine Tail Ford (i.e. Ford B-1). This notion was first published by Godfrey in his *Century Magazine* article, "Custer's Last Battle]. [If Trumpeter Martin] left Custer after he reached Medicine Tail [Coulee] and Custer's firing commenced 20 minutes after Reno left the bottom, there is about an hour to be accounted for." [Note: General Woodruff notes that Custer delayed the fight for roughly one hour from the time Martin left with his message to Benteen until 20 minutes after Reno reached the bluffs, The Participant Timeline supports this time estimate supported by a number of participant accounts].

5. Lieutenant Varnum, remarking that Custer would not be able to wait five minutes under such circumstances, see Graham, *Custer Myth,* 323.

6. Regarding the effectiveness of a 'fix and flank', see Ernest B. Furgurson, *Chancellorsville 1863: The Souls of the Brave,* (Vintage Books, A Division of Random House, 1992), see Chapter 11: "Such a Stampede was Never Seen." In our judgment, this is the most complete and riveting study ever written on this famous Civil War battle. It demonstrates both the risks and effectiveness of a daring and well-executed 'fix and flank' attack.

7. Regarding Custer's success at the Battle of Waynesboro in the Shenandoah Valley, see Gregory, J.W. Urwin, *Custer Victorious: The Civil War Battles of General George Armstrong Custer,* (4 Cornwell Drive, East Brunswick, N.J.: Associated University Press Inc.). See also in that account a quote about Custer from Stevenson's *History of the 1st N.Y. Cavalry,* 329.

8. That Custer's eventual attack was long enough from the Reno battle (perhaps a forty-five-minute delay) to mistakenly believe that Custer's and Reno's troopers were one and the same, see Sitting Bull interview in the *New York Herald,* November 16, 1877.

9. Custer releasing the scouts can be found in Graham, *The Custer Myth,* 15, and Hammer, *Custer in '76,* 175.

10. Boyer telling the Indian scouts he was going to fight with the soldiers, see Hammer, *Custer in '76*, 179.

11. With respect to this quote of Curley not coming back if he had crossed Medicine Tail Coulee, see Thomas B, Marquis: *Memories of a White Crow Indian.* (N.Y. N.Y.: The Century Company, 1928), 35. There are numerous other references to Curley being nowhere near the final stages of the battle waging on the east side of the river. Its staying power can be explained that it is simply a good story. The lore about Curley witnessing the last of the Custer battle has been a subject fraught with drama and fascination. The saga of the controversial Curley fits in with the traditional lone survivor mystique, where one individual miraculously survives while all others perish in a valiant last stand. These romantic accounts help fill in all those exasperating gaps in the story for those so anxious to know what happened, who want answers to the mysteries of the event that no one can answer. But Curley, whose accounts are

often woefully contradictory, is not credible—at least for the most part. In some ways, he is the mirror image of Private Theodore Goldin, who prolifically ladled out all kinds of tales in which he would make himself one of the protagonists. Like Curley, Goldin was very young with the tongue of an incorrigible raconteur. Too many others have Curley elsewhere than in the locale by Last Stand Hill. Black Fox said the young brave was with them when he and others watered their horses and could not have been at the Custer fight. Three Crow scouts from Custer's battalion who were with Curley, Hairy Moccasin, Goes Ahead, and White Man Runs Him swore Curley did not ride with them past Reno Creek, see Hardorff, *Camp, Custer and the Little Big Horn,* 59. Also Graham, *The Custer Myth,* 15, 18. Hairy Moccasin remembered that the nineteen-year-old youth had departed from them at Weir Point and that he cleared out of the country, and we did not see him again for fourteen sleeps—in other words, two weeks, see Hammer, *Custer in '76*, 177. There is also a letter from General Hugh Scott to Luther Hare where he says that he had separate conversations with White Man Runs Him, whose honesty he was impressed with, and the Indian did not agree with Curley's movements. see letter from Scott to Luther Hare, November 28, 1919, quoted in O'Neill, *GaryOwen Tidbits,* Vol. 4, 8–9. Curley may have also ridden off with some Arikara trying to capture Sioux ponies at the time of the denouement of the Custer fight (see Graham, *The Custer Myth,* 24). John Burkman, Custer's orderly, saw Curley driving a herd of Sioux ponies far from where Custer would have been, see Hardin, *Tribune Herald*, April 1, 1932. In truth, it is almost impossible to find an Indian who believed Curley was anywhere near Custer during his final moments. This includes Curley who once admitted that he left the battalion well before Custer reached Medicine Tail Coulee, see Marquis, *Custer on the Little Big Horn*, 53–56. Michael Sheridan, who was an aide de camp to his famous brother, General Philip Sheridan, accompanied by Captain Nowlan, led a party to reinter the remains of the dead at Little Big Horn. After an inspection of the battlefield, a small detachment of Crow scouts, including Curley, joined them. Sheridan asked Curley what he had witnessed, and after hearing his story, he knew immediately that it was mostly the product of a fertile, self-supporting imagination, see Graham, *Custer Myth,* 373-374. Finally, Curley was so drawn to the Little Big Horn that for years afterward, he would live on the battlefield, making himself availabl e to sightseers and curiosity seekers. Who among them would not want to talk to the scout who saw the last of Custer and his five companies? From the start, Curley always had a bit of drama about him. Late in the morning on June 27, 1876, the youthful Crow scout was spotted by the crew of Captain Grant Marsh's *Far West* wearing a red Sioux blanket which made Curley stand out. His hair in a pompadour and carrying a Winchester or carbine, forded his way to the steamer, see Hammer, *Custer in '76*, 241. Once aboard he gave way to the most violent demonstrations of grief. No one could speak Crow, so Curley then engaged in pantomime demonstrating soldiers being shot where they all began to understand that some kind of calamity had taken place, see Joseph Mill Hanson, *The Conquest of the Missouri*. This, however, would not be the last time Curley's histrionics would be on display for an audience. What better stage to take his act than the battlefield itself? His imagination soared, in which he embroidered his accounts in rich detail. It surely made for entertaining mythmaking, but there was little sobering history in it. We believe that Curley likely saw the end of the Reno fight in the valley and possibly the beginning and perhaps even the very end of the Custer fight from a distance. For a good treatment of the Curley episode, see Graham, *Custer Myth*, 21–25, and Hammer, *Custer in '76,* 174–179.

12. For John Martin's exchange with Boston Custer, see Graham, *The Custer Myth,* 290, and Hammer, *Custer in '76*, 104.

13. In a fascinating, unpublished account, "Thompson in Custer's Cavalry: 1875–1880," Susan Thompson Taylor, 224, recounts how her father, Private Peter Thompson, of C. Company, told her about his horse badly tiring and lagging well behind the column and finally having to abandon it deep in enemy territory. How he was chased by a mounted warrior and escaped by going "like a bat out of hell with its wings on fire." She relates how her father encountered Private James Watson, also from C Company, at the foot of the bluff near the river. She also relates how her father finally espied Custer and the tale about an Indian scout—some presumed to be Curley—and a struggling Lakota woman being led by him with a rawhide rope. Among historians, the reliability of Thompson's account is mixed. Camp believed that Thompson was mistaken about Custer; see Hardorff's "*On the Little Big Horn with Walter Camp: A collection of Walter Mason's Camp's Letters, Notes and Opinions on Custer's Last Fight,* (El Secundo, Calif.: Upton and Sons, 2002). Camp also reasoned in a letter dated December 20, 1922, that Thompson and Watson did not speak of their escape because the company was filled up with new men in the fall of 1876, who would not understand, and that the old men never said much. That, at least, was the best Camp could do on behalf of Thompson's tale. Additionally, none of the Crow scouts heard anything about Custer lariating a Sioux squaw (a good story to tell around the campfire). In 1974, Daniel O. Magnussen published an annotated edition of the Thompson account and was not sympathetic to its veracity. On the other hand, in a 2007 edition of the account, he concluded that some important remembrances of Thompson were indeed historically viable. Thompson's odyssey is still robustly debated, with Camp saying it cannot be accepted in its entirety. Kanipe remembered Thompson telling him on Reno Hill how his horse had given out in the low country and that he had to fall out. He left the horse and got lost before backtracking his way back to Reno, but he says nothing about Watson, or seeing Custer and a squaw woman, see Walter Camp, folder 15, Brigham Young University Library. The authors of this work are also partially divided regarding Thompson's tale. While both acknowledge the legitimacy of some of what Thompson has to say, we part company regarding the Custer sighting with Rini inclined to largely accept it, stating that it agrees with the timeline and Guarnieri more tentative to accept it in its entirety.

14. For another account regarding the Thompson–Watson saga, see Marguerite Merington, *The Custer Story,* 307–308.

15. For Thompson's heroics on Reno Hill, he was wounded in the hand and made more than one trip for water on behalf of the thirst-ravaged command. Fifteen soldiers, including Thompson, would be awarded the Medal of Honor, see Donahue, *Where the Rivers Run Red,* 216, and National Park Service publication, "Reno/Benteen Entrenchment Trail Guide," a useful booklet detailing the Reno fight.

16. Regarding the near brawl between Slaper and Thompson, see Brininstool, *Troopers with Custer,* 66. Slaper was particularly vexed by impostors who had the temerity to recount their experiences in that fatal field. He felt the same about the claims of those who said, like Curley, they escaped the Last Stand. As the famous Irish playwright Eugene O'Neill said in his play *The Iceman Cometh,* "Man cannot

live without illusion." Trooper Slaper saw Thompson's tale as part of this "make-believe tendency, either being an illusion or a downright lie which precipitated this confrontation."

17. About shots coming from Springfield carbines, there are several accounts, not least of all from Reno, who said "we heard firing in that direction and knew it could only be from Custer," see Carroll, *The Federal View*, 103. Varnum essentially said the same thing, see Graham, *Reno Court*, 356. Benteen said he just heard shots but no volleys, although he heard officers talking among themselves about hearing volley firing, see Graham, *Reno Court*, 357, 360.

18. Boyer intercepting Custer's column to inform Custer that Reno had been defeated, see "Camp Interview," 1910, published in Hammer, *Custer in '76,* 166. According to Indian scouts, "Boyer intercepted Custer's column as they marched toward the river. It was probably then that Boyer told Custer that Reno had been defeated, for Boyer did a whole lot of talking to Custer when he joined him and kept talking while they were riding side by side." Based on timing sequences and subsequent actions, the authors believe this is a highly believable and likely scenario.

Chapter 9

(Pages 143–148)

1. Regarding the difficulty of calculating the Indian population even in a small tribe, see Gray, *Centennial Campaign*, 346.

2. The officers with high warrior counts are Lieutenant Wallace and Captain Benteen. Wallace based his number of nine thousand warriors not so much on his own empirical observations but on information he later received from Indians. He would subsequently reduce the number to four thousand warriors. See Nichols, *Reno Court,* 81.

3. Benteen, in his testimony before the Reno Court, states that at the time of the battle, he estimated some 2,500 warriors, but now believed they were facing eight thousand to nine thousand. The Court Recorder does not pursue this line of questioning as to why Benteen now thought that the warrior force was more than three times what he initially estimated. See *ibid.*, 411. Earlier in his testimony, Benteen says that there were about 1,800 tepees in the village. It was customarily believed that there were two warriors to every tepee, which would have given a total of 3,600 warriors. See *ibid.*, 409.

4. For Godfrey's response to the Indian exodus from the valley on June 27, see Donovan, *Terrible Glory*, 297.

5. Edgerly's characterization of the Indian exodus, see Nichols, *Reno Court of Inquiry,* 780.

6. Reno's remark to Moylan about the size of the caravan of departing Indians and what they were seeing, see Hardorff, *Camp, Custer and the Little Big Horn,* 83.

7. Author's phone conversation with a Battlefield Ranger in May 2017 regarding Indian numbers.

8. Grinnell talks about Indian tactics in warfare, see George Bird Grinnell, *The Fighting Cheyennes,* (New York: Charles Scribner's Sons, 1915), 357.

9. For Benteen's battle encounter with Sioux warriors in Kansas, see Charles K. Mills, *Harvest of Barren Regrets: The Army Career of William Frederick Benteen, 1834–1898,* (Glendale, CA.: Arthur H. Clark, 1985).

10. Michno's Indian numbers can be found in *Lakota Noon,* 294.

11. For Sheridan's bleak description of Indian reservations, see Jim Harrison, "The Road Home," *Atlantic Monthly Press* (New York, 1998), 9.

12. For a discussion about how the village followed the standard practice of striking the lodges in preparation for the flight of women and children to safety, see Gray, *Centennial Campaign,* 353–354.

13. For the U.S. ultimatum for Indians to return to the reservation by January 31, 1876, see Edward Lazarus, *Black Hills, White Justice*, (New York, Harper Collins, 1991), 83-84 and Gray's *Centennial Campaign*, 23–31.

14. Gray discusses the absence of wikiups housing braves, see *ibid.*, 354.

15. Also discussing the number of warriors, see Robert Hughes, "Campaign Against the Sioux," *Journal of the Military Service Institution* (Governor's Island, Volume 18, 1896), 17–18.

16. For the number revolving around the figure of twenty-five hundred warriors, see Reno's Report, Headquarters Seventh U.S. Cavalry, Camp on Yellowstone River, addressed to Captain E.W. Smith dated July 5, 1876; Also see Reno's article printed in the *New York Herald,* August 8, 1876. As for Captain Benteen, consult his letter to the Adjutant General of the Military Division of Missouri, Chicago, Illinois, via Fort Ellis, dated June 27, 1876.

(Pages 149–154)

17. A discussion of Philo Clark's estimate of warrior numbers, twelve hundred lodges, and two hundred wikiups, see Graham, *Custer Myth,* 116.

18. For a discussion on the Indians' proclivity to move away from the death and stench after a battle, see Dixon, *The Vanishing Race,* 170.

19. For William Cross's estimate on the size of the standing village, see Gray, *Centennial Campaign*, 353.

20. On the erroneous multiplier effect on reporting the number of Indian teepees, see General Nelson Miles, *From the Personal Recollections and Observations of General Nelson Miles,* (Da Copa Press), 289.

21. For Gray's interesting and penetrating analysis on Indian numbers, see Gray, *Centennial Campaign,* 335.

22. That the majority of Indian roamers arrived in the village between June 18 and June 25, see *ibid.*, 337.

23. Mark Kellogg's posthumous dispatch, published in the *New York Herald* on July 11, 1876, contains data on how many warriors Sitting Bull had.

24. For Anson Mills' estimate of twenty thousand inhabitants, see *My Story* (Washington DC: privately published 1918), 409.

25. For Hugh Scott agreeing with Anson Mills, see Scott, *Some Memories of a Soldier* (New York, NY: The Century Company, 1928), 49.

26. For Terry's concerns regarding more significant Indian numbers before the start of the campaign. see Terry's letter to Sheridan dated May 15, 1876, which was quoted in Gray's *Centennial Campaign,* 89.

27. For Sheridan's comments regarding hostile Indians not being able to stay together for very long, see *ibid.*, 90.

28. For information pertaining to Gray's calculations regarding the size of the Indian village, see Gray's chart on tribal distribution in *Centennial Campaign*, 356.

29. Godfrey on fourteen-year-old boys as warriors see his narrative in Graham, *The Custer Myth*, 130.

30. Gerard referencing boys fighting as warriors; see Nichols, *Reno Court,* 100.

31. Custer surmising three fighting men for every ten Indians; see Custer interview, May 22, 1875, *New York Herald*, published in *Fates Change Horses* by Gary Stewart and DJ Richardson. As already noted, unanimity of opinion does not exist on this subject, and this includes the authors themselves. Rini, relying heavily on Philo Clark's data, believes that warrior strength on June 25 approached twenty-six hundred. Guarnieri, more conservative, believes that while warriors exceeded two thousand, it was several hundred less than 2,600, although both authors agree that obtaining exact numbers is not possible.

32. Gall quoted on the great spirit presiding over the field of battle, see *Saint Paul Pioneer Press,* published July 1886.

33. For Pickett's quote about the Yankees, see Gabor S. Boritt, *The Gettysburg Nobody Knows* (New York: Oxford University Press, 1997), 19.

Chapter 10

(pages 155–158)

1. For Benteen's "horns of a dilemma" comment, see Hammer, *Custer in '76*, 80.

2. Benteen orders the troopers to draw their revolvers; see Carroll, *Gibson-Edgerly Narratives,* 10, and Graham, *Custer Myth,* 141.

3. For the appearance of the Crow scouts and Half Yellow Face gesticulating to the bluffs, see Graham, *The Custer Myth,* 219.

4. Benteen's decision to go up the bluffs instead of down in the valley can be found in Carroll, *The Federal View,* 106. Benteen testified that he did not cross the river at Ford A because he thought the whole command had been defeated and Ford A was not a suitable crossing, see Graham, *Reno Court*, 357.

5. Regarding Benteen's estimate of nine hundred Indians in the valley, see Nichols, *Reno Court*, 406, and Carroll, *The Federal View,* 166.

6. For a description of Benteen riding to Reno on the hill, see Nichols, *Reno Court, 405.*

7. Benteen moving at a funeral pace, see Brady, *Indian Fights and Fighters*, 404.

8. For an account of an excited Reno absent a hat and wearing a red handkerchief tied about his head, see Graham, *Custer Myth,* 141.

9. Reno said to Benteen, "For God's sake halt your command," see *ibid.*, 101.

10. Benteen's "Where is Custer?" quote, see Hammer, *Custer in '76,* 181.

11. Regarding Porter at Little Big Horn, see L.G. Walker, Jr., *Henry Porter—the Surgeon who Survived Little Big Horn.* (Jefferson, NC: McFarland & Company, 2008), 54–60.

12. Reno saying to Porter, "That was a charge, sir," see Nichols, *Reno Court,* 19.

13. For a description of Benteen describing events that occurred early on at Reno Hill, see Carroll, *Benteen–Goldin Letters,* 170.

14. For Windolph's remarks on the demoralization of the command, see Hunt, *I Fought With Custer,* 96–97.

15. Red Bear's description of Reno, see Libby, *The Arikara Narrative,* 128. This characterization has to be taken with a grain of salt because Red Bear especially detested Reno.

16. Defenses were belatedly erected on Reno Hill as another indication of how unprepared and disorganized the troops on Reno Hill were. When McDougall from the pack train arrived on the hill said he was shocked that he saw no firing lines, and looking north, he said that if the Indians came over the ridge line, the command would have been caught totally unprepared, see Camp, Brigham Young University, roll 5, box 5, folder 5. When Mathey arrived back at the hill's crest he made a barricade of dead animals and commissary and medical stores from the pack train to help protect the troopers from Indian sniping, see Barnard, *Ten Years with Custer,* 308.

17. The nature of the status of the command structure after Benteen reached Reno on the hilltop has created a great deal of discussion. Was Benteen still obligated to follow Custer's orders after he showed Reno Martin's note to "Be Quick?" Reno, after all, had the higher rank, and no one knew exactly where Custer was. In reviewing the subject, General Nelson Miles said 'yes,' that Custer's orders were not subject to override by Reno, see Gray, *Custer's Last Campaign*, 308. Miles believed Custer was not as undetectable as some of those on the hill would later claim. They heard the gunfire; they saw the dust downstream, see Merington, *The Custer Story,* 325. Others have disputed Miles' assessment regarding the command hierarchy on Reno Hill, arguing that when two separate commands unite, the senior officer present (in this case, Reno) is in de facto and de jure command of the two battalions. Benteen himself noted, "not that I didn't feel free to act in opposition to Reno's wishes, and did so act." See Carroll, *Benteen-Goldin Letters,* 246. To read a brief but solid treatment of this subject. see Liddic, *Vanishing Victory,* 119.

18. An account of Reno's procrastination and indecision on Reno Hill, see Ron H. Nichols, *In Custer's Shadow: Major Marcus Reno,* (Fort Collins, CO, Old Army Press, 1999), 190.

19. The troopers thinking that Custer abandoned them as he supposedly did with Joel Elliot at Washita, see Donovan, *A Terrible Glory,* 260. Wallace hearing swearing about Custer leaving them, see Nichols, *Reno Court*, 37.

20. For Reno firing away at Indians an impossible thousand yards away, see Nichols, *Reno Court of Inquiry*, 189.

21. Varnum shooting at out-of-range Indians because of lack of sleep, anger, and frustration, see Nichols, *Reno Court*, 443; also see John D. Mackintosh, *Custer's Southern Officer: Captain George D. Wallace, 7th U.S. Cavalry,* (Lexington, S.C.: Cloud Creek Press, 2002).

22. Varnum reacting to the gunfire downstream and exclaiming, "Jesus Christ, hear that", see Hardorff, *Indian Views of the Custer Fight,* 51. The first set of volleys likely came from the Luce–Nye Cartright Ridges about the time Reno descended the bluff in search of Hodgson.

23. Varnum dismissing the Custer abandonment talk, see Colonel Charles Francis Bates, *Custer's Indian Battles,* (Bronxville, New York: Privately Printed, 1916), 33.

24. For Herendeen hearing the volleys from the woods, see Hammer, *Custer in '76*, 224; Herendeen's account also can be found in *The New York Herald,* July 7, 1876.

25. Davern stating that Custer must be fighting down at the bottom, see Nichols, *Reno Court,* 352, 365.

26. Benteen commenting on Moylan's temporary breakdown, see Carroll, *Benteen–Goldin Letters,* 243.

27. Moylan on Custer's mistake in dividing the regiment, see E.S. Godfrey, *Custer's Last Battle, 1876,* 26.

(Pages 159–164)

28. Reno hearing firing in that direction, see Nichols, *Reno Court,* 574, 625.

29. For Edgerly stating, "There must be General Custer …," see Nichols, *Reno Court,* 444–445.

30. Hare's horse shot, and he borrows another mount to hurry along ammunition packs, see Hammer, *Custer in '76*, 66. See also Carroll, *Sunshine Magazine Articles*, 21.

31. Reno going down the bluffs to retrieve Hodgson's body elicited Wallace's astonishment, asking why the officer in command would not send someone else to do it, see Graham, *Reno Court*, 50.

32. For a good analysis about Reno's state of mind at the time he descended the bluffs to find Hodgson, see Liddic, *Vanishing Victory,* 521.

33. Reno retrieving only Hodgson's ring and some other personal items, see Nichols, *Reno Court,* 566.

34. Edgerly stating "we ought to go down there," see Nichols, *Reno Court,* 443.

35. Indians plotting a trap by getting the soldiers to leave Reno Hill, see Liddic, *I Buried Custer,* 19.

36. For the packer Churchill hearing faint shots, see Nichols, *Reno Court of Inquiry,* 467.

37. McDougall expresses concern about hearing gunfire, see Nichols, *Reno Court*, 236. But Reno, at this time, is mostly consumed and still fretting about the dead Hodgson. Benteen walks away after talking about hearing faint gunshots, see Hammer, *Custer in '76*, 70. McDougall's story, like Edgerly's, varied a bit. He told Walter Camp that he communicated to Godfrey about hearing gunfire but not Benteen. But a source does not have to be exact in every detail to be viable. Clearly, the firing downriver was heard by many, and both Reno and Benteen were obviously aware of it. Reno's only response about McDougall hearing shots was to painfully tell McDougall, "Captain, I lost your lieutenant [Hodgson]; he is lying down there," see Nichols, *Reno Court,* 529.

38. Edgerly tells Weir about going to Custer, see Nichols, *Reno Court,* 444.

39. Edgerly's exchanges with Captain Weir vary slightly. The upshot, however, is that after Edgerly agreed he would move to the sound of the firing with his commander, he was resolved to ask an immobile Reno for permission to do so, see Hammer, *Custer in '76*, 55.

40. The Edgerly January 17, 1886, letter can be found in the New York Public Library, The Hager Collection.

41. Edgerly watching Weir mount and head to the north, thinking he got permission to make contact with Custer, see George M. Clark, *Scalp Dance: The Edgerly Papers,* (Oswego, New York: Heritage Press, 1985), 19.

42. Weir denies he asked Reno for permission to go to Custer, see Hammer, *Custer in '76,* 55.

43. Private John Fox's account of the Weir–Reno account can be found in *Camp on Custer,* 94–95.

44. Fred Dustin's account in *The Custer Tragedy,* 141-142, wherein he describes the angry exchange between Weir and Reno. While Dustin could be melodramatic, the authors believe the gunpowder, in their words, was by no means fueled by what some might think was Dustin's hyperbole. Tensions on that hill were that high.

45. For Reno's explanation of Weir moving north, see Nichols, *Reno Court of Inquiry,* 582.

46. For Windolph's recollection of the Weir-Reno exchange, see *The Sunshine Magazine,* 21.

47. Goldin's recollection of this altercation can be found in the Benteen-Goldin letters. The complete Goldin account of the Little Big Horn Battle was published in the *Army Magazine* in the June and July 1893 issue and published in Gordon Harper's *Fights on the Little Big Horn Companion,* Full Appendices and Bibliography. But also see Brady, *Indian Fights and Fighters,* 275. Initially, Goldin maintained that he witnessed the Weir-Reno argument, with Weir telling Reno, "if you don't go I will …" and that "some will suffer for this." Then, he later would say that he heard some loud talking between the two; he did not know what was said at the time, but later was told by Wallace that Reno and Weir had hot words because Reno had refused to advance until the packs came up. Hence, he said, Weir's actions were insubordinate, and some suggested that Reno placed Weir under arrest, but Reno declined to do so. To get a complete picture of Goldin's remarks, consult Carroll, *Benteen-Goldin Letters,* 89.

48. John Martin saying it sounded like a big fight. See Taunton, *No Pride in the Little Big Horn,* 21–23.

49. Francis Gibson, in a private letter to George L. Yates, a nephew of Richard Roberts, brother-in-law to Captain George Yates of the Seventh Cavalry, who died with Custer, stated that Weir asked Reno, not Benteen, to go farther. This letter was written in 1915 and was sourced at https://the lbha.proboards.com/thread/1872/.

50. Taylor reflecting on the two hours spent on Reno Hill without moving, see Taylor, *With Custer on the Little Big Horn* 48.

51. For Benteen's comment on Weir (in a fit of bravado) heading in Custer's direction, see Brininstool, *Troopers with Custer*, 81.

52. Companies M, K, and H are moving downstream; see Hammer, *Custer in '76*, 56.

53. Reno's trumpeter ignored, see Brininstool, *Troopers with Custer,* 81.

54. Goldin calls Weir moving north an act of insubordination, see Carroll, *Benteen-Goldin Letters*, 89.

(Pages 165–172)

55. Benteen's comments on Custer being able to take care of himself, see *ibid.*, 81.

56. For Reno's comment about 225 soldiers being able to hold off a great many Indians, see Nichols, *Reno Court*, 28. While Reno's assessment is largely correct, this is only providing that their ammunition holds out (the pack trains were nowhere in proximity to Custer) and that the soldiers are in a properly deployed defensive position. In this case, the ammunition pack train was nowhere near Custer's battalion, and Custer likely deployed his troops not in a strictly defensive, survival mode, but rather in anticipation of the arrival of reinforcements.

57. However strong the Indians were, Reno opined, he believed the 7th cavalry could have whipped them if they were handled properly since he (Reno) had stood them off with half a regiment. See Nichols, *In Custer's Shadow: Major Marcus Reno,* (Fort Collins Co.: Old Army Press, 1999).

58. For Recorder Lee's impressions on the contradictory statements by eyewitnesses, see Camp, Lilly Library, Notes, 81–82, Recorder Jesse Lee to Elizabeth Custer, June 27, 1897, Hagner Collection.

59. De Rudio telling Camp that there was a private understanding between a number of officers that they would do all they could to protect Reno. (See, Hardorff, *On the Little Big Horn)* 24. Also, Camp, IU Notes, 102.

60. For Whittaker's literary output, see A. I. W. Crowell: "A Dedication to the Works of Captain Frederick Whittaker, The Prince of Novelists," *Banner Weekly*, November 30, 1895, quoted in Albert Johansen, *The House of Beadle and Adams and its Dime and Nickels Novels*, 2 (Norman: University of Oklahoma Press, 1950), 300–301.

61. As a sample of what some of the major newspapers were saying in terms of placing the entire blame for the debacle on Custer, see *New York Herald,* July 7, 8, 1876, *and New York Times,* July 17, 1876.

62. Frederick Whittaker's biography, *A Complete Life of General George A. Custer.* (New York: Sheldon and Company, 1876), 608.

63. For Elizabeth Custer's reaction to the news about Little Big Horn, see *Little Big Horn Associates Newsletter*, February 1987.

64. For Whittaker's actions in the Civil War, i.e., Battle of the Wilderness, see Hall, *History of the Sixth New York Cavalry,* 318, 349.

65. Whittaker stating the nation demands the vindication of a dead hero (Custer), see ibid., 1–2.

66. Sitting Bull interview in which he mocks Reno's inaction, see *New York Herald,* November 16, 1877.

67. For Reno's demands for an official inquiry, see Hutchins, *The Army and Naval Journal*, 176.

68. Whittaker's interview of Weir, see Shirley A. Leckie, *Elizabeth Bacon Custer and the Making of a Myth.* (Norman: University of Oklahoma Press, 1993), 210–211.

69. Weir fearful that his testimony at an inquiry would be seen as biased; see letter to Elizabeth Custer, November 28, 1876, Leckie, *Elizabeth Bacon Custer and the Making of a Myth,* (Norman: University of Oklahoma Press, 1993), 210–211.

70. Weir hints at disreputable behavior at Little Big Horn; see Shirley A. Leckie, *Elizabeth Bacon Custer, and the Making of a Myth,* (Norman: University of Oklahoma Press, 1993).

71. It was said that the Weir affidavit was secured by O'Kelly but was entrusted to Whittaker. That is false. No mention of a Weir affidavit was ever made at the Reno Court. But, see Katherine Gibson Fougera, *With Custer's Cavalry* (Caldwell, ID: Caxton Printers, 1940), 272.

72. Whittaker telling Elizabeth Custer that Weir will be with them once the fight begins, letter from Whittaker to Elizabeth Custer dated November 28, 1876. In "Brice C. W. Custer Collection," a private collection.

73. O'Kelly stating that the after-action reports were full of inaccuracies, caused astonishment in those who fought in the battle, see James J. O'Kelly, "A Review of Custer's Fight on the Little Big Horn" published in the September 21, 1876, *New York Herald,* and https://lbha.proboards.com/thread/5705/captain-thomas-weir.

74. O'Kelly stating that Weir's C Company went to the Weir Peaks when they realized Custer's command was being done to death, see *ibid.*, September 20, 1876.

75. For a review of James J. O'Kelly's critical remarks about "the long delayed investigation", see "Fight on the Little Big Horn," published in the September 21, 1876, *New York Herald*, and also at https://lbha.proboards.com/thread/5705/captain-thomas-weir.

76. Benteen stating Weir was insubordinate and had sallied out without orders, see Carroll, *Benteen–Goldin Letters,* 89, 198.

77. For Benteen saying that they could have got to Custer but that they would have all been killed, see Graham, *Reno Court,* 360.

78. Lieutenant Edgerly agrees with Benteen that had Reno advanced at once upon Benteen's junction with him, it would have resulted only in the practical annihilation of the regiment, see Graham, *Custer Myth,* 217.

79. For Evans's thesis that things happened too quickly on the battlefield for the officers to react, see David C. Evans, *Custer's Last Fight: The Story of the Battle of Little Big Horn,* (El Secundo, CA: Upon and Sons, 1999), 306–307.

80. General Winfield Scott Edgerly's letter to Colonel W.A. Graham stating that we then stood the horses ready to mount, awaiting the order to advance, December 5, 1923, see Graham, *Custer Myth,* 217.

(Pages 173–177)

81. McDougall assessing Reno's martial qualities, stating that men are different, see Nichols, *Reno Court,* 534.

82. The authors want to restate that while Reno was imbibing alcohol as early as his appearance at Ford A, he did not become noticeably drunk until he was on the hill, where he began to show increasing signs of inebriation. One could quote from a dozen sources of Reno being seen drinking at one time or another on June 25 prior to the charge. Sometime before 10 p.m. that night, Reno was making his rounds in the pack train area as men were hunkering down for the night, possibly in search of whisky, when he ran into citizen mule packer John Frett. Seemingly making an excuse for rummaging about the pack mules, he asked Frett, "Are the mules tight." Frett, not understanding Reno's remark and thinking that the major meant tied instead of tight, asked, "what do you mean by tight?" Reno lost his cool and spat out, "Tight, God damn you," and slapped Frett across the face as whiskey spilled out from the flask he was carrying. Benjamin Churchill, a packer who was with Frett, held him from striking back at Reno. Reno picked up his carbine and threatened to shoot Frett, and Churchill dragged his fellow mule packer from the ugly and potentially deadly scene, and neither of them saw Reno until the next morning. This definitely strikes one as an alcohol-induced incident by Reno. For the Reno–Frett episode, see Nichols, *Reno Court of Inquiry,* 470–471, 505–506. For an excellent summation of eyewitness accounts of Reno drinking, see Donovan, *Terrible Glory.* Notes, 460–462.

83. Indians firing at objects on the ground, see Hammer, *Custer in '76*, 56, Graham, *Reno Court,* Abstract 161.

84. Indians with guidons, see Hammer, *Custer in '76*, 56, 129. Sergeant Flanagan also saw something else: a soldier galloping south in an attempt to escape until the Indians cut him off and killed the trooper.

85. For the Pigford account, see John Upton Terrell, and Colonel Walton, *Faint the Trumpet Sounds*, 143. A careful vetting of the Pigford accounts is necessary. To use an old saying, 'one must not throw out the baby with the bathwater,' Pigford was another who had a tendency to embellish and wax his accounts. He once described himself and several troopers crawling through some ravines to get close to what he clearly implies is the Custer fight on Last Stand Hill, where through the dust and smoke, he sees a little group of white men fighting where he can hear their death throes. Somehow, Pigford and his companions make an utterly miraculous escape. Pigford must have enrolled in some classes at the Theodore Goldin school of storytelling. Pure balderdash. For this Pigford account, see the *Washington Observer* (Washington, PA, October 8, 1932). On the other hand, it is entirely plausible, as it fits comfortably in the narrative of others, that from the summit of the Weir Peaks, he saw a huge black mass of Indians firing at supine objects on a side hill and ridge, obviously Calhoun Hill and Finley Ridge.

86. Weir telling Libbie, "I have so much to tell you," See Lawrence Frost, *General Custer's Libbie.* (Hesperia, CA: Superior Publishing, 1976), 237.

87. For a good discussion of Custer leaving his command to see his wife and its ensuing consequences, see Jeffrey D. Wert *The Controversial Life of George Armstrong Custer.* (New York: Simon & Schuster, 1996), 262–263, and Leckie, *Elizabeth Bacon Custer and the Making of a Myth,* 97–103.

88. For Benteen quoting Lieutenant Brewster regarding Custer's wife and the attention she was getting from Captain Weir, see John M. Carroll, *Custer from the Civil War to the Little Big Horn.*

89. For Lieutenant Edward Mathey reinforcing the concern about Custer's wife with Weir, see Colonel Edward Mathey's interview with Walter M. Camp, October 19, 1910, Box 6, Folder 5, Brigham Young University.

90. For Weir's comment to Libbie Custer about us being alone in the parlor reference with the curtains closed, see Captain Charles E. Merkel: *Custer's Forgotten Lieutenant,* 191–193.

91. Weir's letter to Elizabeth Custer: Little Big Horn Associates Message Board:https//thelbha.proboards.com/thread/1276/captain-thomas-weir.

92. Captain Weir's letter containing this haunting sentiment about "...the order that never came" was written to the family of Lieutenant W. W. Cooke on October 22, 1876. It can be found in a previously unpublished letter in the private collection of George Kush.

Chapter 11

(Page 179–183)

1. Gall recalling the women and children being told to move hastily downstream to where the Cheyenne were camped, see Graham, *Custer Myth*, 88.

2. Regarding the non-combatants running en masse downstream, see John Neihardt, *Black Elk Speaks,* 110.

3. Iron Hawk watching all the women and children flee down the valley and realized that there was a new threat downstream, see *ibid.*, 122.

4. Soldier Wolf seeing women and children moving toward the Cheyenne circle, see Hardorff, *Cheyenne Memories,* 43.

5. The lack of mobility early and even midway into the fight inhibiting the Indians from pressing their advantage, see DeMallie, *The Sixth Grandfather*, 18.

6. Regarding Crazy Horse's war party leaving the village, see Eleanor H. Hinman: *Interviews on the Life and Death of Crazy Horse,* NP, (Gary Owen Press, 1966), 4.

7. A widespread belief exists among the majority of noteworthy battle scholars and authors of this battle, including Bruce Liddic, Michael Donahue, James Donavan, David Evans, Larry Sklenar, Nathanial Philbrick, and many others that, by the time Custer made his way to Medicine Tail Coulee and perhaps even at his first sighting of the village, his first priority was to capture the non-combatants forcing a surrender of the hostiles. We dissent from this orthodoxy maintaining that Custer's first objective, as has always been the case in a military operation, was to destroy the armed combatants and their means to make war (i.e., their camps).

8. Elliot's death and the criticism that Custer did not do enough continued to polarize some in the 7th all the way up to the battle of Little Big Horn. See Godfrey, *Some Reminiscences including the Washita Battle*; also see *Cavalry Journal* 37, No.153, October 1928, 493.

9. Custer's remark about finishing them up and then go to our station, see Nichols, *Reno Court,* 395; and Graham, *Custer Myth, 290.*

10. Edward Curtis, "The American Indian," published in the *New York Herald*, November 10, 1907. Curtis was a celebrated photographer who compiled a massive twenty-volume work about the American Indian, 115.

11. This reference about dispersing power among your command, whether a Roman legionnaire..., see Clair Conzelman from the Little Big Horn Associates Message Board: https://thelbha.proboards.com/ thread/374/cavalry-tactics-small-units.

(Pages 184–187)

12. Kill Eagle's quote that the soldiers never came within six hundred yards of the river crossing can be found in *The Custer Myth,* 53.

13. For Gall's comments on how close the soldiers came to the crossing (not within a half mile), see Asher L. Burdick, *David F. Barry's Indian Notes on the Custer Battle,* (Baltimore MD: The Proof Press, 1937) 27.

14. For Foolish Elk's account that the soldiers only fired from their horses on the east side of the river and did not cross, see Camp, Folder 54, Brigham Young University.

15. He Dog saying not much firing between the fifteen to twenty Indians on east side of the river battling with the soldiers, see Hammer, *Custer in '76*, 207.

16. Fred Gerard testified that the miry bank was eighteen inches higher than the water. He concluded that he knew it must be miry, or the Indians themselves would have used it. See Nichols, *Reno Court*, 138.

17. For Two Moons account that the soldiers did not come down to the river, see Dixon, *The Vanishing Race*, 181.

18. Respects Nothing also says that the soldiers never reached the riverbank, see Hardorff, *Lakota Recollections of the Custer Fight,* 145. As with virtually every aspect of this battle, other than the Indians won and the 7th lost, there are those who believe Custer crossed the river and was driven back. But the weight of evidence is that Custer did not attempt to cross the river and probably none of his companies got within several hundred yards of the crossing. Thunder Hawk is the only Indian account, out of some 225 extant accounts, that supported a crossing at Ford B and a direct attack on the village. Thunder Hawk told interviewer Eli S. Ricker that "Custer crossed the river at the mouth of Reno Creek [aka Medicine Tail Coulee at Ford B-1] east [actually south] of Calhoun Hill, that he [Custer] swept down with all his command along the river on the west side and passed by his tent [tepee], and he was shooting into the Indian camp, his men doing so wildly. Evidently frightening but not trying to hit the Indians"—which, frankly, does not make much sense. For Eli Ricker's interview, see The Settler and Soldier Interviews of Eli S. Ricker, 1903–1919, published in *Voices of the American West,* 309.

19. Regarding the Indians' perception of soldiers moving down to the crossing as an attack rather than a demonstration, see Richard Hardorff, *Markers, Artifacts, and Indian Testimony: Preliminary Findings on the Custer Battle* (Short Hills, N. J.: Don Horn Publications, 1985), 10.

20. Benteen's July 4, 1876, letter to his wife is published in Graham, *The Custer Myth,* 297–300.

21. Regarding the river crossing by Custer, see John M. Carroll, Editor, *The Federal View,* 105. Benteen furnished two different accounts about whether Custer crossed the river at Ford B. One of Benteen's accounts said that most of Custer's command had crossed the Little Big Horn and had gotten into the village, "but were driven out immediately—flying in great disorder," see Hunt, *I Fought With Custer,* 189. The other account stated that Custer had never reached the ford but did get within three furlongs of the river before turning north, see Brinistool, *Troopers with Custer*, 23.

22. The gathering of additional Indian accounts (225 in total) since Walter Mason Camp stating that none of Custer's troops crossed the river into the village, see Bill Rini's annotated Indian accounts. This is the most extensive recounting of Indian accounts about the Little Big Horn battle extant. See LBHA Message Board at: https://thelbha.proboards.com.

23. Regarding an officer with field glasses and a wooden box, see Brust, "Lieutenant Oscar Long's Early Map Details, Terrain," *Greasy Grass,* Volume 11, May 1995, 8. See also David Humphrey Miller, "Echoes of the Little Big Horn," *American Heritage*, 22, no. 4, June 1971.

24. Regarding the supposition that an officer was shot leading a charge at the river crossing, much of it derived from oral Indian history, can be found in David Humphrey Miller, *Custer's Fall* (NY, NY: Dorsett Press, 1988), 128.

25. That Custer was allegedly killed in action, dying in the waters at Ford B, see Frank B. Linderman, *Pretty Shield,* (NY, NY: The John Day Company, 1972).

26. White Cow Bull account claiming that Custer was not only shot, but that he shot him, see Michno, *Lakota Noon*, 223–224.

27. Perhaps the first to surmise that Lieutenant Jack Sturgis was killed charging across Ford B, see James Donovan, *Terrible Glory,* 266, The authors, however, respectfully disagree for reasons expressed in the chapter.

28. For a description of the colorful Myles Keogh, see the letter from Ed Luce to Charles Kuhlman, January 16, 1939, Box 1, Kuhlman Collection, John P. Langellier, Kurt Hamilton Cox, and Brian C. Pohanka eds. Also see *Myles Keogh: The Life and Legend of an Irish Dragoon in the Seventh Cavalry* (El Secundo, CA: Upton and Sons, 1999); and Leakie, *Elizabeth Bacon Custer*, 35-37.

29. Regarding the shoe and bloody leggings found near the riverbank, see Captain E.A. Garlington Papers, Cromby letter to E.A. Garlington, June 10, 1891, John M. Carroll Collection as referenced in Michael Donahue's *Where the Rivers Ran Red,* 174. Donahue acknowledges the courtesy of George Kush for furnishing the Cromby letter.

(Pages 188–193)

30. John Stands In Timber saying that the firing at Ford B was not very long and few, if any, soldiers were killed, see Liberty, Margot. *Cheyenne Memories*. Lincoln: University of Nebraska Press, 1967, 197-198.

31. For a breakdown of the companies under Keogh and Yates, see John Gray, *Custer's Last Campaign,* 360.

32. McClernand seeing a set of fours meaning two companies going down Medicine Tail Coulee, see Hardorff, *Camp, Custer, and The Little Big Horn,* Undated interview between General Edward J. McClernand and Walter Camp, 99.

33. With regard to the movements of E Company and it being used as Custer's Headquarters that day, it should be noted that it was not always attached, as many believe, to F Company. In truth, E Company was spotted just about everywhere on the Custer battlefield—at Ford B, Luce Ridge, Last Stand Hill, and The Basin area by Cemetery Ridge. But they could not have been at all these places unless they were operating as two platoons, which we believe they were. What is uncertain and subject

to speculation is exactly how and in what strength they were split up. The authors believe that Custer took E Company to reinforce whichever of his two battalions needed support. This would also give him extra support to travel across greater distances than he might otherwise attempt with only Headquarters Staff. If Keogh took his battalion down Medicine Tail Coulee, then it makes perfect sense that Headquarters Staff and E Company would have attached themselves to his (Keogh's) battalion. The unanswered question is whether Custer sent all of E Company down with Keogh or just a platoon, which would amount to half a company, leaving F and L Companies on Luce Ridge in reserve, with possibly a platoon attaching itself to Yates.

34. For an interesting treatment of Private Gustave Korn and his story, see Albert Winkler, "The Case for a Custer Battalion Survivor: Private Gustave Korn's Story," *The Magazine of Western History,* Spring 2013, Montana, 49–50.

35. That Korn was one of the bravest men in the regiment, see Camp, DPL no. 6; See also *National Tribune,* Washington D.C., January 15, 1891.

36. The Korn episode should be looked at through the same prism that Fred Dustin said we should look at the Peter Thompson episode, meaning that the story in all its parts might be unreliable, but it may also furnish corroborative facts that might otherwise not be obtained. See Fred Dustin's letter in *The Benteen–Goldin Letters,* 116.

37. Custer's troops almost certainly decided not to approach Ford B-1 because the west side (where the village was) was heavily wooded and more likely to conceal enemy combatants.

38. Another question that is addressed about Luce Ridge is regarding the firing that was heard in that location by the Benteen/Reno wing. While a look at a map of that area will show that the only place where firing could have been effective, providing the Indians were in force (which we do not think they were) pushing the soldiers up Medicine Tail Coulee, was at the egress of one of the coulees at Medicine Tail Coulee otherwise they were just wasting ammunition. We do not necessarily subscribe to that view since firing at Luce Ridge could have been because they were providing cover for the retreating companies moving back from Ford B, shooting at Wolf Tooth's band in North Medicine Tail Coulee, whose sudden appearance would be a threat to Captain Yates, or it was a distress signal to Benteen, or, if not a distress signal, a signal indicating the rest of the regiment's location. Or the sound of firing could have been for all these reasons.

39. Yellow Nose hearing music, see "Interview of Yellow Nose" appearing in *The Big Horn-Yellowstone Journal* 1, no.5, Summer 1992, taken from an article in *Chicago Record Journal.* The interview was conducted in November 1905 with a then-blind Yellow Nose.

40. For Mrs. Spotted Horn Bull/Pretty White Buffalo Woman hearing the music, see Graham, *Custer Myth*, 86.

41. White Shield calling the gray horses pretty white, see Hardorff, *Cheyenne Memories,* Grinnell interview with White Shield, 1908, 52.

42. Gray Horse Company getting close to the river, see Hardorff, *Cheyenne Memories*, 51–52.

43. John Gray, author of *Centennial Campaign* and *Custer's Last Campaign: Mitch Boyer and the Little Big Horn Reconstructed,* (Lincoln: University of Nebraska Press, 1991), was the first to posit the theory that Custer's noisy descent to Ford B was not a charge but a feint.

44. Short Bull thinking there must be a million soldiers on the hills, see John M. Carroll, *Eleanor Hinman: Interviews on the Life and Death of Crazy Horse,* 19, 41.

45. Runs the Enemy sees soldiers filling the hill, see Dixon, *Vanishing Race,* 174.

46. Red Horse said the warriors turned from Reno to save non-combatants, see Graham, *The Custer Myth,* 60–61.

47. Crow King stated that the only reason warriors confronted Reno was to give women and children time to get to safety, see Judson Walker: *Campaigns of General Custer in the North and West and the Final Surrender of Sitting Bull*, 102.

48. Regarding soldiers moving down to the crossing for a demonstration, see Richard Hardorff, *Markers, Artifacts, and Indian Testimony: Preliminary Findings on the Custer Battlefield*, (Short Hill, New Jersey: Don Horn Publications, 1985), 10.

49. Of course, there is no documentation or source whatsoever on exactly what kind of music Custer's battalion played. It possibly could have been "Garryowen," which Custer played at the last Yellowstone camp and was, after all, Custer's favorite regimental song. So, it would seem "Garryowen" would be as good a guess as any.

50. That the fight at the river was short, see Hardorff, *Lakota Recollections of the Custer Fight,* 145. Also see Liberty, *Cheyenne Memories*, 197–198. Kill Eagle said Custer never got to within six hundred yards of the river, see Graham, *The Custer Myth,* 53; Shave Elk said the fight there was just a little while but not long. See Liddic and Harbough, *Camp on Custer*, 122. Respects Nothing said Custer did not come to the river (meaning the river crossing)—he went from the high ridges to Calhoun Hill, see Hardorff, *Lakota Recollections of the Custer Fight,* 31. Two Moons said the soldiers never got to the river, see Dixon, *The Vanishing Race,* 181. Waterman, an Arapaho, said the troops were trying to cross the river, but the Indians drove them back, see Graham, *Custer Myth,* 72. Godfrey agreed, not because he witnessed the fight at Ford B, but because he saw Custer's trail on the high ridge back of the field, see Graham, *The Custer Myth,* 106.

51. Indians thinking the soldiers were shooting at their own men, see James Brust, *Lt. Oscar Long's Early Map Details Terrain, Battle Positions*, 8.

52. Regarding Custer never turning his battalion away in the face of such weak opposition, see Schoenberger, *End of Custer,* 146.

53. The line of sight from east ridge to the bluffs where the Indians could see Custer's flankers and Custer's flankers could see the Indians on the bluffs, see Michno, *Lakota Noon,* 108. Custer's

objective might have been that by moving downstream, he was forcing the warriors to follow him and to trap them between his column and that which would ultimately consist of reinforcements from Benteen and Reno—a classic 'fix and flank' maneuver.

54. The view that Custer was moving downstream in an attempt to trap the warriors between his own and the anticipated Benteen/Reno reinforcements, see Stewart, *Custer's Luck,* 447.

55. Regarding the encounter with Wolf Tooth and an exchange of fire, see John Stands In Timber, *Cheyenne Memories,* 197–198. See also Fox, "Chapter 11: The Cemetery Ridge Episode," *Archeology, History, and Custer's Last Battle,* 1997.

56. Kate Big Head says that after the attack on Ford B, there was a long period of slow fighting, see Marquis, *Custer on the Little Big Horn,* 85—86.

Chapter 12

(Pages 199–208)

We quote Dr. Charles Kuhlman, from his book *Legend into History*, 175, because it is relevant to the subject matter of this chapter and sheds light upon how a military unit acts and behaves during a battle, and this can provide some insights on the Keogh fight, We have taken the liberty of quoting him here: "If we wish to describe a battle intelligibly, we must begin at the beginning and follow the action step by step to the end. Among civilized nations, battles are fought by troops organized in definite units whose formations and manners of movement are known. The Custer battalion was composed of five troops or companies, each commanded by two commissioned officers assisted by a number of sergeants and corporals. A troop might act as a unit or two separate platoons. The basic unit in all military evolutions was the set of fours, except in actual fighting in the skirmish line. Under no circumstances, except stark panic or the death of every commissioned or non-commissioned officer, would either the larger or smaller units scatter to wander aimlessly. When they went anywhere, they did so for a reason and under orders. In the course of action, they might be crowded back or rush forward without specific orders, but they would continue to act as a unit, not as a rabble."

Using the above as a preface, we continue sourcing our data.

1. While there are no primary sources or accounts that detail the exact movements of Custer's companies, the archeological evidence around Luce Ridge and the timing of the volleys heard by those on Reno Hill and the timber area is enough to draw a conclusion about the heavier firing at this point and at this location. See Hardorff, *Lakota Recollections*, 143, 155.

2. Regarding Dose's death, see Hammer, *Custer in '76*, 134. Private Henry Petring, a friend of Dose and therefore capable of identifying his remains, told Walter Camp the circumstances of his death. Nevertheless, other locations for his death have also been given: See Hardorff, Custer Battle Casualties, 111–112.

3. Sergeant Thomas Harrison is the source in question, and his interview appeared in the *Arkansas Gazette* published in 1905, also appearing in Margot Liberty's *Cheyenne Memories.* The authors synopsize the interview with their annotations:

> **Harrison:** Custer soon saw that he had caught up with a large body of warriors, and about noon the fight between them began. [Note: Harrison's times, like nearly all the estimated times given by enlisted men, are likely local sun times. Headquarters watch time would be about an hour later, in this case, about 1 p.m., which was the general start of the Reno Valley fight.]
>
> **Gazette:** And you were sent back for reinforcements?
>
> **Harrison:** Yes, I was sent back to Benteen, about a mile and a half, to tell him to bring up his battalion to support Custer. [Note: This distance suggests that Harrison left the Custer wing in the vicinity of either Medicine Tail Coulee, Butler Ridge, or Luce Ridge.]
>
> **Gazette:** Why did you fail to follow his instructions?
>
> **Harrison:** When I got to Benteen I found him pretty busy. Reno had been beaten and formed a junction with Benteen, and both of them had their hands full. They could neither advance nor retreat. We could hear the firing on the right, but toward 3 in the afternoon, the firing where Custer had been became infrequent. [Note: It appears that Harrison would have reached Reno Hill sometime between 3:15 and 3:30 p.m. Harrison's reference to hearing the firing from Custer's field— if true, which is likely a reference to the fighting around the Keogh position— becoming infrequent "toward 3 in the afternoon" and this would convert to about 4 p.m. from Headquarters watch time. It is interesting that the last heavy firing reportedly heard by Captain McDougall occurred about 3:45 p.m.]. Not surprisingly, whether Harrison can be believed has been the subject of debate. Even the authors, while generally concurring with one another, find that this question represents one of the few times there lies a difference between them. Rini believes there is enough in the Harrison interview, supplemented with a logical necessity for Custer to gather his reinforcements, to accept Harrison's story. Guarnieri is more dubious, believing the lack of corroborative evidence of Harrison's story from the soldiers on Reno Hill is indicative of its general unreliability. The reader is invited to form his own conclusions on the matter.

4. For Custer's fateful move linking up with Keogh on Calhoun Hill eliminated any possible return by him (Custer) to link up with either Benteen or Reno, see Wagner, *Strategy of Defeat,* 136, Also see Andrew Paul Hutton, *"The Custer Reader"* (Lincoln, NB: University of Nebraska Press, 1992), 209–210.

5. For a good synopsis of the Yellowstone fight in 1873, see Andrew Paul Hutton, Chapter 6, "The Custer Reader" (Lincoln, NB: University of Nebraska Press, 1992), 209–210.

6. Regarding organized military units trumping superior numbers, "whether it is the Roman

Legion versus the Germanic tribe …" is from Claire Conzelman.

7. For Indians retrieving their horses and coming up in large forces, see Hammer, *Custer in '76*, 197-198.

8. For Young Two Moons quote of Indians rushing out like ants, see Michno, *Lakota Noon,* 215.

9. For a solid background on the strategic thinking of securing Calhoun Hill and basing future operations there, see Liddic, *Vanishing Victory,* 113. Also sharing the same view regarding Calhoun Hill was Michno, *Lakota Noon,* 107–108.

10. For the popular and prevailing belief that Captain Keogh chose to keep I Company relatively inactive in a reserve role east of Battle Ridge, see Rini, *Crow's Nest.* The Journal of the Custer Association of Great Britain, Volume 20. January 2020–December 2021.

11. For an account of the conversation regarding the ammunition packs between Reno and Mathey when the latter first reaches the hill, see Nichols, 529.

12. On Lieutenant Mathey encountering Lieutenant Hare, see Nichols, *Reno Court*, 515.

13. Regarding the two ammunition mules that were cut out of the advance pack train to speed up the arrival of the ammunition to Reno Hill, see *ibid.*, 514.

14. Regarding the deployment of I and C Companies as well as the heated fighting taking place in Keogh's Swale, see Margot Liberty, *A Cheyenne Voice–The Complete John Stands In Timber Interviews* (Norman, OK: University of Oklahoma Press, 2006), 444.

15. Professor Charles Kuhlman remains one of the most analytical students of the battle. And even though some of these insights were given more than seventy years ago, encompassing likely misunderstandings (e.g., the existence of the South Skirmish Line, for example) or when we do not wholly agree with his conclusions, his discernments and penetration into the subject matter can still provide valuable information. See Kuhlman, *Legend into History*, 117–119.

16. For Keogh (in Kuhlman's interpretation, Custer) mistaking Weir's intentions to traverse to the east side of the river, thereby breaking up his hedgehog defense. Kuhlman's basic premise, providing one substitutes Keogh for Custer, makes a great deal of sense, and the timing for such a scenario fits well into the timing of events and explains Keogh's decision to send C Company to a position that dangerously extended his perimeter defenses. See our Appendix A or https://thelbha.proboards.com/board/62timelines-little-big-horn-battle. Also see *ibid.*, 181.

17. That there was a direct line of fire on Finley Ridge to lay down enfilade fire into Deep Coulee to drive off heavily concentrated hostile elements at the base of Calhoun Hill, see Michno, *Lakota Noon*, 225–231.

18. For Keogh deploying a detachment of C Company troops on Finley Ridge, see Rini, *Crow's Nest. The Journal of the Custer Association of Great Britain*, Volume 20. January 2020–

December 2021.

(Page 209–215)

19. For Kate Big Head *(*also known as Antelope) saying the warriors were creeping closer and closer to the soldiers, see Marquis, *Custer on the Little Big Horn,* 87.

20. For an excellent account by a primary source, Kate Big Head, who recounted not only what occurred with the non-combatants in the village but also saw swaths of the battle itself. She had begged her elder brother, White Bull, for a horse. She sang for her nephew, Noisy Walking, that he would be brave and spared in the battle. Kate Big Head was incredibly brave; she had also seen and been in other battles, including Washita; she liked to see the men fighting and was often teased about it. See Thomas B. Marquis, Custer on the Little Big Horn: "She Watched Custer's Last Battle," (Lodi, CA: End-Kian Publishing, 1980), 37–38.

21. Yellow Nose yelling charge can be found in the George Bird Grinnell Collection, Grinnell Field Notebook, The Braun Research Library, Southwest Museum in Los Angeles, 1908, 348.

22. For evidence that places Lame White Man's attack at neither the extreme north nor south end of Battle Ridge but closer to the middle of Battle Ridge, leading to a mistaken model of where White Lame Man actually died, see Rini, *Crow's Nest. The Journal of the Custer Association of Great Britain*, Volume 20. January 2020–December 2021. Gregg Michno, who is a deeply analytical scholar of this battle, had a different location for Lame White Man's death. Michno's conclusions are always interesting and sometimes profound. He again does himself credit in analyzing the Keogh fight, though we ultimately depart from some of his deductions. In short, Michno's *Mystery of E Troop* offers a solid analysis with supporting documentation as to why Lame White Man's attack on Battle Ridge likely did not occur along Calhoun Coulee at the extreme south end of Battle Ridge as suggested by Dr. Richard Fox in his book Archeology, History and Custer's Last Battle. After effectively undermining Fox's thesis in this regard, Michno then reverts to an older model that claimed Lame White Man's attack originated from Deep Ravine. So far, we agree. However, Michno would have Lame White Man's charge veer to the left in the upper reaches of Deep Ravine, attacking troops deployed—Michno would identify this troop as E Company. However, that is speculative. This may just as easily have been a detachment of F Company—on the low Basin Ridge, which some have referred to as the South Skirmish line, at the north end of Battle Ridge. There is evidence, however, that Michno did not consider, and that evidence places Lame White Man's attack at neither the extreme north nor south end of Battle Ridge, but rather, closer to the middle of Battle Ridge, albeit closer to Calhoun Hill than Last Stand Hill where his body, along with that of Noisy Walking, were actually found. To reach that point of attack, Lame White Man would have charged out of the upper reaches of Deep Ravine after following the ravine as it bends to the right (i.e. southwards) rather than into The Basin area, as Michno proposes in this model—not considered by Michno in his book—Lame White Man's attack would be directed against elements of Keogh's battalion deployed on the west side of Battle Ridge at a point where two markers from Keogh's men are found today on the west side of Battle Ridge, just above the death sites marked for Lame White Man and Noisy Walking. This would be located at a point west of

Battle Ridge and just north of Harrington Ridge. See Michno, *Lakota Noon*, "Taking Calhoun Hill," 225–229.

22. Lakota Warrior whose lower jaw was shot away, see Margot Liberty, *A Cheyenne Voice*,

400.

23. For Indians, amid the smoke and dust, mistakenly killing each other, see Dixon, *Vanishing Race*, 177.

For the attack that uprooted Keogh's defensive position, see Rini, *Crow's Nest. The Journal of the Custer Association of Great Britain*, Volume 20. January 2020–December 2021.

24. Crazy Horse seen as a leader in the fighting as bullets whizzed harmlessly by him. This source is from the Brulé, Paul Noisy, given to Joseph Masters at the Rosebud Indian Agency, South Dakota, July 25, 1936, published in Richard Hardorff's book, *Indian Views of the Custer Fight.*

25. Red Hawk administering a fatal blow to the soldier who shot at Crazy Horse by crushing his skull with the soldier's own carbine, see Red Hawk interview, July 19, 1907, Edward S. Meany Papers, Box 13, Folder 20, University of Washington Special Collections.

26. For Red Horse witnessing the final destruction of Keogh's battalion in the swale, see Graham, *The Custer Myth,* 62.

27. Two Moons quote, "like water around a stone," see Graham, *The Custer Myth*, 102.

28. Calhoun being identified by the filings in his teeth, see Hammer, *Custer in '76*, 63.

29. Regarding the condition of the corpses of Calhoun and Crittenden, see *ibid.*, 58.

30. For soldiers lying in one great cluster of death, including First Sergeant Edwin Bobo, First Sergeant Frank Varden, and others, see Rini, *Crow's Nest. The Journal of the Custer Association of Great Britain,* Volume 20. January 2020–December 2021.

31. For information on Commanche's fate after the battle, see Donahue, *Where the Rivers Ran Red,* 235.

32. Mention of Keogh's Last Will and Testament, see *New York Herald, July 26. 1876,* quoted in James Wengert, Manhattan, *The Custer Dispatches,* (KS: Sunflower University Press, 1987), 55.

Chapter 13

(Pages 225–231)

1. Calhoun telling Custer when the time of testing comes he would not be found wanting, see James Calhoun's letter to Custer dated April 23, 1871 found in Thom Hatch's *The Custer Companion,* 261.

2. That Custer's men were together at all times and not any part of them became detached can be found in Camp, DPL No. 6. Some Indians did say there were two distinct bodies of soldiers, see Hardorff, *Lakota Recollections of the Custer Fight*, 142–143.

3. John Stands In Timber saying Custer heading toward Willy Bends (this was a cabin owned by a Crow named Willy Bends in the 1920s and has since been referred to as such, August 8, 1956, Folder 15878), see Don Rickey, "John Stands In Timber Interview," Custer Battlefield.

4. Sheridan attributing the disaster to Custer's superabundance of courage can be found in Graham, *The Custer Myth*, 117.

5. Custer, Headquarters Staff, E and F Companies leave Keogh and head north. Fox, *Archeology, History and Custer's Last Battle*, 297.

6. Analysis of Indian accounts and archeological findings strongly suggest that the left squadron under Yates, accompanied by Custer and Headquarters Staff, moved from Calhoun Hill, through the Keogh sector past Last Stand Hill, and continued in a westerly direction toward the river north of the village site can be found in Fox, *Archeology*, 173–194.

7. See also Sioux and Cheyenne accounts from Runs the Enemy, Two Moons, as well as Lieutenant William Philo Clark's interviews and his map of the Little Big Horn Battle. Runs the Enemy: LBHA Proboard: https://thelbha.proboards.com/thread/1806/runs-enemy-kettles/. Two Moons: *Drawing Battle Lines*, by Michael Donahue, 198-204. Philo Clark: *Drawing Battle Lines*, by Michael Donahue, 97-104.

8. Discussing the path Custer took as he moved north away from Keogh, see Douglas Scott and Peter Bleed, *A Good Walk Around the Boundary,* N.P, National Association of Professional Archeologists and the Nebraska Historical Association, 1997.

9. Scott's doubts about Yates's squadron moving west and getting as far as the river can be found, see *ibid.,* 42.

10. Regarding how contemporary writers and historians scrutinize or avoid scrutinizing Custer's hostage-taking, see Donahue, *Rivers Run Red,* 54.

(Pages 232-238)

11. Magpie stating Indians would have attacked the soldiers if not for the women and children, see Hardorff *Washita Memories,* (Norman: University of Oklahoma Press, 2006), 308–309.

12. Moving Behind Woman also recalling the fear of shooting their own women and children, see Theodore A. Ediger and Vinnie Hoffman, "Some Reminiscences of the Battle of Washita," *Chronicles of Oklahoma*, Summer 1955, 139–140.

13. Sergeant John Ryan recalling that when Custer took squaw prisoners and put them on the side of the camp, the Indians would cease shooting at the soldiers, see Barnard, *Ten Years With Custer,* 79.

14. Custer speculating on why the Indians did not shoot at his troops, see Custer, *My Life on the Plains,* 372–373.

15. The risks of small detachments chasing down Indian non-combatants, see Lorne Langley, "The Intertwining Tales of Major Joel Elliott, Clara and Willie Blinn," *Little Big Horn Research Review*, Vol. 30, 2016, 23.

16. On the ordeal and tragedy of Clara H, Blinn, see Judith P. Justus, "The Saga of Clara H. Blinn at the Battle of Washita," *Research Review*: The Journal of Little Big Horn Associates, Volume 14, no.1, 2000, 11.

17. For Elliot's quote 'a brevet or a coffin,' see Godfrey, *Battle of Washita,* November 27, 1868, Collection S-314.

18. Indians contemplating a battle are always anxious to have their women and children removed from danger, see Custer, *My Life,* 497.

19. For Custer stating he is not impetuous or impulsive, see David Nevin, *The Soldiers,* (Time-Life Book), Time-Life Publishers, January 1, 1973, 199.

20. Yellow Nose's comment that the Indians that day also suffered from lack of water can be found in Eli S. Ricker Volume 1, *Voices of the American West, 1903-1909.*

21. General Stanley's last-minute arrival at the Yellowstone, see Hutton, *The Custer Reader,* 218–219. See also *New York Tribune*, September 8, 1873.

22. Custer attempts to do too much with too little, see Gray, *Centennial Campaign,* 183.

Chapter 14

(Pages 239–242)

1. For Lieutenant Gibson saying they could not aid Custer because they could not abandon the wounded, see Katherine Gibson Fougera, *With Custer's Cavalry,* (Caldwell, ID: The Caxton Printers, 1942), 269.

2. For the most controversial of all subjects being the failed attempt of Reno going to the aid of Custer, see Kuhlman, *Legend into History,* 97.

3. Lieutenant Edgerly stating that if Reno had advanced upon Benteen's junction with him, the regiment would have been annihilated, see Graham, *Custer Myth,* 216-217.

4. Mathey's response that they didn't know what they were waiting for on Reno Hill, see Nichols, *Reno Court*, 515.

5. The question of authority on who was actually in command on Reno Hill is critical in terms of obeying commands. This controversy created a forensic menu for battle students who were punctilious in reviewing every nuance on the proper chain of command. While Weir asked Reno for permission to go to the sound of the firing, it is not entirely clear if it was within Reno's authority to grant or deny permission. Weir was part of Benteen's battalion, and Benteen did have a written order to "come on and be quick." Therefore, Weir could properly refuse to obey the orders of Reno, regardless of rank—see Kuhlman, *Legend into History,* 98. In other words, one knows where Benteen's obligation began in initially following the order, but the area is gray on where the obligation ended. Having ignored Reno's repeated trumpet calls to halt the advance, Benteen clearly acted as if he had an independent command, see Carroll, *Benteen–Goldin Letters*, 215.

6. That there were three separate troop movements to the Weir Peaks underscores what is perhaps the most controversial topic of all: Reno and Benteen's attempt, or lack thereof, to go to Custer's aid. The Reno Court of Inquiry's attempt to resolve this issue is fraught with ambiguity and dissimulation. Although it was only Reno's conduct that was under investigation, and it was Reno himself who had demanded an inquest to exculpate himself from charges of incompetency and cowardice, the effect was that the whole regiment was put under the microscope. The sum of the statements and testimony of those who served in the 7th were, not infrequently, concealed behind the most general and indefinite declarations as if to shield the reputation not only of Reno but of the Regiment for not acting more decisively when continuing fire was heard downstream and everyone seemed to know that the engagement that was taking place could involve no one other than Custer. For a look behind some of the psychology involved in making these decisions, see Eric Bacopulos, *Captain Frederick Benteen's Response to Custer's Last Message*, 7th Annual Symposium, Custer Battlefield & Historical Museum Assn., 1993, 39.

7. Ryan testifying that it appeared the Indians were disposing of Custer's wounded men, see Barnard, *Ten Years*, 297.

8. There were initially seven wounded and eventually no more than ten (not all of them seriously) by the time Weir started to make his way to the sound of the firing. McDougall said the same, that seven men were wounded. See Nichols, *Reno Court of Inquiry Proceedings*, 354. In his official report, Reno said seven to ten wounded, see Overfield, *Little Big Horn* 1876, page 45. Varnum also gives the same number, see Nichols, *Reno Court of Inquiry Proceedings,* 143. Lieutenant Wallace echoes these figures several times, see *ibid.*, 28, 30, 53, 58. From all these sources, it can be concluded that the wounded did not exceed ten. But that is not a high percentage considering there were well over 300 able-bodied men who would eventually occupy Reno Hill. From the authors' perspective, it is a lame excuse

for their tardiness in responding to the persistent and sometimes loud gunfire they were hearing, see Graham, *Reno Court*, 505–506.

9. For how formidable a defensive position the Weir Peaks were, see Kuhlman, *Legend into History,* 123. Kuhlman stated, contrary to Lieutenant Wallace's misguided testimony at the Reno Court, that because of Weir Point's two parallel ridges, the position was ideal in defending the ravine or gorge between them on the downstream end. No Indian force, however strong, would have subjected itself to such a heavy, short-range crossfire as they would have received there.

10. That Benteen was at the highest peak on the battlefield with H Company, see Graham, *Reno Court*, 360.

11. For the disposition of the respective company's defenses on Weir Point, see Graham, *Reno Court,* 53-54.

12. Benteen jamming the guidon in a pile of stones on the high point, see Carroll, *Benteen–Goldin Letters,* 215.

13. Lieutenant Hare estimating 1,500 warriors moving toward the Weir Peaks, see Donovan, *A Terrible Glory*, 280.

14. Godfrey speaking of the onslaught of warriors toward the Weir Peaks, see Godfrey, *Custer's Last Battle*, 25. This dovetails with other accounts such as Lieutenant Hare, who stated "The Indians were thick over on Custer Ridge and were firing ..." See also, Nichols, *Reno Court*, 303, 409, Graham, *Custer Myth*, 142-143, Neihardt, *Black Elk Speaks*, 119.

15. As others have pointed out, Benteen was probably referring to Reno being in no-man's-land and still a distance away, and not at the Weir Peaks when he said, "this is a hell of a place to fight Indians." See Hammer, *Custer in '76,* 81.

16. The claim that Reno Hill was the best defensive position Benteen was going to find, see Nathaniel Philbrick, *The Last Stand, Custer, Sitting Bull, And the Battle of Little Big Horn,* (New York: Viking), 228.

(Page 243–247)

17. Gibbon's analysis that Reno Hill was weak defensive ground, see John Gibbon, *Catholic Quarterly*, 1877. For another analysis of the weakness of Reno Hill and its liabilities, see Kuhlman, *Legend into History*, 124. Reno, almost preposterously, testifies before the Reno Court that he had "been impressed with the position when I had first reached the hill." The main reason that he allegedly chose this position as opposed to the Weir Peaks is that it was nearer to the water for resupply than Weir Point—for Reno's testimony regarding this, see Graham, *Reno Court*, 506.

18. Benteen claiming the battlefield was not in sight, see Nichols, *Reno Court*, 423. This statement is not only hard to believe, but it is also hard to accept that even Benteen believed it. If the battlefield was not in sight, then he would not have been able to see Custer's forces. If that is true, then

why did Benteen jam a guidon into a small mound of stones with the expectation that Custer would see it fluttering? Benteen had so convinced himself that the battlefield was not in sight that he even wrote his wife nine days after the battle saying that from the peaks he could see nothing of Custer, see Graham, *The Custer Myth*, 300. Our view is that while Custer's fight on Last Stand Hill was too far to be seen from the Peaks, the last gasps on Calhoun Hill and that of C Company at Finley Ridge were almost certainly witnessed.

19. For white objects, Varnum first mistook for rocks, see Carroll, *I Varnum, Chief of Scouts*, 70.

20. Hare took some soldiers to climb to the high point on the ridge (Sharpshooter Ridge) that dominated the surrounding area in order to protect the company's flank, see E. S. Godfrey to Adjutant General n.d., Godfrey Family Papers printed in Graham, *Custer Myth*, "Godfrey Narrative," 143, reprinted in Tom O'Neill's edition of *Custer Chronicles*, Volumes 1–10, 20, also *Custer's Last Battle*, (Reprint, Olympic Valley, California: Outlook, 1976), 27–28.

21. Weir stating there were enough Indians to eat up his company one hundred times, see Fougera, *With Custer's Cavalry*, 272.

22. Weir, as he peered out from the Peaks into the theatre of hostilities taking place on the southern side of the battlefield, knew that the news was bad although he probably did not know, or at least would not admit to knowing, the magnitude of the defeat but he sensed that the lack of response from Custer's reinforcements, of which he was part, contributed to the defeat, see Graham, *Reno Court*, 435.

23. For Varnum seeing a large assemblage of Indians moving toward them, see Carroll, *Custer's Chief of Scouts*, 143, 174.

24. That it looked as if all the Indians in the country were heading toward them, see Graham, *Reno Court*, 125.

25. Varnum stating by the time the Indians got to Weir Point they were not very numerous and the command was not hard pressed, see Hammer, *Custer in '76*, 62. (Varnum was inconsistent in describing the events at the Weir Peaks.)

26. That the Indians did not drive or force the soldiers off Weir Point, see Nichols, *Reno Court*, 446.

27. That there was heavy fighting at the Weir Peaks before a retreat was ordered, see Kuhlman, *Legend into History*, 105.

28. That the area north became black with Indians after the soldiers left and re-established themselves on Reno Hill, see Hammer, Hammer, *Custer in '76*, 62.

29. Godfrey seeing Weir's troops posted on the high bluffs, posting his men, giving directions about the use of ammunition, orders brought to him by Lieutenant Hare, see Godfrey, *Century Magazine* article published in January 1982, and Graham, *Custer Myth*, 143.

30. French's troop and Weir's troop tearing over the bluff at the Weir Peaks with Indians in pursuit, see Graham, *Custer Myth*, "Godfrey Narrative," 143. Also regarding the panic retreat from Weir Point, see Graham, *Reno Court*, 539; Carroll, *The Federal View*, 103; Carroll, *The Gibson–Edgerly Narrative*, 11; Hammer, *Custer in '76*, 56; and Graham, *Reno Court*, 394.

31. Edgerly's account on how he reluctantly obeyed the order to fall back, see Hammer, *Custer in '76*, 56.

32. Benteen describing how French weakened on the Peaks, see Carroll, *Benteen–Goldin Letters*, 216.

33. Jerome Greene stating that Custer did not believe there were enough Indians to whip his regiment, see Jerome Greene, *Indian War Veterans: Memories of Army Life and Campaigns in the West, 1864–1898*, (N.Y.C.: Sonas Beatie, 2001), 125. This is only one of several examples, which also came out in the Reno Court hearings, see Varnum's testimony (Nichols, *Reno Court*, 162–163) and others, that the prospect of Custer's defeat, much less his annihilation, was at the time unthinkable.

(Pages 248–252)

34. Shooting back at the Indians on Weir Peaks with Colt 45s, see Liddic, *Camp on Custer*, 97–98.

35. An earlier account had a trooper named Charles Sanders as the amused soldier regarding the Indian's poor marksmanship, and not Harrison. See Camp, BYU Notes, Reel 5, 190.

36. For decades, the name of this unfortunate trooper who was shot was not known until Walter Camp finally revealed it to be Vincent Charley. See Douglas Scott et al., *They Died With Custer*, (Norman, OK: University of Oklahoma Press, 1998), 306.

37. Vincent Charley being seriously wounded, see Douglas Scott and Peter Bleed, *A Good Walk Around the Boundary*, 34, and Hammer, *Custer in '76*, 57.

38. A wounded Charley half-crawling about, see Hardorff: *Custer Battle Casualties*, 160.

39. Terry telling Custer to hold on to his wounded, see John Manion, *Last Statement to Custer*, (El Secundo, Calif.: Upton and Sons, 2000), 26.

40. Corporal Wylie had a ball shot through his canteen and the staff of the guidon he was carrying was shot off, see Hammer, *Custer in '76*, 130.

41. Edgerly telling Charley to hide in a ravine, see Liddic, *Camp on Custer*, 98, and Hammer, *Custer in '76*, 57.

42. Edgerly telling Charley they would come back with reinforcements, see Liddic, *Camp on Custer*, 98.

43. The Indians brutally finishing off Charley, see *ibid.*, 98.

44. That a stick was rammed down Trooper Charley's throat, see George M. Clark, Scalp Dance: The Edgerly Papers on the Battle of Little Big Horn, (Oswego, N.Y.: Privately Printed, 1985), 25.

45. Edgerly remorsefully writing to his wife about the man left behind, see George M. Clark, Scalp Dance: the Edgerly Papers on the Battle of Little Big Horn (Oswego, N.Y., privately printed, 1985, 25.

46. Even before Edgerly began to write some of his narratives, he lied to the Reno Court in blaming Captain Weir, now dead nearly six months, for Vincent Charley not being rescued, see Nichols, *Reno Court*, 446.

47. Seeing troops galloping down from the peaks and gunfire coming from the Indians in the rear, see Edward S. Godfrey, "Cavalry Fire Discipline," *Journal of the Military Service Institution*, Volume 19, No. 83, September 1896, 257–258.

48. Hare covering for Reno by saying he ordered the cover fire, see Hammer, *Custer in '76*, 67.

49. Godfrey making the air blue with oaths, threatening to shoot his own men, Godfrey: *Cavalry Fire Discipline*, 258.

50. There was a similar performance by Godfrey at the Battle of Washita, where he heroically formed a rear guard action, see Godfrey, *The Cavalry Journal*, October 1928.

51. Godfrey later tells Walter Camp that the primary reason he stopped and formed a skirmish line was because he saw Vincent Charley had been hit and fell off his horse. See Camp, IU Box 5, Folder 17.

52. McDougall telling Benteen that as the senior captain, he better take charge and take command, see Hammer, *Custer in '76*, 71.

53. As for books dealing with the subject of the Reno Hill fight, Liddic's *Vanishing Victory* is arguably not only the best thing ever written about that particular struggle but is also one of the most insightful analyses of the Battle of the Little Big Horn ever written.

54. Godfrey hearing the post-battle celebrations of the Indians the night of June 25; see Graham, *Custer Myth*, 144.

55. Windolph's quote "A million miles from nowhere …", Hunt, *I Fought With Custer*, 102.

Chapter 15

(Pages 259–262)

1. For the Laurence Barrett quote, "…making history," see Elizabeth Bacon Custer, *Tenting on the Plains or General Custer In Texas and Kansas*, (University of Oklahoma Press, 1994), 220.

2. For Major Reno's description of the carnage of the battlefield, see Hardorff, *Battle Casualties,* 97–98.

3. Dean Hudnutt: Varnum weeping uncontrollably when apprised of the decimation of Custer's column. "New Light on the Little Big Horn," *Field Artillery Journal,* 26, no. 4 (July–August 1936).

4. For Weir's quote, "They look so white," see Graham, *Custer Myth*, 326, also Godfrey letter to John Neihardt, January 6, 1924, about the sentiment's poetic value.

5. An arrow in Crittenden's glass eye, see Hardorff, *Battle Casualties II,* 104.

6. Calhoun is identified by the fillings in his teeth, see *ibid*., 15.

7. Finding Custer and Headquarters Staff on a knoll as well as his brother and nephew, see Graham, *The Custer Myth*, 220.

8. Adjutant Cooke's scalped long sideburns is recorded by Marquis, *Wooden Leg,* 240. 263. Wooden Leg describes how the women used sheath knives and hatchets on Cooke.

9. How soldiers' gaping wounds were filled with a host of blue flies, see Francis B. Taunton, *Custer Field: A Scene of Sickening Ghastly Horror* (London: The Johnson–Taunton Military Press), 26.

10. "All of the bodies we heaped with mounds of dirt, see Greene, *Indian War Veterans,* 136.

11. Regarding Captain Tom Custer's gruesome wounds, see Hardorff, *Battle Casualties, II*, 31–32. Regarding Tom Custer, one of the most mutilated men on the battlefield, see Hardorff, *Battle Casualties II,* 31–32. Regarding Tom Custer's being identified by the initials TWC on his arm along with the seated Goddess of Liberty and an American flag, see Hardorff, *Battle Casualties*, 99.

12. The curious post-mortem of Private Harrington, see Hammer, *Men With Custer,* 31–32.

13. Description of breastwork of six dead sorrel horses, thirty feet in diameter, see Hardorff, *Battle Casualties,* 98, 103.

14. General Custer napping and dying with a smile, see Hardorff, *Battle Casualties,* 103; Liddic, ed., *I Buried Custer: Diary of Private Thomas Coleman, 7th U.S. Cavalry*. A smile is very suggestive of a wink and a nod. However, Custer may have appeared to some; there is no question that, even if not as mutilated as others, Custer died brutally and violently.

15. Bullet entering Custer's left breast, see Hardorff, *Battle Casualties, II*, 26. Memories of the wounds of soldiers on Last Stand Hill naturally vary. For a sampling of just a few, see Hardorff, *The Dead, the Missing, and a Few Survivors* (El Secundo, CA: Upton and Sons, 1999), 15–31. See also Horn, *Fifty Years* (Privately published, West Orange, N. J.: David McMillin, 2010, 48. Regarding the state of Custer's body, Dr. Porter said he was naked and clean as a baby, see Hardorff, *Custer Battle Casualties II*, 27. That an arrow was jammed into Custer's penis see Hardorff, *The Custer Battle Casualties II,* 21–22. Private Windolph, however, said he was dressed in the same buckskin outfit he wore on the morning of the battle, see Hunt, Frazer; Hardorff, *Battle Casualties II,* 24–25. There is a stark contrast between the two eyewitness accounts, and we side with Porter for no other reason than that he was a medical doctor and that most other soldiers were also naked or near naked. For General Custer's remains on Last Stand Hill, see Hardorff, *Custer Battle Casualties,* 98, 103.

16. It should be noted that soldiers were also capable of sadistic mutilations. Twelve years earlier at the Battle of Sand Creek, Colonel Chivington's militia had brutally killed women and children. It is believed that some of the native women who mutilated the dead were at Sand Creek. For brief descriptions of that battle, see Jerome Greene, *Washita: The U.S. Army and Southern Cheyennes, 1876–1879,* 3, 5; Michno, *Encyclopedia of Indian Wars*, 157–159.

17. There were a number of accounts that said Custer had shells nearby, Ryan, Barnard, *Ten Years With Custer,* quoting Sergeant John Ryan, 303; Hardorff, *Custer, Camp and the Little Big Horn,* quoting Private Henry Mechling, 77; Frost, *General Custer's Libby,* quoting Lieutenant John Carland, 246, *New York Herald*, July 14, 1876, quoted in Wengert, *The Custer Dispatches,* 66. The point about the shells from Custer's Remington Sporting rifle is that they were distinctive in that they were brass instead of copper cartridge cases and, therefore, easily identified. It is also good evidence that Custer was fighting hard and was probably not an early casualty even on Last Stand Hill.

18. Kate Big Head regaling about two Southern Cheyenne women puncturing Custer's eardrums with a sewing awl, see Hutton, *The Custer Reader,* 376.

19. De Rudio describes how he, along with Benteen and others, went to inspect the battlefield, see Nichols, *Reno Court of Inquiry,* 322.

20. Sergeant James Butler was considered one of the better shots in the regiment and is generally believed to have bled the enemy before he perished. See Hardorff, *Battle Casualties*, 110.

21. For a description of Keogh just wearing worn-out socks, see Stewart, *Custer's Luck,* 469.

(Pages 263–271)

22. What Benteen actually said when he saw the prone Custer varies, and one must realize that no one was taking dictation for the benefit of posterity, see Mills, *Harvest of Barren Regrets*, 279. Private Jacob Adams, accompanying his company captain, remembered a milder exclamation from Benteen, "By God that is him!" See Richard Hardorff, "Some Recollections of Custer: His Last Battle." *Research Review* 4, no. 2, June 1990.

23. Benteen always said it was a rout and that the only sign of a formal defense was on Last Stand Hill. See Graham, *Reno Court*, 145–146, and Nichols, *Reno Court*, 417. Benteen conclusion that the battle was a rout—"you can take a hand of corn and scatter it over the floor." See Nichols, *Reno Court,* 417.

24. Benteen contradicts himself badly here. At one point, he said that he plunged the guidon into some stones to attract Custer's attention. He was then asked if he believed Custer was alive at that time. He replied in the affirmative, adding that no one thought that Custer was dead or his command destroyed, see Nichols, *Reno Court,* 423. Yet, he later states that before Trumpeter Martin's order reached him, long before he placed the guidon in between those stones on Weir Point to attract Custer who stated he believed his commanding officer was alive, he now claims that both Custer and his command were dead, see Nichols, *Reno Court,* 430.

25. Benteen stated that he knew everyone in Custer's command was dead and the fight was over in thirty minutes, see *ibid.*, 589.

26. Benteen's "we would have been there yet" comment, see Graham, *Reno Court*, 360.

27. The mystery surrounding Custer's fate after trumpeter Martin gallops away with a message from his commander, see Donovan, *Terrible Glory*, Author's Notes, xiv.

28. Two Moons is one among several sources where we can deduce with reasonable accuracy that a platoon of E Company was deployed on Last Stand Hill, which we post for the reader's perusal:

> **Two Moons' account dated February [sic March] 5, 1913, interpreted by Willis T. Rowland, with our annotations in brackets.** W.T. Rowland (interpreter) who sat on the field with his uncle Roan Bear [who was one of the Cheyennes who first met Custer's troops coming down Medicine Tail Coulee in the vicinity of Ford B, who was the brother of his mother and spoke good English (now dead) who was in the battle, says his uncle told him that the Gray Horse Troop [note: or at least one platoon thereof] stood at the monument with their horses, but finally had to let them go, being surrounded and hard pressed.
>
> **Two Moons (Cheyenne) 1909 interview with Joseph Dixon, published in Joseph Dixon's book, *The Vanishing Race,* with our annotations in brackets:** I whipped up my horse and told them to come on … and again I started for the bunch of gray horses on the hilltop. [note: Two Moons leads a charge from the northern end of Keogh's swale heading west towards the Gray Horse troop on Last Stand Hill, based on his map details] The Indians followed me, yelling and firing. I could not break the line at the bunch of gray horses, and I wheeled and went to the left…
>
> **Black Elk – as recorded in the book *Black Elk Speaks* by John Neihardt, with our annotations in brackets:** Then the women came over [on the battlefield] and we went to the top of the [Last Stand] hill. Gray horses were lying dead there, and some of

them were on top. Wasichus (whites) and dead Wasichus were on top of them. [Note: The gray horses seen by Black Elk lying dead on top of Last Stand Hill likely belonged to the 1st Platoon of E Company under Lieutenant Algeron Smith. This platoon was deployed on Last Stand Hill throughout much of the fight, according to Two Moons' Account.]

Young Two Moons [aka John Two Moons] was a nephew of Two Moons [aka Old Two Moons]. This is his account of the Little Big Horn battle given to Walter Campbell in 1929, Walter Campbell Collection, Box 105, Notebook 15, Oklahoma University, published in Richard Hardorff's *Cheyenne Memories of the Custer Fight,* with our annotations in brackets: The Gray Horse Company [i.e., E Company] moved from the south onto the top of the ridge [at Last Stand Hill]… [note: Young Two Moons comments on his hand drawn battle map follows:] 1. [The] Gray Horse Company [came] from [the] S[outh] on top of [Battle] ridge.

Young Two Moons (aka. John Two Moons) account of the Little Big Horn Battle given to George Bird Grinnell in 1908, Grinnell Collection, Grinnell Field Notebook 348, Braun Research Library, Southwest Museum Library, Los Angeles, published in Jerome Greene's *Lakota and Cheyenne: Indian Views of The Great Sioux War, 1876–1877* (1994) pp, 67–72, with our annotations in brackets: After the soldiers turned upon the little ridge [i.e., Battle Ridge], the gray horse company [Note: likely a platoon of E Company] stopped where [the] monument is [at last Stand Hill] … The Indians were trying to drive the three companies on this ridge [Note: Companies I, L, and C] running about north and south over to the gray horse company [note: E Company's platoon deployed on Last Stand Hill] … As they got down [Battle Ridge] part way toward the gray horse company [i.e., platoon] the latter began to fire and drove [the] Indians off, and the [Keogh] soldiers [Note: only about 10 or so] reached the gray horse company [on Last Stand Hill]. Some [of Keogh's] soldiers were killed. However, when they reached [Note: or as they attempted to reach] the Gray Horse Company, the latter shot at [the] Indians so fast that they drove [the] Indians back out of sight over [the] hill toward the [later Crow] agency [note: Beyond Battle Ridge Extension to the north]… When the Indians dismounted, they shot at [the] soldiers to the top of the [Last Stand] hill. Then all [the] Indians mounted and charged [toward Last Stand Hill from the east]. Then the Gray Horse Company [i.e., E Company's platoon] turned their horses loose [i.e., were stampeded], and some of the horses rushed through the Indians and toward the river… The gray horse company [note: likely just the 1st Platoon of E Company under Lieutenant Algernon Smith] was destroyed on the [Last Stand] hill near where they went out of sight [or to the west side of the slope from Young Two Moons' line of sight]. The Indians charged among them.

29. For Indians showing Philo Clark Custer's path after he leaves Calhoun Hill, see Wells, *Little Big Horn Notes,* 17–18. Consult also Fox *Archeology, History, and Custer's Last Battle,* in which he argued that Custer moved farther north than was commonly recognized at the time, 124. John Stand In

Timber not only says that Custer was headed for Ford D, he even indicated the Willy Bends place by name as Custer's objective, see Don Rickey, John Stands In Timber Interview, Custer Battlefield Collection, August 8, 1956, folder 15878. Finally, as part of his summation, Jesse Lee, Recorder at the Reno Court, stated that Custer led his troops in that direction. Though we disagree with Lee that Custer's intention was to cross the river, we believe the general trajectory he had indicated was correct.

30. Indians firing at soldiers by the brush near the river at Ford D and the soldiers moved back to where the national cemetery is. See John Stands In Timber interview, Liberty, *Cheyenne Memories*, 199, Hardorff, *Lakota Reflections*, 80, also records warrior resistance here by the river—see Red Feather's map showing warriors congregating at Squaw Creek, north of the Cheyenne camp. As they waited on Cemetery Ridge, more and more Cheyennes and Sioux kept crossing all the time as the soldiers moved to the top. A number of warriors got up to move across the ridge where the monument stands to join Wolf Tooth and others. These movements, by both parties, set the stage for the final actions of the battle.

31. When found, Kellogg was missing a scalp and one ear. Kellogg was identified by his distinctive-shaped boot heels. Found in his pocket were 37 sheets of narrow, folded paper, an indication that he intended to record a great many impressions about the battle that was now underway. See Hardorff, *The Custer Battle Casualties,* 122, and Barnard, *I Go With Custer,* 142–147.

32. Charles Corn saying "He tried to get our children and wives ...," see Camp C. B. N. M., A312, folder 8, #11404.

33. Once again, just as Custer did when he was at Medicine Tail Coulee, he waited for the situation to develop, notwithstanding the testimony of others (Lieutenant Varnum, for example) that Custer was incapable of waiting five minutes. See John Stands In Timber interview by Dr. John Rickey.

34. The joke circulated among some Cheyenne warriors that before the Last Stand, Custer teased that when he got to the village, he would look for the Sioux girl with the most Elk teeth on her dress and take her along with him. This story was told by Beaver Heart and can be found in Liberty, *Cheyenne Memories,* 199.

35. Regarding the death of John Foley, there are several accounts. See Schoenberger, *End of Custer,* 196; Liberty, *Cheyenne Memories*, 207–208; also see Camp, IU, box 6, Folder 19, and Hammer, *Custer in '76,* 20, 207, 254. There are those who believe that Foley was another messenger. It is also possible that at this stage of the chaos and mayhem, it became more of every man for himself. See Hardorff, *Battle Casualties,* 110–111.

36. That Custer did move into The Basin, at least for a brief time, see "John Stands In Timber interview," Liberty, *Cheyenne Memories,* 200.

(Pages 272–278)

37. How the Suicide Boys pledged to throw away their lives to never return, see Peter J. Powell, *Sweet Medicine,* (Norman: University of Oklahoma Press, 1998), 112.

38. Although belatedly acknowledged, there was evidence that there was a suicide charge by young Cheyenne and Sioux warriors. Male status depended on war honors. It is well-documented in Indian literature that the difficulty involved in restraining a young, unproven group whose rashness was potentially dangerous in its tendency to put personal (war honors) goals, in which Indian warfare is generally oriented, above tribal welfare. See Liberty, *Cheyenne Memories,* 193.

39. For the chaos involved in the Suicide Boys charge, see Liberty, *Cheyenne Memories, 203;* Demallie, *The Sixth Grandfather,* 186, and Hardorff, *Lakota Recollections.*

40. The rush of thirty to forty dismounted troopers speeding down into The Basin, see Dale Schoenberger, *End of Custer: The Death of an American Military Legend* (Blaine, WA: Hancock House), 208.

41. One account states that Boyer's body was supposedly found on the top of a little knoll two hundred yards below Last Stand Hill. He possibly suffered a blunt force trauma to the head, though other reports state he was shot. Fragments of his skull appear to have been discovered near Deep Ravine. (See Hammer, *Custer in '76*, 95). Near Deep Ravine is the more likely location of Boyer's death since it is our belief that Boyer bolted, not from Last Stand Hill, but Cemetery Ridge. Had he been positioned on Last Stand Hill, the distance to Deep Ravine, where his body was found, was too far a distance to traverse on foot. Other than that, he was killed with Custer in the vicinity of Deep Ravine; there is some ambiguity surrounding how he died and exactly where. A widely-attested photograph purporting to be the image of Boyer is almost certainly bogus. Those interested in other testimonies regarding his death, see Scott and Fox, *Archeological Insights,* 39; Hardorff, *Battle Casualties,* 119–120. For a short bio of Mitch Boyer, see Kenneth Hammer's *Men With Custer: Biographies of the 7th Cavalry*, edited by Ronald Nichols (Hardin, MT: Custer Battlefield & Museum Associates, Inc., 1995), 280. One further note regarding Custer's best scout is his name. Many sources spell his last name as Bouyer, but the authors are persuaded by a conversation between battle scholar Frederic C. Wagner and Bill Rini conveying Wagner's research that the actual spelling of his name is Boyer.

42. This quote that Custer "would rather fight a veteran Indian rather than a fourteen-year-old Indian boy" comes from a Custer interview by the *New York Herald* on May 22, 1875, and was also published in *Fates Change Horses* by Gary Stewart and DJ Richardson.

43. The number of men from Keogh's fight that made it to Last Stand Hill has been estimated by Dr. Richard Fox as about twenty. We think, not for any particularly concrete reason, that it was a handful less than this number.

44. Two Moons remembered that the left wing and the survivors of the right wing gathered around Last Stand Hill. While Custer was taking his wing from The Basin to Last Stand Hill, Two Moons said not a shot was fired, leading to the conclusion that Custer arrived there from The Basin without any casualties. See Hardorff, *Indian Views*, 111.

45. On the continuous blowing of notes by the bugler on Last Stand Hill, see Hardorff, *Cheyenne Memories of the Custer Fight*.

46. For a description of Indian armaments used that day, see Scott and Fox, *Archeological Insights in the Custer Battle,* 112.

47. Moving Robe observing the soldiers who were trapped, see Wells, *Little Big Horn Notes,* 12.

48. Kate Big Head saw hundreds of Indians for every soldier left alive, see Hutton, *The Custer Reader,* (University of Nebraska Press, 1992), 370.

49. Lone Bear's comments that Custer's soldiers made a stand on top of the hill (Last Stand Hill) see *Lakota Recollections,* 159.

50. About Custer placing his best marksmen on top of the knoll, see Gray, *Custer's Last Campaign,* 389.

51. Indians getting around and to the east of Custer, see Hammer, *Custer in '76,* 198.

52. That there was heavy fire by the Indians from both sides of Last Stand Hill, see Father Powell, *Sweet Medicine,* 116.

53. Custer ordering six sorrels shot comes from De Rudio remembering that many of these horses used as a makeshift breastwork came from Company C, see Hammer, *Custer in '76,* 87, and Fox, *Archeology: Custer's Last Fight,* 196. See also James S. Hutchins, *Great Western Indian Fights,* (Lincoln, Nebraska: University of Nebraska Press, 1960), 165–174. For Indians disliking nothing more than attacking a foe behind a breastwork, see Custer, *My Life on the Plains.*

54. Despite the firearms the Indians possessed, a number of the soldiers appeared to be killed by arrows being fired from their points of cover. Indians were highly skilled archers firing their arrows in long, high arcs that took a terrible toll on both troopers and horses alike. See comments of Waterman, an Arapaho, quoted in Graham, *The Custer Myth,* who noted most of the soldiers had arrows sticking out of them, 110. See also Jay Smith, *A Hundred Years Later: In Custer and His Times*, edited by Paul Andrew Hutton (El Paso, Texas: Little Big Horn Associates, 1981).

55. Soldiers foiling a charge on the east side, see Donovan, *A Terrible Glory*, 274.

56. Black as evening quote by Chief Gall, see Usher L. Burdick, *David F. Barry's Indian Notes on the Custer Battle,* (Baltimore, MD: The Proof Press, 1937), 27.

57. For soldiers fighting stubbornly and bravely, see Hardorff, *Lakota Recollections,* 149.

58. There are numerous Indian accounts that most of the soldiers, though the courage of a few might have faltered when all was lost, fought bravely. While the Indians later held in captivity may have laid it on a bit thick with the heroics to appease their captors, nevertheless, there are enough of these statements and enough archeological evidence that the fighting was fierce, and such resistance does not emanate from a craven fighting force. Here are some of the statements the Indians made about the fighting character of Custer's men: He Dog said, "the soldiers put up a good fight;" Brave Wolf said, "they held their line of battle all the way up, nearly to where the monument now stands;" Rain In the

Face always thought white men were cowards, "but I had great respect for them after this day;" Lone Bear said "the soldiers were fighting hard and making a good stand on Custer Hill...," even Sitting Bull would say, "I tell no lies about dead men, those men who came with Long Hair were as good men who ever fought," see Graham, *Custer Myth*, 71. See Michno's *Lakota Noon,* 261–265; Hardorff, *Lakota Recollections*, 14.

59. For Wooden Leg saying they only charged the soldiers on the hill when they stopped firing, see Liddic, *Vanishing Victory*, 163.

60. Red Hawk remembering an Indian yell, bone whistles, and a stampede, see Red Hawk interview, July 19, 1907, Edward S. Meany Papers, Box 13, Folder 20, University of Washington Special Collections.

61. Standing Bear comparing the battle to a bad dream, see Neihardt, *ibid.*, 116.

62. Tall Bull saying the fight on top of the hill ended with the soldiers being killed very quickly, see Father Peter J. Powell, *People of the Sacred Mountain.* N.Y., NY: Harper & Row, 1981, 1028.

63. Gall quotable line about a hungry man devouring lunch was taken from an extended interview given to the *St. Paul Pioneer Press,* July 1886, published in the *Custer Myth*, 91, with our annotations in brackets.

> **Question**: How long before all the soldiers were killed?
>
> **Gall** (The chief made the sign of the white man's dinner time, which means noon): [Note: Today's dinner time was actually lunch at that time. What we call dinner today was referred to as supper in the 19th century], and then with his finger cut in half, which would signify half an hour consumed in slaughtering everybody. [Note: this is not necessarily a correct interpretation. Gall's sign language could just as easily, and more likely, meant the battle lasted from its start around noon, sun time, until half the distance to sundown, or roughly 4 hours duration, until 4 p.m. sun time or 5 p.m. Headquarters watch time, which comports with most other Indian time estimates.]

64. George Herendeen stated that they found four severed heads, including Lieutenant Sturgis, in the village. For the narrative of George Glenn, see Godfrey, *Where Custer Fell*, 112. It seems very likely that a few soldiers were tortured to death in the village. Sitting Bull always denied this saying that there were no celebrations of any kind or kill dances but rather they were mourning for our dead and for those brave white men lying up yonder on the hillside, see Vestal, *Sitting Bull*, 174. These softening words may have been more of an attempt to appease any anger that the army might harbor over the destruction of Custer's command. Others remembered differently. Rain In The Face said that there was a big feast and scalp dance that night, see Brady, *Indian Fights and Fighters,* 290. Black Elk had the same recollection as Rain In The Face, see Neihardt, *Black Elk Speaks*, 131, 133. Little Knife said that at least one soldier was tortured to death in the village, see Hardorff, *Dakota Recollections,* 96. Two

Moons remembered that some dead soldiers were dragged into the village, dismembered, and burned at the big dance, see Camp, BYI box, folder 9. Trooper Theodore Goldin, of course, had to put his two cents worth in—stating that several skulls were found at the lower end of the village, whose victims he believed were burned at the stake. He also stated that a piece of blue flannel shirt was discovered nearby with Lieutenant James Sturgis's initials embroidered on it, See Camp, BYU, roll 1, box 1, folder 2. Turtle Rib said no soldiers were taken off the battlefield. That they found a white man's head in the village who was dragged around with a lariat until his head was pulled from his body, see Liddic and Harbaugh, *Custer on* Camp, 93. They were all killed where they laid, see Hammer, *Custer in '76,* 199, 202.

(Pages 276–278)

65. Custer took a big gamble parlaying with the Cheyenne by riding into their village, see Barnard, *Ten Years With Custer*, 90.

66. Benteen offering Custer his pocket derringer, see Whittaker, *A Life of Major General George A. Custer*, (New York: Sheldon, 1876), 461. See also *Custer, My Life*, for Custer's thoughts about becoming his own executioner, 479–480.

Private Jacob Adams examining Custer's body and Godfrey seeing no black powder burns, see Bates Collection, Jacob Adams to Godfrey, January 10, 1927, LBBNM; Donahue, *Where the Rivers Ran Red*, 232.

In a letter to Godfrey, January 20, 1927, Private Jacob Adams saw Custer's body and said that the wound in his left breast had not bled at all, but the wound from the temple had bled and run down his face, see Bates Collection, Jacob Adams to Godfrey, January 20, 1927, LBBNM. Private Goldin examined the bullet holes in Custer's body with others and determined they were possibly made by a gun of smaller caliber than a .45, and we speculated that they were made by a Winchester or Henry. There were no powder marks around the wounds at all," which otherwise would have indicated death by one's own hand. See Carroll, *The Benteen–Goldin Letters,* 26.

67. Dr. Henry Porter, who examined Custer posthumously, stated that Custer's chest wound was not necessarily fatal, see Don Horn, *Fifty Years on Custer's Trail,* 48.

68. Private O'Neil thought that Custer's chest wound, based on the entry point, was fired from a warrior on horseback and the head wound fired from an Indian on foot. Hardorff, *Battle Casualties II,* 27.

69. Private Charles Windolph's save the last cartridge for yourself, see Hunt-Frazier, *I Fought With Custer,* 36.

70. Dr. Thomas Marquis, who communicated with the Indians through hand signs and gathered from his Cheyenne informants, was the only one who spoke of mass suicides among the soldiers, and it was something he promoted a half-century after the battle. See Marquis, *Keep the Last Bullet for Yourself.*

71. For some soldiers making a break from their defenses, see Michno, *Lakota Noon*, 264–265.

72. Regarding the death and locations of the bodies of Boston Custer and Autie Reed, see Lawrence A. Frost, *General Custer's Libbie*, (Seattle, WA: Superior Publishing Company), 246; Hammer, *Custer in '76*, 248; Hardorff, *Custer Battle Casualties,* 104–105.

73. Richard Fox in *Archeology, History and Custer's Last Battle* claimed that there was no Last Stand and that the soldiers panicked and ran.

74. As Gregory Michno rightly points out, this cannot be squared with the facts garnered from the Indian accounts. See Michno, *Lakota Noon,* who waxes both sensibly and sentimentally about this fact, 285.

75. Yates stating that "one day Custer would get us in a hole," see Liddic and Harbaugh, *Camp on Custer,* 50.

76. Gall having no idea he was fighting Custer, see Interview with Herbert J. Slocum: Walter Camp Field Notes, folder 92, Brigham Young University Library. What could be said of Gall's ignorance about Custer being in the battle could be said of almost all the Indians that day. See also Walter Camp Field notes, Brigham Young University.

Chapter 16

(Pages 285–290)

1. Regarding Yellow Nose capturing a battle flag, see Hardorff, *Lakota Recollections,* 140.

2. Only a few horses killed between Deep Ravine and Last Stand Hill, see Hammer, *Custer in '76,* 72.

3. Wooden Leg's account of the Little Big Horn Battle given to Dr. Thomas Marquis is found in *A Warrior Who Fought Custer*(1931), and the account was later published in Colonel William Graham's book, *The Custer Myth, 105.*

4. Regarding the number of troopers who fled from Cemetery Ridge in the direction of Deep Ravine, see Michno, *Lakota Noon*, 272. Michno points out that Two Moon's estimate of forty-and-five men moving down from the hill is not mentioned in any of the other interviews. Dixon, *Vanishing Race.* In a 1901 interview by J. M. Thralls, Two Moons said nineteen men, not forty-five, broke away. In a *Harness Gazette* account published in 1908, the numbers are trimmed even further now, stating that "some men" tried to get away toward the river, but they were all killed. Micho documents all these inconsistencies.

5. That there were few if any, dead cavalry horses between the top of the ridge where E Company was found, see the letter written by Captain Thomas McDougall to General E.S. Godfrey dated Wellsville, New York, May 18, 1909, from the Francis R. Hagner Collection, Box 1, Manuscript Division, New York Public Library.

6. That Gregory Michno expressing doubts that twenty-eight soldiers lay in Deep Ravine, see Michno's article in *The Wild West Magazine* can be found at this link: https://www.historynet.com/little-bighorn-myths/.

7. Stating there were no dead soldiers piled in Deep Ravine, see Lieutenant James Bradley's account can be found at https://www.amazon.com/Montana-Column-March-LittleBigHorn/dp/1519042221/ref=sri13.

8. Ryan discussing the difficulty of burying bodies, see Barnard, *Ten Years With Custer,* 303 and 305.

9. Private Dennis Lynch questioning his own memories as a member of the burial detail, see interviews with Lynch on February 8, 1909, see Hammer, *Custer in '76,* 139. There, Private Lynch gives his own comments on the battle to Walter Camp. Camp, however, likely mistook Private Dennis Lynch for Private Patrick Lynch of I Company. The latter had probably accompanied the steamer *Far West* with Lieutenant Nowlan, whereas Private Dennis Lynch was likely traveling with the pack train on June 25, 1876.

10. Regarding the deep soil deposits in Deep Ravine, stratospheric units, the soil's alkaline quality, etc., see Scott, *Uncovering History: Archeological Investigations at The Little Big Horn,* (Norman: University of Oklahoma Press, 2013), 127. Also see Scott, Fox, Connor, and Harmon, *Archaeological Perspectives on the Battle of Little Big Horn,* 224–242.

11. For Michno's claims that Deep Ravine was really Cemetery Ravine with the buried bodies, see Michno, *Mystery of E Company* 349–350.

12. Remains of enlisted men being buried around a monument in 1881, see Hammer, *Custer in '76,* 250.

13. For Walter Camp accompanying Kanipe, who drew a map marking the locations of the bodies, see Sandy Barnard, *Walter Camp, Greasy Grass Volume II,* Hardin, Montana: Custer Battlefield Museum and Historical Association, May 1995.

14. On Reno ordering Captain McDougall to bury the dead on Custer's Field, see Nichols, *Reno Court of Inquiry,* 535.

15. An excerpt of this letter to General Godfrey from Captain McDougall can be found in Graham, *Custer Myth*, 377.

16. Benteen seeing twenty-two dead in Deep Ravine, see Nichols, *Reno Court of Inquiry,* 418.

(Pages 292–296)

17. Lieutenant Charles De Rudio gives his condensed account of the battle to Walter Camp in an interview given on February 2, 1910. He saw a heap of men in the gully. See Hammer, *Custer in '76,* 87.

18. Private Glenn's account is given to Walter Camp on January 24, 1914, regarding bodies being thick in Deep Ravine and exclamations about the whole company lying in there, see *ibid.*, 136.

19. Hare also counted twenty-eight bodies in Deep Ravine. For his account, see Graham, *Reno Court of Inquiry,* 261.

20. For this quote of Captain Henry Freeman stating that after the "hardest marching I ever made," he saw twenty-eight bodies in Deep Ravine, see *The Freeman Journal* at https://thelbha.proboards.com/thread/2696/capt-henry-freeman/.

21. There are six soldier maps referencing Deep Ravine: Edward Maguire, Frederick Benteen, Henry Freeman, George Gillespie, John Kimm, and Edward Godfrey. There are five additional warrior maps referencing Deep Ravine: Kill Eagle, Sitting Bull, Standing Bear, Red Feather, and White Bull. Not all of these maps indicate that soldiers were buried in the ravine. Donahue, *Where the Rivers Ran Red:* Appendix 6, 276.

22. On the recollections of Chief Engineering Officer Edward McGuire mapping the battlefield of Little Big Horn, see Edward J. McClernand, *On Time for Disaster: The Rescue of Custer's Command.* (Lincoln and London: University of Nebraska Press, 1969), 64–65, Nichols, *Reno Court,* 7–8. For McGuire establishing the first field map, drawing topography, noting the death sites, and taking various measurements on the battlefield, see Michael Donahue, *Drawing Battle Lines*, 27-30.

23. McDougall about bodies of E Company lying on their side, half in the ravine and the other half outside the ravine, see Nichols, *Reno Court,* 535.

24. For Captain Moylan seeing twenty or so odd bodies of E Company in Deep Ravine, see Nichols, *Reno Court of Inquiry,* 237.

25. This was a lengthy letter written by Walter Camp to General Godfrey dated July 17, 1908, to wit, Godfrey responded in a letter dated July 20, 1908. This correspondence is contained in the Godfrey Family Papers at the United States Army War College, Carlisle Barracks.

26. For Captain Owen Sweet overseeing mistaken marker places on graves, see Captain Owen Sweet's Battle Report, dated May 15, 1890. Note that John Kleinschmidt, of Mandan, was in possession of the original report made by Captain Owen J. Sweet, 25th Infantry in his Report on the Custer Battlefield. From the handwritten report [i.e., Major Welch] have copied the following. See: www.welchdakotapapers.com/2012/02/col-welch-notes-on-custer/# captain-sweets-.

27. For Tall Bull speaking of the last soldiers being killed in Deep Ravine, see Jerome Greene, *Lakota and Cheyenne: Indian Views of the Great Sioux War, 1876-1877,* (Norman: University of Oklahoma Press, 1994), 53, Powell, *Sacred Mountains,* 1028.

28. Standing Bear on death of soldiers in Deep Ravine, see Hardorff *Hokahey,* 64–65.

29. Good Voiced Elk's seeing soldiers jump over the deep bank only to be killed can be found in The Robert S. Ellison Collection in the Denver Public Library.

30. Description of Deep Ravine as a ravine over a ravine, see Scott and Fox, *Archeological Insights into the Custer Battle,* 68–71.

31. Regarding the result of the geomorphology tests done in 1984 and its disappointing result, see Scott, Fox, Connor, and Harmon, *Archeological Perspectives on the Battle of Little Big Horn*, 39–48. For conclusion that soldiers were buried in Deep Ravine, see *Archeological Perspectives on the Battle of Little Big Horn,* (Norman: University of Oklahoma Press, 1989), 41. Regarding the investigations using penetrating sonar devices and their findings, see Scott, Fox, Connor, and Harmon, *Uncovering History,* 125, 128.

Afterword

1. Sherman's shocked reaction came at the Centennial Exhibition in Philadelphia, the nation's birthplace, and can be found in *The New York Herald,* July 7, 1876.

2. For John Ryan's description of General Custer and his brother Tom's burial on the battlefield, see Barnard, *Ten Years with Custer*, 304.

3. For General Sturgis's criticism of Custer, see *New York Times,* July 18, 1876.

4. Sheridan's quote about if Custer had only waited for the junction with Terry, see Michael Wayne Sarf, *The Little Big Horn Campaign*, 265.

5. Terry's comments on the aftermath of the battle can be found in *The Little Big Horn 1876: The Official Communications* edited by Lloyd J. Overfield, (Glendale, California: Arthur H. Clark, 1971), 37.

6. President Grant's comments about the sacrifice of Custer's men being totally unnecessary, see *New York Herald,* August 2, 1876.

7. Regarding Reno's largesse and generosity before the Reno Court, see *The New York Sun*, February 26, 1879, Camp I.U. notes 818. For Gerard recounting Reno's uncharacteristic hospitality, see Hammer, *Custer in '76*, 238–239.

8. De Rudio's criticism of the officers' questionable testimony at the Reno Court, see Hardorff, *On the Little Big Horn*, 24.

9. Fred Gerard stating about McDougall not daring to tell the truth see Holley, *Once Their Home,* 246.

10. Charges of testimony being whitewashed at the Reno Court see Liddic, *Vanishing Victory,* 46.

11. Wesley Merritt damned Reno with faint praise see Graham, *Custer Myth,* 347.

12. For the Court of Inquiry's verdict, see Nichols, *In Custer's Shadow,* 285.

13. Benteen recounting years later that he would shield Reno now as much as he did back then can be found in the *Benteen–Goldin Letters,* 212.

14. Possibly one of the things Benteen was alluding to was a conversation he supposedly had on June 28, a day after Terry relieved Reno's battalion, Benteen and Godfrey rode together and had the following conversation:

> Godfrey: Benteen, it's pretty damn bad.
> Benteen: What do you mean?
> Godfrey: Reno's conduct.
> Benteen paused then said: God, I could tell you things that would make your hair stand on end.
> Godfrey: What is it? Tell me.
> Benteen: I can't tell you now.
> Godfrey: Will you tell me sometime?

"Yes," said Benteen, concluding the conversation. It would not be until another five years before Benteen told Godfrey how, on the night of June 25, Reno proposed abandoning the wounded to make their escape (see, Godfrey to J.A. Shoemaker, March 2, 1926, Roll 4, Elizabeth B. Custer Collection.) By the time of Benteen's revelation to Godfrey, two years after the Court of Inquiry's hearings concluded, Reno had been much maligned, but no one other than Benteen ever even hinted that Reno made such a proposal. So, is it true—did Reno suggest abandoning the wounded? Colonel William Graham surmised that, since Benteen is Reno's sole accuser, if Reno did broach the subject, he did so in offering it as an option among many and, in that light, is therefore excusable. At least that is one man's opinion. See Graham, *The Custer Myth,* 437.

15. Benteen calling Custer a liar and a murderer, see Carroll, *Benteen–Goldin Letters,* February 17, 1896, 271-272.

16. Benteen's "out of the whirlpool" remarks, see *ibid.,* May 26, 1896, 302.

17. Custer's sympathy for the Indian's way of life and struggle, see Custer, *My Life on the Plains,*

A Word from the Authors about Sources and Bibliography

Our manuscript is based entirely on primary sources and what we consider to be reliable secondary sources. We have also used maps and photographs as teaching aids which will be found in the body of this work. From the start we have attempted to show the reader not only what we believe happened at Little Big Horn but, even more consequential, why we believe it. For this reason, sources become instructive, if not indispensable, as a tool of explication.

The overriding question for us was how we could cleanly manage this informational task without sacrificing clarity and the reader's attention. end that punctuate each page can act as insidious intruders and not infrequently prove to be an irritant to the reader. Still, citations and parenthetical information are necessary for a book like this, one that is polemical in nature and whose purpose is one of explication to emphasize a point or clinch an argument. These citations can be useful to the reader since one does not want to interrupt the narrative with lengthy explanations that can be distracting and tedious. Once we decided that citations were necessary to supplement the reader's understanding, the question became how to most effectively utilize them in the least intrusive and disruptive manner.

For this reason, we have chosen not to employ these citations throughout the work, usually done at the footing of the page or at the end of each respective chapter, but instead relegate these notes and supplements to the end of the book, sorting them chronologically under the heading of the numbered chapters. Organizing the material in this way will give the reader the opportunity to ignore whatever supplementary information they wish or, alternatively, to pause and peruse the citation at their convenience. Every source we use in this work has been determined to be relevant to the matter at hand and able to withstand the hard glare of scrutiny. We have also made an effort to amass every known extant account we deemed worthy of consideration, including eyewitness accounts, memoirs, correspondence, military reports, and analyses of all those illustrious historians, scholars, and battle students who have weighed in on this saga to get as close as we could to the core of its truth.

And yet, we have also made every effort to limit these notes wherever we reasonably could. Too much supplemental information is like too many musical notes that crowd the ear, supplanting the natural rhythms of sound with an oppressive cacophony of noise diminishing, if not precluding, whatever guidance and discernment the footnote or endnote might provide. Nor did we wish by a plethora of citations to polish the pretentiousness of the scholars' whim for what we could see as serving no other purpose aside from a futile attempt to impress others with the depth of our scholarship and research. In short, we have simply provided the reader with the sources we think are necessary to tell the story convincingly and no more.

Bibliography

For a battle that lasted just a few hours, with its share of fits and starts, the literature on the subject is astonishingly vast and breathtaking. It is an event that has truly captured the American imagination and the sheer amount that has been written about it confirms the story's endless fascination and romance. Perhaps no mortal can plumb its depths, but one of the authors, Bill Rini, has studied the subject for sixty years, reading everything he can find about it, and still shows no signs of slowing down. The following includes books and articles read or consulted.

Books

Adams, Alexander B. *Sitting Bull. An Epic of the Plains.* New York: Barnes and Noble Books, 1992.

Ambrose, Stephen E. *Crazy Horse and Custer: The Parallel Lives of Two American Warriors*. New York: Meridian, 1986.

Anderson, Gary C. *Sitting Bull and the Paradox of the Lakota Nationhood.* New York: Longman, 1996.

Andrist, Ralph K. *The Long Death: The Last Days of the Plains Indian*. New York: MacMillan, 1964.

Arnold Steve and Tim French. *Custer's Forgotten Friend: The Life of W.W. Cooke*. Howell, Michigan: Power River Press, 1993.

Assay, Karol. *Gray Head and Long Hair: The Benteen-Custer Relationship*. New York: Mad Printers of Mattituck, 1983.

Bailey, John W. *Pacifying the Plains: General Terry and the Decline of the Sioux 1866-1890.* Westport Conn..: Greenwood Press, 1979.

Barnard, Sandy. *Custer's First Sergeant John Ryan*. Terre Haute, Indiana: AST Press. 1996..

Barnard, Sandy. *Digging into Custer's Last Stand*. Terre Haute, IN: AST Press, 1998.

Barnard, Sandy. *Speaking about Custer: A Collection of Lectures*. Terre Haute, IN: AST Press, 1991.

Barnard, Sandy. *Ten Years With Custer: A 7th Cavalryman's Memoirs.*, Terre Haute, IN: AST press, 2001.

Barnett, Louise. *Touched by Fire: The Life, Death, and Mythic Afterlife of George Armstrong Custer.* New York: Holt, 1996.

Bates, Col. John Francis. *Custer's Indian Battles*. Bronxville, New York: 1936.

Bates, Col. John Francis. *Terry's Last Order to Custer. 50th Anniversary Celebration* compiled by Tom O'Neil. Brooklyn N.Y: Arrow and Trooper Publishing, 1991.

Beadle, J. H. *Western Wilds and the Men Who Redeem Them*, Cincinnati, Jones Brothers. (1878).

Bookwalter, Thomas E. *Honor Tarnished.* West Carrollton, OH: Little Horn Press, 1979.

Boritt, Gabor S. *The Gettysburg Nobody Knows*: New York: Oxford University Press, 1997.

Boyes, Walter F. *The Cheyenne Tribal Historian John-Stands-in-Timber's Account of the Custer Battle*. N.p., Little Big Horn Associates, 1991.

Boyes, Walter F. *Surgeon's Diary with the Custer Relief Column*. Washington, D.C. South Capitol Press, 1972.

Bradley, James H. *The March of the Montana Column: A Prelude to the Custer Disaster*. Norman: University of Oklahoma Press. 1961.

Brady, Cyrus T. *Indian Fights and Fighters*. New York: McClure, Phillips, 1904.

Bray, Kingsley M. *Crazy Horse: A Lakota Life*. Norman: University of Oklahoma Press, 2006.

Brininstool, E. A. *Troopers with Custer*. Reprint. Lincoln: University of Nebraska Press, 1989.

Brown, Dee. *Bury My Heart at Wounded Knee*.: New York, Chicago, and San Francisco: Holt, Rinehart, and Winston, 1971.

Brown, Jesse, and A.M. Willard. *The Black Hill Trails*. Rapid City Journal Company, 1924.

Brown, Mark H. *The Plainsmen of the Yellowstone*. New York: G.P. Putnam's Sons, 1961.

Buecker, Thomas R. and Eli Paul, eds. *The Crazy Horse Surrender Ledger*. Lincoln: Nebraska State Historical Society, 1994.

Burdick, Usher L. *The Army Life of Charles "Chip" Creighton*. Paris, MD: National Reform Associates," 1937.

Burdick, Usher L. *David F. Barry's Notes on the Custer Battle*. Baltimore: Wirth Brothers, 1949.

Capps, Benjamin. *The Indians*. New York: Time-Life Books, 1973.

Carroll, John M. *The Benteen-Goldin Letters on Custer and His Last Battle*. Lincoln and London: University of Oklahoma Press. 1991.

Carroll, John M. *Cavalry Bits: The Yellowstone Expedition of 1873*. Mattituck, New York and Bryan, Texas: J.M. Carroll and Company, 1986.

Carroll, John M. *A Custer Chrestomathy*. Bryan, Texas.

Carrol, John M. *Custer in Texas*. New York: Sol Lewis Liveright, 1975.

Carroll, John M. *Custer's Chief of Scouts: The Reminiscences of Charles A. Varnum.* Lincoln and London: University of Nebraska Press. 1987.

Carroll, John M. *General Custer and the Battle of Little Big Horn: The Federal View*. New Brunswick, New Jersey: Garry Owen Press, 1976.

Carroll, John M. *General Custer and the Battle of Washita: The Federal View*. Bryan, Texas: Guidon Press, 1978.

Carroll, John M. *The Gibson and Edgerly Narratives*. Bryan, Texas: Privately printed, 1981.

Carroll, John M. *Roll Call on the Little Big Horn*. Fort Collins, CO: Old Army Press, 1974.

Carroll, John M. *The Seventh Cavalry Scrapbook, Nos. 1-13*. Bryan, TX: 1978–1980.

Carroll, John M. ed. *The Sunshine Magazine Articles by John B. Everett*. Bryan TX: Privately printed, 1979.

Carroll, John M. ed. *The Two Battles of Little Big Horn*. Bryan, TX. 1974.

Clark, George M. *Scalp Dance: The Edgerly Papers*. Oswego, NY: Heritage Press, 1985.

Clark, George M. *Killing of Crazy Horse.* Lincoln & London: University of Nebraska Press. 1988

Clark, William Philo. *Indian Sign Language*. Legore Street Press, 2023.

Clausewitz, Carl Von General. *On War*. Ed. and translated by Michael Howard and Peter Paret, Princeton University, 1976.

Coffeen, Herbert. *The Tepee Books, Parts I and II*. New York: Sol Lewis, 1974.

Coleman, Thomas W. *I Buried Custer: The Diary of Pvt. Thomas W. Coleman, 7th U.S. Cavalry.* Ed. By Bruce R. Liddic. College Station, Tex: Creative Publishing, 1979.

Connell, Evan. *Son of the Morning Star: Custer and the Little Big Horn*. New York: Harper Perennial, 1991.

Cozzens, Peter Gould, ed. *Conquering the Southern Plains: Eyewitness to the Indian Wars*, Mechanicsburg Pa.: Stackpole, 2003.

Cozzens, Peter Gould, ed. *Eyewitness to the Indian Wars, 1865 – 1890. Vol. 4, The Long War for the Northern Plains*, Mechanicsburg Pa.: Stackpole, 2004.

Cozzens, Peter Gould, ed. *Eyewitness to the Indian Wars, 1865 – 1890. Vol. 5, The Army and the Indian*. Mechanicsburg Pa.: Stackpole, 2005.

Crook, George. *General George Crook: His Autobiography*. Edited by Martin F. Schmidt. Norman: University of Oklahoma Press, 1946.

Custer, Elizabeth B. *Boots and Saddles*: Norman and London: University of Oklahoma Press, 1961.

Custer, Elizabeth B. *Tenting on the Plains.* Norman and London: University of Oklahoma Press, 1994.

Custer, George Armstrong. *My Life on the Plains*, Reprint. Norman: University of Oklahoma Press, 1962.

Darling, Roger. *Benteen's Scout to the Left*. El Secundo, CA: Upton and Sons, 1987.

Darling, Roger. *A Sad and Terrible Blunder*. Vienna, VA: Potomac-Western Press, 1990.

Davis, E. Elden and Karen I. *That Fatal Day: Eight More With Custer*. Howell MI.: Powder River Press. 1992.

DeMallie, Raymond J. *The Sixth Grandfather*. Lincoln: University of Nebraska Press, 1984.

Dippie, Brian W. *Custer's Last Stand: The Anatomy of an American Myth*. Lincoln and London: University of Nebraska Press, 1994.

Dixon, Dr. Joseph K. *The Vanishing Race*. Garden City, NY: Doubleday, Page, 1913.

Donahue, Michael N. *Drawing Battle Lines: The Map Testimony of Custer's Last Fight.* El Secundo, CA: Upton and Sons, 2008.

Donahue, Michael N. *Where the Rivers Ran Red: The Indian Fights of George Armstrong Custer*. San Juan Publishing Group, 2018.

Donovan, James. *A Terrible Glory: Custer and the Little Big Horn; The Last Great Battle of the American West*. New York: Little Brown, 2008.

Dubois, Charles G. *The Custer Mystery.* El Secundo: Upton and Sons. 1986.

Dubois, Charles G. *Kick the Dead Lion: A Casebook of the Custer Battle*. El Secundo: Upton and Sons, 1987.

Dustin, Fred. *The Custer Tragedy: Events Leading up to and following the Little Big Horn Campaign of 1876*. Ann Arbor, Michigan. Reprint: El Secundo CA: Upton and Sons, 1987.

Eastman, Charles A. *Indian Heroes and Great Chieftains*. Lincoln and London: University of Nebraska Press, 1991.

Ege, Robert J. *Curse Not His Curls*. Fort Collins, Colorado: The Old Army Press, 1974.

Evans, David C. *Custer's Last Fight: The Story of the Battle of Little Big Horn*. El Secundo CA: Upton and Sons, 1999.

Fay, George E. *Military Engagements Between United States Troops and Plains Indians, Part IV, 1872-1890*. Greely. University of Northern Colorado, 1973.

Furgurson, Ernest B. *Chancellorsville 1863: The Souls of the Brave*. New York: Vintage Books, A Division of Random House, Inc.

Forrest, Earle R. *Witnesses at the Battle of Little Big Horn*. Monroe, Mich.: Monroe County Library System, 1986.

Fougera, Katherine Gibson. *With Custer's Cavalry*. Lincoln and London: University of Nebraska Press, 1986.

Fox, Richard Allan, Jr. *Archeology, History and Custer's Last Battle*. Norman: University of Oklahoma Press, 1993.

Freeman, Henry B. *The Freeman Journal: The Infantry in the Sioux Campaign of 1876.* Edited by George A. Schneider. San Rafael CA: Presidio Press, 1977.

Frost Lawrence A. *The Court-Martial of George Armstrong Custer*. Norman: University of Oklahoma Press, 1968.

Frost Lawrence A. *The Custer Album: A Pictorial Biography of George A. Custer.* Norman: University of Oklahoma Press, 1990.

Frost Lawrence A. *Custer Legends*. Bowling Green, Ohio: Bowling Green University Popular Press, 1981.

Frost Lawrence A. *General Custer's Libbie*. Seattle, Washington: Superior Publishing Company, 1976.

Frost Lawrence A. *With Custer In 74: James Calhoun's Diary of the Black Hill's Expedition*. Provo, Utah: Brigham Young University Press, 1979.

Ghent, W.J. *Varnum, Reno and The Little Big Horn: Winners of the West, Volumes 13, 35. April 30, 1936.*

Gibbon, John *Gibbon on the Sioux Campaign of 1876*. Bellevue, NE: Old Army Press, 1970.

Godfrey, Edward *Custer's Last Battle*. Reprint, Olympic Valley, CA: Outbooks, 1976.

Godfrey, Edward. *The Field Diary of Lt. Edward Settle Godfrey*. Portland, OR: Champoeg Press, 1957.

Goldin, Theodore. *With the 7th Cavalry in 1876*. N.p. Privately printed, 1980.

Graham, Col. William T. *The Custer Myth: A Sourcebook of Custerania*. Harrisburg PA: Stackpole, 1953.

Graham, Col. William T. *The Reno Court of Inquiry: Abstract*. Harrisburg, PA: Stackpole, 1954.

Graham, Col. William T. *The Story of the Little Big Horn*. New York: Bonanza Books, 1959.

Grant, Bruce. *The Concise Encyclopedia of the American Indian*. Bonanza Books, 1989.

Gray, John S. *Centennial Campaign: The Sioux War of 1876.* Norman and London: University of Oklahoma Press, 1988.

Gray, John S. *Custer's Last Campaign*. Lincoln and London: University of Nebraska Press, 1991.

Greene, Jerome. *Indian War Veterans: Memories of Army Life and Campaigns in the West, 1864-1898.* Savas Beatie; reprint edition (January 27, 2012).

Greene, Jerome. *Battles and Skirmishes of the Great Sioux War, 1876–1877: The military View.* Norman: University of Oklahoma Press, 1993.

Greene, Jerome. *Evidence and the Custer Enigma*. Golden, CO: Outbooks, 1986.

Greene, Jerome. *Indian War Veterans*. New York: Savas Beatie, 2007.

Greene, Jerome. *Lakota and Cheyenne: Indian Views of the Great Sioux War, 1876-1877.* Norman: University of Oklahoma Press, 1994.

Greene, Jerome. *Washita*. Norman: University of Oklahoma Press, 2004.

Grinnell, George Bird. *The Fighting Cheyennes*. New York: Charles Scribner, 1915.

Hammer, Kenneth. Custer In 76: *Walter Camp's Notes on the Custer Fight*. Provo: Brigham Young University Press, 1981.

Hammer, Kenneth. *Men With Custer: Biographies of the 7th Cavalry.* Huntington Beach CA: Ventana Graphics, 1995.

Hanson, Joseph Mills. *The Conquest of the Missouri*. Chicago: A.C. McClurg, 1910.

Hardorff, Richard G. *Camp, Custer and the Little Big Horn: A Collection of Walter Camp's Research Papers on General George A. Custer's Last Fight*." El Secundo, CA: Upton and Sons, 1989.

Hardorff, Richard G. *Cheyenne, Memories of the Custer Fight*. Spokane, WA: Arthur H. Clark, 1995.

Hardorff, Richard G. *The Custer Battle Casualties: Burials, Exhumations, Reinterments*. El Secundo, CA: Upton and Sons, 1989.

Hardorff, Richard G. *The Custer Battle Casualties ll: The Dead, the Missing, and a Few Survivors.* El Secundo, CA: Upton and Sons, 1989.

Hardorff, Richard G. *Hokahey: A Good Day to Die: The Indian Casualties of the Custer Fight*. Spokane, WA: Arthur H. Clark, 1991.

Hardorff, Richard G. *Markers, Artifacts and Indian Testimony: Preliminary Findings on the Custer Battle*. Short Hills: Don Horn Publications, 1985.

Hardorff, Richard G. *On the Little Big Horn with Walter Camp*. El Secundo, CA: Upton and Sons, 2002.

Hardorff, Richard G. *The Surrender and Death of Crazy Horse*. Spokane, WA: Arthur H. Clark, 2002.

Hardorff, Richard G. *Indian Views of the Custer Fight: A Source Book.* Norman: University of Oklahoma Press, 2005.

Hardorff, Richard G. *Washita Memories*. Norman: University of Oklahoma Press, 2006.

Harper, Gordon. *The Fights on the Little Horn—Unveiling the Mysteries of Custer's Last Stand*. Philadelphia, PA: Casemate Publishing, 2014.

Harrison, Jim. *The Road Home*. New York: Atlantic Monthly Press, 1998.

Hatch, Thom. *The Custer Companion*. Mechanicsburg, Pa.: Stackpole, 2001.

Hedren, Paul L. *The Great Sioux War, 1876 – 1877*. Helena, Montana: Montana Historical Society Press, 1991.

Hillman, Allyn Hall, William B. Beasley, Gilbert Guion Wood. *History of the Sixth New York Cavalry*. Kessinger Publishing, 1908.

Horn, Donald W. *Portrait of a General*. West Orange, NJ: Don Horn Publications, 1998.

Horn, Donald W. *Witnesses For the Defense of George Armstrong Custer*. West Orange, New Jersey: Midland Press, 1981.

Hunt, Frazier, and Robert Hunt, eds. *I Fought with Custer: The Story of Sergeant Windolph, Last Survivor of the Battle of Little Big Horn*. Lincoln and London: University of Nebraska Press, 1987.

Hutchins, James S. *Boots and Saddles at the Little Big Horn*. Fort Collins, Colo.: Old Army Press, 1976.

Hutchins, James S. *Great Western Indian Fights*. Lincoln NB: University of Oklahoma Press, 1960.

Hutchins, James S., ed. *The Papers of Edward S. Curtis Relating to Custer's Last Battle*. El Secundo, Calif.: Upton and Sons, 2000.

Hutton, Paul A. *Custer and His Times*. El Paso, Tex.: Little Big Horn Associates, 1981.

Hutton, Paul A. *The Custer Reader*. Lincoln and London: University of Nebraska Press, 1992.

Innis, Ben. *Bloody Knife: Custer's Favorite Scout*. Fort Collins CO: Old Army Press, 1973.

Jackson, Donald. *Custer's Gold: The United States Cavalry Expedition of 1874.* New Haven, Conn.: Yale University Press, 1966.

Johansen, Albert. *The House of Beadle Adams and its Dime and Nickels Novels, Volume 2.* Norman: University of Oklahoma Press, 1950.

Johnson, Barry. *Custer, Reno, Merrill and Lauffer Case*. London: English Westerner's Society, 1971.

Johnson, Barry. *Merritt and the Indian Wars*. London: Johnson-Taunton Military Press, 1972.

Johnson, Barry and Francis B. Taunton. *More Sidelights of the Sioux Wars*. London: Westerner's Publications, 2004.

King, Charles Captain. *Campaigning With Crook*. Norman: University of Oklahoma Press, 1988.

Kime, Wayne R., ed. *The Black Hills Journals of Colonel Richard Irving Dodge*. Norman: University of Oklahoma Press, 1996.

Koury, Michael. *Diaries of the Little Big Horn*. Fort Collins, CO: Old Army Press, 1968.

Knight, Oliver. *Following the Indian Wars*. Norman: University of Oklahoma Press, 1963.

Kuhlman, Charles. *Legend Into History*. Harrisburg PA.: Stackpole, 1951.

Langellier, John P., Kurt Hamilton Cox, and Brian C. Pohanka, eds. *Myles Keogh: The Life and Legend of an Irish Dragoon in the Seventh Cavalry.* El Secundo, CA: Upton and Sons, 1999.

Lazarus, Edward. *Black Hills, White Justice*. New York: Harper Collins, 1991.

Leckie, Shirley A. *Elizabeth Bacon Custer and the Making of a Myth*. Norman and London: University of Oklahoma Press, 1987.

LeForge, Thomas H., as told by Thomas Marquis. *Memories of a White Crow Indian*. Reprint, Lincoln: University of Nebraska Press, 1962.

Libby, Orrin G., ed. *The Arikara Narrative of Custer's Campaign and the Battle of the Little Big Horn*. Norman: University of Oklahoma Press, 1998.

Liberty, Margot. *A Cheyenne Voice-The Complete John Stands in Timber Interviews*. Norman: University of Oklahoma Press, 2013.

Liberty, Margot. *A Northern Cheyenne Album*. Norman: University of Oklahoma Press, 2006.

Liberty, Margot. *Cheyenne Memories*. Lincoln: University of Nebraska Press, 1967.

Liddic, Bruce R. *I Buried Custer: The Diary of Pvt. Thomas W. Coleman, 7th U.S. Cavalry*. College Station, Texas: Creative Publishing Company, 1979.

Liddic, Bruce R. *Vanishing Victory: Custer's Final March*. El Secundo CA: Upton and Sons, 2004.

Liddic, Bruce R. and Paul Harbaugh. *Camp on Custer: Transcribing the Custer Myth.* Spokane, Washington: The Arthur H. Clark Company, 1995.

Mackintosh, John D. *Custer's Southern Officer: Captain George D. Wallace, 7th U.S. Cavalry.* Lexington, S.C.: Cloud Creek Press, 2002.

Magnussen, Daniel O. *Peter Thompson's Narrative of the Little Big Horn Campaign 1876*. Glendale, CA: The Arthur H. Clark Company, 1974.

Mails, Thomas E. *The Mystic Warriors of the Plains*. New York: Mallard Press, 1991.

Magnum, Neil C. *Battle of the Rosebud: Prelude to the Little Big Horn*. El Secundo, CA: Upton & Sons, 1987.

Manion, John S. *General Terry's Last Statement to Custer*. El Secundo, CA: Upton and Sons 2000.

Marquis, Thomas B. *The Cheyennes of Montana*. Algonac, Michigan: Reference Publications, Inc., 1987.

Marquis, Thomas B. *Custer: Cavalry & Crows*. Fort Collins, CO: Old Army Press, 1975.

Marquis, Thomas B. *Custer on the Little Big Horn*. Algonac, Michigan: Reference Publications Inc., 1985.

Marquis, Thomas B. *Keep the Last Bullet For Yourself: The True Story of Custer's Last Stand.* Algonac, Michigan: Reference Publications Inc., 1985.

Marquis, Thomas B. *Memoirs of a White Crow Indian*. Lincoln and London: University of Oklahoma Press, 1974.

Marquis, Thomas B. *Wooden Leg: A Warrior Who Fought Custer*. Lincoln and London: University of Nebraska Press, 1931.

McLaughlin, James. *My Friend the Indian*. Lincoln and London: University of Nebraska Press, 1989.

McClernand, Edward J. *On Time for Disaster: The Rescue of Custer's Command*. London, NB: University of Oklahoma Press, 1969.

Merkel, Captain Charles E. Jr. *Unraveling the Custer Enigma*. Enterprise, AL: Merkel Press, 1977.

Merkel, Captain Charles E. Jr. *Custer's Forgotten Lieutenant*. ProQuest UMI, 1996.

Merington, Marguerite. *The Custer Story: The Life and Intimate Letters of George A. Custer and His Wife Elizabeth*. Lincoln and London: University of Nebraska Press, 1987.

Michno, Gregory. *Encyclopedia of Indian Wars*. Missoula, MT: Mountain Press Publishing, 2003.

Michno, Gregory. *Lakota Noon*. Missoula, MT: Mountain Press Publishing Company, 1997.

Michno, Gregory. *Mystery of E Troop.* Missoula, MT: Mountain Press Publishing Company, 1994.

Miles, Nelson A. *Personal Recollections & Observations of General Nelson A. Miles Vols. 1 and 2*. Lincoln and London: University of Nebraska Press, 1992.

Miller, David Humphreys. *Custer's Fall: The Indian Side of the Story*. Lincoln and London: University of Nebraska Press, 1985.

Milligan, Edward Archibald. *High Noon on the Greasy Grass*. Bottineau, ND: Bottineau–Courant Printing, 1972.

Mills, Charles K. *Charles C. De Rudio*, Mattituck N.J. J.M. Carroll, 1983.

Mills, Charles K. *Harvest of Barren Regrets*. Glendale, CA: Arthur H. Clark, 1985.

Monaghan, Jay. *Custer: The Life of General George Armstrong Custer*. Lincoln and London. University of Nebraska Press, 1971.

Monnett, John H. *The Battle of Beecher Island and the Indian War of 1867 -1869.* Niwot. University of Colorado Press, 1992.

Mooris, Donald. *The Washing of Spears*. New York: Simon & Schuster, Inc., 1965.

Neihardt, John G. *Black Elk Speaks: Being the Life Story of a Holy Man of the Oglala Sioux*. Lincoln and London: University of Nebraska Press, 1988.

Nevin, David. *The Soldiers* (Old West Time-Life Series). Time Life Publishers, January 1, 1973.

Nichols, Ronald H. *In Custer's Shadow: Major Marcus Reno*. Fort Collins, CO: Old Army Press, 1999.

Nichols, Ronald H., ed. *Men with Custer: Biographies of the 7th Cavalry*. Hardin, MT: *Custer Battlefield* Museum, 2000.

Nichols, Ronald H., ed. *Reno Court of Inquiry*, Hardin, Mt.: Custer Battlefield Museum, 1996.

Nye, Elwood, L. *Marching with Custer*. Glenwood, Calif: Arthur H. Clark, 1964.

Hein, Lt. *Memoirs of Long Ago by an Old Army Officer*. New York: G.P. Putnam and Sons, 1925.

O'Neil, Tom. *Custer Chronicles. Vols. 1-10*. Brooklyn and Arrow Trooper, 1998.

O'Neil, Tom. *GarryOwen Tidbits, Vols. 1-10.* Brooklyn: Arrow and Trooper, 1989.

O'Neil, Tom. *Sagas of the Greasy Grass*. Brooklyn: Arrow and Trooper Publishing Company, 1993.

O'Neil, Tom. *The Gibson–Edgerly Narratives of the Little Big Horn*. Brooklyn: Arrow and Trooper, 1993.

Overfield, Lloyd II. *The Little Big Horn, 1876: The Official Communications, Documents, and Reports, with Rosters of the Officers and Troops of the Campaign*. Glendale, CA: Arthur H. Clark, 1971.

Pennington, Jack. *The Battle of the Little Big Horn: A Comprehensive Study*. El Secundo, CA: Upton and Sons, 2001.

Perkins, J.R. *Trails, Rails, and War: The Life of General G.M. Dodge*. Indianapolis: Bobbs-Merrill, 1929.

Philbrick, Nathaniel. *The Last Stand: Custer, Sitting Bull, and the Battle of Little Big Horn*. New York: Viking, 2010.

Powell, Father Peter J. *People of the Sacred Mountain*. N.Y., N.Y.: Harper & Row, 1981. University of Oklahoma Press, 1976.

Powell, Peter J. *Sweet Medicine: The Continuing Role of the Sacred Arrows, the Sun Dance, and the Sacred Buffalo Hat in Northern Cheyenne History*. United States: University of Oklahoma Press, 1998.

Ricker, Eli S. *The Indian Interviews of Eli S. Ricker, 1903–1919.* Lincoln: University of Nebraska Press, 2005.

Ricker, Eli S. *The Settler and Soldier Interviews of Eli S. Ricker, 1903– 1919*. Lincoln: University of Nebraska Press, 2005.

Rickey, Don. *Forty Miles a Day on Beans and Hay*. Norman: University of Oklahoma Press, 1963.

Roberts, Richard. *Custer's Last Battle: Reminiscences of General Custer*. Monroe, MI: Monroe County Library System, 1978.

Sajna, Mike. *Crazy Horse: The Life Behind the Legend*. New York: Wiley, 2000.

Sarf, Wayne Michael. *The Little Bighorn Campaign*. New York: Combined Publishing, 1993.

Schmidt, Martin, ed. *The Life of General Crook: His Autobiography*. Norman: University of Oklahoma Press, 1946.

Schoenberger, Dale. *The End of Custer: Death of an American Military Legend*. WA: Hancock House, 1995.

Schurtz, James Willard. *William Jackson, Indian Scout*. Boston: Houghton Mifflin, 1926.

Scott, Douglas D., and Peter Bleed. *A Good Walk Around the Boundary*.: Np. National Association of Professional Archaeologists and the Nebraska State Historical Society, 1997.

Scott, Douglas D., and Richard Fox, Jr. *Archeological Insights into the Custer Battle*. Norman: University of Oklahoma Press, 1987.

Scott, Douglas D., Richard Fox, Melissa A. Connor, and Dick Harmon. *Archeological Perspectives on the Battle of Little Big Horn*. Lincoln NB: University of Oklahoma Press, 1989.

Scott, Douglas D., P. Willey, and Melissa A Connor. *They Died With Custer*. Norman: University of Oklahoma Press, 1989.

Scott, Douglas D. *Uncovering History: Archaeological Investigations at the Little Bighorn.* Norman: University of Oklahoma Press, 2013.

Sklenar, Larry. *To Hell with Honor*. Norman: University of Oklahoma Press, 2000.

Smalley, Vern. *Little Big Horn Mysteries*. Bozeman MT: Little Buffalo Press, 2005.

Smalley, Vern. *More Little Big Horn Mysteries*. Bozeman MT: Little Buffalo Press, 2005.

Stands In Timber, John. *Cheyenne Memories*. New Haven, Conn: Yale University Press, 1967.

Stewart, Edgar I. *Custer's Luck*. Norman: University of Oklahoma Press, 1955.

Stewart, Gary W. and D. J. Richardson. *Fates Change Horses*.2023

Taunton, Francis B. *Custer's Field: "A Scene of Sickening, Ghastly Horror."* London. Johnson-Taunton Military Press, 1989.

Taunton, Francis B. *No Pride in the Little Big Horn*. London: The English Westerners Society, 1987.

Taylor, William O. *With Custer on the Little Big Horn*. New York: Viking, 1996.

Stiles. T.J. *Custer's Trials: A Life on the New American Frontier*. Random House, 2016.

Terrell, John Upton and Colonel George Walton. *Faint the Trumpet sounds: The life and trial of Major Reno.* David McCoy Co., 1966.

Thompson, Peter. *Peter Thompson's Account of the Little Big Horn*: The Waddington Typescript. Edited by Michael L. Wydman and Rocky L. Boyd. Self-Published, 2004.

Thompson, Peter. *Peter Thompson's Narrative of the Little Big Horn Campaign, 1876*. Edited by Daniel O. Magnussen. Glendale, California: Arthur H. Clark, 1974.

Thompson, Peter. *Thompson's Narrative of the Little Big Horn.* Edited by Walter Cross. Stillwater, Oklahoma: Cross Publications, 2007.

Unger, Arthur C. *Custer's First Messenger!? Debunking the Story of Sergeant Daniel A. Kanipe.* Washington, NC: Custer 1876 Publisher, 2011.

Upton, Richard. *Custer Catastrophe at the Little Big Horn 1876* El Secundo CA: Upton and Sons, 2012.

Urwin, Gregory. *Custer Victorious*. Edison, NJ: Blue and Grey Press, 1983.

Utley, Robert M. *Cavalier in Buckskin: George Armstrong Custer and the Western Military Frontier*. Norman: University of Oklahoma Press, 1988.

Utley, Robert M. *Frontier Regulars: U.S. Army and the Indian, 1866 -1891*. New York: MacMillan Publishing, 1973.

Utley, Robert M. *Life in Custer's Cavalry: Diaries and Letters of Albert and Jennie Barnitz.* Lincoln: University of Nebraska Press, 1977.

Utley, Robert M. *Little Big Horn Battlefield: A History and Guide to the Battle of Little Big Horn*. Washington DC: National Park Service, 1994.

Utley, Robert M. *The Lance and the Shield*. New York: Henry Holt Company, 1993.

Utley, Robert M. *The Reno Court of Inquiry: The Chicago Times Account*. Fort Collins: Old Army Press, 1992.

Vaughn, J.W. *Indian Fights: New Facts on Seven Encounters.* Norman: University of Oklahoma Press, 1960.

Vestal, Stanley. *New Sources of Indian History, 1850–1891*. Norman: University of Oklahoma Press, 1934.

Vestal, Stanley. *Sitting Bull*. Norman: University of Oklahoma Press, 1989.

Wagner, Frederic C. *Participants in the Battle of Little Big Horn: A Biographical Dictionary of Sioux, Cheyenne and United States Military Personnel*. Second Edition. Jefferson, North Carolina and London: McFarland & Company, Inc., 2016.

Wagner, Frederic C. *The Strategy of Defeat at the Little Big Horn—A Military and Timing Analysis of the Battle*. Jefferson, North Carolina: McFarland & Company, 2011.

Wagner, Glendolin Damon. *Old Neutriment*. Lincoln: University of Nebraska Press, 1989.

Walker, Judson. *Campaigns of General Custer*. Reprint. New York: Promontory Press, 1966.

Walker, L.G., Jr. *Dr. Henry Porter—The Surgeon who Survived Little Bighorn*. Jefferson NC: McFarland and Company, 2008.

Weibert, Henry and Don Weibert. *Sixty-Six Years in Custer's Shadow*. Billings, MT: Falcon Press, 1988.

Wengert, James. *The Custer Dispatches*." Manhattan, Kans.: Sunflower University Press, 1987.

Wengert James and E. Elden Davis, eds. *That Fatal Day: Eight More with Custer*. Howell Mich.: Powder River Press, 1992.

Wert, Jeffrey D. *Custer: The Controversial Life of George Armstrong Custer*. New York: Simon and Schuster, 1996.

Whittaker, Frederick. *A Complete Life of George A. Custer*. New York: Sheldon and Company, 1876.

Willert, James. *Little Big Horn Diary: Chronicle of the 1876 Indian War*. El Secundo CA: Upton and Sons, 1998.

Windolph, Charles. *I Fought With Custer: The Story of Sergeant Windorf: Last Survivor of the Battle of the Little Big Horn as told to Frazier and Robert Hunt.* University of Nebraska Press. Lincoln and London. The Montana Historical Society contributions, 1903.

Wooster, Robert. *The Military and United States Military Policy, 1865 – 1903.* New Haven CT: Yale University Press, 1983.

Articles, Collections, Periodicals, Pamphlets & Proboards

Anderson, Harry H. "Cheyennes at the Little Big Horn; A Study in Statistics." *North Dakota History* 27, no. 3 (Summer 1960).

Arnold, Steve. "Coming to the Aid of Custer?" *Research Review* 27, no. 2, (2013).

Bacopulos, Eric. "Captain Fredrick W. Benteen's Response to Custer's Last Message." Paper presented at the 7th Annual Custer Battlefield and Historical Museum Association Symposium, (June 25, 1993).

Bailly, Edward C. "Echoes from Custer's Last Fight." *Military Affairs* 17, no. 4, (1953).

Banfill. W.H. "The Story of John Burkman's Experience with General George Custer and the 7th Cavalry." Thomas Marquis Collection, National Library of Medicine, Bethesda, Maryland. See Little Big Horn Battlefield National Monument.

Barnard, Sandy. "Walter Camp," Greasy Grass, Volume II, Hardin, Montana: Custer Battlefield and Museum Historical Association 7, (May 1995).

Barnard, Sandy. "I Go with Custer: The Life and Death of Reporter Mark Kellogg." Bismark, ND: *Bismark Tribune*, (1996).

Blake, Bill. "Was Weir a Hero?" *Little Big Horn Associates Newsletter* Vol. XI, No.5, (June 2012).

Braden Charles. "The Yellowstone Expedition." *Journal of the United States Cavalry Association* (Oct. 15, 1904), and a second article, same journal, (October 16, 1905).

Bradley, James. "Journal of the Sioux Campaign of 1876 under the Command General John Gibbon." *Contributions to the Historical Society of Montana* 1, (1903).

Bray, Kingsley M. "Teton Sioux Population History, 1655–1881." *Nebraska History* 75, no. 2, Summer 1994.

Brininstool, E.A., et al. "Chief Crazy Horse: His Career and Death." *Nebraska History* 12, no. 1, January–March 1929.

Broome, Jeff. "In Memory of Lt. James Sturgis." Guidon 3, no. 3, June 2000.

Brust, James. "Lt. Oscar Long's Early Map Details, Terrain, Battle Positions." Volume 11 of *Greasy Grass*, May 1995.

Buecker, Thomas R. "A Surgeon at the Little Big Horn: The Letters of Dr. Holmes O. Paulding." Vol 32, No.4, Montana Historical Society, Autumn 1982.

Buecker, Thomas R. "Lt. William Philo Clark's Sioux War Report." Volume 7 of *Greasy Grass*, May 1991.

Calhoun, Samuel W. "Did Custer Disobey?" Volume 6 of *Greasy Grass*, May 1990. Billings Montana: Fenske Printing Inc.

Caniglia, Salvatore A. "Profile of the Seventh," Volume XI of *Research Review*. El Paso, Texas: (June 1977).

Captain Owen J. Sweet's Battle Report from a handwritten report of Major Welch. See https://www.welchdakotapapers.com/2012/02/col-welch-notes-on-custer/. Select Captain Sweet's Battle Report, May 15, 1890.

Cecil, Jerry. "Lt. Crittenden: Striving for the Soldier's Life," Volume 11 of *Greasy Grass*, (May 1995), Billings Montana: Fenske Printing Inc.

Conzelman, Clair. https://lbha.proboards.com/thread/1154/cavalry-tactics-wichita-village. Pages 1-2.

Coughlan, Colonel T.M. "The Battle of Little Big Horn: A Tactical Study." *Cavalry Journal,* Vol. 43, No. 191, (Jan–Feb. 1994).

Davis, E. Elden. "A Look At The Reno Court of Inquiry and The Defense: A Correspondent's View." Newsletter: Little Big Horn Associates. Vol. XXVIII, (November 1994), Boaz, Alabama: Four Star Printing and Office Supply, Inc., 1994.

Dickson, Ephriam. "Reconstructing the Indian Village on the Little Big Horn." *Research Review* Vol. 22, (May 2006).

Donahue, Michael. "The Battle of Little Big Horn—A New Timeline: 18th Annual Symposium of the Custer Battlefield Association and Museum Association, June 2004.

Donahue, Michael. "The Maps of John Stands In Timber." Paper presented at 27th Annual Symposium of the Custer Battlefield Historical and Museum Association, June 2012.

Donahue, Michael. "Beyond Custer's Hill.", Paper presented at 17th Annual Symposium of the Custer Battlefield Historical & Museum Association, Hardin, Montana, June 2003.

Ediger, Theodore A. and Vinnie Hoffman. "Some Remembrances of the Battle of Washita." *Chronicles of Oklahoma*, Summer 1955.

Fiske, Frank B. "The Taming of the Sioux." *Bismarck Tribune*, 1917.

Foley, James R. "Walter Camp & Ben Clark." *Research Review* 10, no.1, January 1996.

Frost, Lawrence. "The Black Hills Expedition of 1874." *Red River Valley Historical Review* 4, no.4, Fall 1979.

Gibbon, John. "Hunting Sitting Bull." *American Catholic Quarterly Review*, October 1977.

Gibbon John. "Last Summer's Expedition Against the Sioux." *American Catholic Quarterly Review*, April 1877.

Godfrey, Edward S. "Cavalry Fire Discipline." *Journal of the Military Service Institution of the United States,* 19, September, 1896.

Godfrey, Edward S. "Some Reminiscences, Including the Washita Battle, November 27, 1868." *The Cavalry Journal* 37, no.153, October 1928.

Graham, W.A. "Come On—Be Quick: Bring Packs," *The Cavalry Journal*.

Gray, John S. "Arikara Scouts With Custer." *North Dakota History* 35, No. 2, Bismark, North Dakota. State Historical Society of North Dakota, Spring 1968.

Gray, John S. "The Pack Train on George A. Custer's Last Campaign." *Nebraska History* 57, no.1, Spring 1976.

Gray, John S. "Sutler on Custer's Last Campaign." *Nebraska History* 57, no.1, Spring 1976.

Hardorff, Richard G. "Some Recollections on Custer's Last Battle." *Research Review,* June 1990.

Hedren, Paul L. "Carbine Extractor Failure at the Little Big Horn." *Military Collector and Historian,* Summer 1973.

Horn, Donald W. "Custer's Turn to the North at the Little Big Horn: The Reason." *Research Review* 8, no. 1, January 1994.

Horn, Donald W. "The Tainted Testimony of Captain Frederick W. Benteen." *Research Review* 15, no. 2, Summer 2001.

Hudnutt, Dean. "New Light on the Little Big Horn." *Field Artillery Journal* 26, no.4, 1937.

Hughes, Robert "Campaign Against the Sioux in 1876." *Journal of the Military Service Institution of the United States.*

Johnson, Barry C. "A Captain of Chivalric Courage." London: *English Westerners Society,* 1989.

Johnson, Barry. "Dr. Paulding and His Remarkable Diary". In *Sidelights of the Sioux Wars, Issue 2*, edited by Francis Taunton. London. English Westerner's Society, 1971.

Justus, Judith P. "The Saga of Clara Blinn at the Battle of Washita." *Research Review,* Vol.14, No.1, Winter 2000.

Kanipe, Daniel A. "A New Story of Custer's Last Battle; Told by a Messenger Boy Who Survived." *Contributions to the Historical Society of Montana 9,* 1923.

King, Capt. Charles. "Custer's Last Battle." *Harper's,* August 1990.

Kush, George. George Kush Collection, Alberta, Canada.

Krott, Rob. "Was Custer Outgunned." *Military Illustrated, no.139,* 2004.

Langley, Lorne. "Intertwining the Tales of Major Joel Elliott, Clara and Willie Blinn." *Little Big Horn Research Review,* Vol. 30, 2016.

Luce, Edward S. "The Diary and Letters of Dr. James DeWolf." *North Dakota History 25* no.3, April–July 1958.

Magnum, Neil C. "Battle of the Little Big Horn." Volume 1 of *Greasy Grass,* May 1985.

Magnum, Neil C. "Reno's Battalion In the Battle of Little Big Horn." Volume 2 of *Greasy Grass.*

McBlain, John F. "With Gibbon on the Sioux Campaign of 1876." *Journal of the United States Cavalry Association* 9, no. 33, June 1896.

McClernand, Edward J. "With the Indians and the Buffalo in Montana." *The Cavalry Journal 36, no.36, January to April 1927.*

McClernand, Edward J. "Chapter II: History of the 2nd Cavalry," *Washington Observer,* (October 5, 1932).

McKeta, R.T. "The Press and the Little Big Horn." *Research Review* 1, no.1, March 1984.

Michno, Greg: "Little Big Horn Mystery Solved." El Paso, Texas. *Research Review Magazine, January 1992.*

Michno, Gregory. Battle of Little Bighorn Myths. *Wild West Magazine* 10, https://www.historynet.com/little-bighorn-myths/

Millbrook, Minnie D. "Godfrey on Custer" *Research Review* 6, no.4, Winter 1972.

Miller, David Humphreys. "Echoes of the Little Big Horn." *American Heritage* 22 no.4, June 1971.

Myers, Stephen W. "Roster of Known Hostile Indians at the Battle of Little Big Horn." *Research Review Magazine,* Volume 5, No.2, June 1991.

Nettleton, General A.B. "Custer's Characteristics." *Army and Navy Journal: Gazette of the Regular and Volunteer Forces*. New York. Publications Office. Page 818. Online resource: https://babel.hathitrust.org/cgi/pt?id=coo.31924069759953&seq=870

Nichols, Ronald H. "Marcus Albert Reno." Volume 2 of *Greasy Grass,* May 1986.

Noyes, C. Lee. "The Battle of the Little Big Horn: Reno, Terry, and a Variation of a Major Theme." Paper presented at the 13th Annual Symposium of the Custer Battlefield Historical & Museum Association, *Hardin, Montana,* June 25, 1999.

Noyes, C. Lee. "Captain Robert P. Hughes and the Case Against Custer An Early Perspective of the Little Big Horn." *Little Big Horn Associates Newsletter* 33, no. 1, February 1999.

Noyes, C. Lee. "Custer's Surgeon, George Lord, Among the Missing at Little Big Horn Battle." *Greasy Grass* 16, May 2000.

Noyes, C. Lee. "A Dispatch from the Battlefield," *Research Review* 18, no.2, Summer 2004.

Noyes, C. Lee. "The Valley Fight: The Enlisted Man's Perspective." Paper presented at the 20th Annual Symposium of the Custer Battlefield Historical & Museum Association, Hardin, Montana, June 2006.

O'Neil, Tom and Hoyt S. Vandenberg. "A Modern Look at Custer's Orders." *Research Review,* Printers Inc., El Paso, Texas, Vol.8, No.2, June 1994.

Paulding, Thomas O. "Diary with Gibbon Expedition to Meet Custer." *Little Big Horn Battlefield National Monument.*

Pennington, Jack. "The Reno Court—The Second Cover-Up." *Research Review.* Printers Inc., El Paso, Texas, Vol. 9, No. 1, January 1995.

Pohanka, Brian C. "George Yates—Captain of the Band Box Troop." *Greasy Grass,* Vol. 8, May 1992.

Pohanka, Brian C. "A Study of Company Discipline: General Court-Martial of Enlisted Personnel, 7th U.S. Cavalry, 1875 -1876." Paper presented at the 9th Annual Symposium of the Custer Battlefield Historical & Museum Associates, June 23, 1995.

Prickett, Robert C. "The Malfeasance of William Worth Belknap." *North Dakota History* 17, no.1, January 1950.

Rector, William G. "The Reno-Benteen Defense Perimeter." Helena Montana, *Montana Magazine*, Spring 1966.

Reedstrom, Ernest Lisle. "Bloody Knife: Custer's Favorite Scout." *True West,* February 1989. Vol.36, No.2.

Reedstrom, Ernest Lisle. "Tom Custer: In the Shadow of His Brother." *True West,* November 1994, Vol 41, No. 11.

Rickey, Don. "Interview With Stands In Timber." *Little Big Horn Battlefield National Monument,* August 8, 1956.

Rini, Bill. "The Reno Skirmish Line Locations." *Research Review: Journal of the Little Big Horn,* Volume 34, 2020.

Rini, Bill. "Myles Keogh on Battle Ridge." *The Crow's Nest, The Journal of the Custer Association of Great Britain*, Volume 20, 2020–2021.

Roe, Charles Francis. "Custer's Last Battle." Fort Collins CO. *Old Army Press,* No date of publication given.

Ryan, John M. "Ten Years With General Custer Among the Indians." *The Newton Mass. Circuit,* May 28, 1909.

Schneider, Jim. "Rations, Forage and Ammunition: Logistical Problems Facing the Dakota Column in the Summer of 1876." *Research Review.* McNaughton & Gunn, Inc., El Paso Texas, Vol.5, No.2, June 1991.

Schoenberger, Dale T. "A Trooper with Custer: Augustus DeVoto's Account of the Little Big Horn." *Montana* 40 no, 1, Winter 1990.

Schoenberger, Dale T. "Custer's Scouts." Helena, Montana, *Montana Magazine*. Spring 1996.

Scott, Douglas. "Cartridges, Bullets and Bones." *Research Review* 18, May 2002.

"Seven Anonymous Warriors' Accounts of the Little Big Horn Battle," *Chicago Tribune*, August 1, 1876.

Sills, Joe Jr. "The Recruits Controversy: Another Look." Volume 5, *Greasy Grass,* " 1989. Billings, Montana, Fenske Printing Inc.

Sklenar, Larry. "Private Theodore Goldin: Too Early Discredited?" *Research Review.* Printers Inc., El Paso, Texas. Vol. 9. No.1, January 1995.

Sklenar, Larry. "The Wallace Factor at the Reno Court of Inquiry." *Research Review* 14, no. 1, Winter 2000.

Smalley, Vern G. "Lone Tepee along Reno Creek" Paper presented at the 12th Annual Symposium of the Custer Battlefield Historical & Museum Association. Hardin, MT, June 26, 1998.

Smith, Jay. "Custer Didn't Do It." *Little Big Horn Associates Newsletter* 9, no.2, February 1975.

Smith, Jay. "A Hundred Years Later." In *Custer and His Times*, edited by Andrew Paul Hutton. El Paso, Texas: Little Big Horn Associates, 1981.

Stewart, Edgar I., ed. "I Rode With Custer." *Montana* 4, no.3, July, 1952.

Stewart, Edgar I. "The Little Big Horn 90 Years Later." Helena, Montana. *Montana Magazine.* Spring 1966.

Summit, Christopher M. "Apologia Pro 'Custer's Last Stand.'" Volume 5 of *Greasy Grass,* May 1989, billings, Montana, Fenske Printing Inc.

Taunton, Francis B. "The Mystery of Miss Adams." *Brand Book* 28, no.2, Summer 1991.

Taylor, Joseph Henry. "Lonesome Charley." *North Dakota Historical Quarterly* 4, July 1930.

Taylor, Susan Thompson. "Thompson in Custer's Cavalry, 1875–1880." Unpublished.

Trinque, Bruce A. "The Defense of Custer Hill." *Research Review.* Printers Inc., El Paso, Texas, Vol.9, No. 2, June 1994.

Trinque, Bruce A. "Elusive Ridge." *Research Review.* Printers Inc., El Paso, Texas, Vol. 9, No. 1, January 1995.

Trinque, Bruce A. "The Cartridge Case Evidence on Custer Field: An Analysis and Reinterpretation." Paper presented at the 5th Annual Symposium of the Custer Battlefield Historical and Museum Association, Inc., Hardin, Montana, June 21, 1991.

Weir, Thomas. "Little Big Horn Associates Message Board: Captain Thomas Weir." https://lbha.proboards.com/thread/5705/captain-thomas-weir Wells, Wayne.. "Little Big Horn Notes," *Greasy Grass,* Volume 5, May 1989.

Whittaker, Frederick. "George A. Custer." *Galaxy,* September 1876.

Willert, James. "Does Anomaly Contain Sturgis's Body?" *Research Review* 11, no. 2, Summer 1997.

Winkler, Albert. "The Case for a Custer Battalion Survivor: Private Gustave Korn's Story." Montana, *The Magazine of Western History,* Spring, 2013.

Wolseley, Garnet. "The Story of a Soldier's Life." Volumes 1 & 2. Westminster, Constable & Co., Ltd. 1903–1904.

Yellow Nose. "The Big Horn." *Yellowstone Journal* 1, no.5, Summer 1992.

Newspapers and other Periodicals

Anaconda Standard

Annual Symposium: Custer Battlefield Historical and Museum Association

Arkansas Gazette

Bibliography

Army and Naval Journal

Battlefield Dispatch

Bismark Tribune

Brand Book

Chicago Daily News

Chicago Daily Tribune

Chicago Evening Journal

Chicago Times

Chicago Tribune

Crow's Nest, Journal of the Custer Association of Great Britain

Greasy Grass, Journal of the Custer Battlefield Historical & Museum Association.

Guidon

Harness Gazette

Janesville Gazette

Little Big Horn Associates Newsletter

National Tribune (Washington D.C)

New York Evening Post

New York Herald

New York Sun

New York Times

Philadelphia Times

Research Review, Journal of the Little Big Horn Associates

St. Paul Pioneer Press

Washington Observer

Washington Post

Collections and Archives

Camp, Walter M. Papers. Denver Public Library.

Camp, Walter M. Papers. Harold B. Lee Library, Brigham Young University, Provo, UT.

Campbell, Walter Stanley Collection. Western History Collection, University of Oklahoma, Norman.

Custer, Elizabeth B. Collection: Little Bighorn Battlefield National Monument, Crow Agency MT.

Custer Collection. Monroe County Historical Museum, Monroe MI.

Custer File and Scrapbooks. Montana Room, Billings Public Library, Billings, MT.

Frost, Lawrence A. Collection: Monroe County Historical Museum, Monroe, MI.

Ghent, William J. Papers. Library of Congress, Washington D.C.

Godfrey, Edward S. Papers. Library of Congress, Washington D.C.

Godfrey Family Papers. U.S. Army Military Institute, Carlisle, PA.

Grinnell, George Bird Papers. Braun Research Library, Institute for the Study of the American West, Autry National Center, Los Angeles.

Hagner, Francis R. Collection. New York Public Library.

Kuhlman, Charles Collection . Harold B. Lee Library, Brigham Young University, Provo UT.

Kush, George Collection. Privately held collection of George Kush in Canada.

Merington, Marguerite Papers. New York Public Library.

National Archives. Washington D.C.

Order of the Indian Wars Papers. U/S. Military Institute, Carlisle, PA.

Perry, L. Tom Special Collections; MSS P 16. Brigham Young University Harold B. Lee Library.

Ricker, Eli Collection. Nebraska State Historical Society, Lincoln.

Acknowledgments

It would be an act of impertinence to maintain that a work of this size and scope is purely one's own. This work has relied on the many authors, scholars, and battle participants who have written books, examined every clue, dug for every scrap of evidence, and, to this present day, have tenaciously sleuthed after the many tantalizing mysteries of the Battle of the Little Big Horn. The authors stand on the shoulders of those who have made immeasurable contributions to our understanding and to the writing of this book. Little Big Horn online sites, where multiple ideas and theories have been discussed, debated, and vetted, have proven to be our own personal battleground, where we have gained a new and deeper understanding of the facts and where their implications can be drawn out to the fullest. Broadening our perspective on the battle were countless conversations that have been enormously fruitful over the years, in person, by phone, by e-mail, and at yearly symposia with other battle students.

While it is impossible to name everyone who has contributed to this work, we would like to extend special thanks to the following individuals, some of whom are no longer with us but are here affectionately and gratefully remembered: Joan Croy, the Dean of Western Historians Paul Andrew Hutton, Sandy Barnard, author Ephriam Dickson, author Kingsley Bray, former Little Big Horn Battlefield Chief Historian John Doerner, James Donovan author of perhaps the best overall treatment of the battle, Pat Haddock for proofreading and editorial suggestions, Diane Windsor for the cover work, Reverend Vince Heier and his unparalleled Custer collection, Ray Hillyer, Tom Heski, Don Horn Sr., Lowell Smith, NPS Ranger Jerry Jasmer, Morris Knorr, Dale Kosman, Western artist George Kush and his unique collection of Little Big Horn sources, cartographer Geoff Lackey who blessed this work with a series of extraordinary maps, Little Big Horn historian nonpareil Bruce Liddic, author/historian Eric Wittenberg, Scott Lundin, author Greg Michno, and C. Lee Noyes. Tom Van Riper and David Taub for proofreading and editorial suggestions; Nancy Pinson; Western artist Joe Grandee, whose painting of the battle graces the cover of this book; and the irrepressible Frederic Wagner, III, for his friendship and being the best correspondent one could possibly ask for.

We would like to recognize and salute all the fellow Little Big Horn enthusiasts, Historical Reenactors, and Living Historians who gather several times a year at conferences, field trips, museums, research centers, roundtables, reenactments, etc. The names are too many for Bill to list but among them are: Civil War/Indian War reenactor Peter Meyer, Foremost Custer Living Historian Steve Alexander, author/historian Siobhan Fallon, Nicola Scro (aka Lt. Carlo De Rudio) and his talented photographer wife Stefania Pelfini, Crow warriors and reenactors Casey Birdinground, Kevin Shane and Kyle C. Reed, as well as Ken Woody and Michael Badhand for their unparalleled knowledge on Plains Indians culture and history, noted Crow poet Henry Real Bird, Living Historian and Reenactors Gary Stewart (aka Tom Custer), Ron Glasgow (aka Lt. W. W. Cooke), Melani Van Petten, historian Jeff

Wahl, and Western artist Jerry Thomas for our many discussions over the years while riding the battlefield on horseback—by far the best way to get a sense of the timing and distance factors of the Battlefield, Colonel Rodney G. Thomas, Lieutenant Colonel Mark Hoyt, Lieutenant Colonel Kevin Galvin, Captain Clair Conzelman, Joe Creadon, Joe Kelly, Western artist and reenactor Jim Hatzell, author/historian Louis Kraft, author/historian Gregory Urwin, author/reenactor Lorne Langley, Living Historian and Battle Scholar Kevin Connelly for his keen eye, input and invaluable suggestions, and James "Putt" Thompson and his wonderful Custer Battlefield Trading Post & Café, the scene of many memorable discussions on the battle. And to the Little Big Horn Associates, the Custer Battlefield Historical and Museum Association, and the Custer Association of Great Britain for keeping the history of this unique period of our history alive.

Personal note from Bill Rini: I would especially like to acknowledge the contribution of Navajo-Ute warrior/professional jockey Jarell James, one of the fiercest competitors I know, for his interest and encouragement in my pursuing this particular project to its completion. Horses and racing are a particular passion of mine, along with a love of western history and Plains Indian/Indigenous culture.

Personal note from Phil Guarnieri|: The patience and understanding of family are indispensable in undertaking a project of this magnitude. For this, I especially want to thank my beautiful bride and supportive wife of 33 years, Sonia Guarnieri, to whom I have dedicated this work, and for allowing me to turn our immaculately kept home into a storm of books, maps, and reams of foolscap.

Finally, any errors found herein are due to the authors and the authors alone.

AUTHOR BIOGRAPHIES

Phil Guarnieri

Phil Guarnieri is an author of two previous books and was a weekly columnist for local publications, writing on a variety of subjects. He has written for a national sports magazine. He developed an early and consuming interest in the Civil War and the Battle of Little Big Horn. A proud member of the Little Big Horn Association, he lives on Long Island, New York.

Bill Rini

Bill Rini is a retired History teacher with over three decades of classroom teaching experience, currently living in the New York City suburbs. He has a Bachelor of Arts degree in History along with a minor in Military Science from Washington & Lee University and a Master of Science degree in Education from CUNY.

As a trained historian, Bill has been an avid student of Western Military History and the Little Big Horn Battle for over four decades, having first joined the Little Big Horn Associates in 1975. Bill is also a member of the Custer Battlefield Historical Museum Association and the Custer Association of Great Britain.

In addition to field work in history, Bill is an avid Historical Reenactor and Living Historian and has served as the mounted Cavalry Field Coordinator and associate producer for several years -- including the 125th Anniversary Reenactment -- at the annual Custer's Last Stand Reenactment held at the Real Bird Ranch in Crow Agency, MT, as well as serving for some years on the Federal Headquarters Staff at Gettysburg Reenactments and in other Civil War and Living History events around the country. Bill has portrayed 7th Cavalry Captain Myles Keogh in several documentary productions, including the award-winning docudrama 'Custer's Strategy of Defeat.' Bill has a page listing his acting work at http://WWW.IMDB.COM

Bill has published several articles on the Little Big Horn Battle, including "The Reno Valley Skirmish Line Locations," published in the *Research Review Journal* of the Little Big Horn Associates, and "Myles Keogh on Battle Ridge," published in *The Crow's Nest Journal* of the Custer Association of Great Britain. Bill has served as the moderator/administrator for the official Message Board of the Little Big Horn Associates since 2007.

AUTHOR BIOGRAPHIES

Index

Note 1: General George Armstrong Custer appears on almost every page making it impractical to note every entry. However, an attempt was made to select those pages that described the General and some of his more significant decisions.

Note 2: Ranks may have changed over time. For the most part, the rank at the time of the Little Big Horn Battle was used for the Index.

www.ingramcontent.com/pod-product-compliance
Lightning Source LLC
LaVergne TN
LVHW081948200926
848484LV00009B/2525